Some Useful Constants and Physical Measurements*

astronomical unit	1 A.U. = 1.496×10^8 km (1.5×10^8 km)
light year	1 ly = 9.46×10^{12} km (10^{13} km; 6 trillion miles)
parsec	1 pc = 3.09×10^{13} km = 3.3 ly
speed of light	c = 299,792.458 km/s (3×10^5 km/s)
Stefan-Boltzmann constant	σ [Greek sigma] = 5.67×10^{-8} W/m$^2 \cdot$ K^4
Planck's constant	$h = 6.63 \times 10^{-34}$ Js
gravitational constant	$G = 6.67 \times 10^{-11}$ Nm^2/kg^2
mass of the Earth	$M_\oplus = 5.97 \times 10^{24}$ kg (6×10^{24} kg; about 6000 billion billion tons)
radius of the Earth	$R_\oplus$ = 6378 km (6500 km)
mass of the Sun	$M_\odot = 1.99 \times 10^{30}$ kg (2×10^{30} kg)
radius of the Sun	$R_\odot = 6.96 \times 10^5$ km (7×10^5 km)
luminosity of the Sun	$L_\odot = 3.90 \times 10^{26}$ W
effective temperature of the Sun	$T_\odot$ = 5778 K (5800 K)
Hubble constant	$H_0 \approx$ 75 km/s/Mpc
mass of the electron	$m_e = 9.11 \times 10^{-31}$ kg
mass of the proton	$m_p = 1.67 \times 10^{-27}$ kg

**The rounded-off values used in the text are shown in parentheses.*

Conversions between English and Metric Units

1 inch	=	2.54 centimeters (cm)	1 mile	=	1.609 kilometers (km)
1 foot (ft)	=	0.3048 meters (m)	1 pound (lb)	=	453.6 grams (g) or .4536 kilograms (kg) (on Earth)

The Entire Electromagnetic Spectrum

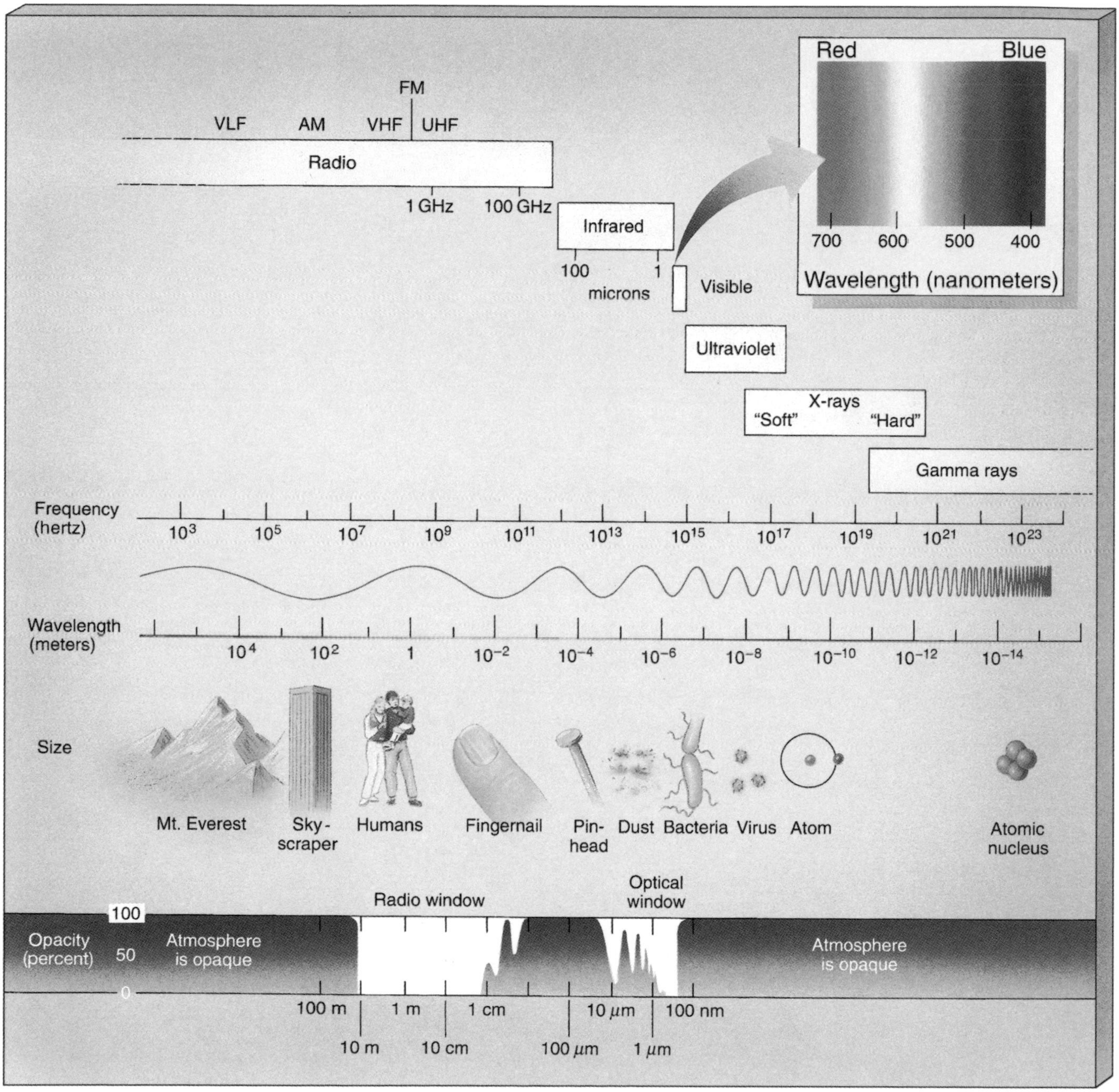

ASTRONOMY

ABOUT THE AUTHORS

Eric Chaisson Eric holds a doctorate in Astrophysics from Harvard University, where he spent ten years on the faculty of Arts and Sciences. For five years, Eric was a Senior Scientist and Director of Educational Programs at the Space Telescope Science Institute and Adjunct Professor of Physics at Johns Hopkins University. He joined Tufts University, where he is now Professor of Physics, Professor of Education, and Director of the Wright Center for Innovative Science Education. He has written nine books on astronomy, which have received such literary awards as the Phi Beta Kappa Prize, two American Institute of Physics Awards, and Harvard's Smith Prize for Literary Merit. He has published more than 100 scientific papers in professional journals, and has also received Harvard's Bok Prize for original contributions to astrophysics.

Steve McMillan Steve holds a bachelor's and master's degree in Mathematics from Cambridge University and a doctorate in Astronomy from Harvard University. He held post-doctoral positions at the University of Illinois and Northwestern University, where he continued his research in theoretical astrophysics, star clusters, and numerical modeling. Steve is currently Distinguished Professor of Physics at Drexel University and a frequent visiting researcher at Princeton's Institute for Advanced Study and the University of Tokyo. He has published over 50 scientific papers in professional journals.

ASTRONOMY

A Beginner's Guide to the Universe

THIRD EDITION

Eric Chaisson

Tufts University

Steve McMillan

Drexel University

Prentice Hall

PRENTICE HALL

Upper Saddle River, New Jersey 07458

Library of Congress Cataloging-in-Publication Data
Chaisson, Eric.
Astronomy: a beginner's guide to the universe/Eric Chaisson, Steve McMillan.—3rd ed.
p. cm.
Includes index.
ISBN 0-13-087307-1
1. Astronomy. I. McMillan, S. (Stephen). II. Title
QB43.2.C43 2001
520—dc21 00-038488
CIP

Executive Editor: Alison Reeves
Editorial Director: Paul F. Corey
Development Editor: Donald Gecewicz
Editor in Chief, Development: Ray Mullaney
Production Editor: Joanne Hakim
Assistant Vice President of Production and Manufacturing: David W. Riccardi
Executive Managing Editor: Kathleen Schiaparelli
Assistant Managing Editor, Science Media: Alison Lorber
Manager of Formatting: Jim Sullivan
Electronic Production Specialist/Electronic Page Makeup: Joanne Del Ben
Creative Director: Paul Belfanti
Art Director: Joseph Sengotta
Illustrations: Imagineering Art
Photo Editor: Beth Boyd
Editorial Assistant: Christian Botting
Interior and Cover Designer: Judy Matz-Coniglio
Manufacturing Manager: Trudy Pisciotti
Manufacturing Buyer: Michael Bell
Art Manager: Gus Vibal
Art Editor: Karen Branson
Marketing Manager: Eric Fahlgren
Photo Researcher: Yvonne Gerin
Photo Coordinator: Anthony Arabia
CD Rom Project Manager: Barbara Booth
CD Rom Production: William Johnson, Lawrence La Raia, Terrence Cummings
Associate Media Editor: Michael Banino
Copy Editor: Jocelyn Phillips
Project Manager, Physics and Astronomy: Liz Kell
Testing Supervisor: Michael Hornbostel

Prentice Hall

Upper Saddle River, NJ 07458

Printed in the United States of America

10 9 8 7 6 5 4 3 2

ISBN 0-13-087307-1

Prentice-Hall International (UK) Limited, *London*
Prentice-Hall of Australia Pty. Limited, *Sydney*
Prentice-Hall Canada Inc., *Toronto*
Prentice-Hall Hispanoamericana, S.A., *Mexico*
Prentice-Hall of India Private Limited, *New Delhi*
Prentice-Hall of Japan, Inc., *Tokyo*
Pearson Education Asia Pte. Ltd.
Editora Prentice-Hall do Brasil, Ltda., *Rio de Janeiro*

BRIEF CONTENTS

CONTENTS

(NASA)

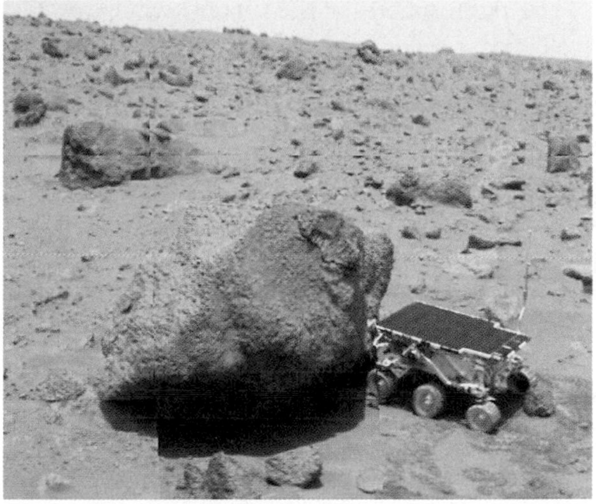

(NASA)

(NASA)

(Sky Publishing Corporation)

12 Stellar Evolution

13 Neutron Stars and Black Holes

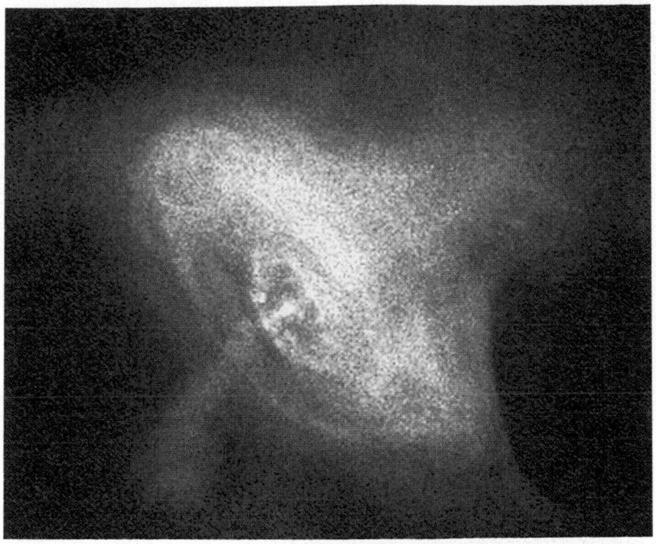

(NASA)

PART FOUR
GALAXIES AND COSMOLOGY

14 The Milky Way Galaxy

15 Normal Galaxies

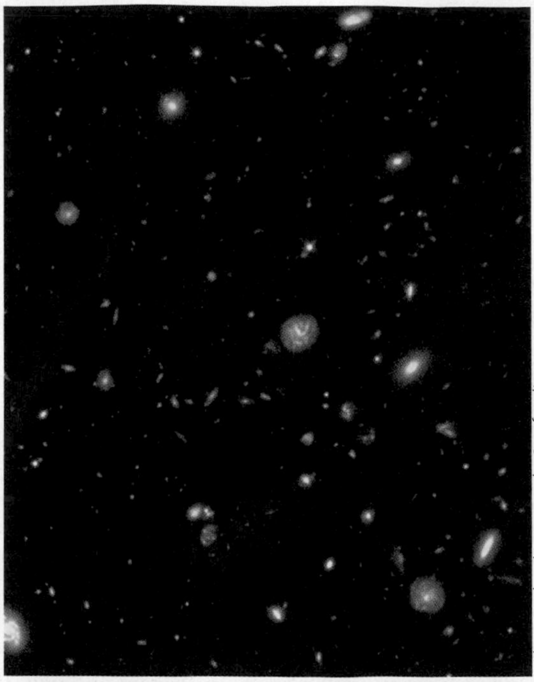
(Chris Butler/Astrostock-Sanford)

16 Active Galaxies and Quasars

17 Cosmology

18 Life in the Universe

(NASA)

PREFACE

Astronomy continues to enjoy a golden age of exploration and discovery. Fueled by new technologies and novel theoretical insights, the study of the cosmos has never been more exciting. We are pleased to have the opportunity to present in this book a representative sample of the known facts, evolving ideas, and frontier discoveries in astronomy today.

Astronomy: A Beginner's Guide to the Universe has been written for students who have taken no previous college science courses and who will likely not major in physics or astronomy. It is intended primarily for use in a one-semester, non-technical astronomy course. We present a broad view of astronomy, straightforwardly descriptive and without complex mathematics. The absence of sophisticated mathematics, however, in no way prevents discussion of important concepts. Rather, we rely on qualitative reasoning as well as analogies with objects and phenomena familiar to the student to explain the complexities of the subject without oversimplification. We have tried to communicate the excitement that we feel about astronomy and to awaken students to the marvelous universe around us.

We are very gratified that the first two editions of this text have been so well received by many in the astronomy education community. In using those earlier texts, many of you—teachers and students alike—have given us helpful feedback and constructive criticisms. From these, we have learned to communicate better both the fundamentals and the excitement of astronomy. Many improvements inspired by your comments, as well as numerous innovations and popular new features from our companion hardback text *Astronomy Today*, have been incorporated into this new edition.

Organization and Approach

As in the first two editions, our organization follows the popular and effective "Earth-out" progression. We have found that most students, especially those with little scientific background, are much more comfortable studying the relatively familiar solar system before tackling stars and galaxies. Thus, Earth is the first object we discuss in detail. With Earth and Moon as our initial planetary models, we move through the solar system. Integral to our coverage of the solar system is a discussion of its formation. This line of investigation leads directly into a study of the Sun.

With the Sun as our model star, we broaden the scope of our discussion to include stars in general—their properties, their evolutionary histories, and their varied fates. This journey naturally leads us to coverage of the Milky Way Galaxy, which in turn serves as an introduction to our treatment of other galaxies, both normal and active. Finally, we reach the subject of cosmology and the large-scale structure and dynamics of the universe as a whole. Throughout, we strive to emphasize the dynamic nature of the cosmos—virtually every major topic, from planets to quasars, includes a discussion of how those objects formed and how they evolve.

We continue to place much of the needed physics in the early chapters—an approach derived from years of experience teaching thousands of students. Additional physical principles are developed as needed later, both in the text narrative and in the boxed *More Precisely* features (described on p. xiv). We feel strongly that this is the most economical and efficient means of presentation. However, we acknowledge that not all instructors feel the same way. Accordingly, we have made the treatment of physics, as well as the more quantitative discussions, as modular as possible, so that these topics can be deferred to later stages of an astronomy course if desired. Instructors presenting this material in a 1-quarter course, who wish to (or have time to) cover only the essentials of the solar system before proceeding on to the study of stars and the rest of the universe, may want to teach only Chapter 4, and then move directly to Chapter 9 (the Sun).

New and Revised Material

Astronomy is a rapidly evolving field, and the three years since the publication of the second edition of *Astronomy: A Beginner's Guide to the Universe* have seen many new discoveries covering the entire spectrum of astronomical research. Almost every chapter in the third edition has been substantially updated with new and late-breaking information. Several chapters have also seen significant internal reorganization in order to streamline the overall presentation. Among the many changes are:

- Addition of more quantitative material and worked examples to both the text and *More Precisely* boxes throughout the book.

- Updates to material on ground-based adaptive optics and interferometry, the present status of the *Hubble Space Telescope* and the *Compton Gamma-Ray Observatory*, and new coverage of the *Chandra Advanced X-ray Astrophysics Facility* (Chapter 3).
- Discussion of the *NEAR* mission to asteroid Eros (Chapter 4).
- Expanded discussion of the Kuiper belt and Oort cloud (Chapters 4 and 8).
- Further updates on the search for extrasolar planets (Chapter 4).
- Expanded discussion of tidal forces (Chapter 5).
- Updates on lunar exploration by *Clementine* and *Lunar Prospector* (Chapter 5).
- Updated discussion of recent spacecraft missions to Mars—*Pathfinder, Sojourner, Polar Lander*, and *Global Surveyor* (Chapter 6).
- Continuing coverage of the *Galileo* mission to Jupiter and its main findings regarding that planet's Galilean satellites (Chapters 7 and 8).
- The reported detection of neutrino oscillations, and its relevance to the solar neutrino problem (Chapter 9).
- New H–R diagram based on *Hipparcos* data on nearby stars (Chapter 10).
- Expanded material on the search for brown dwarfs (Chapter 11).
- New composite H–R diagram for the oldest globular clusters (Chapter 12).
- Greatly expanded discussion of cosmic gamma-ray bursts (Chapter 13).
- Update on the *LIGO* project (Chapter 13).
- Expanded coverage of conditions near the Galactic center (Chapter 14).
- Updated Hubble's constant of 65 km/s/Mpc used consistently throughout the text (Chapter 15).
- New imagery and discussion of galaxy collisions (Chapter 15).
- Updated observations of quasar host galaxies (Chapter 16).
- Incorporation of larger-scale (Las Campanas) redshift surveys into the text discussion of the cosmological principle (Chapter 17).
- Extensive discussion of the "accelerating universe," the cosmological constant, and their possible ramifications for the structure and large-scale geometry of the cosmos (Chapter 17).

The Illustration Program

Visualization plays an important role in both the teaching and the practice of astronomy, and we continue to place strong emphasis on this aspect of our book. We have tried to combine aesthetic beauty with scientific accuracy in the artist's conceptions that adorn the text, and we have sought to present the best and latest imagery of a wide range of cosmic objects. Each illustration has been carefully crafted to enhance student learning; each is pedagogically sound and tightly tied to nearby discussion of important scientific facts and ideas.

Full-Spectrum Coverage and Spectrum Icons. 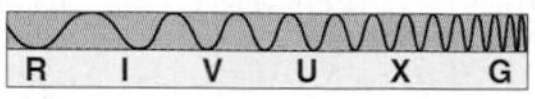

Increasingly, astronomers are exploiting the full range of the electromagnetic spectrum to gather information about the cosmos. Throughout this book, images taken at radio, infrared, ultraviolet, X ray, or gamma ray wavelengths are used to supplement visible-light images. As it is sometimes difficult (even for a professional) to tell at a glance which images are visible-light photographs and which are false-color images created with other wavelengths, each photo in the text is provided with an icon that identifies the wavelength of electromagnetic radiation used to capture the image.

Other Pedagogical Features

As with many other parts of our text, adopting instructors have helped guide us toward what is most helpful for effective student learning. With their assistance, we have revised both our in-chapter and end-of-chapter pedagogical apparatus to increase its utility to students.

Learning Goals. Studies indicate that beginning students often have trouble prioritizing textual material. For this reason, a few (typically 5 or 6) well-defined Learning Goals are provided at the start of each chapter. These help students to structure their reading of the chapter and then test their mastery of key facts and concepts. The Goals are numbered and cross-referenced to key sections in the body of each chapter. This in-text highlighting of the most important aspects of the chapter also helps students to review. The Goals are organized and phrased in such a way as to make them objectively testable, affording students a means of gauging their own progress.

Compound Art ▶

It is rare that a single image, be it a photograph or an artist's conception, can capture all aspects of a complex subject. Wherever possible, multiple-part figures are used in an attempt to convey the greatest amount of information in the most vivid way:

- Visible images are often presented along with their counterparts captured at other wavelengths.
- Interpretive line drawings are often superimposed on or juxtaposed with real astronomical photographs, helping students to really "see" what the photographs reveal.
- Breakouts—often multiple ones—are used to zoom in from wide-field shots to close-ups, so that detailed images can be understood in their larger context.

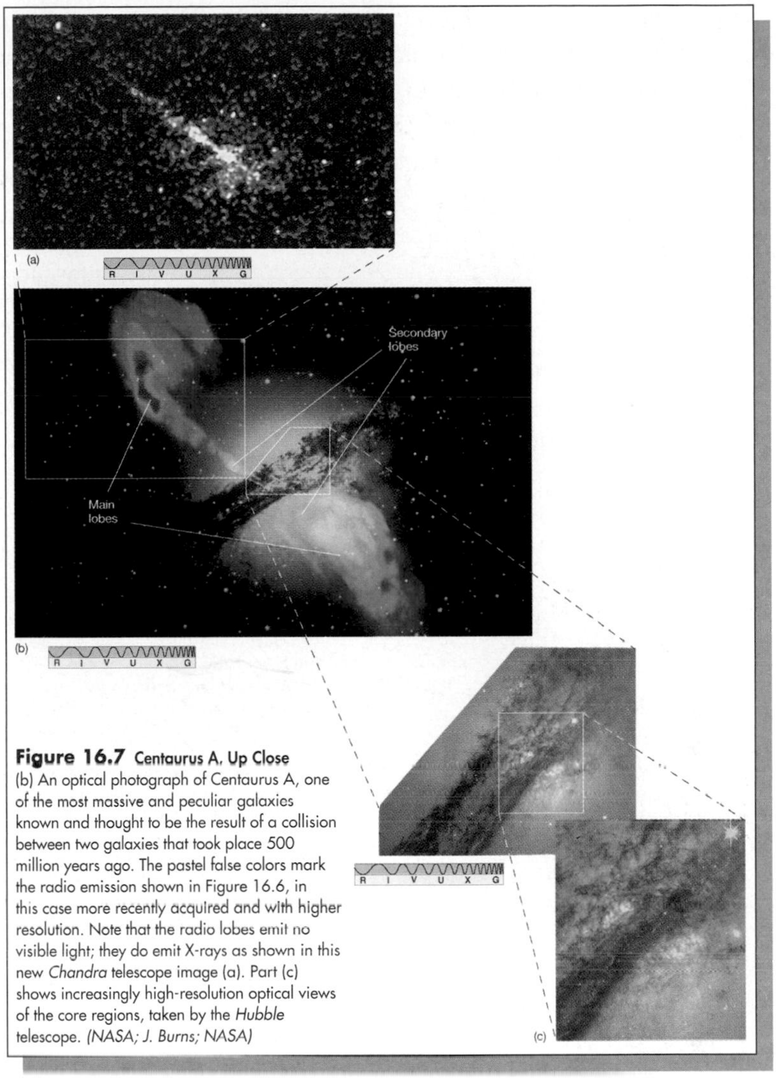

Figure 16.7 **Centaurus A, Up Close**
(b) An optical photograph of Centaurus A, one of the most massive and peculiar galaxies known and thought to be the result of a collision between two galaxies that took place 500 million years ago. The pastel false colors mark the radio emission shown in Figure 16.6, in this case more recently acquired and with higher resolution. Note that the radio lobes emit no visible light; they do emit X-rays as shown in this new *Chandra* telescope image (a). Part (c) shows increasingly high-resolution optical views of the core regions, taken by the *Hubble* telescope. *(NASA; J. Burns; NASA)*

Explanatory Captions ▶

Students often review a chapter by "looking at the pictures." For this reason, the captions in this book are often a bit longer and more detailed than those in other texts.

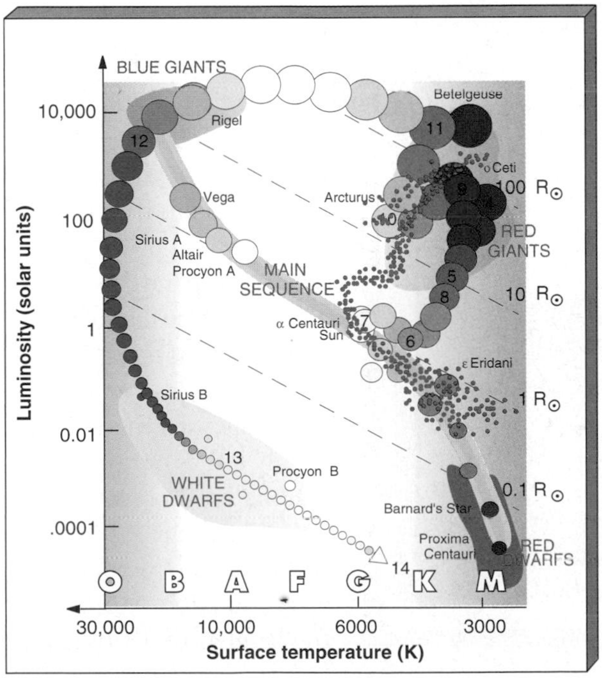

◀ H–R Diagrams and Acetate Overlays

All of the book's H–R diagrams are drawn in a uniform format, using real data. In addition, a unique set of transparent acetate overlays dramatically demonstrate to students how the H–R diagram helps us to organize our information about the stars and track their evolutionary histories.

Key Terms. Like all subjects, astronomy has its own specialized vocabulary. To aid student learning, the most important astronomical terms are boldfaced at their first appearance in the text. Each boldfaced Key Term is also incorporated in the appropriate chapter summary, together with the page number where it was defined. In addition, a full alphabetical glossary, defining each Key Term and locating its first use in the text, appears at the end of the book.

Interludes explore a ▶ variety of interesting supplementary topics, such as *The Hubble Space Telescope*, *Life on Mars?*, *A Cometary Impact*, *Supernova 1987A*, *Colliding Galaxies*, *The Cosmological Constant*, and *What Killed the Dinosaurs?*

INTERLUDE 4-1

Planets Beyond the Solar System

The past few years have seen enormous strides in the search for extrasolar planets. These advances have come not through any dramatic scientific or technical breakthrough, but rather through steady improvements in both telescope and detector technology and computerized data analysis. The count of potential extrasolar planets currently stands at 28. However, it is not yet possible to image any of these newly discovered worlds. The techniques used to find them are indirect, based on analysis of light from the parent star, not from the unseen planet.

The accompanying figure shows two sets of data that betray the presence of planets. As a planet orbits a star, gravitationally pulling one way and then the other, the star wobbles slightly. The more massive the planet or the closer its orbit to the star, the greater its gravitational pull and hence the star's movement.

If the wobble happens to occur along our line of sight to the star (as illustrated in the upper graph, which shows the line-of-sight velocity of the star 51 Pegasi, a near-twin of our Sun lying some 40 light-years away), then we see small fluctuations in the star's radial velocity, which can be measured using the Doppler effect. ∞ (*More Precisely 2-3*) These data were acquired in 1994 by Swiss astronomers using the 1.9-m telescope at Haute-Provence Observatory in France, and imply that a planet at least half the mass of Jupiter orbits 51 Peg with a period of just 4.2 days. Alter-

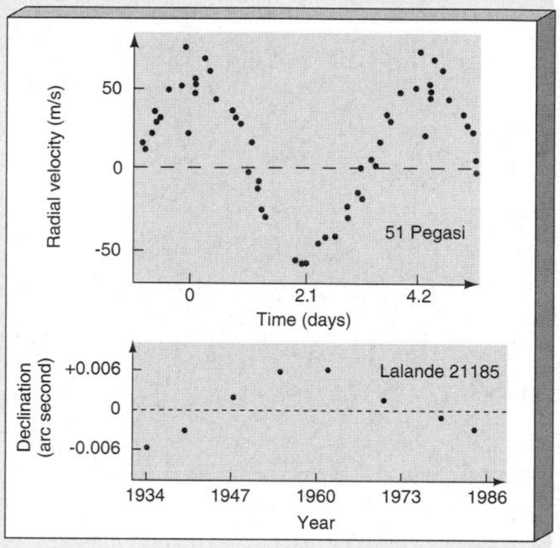

natively, if the wobble is predominantly perpendicular to our line of sight (as indicated in the lower graph, which plots the declination of the star Lalande 21185 over the last half century), then the star's position in the sky changes slightly from night to night. ∞ (Sec. P.2) The 30-year-

MORE PRECISELY 1-1

Some Properties of Planetary Orbits

Two numbers—semi-major axis and eccentricity—are all that are needed to describe the size and shape of a planet's orbital path. From them we can derive many other useful quantities. Two of the most important are the planet's *perihelion* (its point of closest approach to the Sun) and its *aphelion* (greatest distance from the Sun). From the definitions presented in the text, it follows that if the planet's orbit has semi-major axis a and eccentricity e, its perihelion is at a distance $a(1 + e)$ from the Sun, while its aphelion is at $a(1 - e)$. These points and distances are illustrated in the accompanying figure.

Note that, while the Sun resides at one focus, the other focus is empty and has no particular significance. Thus, for example, a hypothetical planet with a semi-major axis of 400 million km and an eccentricity of 0.5 (the eccentricity of the ellipse shown in the diagram) would range between $400 \times (1 + 0.5) = 200$ million km and $400 \times (1 - 0.5) = 600$ million km from the Sun over the course of one complete orbit. With $e = 0.9$, the range would be 40–760 million km, and so on.

No planet has an orbital eccentricity as large as 0.5—the planet with the most eccentric orbit is Pluto, with $e = 0.248$ (see Table 1.1). However, many meteoroids, and all comets (see Chapter 4) have eccentricities considerably greater than this. In fact, most comets visible from Earth have eccentricities very close to one. Their highly elongated orbits approach within a few A.U. of the Sun at perihelion, yet these tiny frozen worlds spend most of their time far beyond the orbit of Pluto.

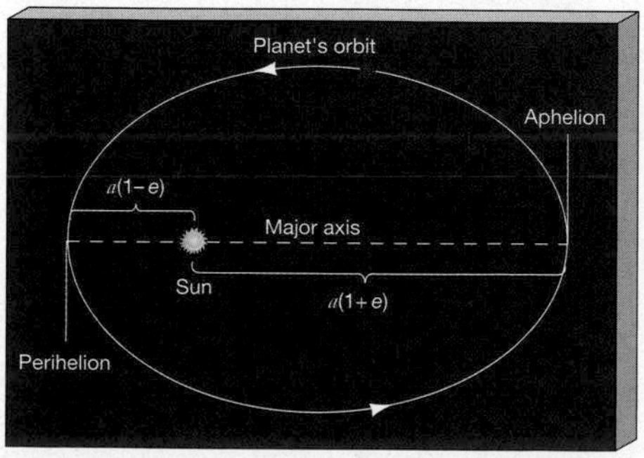

◀ **More Precisely** boxes provide more quantitative treatments of subjects discussed qualitatively in the text. Removing these more challenging topics from the main flow of the narrative and placing them within a separate modular element of the chapter design (so that they can be covered in class, assigned as supplementary material, or simply left as optional reading for those students who find them of interest) will allow instructors greater flexibility in setting the level of their coverage.

> ✓ Concept Check
> ■ How do Newton's and Einstein's theories differ in their descriptions of gravity?

◀ **Concept Checks.** New to this edition, we have incorporated into each chapter a number of "Concept Checks"—key questions that require the reader to reconsider some of the material just presented or attempt to place it into a broader context.

∞ **Cross-Links.** In astronomy, as in many ▶ scientific disciplines, almost every topic seems to have some bearing on almost every other. In particular, the connection between the specifically astronomical material and the physical principles set forth early in the text is crucial. It is important that students, when they encounter, say, Hubble's Law in Chapter 16, recall what they learned about spectral lines and the Doppler shift in Chapter 2. Similarly, the discussions of the masses of binary star components (Chapter 10) and of galactic rotation (Chapter 14) both depend on the discussion of Kepler's and Newton's laws in Chapter 1. Throughout, discussions of new astronomical objects and concepts rely heavily on comparison with topics introduced earlier in the text.

It is important to remind students of these links so that they can recall the principles on which later discussions rest and, if necessary, review them. To this end, we have inserted "cross-links" throughout the text—symbols that mark key intellectual bridges between material in different chapters. The links, denoted by the symbol ∞ together with a section reference (and a hyperlink on the accompanying CD-ROM), signal to students that the topic under discussion is related in some significant way to ideas developed earlier, and direct them to material that they might wish to review before proceeding.

> **17.4 The Geometry of Space**
>
> **4** The idea of the entire universe expanding from a point—with *nothing*, not even space and time, outside—takes a lot of getting used to. Nevertheless, it lies at the heart of modern cosmology, and few modern astronomers seriously doubt it. But this description of the universe itself (not just its contents) as a dynamic, evolving object is far beyond the capabilities of Newtonian mechanics, which we have used almost everywhere in this book. ∞ (Sec. 1.4) Instead, the more powerful techniques of Einstein's theory of general relativity, with its built-in notions of warped space and dynamical spacetime, are needed. ∞ (Sec. 13.5)
>
> The theory of general relativity states that mass (or, equivalently, energy) curves, or "warps" space in its vicinity. The greater the *total* density of the cosmos—including not just matter, but also radiation and dark energy (if any), suitably converted to mass units using the relation $E = mc^2$—the greater the curvature. ∞ (Sec. 9.5) In the universe, the curvature must be the same everywhere (assuming homogeneity), so there are really only three possibilities for the large-scale geometry of space. For more information on the different types of geometry involved, see *More Precisely 17-1*.

Chapter Summaries. The Chapter Summaries, a primary review tool for the student, have been expanded and improved for the second edition. All Key Terms introduced in each chapter are listed again, in context and in boldface, in these Summaries, along with page references to the text discussion.

Questions, Problems, and Projects. Many elements of the end-of-chapter material have seen substantial reorganization:

- Each chapter now incorporates 30 Self-Test Questions, equally divided between "true/false" and "fill-in-the-blank" formats, designed to allow students to assess their understanding of the chapter material. Answers to all these questions appear at the end of the book.
- Each chapter also has 15 Review and Discussion Questions, which may be used for in-class review or for assignment. As with the Self-Test Questions, the material needed to answer Review Questions may be found within the chapter. The Discussion Questions explore particular topics more deeply, often asking for opinions, not just facts. As with all discussions, these questions usually have no single "correct" answer.

- The end of chapter material includes a number of Problems, based on the chapter contents and entailing some numerical calculation. In this edition we have increased the number of problems to 10, have significantly broadened their range of difficulty, and in many cases have tied their contents directly to quantitative statements made (but not worked out in detail) in the text. The solutions to the problems are not contained verbatim within the chapter, but the information necessary to solve them has been presented in the text. Answers appear at the end of the book.
- Each chapter ends with a few (2–4) Projects meant to get the student out of the classroom and looking at the sky, although some entail research in libraries or other extracurricular activities.

CD-ROM

The free CD for *Astronomy: A Beginner's Guide 3/e* is included at the end of the text and contains a fully hyperlinked electronic version of the text to help the reader quickly find related information and assist in review. It also contains integrated animations and videos to bring text figures to life, and links to our companion website, which is organized by text chapter and updated monthly. The CD for this edition has been redesigned for easier and clearer navigation, and to include larger, higher-resolution videos with voiceovers. We are excited about the innovative use of media to complement the text and look forward to your response to it.

The CD-ROM material can be used on both Macintosh and PC computers using any standard browser (such as Netscape Navigator or Microsoft Explorer). For those students who do not already have a browser, Netscape Navigator 4.08 is included on the CD. A script to facilitate use of the CD under UNIX is available at

ftp://ftp.prenhall.com/pub/esm/physics.s-085/chaissonbg/

Web Site

For both teachers and students, we have created a companion website specifically for *Astronomy: A Beginner's Guide 3/e* at

http://www.prenhall.com/chaisson/bg

This powerful resource organizes material from a variety of sources on the web on a chapter-by-chapter basis, is updated monthly, and provides interactive online exercises for each chapter.

Each chapter of the website for *Astronomy: A Beginner's Guide 3/e* has the following four categories of materials:

- *Online Exercises*—interactive questions for students to answer on-line; scoring and feedback are provided immediately. The new edition features a significantly increased number of true/false and multiple choice questions.
- *Online Archives*—annotated images, videos, animations, and free downloadable software
- *Online Destinations*—annotated links to relevant websites that are regularly updated for currency and functionality
- *Multimedia Study Guide*—twenty-five questions per chapter that focus on both quantitative and conceptual understanding; many chapters also include graphical labeling exercises

Supplementary Material

This edition is accompanied by an outstanding set of instructional aids.

Comets. This is an annual update kit for *Astronomy: A Beginner's Guide, 3/e* containing videos, slides, and *New York Times* articles. The VHS tape in the Fall 2000 Comets kit includes 27 custom animations prepared by the Wright Center for Science Visualization to accompany *Astronomy: A Beginner's Guide to the Universe 3/e* and *Astronomy Today, 3/e 2000 Media Edition*. It also contains many other videos of new discoveries and animations from various sources: seven videos from the Space Telescope Science Institute including an imaginary look at a Saturn-like extrasolar planet; seven from the Jet Propulsion Laboratories, including a computer simulated fly-over of the Martian north pole; and four from the Applied Physics Laboratory including an image sequence of one rotation of the asteroid Eros. The slides, videos, and animations can be shown in class; the collection of *New York Times* articles, called *Themes of the Times*, is published twice yearly and is available free in quantity for your students using either text. A newsletter in the Comets kit provides descriptions of each video and slide. In addition, all slides, videos, and *Times* articles are cross-referenced to the appropriate chapters in both Chaisson/McMillan texts. (ISBN: 0-13-089233-5)

Instructor's Manual, by Leo Connolly (California State University at San Bernardino). This manual provides an overview of each chapter; pedagogical tips, useful analogies, and suggestions for classroom demonstrations; answers to the end-of-chapter review and discussion questions and problems; and a list of selected readings. (ISBN: 0-13-089212-2)

Presentation Manager CD. This flexible, easy-to-use tool contains a wealth of photographs, line art, animations, and videos to use in class lectures. With the *Presentation Manager* system, instructors can easily search, access, and organize the materials according to their lecture outlines and add their own visuals and lecture notes. The Presentation Manager CD contains all of the art and tables from *Astronomy: A Beginner's Guide 3/e*, as well as all animations and videos from the CD that ships with the student text. In addition, the *Presentation Manager* incorporates over 80 slides from the past four editions of Comets. [Available on one dual-platform CD (Macintosh/Windows) ISBN: 0-13-089216-5]

Acetates and Slides. A set of nearly 150 images from the text are available as a package of color acetates or 35-mm slides and are available free to qualified adopters. [(Slide set) ISBN: 0-13-089211-4; (Transparency pack) ISBN: 0-13-089200-9]

Test Item File. An extensive file of test questions, newly compiled for the third edition is offered free upon adoption. Available in both printed and electronic formats (Macintosh or IBM-compatible formats). (ISBN: 0-13-089213-0)

Prentice Hall Custom Test. Prentice Hall Custom Test is based on the powerful testing technology developed by Engineering Software Associates, Inc. (ESA). Available for Windows, Macintosh, and DOS, Prentice Hall Custom Test allows educators to create and tailor the exam to their own needs. With the Online Testing option, exams can also be administered online and data can then be automatically transferred for evaluation. A comprehensive desk reference guide is included, along with on-line assistance. [(Mac) ISBN: 0-13-089214-9; (Win) ISBN: 0-13-089215-7]

Science on the Internet: A Student's Guide, by Andrew Stull and Harry Nickla. A guide to general science resources on the Internet. Everything you need to know to get yourself online and browsing the World Wide Web! (ISBN: 0-13-028253-7)

Acknowledgments

Throughout the many drafts that have led to this book, we have relied on the critical analysis of many colleagues. Their suggestions ranged from the macroscopic issue of the book's overall organization to the minutiae of the technical accuracy of each and every sentence. We have also benefited from much good advice and feedback from users of the first edition of the text and our longer book, *Astronomy Today, Media Edition*. To these many helpful colleagues, we offer our sincerest thanks.

Reviewers of the Third Edition

John Dykla, *Loyola University–Chicago*
David J. Griffiths, *Oregon State University*
Susan Hartley, *University of Minnesota–Duluth*
Marvin Kemple, *Indiana University–Purdue University–Indianapolis*
Mario Klarić, *Midlands Technical College*
Andrew R. Lazarewicz, *Boston College*
M.A.K. Lodhi, *Texas Tech University*
Fred Marschak, *Santa Barbara College*
Edward Oberhofer, *University of North Carolina–Charlotte*
Andreas Quirrenbach, *University of California–San Diego*
Gerald Royce, *Mary Washington College*

Previous Edition Reviewers

Stephen G. Alexander, *Miami University*
Martin Goodson, *Delta College*
David J. Griffiths, *Oregon State University*
Andrew R. Lazarewicz, *Boston College*
Richard Nolthenius, *Cabrillo College*
Robert S. Patterson, *Southwest Missouri State University*
John C. Schneider, *Catonsville Community College*
Don Sparks, *Los Angeles Pierce College*
Jack W. Sulentic, *University of Alabama*

We would also like to acknowledge our gratitude to Leo Connolly for preparing the end-of-chapter questions and problems; to Ray Villard of the Space Telescope Science Institute for compiling the Comets supplement; and to Alan Sill of Texas Tech University for creating and Wesley Phillips, also of Texas Tech, for maintaining our Web site.

The publishing team at Prentice Hall has assisted us at every step along the way in creating this text.

Much of the credit for getting the project completed on time goes to our Executive Editor, Alison Reeves, who has successfully navigated us through the twists, turns, and "absolute final deadlines" of the publishing world, all the while managing the many variables that go into a multifaceted publication such as this. Don Gecewicz, our Development Editor, has skillfully helped us update and trim the manuscript. Production Editor Joanne Hakim has done a remarkable job of tying together the threads of this very complex project, made all the more complex by the necessity of combining text, art, and electronic media into a coherent whole. Mike Banino has again displayed his technical skill in managing this book's accompanying CD-ROM project. Finally, we would like to express our gratitude to renowned space artist Dana Berry for allowing us to use many of his beautiful renditions of astronomical scenes, and to Lola Judith Chaisson for assembling and drawing all the H–R diagrams (including the acetate overlays) for this edition.

MEDIA RESOURCES

While almost all books in astronomy now have web sites, CD-ROMs, videos, and animations, our mission is to integrate all of these together in a more meaningful way for the student.

The videos and other material are intended to be viewed by the student in the context of text discussions. Web resources are therefore organized by chapter, both to expand students' knowledge of a given topic and to save time in locating relevant material. The text is fully hyperlinked, to encourage students to follow their natural curiosity about astronomy and to help them explore this fascinating subject.

The following is a brief tour of how the Media Edition CD-ROM in the back of the text pulls together text, web, videos, and animations to help students take full advantage of these resources.

The navigation window on the left provides an icon for each part of the text. Clicking on one of the icons brings up a list of the chapters in the right-hand frame.

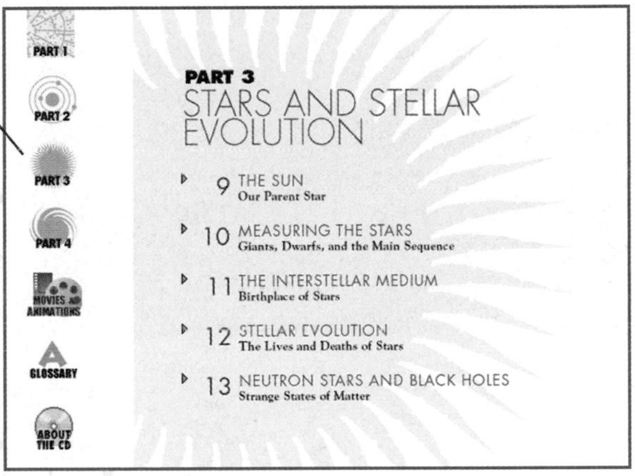

When you click on a given chapter, the complete chapter text loads in the right window and the section-level table of contents for the chapter expands in the left column. The part icons and chapters for other parts of the book remain resident in the left-hand frame so you can navigate easily within the chapter and/or throughout the book.

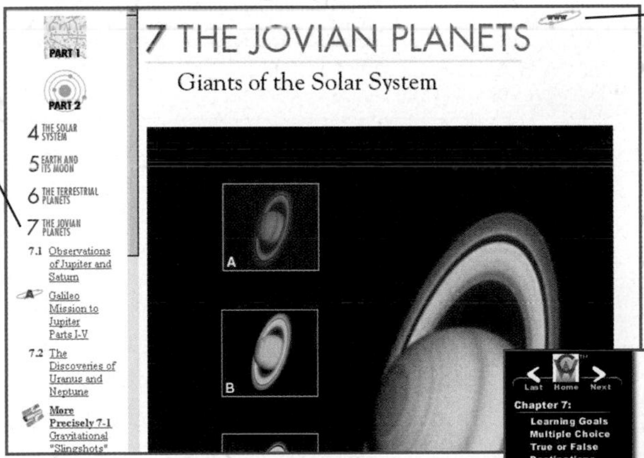

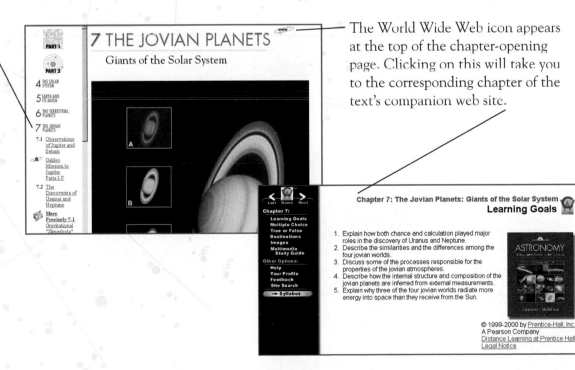

The World Wide Web icon appears at the top of the chapter-opening page. Clicking on this will take you to the corresponding chapter of the text's companion web site.

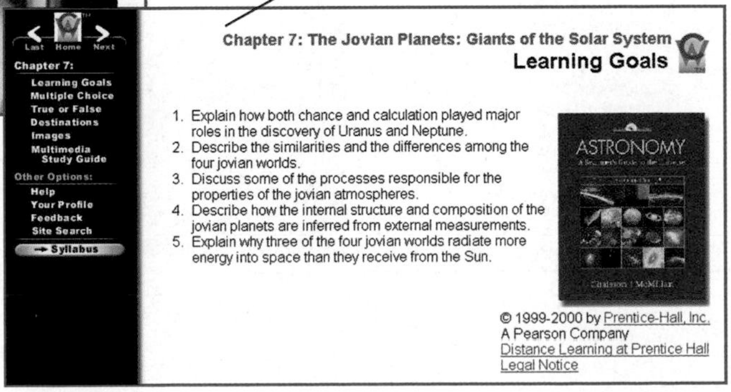

For each text chapter, the web site has self-scoring multiple choice and true/false questions; links to associated web sites (Destinations); annotated images and animations; and a multimedia study guide. The Destinations, Images, and Animations modules are updated on a monthly basis both to repair broken links and to add material on new events and discoveries.

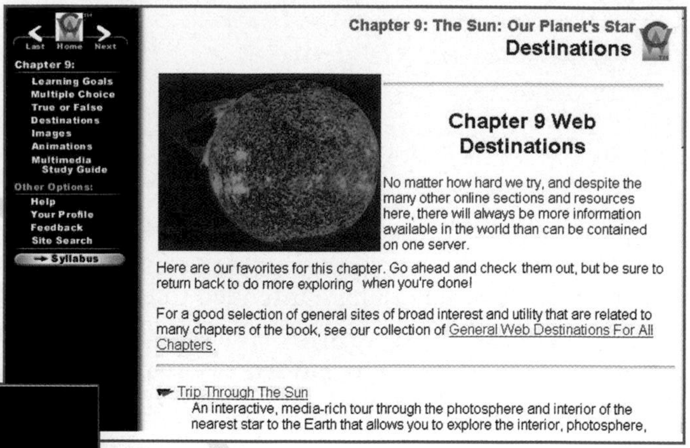

Destinations

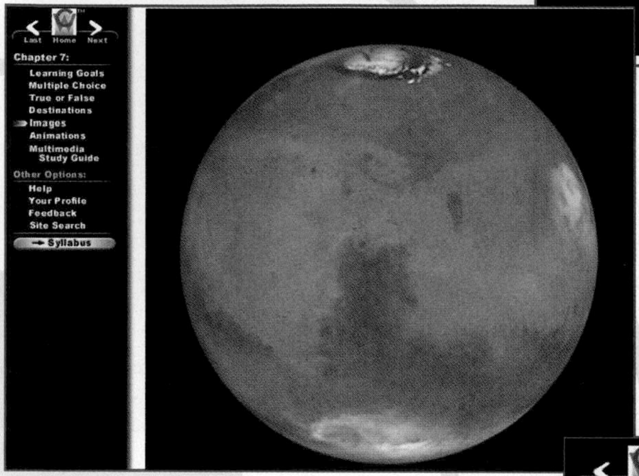

Images

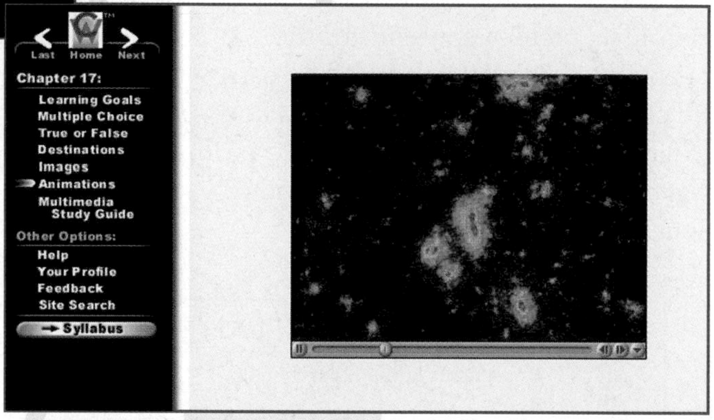

Animations

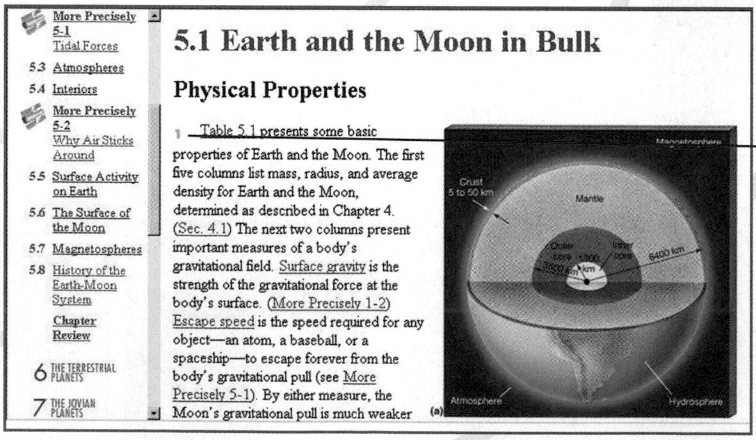

Each text chapter begins with learning objectives. In the printed text, discussions related to that objective are indicated by the corresponding learning objective number in blue type. In the CD-ROM version of the text, the chapter-opening learning objectives are hyperlinked so that you can go directly from one to the other.

Video icons in the printed text indicate the availability of a video or animation to help explain text discussions and illustrations. In the CD-ROM version of the text, these videos are listed in the contents for each chapter. Clicking on the video name will bring you to the corresponding place in the text where you can look at the text figure and watch the animation at the same time. All videos and animations now have audio narration.

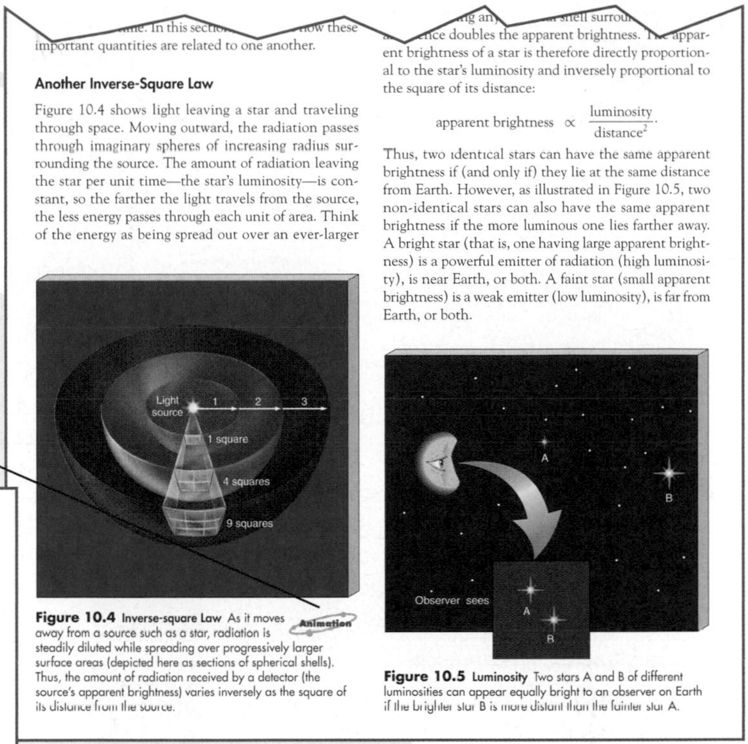

important quantities are related to one another.

Another Inverse-Square Law

Figure 10.4 shows light leaving a star and traveling through space. Moving outward, the radiation passes through imaginary spheres of increasing radius surrounding the source. The amount of radiation leaving the star per unit time—the star's luminosity—is constant, so the farther the light travels from the source, the less energy passes through each unit of area. Think of the energy as being spread out over an ever-larger

...ence doubles the apparent brightness. The apparent brightness of a star is therefore directly proportional to the star's luminosity and inversely proportional to the square of its distance:

$$\text{apparent brightness} \propto \frac{\text{luminosity}}{\text{distance}^2}.$$

Thus, two identical stars can have the same apparent brightness if (and only if) they lie at the same distance from Earth. However, as illustrated in Figure 10.5, two non-identical stars can also have the same apparent brightness if the more luminous one lies farther away. A bright star (that is, one having large apparent brightness) is a powerful emitter of radiation (high luminosity), is near Earth, or both. A faint star (small apparent brightness) is a weak emitter (low luminosity), is far from Earth, or both.

Figure 10.4 **Inverse-square Law** As it moves away from a source such as a star, radiation is steadily diluted while spreading over progressively larger surface areas (depicted here as sections of spherical shells). Thus, the amount of radiation received by a detector (the source's apparent brightness) varies inversely as the square of its distance from the source.

Figure 10.5 **Luminosity** Two stars A and B of different luminosities can appear equally bright to an observer on Earth if the brighter star B is more distant than the fainter star A.

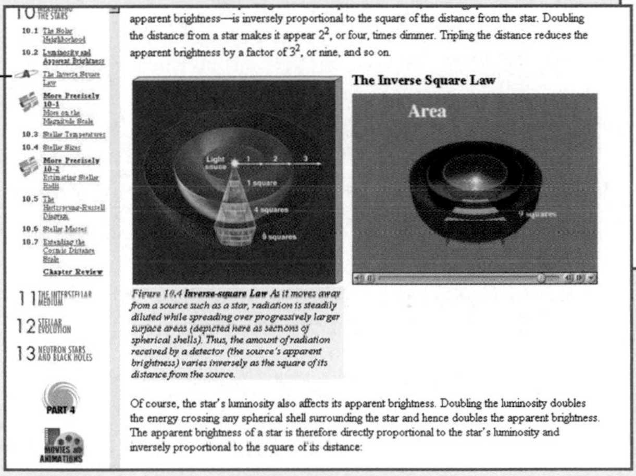

apparent brightness—is inversely proportional to the square of the distance from the star. Doubling the distance from a star makes it appear 2^2, or four, times dimmer. Tripling the distance reduces the apparent brightness by a factor of 3^2, or nine, and so on.

The Inverse Square Law

*Figure 10.4 **Inverse-square Law** As it moves away from a source such as a star, radiation is steadily diluted while spreading over progressively larger surface areas (depicted here as sections of spherical shells). Thus, the amount of radiation received by a detector (the source's apparent brightness) varies inversely as the square of its distance from the source.*

Of course, the star's luminosity also affects its apparent brightness. Doubling the luminosity doubles the energy crossing any spherical shell surrounding the star and hence doubles the apparent brightness. The apparent brightness of a star is therefore directly proportional to the star's luminosity and inversely proportional to the square of its distance:

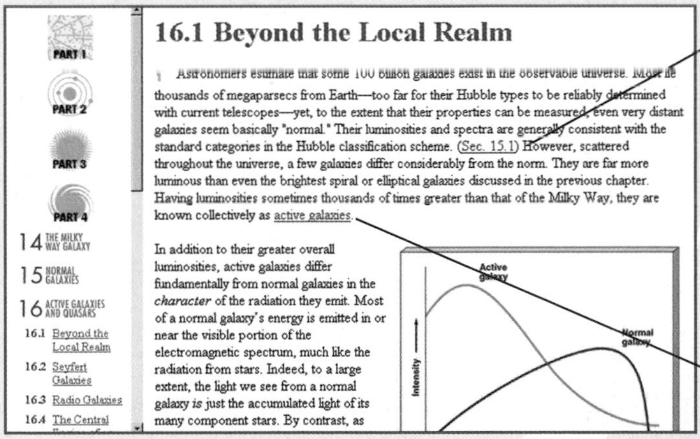

16.1 Beyond the Local Realm

Astronomers estimate that some 100 billion galaxies exist in the observable universe. Most lie thousands of megaparsecs from Earth—too far for their Hubble types to be reliably determined with current telescopes—yet, to the extent that their properties can be measured, even very distant galaxies seem basically "normal." Their luminosities and spectra are generally consistent with the standard categories in the Hubble classification scheme. (Sec. 15.1) However, scattered throughout the universe, a few galaxies differ considerably from the norm. They are far more luminous than even the brightest spiral or elliptical galaxies discussed in the previous chapter. Having luminosities sometimes thousands of times greater than that of the Milky Way, they are known collectively as active galaxies.

In addition to their greater overall luminosities, active galaxies differ fundamentally from normal galaxies in the *character* of the radiation they emit. Most of a normal galaxy's energy is emitted in or near the visible portion of the electromagnetic spectrum, much like the radiation from stars. Indeed, to a large extent, the light we see from a normal galaxy *is* just the accumulated light of its many component stars. By contrast, as

Cross-link icons tell you that a related text discussion is located in another part of the text. Often these link materials in later chapters to the underlying physics principles discussed earlier in the text. In the CD-ROM version of the text, the cross-links are hyperlinked so that you can go directly to the relevant discussion.

All glossary terms in the chapter narrative are linked to the end-of-book glossary definition.

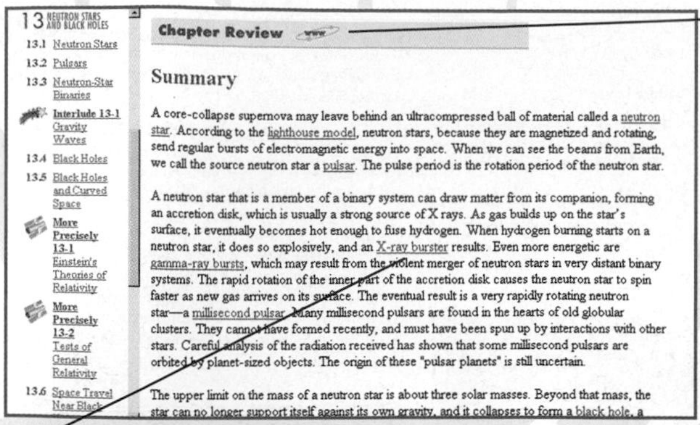

At each end-of-chapter review, clicking on the WWW icon again brings you to the corresponding web chapter. At this point, students may want to take advantage of the self-scoring multiple-choice and true/false quizzes. There are approximately 15 multiple choice and 15 true/false questions for each chapter.

Clicking on the glossary term in the chapter summary will take you to the place in the text where they are discussed.

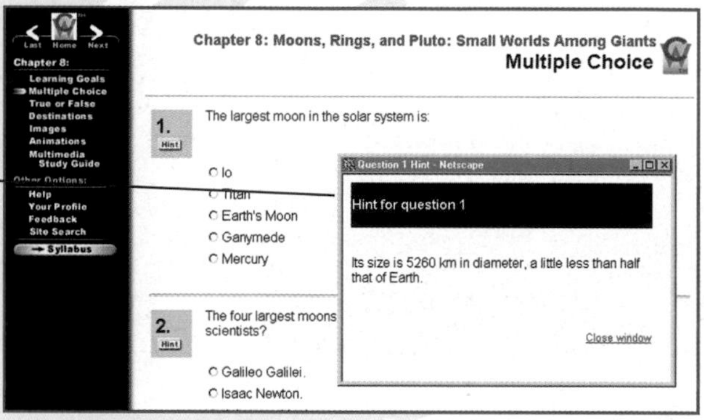

Hints are provided for each question, but if you still get the answer wrong, the *wrong answer feedback* in the *scored results reporter* will tell you what section in the text you need to review. In addition, each chapter has a Multimedia Study Guide that includes 25 detailed multiple-choice questions, and, in many chapters, labeling exercises that help you visually test your knowledge.

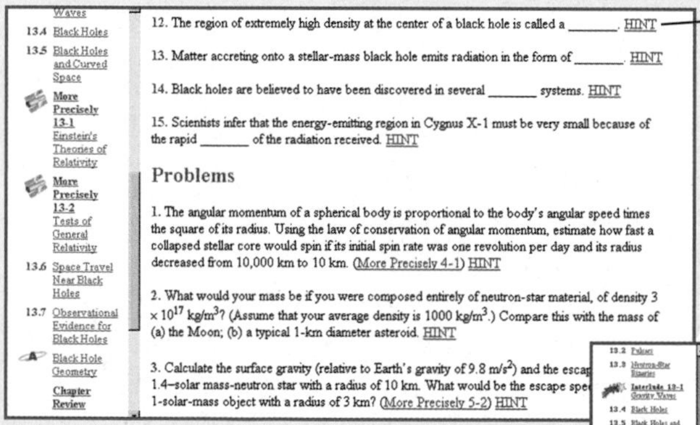

Finally, all end-of-chapter questions and problems have hint buttons which take you to the place in the text where the topic for that question is discussed.

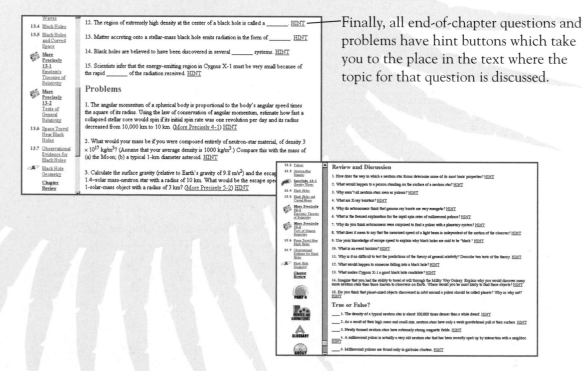

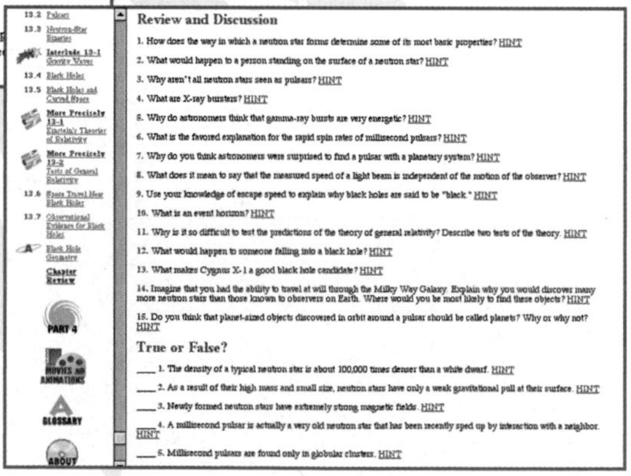

ASTRONOMY

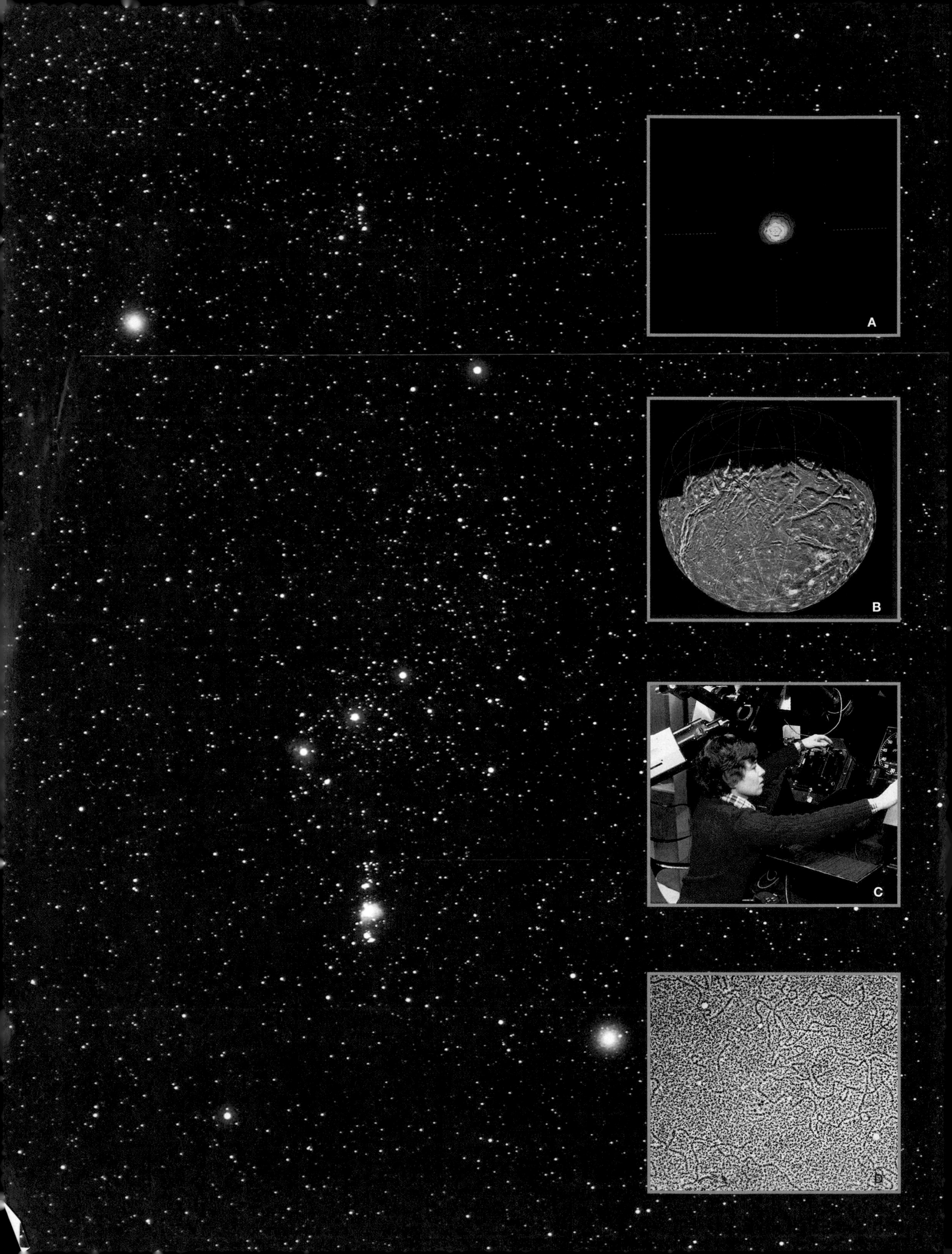
A
B
C
D

PROLOGUE

Charting the Heavens

LEARNING GOALS

Studying this prologue will enable you to:

1 Describe the concept of the celestial sphere and the conventions of angular measurement that enable us to locate objects in the sky.

2 Account for the apparent motions of the Sun, Moon, and stars in terms of the actual motions of Earth and the Moon.

3 Show how the relative motions of Earth, the Sun, and the Moon lead to eclipses.

4 Explain the simple geometric reasoning that allows astronomers to measure the distances and sizes of faraway objects.

(Opposite page, background) One of the most easily recognizable star fields in the winter nighttime sky, the familiar constellation Orion. This field of view spans roughly 100 light-years, or 10^{15} kilometers. (See also Figure P.6.) *(J. Sanford/Astrostock-Sanford)*

(Inset A) If we magnify the wide view of the Orion constellation, shown at left, by a million times, we enter into the realm of the largest stars, with diameters of about a billion kilometers. Such a star is seen in this false-color image of the red-giant star Betelgeuse (which is actually the bright star at the upper left of the Orion constellation). *(AURA)*

(Inset B) Another magnification of a million brings us to the scale of typical moons—roughly 1000 kilometers—represented here by Ariel, one of the many moons of Uranus. *(NASA)*

(Inset C) With yet another million-times magnification, we reach scales of meters, represented here by an astronomer at the controls of her telescope. *(AURA)*

(Inset D) At a final magnification of an additional million, we reach the scale of molecules (about 10^{-6} meter), represented by this coiled DNA molecule of a rat's liver. *(Harvard Medical School)*

Nature offers no greater splendor than the starry sky on a clear, dark night. Silent, jeweled with the constellations of ancient myth and legend, the night sky has inspired wonder throughout the ages—a wonder that leads our imaginations far from the confines of Earth and out into the distant reaches of space and time. Astronomy, born in response to that wonder, is built on two basic traits of human nature: the need to explore and the need to understand. Through the interplay of curiosity, discovery, and analysis, people have sought answers to questions about the universe since the earliest times. Astronomy is the oldest of all the sciences, yet never has it been more exciting than it is today.

P.1 Our Place in Space

Of all the scientific insights achieved to date, one stands out boldly: Earth is neither central nor special. We inhabit no unique place in the universe. Astronomical research, especially within the past few decades, strongly suggests that we live on an ordinary rocky *planet* called Earth, one of nine known planets orbiting an average *star* called the Sun, a middle-aged star near the edge of a huge collection of stars called the Milky Way *galaxy*, one galaxy among countless billions of others spread throughout the observable *universe*. To get a feel for these relationships, consult Figures P.1 through P.4; put them in perspective by studying Figure P.5.

Simply put, the **universe** is the totality of all space, time, matter, and energy. **Astronomy** is the study of the universe. It is a subject unlike any other, for it requires us to change profoundly our view of the cosmos and to consider matter on scales totally unfamiliar from everyday experience. Look again at the galaxy shown in Figure P.3. It is a swarm of about a hundred billion stars—more stars than the number of people who have ever lived on Earth. The entire assemblage is spread across a vast expanse of space some 100,000 **light-years** across. What is a light-year? It is the *distance* traveled by light, moving at a speed of about 300,000 kilometers per second, in a year. One light-year equals about 10 trillion kilometers (around 6 trillion miles)—astronomical systems are truly "astronomical" in size! The light-year is a convenient unit introduced by astronomers to help them describe immense distances. We will encounter many more such "custom" units in our studies (see Appendix 1).

A thousand (1000), a million (1,000,000), a billion (1,000,000,000), a trillion (1,000,000,000,000)—let's take a moment to understand the magnitude of these numbers. One thousand is easy enough to understand: At the rate of one number per second, you could count to a thousand in a little over 16 minutes. However, if you wanted to count to a million, you would need more than two weeks of counting at the rate of one number per second, 16 hours per day (allowing eight hours per day for sleep). To count from one to a billion at the same rate of one number per second for 16 hours per day would take nearly 50 years. In this text we consider spatial domains spanning not just billions of kilometers but billions of light-years, objects containing not just trillions of atoms but trillions of stars, time intervals of not just billions of seconds or hours but billions of years. You will need to be-

Figure P.1 Earth Earth is a planet, a mostly solid object, though it has some liquid in its oceans and its core, and gas in its atmosphere. (In this view, you can clearly see the North and South American continents.) *(NASA)*

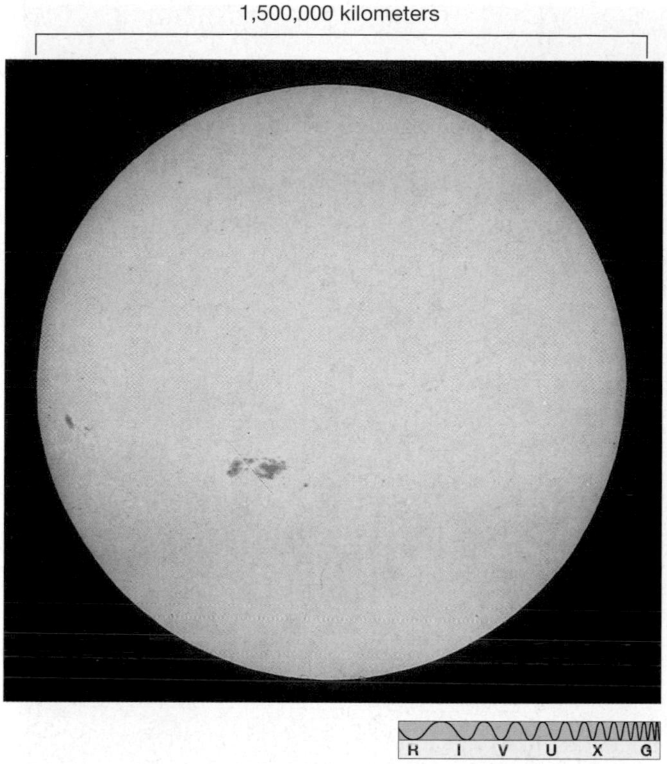

Figure P.2 The Sun Much bigger than Earth, the Sun is a very hot ball of gas held together by its own gravity. *(AURA)*

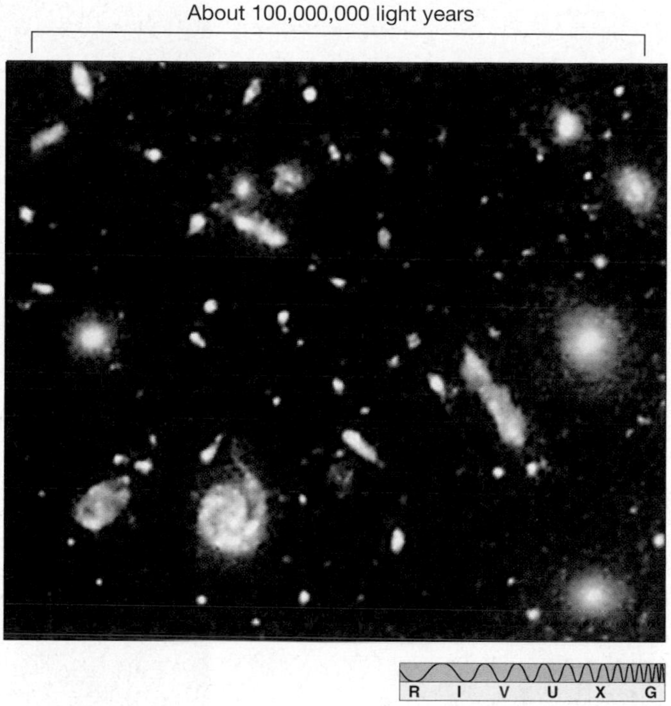

Figure P.4 Galaxy Cluster A portion of a typical cluster of galaxies, roughly a billion light-years from Earth. Each galaxy contains hundreds of billions of stars, probably a comparable number of planets, and possibly living creatures. *(R. Williams/NASA)*

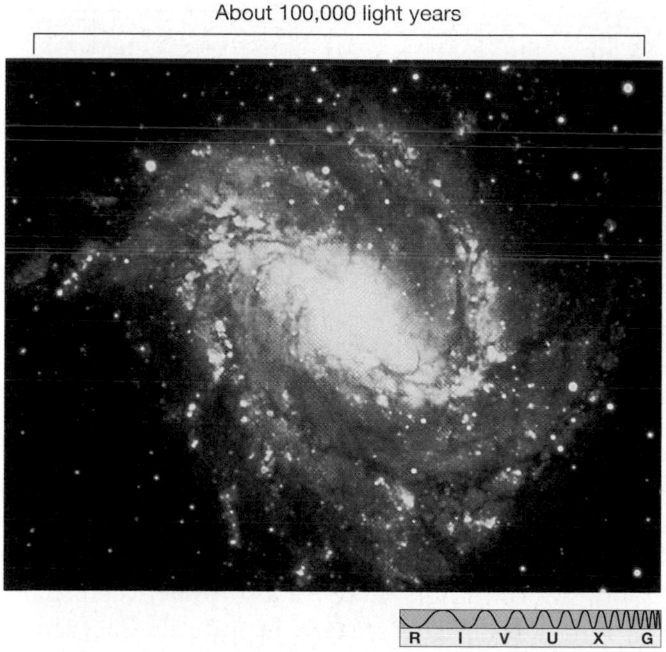

Figure P.3 Galaxy A typical galaxy is a collection of a few hundred billion stars, each separated by vast regions of nearly empty space. This galaxy is M83, the 83rd entry in the catalog compiled by the eighteenth-century French astronomer Charles Messier. Our Sun is an undistinguished star near the edge of another galaxy, the Milky Way. *(R. DuFour/Rice University)*

come familiar with—and comfortable with—such enormous numbers. A good way to start is to recognize just how much larger than a thousand is a million and how much larger still is a billion. (Appendix 2 explains the convenient method used by scientists for writing and manipulating very large and very small numbers.)

P.2 The "Obvious" View

How have we come to know the universe around us? How have astronomers determined the perspective sketched in Figure P.5? Our study of the universe, the science of astronomy, begins by examining the sky.

Constellations in the Sky

Between sunset and sunrise on a clear night, we can see some 3000 points of light. Include the view from the opposite side of Earth, and nearly 6000 stars are visible to the unaided eye. A natural human tendency is to see patterns and relationships between objects even when

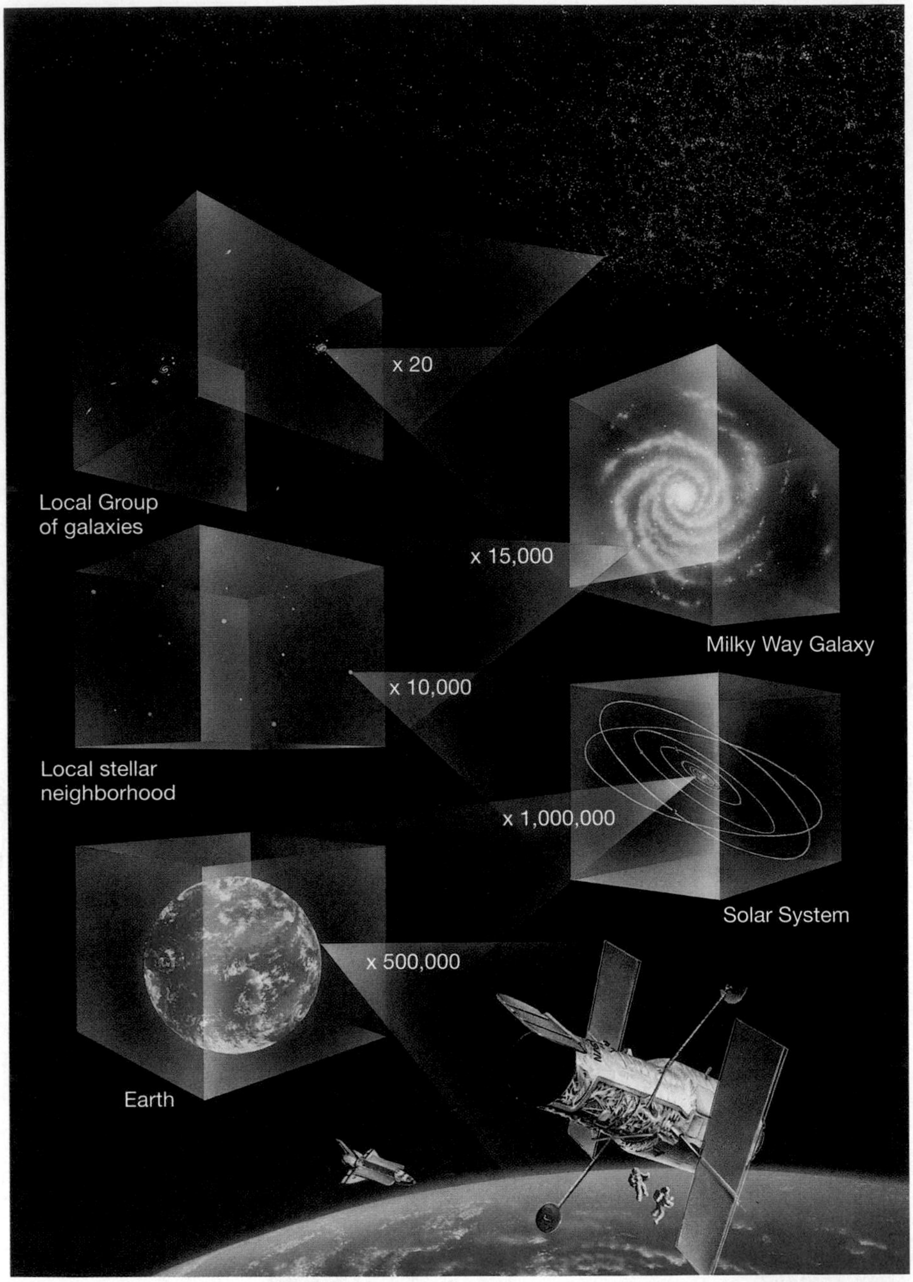

Figure P.5 Size and Scale This artist's conception puts the previous four illustrations in perspective. The bottom shows spacecraft (and astronauts) in Earth orbit, a view that widens progressively from bottom to top in the next five cubes: Earth, the Solar System, the Local Stellar Neighborhood, the Milky Way Galaxy, and the Local Group of Galaxies. The numbers indicate approximately the increase in scale between successive images. *(D. Berry)*

no true connection exists, and people long ago connected the brightest stars into configurations called **constellations**, which ancient astronomers named after mythological heroes and animals. Figure P.6 shows a constellation especially prominent in the northern night sky from October through March: the hunter Orion, named for a mythical Greek hero famed, among other things, for his amorous pursuit of the Pleiades, the seven daughters of the giant Atlas. According to Greek mythology, in order to protect the Pleiades from Orion, the gods placed them among the stars, where Orion now nightly stalks them across the sky. Many other constellations have similarly fabulous connections with ancient cultures.

The stars making up a particular constellation are generally not close together in space. They merely are bright enough to observe with the naked eye and happen to lie in the same direction in the sky as seen from Earth. Figure P.6(c) illustrates this point for Orion, showing the true relationships between that constellation's brightest stars. Despite the fact that constellation patterns have no real significance, the terminology is still used today. Constellations provide a convenient means for astronomers to specify large areas of the sky, much as geologists use continents or politicians use voting precincts to identify certain localities on Earth. In all, there are 88 constellations,

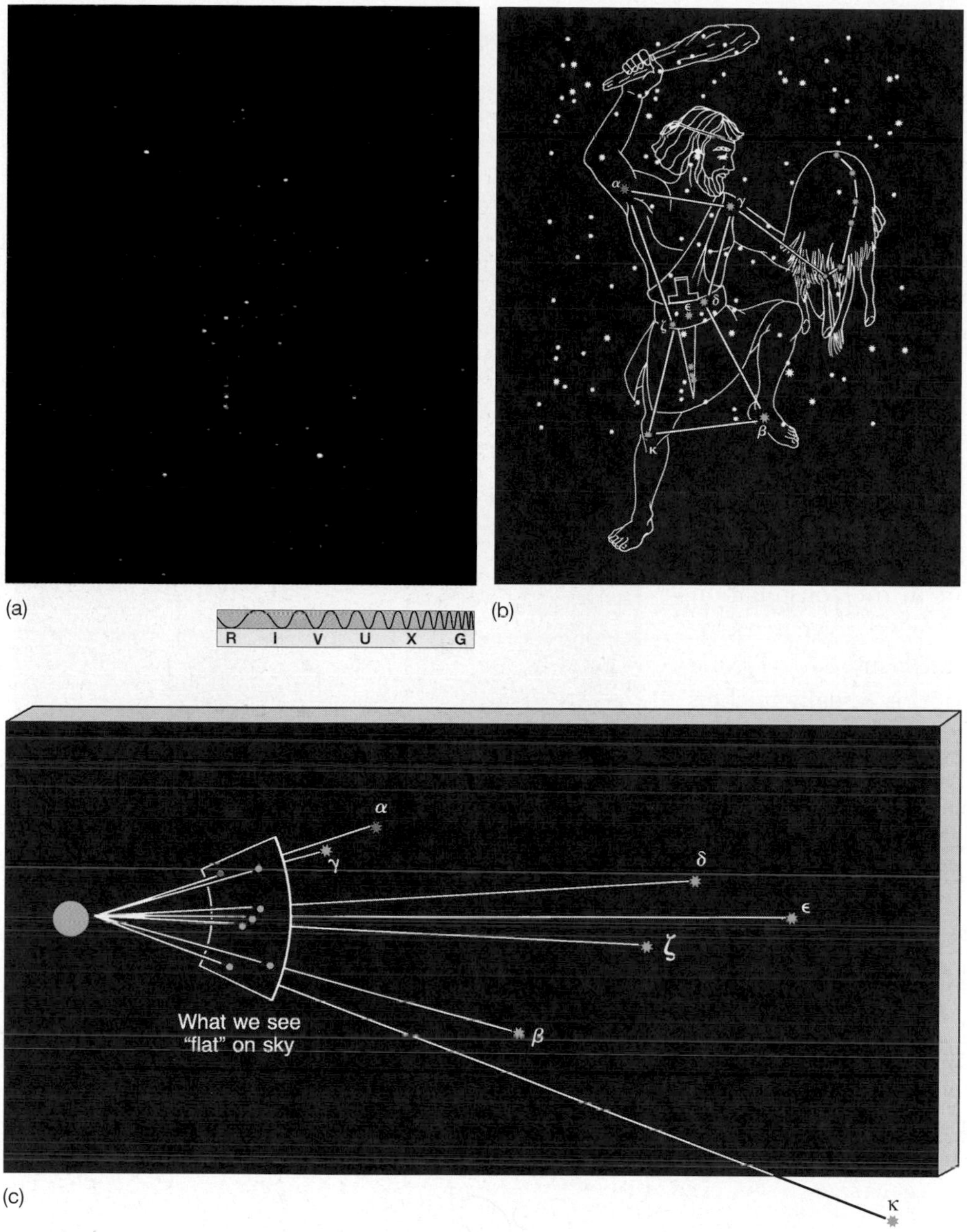

Figure P.6 The Constellation Orion (a) A photograph of the group of bright stars that make up the constellation Orion. (b) The stars connected to show the pattern visualized by the Greeks: the outline of a hunter. You can easily find this constellation in the northern winter sky by identifying the line of three bright stars in the hunter's belt. (c) The three-dimensional relationships for the prominent stars in Orion. The Greek letters are astronomical notations indicating brightness. *(S. Westphal)*

most of them visible from North America at some time during the year.

The Celestial Sphere

1 Over the course of a night, the constellations appear to move across the sky from east to west. However, ancient sky-watchers noted that the *relative* positions of stars remained unchanged as this nightly march took place. It was natural for those first astronomers to conclude that the stars were attached to a **celestial sphere** surrounding Earth—a canopy of stars resembling an astronomical painting on a heavenly ceiling. Figure P.7 shows how early astronomers pictured the stars as moving with this celestial sphere as it turned around a fixed, unmoving Earth. Figure P.8 shows how stars appear to move in circles around a point in the sky very close to the star Polaris (better known as the Pole Star or the North Star). To the ancients, this point represented the axis around which the celestial sphere turned.

From our modern standpoint, the apparent motion of the stars is the result of the spin, or **rotation**, not of the celestial sphere but of Earth. Polaris indicates the direction—due north—in which Earth's rotation axis points. Even though we now know that a revolving celestial sphere is an incorrect description of the heavens, we still use the idea as a convenient fiction that helps us visualize the positions of stars in the sky. The point where Earth's axis intersects the celestial sphere in the Northern Hemisphere is known as the **north celestial pole**; it is directly above Earth's North Pole. In the Southern Hemisphere, the extension of Earth's axis in the opposite direction defines the **south celestial pole**. Midway between the north and south celestial poles lies the **celestial equator**, representing the intersection of Earth's equatorial plane with the celestial sphere.

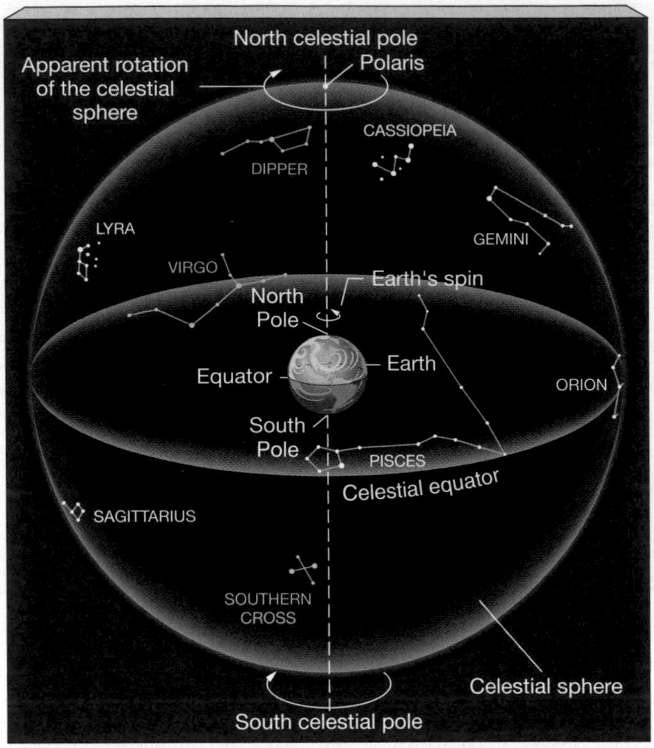

Figure P.7 The Celestial Sphere Planet Earth sits fixed at the hub of the celestial sphere, which contains all the stars. This is one of the simplest possible models of the universe, but it doesn't agree with the facts that astronomers now know about the cosmos.

Celestial Coordinates

The simplest method of locating stars in the sky is to specify their constellation and then rank the stars in it in order of brightness. The brightest star is denoted by the Greek letter α (alpha), the second brightest by β (beta), and so on. Thus, the two brightest stars in the constellation Orion—Betelgeuse and Rigel—are also known as α Orionis and β Orionis, respectively (see Figure P.6). (Precise recent observations show that Rigel is actually brighter than Betelgeuse, but the names are now permanent.) Because there are many more stars in any given constellation than there are letters in the Greek alphabet, this method is of limited utility. However, for naked-eye astronomy, where only bright stars are involved, it is quite satisfactory.

For more precise measurements, astronomers find it helpful to lay down a system of **celestial coordinates** on the sky. If we think of the stars as being attached to the celestial sphere centered on Earth, then the familiar system of latitude and longitude on Earth's surface extends quite naturally to the sky. The celestial analogs of latitude and longitude are called **declination** and **right ascension**, respectively (Figure P.9). Just as latitude and longitude are tied to Earth, right ascension and declination are fixed on the celestial sphere. Although the stars appear to move across the sky because of Earth's rotation, their celestial coordinates remain constant over the course of a night.

Declination (dec) is measured in *degrees* (°) north or south of the celestial equator, just as latitude is mea-

Figure P.8 The Northern Sky A time-lapse photograph of the northern sky. Each trail is the path of a single star across the night sky. The duration of the exposure is about five hours. (How can you tell that this is so?) The center of the concentric circles is near the North Star, Polaris, whose short, bright arc is prominently visible. *(AURA)*

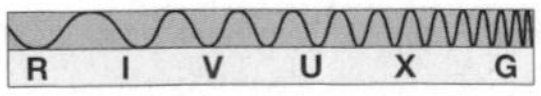

sured in degrees north or south of Earth's equator (see *More Precisely P-1*). Thus, the celestial equator is at a declination of 0°, the north celestial pole is at +90°, and the south celestial pole is at −90° (the minus sign here just means "south of the celestial equator"). Right ascension (RA) is measured in angular units called *hours*, *minutes*, and *seconds*, and it increases in the eastward direction. Like the choice of the Greenwich Meridian as the zero-point of longitude on Earth, the choice of zero right ascension is quite arbitrary—it is conventionally taken to be the position of the Sun in the sky at the instant of the vernal equinox (to be discussed in the next section).

✓ Concept Check

- If Earth is not really enclosed in a sphere with stars attached, why do astronomers find it convenient to retain the fiction of the celestial sphere when describing the sky?

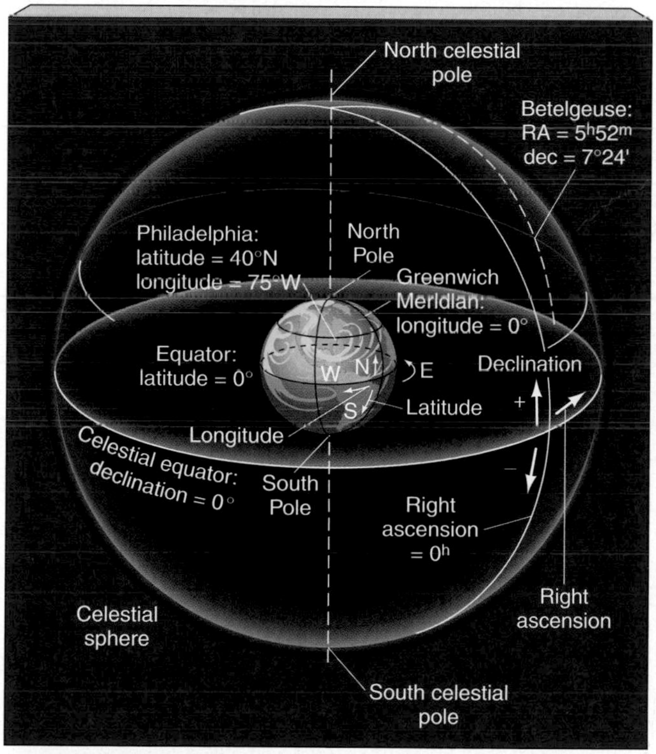

Figure P.9 Right Ascension and Declination Just as longitude and latitude allow us to locate a point on the surface of Earth, right ascension and declination specify locations on the sky. For example, to find Philadelphia on Earth, look 75° west of the Greenwich Meridian and 40° north of the Equator. Similarly, to locate the star Betelgeuse on the celestial sphere, look 5^h52^m east of the vernal equinox (the line on the sky with a right ascension of zero) and 7°24′ north of the celestial equator.

P.3 Earth's Orbital Motion

Day-to-Day Changes

2 We measure time by the Sun. The rhythm of day and night is central to our lives, so it is not surprising that the period of time from one sunrise (or noon, or sunset) to the next, the 24-hour **solar day**, is our basic social time unit. As we have just seen, this apparent daily progress of the Sun and other stars across the sky (known as *diurnal motion*) is a consequence of Earth's rotation. But the stars' positions in the sky do not repeat themselves exactly from one night to the next. Each night, the whole celestial sphere appears shifted a little compared with the night before—you can confirm this for yourself by noting over the course of a week or two which stars are visible near the horizon just after sunset or just before dawn. Because of this shift, a day measured by the stars—called a **sidereal day** after the Latin word *sidus*, meaning "star"—differs in length from a solar day.

The reason for the difference in length between a solar day and a sidereal day is sketched in Figure P.10. It is a result of the fact that Earth moves in two ways simultaneously: It rotates on its central axis while at the

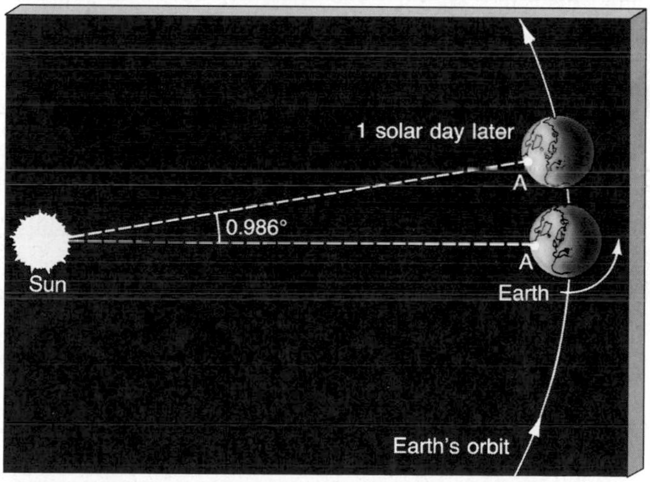

Figure P.10 Solar and Sidereal Days The difference in length between a solar and a sidereal day can be easily explained once we understand that Earth revolves around the Sun at the same time as it rotates on its axis. A solar day is the time from one noon to the next. In that time, Earth also moves a little in its solar orbit. Because Earth completes one circuit (360°) around the Sun in one year (365 days), it moves through nearly 1° in one day. Thus, between noon at point A on one day and noon at the same point the next day, Earth rotates through about 361°. Consequently, the solar day exceeds the sidereal day (360° rotation) by about four minutes. Note that the diagram is not drawn to scale; the true 1° angle is in reality much smaller than shown here.

MORE PRECISELY P-1

Angular Measure

The size and scale of things are often specified by measuring lengths and angles. The concept of length measurement is fairly intuitive. The concept of angular measurement may be less familiar, but it too can become second nature if you remember a few simple facts:

- A full circle contains 360 arc degrees (or just 360°). Therefore, the half-circle that stretches from horizon to horizon, passing directly overhead and spanning the portion of the sky visible to one person at any one time, contains 180°.
- Each 1° increment can be further subdivided into fractions of an arc degree, called arc minutes; there are 60 arc minutes (60′) in one arc degree. Both the Sun and the Moon project an angular size of 30 arc minutes on the sky. Your little finger, held at arm's length, does about the same, covering about a 40-arc-minute slice of the 180° horizon-to-horizon arc.
- An arc minute can be divided into 60 arc seconds (60″). Put another way, an arc minute is 1/60 of an arc degree, and an arc second is $1/60 \times 1/60 = 1/3600$ of an arc degree. An arc second is an extremely small unit of angular measure—it is the angular size of a centimeter-sized object (a dime, say) at a distance of about 2 kilometers (a little over a mile).

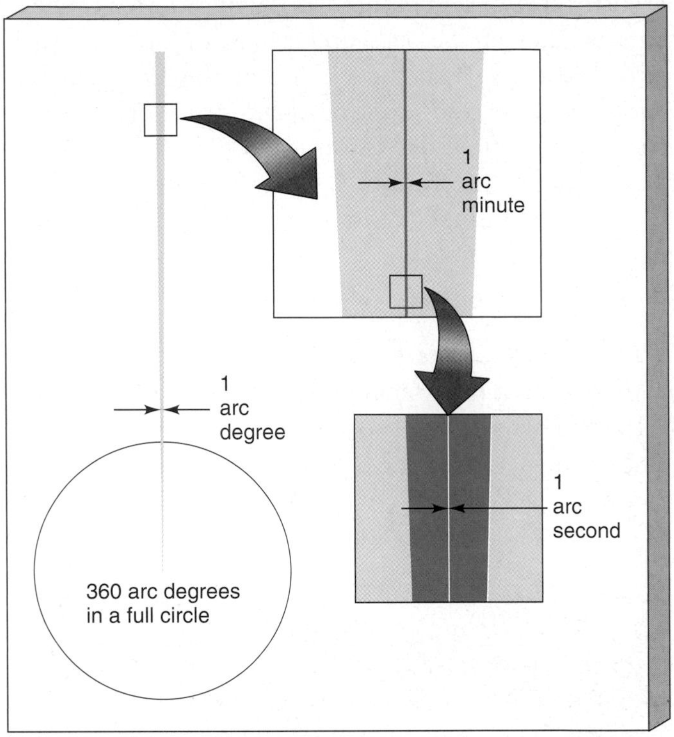

The accompanying figure illustrates this subdivision of the circle into progressively smaller units.

One final note: Arc degrees have nothing to do with temperature, and arc minutes and arc seconds have nothing to do with time. However, the angular units used to measure right ascension (and only right ascension)—hours (^{h}), minutes (^{m}), and seconds (^{s})—*are* constructed to parallel the units of time, the connection being provided by Earth's rotation. In 24 hours, Earth rotates once on its axis, or through 360°. Thus, in a time interval of one hour, Earth rotates through $360°/24 = 15°$, or 1^h. In one minute of time, Earth rotates through 1^m; in one second, Earth rotates through 1^s. Just bear in mind that these units refer only to right ascension and you should avoid undue confusion.

same time **revolving** around the Sun. Each time Earth rotates once on its axis, it also moves a small distance along its orbit. Each day, therefore, Earth has to rotate through slightly more than 360° in order for the Sun to return to the same apparent location in the sky. Thus, the interval of time between noon one day and noon the next (a solar day) is slightly greater than the true rotation period (one sidereal day). Our planet takes 365 days to orbit the Sun, so the additional angle is $360°/365 = 0.986°$. Because Earth takes about 3.9 minutes to rotate through this angle, the solar day is 3.9 minutes longer than the sidereal day.

Seasonal Changes

Because Earth revolves around the Sun, our planet's darkened hemisphere faces in a slightly different direction each night. The change is only about 1° per night (Figure P.10)—too small to be easily discerned with the naked eye from one evening to the next, but clearly noticeable over the course of weeks and months, as illustrated in Figure P.11. In six months, Earth moves to the opposite side of its orbit, and we face an entirely different group of stars and constellations at night. Because of this motion, the Sun appears, to an observer on Earth,

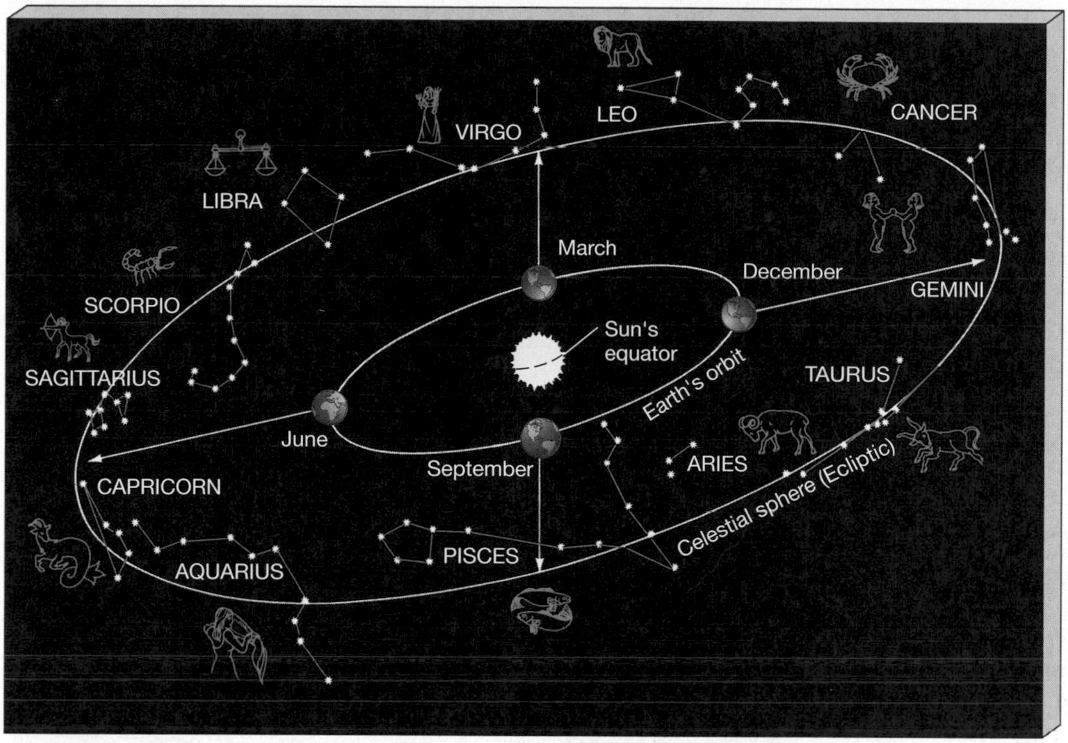

Figure P.11 The Zodiac The view of the night sky changes as Earth moves in its orbit about the Sun. As drawn here, the night side of Earth faces a different set of constellations at different times of the year. The twelve constellations named here make up the astrological zodiac.

to move relative to the background stars over the course of a year. This apparent motion of the Sun on the sky traces out a path on the celestial sphere known as the **ecliptic**. The 12 constellations through which the Sun passes as it moves along the ecliptic—that is, the constellations we would see looking in the direction of the Sun if they weren't overwhelmed by the Sun's light—had special significance for astrologers of old. They are collectively known as the **zodiac**.

As illustrated in Figure P.12(a), the ecliptic forms a great circle on the celestial sphere, inclined at an angle of about 23.5° to the celestial equator. In reality, as illustrated in Figure P.12(b), the plane defined by the ecliptic is *the plane of Earth's orbit around the Sun*. Its tilt is a consequence of the *inclination* of our planet's rotation axis to its orbital plane.

The point on the ecliptic where the Sun is at its northernmost point above the celestial equator is known as the **summer solstice** (from the Latin words *sol*, meaning "sun," and *stare*, "to stand"). As shown in Figure P.12, it represents the point in Earth's orbit where our planet's North Pole points closest to the Sun. This occurs on or near June 21—the exact date varies slightly from year to year because the actual length of a year is not a whole number of days. As Earth rotates on that date, points north of the equator spend the greatest fraction of their time in sunlight, so the summer solstice corresponds to the longest day of the year in Earth's Northern Hemisphere and the shortest day in Earth's Southern Hemisphere.

Six months later, the Sun is at its southernmost point below the celestial equator, and we have reached the **winter solstice** (December 21)—the shortest day in Earth's Northern Hemisphere and the longest in the Southern Hemisphere. These two effects—the height of the Sun above the celestial equator and the length of the day—combine to account for the **seasons** we experience. In northern summer, the Sun is high in the sky and the days are long, with the result that temperatures are generally much higher than in winter, when the Sun is low and the days are short.

The two points where the ecliptic intersects the celestial equator are known as **equinoxes**. On those dates, day and night are of equal duration. (The word *equinox* derives from the Latin for "equal night.") In the fall (in Earth's northern hemisphere), as the Sun crosses from the northern into the southern celestial hemisphere, we have the **autumnal equinox** (on September 21). The **vernal equinox** occurs in spring, on or near March 21, as the Sun crosses the celestial equator moving north. The vernal equinox plays an important role in human timekeeping. The interval of time from one vernal equinox to the next—365.242 solar days—is known as one **tropical year**.

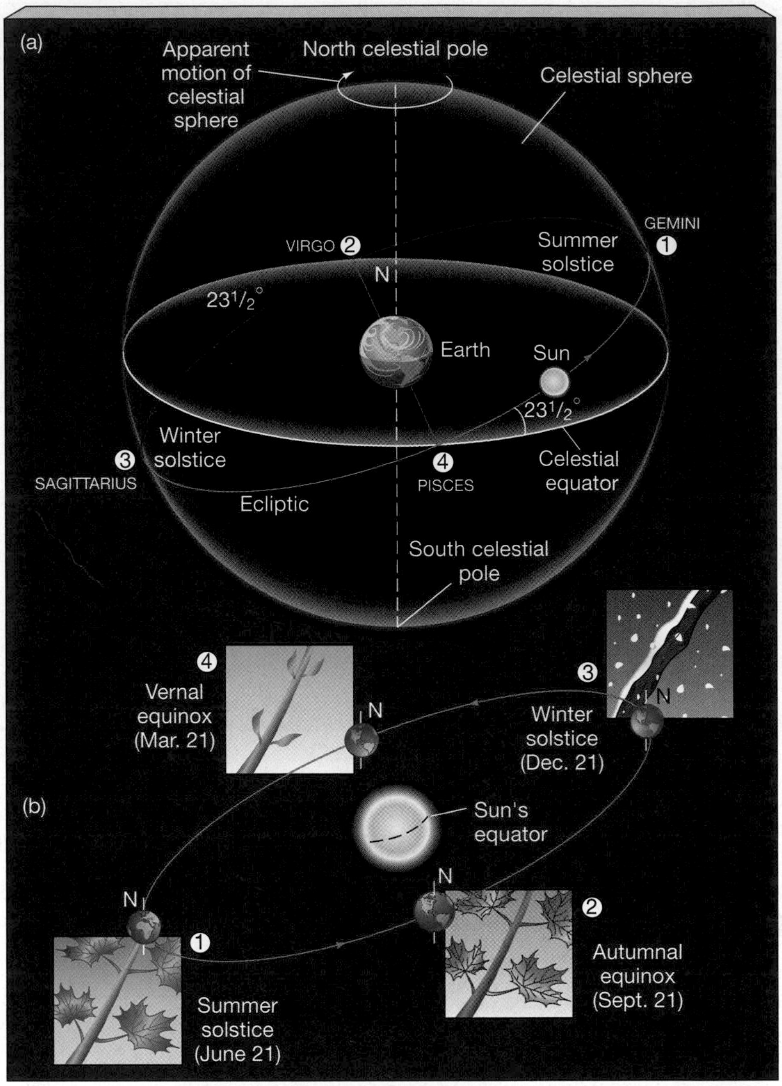

Animation

Figure P.12 **Seasons** (a) The apparent path of the Sun on the celestial sphere and (b) the Sun's actual relationship to Earth's rotation and revolution. The seasons result from the changing height of the Sun above the horizon. At the summer solstice (the points marked 1), the Sun is highest in the sky, as seen from Earth's northern hemisphere, and the days are longest. In the "celestial sphere" picture (a), the Sun at this time is at its northernmost point on its path around the ecliptic; in reality (b), the summer solstice corresponds to the point on Earth's orbit where our planet's North Pole points most nearly toward the Sun. The reverse is true at the winter solstice (points 3). At the vernal and autumnal equinoxes, day and night are of equal length. These are the times when, as seen from Earth (a), the Sun crosses the celestial equator. These times correspond to the points in Earth's orbit when our planet's axis is perpendicular to the line joining Earth and the Sun (b).

Long-Term Changes

The time required for Earth to complete exactly one orbit around the Sun, relative to the stars, is called a **sidereal year**. One sidereal year is 365.256 solar days long, about 20 minutes longer than a tropical year. The reason for this slight difference is a phenomenon known as **precession**. Like a spinning top that rotates rapidly on its own axis while that axis slowly revolves about the vertical, Earth's axis changes its direction over the course of time (although the angle between the axis and a line perpendicular to the plane of the ecliptic remains close to 23.5°). Figure P.13 illustrates Earth's precession, which is caused by the combined gravitational pulls of the Moon and the Sun. During a complete cycle of precession, taking about 26,000 years, Earth's axis traces out a cone. Because of this slow shift in the orientation of

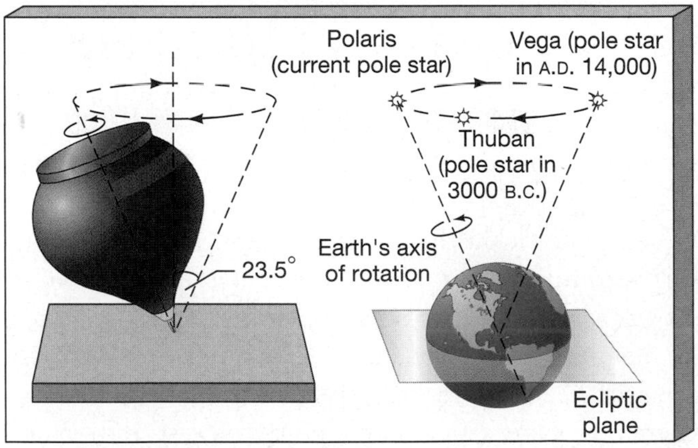

Figure P.13 **Precession** Earth's axis currently points nearly toward the star Polaris. Some 12,000 years from now—nearly halfway through one cycle of precession—Earth's axis will point toward a star called Vega, which will then be the "North Star." Five thousand years ago, the North Star was a star named Thuban in the constellation Draco.

Earth's rotation axis, the vernal equinox, which defines the tropical year, drifts slowly around the ecliptic over the course of the precession cycle.

The tropical year is the year our calendars measure. If our timekeeping were tied to the sidereal year, the seasons would slowly march around the calendar as Earth precessed—13,000 years from now, summer in the Northern Hemisphere would be at its height in mid-February! By using the tropical year instead, we ensure that July and August will always be (northern) summer months. However, in 13,000 years' time, Orion, now a prominent feature of the northern winter sky, will be a summer constellation.

✓ Concept Check

- What is precession? Explain in your own words what changes you would see on the celestial sphere over the course of one precession cycle.

P.4 The Motion of the Moon

The Moon is our nearest neighbor in space. Apart from the Sun, it is by far the brightest object in the sky. Like the Sun, the Moon appears to move relative to the background stars. Unlike the Sun, however, the explanation for this motion is the obvious one—the Moon really does revolve around Earth.

Lunar Phases

2 The Moon's appearance undergoes a regular cycle of changes, or **phases**, taking a little more than 29 days to complete. (The word *month* is derived from the word Moon.) Figure P.14 illustrates the appearance of the Moon at different times in this monthly cycle. Starting from the **new Moon**, which is all but invisible in the sky, the Moon appears to *wax* (grow) a little each night and is visible as a growing *crescent* (panel 1 of Figure P.14). One week after new Moon, half of the lunar disk can be seen (panel 2). This phase is known as a **quarter Moon**. During the next week, the Moon continues to wax, passing through the *gibbous* phase (more than half of the lunar disk visible, panel 3) until, two weeks after new Moon, the **full Moon** (panel 4) is visible. During the next 2 weeks, the Moon *wanes* (shrinks), passing in turn through the gibbous, quarter, and crescent phases (panels 5–7), eventually becoming new again.

Unlike the Sun and the other stars, the Moon emits no light of its own. Instead, it shines by reflected sunlight, giving rise to the phases we see. As illustrated in Figure P.14, half of the Moon's surface is illuminated by the Sun at any instant. However, not all of the Moon's sunlit face can be seen because of the Moon's position with respect to Earth and the Sun. When the Moon is full, we see the entire "daylit" face because the Sun and the Moon are in opposite directions from Earth in the sky. In the case of a new Moon, the Moon and the Sun are in almost the same part of the sky, and the sunlit side of the Moon is oriented away from us. At new Moon, the Sun is almost behind the Moon, from our perspective.

As it revolves around Earth, the Moon's position in the sky changes with respect to the stars. In one **sidereal month** (27.3 days), the Moon completes one revolution and returns to its starting point on the celestial sphere, having traced out a great circle in the sky. The time required for the Moon to complete a full cycle of phases, one **synodic month**, is a little longer—about 29.5 days. The synodic month is a little longer than the sidereal month for the same basic reason that a solar day is slightly longer than a sidereal day: Because of Earth's motion around the Sun, the Moon must complete slightly more than one full revolution to return to the same phase in its orbit.

Eclipses

3 From time to time—but only at new or full Moon—the Sun, Earth, and the Moon line up precisely and we observe the spectacular phenomenon known as an **eclipse**. When the Sun and the Moon are in exactly *opposite* directions as seen from Earth, Earth's shadow sweeps across the Moon, temporarily blocking the Sun's light and darkening the Moon in a **lunar eclipse**, as illustrated in Figure P.15. From Earth, we see the curved edge of Earth's shadow begin to cut across the face of the full Moon and slowly eat its way into the lunar disk.

Usually, the alignment of the Sun, Earth, and the Moon is imperfect, so the shadow never completely covers the Moon. Such an occurrence is known as a **partial eclipse**. Occasionally, however, the entire lunar surface is obscured in a **total eclipse** (such as that shown in the inset in Figure P.15). Total lunar eclipses last only as long as is needed for the Moon to pass through Earth's shadow—no more than about 100 minutes. During that

7 Waning crescent (26 days old)

6 Third quarter (22 days old)

5 Waning gibbous (18 days old)

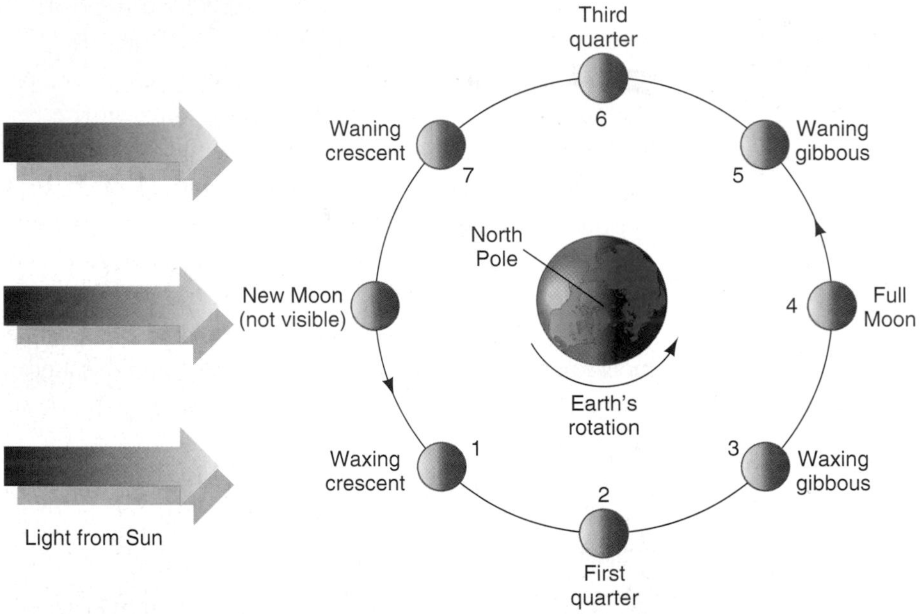

4 Full Moon (14 days old)

1 Waxing crescent (4 days old)

2 First quarter (7 days old)

3 Waxing gibbous (10 days old)

R I V U X G

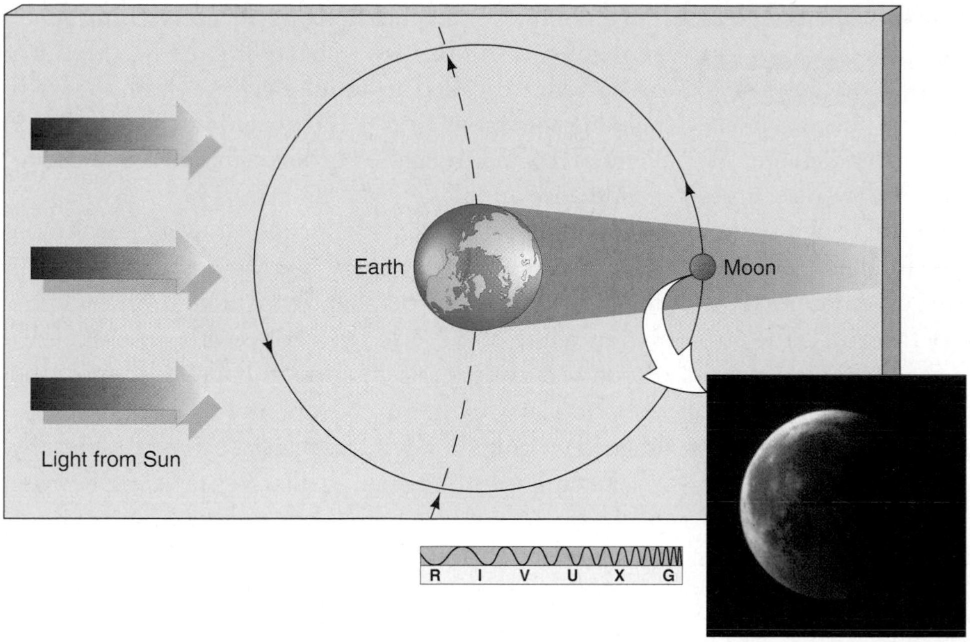

Figure P.15 **Lunar Eclipse** A lunar eclipse occurs when the Moon passes through Earth's shadow. At these times we see a darkened, copper-colored Moon, as shown in the inset photograph. The coloration is caused by sunlight deflected by Earth's atmosphere onto the Moon's surface. (Note that this figure is not drawn to scale.) *(Inset: G. Schneider)*

time, the Moon often acquires an eerie, deep red coloration, the result of a small amount of sunlight being refracted (bent) by Earth's atmosphere onto the lunar surface, preventing the shadow from being completely black.

When the Moon and the Sun are in exactly the *same* direction as seen from Earth, an even more awe-inspiring event occurs. The Moon passes directly in front of the Sun, briefly turning day into night in a **solar eclipse**. In a *total solar eclipse*, when the alignment is perfect, planets and some stars become visible in the daytime as the Sun's light is blocked. By pure chance, the Sun and Moon have almost exactly the same angular size as seen from Earth (the Sun is much bigger than the Moon, but it also lies much farther away—see *More Precisely P-2*). Consequently, during a total solar eclipse, we can often see the Sun's ghostly outer atmosphere, or *corona* (Figure P.16), its faint glow becoming temporarily visible when the rest of the Sun's glare is obscured. In a *partial solar eclipse*, the Moon's path is slightly "off center," and only a portion of the Sun's face is covered.

Figure P.14 **Lunar Phases** Because the Moon orbits Earth, the visible fraction of the lunar sunlit face varies from night to night. The complete cycle of lunar phases takes about 29 days to complete. *(UC/Lick Observatory)*

Figure P.16 **Solar Eclipse** During a total solar eclipse, the Sun's corona becomes visible as an irregularly shaped halo surrounding the blotted-out disk of the Sun. This was the August 1999 eclipse, as seen from the banks of the Danube River near Sofia, Bulgaria. *(Magdalena Tsavkova)*

Unlike a lunar eclipse, which is simultaneously visible from all locations on Earth's night side, a total solar eclipse can be seen from only a small portion of the daytime side. The Moon's shadow on Earth's surface is about 7000 kilometers wide—roughly twice the diameter of the Moon (Figure P.17). Outside that shadow, no eclipse is seen. However, only within the central region of the shadow, in the **umbra**, is the eclipse total. Within the shadow but outside the umbra, in the **penumbra**, the eclipse is partial, with less and less of the Sun being obscured the farther one travels from the shadow's center. The connections between the umbra, the penumbra, and the relative locations of Earth, Sun, and Moon during eclipses of various types are illustrated in Figure P.18. The umbra is always very small—even under the most favorable circumstances, its diameter never exceeds 270 kilometers. Because the Moon's shadow sweeps across Earth's surface at a speed of more than 1700 kilometers per hour, the duration of a total solar eclipse at any given point can never exceed 7.5 minutes.

The Moon's orbit around Earth is not exactly circular. Thus, the Moon may be far enough from Earth at the moment of an eclipse that its disk fails to cover the disk of the Sun completely, even though their centers coincide. In that case, there is no region of totality—the umbra never reaches Earth at all, and a thin ring of sunlight can still be seen surrounding the Moon. Such an occurrence, called an **annular eclipse** (from the word *annulus*, meaning "ring"), is depicted at the bottom of Figure P.18 and in Figure P.19. Roughly half of all solar eclipses are annular.

Why isn't there a solar eclipse at every new Moon and a lunar eclipse at every full Moon? The answer is that the Moon's orbit is slightly inclined to the ecliptic (at an angle of 5.2°) so the chance of a new (or full) Moon occurring just as the Moon happens to cross the ecliptic plane (so Earth, Moon, and Sun are perfectly aligned) is quite low. For this reason, eclipses, especially solar eclipses, are quite rare events. Nevertheless, because we know the orbits of both Earth and the Moon to great accuracy, we can predict eclipses far into the future. Figure P.20 shows the location and duration of all total and annular eclipses of the Sun between 1995 and 2005.

Concept Check

- If Earth's distance from the Sun were to double, would you still expect to see total solar eclipses? What if the distance became half its present value?

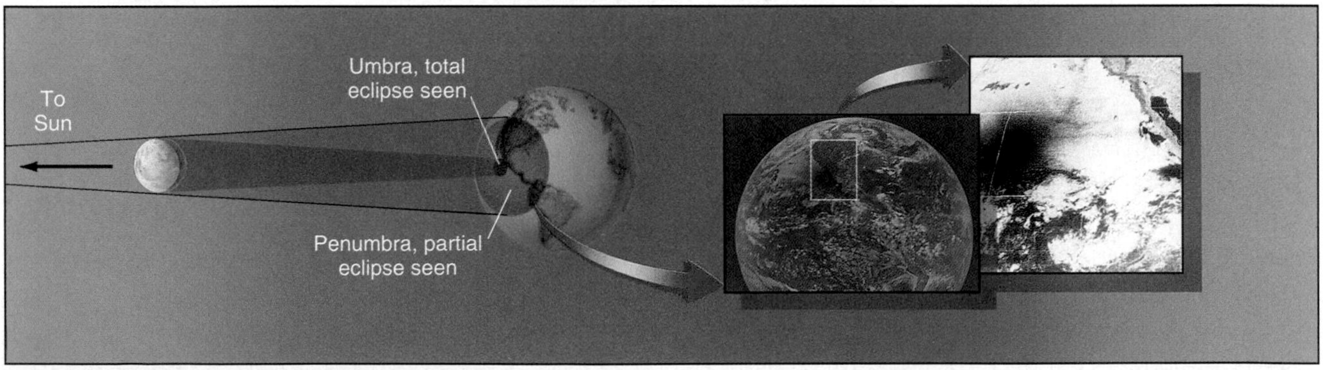

Figure P.17 Lunar Shadow The Moon's shadow on Earth during a solar eclipse consists of the umbra, where the eclipse is total, and the penumbra, where the Sun is only partially obscured. The insets show photographs of the Moon's shadow projected onto Earth's surface (near Baja California) during the total solar eclipse of July 11, 1991. The photographs were taken by an Earth-orbiting weather satellite. *(Insets: NOAA)*

R I V U X G

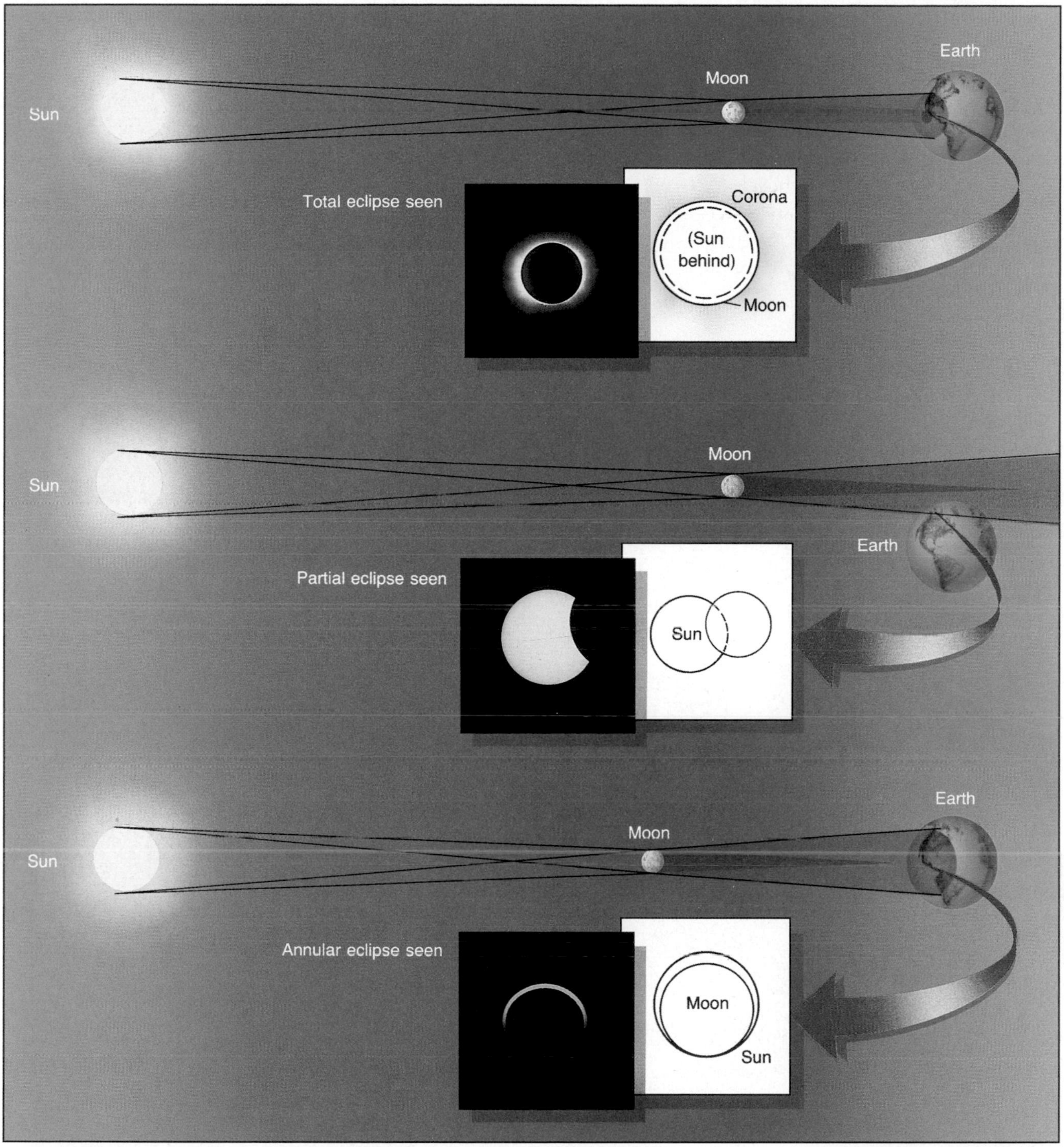

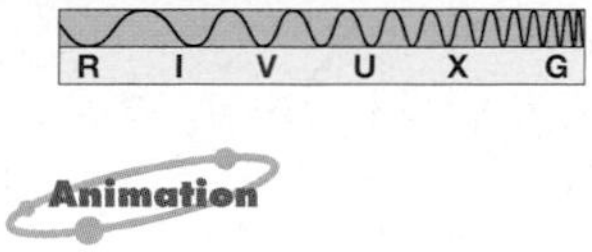

Figure P.18 Solar Eclipse Types Whether we see a total or partial eclipse depends both on our location on Earth and on the precise alignment of Earth, the Moon, and the Sun. If the Moon is too far from Earth at the moment of the eclipse, there is no region of totality; instead, an annular eclipse is seen. *(Insets: G. Schneider)*

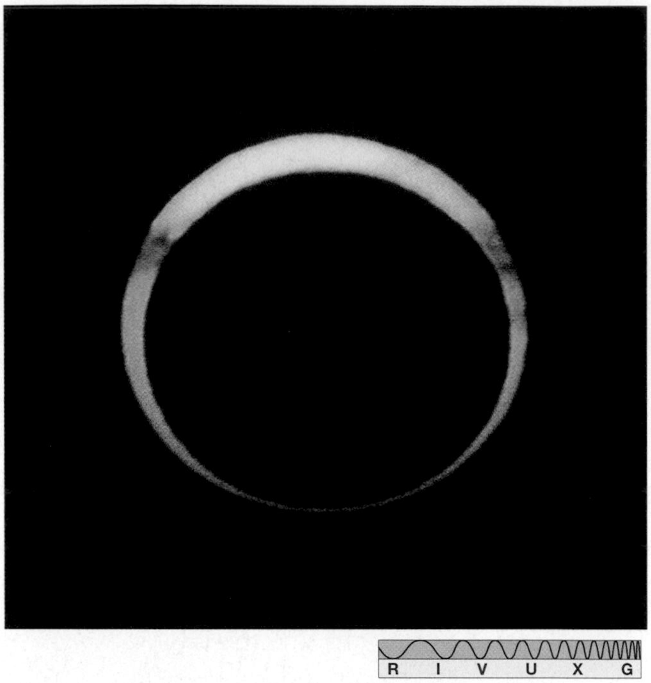

Figure P.19 Annular Eclipse During an annular eclipse, the Moon fails to completely hide the Sun, with the result that a thin ring of light remains. No corona is seen in this case because even the small amount of the Sun still visible completely overwhelms the corona's faint glow. This was the December 1973 eclipse, as seen from Algiers. (The gray fuzzy areas at top left and right of the ring are clouds in Earth's atmosphere.) *(G. Schneider)*

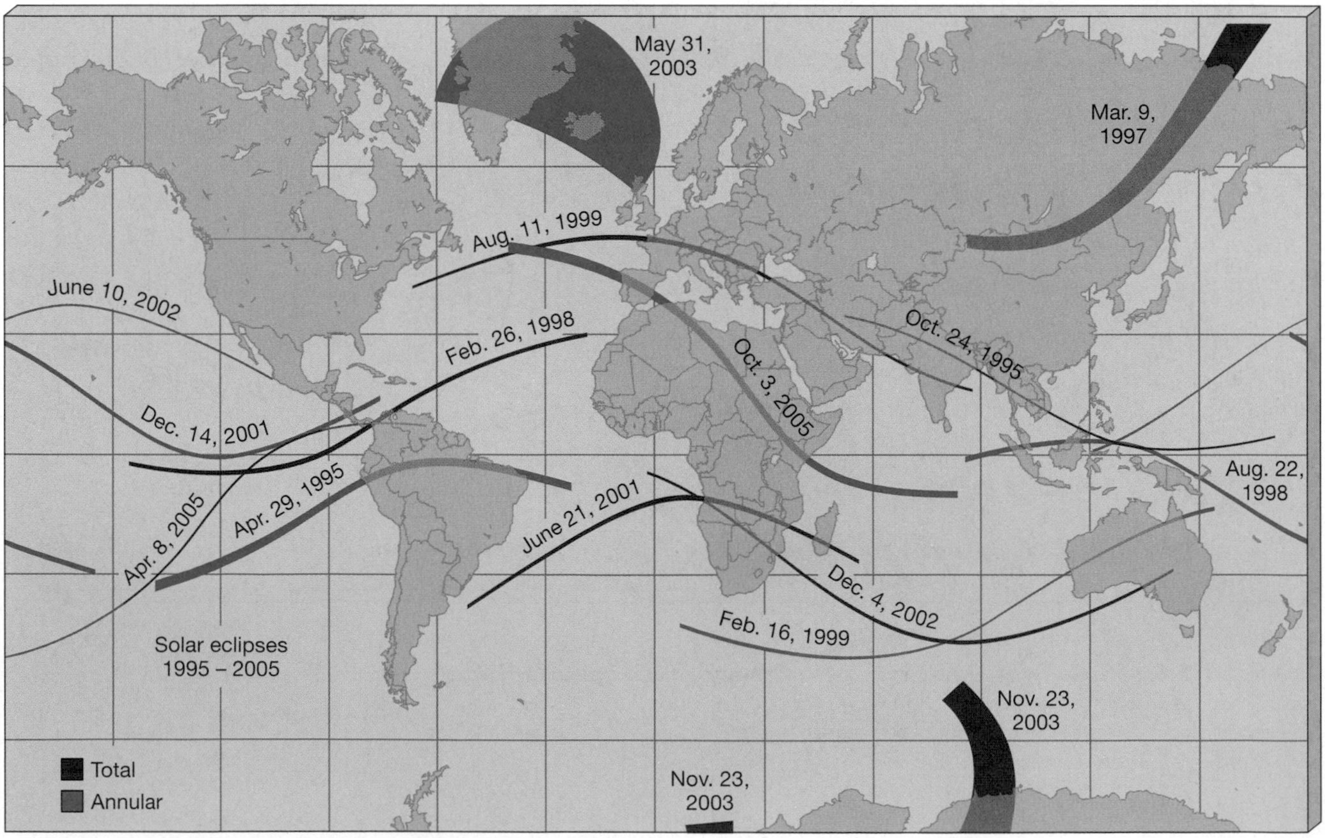

Figure P.20 Eclipse Tracks Regions of Earth that saw or will see total or annular solar eclipses between the years 1995 and 2005. Each track represents the path of the Moon's umbra across Earth's surface during an eclipse. *(Sinnott/Martinez/Sky Publishing Corporation)*

P.5 The Measurement of Distance

4 So far we have considered only the *directions* to the Sun, the Moon, and the stars, as seen from Earth. But knowing the direction in which an object lies is only part of the information needed to locate it in space. Before we can make a systematic study of the heavens, we must find a way of measuring *distances*, too. One distance-measurement method, called **triangulation**, is based on the principles of Euclidean geometry and finds widespread application today in both terrestrial and astronomical settings. Modern engineers, especially surveyors, use this age-old geometrical idea to measure indirectly the distance to faraway objects. In astronomy, it forms the foundation of the family of distance-measurement techniques that together make up the **cosmic distance scale**.

Imagine trying to measure the distance to a tree on the other side of a river. The most direct method is to lay a tape across the river, but that's not the simplest way. A smart surveyor would make the measurement by visualizing an imaginary triangle (hence the term *triangulation*), sighting the tree on the far side of the river from two positions on the near side, as illustrated in Figure P.21. The simplest possible triangle is a right triangle, in which one of the angles is exactly 90°, so it is often convenient to set up one observation position directly opposite the object, as at point A, although this isn't necessary for the method to work. The surveyor then moves to another observation position at point B, noting the distance covered between A and B. This distance is called the **baseline** of the imaginary triangle. Finally, the surveyor, standing at point B, sights toward the tree and notes the angle formed at point B by the intersection of this sightline and the baseline. No further observations are required. Knowing the length of one side (AB) and two angles (the right angle at A and the angle at B) of the triangle, and having some knowledge of elementary geometry and trigonometry, the surveyor can construct the remaining sides and angles and so establish the distance from A to the tree.

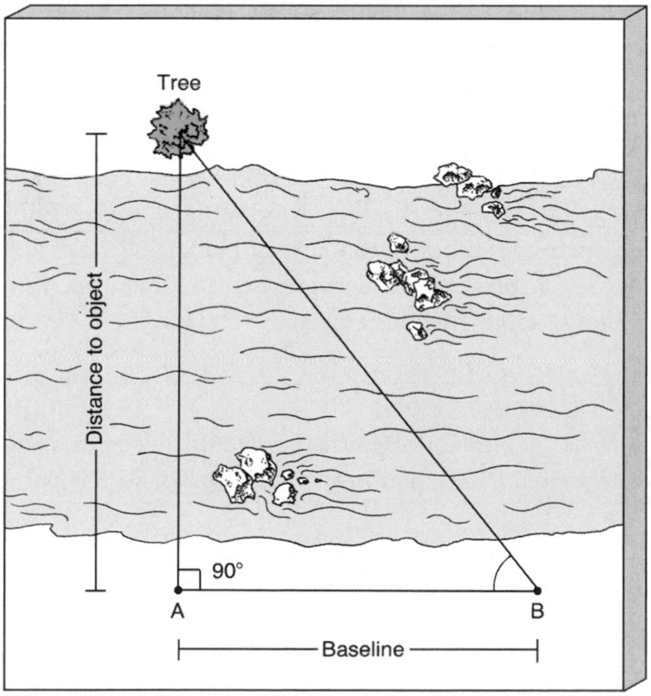

Figure P.21 Triangulation Surveyors often use simple geometry and trigonometry to estimate the distance to a faraway object by triangulation.

Obviously, for a fixed baseline, the triangle becomes longer and narrower as the tree's distance from A increases. Narrow triangles cause problems because it becomes hard to measure the angles at A and B with sufficient accuracy. The measurements can be made easier by "fattening" the triangle—that is, by lengthening the baseline—but in astronomy there are limits on how long a baseline we can choose. For example, consider an imaginary triangle extending from Earth to a nearby object in space, perhaps a neighboring planet. The triangle is now extremely long and narrow, even for a relatively nearby object (by cosmic standards). Figure P.22(a) illustrates a case in which the longest baseline possible on Earth—Earth's diameter, measured from point A to point B—is used. In principle, two observers could sight the planet from opposite sides of Earth, measuring the triangle's angles at A and B. However, in practice it is easier to measure the third angle of the imaginary triangle. Here's how.

The observers each sight toward the planet, taking note of its position *relative to some distant stars* seen on the plane of the sky. The observer at A sees the planet at apparent location A′ relative to those stars, as indicated in Figure P.22(a). The observer at B sees the planet at location B′. If each observer takes a photograph of the same region of the sky, the planet will appear at slightly different places in the two images, as shown in Figure P.22(b). (The background stars appear undisplaced because of their much greater distance from the observer.) This apparent displacement of a foreground object relative to the background as the observer's

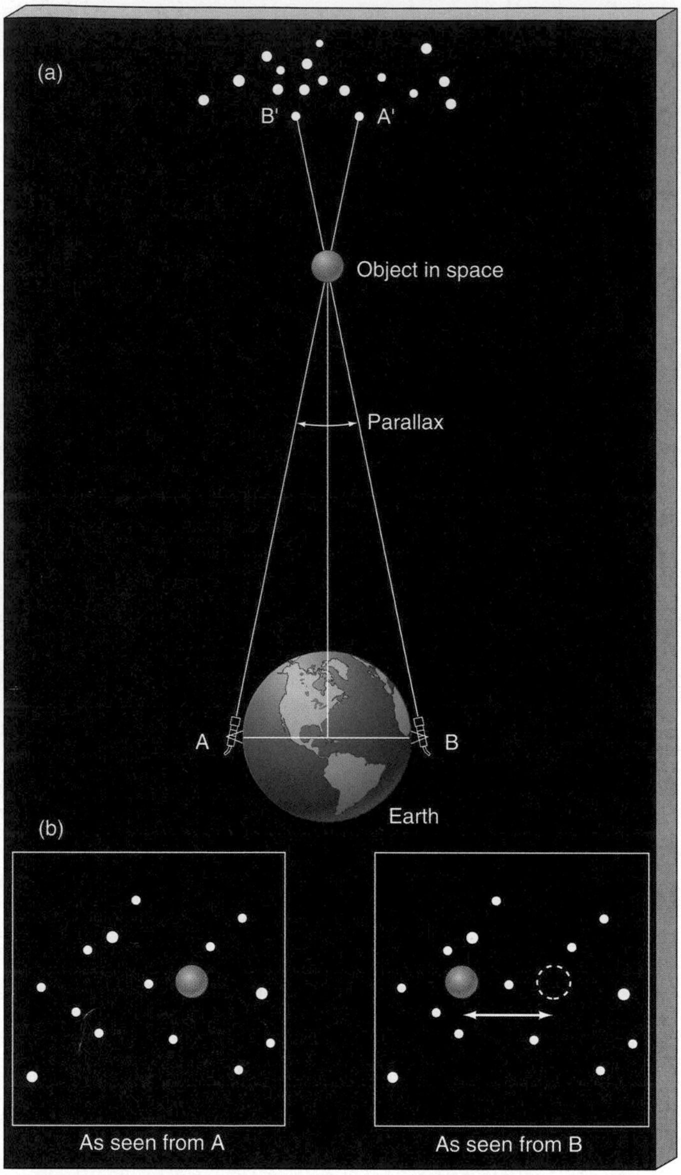

Figure P.22 Parallax (a) This imaginary triangle extends from Earth to a nearby object in space (such as a planet). The group of stars at the top represents a background field of very distant stars. (b) Hypothetical photographs of the same star field showing the nearby object's apparent displacement, or shift, relative to the distant undisplaced stars.

location changes is known as **parallax**. The size of the shift in Figure P.22(b), measured as an angle on the celestial sphere, is equal to the third angle of the imaginary triangle in Figure P.22(a).

The closer an object is to the observer, the larger the parallax. To see this for yourself, hold a pencil vertically just in front of your nose. Look at some far-off object—a distant wall, say. Close one eye, then open it while closing the other. You should see a large shift in the apparent position of the pencil relative to the wall—a large parallax. In this example, one eye corresponds to point A in Figure P.22, the other eye to point B, the distance between your eyeballs to the baseline, the pencil to the planet, and the distant wall to the remote field of stars. Now hold the pencil at arm's length, corresponding to a more distant object (but still not as far away as the distant stars). The apparent shift of the pencil will be smaller. By moving the pencil farther away, you are narrowing the triangle and decreasing the parallax. If you were to paste the pencil to the wall, corresponding to the case where the object of interest is as far away as the background star field, blinking would produce no apparent shift of the pencil at all.

The amount of parallax is inversely proportional to an object's distance. Small parallax implies large distance. Conversely, large parallax implies small distance. Knowing the amount of parallax (as an angle) and the length of the baseline, we can easily derive the distance through triangulation. *More Precisely P-2* explores the connection between angular measure and distance in a little more detail.

Surveyors of the land routinely use these simple geometric techniques to map out planet Earth. As surveyors of the sky, astronomers use the same basic principles to chart the universe.

Concept Check

- What practical difficulties can you think of in using triangulation to determine the distance to the Sun?

MORE PRECISELY P-2

Measuring Distances with Geometry

We can convert baselines and parallaxes into distances using a simple argument made by the Greek geometer Euclid. The figure below represents Figure P.22(a), but we have changed the scale and added the circle centered on the target planet and passing through our baseline on Earth.

To compute the planet's distance, we note that the ratio of the baseline AB to the circumference of the large circle shown in the figure (2π times the distance to the planet) must be equal to the ratio of the parallax to one full revolution, 360°:

$$\frac{\text{baseline}}{2\pi \times \text{distance}} = \frac{\text{parallax}}{360°},$$

from which we find

$$\text{distance} = \text{baseline} \times \frac{(360°/2\pi)}{\text{parallax}}.$$

(The angle $360°/2\pi \approx 57.3°$ in the above equation is called 1 *radian*.)

Thus, for example, two observers 1000 km apart looking at the Moon might measure (using the photographic technique described in the text) a parallax of 9.0 arc minutes—that is, 0.15°. It then follows that the distance to the Moon is 1000 km × (57.3 / 0.15) ≈ 380,000 km. (More accurate measurements, based on laser ranging using equipment left on the lunar surface by *Apollo* astronauts, yield a mean distance of 384,000 km.)

Knowing the distance to an object, we can then determine many other properties. For example, by measuring the object's *angular diameter*, we can compute its size. The figure below illustrates the geometry involved. Notice that this is basically the same diagram as the previous one, except that now the angle (the angular diameter) and the distance are known, instead of the angle (the parallax) and the baseline. The same reasoning as before then allows us to calculate the diameter:

$$\frac{\text{diameter}}{2\pi \times \text{distance}} = \frac{\text{angular diameter}}{360°},$$

so

$$\text{diameter} = \text{distance} \times \frac{\text{angular diameter}}{57.3°}.$$

Continuing our previous example, the Moon's angular diameter is measured to be about 31 arc minutes—a little over half a degree. From the preceding discussion, it follows that the Moon's actual diameter is 380,000 km × (0.52° / 57.3°) ≈ 3450 km. A more precise measurement gives 3476 km.

You should study the above reasoning carefully. Although the observations are straightforward and the geometrical reasoning elementary, simple measurements such as these form the basis for almost every statement made in this book about size and scale in the universe.

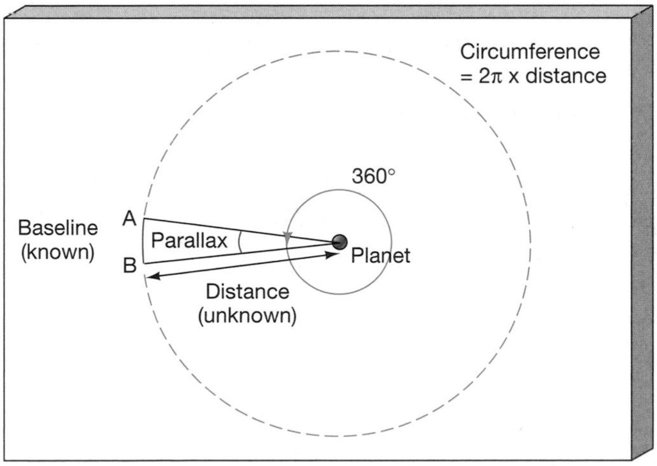

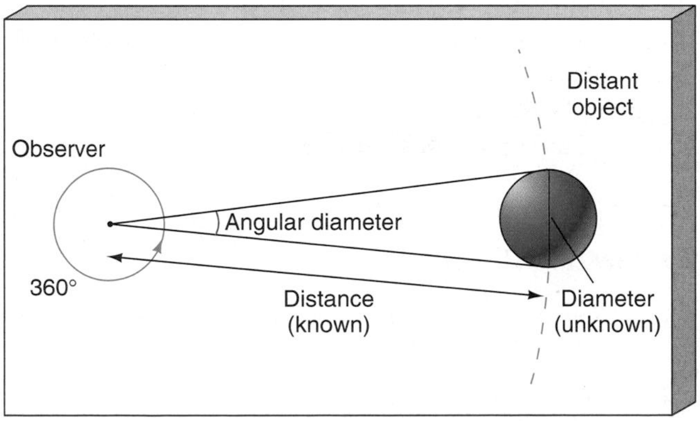

Chapter Review

Summary

The **universe** (p. 2) is the totality of all space, time, matter, and energy. **Astronomy** (p. 2) is the study of the universe. A widely used unit of distance in astronomy is the **light-year** (p. 2), the distance traveled by light in one year.

Early observers grouped the thousands of stars visible to the naked eye into patterns called **constellations** (p. 4). These patterns have no physical significance, although they are a useful means of labeling regions of the sky.

The nightly motion of the stars across the sky is the result of Earth's **rotation** (p. 5) on its axis. Early astronomers imagined that the stars were attached to a vast **celestial sphere** (p. 5) centered on Earth. The points where Earth's rotation axis intersects the celestial sphere are called the **north and south celestial poles** (p. 6). The line where Earth's equatorial plane cuts the celestial sphere is the **celestial equator** (p. 6). **Celestial coordinates** (p. 6) specify a star's location on the celestial sphere. **Declination** and **right ascension** (p. 6) are the celestial equivalents of latitude and longitude on Earth's surface.

The time from one sunrise to the next is a **solar day** (p. 7), and that between successive risings of any given star is one **sidereal day** (p. 7). Because of Earth's **revolution** (p. 8) around the Sun, the solar day is a few minutes longer than the sidereal day. The Sun's apparent yearly path around the celestial sphere, or the plane of Earth's orbit around the Sun, is called the **ecliptic** (p. 9). The constellations lying along the ecliptic are collectively called the **zodiac** (p. 9).

Because Earth's axis is inclined to the ecliptic plane, we experience **seasons** (p. 9). At the **summer solstice** (p. 9), the Sun is highest in the sky and the length of the day is greatest. At the **winter solstice** (p. 9), the Sun is lowest and the day is shortest. At the **vernal and autumnal equinoxes** (p. 9), Earth's rotation axis is perpendicular to the line joining Earth to the Sun, so day and night are of equal length. The time interval between one vernal equinox and the next is one **tropical year** (p. 9).The time required for the same zodiac constellations to reappear at the same location in the sky, as viewed from a given point on Earth, is one **sidereal year** (p. 10).

In addition to its rotation about its axis and its revolution around the Sun, Earth undergoes a motion called **precession** (p. 10), where the influence of the Moon and Sun causes Earth's axis to wobble slightly. As a result, the particular constellations that happen to be visible on any given night change slowly over the course of thousands of years.

As the Moon orbits Earth, we see lunar **phases** (p. 11) as the fraction of the Moon's sunlit face visible to us varies. At **full Moon** (p. 11), we see the entire illuminated side. At **quarter Moon** (p. 11), only half the sunlit side can be seen. At **new Moon** (p. 11), the sunlit face points away from us, and the Moon is all but invisible from Earth. The time between successive full Moons is one **synodic month** (p. 11). The time taken for the Moon to return to the same position in the sky relative to the stars is one **sidereal month** (p. 11). Because of Earth's motion around the Sun, the synodic month is about two days longer than the sidereal month.

A **lunar eclipse** (p. 11) occurs when the Moon enters Earth's shadow. The eclipse may be **total** (p. 11)—the entire Moon is (temporarily) darkened—or **partial** (p. 11)—only a portion of the Moon's surface is affected. A **solar eclipse** (p. 13) occurs when the Moon passes between Earth and the Sun and a small part of Earth's surface is plunged into shadow. For observers in the **umbra** (p. 14), the entire Sun is obscured and the solar eclipse is total. In the **penumbra** (p. 14), a **partial solar eclipse** (p. 13) is seen. If the Moon happens to be too far from Earth for its disk to completely hide the Sun, an **annular eclipse** (p. 14) occurs.

Surveyors on Earth use **triangulation** (p. 17) to determine the distances to remote objects. Astronomers use the same technique to measure the distances to planets and stars. The **cosmic distance scale** (p. 17) is the family of distance-measurement techniques by which astronomers chart the universe. **Parallax** (p. 18) is the apparent motion of a foreground object relative to a distant background as the observer's position changes. The larger the **baseline** (p. 17), the distance between the two observation points, the greater the parallax.

Review and Discussion

1. Compare the size of Earth with that of the Sun, the Milky Way, and the entire observable universe.
2. What is a constellation?
3. How and why does a day measured by the Sun differ from a day measured by the stars?
4. Why do we see different stars in summer than in winter?
5. Why are there seasons on Earth?
6. Why do we see different phases of the Moon?
7. Why aren't there lunar and solar eclipses every month?
8. What is precession, and what is its cause?
9. What is parallax? Give an everyday example.
10. Why is it necessary to have a long baseline when using triangulation to measure the distances to objects in space?

True or False?

____ **1.** The light-year is a measure of distance.

____ **2.** The stars in a constellation are physically close to one another.

____ **3.** The solar day is longer than the sidereal day.

____ **4.** The constellations lying immediately adjacent to the north celestial pole are collectively referred to as the zodiac.

____ **5.** The vernal equinox marks the beginning of fall.

____ **6.** The new phase of the Moon cannot be seen because it always occurs during the daytime.

____ **7.** A lunar eclipse can occur only during the full phase.

____ **8.** Solar eclipses are possible during any phase of the Moon.

____ **9.** The parallax of an object is directly proportional to its distance from the observer.

____ **10.** Parallax is useful as a distance-measuring technique only on Earth's surface.

Fill in the Blank

1. Earth and the Sun are part of a vast collection of matter called the Milky Way _______.

2. The apparent path of the Sun across the sky is known as the _______.

3. On December 21, known as the _______, the Sun is at its _______ point on the celestial sphere.

4. The solar day is measured relative to the Sun; the sidereal day is measured relative to the _______.

5. To explain the daily and yearly motions of the heavens, ancient astronomers imagined that the Sun, Moon, stars, and planets were attached to a rotating _______.

6. A tropical year is the time between one _______ and the next.

7. Declination measures the position of an object north or south of the _______.

8. A _______ eclipse can be seen by about half of Earth at once.

9. The distance to an object can be determined, for a known baseline, by measuring its _______.

10. The size of an object can be determined, if we know its distance, by measuring its _______.

Problems

1. (a) Write the following in scientific notation (see Appendix 1): 1000; 0.000001; 1001; 1,000,000,000,000,000; 123,000; 0.000456. (b) Write the following in "normal" numerical form: 3.16×10^7; 2.998×10^5; 6.67×10^{-11}; 2×10^0. (c) Calculate: $(2 \times 10^3) + 10^{-2}$; $(1.99 \times 10^{30}) \div (5.97 \times 10^{24})$; $(3.16 \times 10^7) \times (2.998 \times 10^5)$.

2. In one second, light leaving Los Angeles will reach approximately as far as (a) San Francisco (about 500 km), (b) London (roughly 10,000 km), (c) the Moon (400,000 km), (d) Venus (roughly 45,000,000 km from Earth at closest approach), or (e) the nearest star (about 3 light-years from Earth). Which is correct?

3. How would the length of the solar day change if Earth's rotation were suddenly to reverse direction?

4. The vernal equinox is now just entering the constellation Aquarius. In what constellation will it lie in the year A.D. 15,000?

5. Through how many degrees, arc minutes, or arc seconds does the Moon move in (a) one hour of time, (b) one minute, (c) one second? How long does it take for the Moon to move a distance equal to its own diameter?

6. A surveyor wishes to measure the distance between two points, as illustrated in Figure P.21. She measures the distance AB to be 250 m and the angle at B to be 30°. What is the distance between the two points?

7. At what distance is an object if its parallax, as measured from either end of a 1000-km baseline, is (a) 1°, (b) 1′, (c) 1″?

8. Given that the angular size of Venus is 55″ when the planet is 45,000,000 km from Earth, calculate Venus's diameter.

9. The Moon lies 384,000 km from Earth and the Sun lies 150,000,000 km away. If both have the same angular size, how many times larger than the Moon is the Sun?

10. Given that the distance from Earth to the Sun is 150,000,000 km, through what distance does Earth move in an hour? A day?

A
B
C
D

1 THE COPERNICAN REVOLUTION

The Birth of Modern Science

Learning Goals

Studying this chapter will enable you to:

1 Explain how the observed motions of the planets led to our modern view of a Sun-centered solar system.

2 Sketch the major contributions of Galileo and Kepler to the development of our understanding of the solar system.

3 State Kepler's laws of planetary motion.

4 Explain how astronomers have measured the true size of the solar system.

5 State Newton's laws of motion and his law of universal gravitation, and explain how the latter permits us to measure the masses of astronomical bodies.

(Opposite page, background) In an early twentieth-century classroom at Radcliffe College, as at other schools, astronomy came into its own as a viable and improtant subject. (*Harvard College Observatory*)

(Inset A) Harlow Shapley (1885–1972) discovered our place in the "suburbs" of the Milky Way, dispelling the notion that the Sun resides at the center of the universe. (*Harvard College Observatory*)

(Inset B) Annie Cannon (1863–1941), one of the greatest astronomical catalogers of all time, carefully analyzed photographic plates to classify nearly a million stars over the course of fifty years of work at Harvard University. (*Harvard College Observatory*)

(Inset C) Maria Mitchell (1818–1889) in her observatory on Nantucket Island, made important contributions to several areas of astronomy. (*Special Collections, Vassar College Libraries, Poughkeepsie, NY*)

(Inset D) Edwin Hubble (1889–1953), here posing in front of one of the telescopes on Mount Wilson, in California, is often credited with having discovered the expansion of the universe. (*California Institute of Technology/Palomar Hall/Hale Observatory/Caltech*)

Living in the Space Age, we have become accustomed to the modern view of our place in the universe. Images of our planet taken from space leave little doubt that Earth is round, and no one seriously questions the idea that we orbit the Sun. Yet there was a time, not so long ago, when our ancestors maintained that Earth was flat and lay at the center of all things. Our view of the universe—and of ourselves—has undergone a radical transformation since those early days. In this transformed view, Earth is a planet like many others. Humankind has been torn from its throne at the center of the cosmos and relegated to an unremarkable position on the periphery of the Milky Way Galaxy. But we have been amply compensated for our loss of prominence, for we have gained a wealth of scientific knowledge in the process. The story of how all this came about is the story of the rise of science and the genesis of modern astronomy.

1.1 The Motions of the Planets

Over the course of a night, the stars slide smoothly across the sky. Over the course of a month, the Moon moves smoothly and steadily along its path on the sky relative to the stars, passing through its familiar cycle of phases. Over the course of a year, the Sun progresses along the ecliptic at an almost constant rate, varying little in brightness from day to day. In short, the behavior of the Sun, the Moon, and the stars seems fairly simple and orderly. But ancient astronomers were also aware of five other bodies in the sky—the planets Mercury, Venus, Mars, Jupiter, and Saturn—whose behavior was not so easy to grasp.

Planets do not behave in as regular a fashion as do the Sun, Moon, and stars. They vary in brightness and don't maintain a fixed position in the sky. Unlike the Sun and the Moon, they seem to wander around the celestial sphere—indeed, the word *planet* derives from the Greek word *planetes*, meaning "wanderer." Planets never stray far from the ecliptic and generally traverse the celestial sphere from west to east, as the Sun does. However, they seem to speed up and slow down during their journeys, and at times they even appear to loop back and forth relative to the stars, as shown in Figure 1.1. Motion in the eastward sense is usually referred to as *direct*, or *prograde*, motion. The backward (westward) loops are known as **retrograde motion**.

Like the Moon, the planets produce no light of their own. Instead, they shine by reflected sunlight. Ancient astronomers correctly reasoned that the apparent brightness of a planet in the night sky is related to its distance from Earth—planets appear brightest when closest to us. However, the planets Mars, Jupiter, and Saturn are always brightest during the retrograde portions of their orbits. The challenge facing astronomers was to explain the observed motions of the planets and to relate those motions to the variations in planetary brightness.

The Geocentric Universe

The earliest models of the solar system followed the teachings of the Greek philosopher Aristotle (384–322 B.C.) and were **geocentric**, meaning that Earth lay at the center of the universe and all other bodies moved around it. (Figures P.7 and P.12a illustrate the basic geocentric view.) ∞ (Sec. P.2) These models employed what Aristotle had taught was the perfect form: the circle. The simplest possible description—uniform motion around a circle having Earth at its center—provided a fairly good approximation to the orbits of the Sun and the Moon, but it could not account for the observed variations in planetary brightness, nor for their retrograde motion. A better model was needed to describe the planets.

In the first step toward this new model, each planet was taken to move uniformly around a small circle, called an **epicycle**, whose *center* moved uniformly around Earth on a second and larger circle, known as the **deferent** (Figure 1.2). The motion was now composed of two separate circular orbits, creating the possibility that, at some times, the planet's apparent motion could be retrograde. Also, the distance from the planet to Earth would vary, accounting for changes in brightness. By tinkering with the relative sizes of epicycle and deferent, with the planet's speed on the epicycle, and with the epicycle's speed along the deferent, early astronomers were able to bring this "epicyclic" motion into fairly good agreement with the observed paths of the planets in the sky. Moreover, this model had good predictive power, at least to the accuracy of observations at the time.

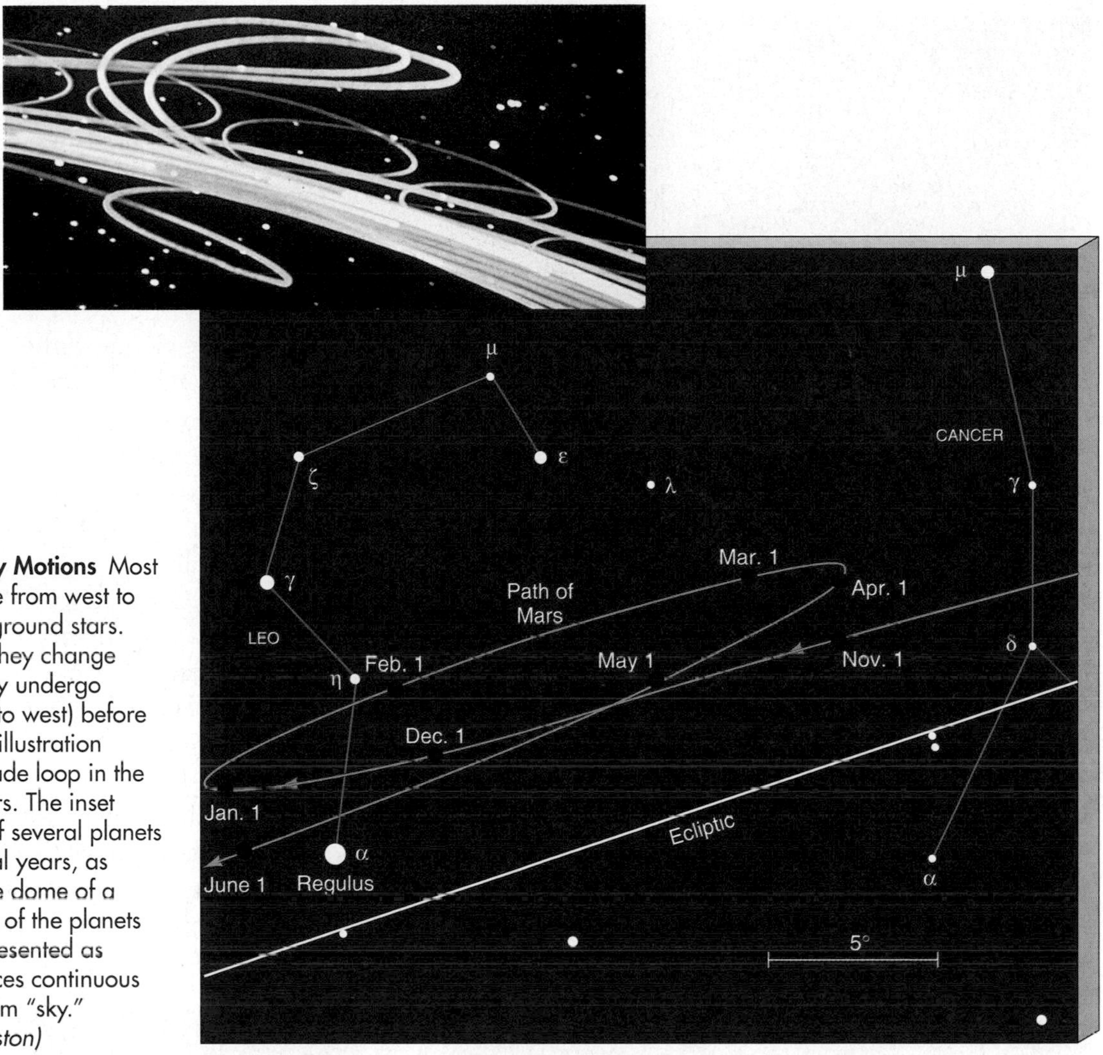

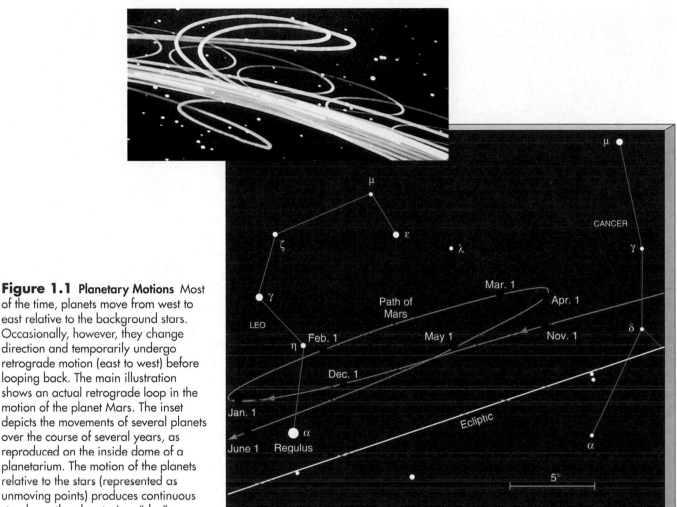

Figure 1.1 **Planetary Motions** Most of the time, planets move from west to east relative to the background stars. Occasionally, however, they change direction and temporarily undergo retrograde motion (east to west) before looping back. The main illustration shows an actual retrograde loop in the motion of the planet Mars. The inset depicts the movements of several planets over the course of several years, as reproduced on the inside dome of a planetarium. The motion of the planets relative to the stars (represented as unmoving points) produces continuous streaks on the planetarium "sky." *(Museum of Science, Boston)*

However, as the number and the quality of observations increased, small corrections had to be introduced into the simple epicyclic model to bring it into line with new observations. The center of the deferents had to be shifted slightly from Earth's center, and the motion of the epicycles had to be imagined uniform with respect not to Earth but to yet another point in space. Around A.D. 140, a Greek astronomer named Claudius Ptolemaeus (known today as Ptolemy) constructed perhaps the best geocentric model of all time. Illustrated in simplified form in Figure 1.3, it explained remarkably well the observed paths of the five planets then known, as well as the paths of the Sun and the Moon. However, to achieve its explanatory and predictive power, the full **Ptolemaic model** required a series of no fewer than 80 circles. To account for the paths of the Sun, the Moon, and all the nine planets (and their moons) that we know today would require a vastly more complex set.

Today, our scientific training leads us to seek simplicity, because simplicity in the physical sciences has so often proved to be an indicator of truth. We would regard the intricacy of a model as complicated as the Ptolemaic system as a clear sign of a fundamentally flawed theory. With the benefit of hindsight, we now recognize that the major error lay in the assumption of a geocentric universe, compounded by the insistence on uniform circular motion, the basis of which was largely philosophical, rather than scientific, in nature.

Actually, history records that some ancient Greek astronomers reasoned differently about the motions of heavenly bodies. Foremost among them was Aristarchus of Samos (310–230 B.C.), who proposed that all the planets, including Earth, revolve around the Sun and, furthermore, that Earth rotates on its axis once each day. This, he argued, would create an apparent motion of the sky—a simple idea that is familiar to anyone who

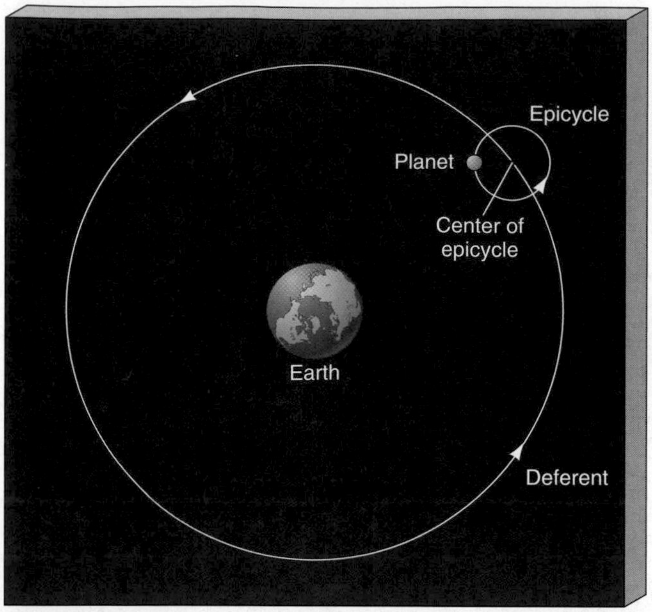

Figure 1.2 Geocentric Model In the geocentric model of the solar system, the observed motions of the planets made it impossible to assume that they moved on simple circular paths around Earth. Instead, each planet was thought to follow a small circular orbit (the epicycle) about an imaginary point that itself traveled in a large, circular orbit (the deferent) about Earth.

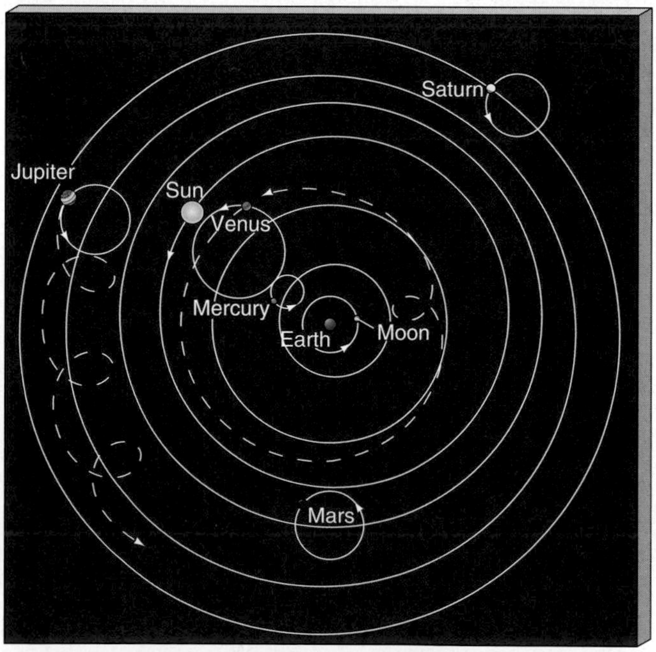

Figure 1.3 Ptolemy's Model The basic features, drawn roughly to scale, of Ptolemy's geocentric model of the inner solar system, a model that enjoyed widespread popularity prior to the Renaissance. To avoid confusion, partial paths (dashed) of only two planets, Venus and Jupiter, are drawn here.

has ridden on a merry-go-round and watched the landscape appear to move past in the opposite direction. However, Aristarchus's description of the heavens, though essentially correct, did not gain widespread acceptance during his lifetime. Aristotle's influence was too strong, his followers too numerous, his writings too comprehensive.

The Heliocentric Model of the Solar System

1 The Ptolemaic picture of the universe survived, more or less intact, for almost 13 centuries, until a sixteenth-century Polish cleric, Nicholas Copernicus (Figure 1.4), rediscovered Aristarchus's **heliocentric** (Sun-centered) model. Copernicus asserted that Earth spins on its axis and, like all other planets, orbits the Sun. Not only does this model explain the observed daily and seasonal changes in the heavens, as we have seen, but it also naturally accounts for planetary retrograde motion and brightness variations. The critical realization that Earth is not at the center of the universe is now known as the **Copernican revolution**.

Figure 1.5 shows how the Copernican view explains both the changing brightness of a planet (in this case, Mars) and its apparent looping motions. If we suppose that Earth moves faster than Mars, then every so often Earth "overtakes" that planet. Each time this hap-

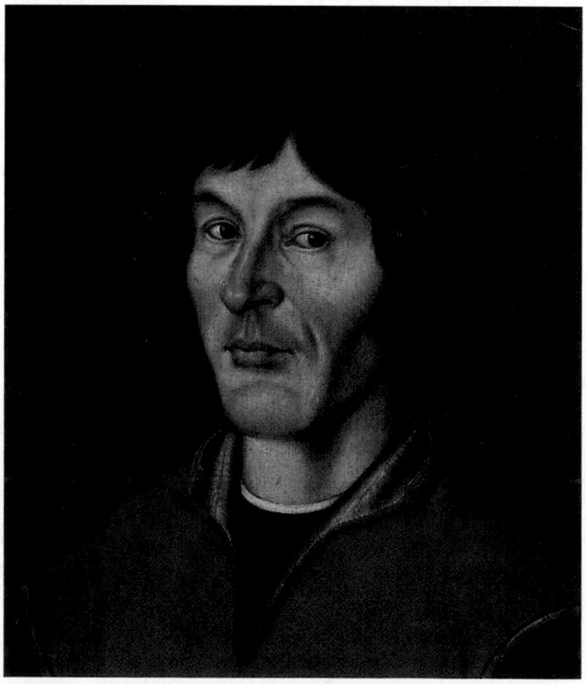

Figure 1.4 Nicholas Copernicus (1473–1543). *(E. Lessing/Art Resource, NY)*

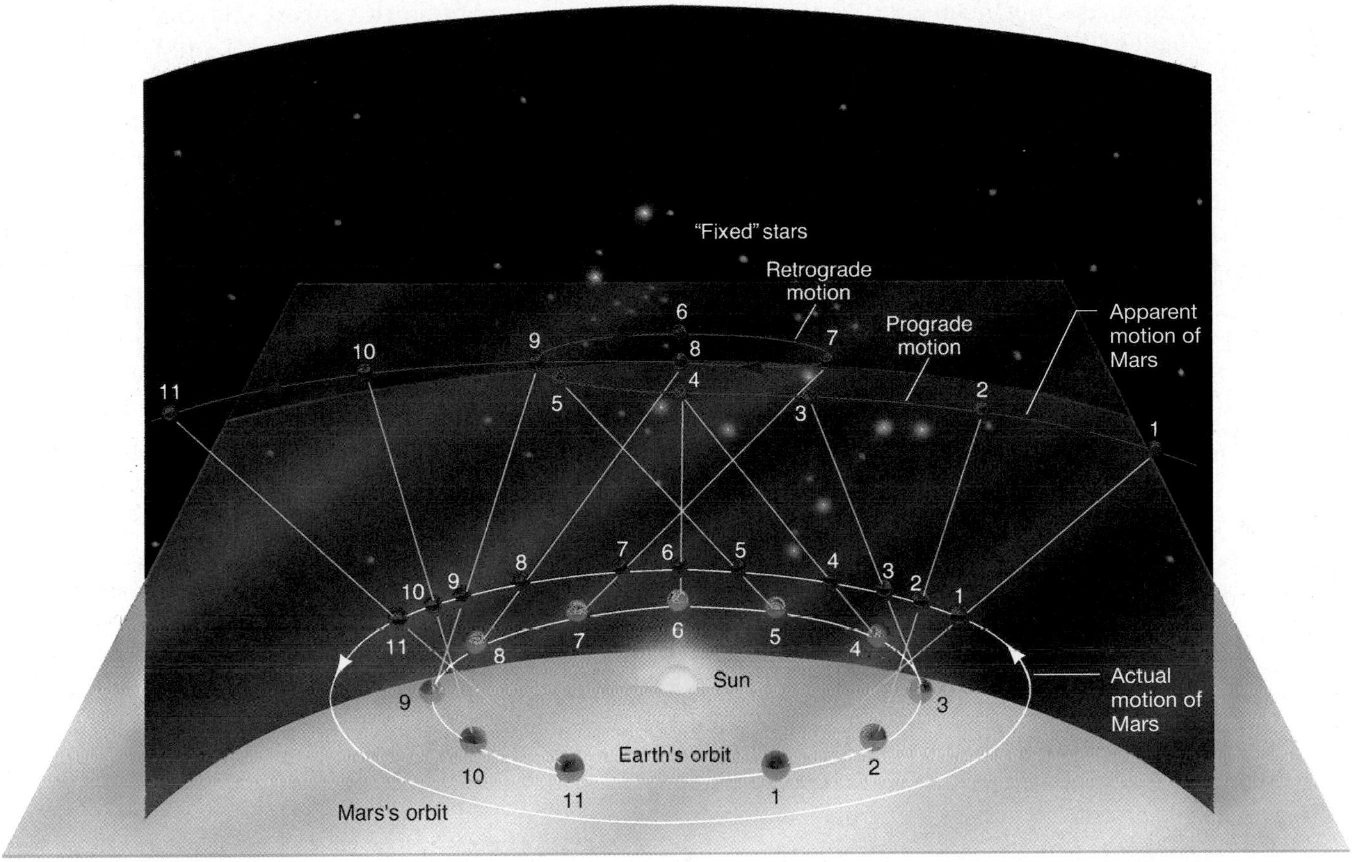

Figure 1.5 Retrograde Motion The Copernican model of the solar system explains both the varying brightnesses of the planets and the phenomenon of retrograde motion. Here, for example, when Earth and Mars are relatively close to one another in their respective orbits (as at position 6), Mars seems brighter. When they are farther apart (as at position 1), Mars seems dimmer. Also, because the line of sight from Earth to Mars changes as the two planets orbit the Sun, Mars appears to loop back and forth in retrograde motion. The line of sight changes because Earth, on the inside track, moves faster in its orbit than does Mars. The white curves are the actual planetary orbits. The apparent motion of Mars, as seen from Earth, is shown as the red curve.

pens, Mars appears to move backwards in the sky, in much the same way as a car we overtake on the highway seems to slip backwards relative to us. Furthermore, at these times, Earth is closest to Mars, so Mars appears brightest, in agreement with observations. Notice that in the Copernican picture the planet's looping motions are only apparent—in the Ptolemaic view, they were real.

Copernicus's major motivation for introducing the heliocentric model was simplicity. Even so, he was still influenced by Greek thinking and clung to the idea of circles to model the planets' motions. To bring his theory into agreement with observations, he was forced to retain the idea of epicyclic motion, though with the deferent centered on the Sun rather than on Earth, and with the epicycles being smaller than in the Ptolemaic picture. Thus, he retained unnecessary complexity and actually gained little in accuracy over the geocentric model. The heliocentric model did rectify some small discrepancies and inconsistencies in the Ptolemaic system, but for Copernicus, the primary attraction of heliocentricity was its simplicity, its being "more pleasing to the mind." To this day, scientists still are guided by simplicity, symmetry, and beauty in modeling all aspects of the universe.

Copernicus's ideas were never widely accepted during his lifetime. By relegating Earth to a non-central and undistinguished place within the solar system, heliocentricity contradicted the conventional wisdom of the time, and violated the religious doctrine of the Roman Catholic Church. Possibly because Copernicus wished to avoid direct conflict with the Church, his book *On the Revolution of the Celestial Spheres* was not published until

1543, the year he died. Only much later, when others—notably Galileo Galilei—popularized the heliocentric model, did the Copernican theory gain widespread recognition.

✓ Concept Check

- How do the geocentric and heliocentric models of the solar system differ in their explanations of planetary retrograde motion?

1.2 The Birth of Modern Astronomy

2 In the century following the death of Copernicus and the publication of his theory of the solar system, two scientists—Galileo Galilei and Johannes Kepler—made indelible imprints on the study of astronomy. Each achieved fame for his discoveries and made great strides in popularizing the Copernican viewpoint.

Galileo Galilei (Figure 1.6) was an Italian mathematician and philosopher. By his willingness to perform experiments to test his ideas—a radical approach in those days (see *Interlude 1-1*)—and by embracing the brand-new technology of the telescope, he revolutionized the way science was done, so much so that he is now widely regarded as the father of experimental science. The telescope was invented in Holland in the early seventeenth century. Having heard of the invention (but without having seen one), Galileo built a telescope for himself in 1609 and aimed it at the sky. What he saw conflicted greatly with the philosophy of Aristotle and provided much new data to support the ideas of Copernicus.

Figure 1.6 Galileo Galilei (1564–1642). *(Art Resource, NY)*

Using his telescope, Galileo discovered that the Moon had mountains, valleys, and craters—terrain in many ways reminiscent of that on Earth. He found that the Sun had imperfections—dark blemishes now known as *sunspots*. By noting the changing appearance of these sunspots from day to day, he inferred that the Sun *rotates*, approximately once per month, around an axis roughly perpendicular to the ecliptic plane. These observations ran directly counter to accepted scientific beliefs. Galileo also saw four small points of light, invisible to the naked eye, orbiting the planet Jupiter and realized that they were moons. To Galileo, the fact that another planet had moons provided the strongest support for the Copernican model. Clearly, Earth was not the center of all things. He also found that Venus showed a complete cycle of phases, like those of our Moon (Figure 1.7), a finding that could be explained only by the planet's motion around the Sun. These observations were further strong evidence that Earth is not the center of things and that at least one planet orbited the Sun.

In 1610, Galileo published *The Starry Messenger*, detailing his observational findings and his controversial conclusions supporting the Copernican theory, and challenging both scientific orthodoxy and the religious dogma of the day. In 1616, his ideas were judged heretical, both his works and those of Copernicus were banned by the Church, and Galileo was instructed to abandon his astronomical pursuits. This he refused to do, and instead continued to amass data in support of the heliocentric view. In 1632, he raised the stakes further by publishing *Dialogue Concerning the Two Chief World Systems*, a side-by-side comparison of the Ptolemaic and Copernican models in the form of a discussion among three people, one of them a dull-witted Aristotelian, whose views time and again were roundly defeated by the arguments of one of his two companions, an articulate proponent of the heliocentric system. To make the book accessible to a wider audience, Galileo wrote it in Italian rather than Latin (the standard language of academic discourse at the time).

These actions brought Galileo into direct conflict with the Church. The Inquisition forced him, under threat of torture, to retract his claim that Earth orbits the

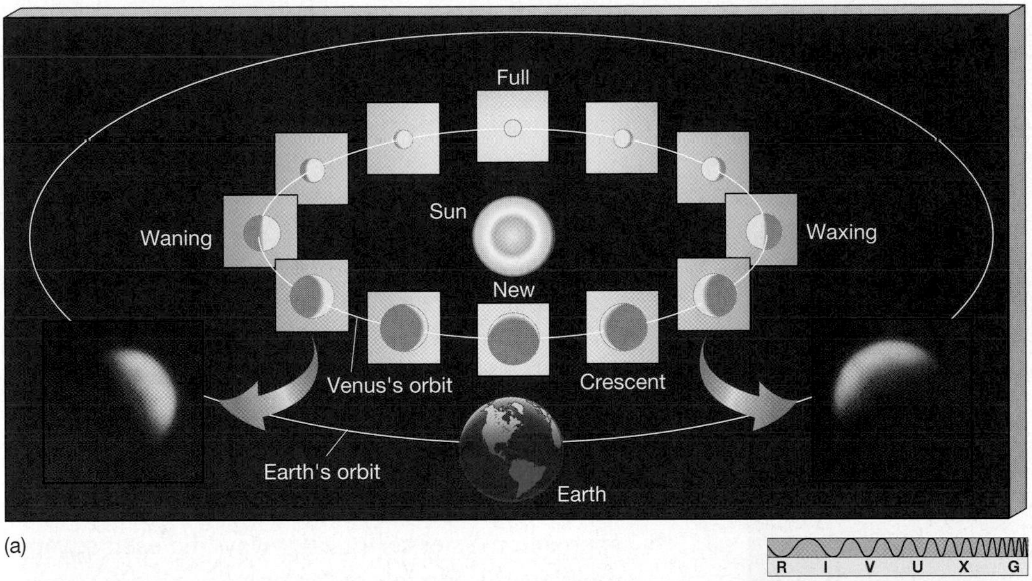

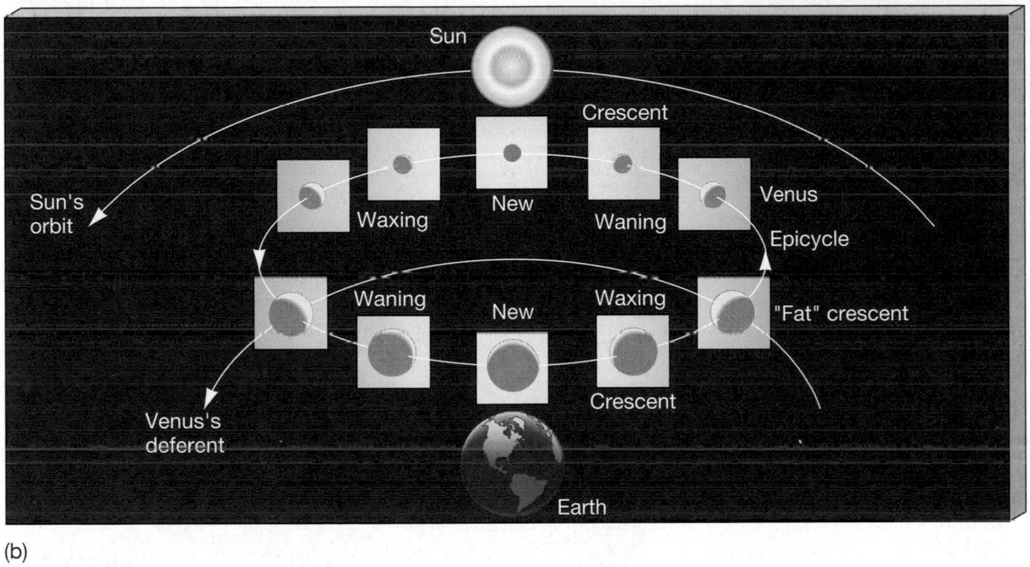

Figure 1.7 Venus Phases (a) The phases of Venus, at different points in the planet's orbit. If Venus orbits the Sun and is closer to the Sun than is Earth, as Copernicus maintained, then Venus should display phases, much as our Moon does. When directly between Earth and the Sun, Venus's unlit side faces us, and the planet is invisible to us. As Venus moves in its orbit, progressively more of its illuminated face is visible from Earth. Note the connection between orbital phase and the apparent size of the planet: Venus seems much larger in its crescent phase than when it is full because it is much closer to us during its crescent phase. (The insets at bottom left and right are actual photographs of Venus at two of its crescent phases.) *(Rita Beebee)* (b) The Ptolemaic model (see also Figure 1.3) is unable to account for these observations. In particular, the full phase of the planet cannot be explained. Seen from Earth, Venus reaches only a "fat crescent" phase, then begins to wane as it nears the Sun. *(Rita Beebe)*

Sun, and he was placed under house arrest in 1633. He remained imprisoned for the rest of his life. Not until 1992 did the Church publicly forgive Galileo's "crimes." But the damage to the orthodox view of the universe was done, and the Copernican genie was out of the bottle once and for all.

1.3 The Laws of Planetary Motion

2 At about the time Galileo was becoming famous for his telescopic observations, Johannes Kepler (Figure 1.8), a German mathematician and astronomer, announced his discovery of a set of simple empirical (that is, based on observation) "laws" that accurately described the motions of the planets. While Galileo was the first "modern" observer, Kepler was a pure theorist. He based his work almost entirely on the observations of another scientist (in part because of his own poor eyesight). Those observations, which predated the telescope by several decades, had been made by Kepler's employer, Tycho Brahe (1546–1601), arguably one of the greatest observational astronomers who ever lived.

Brahe's Complex Data

Tycho, as he is often called, was an eccentric aristocrat and a skillful observer. Born in Denmark, he was educated at some of the best universities in Europe, where

Figure 1.8 Johannes Kepler (1571–1630). *(E. Lessing/Art Resource, NY)*

he studied astrology, alchemy, and medicine. Most of his observations, which predated the invention of the telescope by several decades, were made at his own observatory, named Uraniborg, in Denmark (Figure 1.9). There, using instruments of his own design, Tycho maintained meticulous and accurate records of the stars, the planets, and noteworthy celestial events.

In 1597, having fallen out of favor with the Danish court, Tycho moved to Prague, which happens to be fairly close to Graz, in Austria, where Kepler lived. Kepler joined Tycho in Prague in 1600, and was put to work trying to find a theory that could explain Brahe's planetary data. When Tycho died a year later, Kepler inherited not only Brahe's position as Imperial Mathematician of the Holy Roman Empire but also his most priceless possession: the accumulated observations of the planets, spanning several decades. Tycho's observations, though made with the naked eye, were nevertheless of very high quality. Kepler set to work seeking a unifying principle to explain the motions of the planets without the need for epicycles. The effort was to occupy much of the remaining 29 years of his life.

Kepler's goal was to find a simple description of the solar system, within the basic framework of the Copernican model, that fit Tycho's complex mass of detailed observations. In the end, he had to abandon Copernicus's original simple notion of circular planetary orbits, but even greater simplicity emerged as a result. Kepler determined the shapes and relative sizes of each planet's orbit by triangulation not from different points on Earth but from different points on Earth's orbit, using observations made at many different times of the year. ∞ (Sec. P.5) Noting where the planets were on successive nights, he was able to infer the speeds at which they moved. After long years working with Brahe's planetary data and after many false starts and blind alleys, Kepler succeeded in summarizing the motions of all the known planets, including Earth, in the three **laws of planetary motion** that now bear his name.

✓ Concept Check

- In what ways did Galileo and Kepler differ in their approach to science? In what ways did each advance the Copernican view of the universe?

Figure 1.9 Tycho Brahe in his observatory Uraniborg, on the island of Hveen, in Denmark. *(The Royal Ontario Museum © ROM)*

INTERLUDE 1-1

The Scientific Method

The earliest known models of the universe were based largely on imagination and mythology, and made little attempt to explain the workings of the heavens in terms of known earthly experience. However, history shows that some early scientists did come to realize the importance of careful observation and testing to the formulation of their theories. The success of their approach changed, slowly but surely, the way science was done and opened the door to a fuller understanding of nature.

As knowledge from all sources was sought and embraced for its own sake, the influence of logic and reasoned argument grew and the power of myth diminished. People began to inquire more critically about themselves and the universe. They realized that thinking about nature was no longer sufficient, that looking at it was also necessary. Experiments and observations became a central part of the process of inquiry. To be effective, a *theory*—the framework of ideas and assumptions used to explain some set of observations and make predictions about the real world—must be continually tested. If experiments and observations favor it, a theory can be further developed and refined. If they do not, the theory must be reformulated or rejected, no matter how appealing it originally seemed.

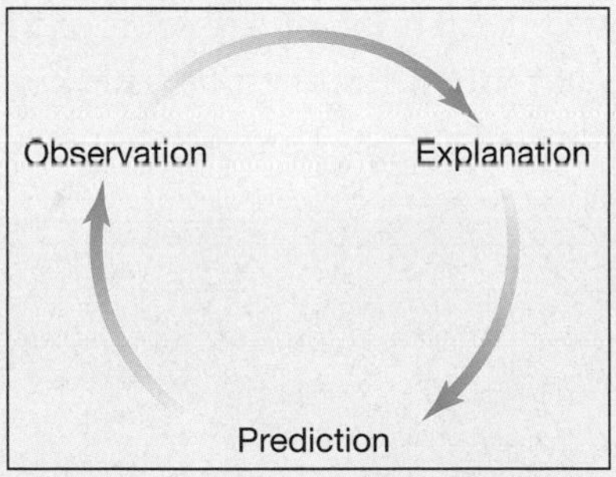

The process is illustrated schematically in the accompanying figure. This new approach to investigation, combining thinking and doing—that is, theory and experiment—is known as the *scientific method*. It lies at the heart of modern science.

Notice that there is no "end-point" to the process depicted in the figure. A theory can be invalidated by a single wrong prediction, but no amount of observation or experimentation can ever prove it correct. Theories simply become more and more widely accepted as their predictions are repeatedly confirmed.

In astronomy, we are rarely afforded the luxury of performing experiments to test our theories, so observation is vitally important. One of the first documented uses of the scientific method in an astronomical context was by Aristotle (384–322 B.C.) nearly 25 centuries ago. He noted that during a lunar eclipse, Earth casts a curved shadow onto the surface of the Moon, as shown in the accompanying figure. Earth's shadow, projected onto the Moon's surface, is indeed slightly curved. This is what Aristotle must have seen and recorded so long ago.

Because the shadow seemed always to be an arc of the same circle, Aristotle theorized that Earth, the cause of the shadow, must be round. On the basis of this *hypothesis*—this possible explanation of the observed facts—he then predicted that any and all future lunar eclipses would show Earth's shadow to be curved, regardless of our planet's orientation. That prediction has been tested every time a lunar eclipse has occurred. It has yet to be proved wrong. Aristotle was not the first person to argue that Earth is round, but he was apparently the first to offer proof using the lunar-eclipse method.

The reasoning Aristotle used forms the basis of all scientific inquiry today. He first made an observation. He then formulated a hypothesis to explain that observation. Then he tested the validity of his hypothesis by making predictions that could be confirmed or refuted by further observations. Observation, theory, and testing—these are the cornerstones of the scientific method, a technique whose power is demonstrated again and again throughout our text.

Used properly over a period of time, this rational, methodical approach enables us to arrive at conclusions that are mostly free of the personal bias and human values of any one scientist. The scientific method is designed to yield an objective view of the universe we inhabit.

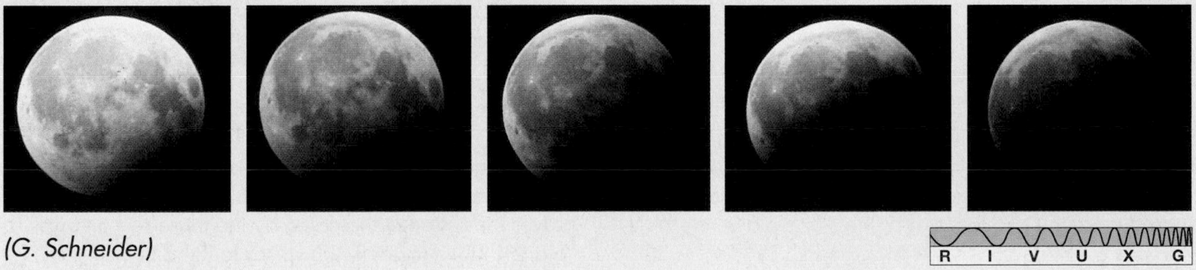

(G. Schneider)

Kepler's Simple Laws

3 *Kepler's first law* has to do with the *shapes* of the planetary orbits:

> The orbital paths of the planets are elliptical (*not* circular), with the Sun at one focus.

An **ellipse** is simply a flattened circle. Figure 1.10 illustrates a means of constructing an ellipse using a piece of string and two thumbtacks. Each point at which the string is pinned is called a **focus** (plural: foci) of the ellipse. The long axis of the ellipse, containing the two foci, is known as the *major axis*. Half the length of this long axis is referred to as the **semi-major axis**, a conventional measure of the ellipse's size. The **eccentricity** of the ellipse is the ratio of the distance between the foci to the length of the major axis. The length of the semi-major axis and the eccentricity are all we need to describe the size and shape of a planet's orbital path. (A circle is a special kind of ellipse in which the two foci happen to coincide, so the eccentricity is zero. The semi-major axis of a circle is simply its radius. See *More Precisely 1-1*.)

In fact, no planet's elliptical orbit is nearly as elongated as the one shown in Figure 1.10. With two exceptions (the paths of Mercury and Pluto), planetary orbits have such small eccentricities that our eyes would have trouble distinguishing them from true circles. Only because the orbits are so nearly circular were the Ptolemaic and Copernican models able to come as close as they did to describing reality.

Kepler's second law, illustrated in Figure 1.11, addresses the *speed* at which a planet traverses different parts of its orbit:

> An imaginary line connecting the Sun to any planet sweeps out equal areas of the ellipse in equal intervals of time.

While orbiting the Sun, a planet traces the arcs labeled A, B, and C in Figure 1.11 in equal times. Notice, however, that the distance traveled along arc C is greater than the distance traveled along arc A or arc B. Because the time is the same and the distance is different, the speed must vary: When a planet is close to the Sun, as in sector C, it moves much faster than when farther away, as in sector A.

These laws are not restricted to planets. They apply to *any* orbiting object. Spy satellites, for example, move very rapidly as they swoop close to Earth's surface not because they are propelled by powerful on-board rockets but because their highly eccentric orbits are governed by Kepler's laws.

Kepler published his first two laws in 1609, stating that he had proved them only for the orbit of Mars. Ten years later, he extended them to all the known

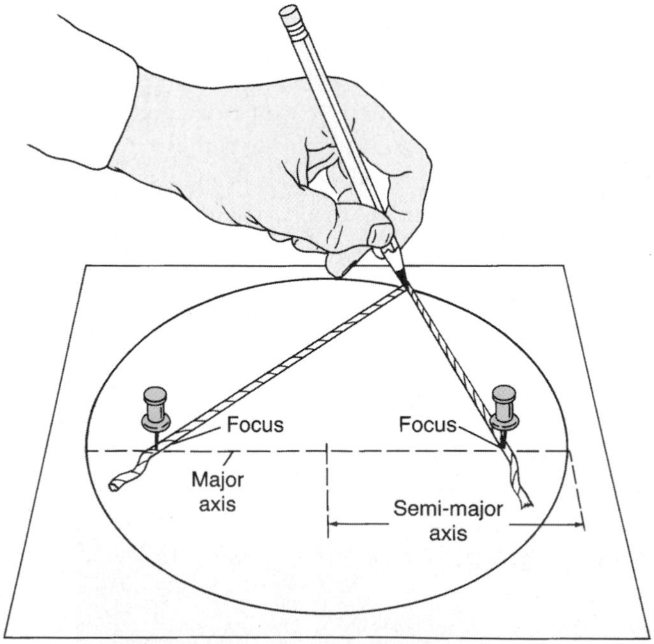

Figure 1.10 Ellipse An ellipse can be drawn with the aid of a string, a pencil, and two thumbtacks. The wider the separation of the foci, the more elongated, or eccentric, the ellipse. In the special case where the two foci are at the same place, the drawn curve is a circle.

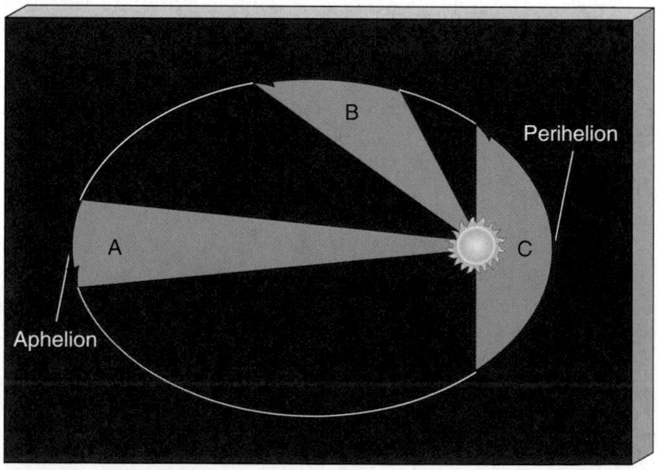

Figure 1.11 Kepler's Second Law A line joining a planet to the Sun sweeps out equal areas in equal intervals of time. The three shaded areas A, B, and C are equal. Any object traveling along the elliptical path would take the same amount of time to cover the distance indicated by the three red arrows. Therefore, planets move faster when closer to the Sun.

MORE PRECISELY 1-1

Some Properties of Planetary Orbits

Two numbers—semi-major axis and eccentricity—are all that are needed to describe the size and shape of a planet's orbital path. From them we can derive many other useful quantities. Two of the most important are the planet's *perihelion* (its point of closest approach to the Sun) and its *aphelion* (greatest distance from the Sun). From the definitions presented in the text, it follows that if the planet's orbit has semi-major axis a and eccentricity e, its perihelion is at a distance $a(1 + e)$ from the Sun, while its aphelion is at $a(1 - e)$. These points and distances are illustrated in the accompanying figure.

Note that, while the Sun resides at one focus, the other focus is empty and has no particular significance. Thus, for example, a hypothetical planet with a semi-major axis of 400 million km and an eccentricity of 0.5 (the eccentricity of the ellipse shown in the diagram) would range between $400 \times (1 + 0.5) = 200$ million km and $400 \times (1 - 0.5) = 600$ million km from the Sun over the course of one complete orbit. With $e = 0.9$, the range would be 40–760 million km, and so on.

No planet has an orbital eccentricity as large as 0.5—the planet with the most eccentric orbit is Pluto, with $e = 0.248$ (see Table 1.1). However, many meteoroids, and all comets (see Chapter 4) have eccentricities considerably greater than this. In fact, most comets visible from Earth have eccentricities very close to one. Their highly elongated orbits approach within a few A.U. of the Sun at perihelion, yet these tiny frozen worlds spend most of their time far beyond the orbit of Pluto.

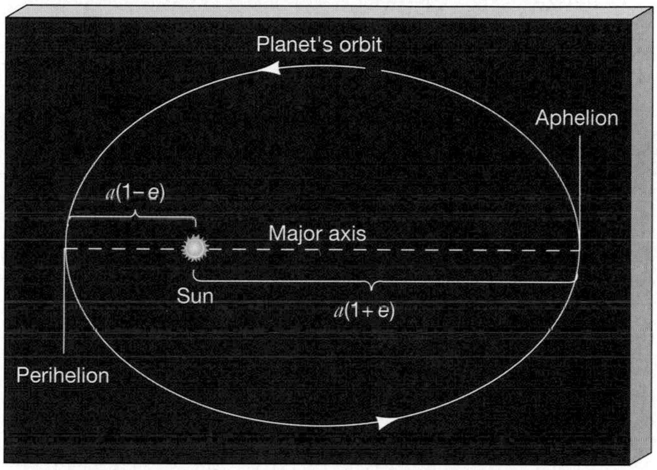

planets (Mercury, Venus, Earth, Mars, Jupiter, and Saturn) and added a third law relating the size of a planet's orbit to its sidereal orbital **period**, defined as the time needed for the planet to complete one circuit around the Sun.

Kepler's third law states that

> The square of a planet's orbital period is proportional to the cube of its semi-major axis.

Kepler's third law becomes particularly simple when we choose the (Earth) year as our unit of time and the *astronomical unit* as our unit of length. One **astronomical unit** (A.U.) is the semi-major axis of Earth's orbit around the Sun—the average distance between Earth and the Sun. Like the light-year, the astronomical unit is a unit custom-made for the vast distances encountered in astronomy. Using these units for time and distance, we can conveniently write Kepler's third law for any planet in the form

$$P^2 \text{ (in Earth years)} = a^3 \text{ (in astronomical units)},$$

where P is the planet's sidereal orbital period and a is its semi-major axis.

Table 1.1 presents some basic data describing the orbits of the nine planets now known. Renaissance astronomers knew these properties for the innermost six planets and used them to construct the heliocentric model of the solar system. However, Kepler's laws are obeyed by *all* the known planets, and *not just by the six on which he based his conclusions*. For purposes of verifying Kepler's third law, the rightmost column lists the ratio P^2/a^3. As we have just seen, in the units used in the table, this number should equal one in all cases. The small but significant deviations of P^2/a^3 from one in the cases of Uranus and Neptune are caused by the gravitational attraction between those two planets (see Chapter 7).

Concept Check

- Do Kepler's laws apply to comets orbiting the Sun? Do they apply to the moons of Jupiter?

TABLE 1.1 Some Planetary Properties

PLANET	ORBITAL SEMI-MAJOR AXIS, a (astronomical units)	ORBITAL PERIOD, P (Earth years)	ORBITAL ECCENTRICITY, e	P^2/a^3
Mercury	0.387	0.241	0.206	1.002
Venus	0.723	0.615	0.007	1.001
Earth	1.000	1.000	0.017	1.000
Mars	1.524	1.881	0.093	1.000
Jupiter	5.203	11.86	0.048	0.999
Saturn	9.537	29.42	0.054	0.998
Uranus	19.19	83.75	0.047	0.993
Neptune	30.07	163.7	0.009	0.986
Pluto	39.48	248.0	0.249	0.999

The Dimensions of the Solar System

4 Kepler's laws allow us to construct a scale model of the solar system, with the correct shapes and *relative* sizes of all the planetary orbits, but they do not tell us the *actual* size of any orbit. We can express the distance to each planet only in terms of the distance from Earth to the Sun—in other words, only in astronomical units. Why is this? Because Kepler's measurements were based on triangulation using a portion of Earth's orbit as a baseline, so his distances could be expressed only relative to the size of that orbit, which was not itself determined.

Our model of the solar system would be analogous to a road map of the United States showing the *relative* positions of cities and towns but lacking the all-important scale marker indicating distances in kilometers or miles. For example, we would know that Kansas City is about three times more distant from New York than it is from Chicago, but we would not know the actual mileage between any two points on the map. If we could somehow determine the value of the astronomical unit—in kilometers, say—we would be able to add the vital scale marker to our map of the solar system and compute the exact distances between the Sun and each of the planets.

The modern method for deriving the absolute scale of the solar system uses a technique called *radar ranging.* The word **radar** is an acronym for **ra**dio **d**etection **a**nd **r**anging. Radio waves are transmitted toward an astronomical body, such as a planet. Their returning echo indicates the body's direction and range, or distance, in absolute terms—in other words, in kilometers rather than in astronomical units. Multiplying the round-trip travel time of the radar signal (the time elapsed between transmission of the signal and reception of the echo) by the speed of light (300,000 km/s, which is also the speed of radio waves), we obtain twice the distance to the target planet.

We cannot use radar ranging to measure the distance to the Sun directly because radio signals are absorbed at the solar surface and are not reflected back to Earth. Instead, the planet Venus, whose orbit periodically brings it closest to Earth, is the most common target for this technique. Figure 1.12 is an idealized diagram of the Sun–Earth–Venus orbital geometry. Neglecting for the sake of simplicity the small eccentricities of the planets' orbits, we see from the figure that the distance from Earth to Venus at the point of closest approach is approximately 0.3 A.U. Radar signals bounced off Venus at that instant return to Earth in about 300 seconds, indicating that Venus lies 45,000,000 km from Earth. With this information, we can compute the magnitude of the astronomical unit: 0.3 A.U. is 45,000,000 km, so 1 A.U. is 45,000,000 km/0.3, or 150,000,000 km.

Through precise radar ranging, the astronomical unit is now known to be 149,597,870 km. In this text, we will use the rounded-off value of 1.5×10^8 km. Having determined the value of the astronomical unit, we can re-express the sizes of the other planetary orbits in more familiar units, such as miles or kilometers. The entire scale of the solar system can then be calibrated to high precision.

Concept Check

- Why don't Kepler's laws tell us the value of the astronomical unit?

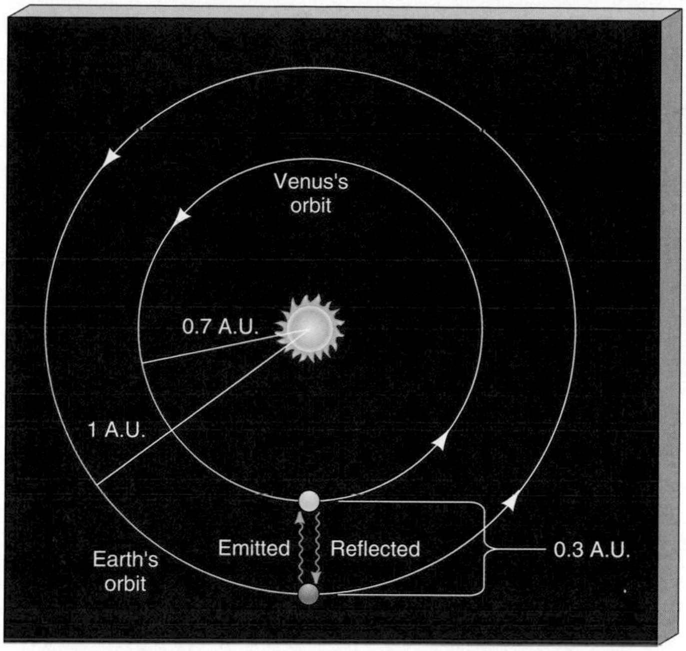

Figure 1.12 **Astronomical Unit** Simplified geometry of the orbits of Earth and Venus as they move around the Sun. The wavy lines represent the paths along which radar signals might be transmitted toward Venus and received back at Earth at the moment when Venus is at its minimum distance from Earth. Because the radius of Earth's orbit is 1 A.U. and that of Venus is about 0.7 A.U., we know that the one-way distance covered by the signal is 0.3 A.U. Thus, radar measurements allow us to determine the astronomical unit in kilometers.

1.4 Newton's Laws

What prevents the planets from flying off into space or falling into the Sun? What causes them to revolve about the Sun, apparently endlessly? Of course, the motions of the planets obey Kepler's three laws, but only by considering something more fundamental can we truly understand planetary motion.

The Laws of Motion

4 In the seventeenth century, the British mathematician Isaac Newton (Figure 1.13) developed a deeper understanding of the way *all* objects move and interact with one another. Newton's theories form the basis for what today is known as **Newtonian mechanics**. Three basic laws of motion, the law of universal gravitation, and a little calculus (which Newton also invented) are sufficient to explain and quantify virtually all of the complex dynamic behavior we see on Earth and throughout the universe. Only in extreme circumstances do Newton's laws break down, and this fact was not realized until the twentieth century, when Albert Einstein's theories of relativity once again revolutionized our view of the universe (see Chapter 13). Most of the time, however, Newtonian mechanics provides an excellent description of the motion of planets, stars, and galaxies throughout the cosmos.

Figure 1.14 illustrates *Newton's first law of motion*, which states that, unless some external **force** changes its state of motion, an object at rest remains at rest and a moving object continues to move forever in a straight line with constant speed. An example of an external force would be the force exerted by, say, a brick wall when a rolling ball glances off it, or the force exerted on a pitched ball by a baseball bat. In either case, a force changes the original motion of the object. The tendency of an object to keep moving at the same speed and in the same direction unless acted upon by a force is known as **inertia**. A familiar measure of an object's inertia is its **mass**—loosely speaking, the total amount of matter the object contains. The greater an object's mass, the more inertia it has and the greater is the force needed to change its motion.

Newton's first law contrasts sharply with the view of Aristotle, who maintained (incorrectly) that the natural state of an object was to be *at rest*—most probably an opinion based on Aristotle's observations of the effect of friction. To simplify our discussion, we will neglect

Figure 1.13 **Isaac Newton** (1642–1727). *(The Granger Collection)*

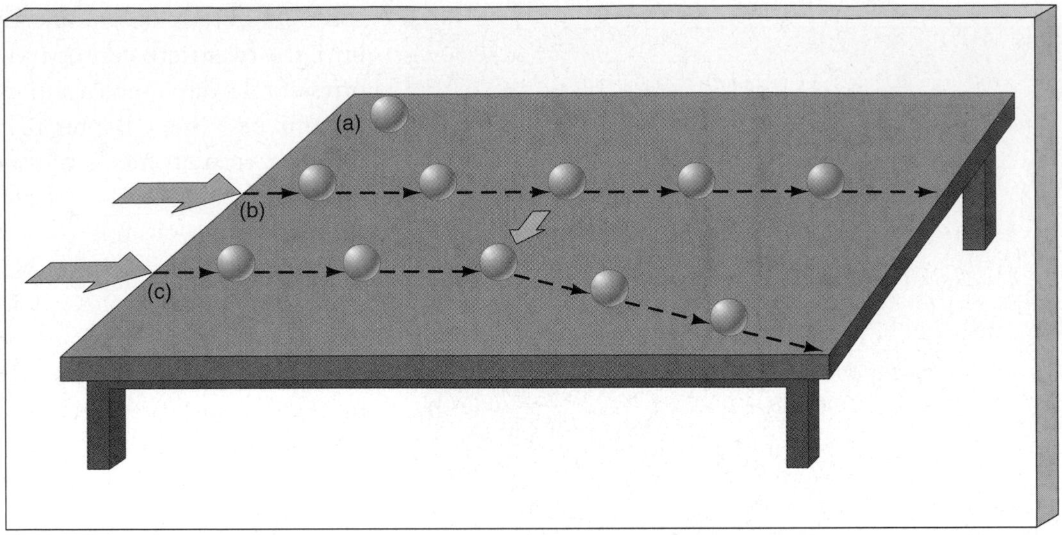

Figure 1.14 **Newton's First Law** An object at rest will remain at rest (a) until some force acts on it (b). Thereafter it will remain in that state of uniform motion until another force acts on it. The arrow in (c) shows a second force acting at a direction different from the first, which causes the object to change its direction of motion.

friction—the force that slows balls rolling along the ground, blocks sliding across tabletops, and baseballs moving through the air. In any case, this is not an issue for the planets because there is no appreciable friction in outer space. The fallacy in Aristotle's argument was first exposed by Galileo, who conceived of the notion of inertia long before Newton formalized it into a law.

The rate of change of the velocity of an object—speeding up, slowing down, or simply changing direction—is called its **acceleration**. *Newton's second law* states that the acceleration of an object is directly proportional to the applied force and inversely proportional to the object's mass—that is, the greater the force acting on the object, or the smaller the mass of the object, the greater its acceleration. Thus, if two objects are pulled with the same force, the more massive one will accelerate less. If two identical objects are pulled with different forces, the one experiencing the greater force will accelerate more.

Finally, *Newton's third law* tells us that forces cannot occur in isolation—if body A exerts a force on body B, then body B necessarily exerts a force on body A that is equal in magnitude but oppositely directed.

Gravity

5 Forces can act either *instantaneously* or *continuously*. To a good approximation, the force from a baseball bat that hits a home run can be thought of as instantaneous in nature. A good example of a continuous force is the one that prevents the baseball from zooming off into space—**gravity**, the phenomenon that started Newton on the path to the discovery of his laws. Newton hypothesized that any object having mass exerts an attractive **gravitational force** on all other massive objects. The more massive an object, the stronger its gravitational pull.

Consider a baseball thrown upward from Earth's surface, as illustrated in Figure 1.15. In accordance with Newton's first law, the downward force of Earth's gravity steadily modifies the baseball's velocity, slowing its initial upward motion and eventually causing the ball

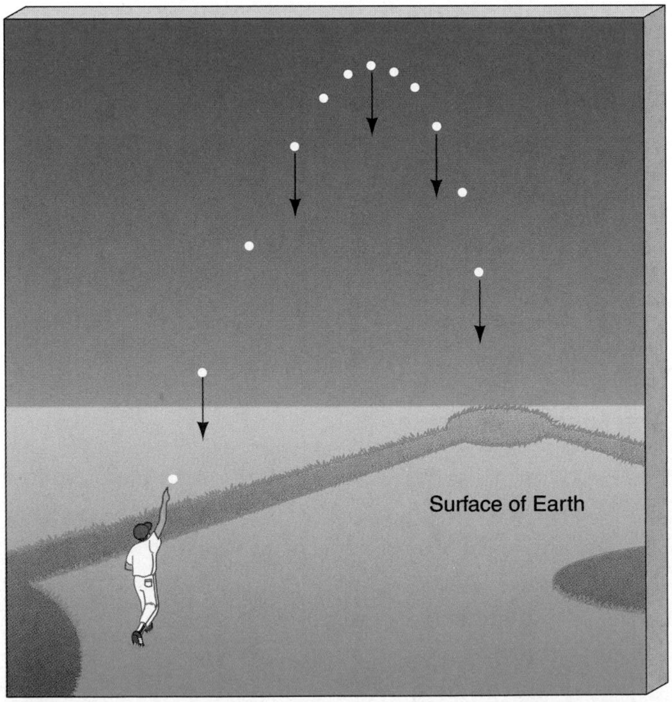

Figure 1.15 **Gravity** A ball thrown upward from the surface of a massive object, such as a planet, is pulled continuously by the gravity of that planet (and, conversely, the gravity of the ball continuously pulls the planet).

to fall back to the ground. Of course, the baseball, having some mass of its own, also exerts a gravitational pull on Earth. By Newton's third law, this force is equal in magnitude to the weight of the ball (the weight of an object is a measure of the force with which Earth attracts that object), but oppositely directed. By Newton's second law, however, Earth has a much greater effect on the light baseball than the baseball has on the much more massive Earth. The ball and Earth each feel the same gravitational force, but Earth's *acceleration* as a result of this force is much smaller.

Now consider the trajectory of a baseball batted from the surface of the Moon. The pull of gravity is about one-sixth as great on the Moon as on Earth, so the baseball's velocity changes more slowly—a typical home run in a ballpark on Earth would travel nearly half a mile on the Moon. The Moon, less massive than Earth, has less gravitational influence on the baseball. The magnitude of the gravitational force, then, depends on the *masses* of the attracting bodies. In fact, it is directly proportional to the product of the two masses.

Studying the motions of the planets reveals a second aspect of the gravitational force. At locations equidistant from the Sun's center, the gravitational force has the same strength, and is always directed toward the Sun. Furthermore, detailed calculation of the planets' accelerations as they orbit the Sun reveals that the strength of the Sun's gravitational pull decreases in proportion to the *square* of the distance from the Sun. (Newton is said to have first realized this fact by comparing the accelerations not of the planets but of the Moon and an apple falling to the ground—the basic reasoning is the same in either case.) The force of gravity is said to obey an **inverse-square law** (see *More Precisely 1-2*).

As shown in Figure 1.16, inverse-square forces decrease rapidly with distance from their source. For example, tripling the distance makes the force $3^2 = 9$ times weaker, while multiplying the distance by five results in a force that is $5^2 = 25$ times weaker. Despite this rapid decrease, the force never quite reaches zero. The gravitational pull of an object having some mass can never be completely extinguished.

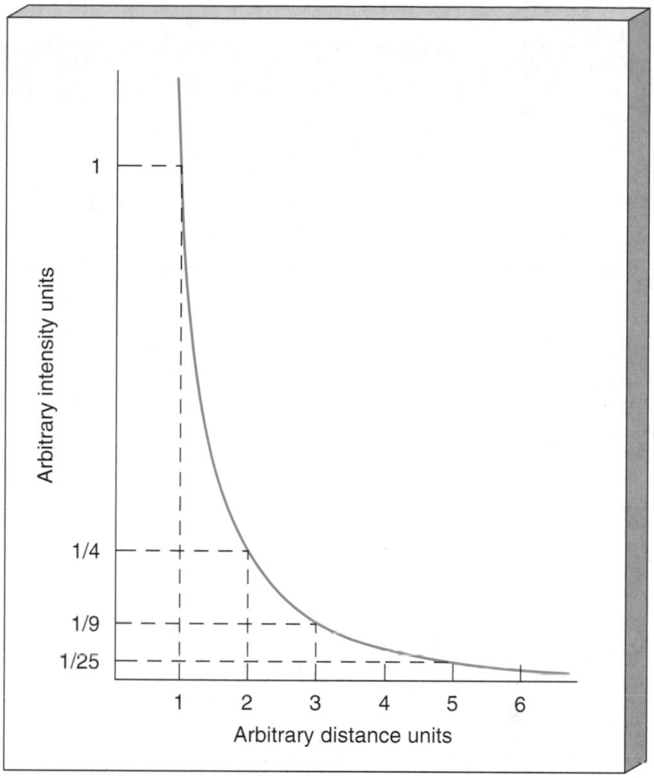

Figure 1.16 **Gravitational Force** Inverse-square forces rapidly weaken with distance from their source. The strength of the Sun's gravitational force decreases with the square of the distance from the Sun. The force never quite diminishes to zero, however, no matter how far away from the Sun we go.

Planetary Motion

The mutual gravitational attraction of the Sun and the planets, as expressed by Newton's law of gravity, is responsible for the observed planetary orbits. As depicted in Figure 1.17, this gravitational force continuously pulls each planet toward the Sun, deflecting its forward

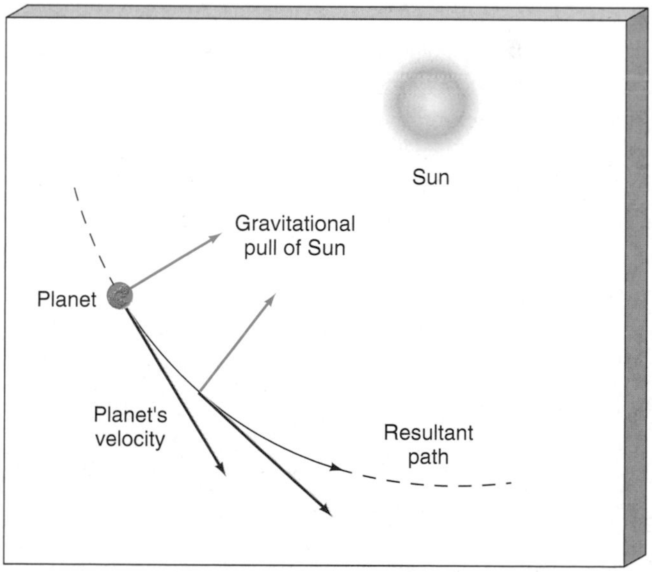

Figure 1.17 **Sun's Gravity** The Sun's inward pull of gravity on a planet competes with the planet's tendency to continue moving in a straight line. These two effects combine, causing the planet to move smoothly along an intermediate path, which continuously "falls around" the Sun. This unending tug-of-war between the Sun's gravity and the planet's inertia results in a stable orbit.

MORE PRECISELY 1-2

Weighing the Sun

Every particle of matter in the universe attracts every other particle with a force that is directly proportional to the product of the masses of the particles and inversely proportional to the square of the distance between them. In other words, two bodies of masses m_1 and m_2, separated by a distance r, attract each other with a force F that is proportional to $(m_1 \times m_2)/r^2$. The constant of proportionality is known as the gravitational constant, or often simply as *Newton's constant*, and is always denoted by the letter G. We can then express the law of gravity as

$$F = \frac{G\, m_1 m_2}{r^2}.$$

The value of G has been measured on Earth in extremely delicate laboratory experiments. In SI units, its value is 6.67×10^{-11} newton meter²/kilogram² (N m²/kg²). (The newton is the SI unit of force, defined as the force required to accelerate a mass of 1 kg at a rate of 1 m/s². One newton is approximately 0.22 pounds.)

Knowing the value of G, we can use Newtonian mechanics to calculate the relationship between the distance (r) and the speed (v) of a planet moving in a circular orbit around the Sun (of mass M). By calculating the force required to keep the planet moving in a circle and comparing it with the gravitational force due to the Sun, it can be shown that the circular speed is

$$v = \sqrt{\frac{GM}{r}}.$$

Dividing this speed into the circumference of the orbit ($2\pi r$) to give the orbital period $P = 2\pi r/v$, we obtain a version of Kepler's third law equivalent to the formula presented in the text:

$$P = 2\pi\sqrt{\frac{r^3}{GM}}.$$

We can then use Newtonian mechanics to *weigh* the Sun. Inserting the known values of v = 30 km/s, r = 1 A.U. = 1.5×10^{11} m, and G (given above), we can calculate the mass of the Sun. The result is M = 2.0×10^{30} kg—an enormous mass by terrestrial standards. Similarly, knowing the distance to the Moon and the length of the (sidereal) month, we can measure the mass of Earth to be 6.0×10^{24} kg.

In fact, this is how *all* masses are measured in astronomy. Because we can't just go out and weigh an astronomical object when we need to know its mass, we must look for its gravitational influence on something else. This principle applies to planets, stars, galaxies, and even clusters of galaxies—very different objects but all subject to the same physical laws.

---

motion into a curved orbital path. Because the Sun is much more massive than any of the planets, it dominates the interaction. We might say that the Sun "controls" the planets, not the other way around.

The planet–Sun interaction sketched here is analogous to what occurs when you whirl a rock at the end of a string above your head. The Sun's gravitational pull is your hand and the string, and the planet is the rock at the end of that string. The tension in the string provides the force necessary for the rock to move in a circular path. If you were suddenly to release the string—which would be like eliminating the Sun's gravity—the rock would fly away along a tangent to the circle, in accordance with Newton's first law.

In the solar system, at this very moment, Earth is moving under the combined influence of these two effects: the competition between gravity and inertia. The net result is a stable orbit, despite our continuous rapid motion through space. In fact, Earth orbits the Sun at a speed of about 30 km/s, or some 70,000 mph. (Verify this for yourself by calculating how fast Earth must move to complete a circle of radius 1 A.U.—and hence of circumference 2π A.U., or 940 million km—in one year, or 3.2×10^7 seconds.) *More Precisely 1-2* describes how astronomers can use Newtonian mechanics and the law of gravity to measure the masses of Earth, the Sun, and many other astronomical objects by studying the orbits of objects near them.

Kepler's Laws Reconsidered

Newton's laws of motion and his law of universal gravitation provided a theoretical explanation for Kepler's empirical laws of planetary motion. Just as Kepler modified the Copernican model by introducing ellipses in place of circles, so too did Newton make corrections to Kepler's first and third laws. Because the Sun and a planet feel equal and opposite gravitational forces (by Newton's third law), the Sun must also move (by Newton's first law), due to the planet's gravitational pull. As a result, the planet does not orbit the exact center of the Sun, but instead, both the planet and the Sun orbit their common **center of mass**, and Kepler's first law becomes:

> The orbit of a planet around the Sun is an ellipse having the center of mass of the planet–Sun system at one focus.

As shown in Figure 1.18, the center of mass for a system consisting of two objects of comparable mass does not lie within either object. For identical masses (Figure 1.18a), the orbits are identical ellipses with a common focus located midway between the two objects. For unequal masses (as in Figure 1.18b), the elliptical orbits still share a focus and have the same eccentricity, but the more massive object moves more slowly and on a tighter orbit. (Note that Kepler's second law, as stated earlier, continues to apply without modification to each orbit separately, but the rates at which the two orbits sweep out area are different.) In the extreme case of a planet orbiting the much more massive Sun (Figure 1.18c), the path traced out by the Sun's center lies entirely within the Sun.

The change to Kepler's third law is small in the case of a planet orbiting the Sun, but may be very important in other circumstances, such as the orbits of two stars that are gravitationally bound to one another. Following through the mathematics of Newton's theory, we find that the true relationship between the semi-major axis a (measured in astronomical units) of the planet's orbit relative to the Sun and its orbital period P (in Earth years) is

$$P^2 \text{ (in Earth years)} = \frac{a^3 \text{ (in astronomical units)}}{M_{\text{total}} \text{ (in solar units)}},$$

where M_{total} is the combined mass of the two objects expressed in terms of the mass of the Sun (see also *More Precisely 1-2*). Notice that this restatement of Kepler's third law preserves the proportionality between P^2 and a^3, but now the proportionality also includes M_{total}, so it is *not* quite the same for all the planets. The Sun's mass is so great, however, that the differences in M_{total} among the various combinations of the Sun and the other planets are almost unnoticeable. Therefore Kepler's third law, as originally stated, is a very good approximation. This modified form of Kepler's third law is true in *all* circumstances, inside or outside the solar system.

Concept Check

- Explain, in terms of Newton's laws of motion and gravity, why planets orbit the Sun.

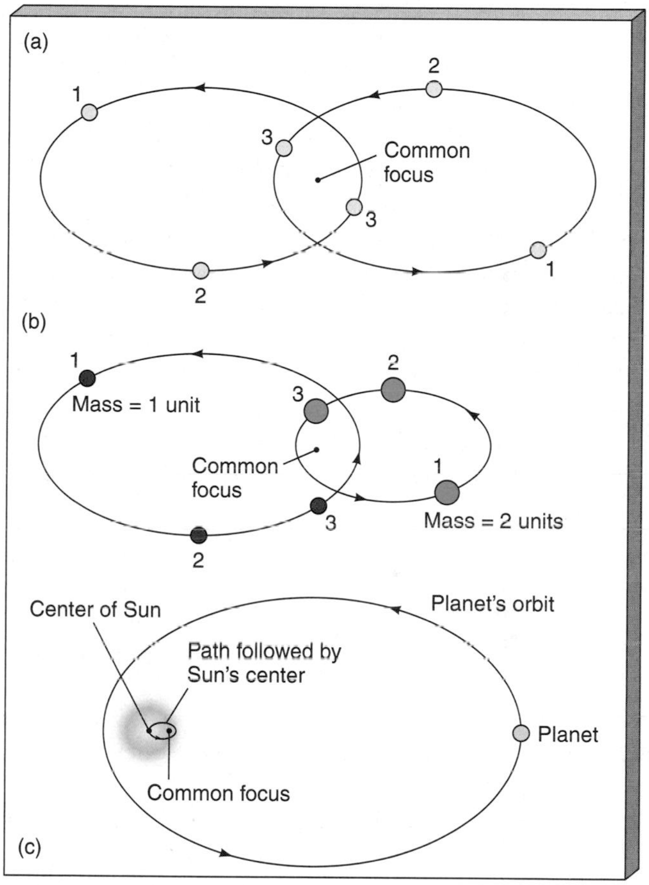

Figure 1.18 Orbits (a) The orbits of two bodies (stars, for example) with equal masses, under the influence of their mutual gravity, are identical ellipses with a common focus. That focus is not at the center of either star but instead at their center of mass, located midway between them. The pairs of numbers indicate the positions of the two bodies at three times. (Note that the line joining the bodies at any given time always passes through the common focus.) (b) The orbits of two bodies when one body is twice as massive as the other. The elliptical orbits again have a common focus (at the center of mass of the two-body system), and the two ellipses have the same eccentricity, but the more massive body moves more slowly and in a smaller orbit. The larger ellipse is twice the size of the smaller one. (c) In the extreme case of a hypothetical planet orbiting the Sun, the common focus of the two orbits lies inside the Sun.

Chapter Review

Summary

Planets sometimes appear to temporarily reverse their direction of motion and undergo **retrograde motion** (p. 24). **Geocentric** (p. 24) models of the universe, such as the **Ptolemaic model** (p. 25) have the Sun, the Moon, and all the other planets orbiting Earth. To account for retrograde motion within the geocentric picture, it was necessary to suppose that planets moved on small circles called **epicycles** (p. 24), whose centers orbited Earth on larger circles called **deferents** (p. 24). Copernicus's **heliocentric** (p. 26) view of the solar system holds that Earth, like all the other planets, orbits the Sun. This model naturally explains retrograde motion and the observed brightness variations of the planets. The widespread realization that the solar system is Sun-centered and not Earth-centered is known as the **Copernican revolution** (p. 26).

Kepler's three **laws of planetary motion** (p. 30) state that (1) planetary orbits are **ellipses** (p. 32) having the Sun as one **focus** (p. 32), (2) a planet moves faster as its orbit takes it closer to the Sun, and (3) the **semi-major axis** (p. 32) of the orbit is related in a simple way to the planet's orbit **period** (p. 33). Most planets move on orbits whose **eccentricities** (p. 32) are quite small, so their paths differ only slightly from perfect circles.

The average distance from Earth to the Sun is one **astronomical unit** (p. 33), today precisely determined by bouncing **radar** (p. 34) signals off the planet Venus.

According to the principles of **Newtonian mechanics** (p. 35), the tendency of a body to keep moving at constant velocity is the body's **inertia** (p. 35). The greater the body's **mass** (p. 35), the greater its inertia. To change the velocity, a **force** (p. 35) must be applied to the body. The rate of change of velocity, called the **acceleration** (p. 36), is equal to the applied force divided by the body's mass. To explain planetary orbits, Newton postulated that **gravity** (p. 36) attracts the planets to the Sun. Every object with any mass exerts a **gravitational force** (p. 36) on all other objects, and the strength of this force decreases with distance according to an **inverse-square law** (p. 37). Newton's laws imply that a planet does not orbit the precise center of the Sun. Instead the planet and the Sun orbit the **center of mass** (p. 39) of the planet–Sun system.

Review and Discussion

1. Briefly describe the geocentric model of the universe.
2. What was the basic flaw?
3. What was the great contribution of Copernicus to our knowledge of the solar system?
4. What was the Copernican Revolution?
5. What discoveries of Galileo helped confirm the views of Copernicus?
6. Why is Galileo often referred to as the father of experimental science?
7. State Kepler's three laws of orbital motion.
8. What is meant by the statement that Kepler's laws are empirical in nature?
9. If radio waves cannot be reflected from the Sun, how can radar be used to find the distance from Earth to the Sun?
10. List the two modifications made by Newton to Kepler's laws.
11. Why would a baseball thrown upward from the surface of the Moon go higher than one thrown with the same velocity from the surface of Earth?
12. What would happen to Earth if the Sun's gravity were suddenly "turned off"?

True or False?

____ **1.** Aristotle was first to propose that all planets revolve around the Sun.

____ **2.** Ptolemy's geocentric model accurately prediced the positions of the planets, Moon, and the Sun.

____ **3.** The heliocentric model of the universe holds that Earth is at the center of the universe.

____ **4.** Copernicus's theories gained widespread scientific acceptance during his own lifetime.

____ **5.** Galileo's telescopic observations provided direct evidence against the geocentric model of the universe.

____ **6.** Kepler's discoveries regarding the orbital motion of the planets were based on his own observations.

____ **7.** A circle has an eccentricity of zero.

____ **8.** The astronomical unit is a distance equal to the semi-major axis of Earth's orbit around the Sun.

____ **9.** The speed of a planet orbiting the Sun is independent of the planet's position in its orbit.

____ **10.** Kepler's laws hold only for the six planets known in his time.

____ **11.** Kepler never determined the true distances between the planets and the Sun, only their relative distances.

____ **12.** Using his laws of motion and gravity, Newton was able to prove Kepler's laws.

Fill in the Blank

1. When the planets Mars, Jupiter, and Saturn appear to move "backwards" (westward) in the sky relative to the stars, this is known as _______ motion.
2. The geocentric model holds that _______ is at the center of the solar system.
3. Observation, theory, and testing are the cornerstones of the _______.
4. Galileo discovered _______ of Jupiter, the _______ of Venus, and the Sun's rotation from observations of _______.
5. Kepler's laws were based on _______'s observations.
6. Kepler discovered that the shape of an orbit is an _______, not a _______ as had previously been believed.
7. According to Kepler's first law, the Sun lies at the _______ of a planet's orbit.
8. Kepler's third law relates the _______ of the orbital period to the _______ of the semi-major axis.
9. The modern method of measuring the astronomical unit uses _______ measurements of Venus.
10. One astronomical unit is approximately _______ km.
11. Newton's first law states that a moving object will continue to move in a straight line with constant speed unless acted upon by a _______.
12. Newton's law of gravity states that the gravitational force between two objects depends on the _______ of their masses and inversely on the _______ of their separation.

Problems

1. Tycho Brahe's observations of the stars and planets were accurate to about 1 arc minute (1′). To what distance does this angle correspond (a) at the distance of the Moon, (b) of the Sun, (c) of Saturn (at closest approach)?
2. To an observer on Earth, through what angle will Mars appear to move relative to the stars over the course of 24 hours, when the two planets are at closest approach? Take Earth and Mars to move on circular orbits of radii 1.0 A.U. and 1.5 A.U., respectively, in exactly the same plane. Will the apparent motion be prograde or retrograde?
3. How long would a radar signal take to complete a round trip between Earth and Mars when the two planets are 0.7 A.U. apart?
4. An asteroid has a perihelion distance of 2.0 A.U. and an aphelion distance of 4.0 A.U. Calculate its orbital semi-major axis, eccentricity, and period.
5. Halley's comet has a perihelion distance of 0.6 A.U. and an orbital period of 76 years. What is its aphelion distance from the Sun?
6. Using the data in Table 1.1, show that Pluto is closer to the Sun at perihelion (the point of closest approach to the Sun in its orbit) than Neptune is at any point in its orbit.
7. Jupiter's moon Callisto orbits the planet at a distance of 1.88 million km. Callisto's orbital period about Jupiter is 16.7 days. What is the mass of Jupiter? ∞ (Sec. 1.6)
8. The acceleration due to gravity at Earth's surface is 9.80 m/s^2. What is the acceleration at altitudes of (a) 100 km? (b) 1000 km? (c) 10,000 km? Take Earth's radius to be 6400 km.
9. Use Newton's law of gravity to calculate the force of gravity between you and Earth. Convert your answer, which will be in newtons, to pounds using the conversion 4.45 N equals 1 pound.
10. The Moon's mass is 7.3×10^{22} kg and its radius is 1700 km. What is the speed of a spacecraft moving in a circular orbit just above the lunar surface?

Projects

1. Look in an almanac for the date of opposition of one or all of these bright planets: Mars, Jupiter, and Saturn. At opposition, these planets are at their closest points to Earth and therefore are at their largest and brightest in the night sky. Observe these planets. How long before opposition does each planet's retrograde motion begin? How long afterwards does it end?
2. Use a small telescope to observe Jupiter's four largest moons. Note their brightnesses and locations relative to Jupiter. If you watch over a period of several nights, draw what you see. You'll notice that these moons change their positions as they orbit the gigantic planet. Check the charts given monthly in *Astronomy* or *Sky & Telescope* magazine to identify each moon you see.

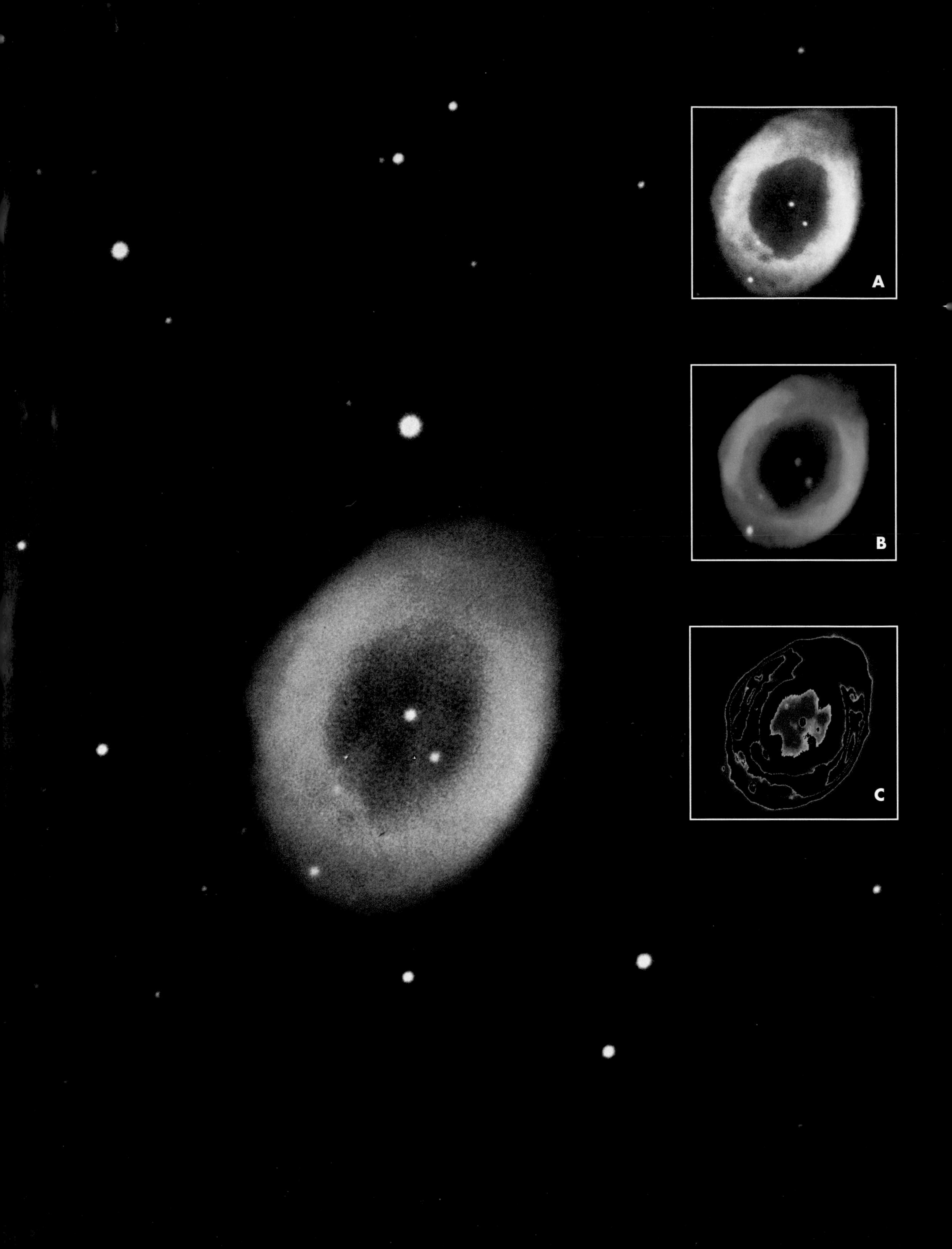
A
B
C

2 LIGHT AND MATTER

The Inner Workings of the Cosmos

LEARNING GOALS

Studying this chapter will enable you to:

1 Discuss the nature of electromagnetic radiation, and tell how that radiation transfers energy and information through interstellar space.

2 Describe the major regions of the electromagnetic spectrum.

3 Explain how we can determine an object's temperature by observing the radiation it emits.

4 Describe the characteristics of continuous, emission, and absorption spectra, and the conditions under which each is produced.

5 Specify the basic components of the atom and describe our modern conception of its structure.

6 Explain how electron transitions within atoms produce unique emission and absorption spectra.

(Opposite page, background) The Ring Nebula in the constellation Lyra is one of the most magnificent sights in the nighttime sky. Seen here glowing in the light of its own emitted radiation, the nebula is actually the expanding outer atmosphere of a nearly dead star. *(AURA)*

(Inset A) This nearly true color view clearly shows the dying dwarf star and the expanding gas cloud, which is really a three-dimensional shell and not a ring. *(Smithsonian Astrophysical Observatory)*

(Inset B) False-color images can sometimes enhance certain features. Here some of the fine structure in the shell of gas can be seen more clearly. *(Smithsonian Astrophysical Observatory)*

(Inset C) Contour images, like this one derived from red light emitted by hydrogen gas in the nebula's shell and star's lingering atmosphere, can be used to map regions of relative brightness. *(Smithsonian Astrophysical Observatory)*

Astronomical objects are more than just things of beauty in the night sky. Planets, stars, and galaxies are of vital significance if we are to understand fully the big picture—the grand design of the universe. Each object is a source of information about the universe—its temperature, its chemical composition, its state of motion, its past history. The starlight we see tonight began its journey to Earth decades, centuries—even millennia—ago. The faint rays from the most distant galaxies have taken billions of years to reach us. The stars and galaxies in the night sky show us not just the far away but also the long ago. In this chapter, we begin our study of how astronomers extract information from the light emitted by astronomical objects. The observational and theoretical techniques that enable researchers to determine the nature of distant atoms by the way they emit and absorb light are the indispensable foundation of modern astronomy.

2.1 Information from the Skies

Figure 2.1 shows a galaxy in the constellation Andromeda. On a dark, clear night, far from cities or other sources of light, the Andromeda Galaxy, as it is generally called, can be seen with the naked eye as a faint, fuzzy patch on the sky, comparable in diameter to the full Moon. Yet the fact that it is visible from Earth belies this galaxy's enormous distance from us. It lies roughly three million light-years away. An object at such a distance is truly inaccessible in any realistic human sense. Even if a space probe could miraculously travel at the speed of light, it would need three million years to reach this galaxy and three million more to return with its findings. Considering that civilization has existed on Earth for fewer than 10,000 years (and its prospects for the next 10,000 are far from certain), even this unattainable technological feat would not provide us with a practical means of exploring other galaxies—or even the farthest reaches of our own galaxy, several tens of thousands of light-years away.

Light and Radiation

How do astronomers know anything about objects far from Earth? How can we obtain detailed information about any planet, star, or galaxy too distant for a personal visit or any kind of controlled experiment? The answer is that we use the laws of physics, as we know them here on Earth, to interpret the **electromagnetic radiation** emitted by these objects. *Radiation* is any way in which energy is transmitted through space from one point to another without the need for any physical con-

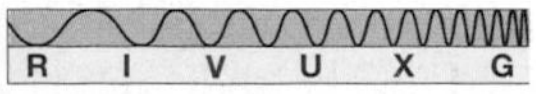

Figure 2.1 Andromeda Galaxy The pancake-shaped Andromeda Galaxy is about three million light-years away and contains a few hundred billion stars. (*NASA*)

nection between those two locations. The term *electromagnetic* means that the energy is carried in the form of rapidly fluctuating *electric* and *magnetic fields* (see Section 2.2). Virtually all we know about the universe beyond Earth's atmosphere has been gleaned from analysis of electromagnetic radiation received from afar.

Visible light is the particular type of electromagnetic radiation to which the human eye happens to be sensitive. But there is also *invisible* electromagnetic radiation, which goes completely undetected by our eyes. **Radio**, **infrared**, and **ultraviolet** waves, as well as **X rays** and **gamma rays**, all fall into this category. Recognize that, despite the different names, the words *light*, *rays*, *electromagnetic radiation*, and *waves* really all refer to the same thing. The names are just historical accidents, reflecting the fact that it took many years for scientists to realize that these apparently very different types of radiation are in reality one and the same physical phenomenon. Throughout this text, we will use the general terms "light" and "electromagnetic radiation" more or less interchangeably.

Wave Motion

All types of electromagnetic radiation travel through space in the form of **waves**. To understand the behavior of light, then, we must know a little about this kind of motion. Simply stated, a wave is a way in which energy is transferred from place to place without physical movement of material from one location to another. In wave motion, the energy is carried by a *disturbance* of some sort that occurs in a distinctive, repeating pattern. Ripples on the surface of a pond, sound waves in air, and electromagnetic waves in space, despite their many obvious differences, all share this basic defining property.

As a familiar example, imagine a twig floating in a pond (Figure 2.2). A pebble thrown into the pond at some distance from the twig disturbs the surface of the water, setting it into up-and-down motion. This disturbance propagates outward from the point of impact in the form of waves. When the waves reach the twig, some of the pebble's energy is imparted to it, causing the twig to bob up and down. In this way,

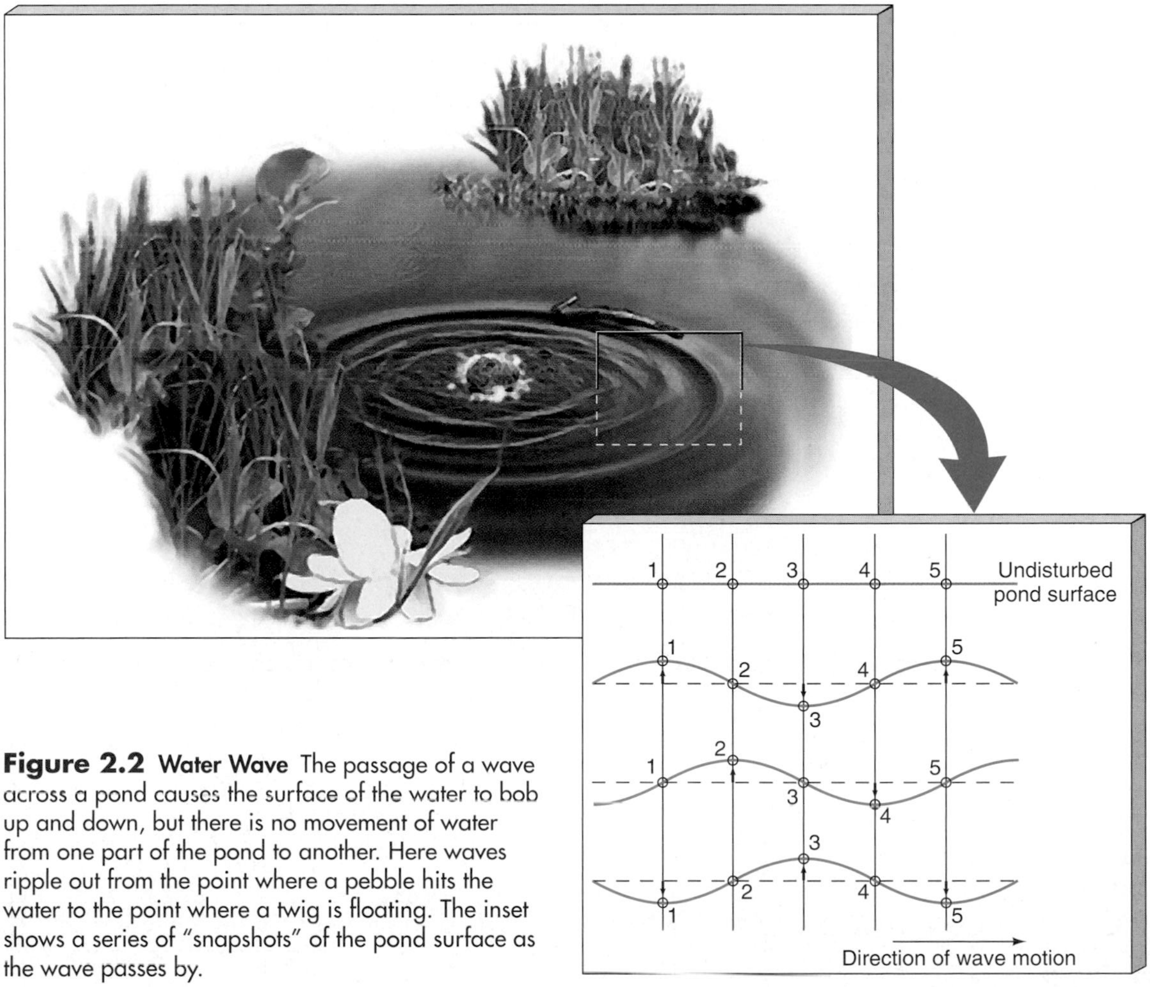

Figure 2.2 Water Wave The passage of a wave across a pond causes the surface of the water to bob up and down, but there is no movement of water from one part of the pond to another. Here waves ripple out from the point where a pebble hits the water to the point where a twig is floating. The inset shows a series of "snapshots" of the pond surface as the wave passes by.

both energy and *information*—the fact that the pebble entered the water—are transferred from the place where the pebble landed to the location of the twig. We could tell just by observing the twig that a pebble (or some object) had entered the water. With a little additional physics, we could even estimate the pebble's energy.

A wave is *not* a physical object. No water traveled from the point of impact of the pebble to the twig—at any location on the surface, the water surface simply moved up and down as the wave passed. What, then, *does* move across the pond surface? The answer is that the wave is the *pattern* of up-and-down motion, and it is this pattern that is transmitted from one point to the next as the disturbance moves across the water.

Figure 2.3 shows how wave properties are quantified. The **wave period** is the number of seconds needed for the wave to repeat itself at some point in space. The **wavelength** is the number of meters needed for the wave to repeat itself at a given moment in time. It can be measured as the distance between two adjacent wave *crests*, two adjacent wave *troughs*, or any other two similar points on adjacent wave cycles (for example, the points marked "X" in Figure 2.3). The maximum departure of the wave from the undisturbed state—still air, say, or a flat pond surface—is called its **amplitude**.

The number of wave crests passing any given point per unit time is called the wave's **frequency**. If a wave of a given wavelength moves at high speed, then many crests pass by per second and the frequency is high. Conversely, if the same wave moves slowly, then its frequency will be low. The frequency of a wave is just one divided by the wave's period:

$$\text{wave frequency} = \frac{1}{\text{wave period}}.$$

Frequency is expressed in units of inverse time (cycles per second), called hertz (Hz) in honor of the nineteenth-century German scientist Heinrich Hertz, who studied the properties of radio waves. A wave with a period of five seconds has a frequency of $(1/5)$ cycles/second $= 0.2$ Hz.

A wave moves a distance equal to one wavelength in one wave period. The product of wavelength and frequency therefore equals the *wave velocity*:

$$\text{wavelength} \times \text{frequency} = \text{velocity}.$$

Thus, if the wave in our earlier example had a wavelength of 0.5 m, its velocity is $(0.5\text{ m}) \times (0.2\text{ Hz}) = 0.1$ m/s. Wavelength and wave frequency are *inversely* related—doubling one halves the other.

2.2 Waves in What?

Waves of radiation differ in one fundamental respect from water waves, sound waves, or any other waves that travel through a material medium—radiation needs *no* such medium. When light travels from a distant cosmic object, it moves through the virtual vacuum of space. Sound waves, by contrast, cannot do this. If we were to remove all the air from a room, conversation would be impossible because sound waves cannot exist without air or some other physical medium to support them. Communication by flashlight or radio, however, would be entirely feasible.

The ability of light to travel through empty space was once a great mystery. The idea that light, or any other kind of radiation, could move as a wave through nothing at all seemed to violate common sense, yet it is now a cornerstone of modern physics.

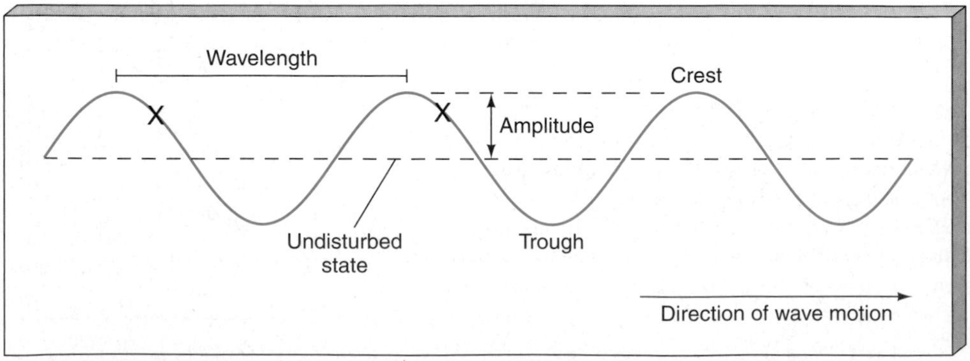

Figure 2.3 Electromagnetic Wave Representation of a typical wave, showing its direction of motion, wavelength, and amplitude. In one wave period, the entire pattern shown moves one wavelength to the right.

Interactions Between Charged Particles

1 To understand more about the nature of light, consider an *electrically charged* particle, such as an **electron** or a **proton**. Electrons and protons are elementary particles—fundamental components of matter—that carry the basic unit of charge. Electrons are said to carry a *negative* charge, while protons carry an equal and opposite *positive* charge. Just as a massive object exerts a gravitational force on any other massive object, an electrically charged particle exerts an *electrical* force on every other charged particle in the universe. ∞ (Sec. 1.4) Unlike gravity however, which is always attractive, electrical forces can be either attractive or repulsive. Particles having like charges (both negative or both positive) repel one another; particles having unlike charges attract (Figure 2.4a).

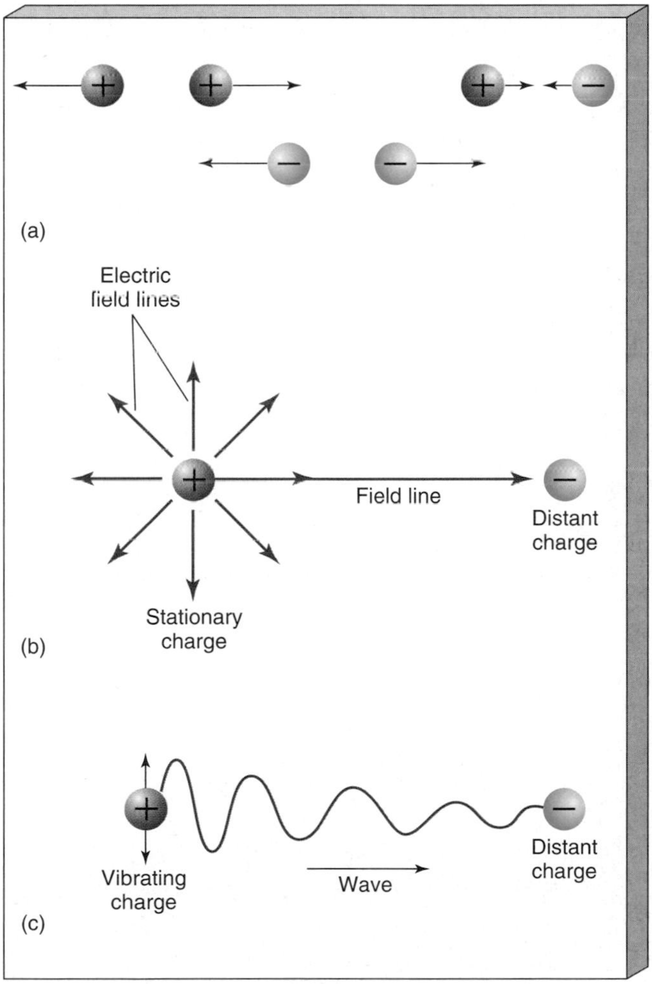

Figure 2.4 Charged Particles (a) Particles carrying like electrical charges repel one another; particles with unlike charges attract. (b) A charged particle is surrounded by an electric field, which determines the particle's influence on other charged particles. We represent the field by a series of field lines. (c) If a charged particle begins to vibrate, its electric field changes. The resulting disturbance travels through space as a wave.

Extending outward in all directions from our charged particle is an **electric field**, which determines the electric force exerted by the particle on other charged particles (Figure 2.4b). The strength of the electric field, like that of the gravitational field, decreases with increasing distance from the source according to an inverse-square law. By means of the electric field, the particle's presence is "felt" by other charged particles, near and far.

Now suppose our particle begins to vibrate, perhaps because it becomes heated or collides with some other particle. Its changing position causes its associated electric field to change, and this changing field in turn causes the electrical force exerted on other charges to vary (Figure 2.4c). If we measure the changes in the forces on these other charges, we learn about our original particle. Thus, *information about our particle's motion is transmitted through space via a changing electric field.* This *disturbance* in the particle's electric field travels through space as a wave.

Electromagnetic Waves

The laws of physics tell us that a **magnetic field** must accompany every changing electric field. Magnetic fields govern the influence of *magnetized* objects on one another, much as electric fields govern interactions between charged particles. The fact that a compass needle always points to magnetic north is the result of the interaction between the magnetized needle and Earth's magnetic field (Figure 2.5). Magnetic fields also exert forces on moving electric charges

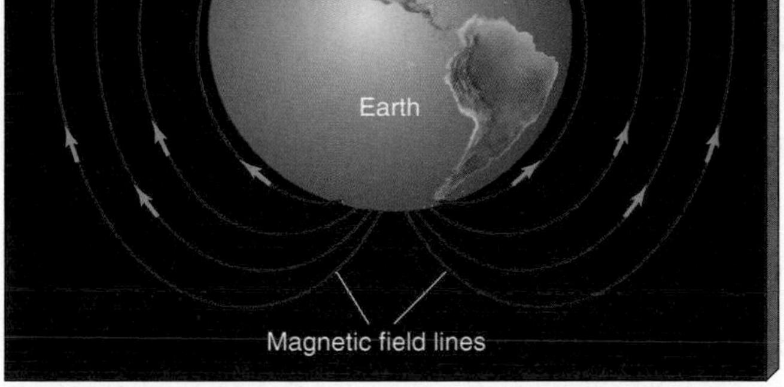

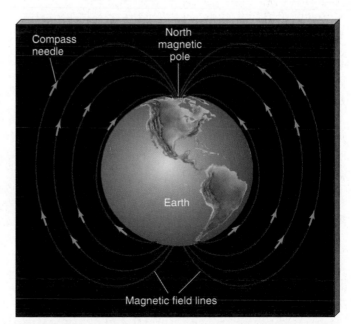

Figure 2.5 Magnetism Earth's magnetic field interacts with a magnetic compass needle, causing the needle to become aligned with the field—that is, to point toward Earth's north (magnetic) pole.

(that is, electric currents)—electric meters and motors rely on this basic fact. Conversely, moving charges create magnetic fields (electromagnets are a familiar example).

Electric and magnetic fields are inextricably linked to one another. A change in either one *necessarily* creates the other. For this reason, the disturbance produced by our moving charge actually consists of oscillating electric *and* magnetic fields, always oriented perpendicular to one another and moving together through space (Figure 2.6). These fields do not exist as independent entities. Rather, they are different aspects of a single physical phenomenon: **electromagnetism**. Together, they constitute an *electromagnetic wave* that carries energy and information from one part of the universe to another.

Now consider a distant cosmic object—a star. It is made up of charged particles, mainly protons and electrons, in constant motion. As these charged contents move around, their electric fields change, and electromagnetic waves are produced. These waves travel outward into space, and eventually some reach Earth. Other charged particles, either in our eyes or in our experimental apparatus, respond to the electromagnetic field changes by vibrating in tune with the received radiation. This response is how we "see" the radiation—with our eyes or with our detectors.

All electromagnetic waves move at a very specific speed—the **speed of light** (always denoted by the letter *c*). Its value is 299,792.458 km/s in a vacuum (and somewhat less in material substances, such as air or water). In this text, we round this value off to $c = 3.00 \times 10^5$ km/s. This is an extremely high speed. In the time needed to snap a finger—about a tenth of a second—light can travel three quarters of the way around our planet! If the currently known laws of physics are correct, then the speed of light is the fastest speed possible (see *More Precisely 13-1*).

✓ Concept Check

- What is light? List some similarities and differences between light waves and waves on water or in air.

2.3 The Electromagnetic Spectrum

White light is a mixture of colors, which we conventionally divide into six major hues—red, orange, yellow, green, blue, and violet. As shown in Figure 2.7, we can separate a beam of white light into a rainbow of these basic colors—called a *spectrum* (plural: spectra)—by passing it through a prism. This experiment was first reported by Isaac Newton over 300 years ago. In principle, the original beam of white light could be recovered by passing the spectrum through a second prism to recombine the colored beams.

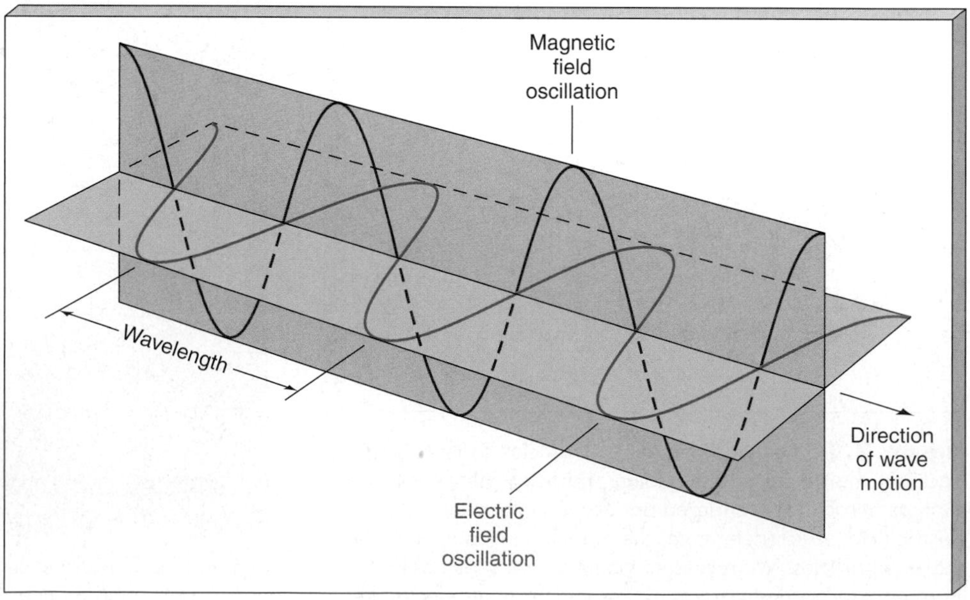

Figure 2.6 Wave Oscillation Electric and magnetic fields oscillate perpendicular to each other. Together they form an electromagnetic wave that moves through space at the speed of light.

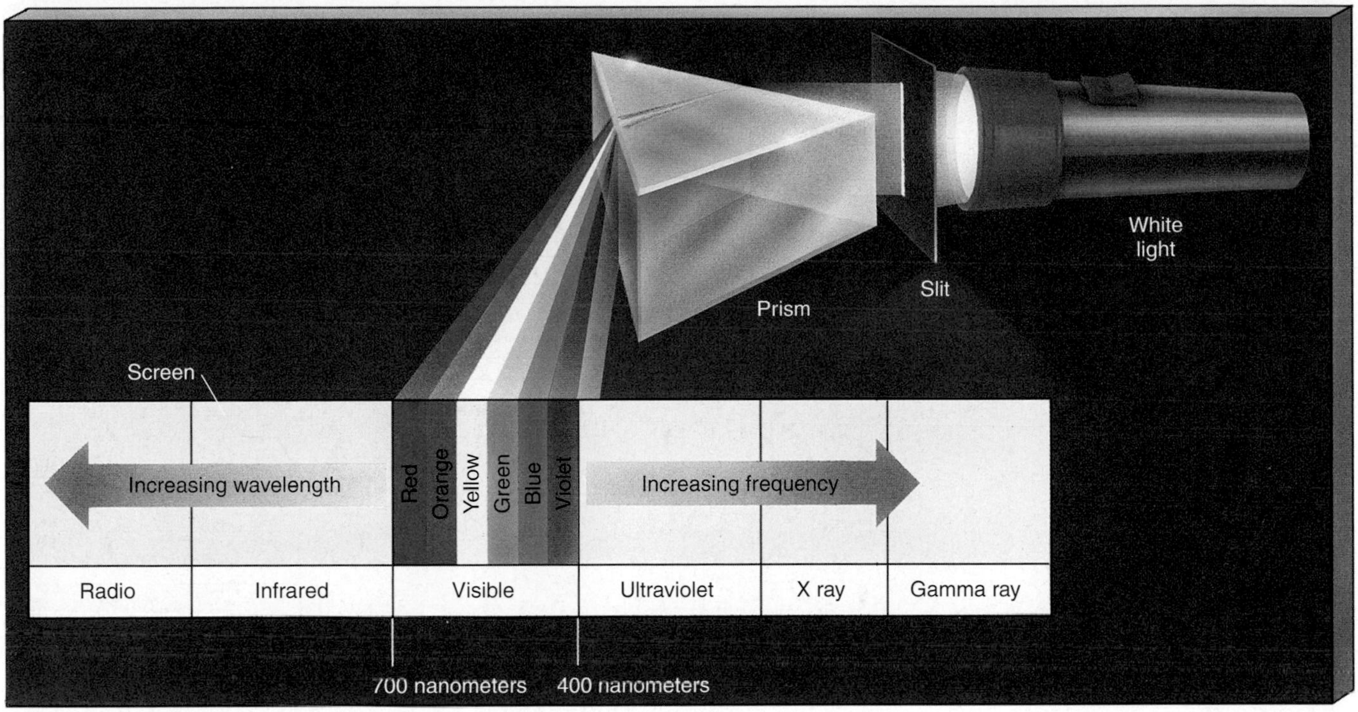

Figure 2.7 Visible Spectrum When passed through a prism, white light splits into its component colors, spanning red to violet in the visible part of the electromagnetic spectrum. The slit immediately in front of the flashlight narrows the beam of radiation. The image on the screen is a series of colored images of the slit. Human eyes are insensitive to radiation of wavelength shorter than 400 nm or longer than 700 nm.

The Components of Visible Light

What determines the color of a beam of light? The answer is its frequency (or, equivalently, its wavelength)—we see different colors because our eyes react differently to electromagnetic waves of different frequencies. Red light has a frequency of roughly 4.3×10^{14} Hz, corresponding to a wavelength of about 7.0×10^{-7} m. Violet light, at the other end of the visible range, has nearly double the frequency—7.5×10^{14} Hz—and (since the speed of light is the same in either case) just over half the wavelength—4.0×10^{-7} m. The other colors we see have frequencies and wavelengths intermediate between these two extremes.

Scientists often use a unit called the *nanometer* (nm) when describing the wavelength of light (see Appendix 2). There are 10^9 nanometers in one meter. An older unit called the *angstrom* ($1\text{Å} = 10^{-10}\,\text{m} = 0.1$ nm) is also widely used, although the nanometer is now preferred. Thus, the visible spectrum covers the wavelength range from 400 to 700 nm (4000 to 7000 Å). The radiation to which our eyes are most sensitive has a wavelength near the middle of this range, at about 550 nm (5500 Å), in the yellow-green region of the spectrum.

The Full Range of Radiation

2 Figure 2.8 plots the entire range of electromagnetic radiation. To the low-frequency, long-wavelength side of visible light lies radio and infrared radiation. Radio frequencies include radar, microwave radiation, and the familiar AM, FM, and TV bands. We perceive infrared radiation as heat. To the high-frequency, short-wavelength side of visible light lies ultraviolet, X-ray, and gamma-ray radiation. Ultraviolet radiation, lying just beyond the violet end of the visible spectrum, is responsible for suntans and sunburns. X rays are perhaps best known for their ability to penetrate human tissue and reveal the state of our insides without our resorting to surgery. Gamma rays are the shortest-wavelength radiation. They are often associated with radioactivity and are invariably damaging to any living cells they encounter.

All these spectral regions, including the visible, collectively make up the **electromagnetic spectrum**. Remember that, despite their greatly differing wavelengths and the very different roles they play in everyday life on Earth, all types of electromagnetic radiation

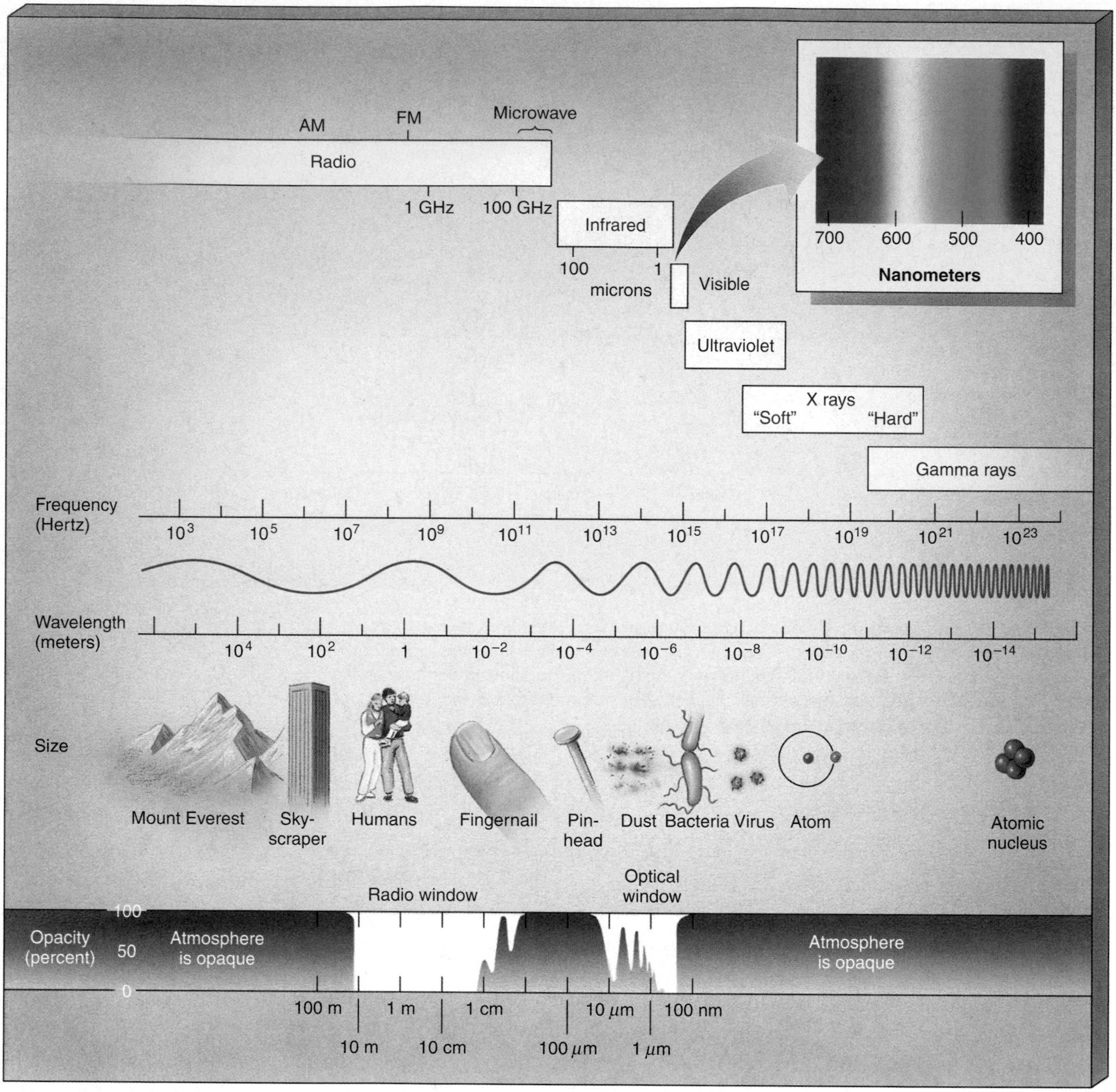

Figure 2.8 Electromagnetic Spectrum The entire electromagnetic spectrum.

are basically the same phenomenon, and all move at the same speed—the speed of light c.

Figure 2.8 is worth studying carefully, as it contains a great deal of information. Note that wave frequency (in hertz) increases from left to right and wavelength (in meters) increases from right to left. Scientists often disagree on the "correct" way to display wavelengths and frequencies in diagrams of this type. This book consistently adheres to the convention that frequency increases toward the *right*. Notice that the wavelength and frequency scales in Figure 2.8 do not increase by equal increments of 10. Instead, successive values marked on the horizontal axis differ by factors of 10—each successive value is 10 times greater than its neighbor. This type of scale (called a *logarithmic* scale) is often used in science in order to condense a very large range of some quantity into a manageable size. Throughout the text we will often find it convenient to use such a scale in order to compress a wide range of some quantity onto a single easy-to-view plot.

Figure 2.8 shows wavelengths extending from the height of mountains for radio radiation to the diameter of an atomic nucleus for gamma-ray radiation. The box at the upper right emphasizes how small the visible portion of the electromagnetic spectrum is. Most objects in the universe emit large amounts of invisible radiation. Indeed, many objects emit only a tiny fraction of their total energy in the visible range. A wealth of extra knowledge can be gained by studying the invisible regions of the electromagnetic spectrum. To remind you of this important fact and to identify the region of the electromagnetic spectrum in which a particular observation was made, we have attached a spectrum icon—an idealized version of the wavelength scale in Figure 2.8—to every astronomical image presented in this text.

Only a small fraction of the radiation arriving at our planet actually reaches Earth's surface because of the *opacity* of Earth's atmosphere. **Opacity** is the extent to which radiation is blocked by the material through which it is passing—in this case, air. The more opaque an object is, the less radiation gets through. Earth's atmospheric opacity is plotted along the wavelength and frequency scales at the bottom of Figure 2.8. Where the shading is greatest, no radiation can get in or out—the energy is completely absorbed by atmospheric gases. Where there is no shading at all, our atmosphere is almost totally transparent. Note that there are just a few *windows*, at well-defined locations in the electromagnetic spectrum, where Earth's atmosphere is transparent. In much of the radio and in the visible portions of the spectrum, the opacity is low, and we can study the universe at those wavelengths from ground level. In parts of the infrared range, the atmosphere is partially transparent, so we can make certain infrared observations from the ground. Over the rest of the spectrum, however, the atmosphere is opaque. As a result, ultraviolet, X-ray, and gamma-ray observations can be made only from above the atmosphere, from orbiting satellites.

✓ Concept Check

- In what sense are radio waves, visible light, and X rays one and the same phenomenon?

2.4 The Distribution of Radiation

All macroscopic objects—fires, ice cubes, people, stars—emit radiation at all times. They radiate because the microscopic charged particles in them are in constant random motion, and whenever charges change their state of motion, electromagnetic radiation is emitted. The **temperature** of an object is a direct measure of the amount of microscopic motion within it (see *More Precisely 2-1*). The hotter the object—the higher its temperature—the faster its constituent particles move and the more energy they radiate.

The Blackbody Spectrum

Intensity is a term often used to specify the amount or strength of radiation at any point in space. Like frequency and wavelength, intensity is a basic property of radiation. No natural object emits all of its radiation at just one frequency. Instead, the energy is often spread out over a range of frequencies. By studying the way in which the intensity of this radiation is distributed across the electromagnetic spectrum, we can learn much about the object's properties.

Figure 2.9 illustrates schematically the distribution of radiation emitted by any object. Note that the curve peaks at a single, well-defined frequency, and falls off to lesser values above and below that frequency, but it is not symmetrical about the peak—the falloff is more rapid on the high-frequency side of the peak than it is toward lower frequencies. This overall shape is characteristic of

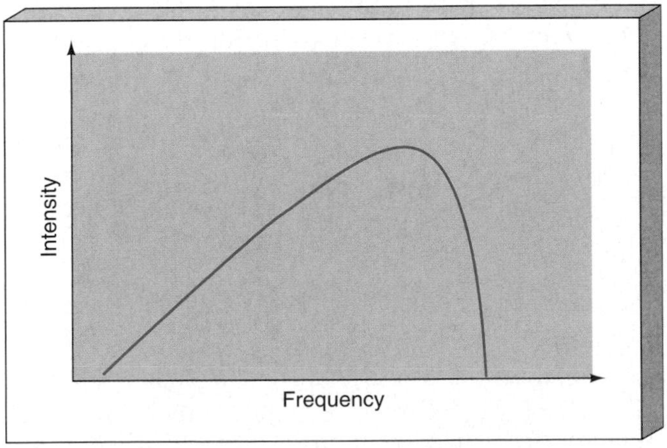

Figure 2.9 Ideal Blackbody Curve The blackbody curve represents the distribution of the intensity of the radiation emitted by any object.

the radiation emitted by *any* object, regardless of its size, shape, composition, or temperature.

The curve drawn in Figure 2.9 refers to a mathematical idealization known as a *blackbody*—an object that absorbs all radiation falling upon it. In a steady state, a blackbody must reemit the same amount of energy as it absorbs. The **blackbody curve** shown in the figure describes the distribution of that reemitted radiation. No real object absorbs or radiates as a perfect blackbody. However, in many cases the blackbody curve is a very good approximation to reality, and the properties of blackbodies provide important insights into the behavior of real objects.

The Radiation Laws

3 The blackbody curve shifts toward higher frequencies (shorter wavelengths) and greater intensities as an object's temperature increases. Even so, the *shape* of the curve remains the same (see Figure 2.10). This shifting of radiation's peak frequency with temperature is familiar to us all. Very hot glowing objects, such as light-bulb filaments or stars, emit visible light because their blackbody curves peak in or near the visible range. Cooler objects, such as warm rocks or household radiators, produce invisible radiation—they are warm to the touch but are not glowing hot to the eye. These latter objects emit most of their radiation in the lower-frequency infrared portion of the electromagnetic spectrum.

There is a very simple connection between the frequency or wavelength at which most radiation is emitted and the absolute temperature (that is, temperature measured in kelvins—see *More Precisely 2-1*) of the emitting object: The peak frequency is *directly proportional* to the temperature. This relationship is more conventionally stated in terms of wavelength:

$$\text{wavelength of peak emission} \propto \frac{1}{\text{temperature}},$$

and is known as **Wien's law** (after Wilhelm Wien, the German scientist who formulated it in 1897). In essence, it tells us that the hotter the object, the bluer its radiation. For example, an object with a temperature of 6000 K (Figure 2.10) emits most of its energy in the visible part of the spectrum, with a peak wavelength of 480 nm. At 600 K, the object's emission would peak at 4800 nm, well into the infrared. At a temperature of 60,000 K, the peak would move all the way through the visible range to a wavelength of 48 nm, in the ultraviolet.

It is also a matter of everyday experience that as the temperature of an object increases, the *total* amount of energy it radiates (summed over all frequencies) increases rapidly. For example, the heat given off by an electric heater increases sharply as the heater warms up and begins to emit visible light. In fact, the total amount of energy radiated per unit time is proportional to the fourth power of an object's temperature:

$$\text{total energy radiated} \propto \text{temperature}^4.$$

This relationship is called **Stefan's law**, after the nineteenth-century Austrian physicist Josef Stefan. It implies that the energy emitted by a body rises dramatically as the body's temperature increases. Doubling the temperature, for example, causes the total energy radiated to increase by a factor of 16. The radiation laws are presented in more detail in *More Precisely 2-2*.

Astronomical Applications

Astronomers use blackbody curves as thermometers to determine the temperatures of distant objects. For example, study of the solar spectrum makes it possible to measure the temperature of the Sun's surface. Observations of the radiation from the Sun at many frequencies yield a curve shaped somewhat like that shown in Figure 2.9. The Sun's curve peaks in the visible part of the electromagnetic spectrum. The Sun also emits a lot of infrared and a little ultraviolet radiation. Applying Wien's law to the blackbody curve that best fits the solar

Figure 2.10 Blackbody Curves Comparison of blackbody curves for four cosmic objects. (a) A cool, invisible galactic gas cloud called Rho Ophiuchi. At a temperature of 60 K, it emits mostly low-frequency radio radiation. (b) A dim, young star (shown red in the inset photograph) near the center of the Orion Nebula. The star's atmosphere, at 600 K, radiates primarily in the infrared. (c) The Sun's surface, at approximately 6000 K, is brightest in the visible region of the electromagnetic spectrum. (d) Some very bright stars in a cluster called Omega Centauri, as observed by a telescope aboard the space shuttle. At a temperature of 60,000 K, these stars radiate strongly in the ultraviolet. *(NASA)*

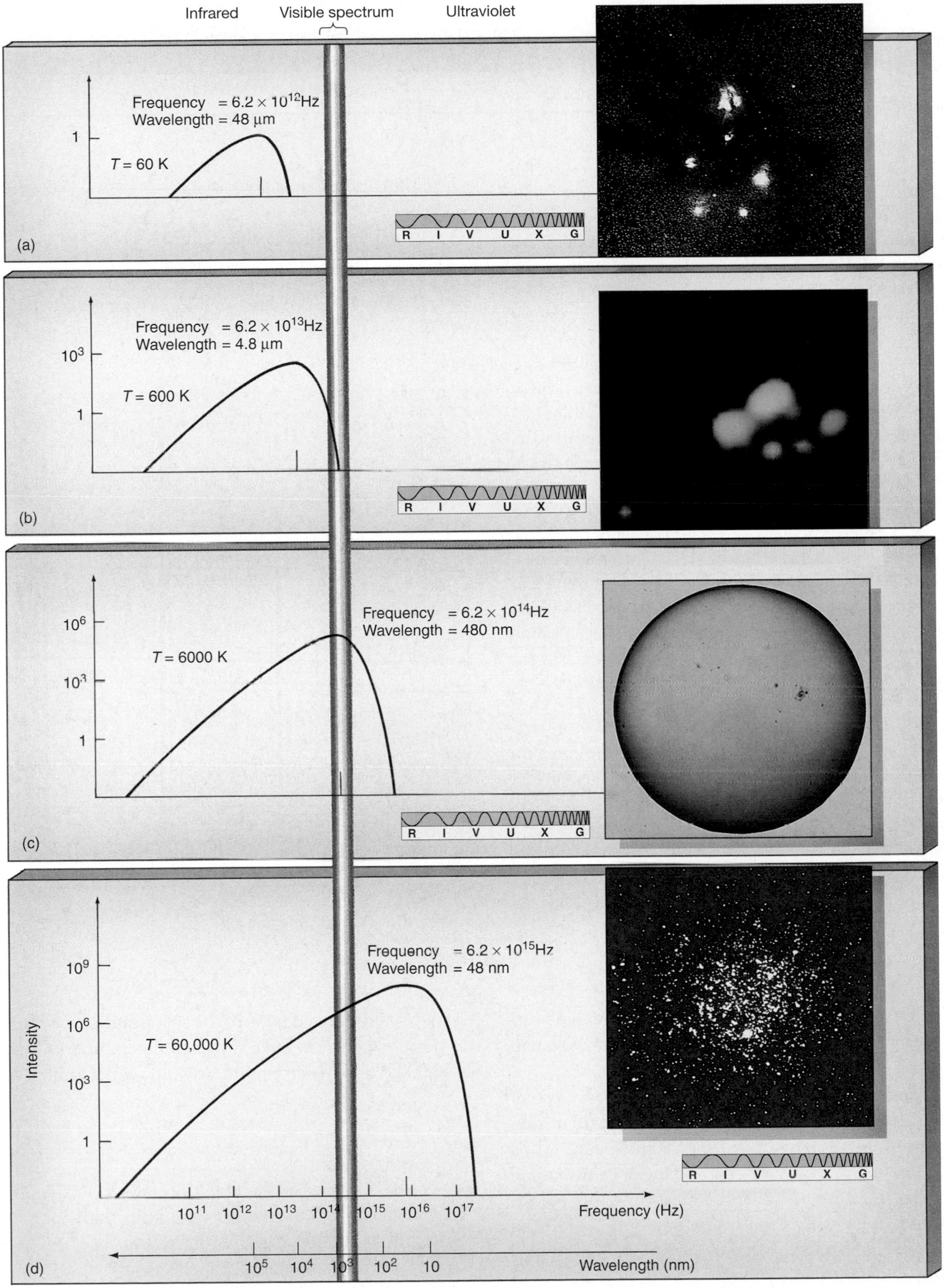

Infrared
Visible spectrum
Ultraviolet
Frequency = 6.2 × 10^12 Hz
Wavelength = 48 µm
T = 60 K
(a)
Frequency = 6.2 × 10^13 Hz
Wavelength = 4.8 µm
T = 600 K
(b)
Frequency = 6.2 × 10^14 Hz
Wavelength = 480 nm
T = 6000 K
(c)
Frequency = 6.2 × 10^15 Hz
Wavelength = 48 nm
T = 60,000 K
Intensity
(d)
R I V U X G
10^11 10^12 10^13 10^14 10^15 10^16 10^17
Frequency (Hz)
10^5 10^4 10^3 10^2 10
Wavelength (nm)

MORE PRECISELY 2-1

The Kelvin Temperature Scale

The atoms and molecules that make up any piece of matter are in constant random motion. This motion represents a form of energy known as thermal energy or, more colloquially, *heat*. The quantity we call temperature is a direct measure of this internal motion: The higher an object's temperature, the faster the random motion of its constituent particles. More precisely, the temperature of a piece of matter specifies the average thermal energy of the particles it contains.

The temperature scale probably most familiar to you, the Fahrenheit scale, is now a peculiarity of American society. Most of the rest of the world uses the Celsius temperature scale, in which water freezes at 0 degrees (0°C) and boils at 100 degrees (100°C), as illustrated in the accompanying figure.

There are, of course, temperatures below the freezing point of water. Although we know of no matter anywhere in the universe that is actually this cold, temperatures can in principle reach as low as −273.15°C. This is the temperature at which, theoretically, all atomic and molecular motion ceases. It is convenient to construct a temperature scale based on this lowest possible temperature, which is called *absolute zero*. Scientists commonly use such a scale, called the Kelvin scale in honor of the nineteenth-century British physicist Lord Kelvin. Since it takes absolute zero as its starting point, the Kelvin scale differs from the Celsius scale by 273.15°. In this book, we round off the decimal places and simply use the relationship

$$\text{kelvins} = \text{degrees Celsius} + 273.$$

Thus,

- Motion of atoms and molecules ceases at 0 kelvins (0 K).
- Water freezes at 273 kelvins (273 K).
- Water boils at 373 kelvins (373 K).

Note that the unit is "kelvins," or "K," not "degrees kelvin" or "°K."

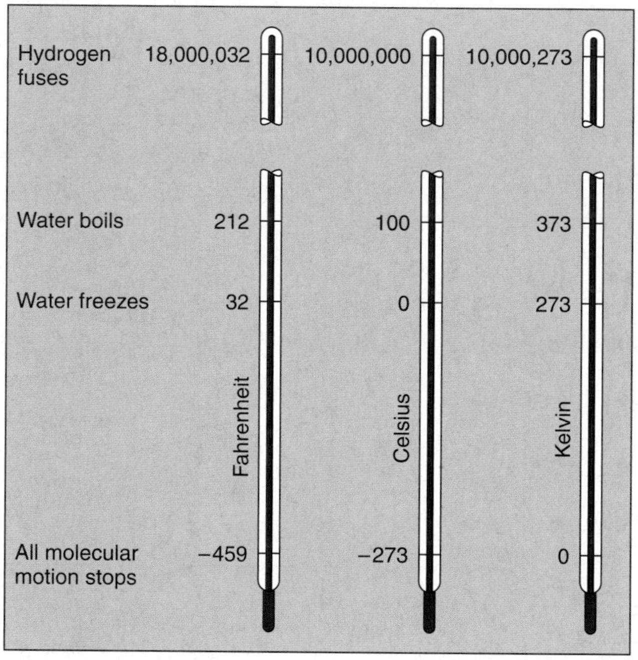

spectrum, we find that the temperature of the Sun's surface is approximately 6000 K. (A more precise measurement yields a temperature of 5800 K.)

Other cosmic objects have surfaces very much cooler or hotter than the Sun's, emitting most of their radiation in invisible parts of the spectrum (Figure 2.10). For example, the relatively cool surface of a very young star might measure 600 K and emit mostly infrared radiation. Cooler still is the interstellar gas cloud from which the star formed. At a temperature of 60 K, such a cloud would emit mainly long-wavelength radiation in the radio and infrared parts of the spectrum. The brightest stars, by contrast, have surface temperatures as high as 60,000 K and hence emit mostly ultraviolet radiation.

✓ Concept Check

- Describe, in terms of the radiation laws, how the appearance of an incandescent light bulb changes as you turn a dimmer switch to increase its brightness from "off" to "maximum."

2.5 Spectral Lines

Radiation can be analyzed with an instrument known as a **spectroscope**. In its most basic form, this device consists of an opaque barrier with a slit in it (to form a narrow beam of light), a prism (to split the beam into its component colors), and either a detector or a screen (to allow the user to view the resulting spectrum). Figure 2.11 shows such an arrangement.

Emission Lines

4 The spectra encountered in the previous section are examples of **continuous spectra**. A light bulb, for instance, emits radiation of all wavelengths (mostly in the visible and near-infrared ranges), with an intensity distribution that is well described by the blackbody curve corresponding to the bulb's temperature. Viewed through a spectroscope, the spectrum of the light from the bulb would show the familiar rainbow running from red to violet without interruption, as presented in Figure 2.11.

Not all spectra are continuous, however. For instance, if we took a glass jar containing pure hydrogen gas and passed an electrical discharge through it (a little like a lightning bolt arcing through Earth's atmosphere), the gas would begin to glow—that is, it would emit radiation. If we were to examine that radiation with our spectroscope, we would find that its spectrum consisted of only a few bright lines on an otherwise dark background, quite unlike the continuous spectrum described for the light bulb. Figure 2.12 shows this schematically (the lenses have been removed for clarity), and a more detailed rendering of the spectrum of hydrogen appears in the top panel of Figure 2.13. The light produced by the hydrogen in this experiment does *not* consist of all possible colors, but instead includes only a few narrow, well-defined **emission lines**, narrow "slices" of the continuous spectrum. The black background represents all the wavelengths *not* emitted by hydrogen.

After some experimentation, we would also find that although we could alter the intensity of the lines (for example, by changing the amount of hydrogen in the jar or the strength of the electrical discharge), we could not alter their color (in other words, their frequency or wavelength). This particular pattern of spectral emission lines is a property of the element hydrogen—whenever we perform this experiment, the same characteristic **emission spectrum** is the result. Other elements yield different emission spectra. Depending on which element is involved, the pattern of lines can be fairly simple or very complex. Always, though, it is *unique* to that element. The emission spectrum of a gas thus provides a "fingerprint" that allows scientists to deduce its presence by spectroscopic means.

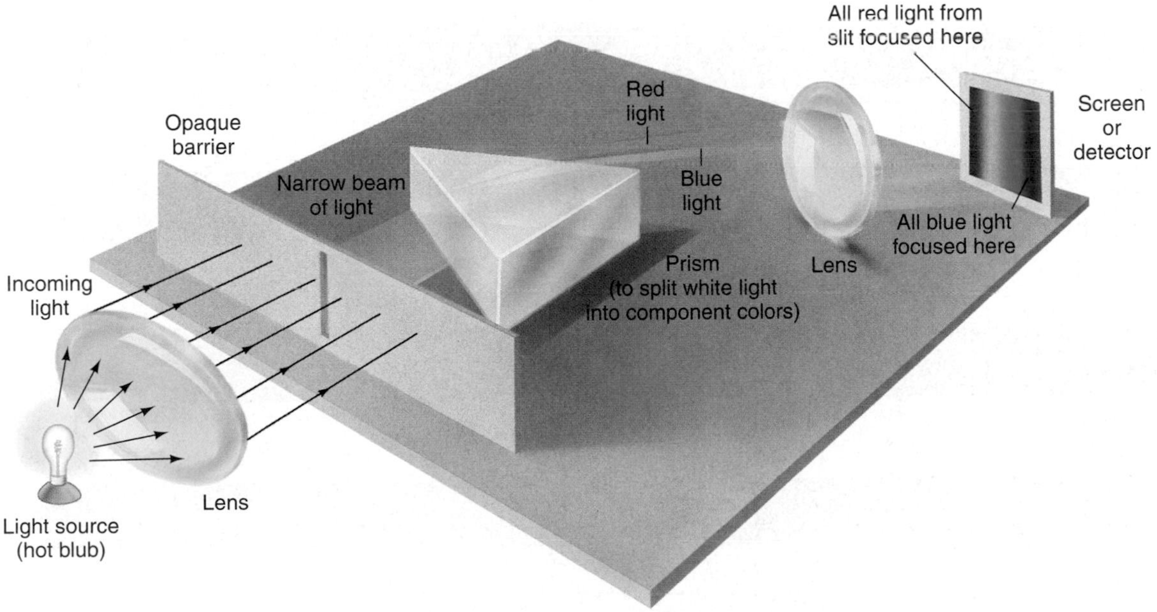

Figure 2.11 Spectroscope Diagram of a simple spectroscope. A slit in the barrier allows a narrow beam of light to pass. The beam passes through a prism and is split into its component colors. A lens then focuses the light into a sharp image that is either projected onto a screen, as shown here, or analyzed as it is passed through a detector.

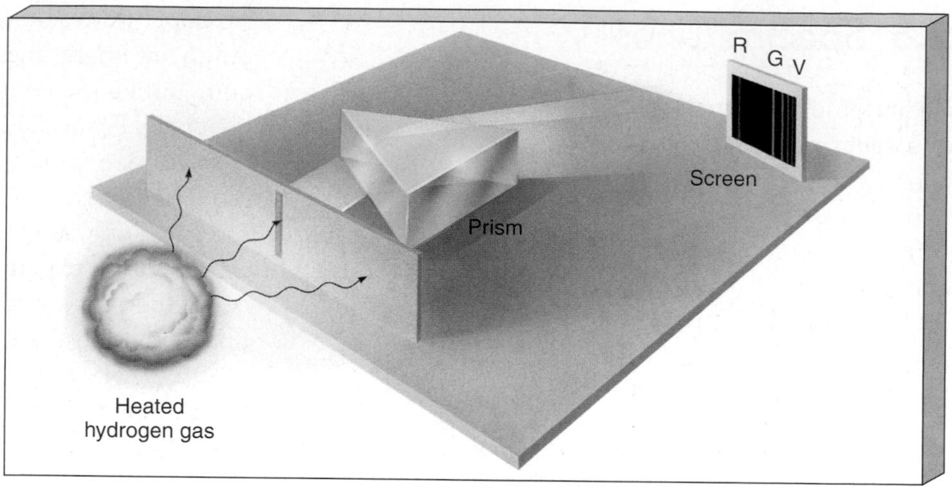

Figure 2.12 Emission Spectrum Instead of a continuous spectrum, the light from excited (heated) hydrogen gas consists of a series of distinct spectral lines. (For simplicity, the focusing lenses have been omitted.)

Examples of the emission spectra of some common substances are shown in Figure 2.13.

Absorption Lines

When sunlight is split by a prism, at first glance it appears to produce a continuous spectrum. However, closer scrutiny shows that the solar spectrum is interrupted by a large number of narrow dark lines, as shown in Figure 2.14. These lines represent wavelengths of light that have been removed (absorbed) by gases present either in the outer layers of the Sun or in Earth's atmosphere. These gaps in the spectrum are called **absorption lines**. The absorption lines in the solar spectrum are referred to collectively as *Fraunhofer lines*, after the nineteenth-

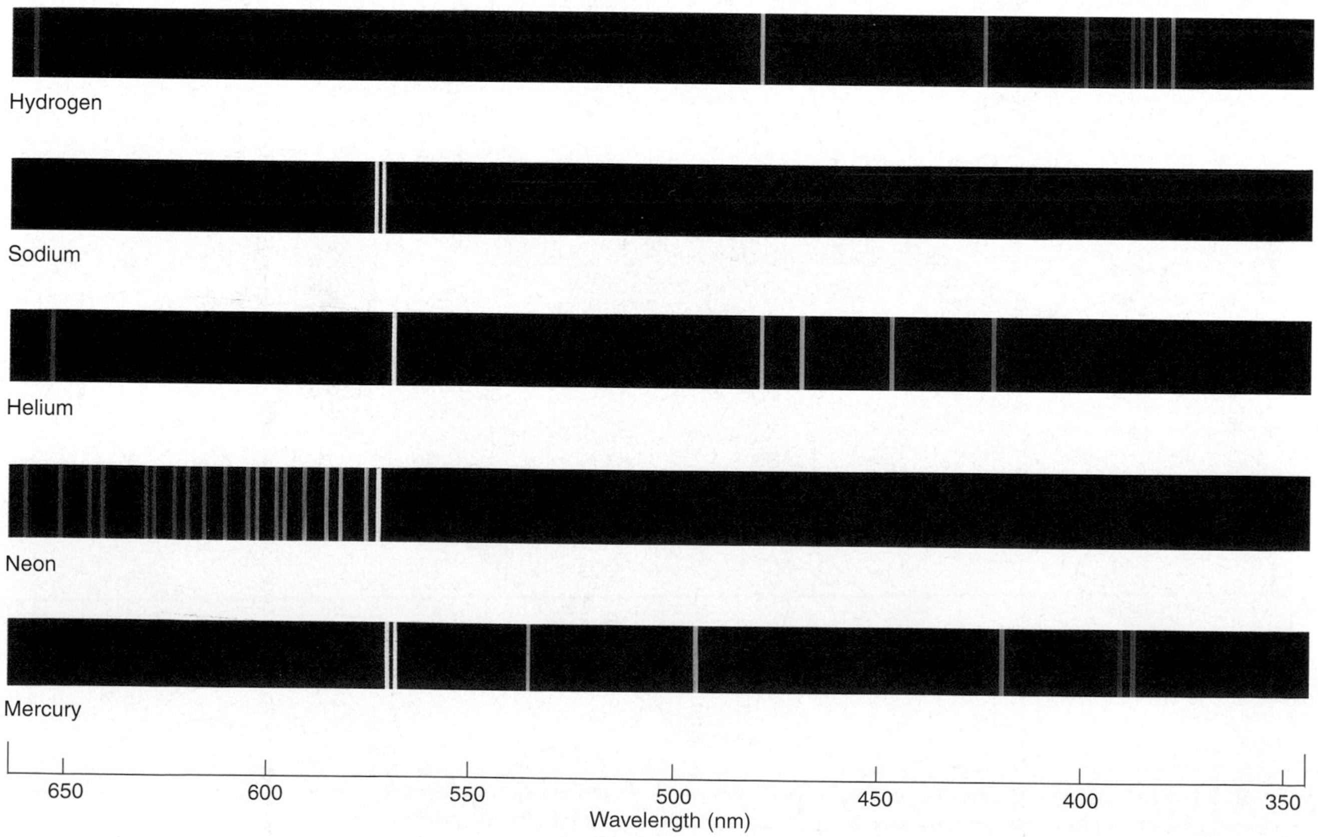

Figure 2.13 Elemental Emission The emission spectra of some well-known elements. *(Wabash Instrument Corporation)*

Figure 2.14 **Solar Spectrum** This visible spectrum of the Sun shows hundreds of dark absorption lines superimposed on a bright continuous spectrum. Here, the scale extends from long wavelengths (red) at the upper left to short wavelengths (blue) at the lower right. *(AURA)*

century German physicist Joseph Fraunhofer, who measured and cataloged more than 600 of them.

At around the time solar absorption lines were discovered, scientists found that absorption lines could also be produced in the laboratory by passing a beam of light from a continuous source through a cool gas, as shown in Figure 2.15. They quickly observed an intriguing connection between emission and absorption lines: The absorption lines associated with a given gas occur at precisely the *same* wavelengths as the emission lines produced when the gas is heated. Both sets of lines therefore contain the *same* information about the composition of the gas.

The analysis of the ways in which matter emits and absorbs radiation is called **spectroscopy**. The observed relationships between the three types of spectra—continuous, emission line, and absorption line—may be summarized as follows:

1. A luminous solid or liquid, or a sufficiently dense gas, emits light of all wavelengths and so produces a continuous spectrum of radiation (Figure 2.11).
2. A low-density hot gas emits light whose spectrum consists of a series of bright emission lines. These lines are characteristic of the chemical composition of the gas (Figure 2.12).
3. A low-density cool gas absorbs certain wavelengths from a continuous spectrum, leaving dark absorption lines in their place, superimposed on the continuous spectrum. These lines are characteristic of the composition of the intervening gas. They occur at precisely the same wavelengths as the emission lines produced by the gas at higher temperatures (Figure 2.15).

These rules are known as **Kirchhoff's laws**, after the German physicist Gustav Kirchhoff, who published them in 1859.

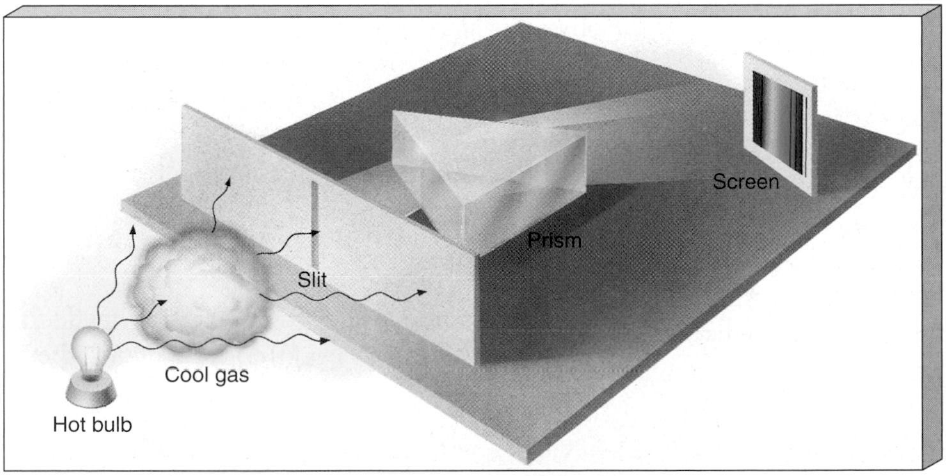

Figure 2.15 **Absorption Spectrum** When a cool gas is placed between a source of continuous radiation (a light bulb) and the detector/screen of a spectroscope, the resulting spectrum is crossed by a series of dark absorption lines. These lines are formed when the cool gas absorbs certain wavelengths (colors) from the light radiating from the light bulb. The absorption lines appear at precisely the same wavelengths as the emission lines that would be produced if the gas were heated to high temperatures (see Figure 2.12).

MORE PRECISELY 2-2

More on the Radiation Laws

Wien's law relates the temperature T of an object to the wavelength λ_{max} at which the object's blackbody radiation spectrum peaks. (The Greek letter λ—lambda—is conventionally used to denote wavelength.) Mathematically, if we measure T in kelvins, we find that

$$\lambda_{max} = \frac{0.29 \text{ cm}}{T}.$$

Thus at 6000 K (the approximate surface temperature of the Sun), the wavelength of maximum intensity is 0.29/6000 cm, or 480 nm, corresponding to the yellow-green part of the visible spectrum.

We can also give Stefan's law a more precise mathematical formulation. With T again measured in kelvins, the total amount of energy emitted per square meter of surface per second (a quantity known as the *energy flux* F) is given by

$$F = \sigma T^4.$$

This equation is usually referred to as the *Stefan-Boltzmann* equation. (Stefan's student, Ludwig Boltzmann, was an Austrian physicist who played a central role in the development of the laws of thermodynamics during the late nineteenth and early twentieth centuries.) The constant σ (the Greek letter sigma) is known as the Stefan-Boltzmann constant.

The SI unit of energy is the joule (J). Probably more familiar is the closely related unit called the watt (W), which measures power—the rate at which energy is emitted or expended by an object. One watt is the emission of one joule per second; for example, a 100-W light bulb emits energy (mostly in the form of infrared and visible light) at a rate of 100 J/s. In these units, the Stefan-Boltzmann constant has the value $\sigma = 5.67 \times 10^{-8}$ W/m^2·K^4.

Notice just how rapidly the energy flux increases with increasing temperature. A piece of metal in a furnace, when at a temperature of 3500 K, radiates energy at a rate of about 850 W for every square centimeter of its surface area. Doubling its temperature to 7000 K (so that it becomes yellow to white hot, by Wien's law) increases the energy emitted by a factor of 16 (four "doublings"), to 13.6 kilowatts (kW) (13,600 W) per square centimeter.

Astronomical Applications

Once astronomers realized that spectral lines are indicators of chemical composition, they set about identifying the observed lines in the Sun's spectrum. Almost all of the lines observed in the light received from extraterrestrial sources could be attributed to known elements (for example, many of the Fraunhofer lines in sunlight are associated with the element iron). However, some new lines also appeared in the solar spectrum. In 1868, astronomers realized that those lines must correspond to a previously unknown element. It was given the name helium, after the Greek word *helios*, meaning "sun." Only in 1895, almost three decades after its detection in sunlight, was helium discovered on Earth.

Yet for all the information that nineteenth-century astronomers could extract from observations of stellar spectra, they still lacked a theory explaining *how* those spectra arose. Despite their sophisticated spectroscopic equipment, they knew scarcely any more about the physics of stars than did Galileo or Newton. To understand how spectroscopy can be used to extract detailed information about astronomical objects from the light they emit, we must delve more deeply into the processes that produce line spectra.

Concept Check

- What are absorption and emission lines? What do they tell us about the properties of the gas producing them?

2.6 The Formation of Spectral Lines

By the start of the twentieth century, physicists had accumulated substantial evidence that light sometimes behaves in a manner that simply cannot be explained by the wave theory of radiation. As we have just seen, the production of absorption and emission lines involves only certain very specific wavelengths of light. This result would not be expected if light behaved only as a continuous wave and matter always obeyed the laws of Newtonian mechanics. It became clear that, when light interacts with matter on very small scales, it does so not in a smooth, continuous way but in a discontinuous, stepwise manner. The challenge was to find an explanation for this unexpected behavior. The solution revolutionized our view of nature and now forms the foundation not just for physics and astronomy, but for virtually all of modern science.

Atomic Structure

5 To explain the formation of spectral lines, we must understand not just the nature of light but also something of the structure of **atoms**—the microscopic building blocks from which all matter is constructed. Let us start with the simplest atom, hydrogen, which consists of an electron, with a negative electrical charge, orbiting a proton, which carries a positive charge. The proton forms the central **nucleus** (plural: nuclei) of the atom. Because the positive charge on the proton exactly cancels the negative charge on the electron, the hydrogen atom as a whole is electrically neutral.

How does this picture of the hydrogen atom relate to the characteristic emission and absorption lines associated with hydrogen gas? If an atom emits some energy in the form of radiation, that energy has to come from somewhere within the atom. Similarly, if energy is absorbed by the atom, that energy must cause some internal change. The energy emitted or absorbed by the atom is associated with changes in the motion of the orbiting electron.

The first theory of the atom to provide an explanation of hydrogen's observed spectral lines was propounded by the Danish physicist Niels Bohr. This theory is now known simply as the **Bohr model** of the atom. Its essential features are as follows. First, there is a state of lowest energy—the **ground state**—which represents the "normal" condition of the electron as it orbits the nucleus. Second, there is a maximum energy that the electron can have and still be part of the atom. Once the electron acquires more than that maximum energy, it is no longer bound to the nucleus, and the atom is said to be *ionized*. An atom having fewer (or more) than its normal complement of electrons, and hence a net electrical charge, is called an **ion**. Third, and most important (and also least intuitive), between those two energy levels, the electron can exist only in certain sharply defined energy states, often referred to as *orbitals*.

An atom is said to be in an **excited state** when an electron occupies an orbital other than the ground state. The electron then lies at a greater than normal distance from its parent nucleus, and the atom has a greater than normal amount of energy. The excited state with the lowest energy (that is, the one closest to the ground state) is called the *first excited state*, that with the second-lowest energy the *second excited state*, and so on.

In Bohr's model, each electron orbital was pictured as having a specific radius, much like a planetary orbit in the solar system, as shown in Figure 2.16. However, the modern view is not so simple. Although each orbital *does* have a precise energy, the electron is now envisioned as being smeared out in an *electron cloud* surrounding the nucleus, as illustrated in Figure 2.17. It is common to speak of the average distance from the cloud to the nucleus as the radius of the electron's orbital. When a hydrogen atom is in its ground state, the radius of the orbital is about 0.05 nm (0.5 Å). As the orbital energy increases, the radius increases, too. For the sake of clarity in the diagrams that follow, we represent electron orbitals by solid lines, but bear in mind always that the fuzziness shown in part (b) of the figure is a more accurate depiction of reality.

An atom can become excited by absorbing some light energy from a source of electromagnetic radiation, or by colliding with some other particle—another atom, for example. However, the atom cannot stay in that state forever. After about 10^{-8} s, it must return to its ground state.

✓ Concept Check

- In what ways do electron orbits in the Bohr atom differ from planetary orbits around the Sun?

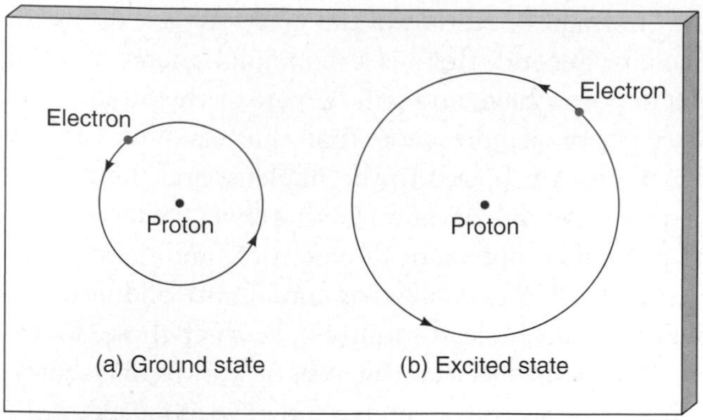

Figure 2.16 Classical Atom An early conception of the hydrogen atom—the Bohr model—pictured its electron orbiting the central proton in a well-defined orbital, like a planet orbiting the Sun. Two electron orbitals of different energies are shown: (a) the ground state and (b) an excited state.

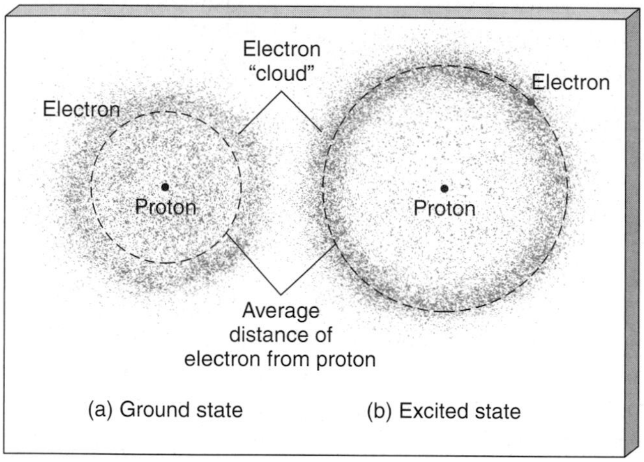

Figure 2.17 Modern Atom The modern view of the hydrogen atom sees the electron as a "cloud" surrounding the nucleus. The same two energy states are shown as in Figure 2.16.

The Particle Nature of Radiation

Here now is the crucial point that links atoms to radiation and allows us to interpret atomic spectra. Because electrons may exist only in orbitals having specific energies, atoms can absorb only specific amounts of energy as their electrons are boosted into excited states. Likewise, atoms can emit only specific amounts of energy as their electrons fall back to lower energy states. Thus, the amount of light energy absorbed or emitted in these processes *must correspond precisely to the energy difference between two orbitals*. This requires that light must be absorbed and emitted in the form of little "packets" of electromagnetic radiation, each carrying a very specific amount of energy. We call these packets **photons**. A photon is, in effect, a "particle" of electromagnetic radiation.

The idea that light sometimes behaves not as a continuous wave but as a stream of particles was proposed by Albert Einstein in 1905. To explain all experimental results then known, Einstein found that the energy contained within a photon had to be proportional to the frequency of the radiation:

$$\text{photon energy} \propto \text{radiation frequency}.$$

Thus, for example, a "red" photon having a frequency of 4×10^{14} Hz (corresponding to a wavelength of about 750 nm, or 7500 Å) has 4/7 the energy of a "blue" photon, of frequency of 7×10^{14} Hz. Because it connects the *energy* of a photon with the *color* of the light it represents, this relationship is the final piece in the puzzle of how to understand the spectra we see.

Environmental conditions ultimately determine which description—wave or stream of particles—better fits the behavior of electromagnetic radiation. As a general rule of thumb, in the macroscopic realm of everyday experience, radiation is more usefully described as a wave, and in the microscopic domain of atoms, it is best characterized as a stream of particles.

The Spectrum of Hydrogen

Absorption and emission of photons by a hydrogen atom are illustrated in Figure 2.18. Figure 2.18(a) shows the atom absorbing a photon of radiation and making a transition from the ground state to the first excited state, then emitting a photon of precisely the same energy and dropping back to the ground state. The energy difference between the two states corresponds to an ultraviolet photon, of wavelength 121.6 nm (1216 Å).

Figure 2.18(b) depicts the absorption of a more energetic (higher-frequency, shorter-wavelength) ultraviolet photon, this one having a wavelength of 102.6 nm (1026 Å), causing the atom to jump to the *second* excited state. From that state, the electron may return to the ground state via either one of two alternate paths.

1. It can proceed directly back to the ground state, in the process emitting an ultraviolet 102.6 nm photon identical to the one that excited the atom in the first place.
2. Alternatively, it can *cascade* down one orbital at a time, emitting *two* photons: one having an energy equal to the difference between the second and first excited states, and the other having an energy equal to the difference between the first excited state and the ground state.

The second step in the cascade produces a 121.6-nm ultraviolet photon, just as in Figure 2.18(a). However, the first part of the cascade—the one from the second to the first excited state—produces a photon of wavelength 656.3 nm (6563 Å), which is in the visible part of the electromagnetic spectrum. This photon is seen as red light. An individual atom—if one could be isolated—would emit a momentary red flash. The inset in Figure 2.18 shows an astronomical object whose red coloration is the result of precisely this process.

Absorption of more energy can boost the electron to even higher orbitals within the atom. As the excited electron cascades back down to the ground state, the atom may emit many photons, each with a different energy and hence a different color, and the resulting spectrum shows many distinct spectral lines. In the case of hydrogen, all transitions ending at the ground state produce ultraviolet photons. However, downward transitions ending at the *first* excited state give rise to spectral lines in or near the visible portion

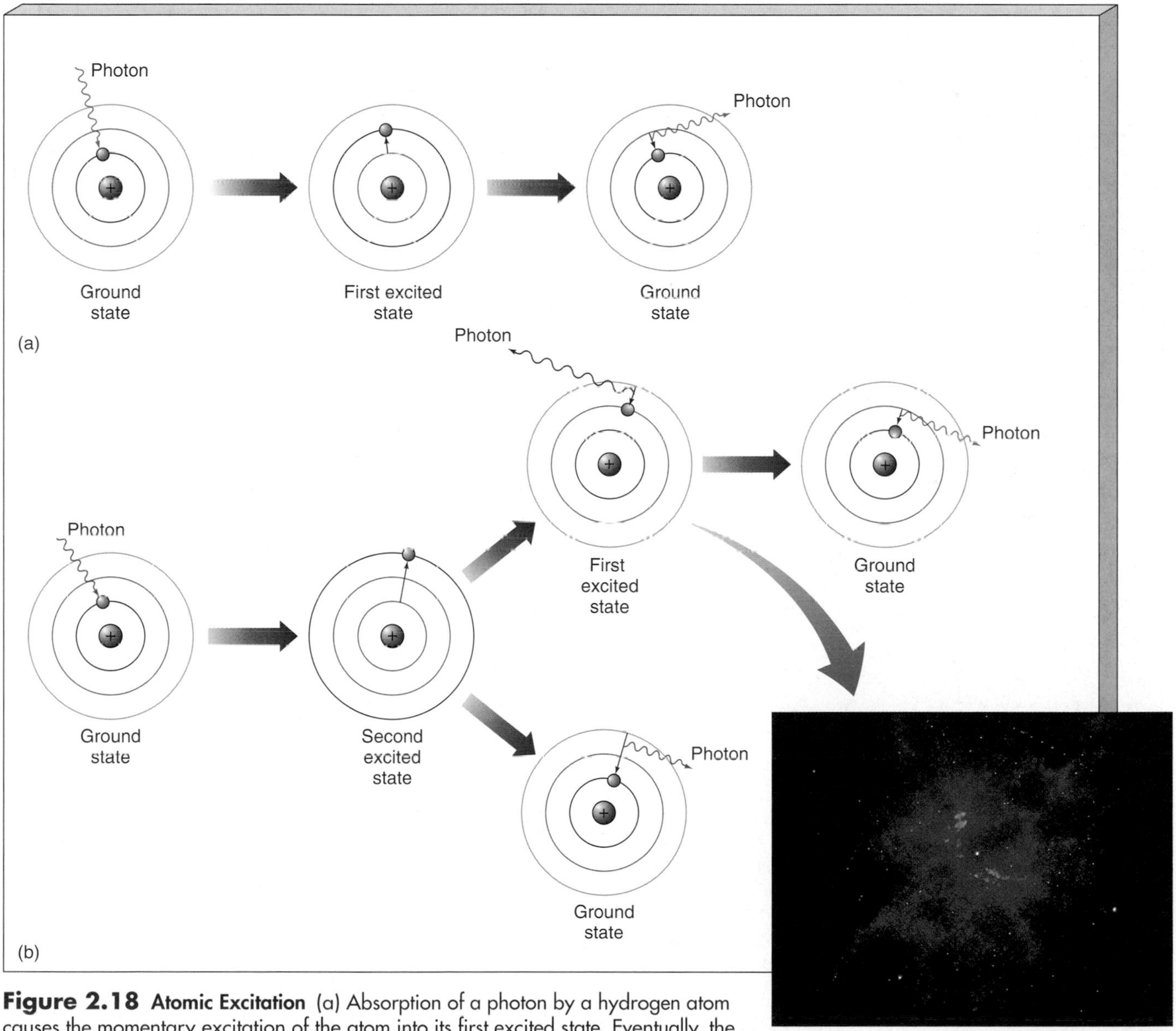

R I V U X G

Figure 2.18 Atomic Excitation (a) Absorption of a photon by a hydrogen atom causes the momentary excitation of the atom into its first excited state. Eventually, the atom returns to its ground state, in the process emitting a photon having the same energy as the original photon. (b) Absorption of a photon might also boost the atom into a higher excited state, from which there are several possible paths back to the ground state. The object shown in the inset, designated NGC 2440, is an emission nebula: an interstellar cloud consisting largely of hydrogen gas excited by an extremely hot star (the white dot in the center). *(Inset: S. Heap/NASA)*

Animation Animation

of the electromagnetic spectrum (Figure 2.13). Because they form the most easily observable part of the hydrogen spectrum, and were the first to be discovered, these lines (also known as *Balmer lines*) are often referred to simply as the "Hydrogen series" and are denoted by the letter H. Individual transitions are labeled with Greek letters, in order of increasing energy (decreasing wavelength): the Hα line corresponds to the transition from the second to the first excited state, and has a wavelength of 656.3 nm (red); H β (third to first) has wavelength 486.1 nm (green); Hγ (fourth to first) has wavelength 434.1 nm (blue), and so on.

Kirchhoff's Laws Explained

6 Let's reconsider our earlier discussion of emission and absorption lines in terms of the model just presented. In Figure 2.15, a beam of continuous radiation shines through a cloud of cool gas. The beam contains photons of all energies, but most of them cannot interact with the gas because the gas can absorb only photons having precisely the right energy to cause an electron to jump from one orbital to another. Photons having energies that cannot produce such a jump do not interact with the gas at all. They pass through it unhindered. Photons having the right energies are absorbed, excite the gas, and are removed from the beam. This is the cause of the dark absorption lines in the spectrum of Figure 2.15. These lines are direct indicators of the energy differences between orbitals in the atoms making up the gas.

The excited gas atoms rapidly return to their original states, each emitting one or more photons in the process. Most of these reemitted photons leave at angles that do *not* take them through the slit and on to the detector. A second detector looking at the cloud from the side would record the reemitted energy as an emission spectrum. (This is what we are seeing in the inset to Figure 2.18.) Like the absorption spectrum, the emission spectrum is characteristic of the gas, not of the original beam.

✓ Concept Check

- How does the structure of an atom determine the atom's emission and absorption spectra?

More Complex Spectra

All hydrogen atoms have the same structure—a single electron orbiting a single proton—but, of course, there are many other kinds of atoms, each having a unique internal structure. The number of protons in the nucleus of an atom determines the **element** that the atom represents. That is, just as all hydrogen atoms have a single proton, all oxygen atoms have eight protons, all iron atoms have 26 protons, and so on.

The next simplest element after hydrogen is helium. The central nucleus of the most common form of helium is made up of two protons and two **neutrons** (another kind of elementary particle having a mass slightly larger than that of a proton but carrying no electrical charge). About this nucleus orbit two electrons. As with hydrogen and all other atoms, the "normal" condition for helium is to be electrically neutral, with the negative charge of the orbiting electrons exactly canceling the positive charge of the nucleus (Figure 2.19a).

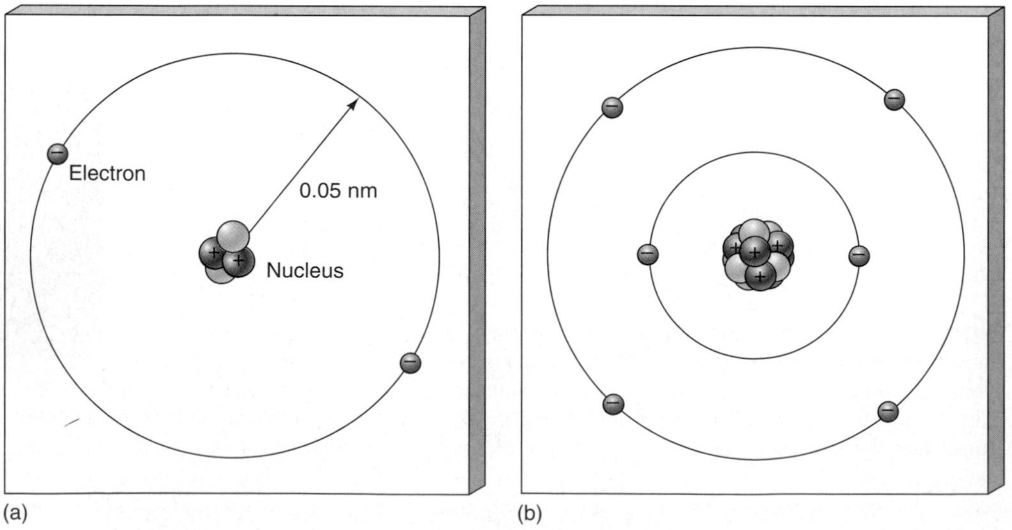

Figure 2.19 Helium and Carbon (a) A helium atom in its ground state. Two electrons occupy the lowest-energy orbital around a nucleus containing two protons and two neutrons. (b) A carbon atom in its ground state. Six electrons orbit a six-proton, six-neutron nucleus, two of the electrons in an inner orbital, the other four at a greater distance from the center.

More complex atoms contain more protons (and neutrons) in the nucleus and have correspondingly more orbiting electrons. For example, an atom of carbon (Figure 2.19b) consists of six electrons orbiting a nucleus containing six protons and six neutrons. As we progress to heavier and heavier elements, the number of orbiting electrons increases, and consequently the number of possible electronic transitions rises rapidly. The result is that very complicated spectra can be produced. The complexity of atomic spectra generally reflects the complexity of the source atoms. A good example is the element iron, which contributes several hundred of the Fraunhofer absorption lines seen in the solar spectrum. The many possible transitions of its 26 orbiting electrons yield an extremely rich line spectrum.

Even more complex spectra are produced by **molecules**. A molecule is a tightly bound group of atoms held together by interactions among their orbiting electrons—interactions called *chemical bonds*. Much like atoms, molecules can exist only in certain well-defined energy states, and again like atoms, molecules produce emission or absorption spectral lines when they make a transition from one state to another. Because molecules are more complex than atoms, the rules of molecular physics are also much more complex. Nevertheless, as with atomic spectral lines, painstaking experimental work over many decades has determined the precise frequencies at which millions of molecules emit and absorb radiation. These lines are molecular fingerprints, just like their atomic counterparts, enabling researchers to identify and study one kind of molecule to the exclusion of all others.

Molecular lines usually bear little resemblance to the spectral lines associated with their component atoms. For example, Figure 2.20(a) shows the emission spectrum of the simplest molecule known—molecular hydrogen. Notice how different it is from the spectrum of atomic hydrogen shown in part (b).

Spectral-Line Analysis

Astronomers apply the laws of spectroscopy in analyzing radiation from beyond Earth. A nearby star or a distant galaxy takes the place of the light bulb in our previous examples, an interstellar cloud or a stellar (or even planetary) atmosphere plays the role of the intervening cool gas, and a spectrograph attached to a telescope replaces our simple prism and detector. We list below a few of the properties of emitters and absorbers that can be determined by careful analysis of radiation received on (or near) Earth. We will encounter other important examples as our study of the cosmos unfolds.

1. The *composition* of an object is determined by matching its spectral lines with the laboratory spectra of known atoms and molecules.
2. The *temperature* of an object emitting a continuous spectrum can be measured by matching the overall distribution of radiation with a blackbody curve.
3. The *(line-of-sight) velocity* of an object is measured by determining the Doppler shift of its spectral lines (see *More Precisely 2-3*).
4. An object's *rotation rate* can be determined by measuring the broadening (smearing out over a range of wavelengths) produced by the Doppler effect in emitted or reflected spectral lines.
5. The *pressure* of the gas in the emitting region of an object can be measured by its tendency to smear out, or broaden, spectral lines. The greater the pressure, the broader the line.
6. The *magnetic field* of an object can be inferred from a characteristic splitting it produces in many spectral lines, when a single line divides into two. (This is known as the *Zeeman effect*.)

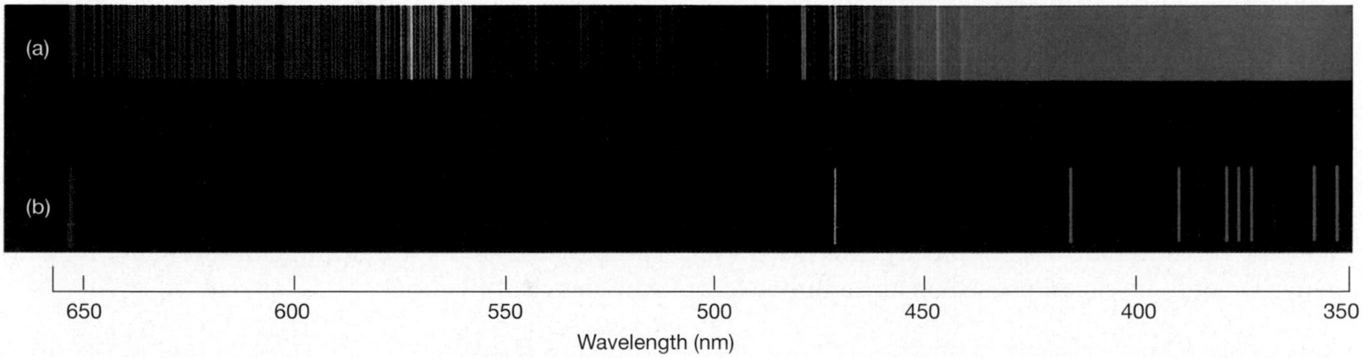

Figure 2.20 Hydrogen Spectrum The emission spectra of (a) molecular hydrogen and (b) atomic hydrogen. *(Bausch & Lomb Inc.)*

MORE PRECISELY 2-3

The Doppler Effect

Most of us have had the experience of hearing the pitch of a train whistle change from high shrill (high frequency, short wavelength) to low blare (low frequency, long wavelength) as the train approaches and then recedes. This motion-induced change in the observed frequency of a wave is known as the *Doppler effect*, in honor of Christian Doppler, the nineteenth-century Austrian physicist who first explained it. Applied to cosmic sources of electromagnetic radiation, it has become one of the most important observational tools in all of twentieth-century astronomy. Here's how it works.

Imagine a wave moving from the place where it is generated toward an observer who is not moving with respect to the wave source, as shown in the accompanying figure. By noting the distances between successive wave crests, the observer can determine the wavelength of the emitted wave. Now suppose that the wave source begins to move. Because the source moves between the times of emission of one wave crest and the next, wave crests in the direction of motion of the source will be seen to be *closer together* than normal, while crests behind the source will be more widely spaced. Thus, an observer in front of the source will measure a *shorter* wavelength than normal, while one behind will see a *longer* wavelength. (The numbers in the diagram indicate successive wave crests emitted by the source and the location of the source at the instant each wave crest was emitted.)

The greater the relative speed of source and observer, the greater the observed shift. In terms of the net velocity of *recession* between source and observer (so a positive velocity means that the two are moving apart, a negative value that they are approaching), the apparent wavelength and frequency (measured by the observer) are related to the true quantities (emitted by the source) by

$$\frac{\text{apparent wavelength}}{\text{true wavelength}} = \frac{\text{true frequency}}{\text{apparent frequency}}$$

$$= 1 + \frac{\text{recession velocity}}{\text{wave speed}},$$

where the "wave speed" is the speed of light in the case of electromagnetic radiation. In the figure, the source is shown in motion. However, the same general statements hold whenever there is any *relative* motion between source and observer. Note also that only motion along the line joining source and observer, known as *radial* motion, appears in the above equation. Motion *transverse* (perpendicular) to the line of sight has no significant effect.

In astronomical parlance, the wave measured by an observer situated in front of a moving source is said to be *blueshifted*, because blue light has a shorter wavelength than red light. Similarly, an observer situated behind the source will measure a longer-than-normal wavelength—the radiation is said to be *redshifted*. This terminology holds even for invisible radiation, for which "red" and "blue" have no meaning. Any shift toward shorter wavelengths is called a blueshift, and any shift toward longer wavelengths is called a redshift.

The Doppler effect is not normally noticeable in visible light—the speed of light is so large that the wavelength change is far too small to be noticeable for everyday terrestrial velocities. However, using spectroscopic techniques, astronomers routinely use the Doppler effect to measure the line-of-sight velocity of cosmic objects, by determining the extent to which known spectral lines are shifted to longer or

---

Given sufficiently sensitive equipment, there is almost no end to the wealth of data contained in starlight. However, deciphering the extent to which each of many competing factors influences a spectrum can be a very difficult task. Typically, the spectra of many elements are superimposed on one another, and several physical processes are occurring simultaneously, each modifying the spectrum in its own way. The challenge facing astronomers is to unravel the extent to which each mechanism contributes to spectral-line profiles and so obtain meaningful information about the source of the lines.

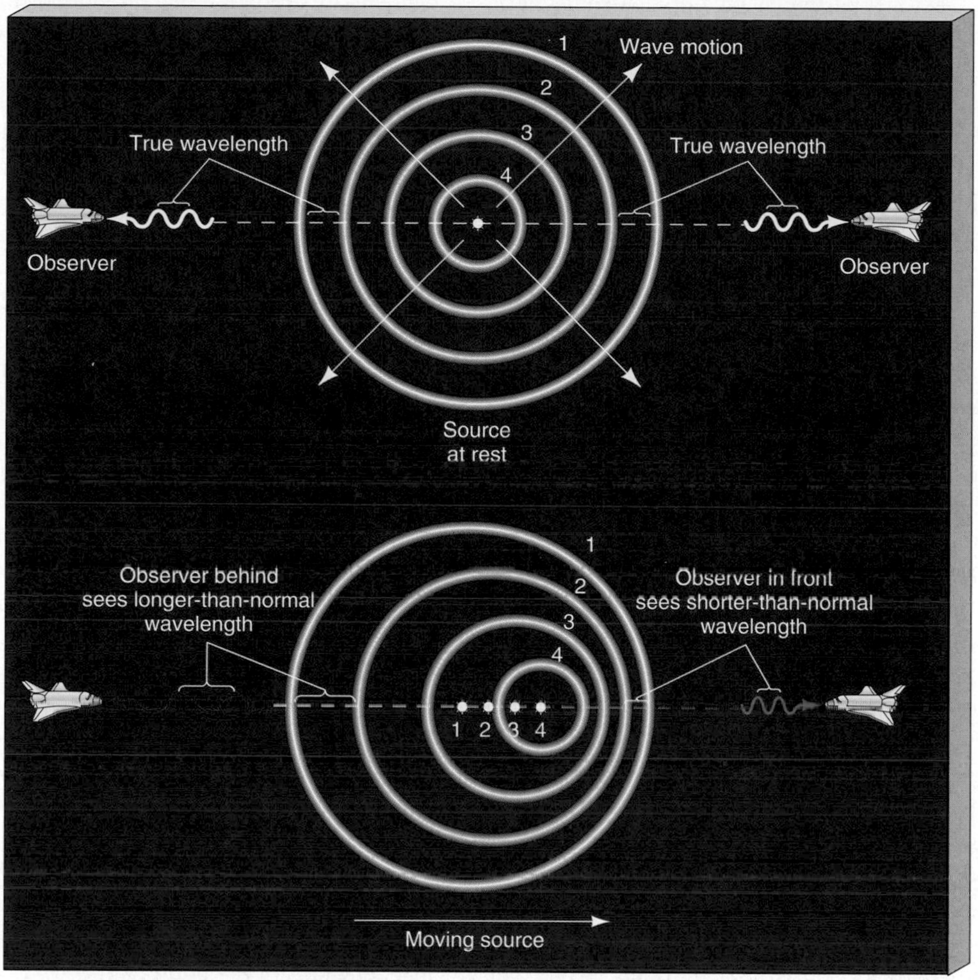

shorter wavelengths. For example, suppose that an astronomer observes the red Hα line in the spectrum of a star to have a wavelength of 657 nm, instead of the 656.3 nm measured in the lab. (How does she know it is the same line? Because she realizes that *all* the hydrogen lines are shifted by the same fractional amount—the characteristic pattern of lines identifies hydrogen as the source.) Using the above equation, she calculates that the star's radial velocity is $657/656.3 - 1 = 0.0056$ times the speed of light. In other words, the star is receding from Earth at a rate of 320 km/s. The motions of nearby stars and distant galaxies—even the expansion of the universe itself—have all been measured in this way.

Chapter Review

Summary

Visible light (p. 45) is a particular type of **electromagnetic radiation** (p. 44). It travels through space in the form of a **wave** (p. 45). A wave is characterized by the **wave period** (p. 46), the length of time taken for one complete cycle; the **wavelength** (p. 46), the distance between successive wave crests; and the wave **amplitude** (p. 46), which measures the size of the disturbance associated with the wave. The wave **frequency** (p. 46) counts the number of wave crests that pass a given point in one second.

Electrons (p. 47) and **protons** (p. 47) are elementary particles that carry equal and opposite electrical charges. Any electrically charged object is surrounded by an **electric field** (p. 47) that determines the force it exerts on other charged objects. Like gravitational fields, electric fields vary as the inverse square of the distance from their source. When a charged particle moves, information about that motion is transmitted throughout the universe by the particle's changing electric and **magnetic fields** (p. 47). The information travels at the **speed of light** (p. 48) in the form of a wave. The phenomenon is known as **electromagnetism** (p. 48).

The **electromagnetic spectrum** (p. 49) consists of (in order of increasing frequency) **radio waves**, **infrared radiation**, **visible light**, **ultraviolet radiation**, **X rays**, and **gamma rays** (p. 45). The **opacity** (p. 51) of Earth's atmosphere—the extent to which it absorbs radiation—varies greatly with the wavelength of the radiation. Only radio waves, some infrared wavelengths, and visible light can penetrate the atmosphere and reach the ground.

The **temperature** (p. 51) of an object is a measure of the speed with which its constituent particles move. The **intensity** (p. 51) of radiation emitted by an object has a characteristic distribution, called a **blackbody curve** (p. 52), that depends only on the temperature of the object. **Wien's law** (p. 52) tells us that the wavelength at which the object's radiation peaks is directly proportional to its temperature. **Stefan's law** (p. 52) states that the total amount of energy radiated is proportional to the fourth power of the temperature.

Many hot objects emit a **continuous spectrum** (p. 55) of radiation, containing light of all wavelengths. A hot gas may instead produce an **emission spectrum** (p. 55), consisting only of a few well-defined **emission lines** (p. 55) of specific frequencies, or colors. Passing a continuous beam of radiation through cool gas will produce **absorption lines** (p. 56) at precisely the same frequencies as would be present in the gas's emission spectrum. **Kirchhoff's laws** (p. 57) describe the relationships between these different types of spectra. The emission and absorption lines produced by each element are unique—they provide a "fingerprint" of that element in the light it emits or absorbs. A **spectroscope** (p. 55) splits a beam of radiation into its component frequencies and delivers them to a detector as a series of spectral lines. **Spectroscopy** (p. 57) is the study of these spectral lines.

Atoms (p. 59) are made up of negatively charged electrons orbiting a positively charged **nucleus** (p. 59) consisting of positively charged protons and, with the exception of the hydrogen nucleus, electrically neutral **neutrons** (p. 62). In normal circumstances, the number of orbiting electrons equals the number of protons in the nucleus, and the atom is electrically neutral. The number of protons in the nucleus determines the type of **element** (p. 62) the atom represents.

The **Bohr model** (p. 59) of the hydrogen atom was an early attempt to explain how atoms produce emission and absorption line spectra. An atom has a minimum-energy **ground state** (p. 59), representing its "normal" condition. In this state, the orbiting electron is at its normal distance from the nucleus. An atom is in an **excited state** (p. 59) when the electron occupies an orbital located at a greater than normal distance from the nucleus. When an electron in an atom acquires sufficient energy, it is no longer bound to the atom. An atom that has gained or lost one or more electrons is called an **ion** (p. 59). In the modern view, the electron is envisaged as being spread out in a "cloud" around the nucleus but still having a sharply defined energy.

As an electron moves from one energy level to another in an atom, the difference in the energy between the states is emitted or absorbed in the form of "packets" of electromagnetic radiation—**photons** (p. 60). Because the energy levels have definite energies, the photons also have definite energies that are characteristic of the type of atom involved. The energy of a photon determines the frequency, and hence the color, of the light emitted or absorbed.

Molecules (p. 63) are groups of two or more atoms bound together by interactions known as chemical bonds. Molecules can exist only in certain well-defined energy states that obey rules similar to those governing the internal structure of atoms. As with atoms, molecules moving from one energy state to another emit or absorb a characteristic spectrum of radiation that identifies them uniquely.

The Doppler effect is a motion-induced change in the observed wavelength or frequency of a wave. Spectroscopy allows astronomers to determine many physical properties of distant objects. These properties include composition, temperature, line-of-sight velocity (from the Doppler effect), pressure, and magnetic field.

Review and Discussion

1. Define the following wave properties: period, wavelength, amplitude, frequency.
2. What is the relationship between wavelength, wave frequency, and wave velocity?
3. Compare the gravitational force with the electric force.
4. Describe the way in which light radiation leaves a star, travels through the vacuum of space, and finally is seen by someone on Earth.
5. What do radio waves, infrared radiation, visible light, ultraviolet radiation, X rays, and gamma rays have in common? How do they differ?
6. In what regions of the electromagnetic spectrum is the atmosphere transparent enough to allow observations from the ground?
7. What is a blackbody? What are the characteristics of the radiation it emits?
8. If Earth were completely blanketed with clouds and we couldn't see the sky, could we learn about the realm beyond the clouds? What forms of radiation might penetrate the clouds and reach the ground?
9. Describe how its blackbody curve changes as a red-hot glowing coal cools off.
10. What is spectroscopy? Why is it so important to astronomers?
11. Describe the basic components of a spectroscope.
12. What is a continuous spectrum? An absorption spectrum?
13. What is the normal condition for atoms? What is an excited atom? What are orbitals?
14. Why do excited atoms absorb and reemit radiation at characteristic frequencies?
15. Suppose a luminous cloud of gas is discovered emitting an emission spectrum. What can be learned about the cloud from this observation?

True or False?

_____ 1. Light, radio, ultraviolet, and gamma rays are all forms of electromagnetic radiation.

_____ 2. Sound is a familiar form of electromagnetic wave.

_____ 3. Electromagnetic waves cannot travel through empty space.

_____ 4. In a vacuum, electromagnetic waves all travel at the same speed, the speed of light.

_____ 5. Visible light makes up the greatest part of the electromagnetic spectrum.

_____ 6. Ultraviolet light has the shortest wavelength of any electromagnetic wave.

_____ 7. A blackbody emits all its radiation at a single wavelength or frequency.

_____ 8. Emission spectra are characterized by narrow, bright lines of different colors.

_____ 9. For an emission spectrum produced by a container of hydrogen gas, changing the amount of hydrogen in the container will change the color of the lines in the spectrum.

_____ 10. In the previous question, changing the gas in the container from hydrogen to helium will change the colors of the lines in the spectrum.

_____ 11. An absorption spectrum appears as a continuous spectrum interrupted by a series of dark lines.

_____ 12. The wavelengths of the emission lines produced by an element are different from the wavelengths of the absorption lines produced by the same element.

_____ 13. The energy of a photon is inversely proportional to the wavelength of the radiation.

_____ 14. An electron can have any energy within an atom, so long as that energy is above the ground state energy.

_____ 15. An atom can remain in an excited state indefinitely.

Fill in the Blank

1. The _______ of a wave is the distance between any two adjacent wave crests.
2. The hertz (Hz) is a unit used to measure the _______ of a wave.
3. When a charged particle moves, information about this motion is transmitted through space via the particle's changing _______ and _______ fields.
4. The visible spectrum ranges from _______ nm to _______ nm in wavelength.
5. Earth's atmosphere has low opacity for three forms of electromagnetic radiation. They are _______, _______, and _______.
6. The peak of an object's emitted radiation occurs at a frequency or wavelength determined by the object's _______.

7. Two identical objects have temperatures of 1000 K and 1200 K. It is observed that one of the objects emits roughly twice as much radiation as the other. Which one is it? _______.

8. Blackbody radiation is an example of a _______ spectrum.

9. Fraunhofer discovered absorption lines in the spectrum of _______.

10. A continuous spectrum can be produced by a luminous solid, a liquid, or a _______ gas.

11. An absorption spectrum is produced when a _______ gas lies in front of a continuous source.

12. Protons carry a _______ charge; electrons carry a _______ charge.

13. When moving to a higher energy level in an atom, an electron _______ a photon of a specific energy.

14. When moving to a lower energy level in an atom, an electron _______ a photon of a specific energy.

15. The "specific energy" referred to in the preceding two questions is exactly equal to the energy _______ between the two energy levels the electron moves.

Problems

1. A sound wave moving through water has a frequency of 256 Hz and a wavelength of 5.77 m. What is the speed of sound in water?

2. What is the wavelength of a 100-MHz ("FM 100") radio signal?

3. What would be the frequency of an electromagnetic wave having a wavelength equal to Earth's diameter? In what part of the electromagnetic spectrum would such a wave lie?

4. The blackbody emission spectrum of object A peaks in the ultraviolet region of the electromagnetic spectrum, at a wavelength of 200 nm. That of object B peaks in the red region, at 650 nm. According to Wien's law, how many times hotter than B is A? According to Stefan's law, how many times more energy per unit area does A radiate per second?

5. Normal human body temperature is about 37°C. What is this temperature in kelvins? What is the peak wavelength emitted by a person with this temperature? In what part of the spectrum does this lie?

6. According to the Stefan-Boltzmann law, how much energy is radiated into space per unit time by each square meter of the Sun's surface (see *More Precisely 2-2*)? If the Sun's radius is 696,000 km, what is the total power output of the Sun?

7. A 1-nm gamma ray photon has how many times more energy than a 10-MHz radio photon?

8. How many different photons (that is, photons of different frequencies) can be emitted as a hydrogen atom in the second excited state falls back, directly or indirectly, to the ground state? What about a hydrogen atom in the third excited state?

9. The Hα line (Sec. 2.6) of a certain star is received on Earth at a wavelength of 656 nm. What is the star's radial velocity with respect to Earth?

10. Imagine you are observing a spacecraft moving in a circular orbit of radius 100,000 km around a distant planet. You happen to be located in the plane of the spacecraft's orbit. You find that the spacecraft's radio signal varies periodically in wavelength between 2.99964 m and 3.00036 m. Assuming that the radio is broadcasting normally, at a constant wavelength, what is the mass of the planet?

Projects

1. Locate the constellation Orion. Its two brightest stars are Betelgeuse and Rigel. Which is hotter? Which is cooler? How can you tell? Which of the other stars scattered across the night sky are hot, and which are cool?

2. Stand near (but not too near!) a train track or busy highway and wait for a train or traffic to pass by. Can you notice the Doppler effect in the pitch of the engine noise or whistle blowing? How does the sound frequency depend on (a) speed? (b) the motion toward or away from you?

3. Use a hand-held spectroscope (available through Learning Technologies, Inc.). While in the shade, point the spectroscope at a white cloud or white piece of paper that is in direct sunlight. Look for the absorption lines in the Sun's spectrum. Note their wavelength from the scale inside the spectroscope. Compare your list to the Fraunhofer lines given in many physics, astronomy, or chemistry reference books.

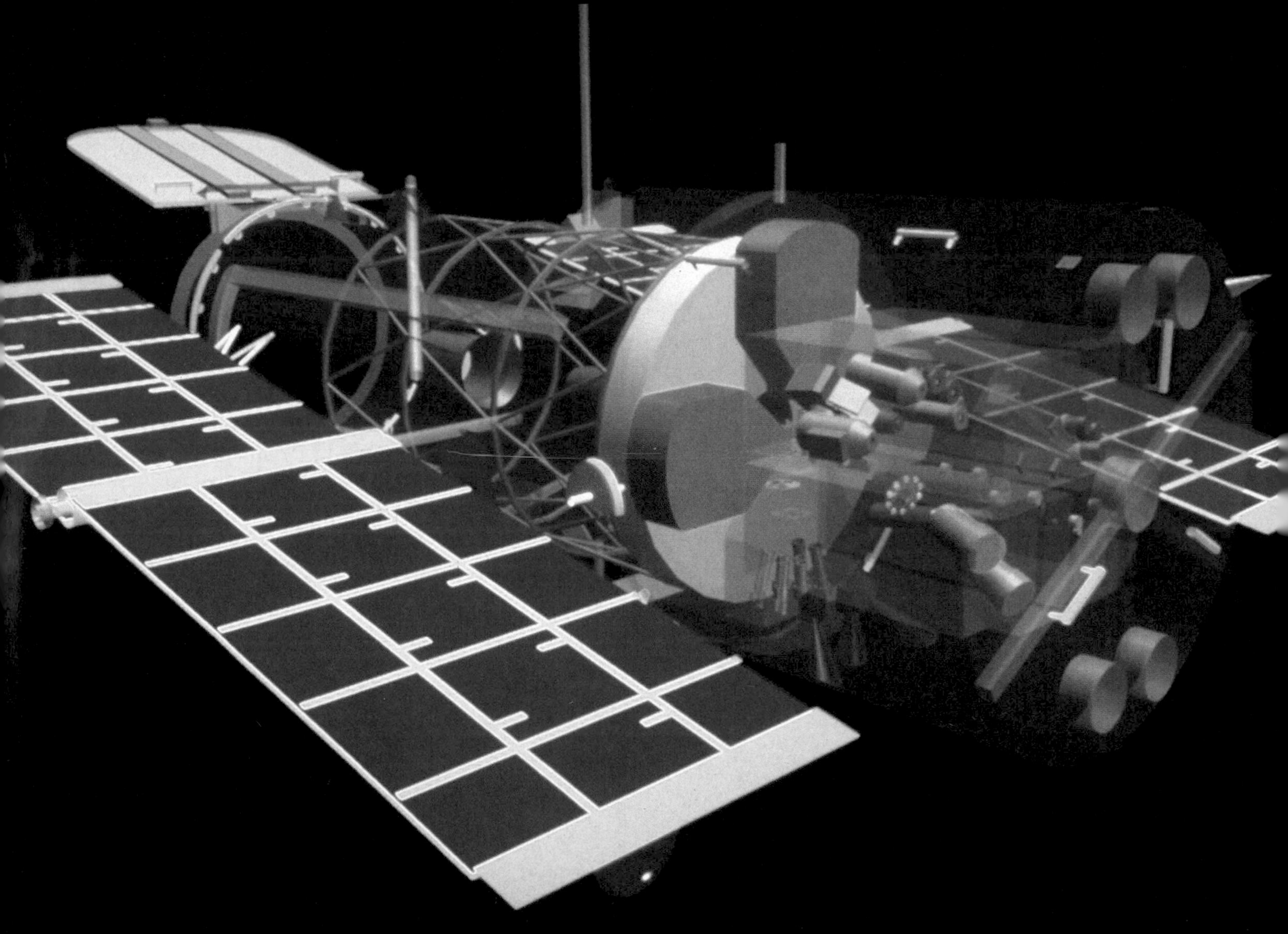

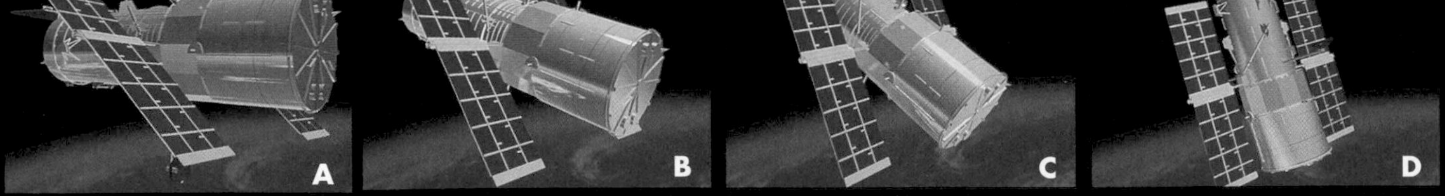
A
B
C
D

3 TELESCOPES

The Tools of Astronomy

LEARNING GOALS

Studying this chapter will enable you to:

1. Sketch and describe the basic designs of the major types of optical telescopes.
2. Explain why very large telescopes are needed for most astronomical studies, and specify the advantages of reflecting telescopes for astronomical use.
3. Describe how Earth's atmosphere affects astronomical observations, and discuss some of the current efforts to improve ground-based astronomy.
4. Discuss the advantages and disadvantages of radio astronomy.
5. Explain how interferometry can enhance the usefulness of radio and other observations.
6. Discuss the advantages, limitations, and chief uses of infrared, ultraviolet, and high-energy astronomies.
7. Explain why it is important to make astronomical observations in many different regions of the electromagnetic spectrum.

(Opposite page, background) This semi-transparent illustration shows some of the main features of the *Hubble Space Telescope*. The large blue disk at center of the spacecraft is the primary mirror; the red gadgets to its rear are the sensors that guide the pointing of the telescope. The open aperture door is at upper left. The huge solar panels are shown in yellow-checkered blue at left and partly obscured at right. Looking inside the aft bay of the vehicle, we can see key components of each of the science instruments—the spectrometers are shown in copper and blue (in the foreground), the cameras in pink, green, and lavender (mostly in the background). (*D. Berry*)

(Insets A, B, C, and D) The four small insets are computer-rendered views of *Hubble* in orbit. Despite having the size of a city bus, *Hubble* is designed to move in space with the grace of a prima ballerina. All these illustrations are taken from video animations made by the astronomy artist and animator Dana Berry.

At its heart, astronomy is an observational science. More often than not, observations of cosmic phenomena precede any clear theoretical understanding of their nature. As a result, our detecting instruments—our telescopes—have evolved to observe as broad a range of wavelengths as possible. Until the middle of the twentieth century, telescopes were limited to visible light. Since then, however, technological advances have expanded our view of the universe to all regions of the electromagnetic spectrum. Some telescopes are sited on Earth, others must be placed in space, and design considerations vary widely from one part of the spectrum to another. Whatever the details of its construction, however, the basic purpose of a telescope is to collect electromagnetic radiation and deliver it to a detector for detailed study.

3.1 Optical Telescopes

1 In essence, a **telescope** is a "light bucket"—a device whose primary function is to capture as much radiation as possible from a given region of the sky and concentrate it into a focused beam for analysis. We begin our study of astronomical hardware with *optical* telescopes, designed to collect wavelengths visible to the human eye. Later we will turn our attention to telescopes designed to capture and analyze radiation in other, *invisible* regions of the electromagnetic spectrum.

Reflecting and Refracting Telescopes

Optical telescopes fall into two basic categories—*reflectors* and *refractors*. Figure 3.1 shows how a **reflecting telescope** uses a curved mirror to gather and concentrate a beam of light. The mirror, usually called the *primary mirror* because telescopes generally contain more than one mirror, is constructed so that all light rays arriving parallel to its axis (the imaginary line through the center of and perpendicular to the mirror), regardless of their distance from that axis, are reflected to pass through a single point, called the *focus*. The distance between the primary mirror and the focus is the *focal length*. In astronomical contexts, the focus of the primary mirror is referred to as the **prime focus**.

A **refracting telescope** uses a lens to focus the incoming light. **Refraction** is the bending of a beam of light as it passes from one transparent medium (for example, air) into another (such as glass). Figure 3.2(a) illustrates the process and shows how a prism can be used to change the direction of a beam of light. As illustrated in Figure 3.2(b), we can think of a lens as a series of prisms combined in such a way that all light rays striking the lens parallel to the axis are refracted to pass through the focus.

Optical telescopes are often used to make **images** of their fields of view. Figure 3.3 illustrates how this is accomplished, in this case by the primary mirror in a reflecting telescope. Light from a distant object reaches the telescope as nearly parallel rays. A ray of light entering the instrument parallel to the mirror axis is reflected through the focus. Light from a slightly different direction—that is, at a slight angle to the axis—is focused to a slightly different point. In this way, an image is formed near the focus. Each point in the image corresponds to a different angle in the field of view. Often, the image is magnified with a lens known as an *eyepiece*. Figure 3.4(a) shows the basic design of a simple reflecting telescope, illustrating how a small secondary mirror and an eyepiece are used to view the image. Figure 3.4(b) shows how a refracting telescope accomplishes the same function.

The two telescope designs shown in Figure 3.4 achieve the same result—light from a distant object is cap-

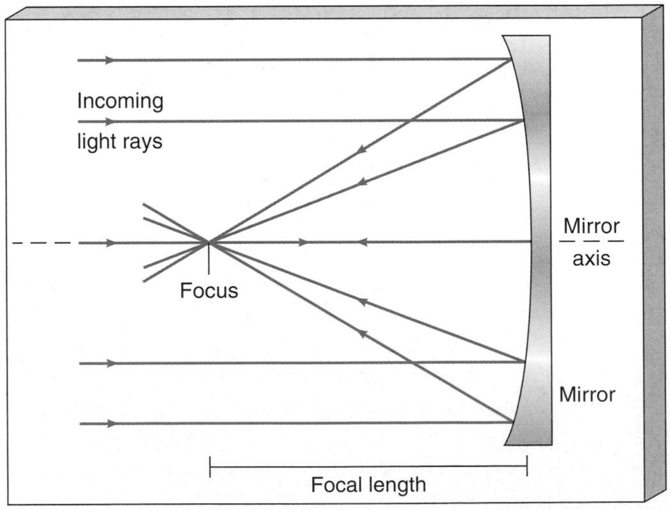

Figure 3.1 Reflecting Mirror A curved mirror can be used to focus to a single point all rays of light arriving parallel to the mirror axis. Light rays traveling along the axis are reflected back along the axis, as indicated by the arrowheads pointing in both directions. Off-axis rays are reflected through greater and greater angles the farther they are from the axis, so that they all pass through the same point—the focus.

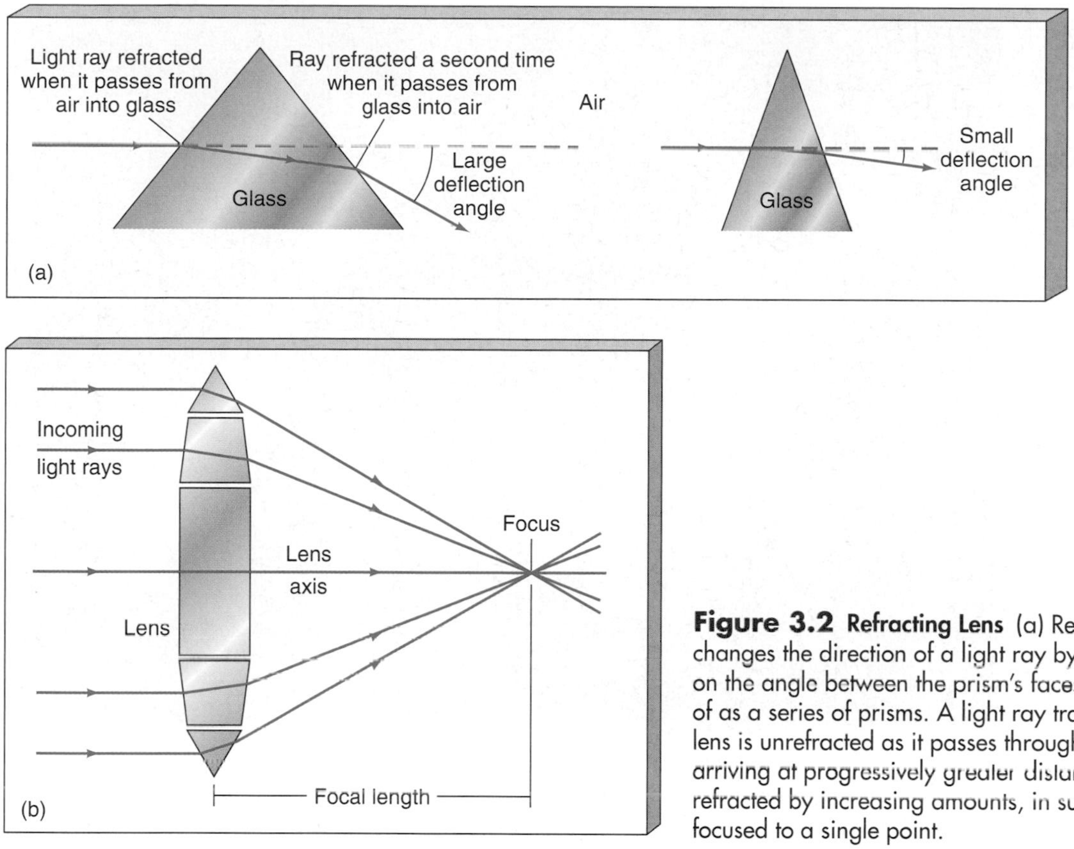

Figure 3.2 Refracting Lens (a) Refraction by a prism changes the direction of a light ray by an amount that depends on the angle between the prism's faces. (b) A lens can be thought of as a series of prisms. A light ray traveling along the axis of a lens is unrefracted as it passes through the lens. Parallel rays arriving at progressively greater distances from the axis are refracted by increasing amounts, in such a way that all are focused to a single point.

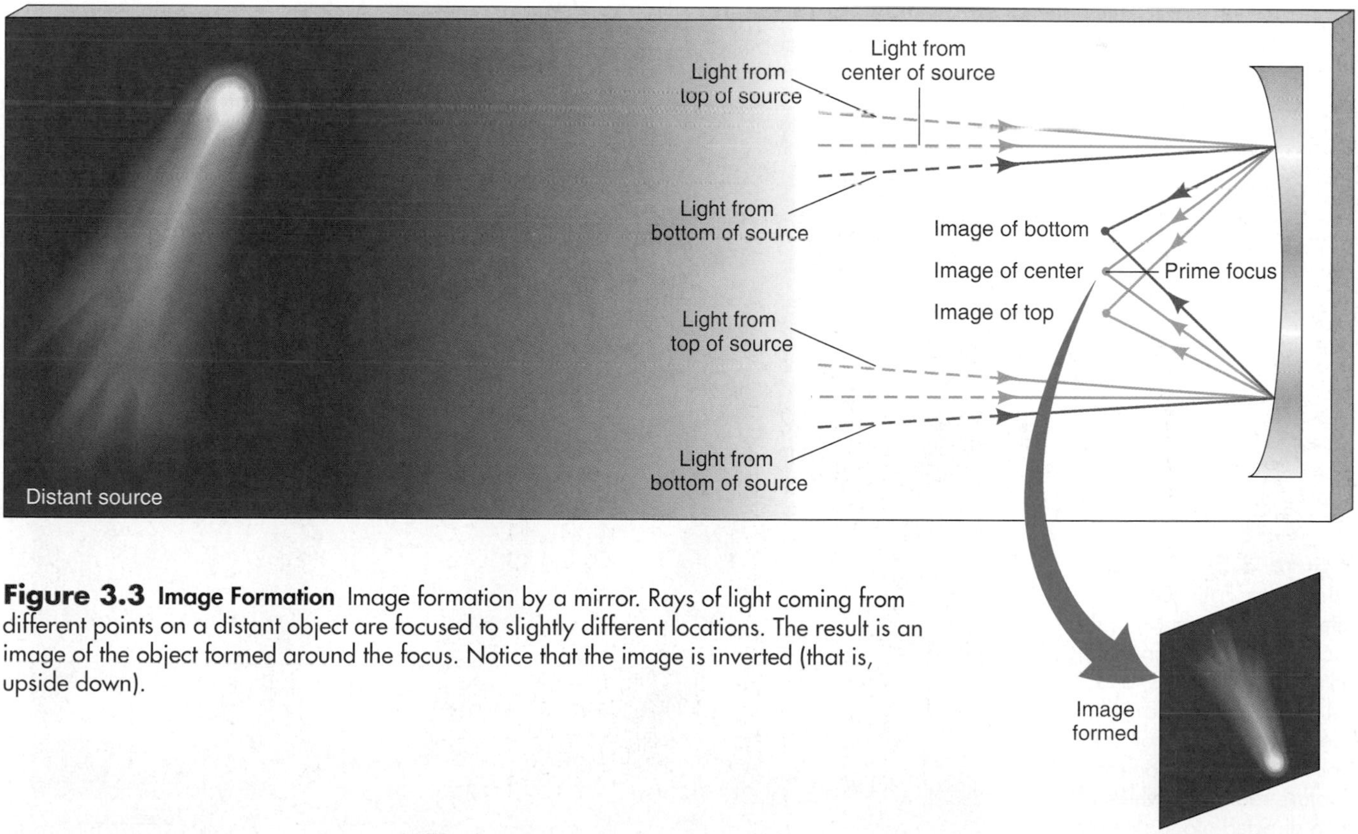

Figure 3.3 Image Formation Image formation by a mirror. Rays of light coming from different points on a distant object are focused to slightly different locations. The result is an image of the object formed around the focus. Notice that the image is inverted (that is, upside down).

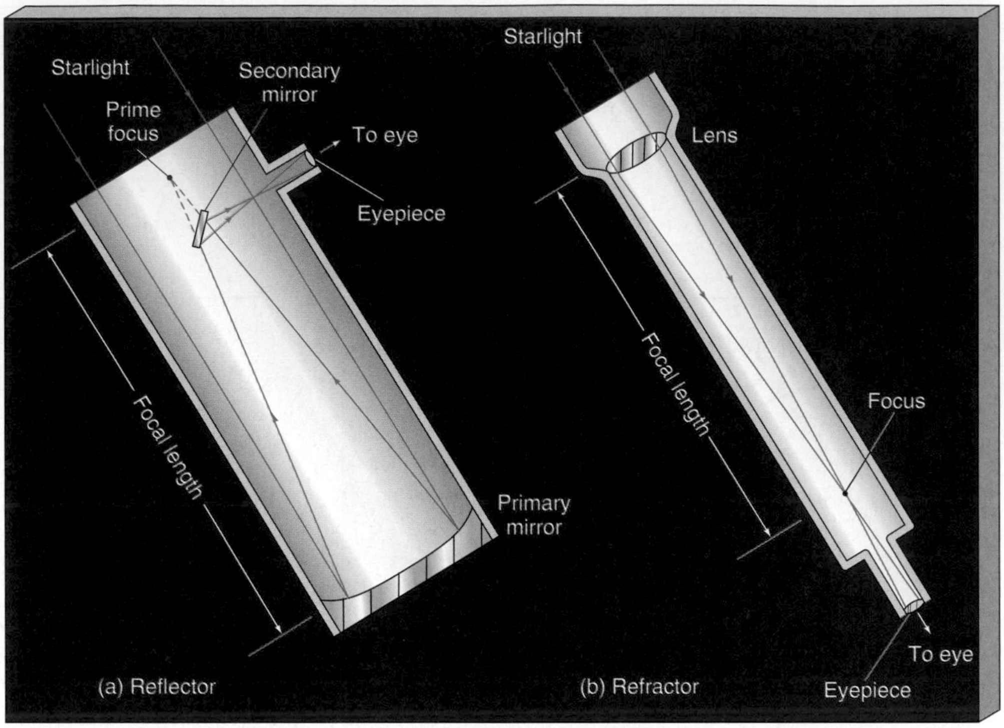

Figure 3.4 Reflectors and Refractors Comparison of (a) reflecting and (b) refracting telescopes. Both types are used to gather and focus cosmic radiation that is either observed by human eyes or recorded on photographs or in computers. In both types, the image formed at the focus is viewed with a small magnifying lens called an eyepiece.

tured and focused to form an image. However, as telescope *size* has steadily increased over the years (for reasons to be discussed in Section 3.2), a number of important factors have tended to favor reflecting instruments over refractors:

- Just as a prism separates white light into its component colors, the lens in a refracting telescope focuses red and blue light differently. This deficiency is known as *chromatic aberration*. Careful design and choice of materials can largely correct this problem, but it is difficult to eliminate it entirely. Mirrors do not suffer from this defect.
- As light passes through the lens, some of it is absorbed by the glass. This absorption is a relatively

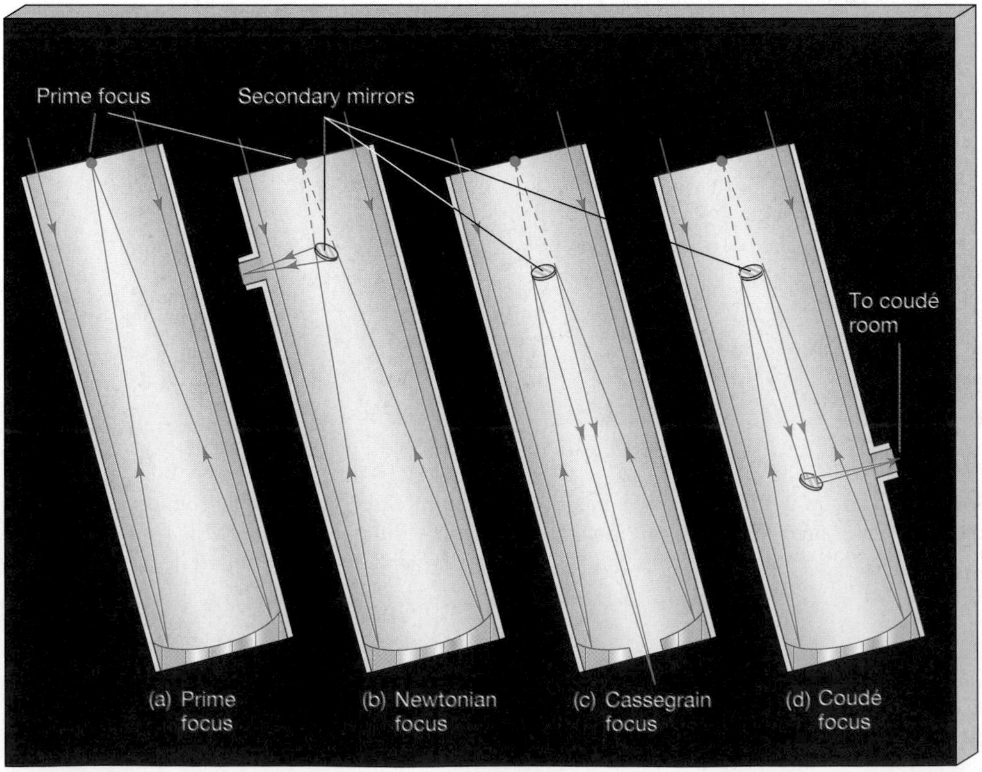

Figure 3.5 Reflecting Telescopes Four reflecting telescope designs: (a) prime focus, (b) Newtonian focus, (c) Cassegrain focus, and (d) coudé focus. Each design uses a primary mirror at the bottom of the telescope to capture radiation, which is then directed along different paths for analysis.

minor problem for visible radiation, but it can be severe for infrared and ultraviolet observations because glass blocks most of the radiation coming from those regions of the electromagnetic spectrum. This problem obviously does not affect mirrors.

- A large lens can be quite heavy. Because it can be supported only around its edge (so as not to block the incoming radiation), the lens tends to deform under its own weight. A mirror does not have this drawback because it can be supported over its entire back surface.
- A lens has two surfaces that must be accurately machined and polished, which can be a difficult task. A mirror has only one surface.

For these reasons, *all* large modern optical telescopes are reflectors.

Telescope Design

Figure 3.5 shows some basic reflecting telescope designs. Radiation from a star enters the instrument, passes down the main tube, strikes the primary mirror, and is reflected back toward the prime focus, near the top of the tube. Sometimes astronomers place their recording instruments at the prime focus. However, it can be very inconvenient, or even impossible, to suspend bulky pieces of equipment there. More often, the light is intercepted on its path to the focus by a secondary mirror and redirected to a more convenient location, as in Figures 3.5(b) through (d).

In a **Newtonian telescope** (named after Isaac Newton, who invented this design), the light is intercepted by a flat secondary mirror before it reaches the prime focus and deflected 90°, usually to an eyepiece at the side of the instrument (Figure 3.5b). This is a popular design for smaller reflecting telescopes, such as those used by amateur astronomers.

Alternatively, astronomers may choose to work on a rear platform where they can use detecting equipment too heavy or delicate to hoist to the prime focus. In this case, the light reflected by the primary mirror toward the prime focus is intercepted by a convex secondary mirror, which reflects the light back down the tube and through a small hole at the center of the primary mirror (Figure 3.5c). This arrangement is known as a **Cassegrain telescope** (after Guillaume Cassegrain, a French lensmaker). The point behind the primary mirror where the light from the star finally converges is called the *Cassegrain focus*.

In a variant on the Cassegrain design, light is first reflected by the primary mirror toward the prime focus and reflected back down the tube by a secondary mirror. A third, much smaller mirror then reflects the light into an environmentally controlled laboratory (Figure 3.5d). Known as the *coudé* room (from the French word for "bent"), this laboratory is separate from the telescope, enabling astronomers to use very heavy and finely tuned equipment that could not possibly be lifted to either the prime focus or the Cassegrain focus.

To illustrate some of these points, Figure 3.6 depicts the Hale 5-m-diameter optical telescope on

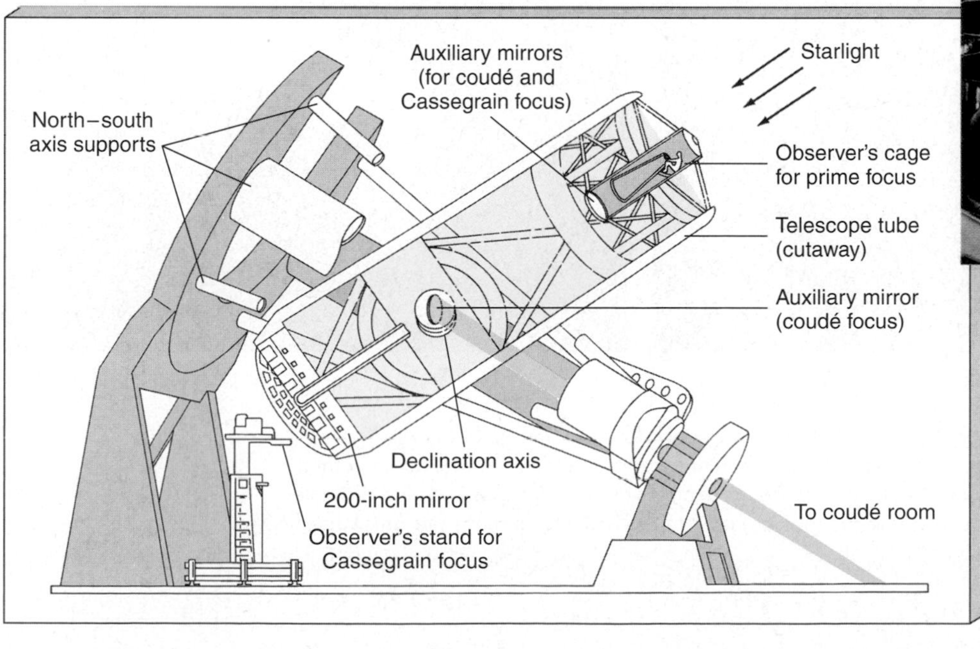

Figure 3.6 Palomar Telescope (a) An artist's illustration of the 5-m-diameter Hale reflecting telescope on Palomar Mountain in California. (b) Astronomer Edwin Hubble in the observer's cage at the Hale prime focus. *(Palomar Observatory/California Institute of Technology)*

California's Palomar Mountain. For almost three decades after its dedication in 1948, the Hale telescope was the largest in the world. It has been at or near the forefront of astronomical research for much of the last half century. Observations can be made at the prime, the Cassegrain, or the coudé focus, depending on the needs of the user.

✓ Concept Check

- Why do all modern telescopes use mirrors to gather and focus light?

3.2 Telescope Size

2 Astronomers generally prefer large telescopes over small ones, for two main reasons. The first has to do with the amount of light a telescope can gather—its *light-gathering power*. The second is related to the amount of detail that can be seen—the telescope's *resolving power*.

Light-Gathering Power

One important reason for using a larger telescope is that it has a greater **collecting area**, which is the area capable of gathering radiation. The larger the telescope's reflecting mirror (or refracting lens), the more light it collects and the easier it is to measure and study an object's radiative properties. Astronomers spend much of their time observing very distant—and hence very *faint*—cosmic sources. In order to make detailed observations of such objects, very large telescopes are essential.

The observed brightness of an astronomical object is directly proportional to the area of the telescope's mirror and hence to the *square* of the mirror diameter (Figure 3.7). Thus, for example, a 5-m telescope produces an image 25 times brighter than a 1-m instrument because a 5-m mirror has $5^2 = 25$ times the collecting area of a 1-m mirror. We can also think of this relationship in terms of the *time* required for a telescope to collect enough energy to create a recognizable image on a photographic plate. A 5-m telescope produces an image 25 times faster than a 1-m device because it gathers energy at a rate 25 times greater. Put another way, a one-hour time exposure with a 1-m telescope is roughly equivalent to a 2.4-minute time exposure with a 5-m instrument.

Currently, the largest optical telescopes are the twin Keck instruments atop Mauna Kea in Hawaii (Figure 3.8), operated jointly by the California Institute of

Figure 3.7 Sensitivity Effect of increasing light-gathering power on an image of the Andromeda Galaxy. Both photographs had the same exposure time, but the image on the right was taken with a telescope twice the size of that used to make the left image. Fainter detail can be seen as the diameter of the telescope mirror increases because larger telescopes are able to collect more photons per unit time. *(AURA)*

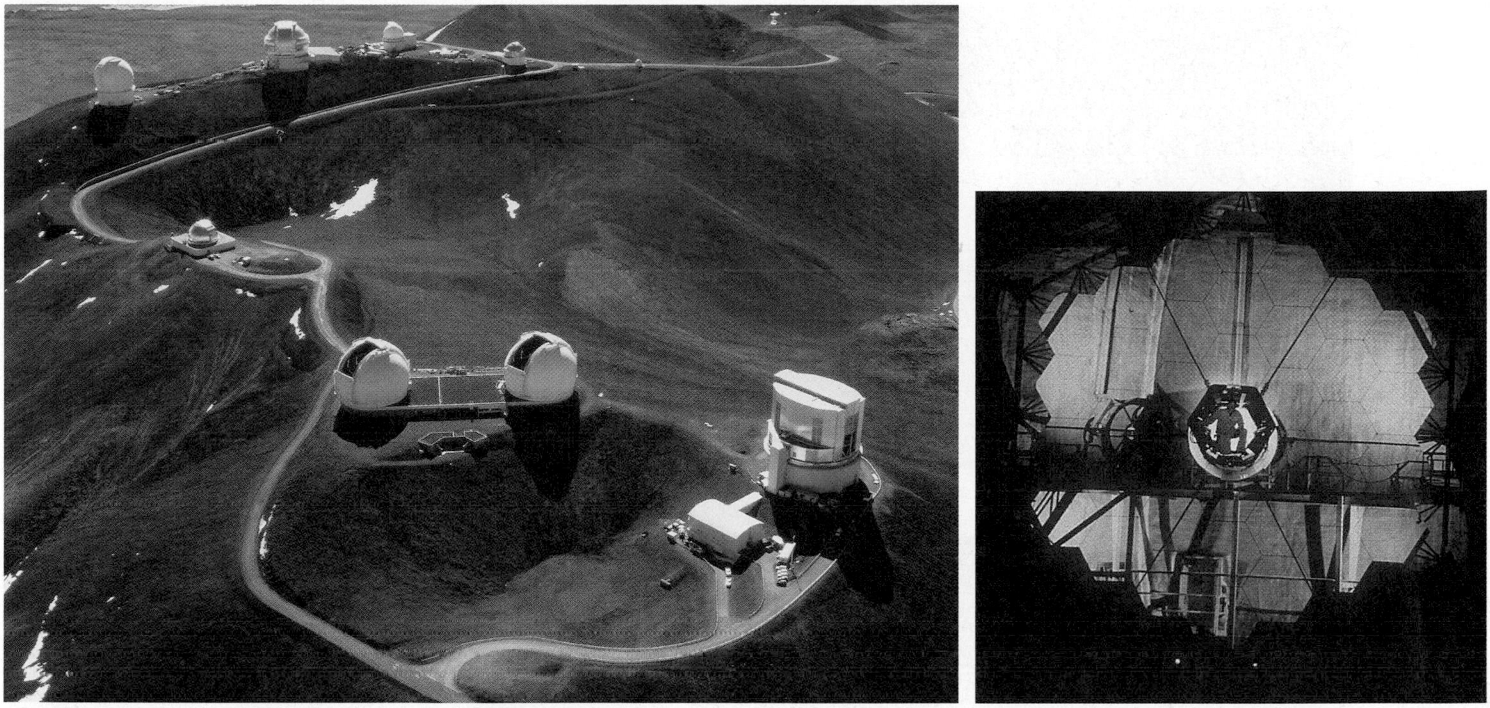

(a) (b)

Figure 3.8 Mauna Kea Observatory (a) The world's highest ground-based observatory, at Mauna Kea, Hawaii, is perched atop an extinct volcano more than 4 km above sea level. Among the domes visible in the picture are those that house the Canada–France–Hawaii 3.6-m telescope, the 2.2-m telescope of the University of Hawaii, Britain's 3.8-m infrared facility, and the twin 10-m Keck telescopes. To the right of the twin Kecks is the Japanese 8.3-m Subaru telescope. The thin air at this high-altitude site guarantees less atmospheric absorption of incoming radiation and hence a clearer view than at sea level, but the air is so thin that astronomers must occasionally wear oxygen masks while working. (b) The 10-m mirror in the first Keck telescope. Note the technician in orange coveralls at center. *(R. Wainscoat; R. Underwood/ W. M. Keck Observatory)*

Technology and the University of California. Each telescope combines 36 hexagonal 1.8-m mirrors into the equivalent collecting area of a single 10-m reflector. The first Keck telescope became fully operational in 1992; the second was completed in 1996. The high altitude and large size of these telescopes makes them particularly well suited for detailed spectroscopic studies of very faint objects. The Very Large Telescope (VLT), now under construction at the European Southern Observatory at Cerro Paranal, in Chile, will consist of four separate 8.2-m mirrors that will function as a single instrument. The first two mirrors became operational in 1998 and 1999, the third in 2000.

Resolving Power

A second advantage of large telescopes is their superior **angular resolution**. In general, *resolution* refers to the ability of any device, such as a camera or a telescope, to form distinct, separate images of objects lying close together in the field of view. The finer the resolution, the better we can distinguish the objects and the more detail we can see. In astronomy, where we are always concerned with angular measurement, "close together" means "separated by a small angle on the sky," so angular resolution is the factor that determines our ability to see fine structure. ∞ (Sec. P.5) Figure 3.9 illustrates the result of increasing resolving power with views of the Andromeda Galaxy at several different resolutions.

One important factor limiting a telescope's resolution is **diffraction**, the tendency of light to "bend" around corners (Figure 3.10). Because of diffraction, when a parallel beam of light enters a telescope, the rays spread out slightly, making it impossible to focus the beam to a

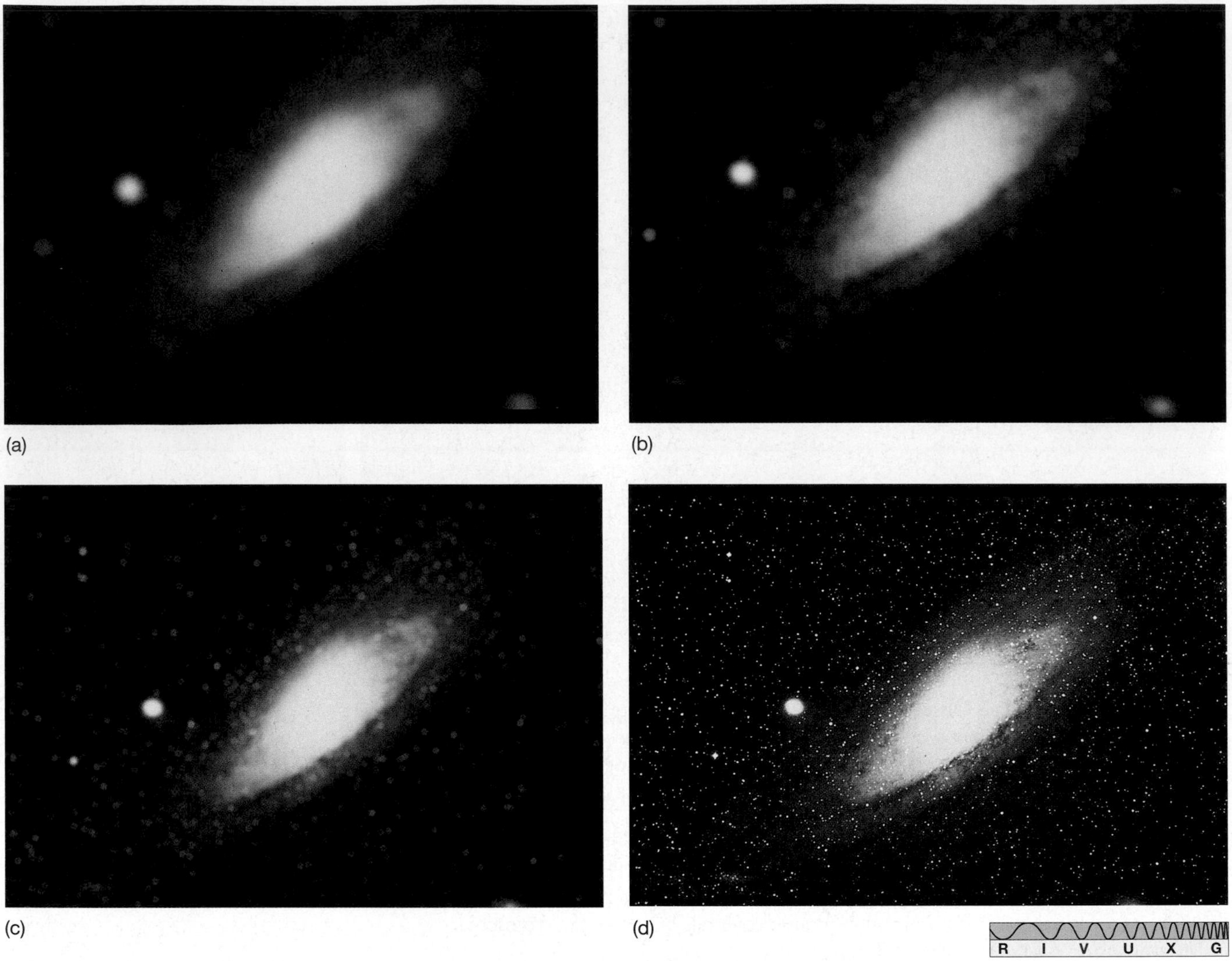

Figure 3.9 Resolution Detail becomes clearer in the Andromeda Galaxy as the angular resolution is improved some 600 times, from (a) 10′, to (b) 1′, (c) 5″, and (d) 1″. *(AURA)*

sharp point, even with a perfectly constructed mirror. Diffraction introduces a certain "fuzziness," or loss of resolution, into the system. The degree of fuzziness—the minimum angular separation that can be distinguished—determines the angular resolution of the telescope. The amount of diffraction is proportional to the wavelength of the radiation divided by the diameter of the telescope mirror. As a result, we can write, in convenient units:

$$\text{angular resolution (arc seconds)} = 0.25\,\frac{\text{wavelength }(\mu\text{m})}{\text{telescope diameter (m)}},$$

where 1 μm (micrometer) = 1000 nm.

For a given telescope size, the blurring effects of diffraction increase in proportion to the wavelength used. Observations in the infrared or radio range are often severely limited by this effect. For example, in an otherwise perfect observing environment, the best possible angular resolution of blue light (wavelength 400 nm = 0.4 μm) using a 1-m telescope is about 0.1″. This quantity is known as the *diffraction-limited resolution* of the telescope. For the same telescope used in the near infrared range, at a wavelength of 10,000 nm = 10 μm, the best resolution obtainable is only 2.5″. For light of a given wavelength, large telescopes produce less diffraction than small ones. A 5-m telescope observing in blue light has a diffraction-limited resolution of 0.02″, five times finer than that of the 1-m telescope just discussed. A 0.1-m telescope has a diffraction limit of 1″, and so on. For comparison, the angular resolution of the human eye in blue light is about 0.5′.

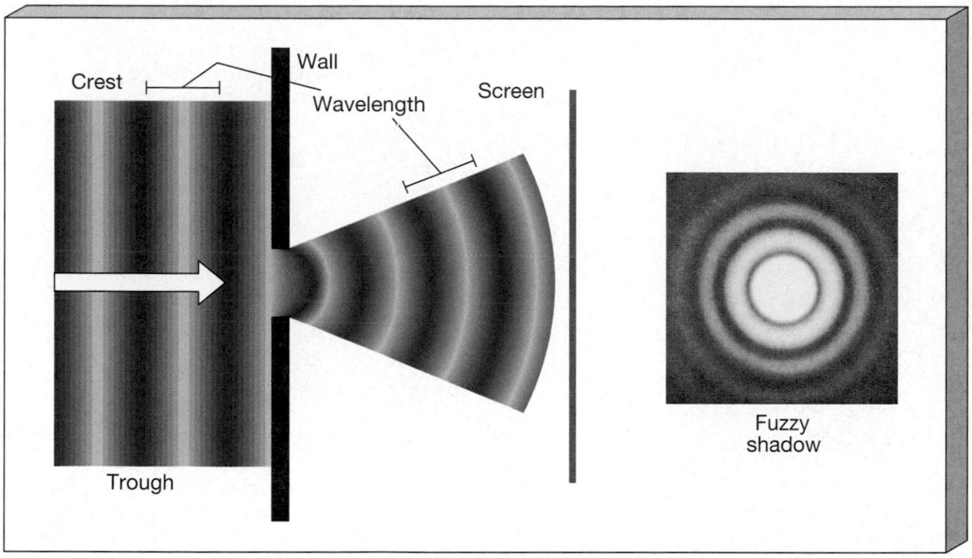

Figure 3.10 Diffraction A wave passing through a gap is diffracted through an angle that depends on the ratio of the wavelength of the wave to the size of the gap, leading to a "fuzzy" shadow on the screen. The longer the wavelength and/or the smaller the gap, the greater the angle through which the wave is diffracted. In the absence of diffraction, the shadow would be perfectly sharp. *(© Springer–Verlag GMbH & Co. KG, M. Cagnet et al, Atlas of Optical Phenomena, 1962)*

✓ Concept Check

- Give two reasons why astronomers need to build very large telescopes.

3.3 High-Resolution Astronomy

3 Even large telescopes have their limitations. For example, according to the discussion in the preceding section, the 5-m Hale telescope should have an angular resolution of around 0.02″. In practice, however, it cannot do better than about 1″. In fact, apart from instruments using special techniques developed to examine some particularly bright stars, *no* ground-based optical telescope built before 1990 can resolve astronomical objects to much better than about 1″. The reason is Earth's turbulent atmosphere, which blurs the image even before the light reaches our instruments.

Atmospheric Blurring

As we observe a star, atmospheric turbulence produces continuous small changes in the optical properties of the air between the star and our telescope (or eye). The light from the star is refracted slightly, again and again, and the stellar image dances around on the detector (or retina). This is the cause of the well-known "twinkling" of stars. It occurs for the same basic reason that objects appear to shimmer when viewed across a hot roadway on a summer day.

On a good night at the best observing sites, the maximum deflection produced by the atmosphere is slightly less than 1″ (i.e. 1 arc second). Consider taking a photograph of a star under such conditions. After a few minutes' exposure time (long enough for the intervening atmosphere to have undergone many small, random changes), the dancing sharp image of the star has been smeared out over a roughly circular region 1″ or so in diameter (Figure 3.11). Astronomers use the term **seeing** to describe the effects of atmospheric turbulence. The circle over which a star's light is spread is called the **seeing disk**.

To achieve the best possible seeing, telescopes are sited on mountaintops (to get above as much of the atmosphere as possible) in locations where the atmosphere is known to be fairly stable and relatively free of dust, moisture, and light pollution from cities. In the continental United States, these sites tend to be in the desert Southwest. The U.S. National Observatory for optical astronomy in the Northern Hemisphere, completed in 1973, is located high on Kitt Peak near Tucson, Arizona. The site was chosen because of its many dry, clear nights. Seeing of 1″ from such a location is regarded as good, and seeing of a few arc seconds is tolerable for many purposes. Even better conditions are found on Mauna Kea in Hawaii (Figure 3.8) and at La Silla and Cerro Paranal in the Andes Mountains of Chile (Figure 3.12), which is why many large telescopes have recently been constructed at those two exceptionally clear sites.

A telescope placed above the atmosphere, in Earth orbit, can achieve resolution close to the

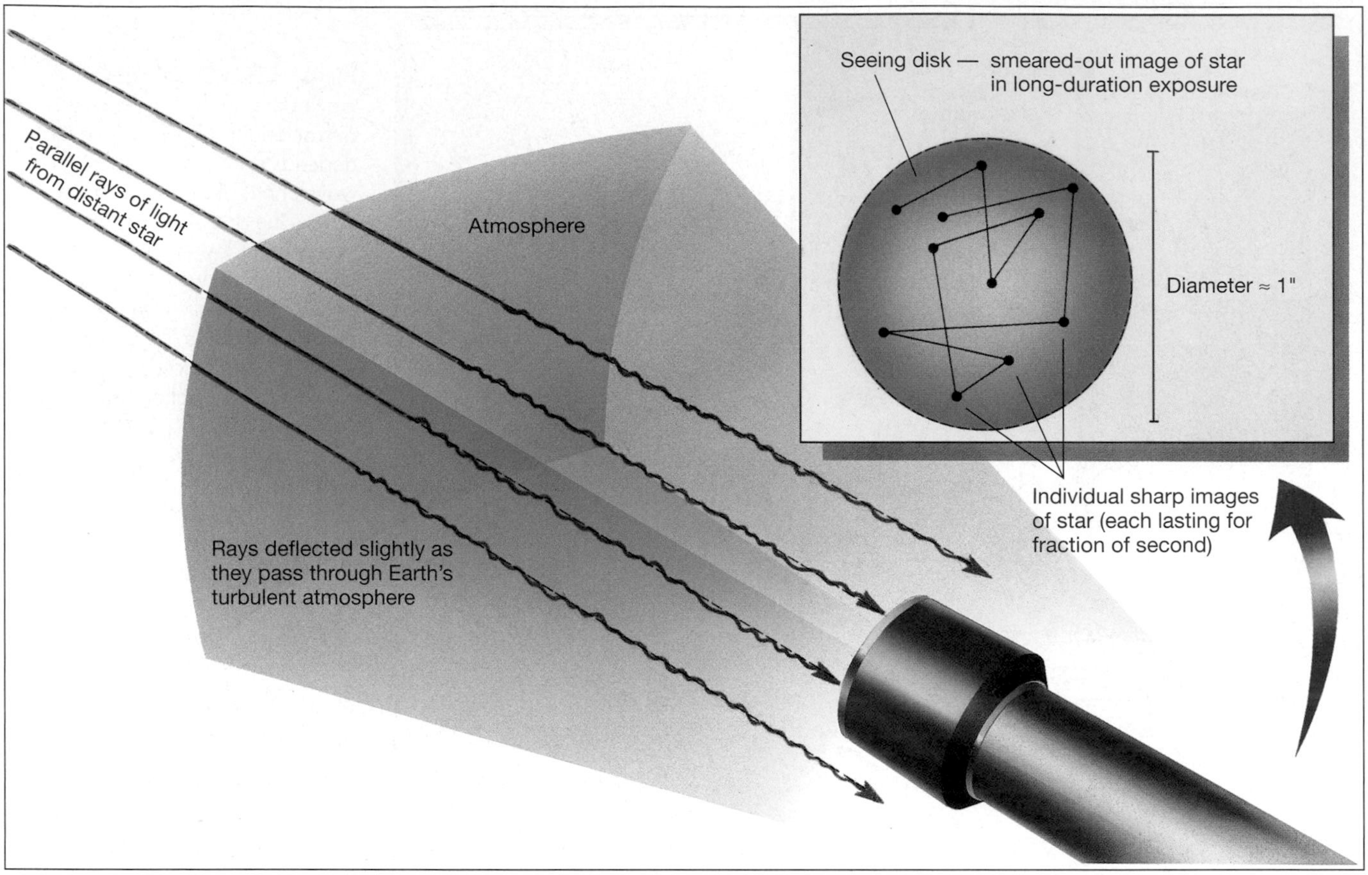

Figure 3.11 Atmospheric Turbulence Individual photons from a distant star strike a telescope detector at slightly different locations because of turbulence in Earth's atmosphere. Over time, the photons cover a roughly circular region on the detector, and even the pointlike image of a star is recorded as a small disk, called the seeing disk.

diffraction limit, subject only to the engineering restrictions of building or placing large structures in space. The *Hubble Space Telescope* (*HST*), named for one of America's most notable astronomers, Edwin Hubble (Figure 3.6b), has a 2.4-m mirror and a diffraction limit of 0.05″, giving astronomers a view of the universe as much as 20 times sharper than that normally available from even much larger ground-based instruments. (See *Interlude 3-1*.)

Image Processing

It is becoming rare for photographic equipment to be used as the primary means of data acquisition at large observatories. Instead, electronic detectors known as

Figure 3.12 Southern Hemisphere Telescope Located in the Andes Mountains of Chile, the European Southern Observatory at La Silla is run by a consortium of European nations. Numerous domes house optical telescopes of different sizes, each carrying various pieces of support equipment, making this one of the most versatile observatories south of the equator. The New Technology Telescope is inside the square dome at center. *(ESO)*

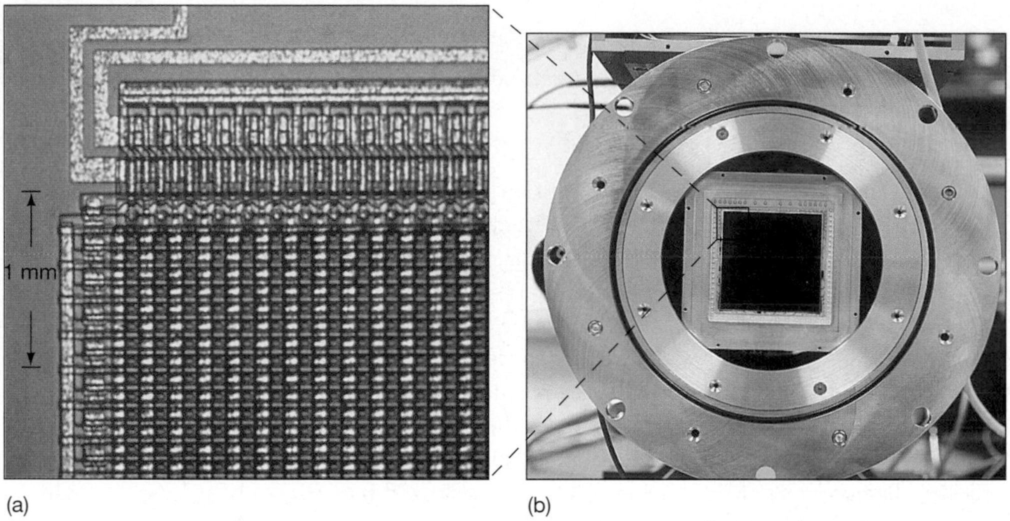

Figure 3.13 **CCD Chip** (a) A charge-coupled device consists of hundreds of thousands, or even millions, of tiny light-sensitive cells called pixels, usually arranged in a square array. Light striking a pixel causes an electrical charge to build up on it. By electronically reading out the charge on each pixel, a computer can reconstruct the pattern of light—the image—falling on the chip. (b) A CCD chip mounted at the focus of a telescope. *(AURA; R. Wainscoat/Peter Arnold)*

charge-coupled devices, or **CCDs**, are in widespread use. A CCD (Figure 3.13) consists of a wafer of silicon divided into a two-dimensional array of many tiny picture elements, known as *pixels*. When light strikes a pixel, an electric charge builds up on the device. The amount of charge is directly proportional to the number of photons striking each pixel—in other words, to the intensity of the light at that point. The charge buildup is monitored electronically, and a two-dimensional image is obtained. A CCD is typically a few square centimeters in area and may contain several million pixels, generally arranged on a square grid. As the technology improves, both the areas of CCDs and the number of pixels they contain are steadily increasing.

CCDs have two important advantages over photographic plates, which were the staple of astronomers for over a century. First, CCDs are much more *efficient* than photographic plates, recording as many as 75 percent of the photons striking them, compared with less than five percent for photographic methods. This means that a CCD instrument can image objects 10 to 20 times fainter—or the same object 10 to 20 times faster—than can a photographic plate. Second, CCDs produce a faithful representation of an image in a digital format that can be placed directly on magnetic tape or disk, or even sent across a computer network to an observer's home institution for analysis.

Using computer processing, astronomers can compensate for known instrumental defects and even correct some effects of bad seeing. In addition, the computer can often carry out many of the tedious and time-consuming chores that must be performed before an image or spectrum reaches its final form. Figure 3.14 illustrates how

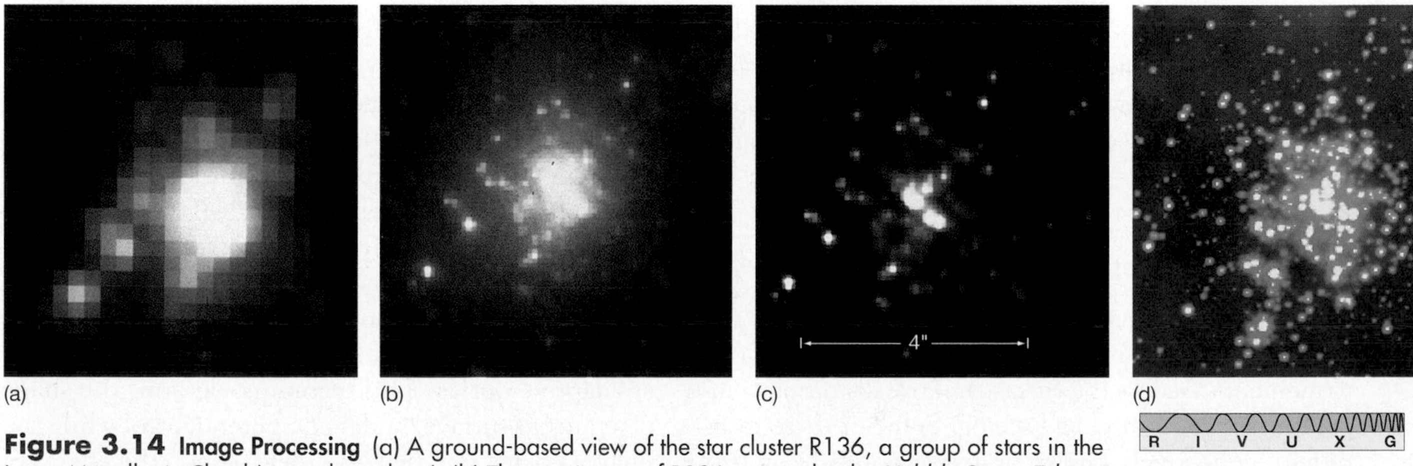

Figure 3.14 **Image Processing** (a) A ground-based view of the star cluster R136, a group of stars in the Large Magellanic Cloud (a nearby galaxy). (b) The raw image of R136 as seen by the *Hubble Space Telescope* in 1990, before the repair mission. (c) The same image after computer processing that partly compensated for imperfections in the mirror. (d) The same region as seen by the repaired *HST* in 1994. *(AURA; NASA)*

INTERLUDE 3-1
The Hubble Space Telescope

The *Hubble Space Telescope* (*HST*) is the largest, most complex, most sensitive observatory ever deployed in space. At over $4.5 billion (including the cost of three missions to service and refurbish the system), it is also the most expensive scientific instrument ever constructed. Built jointly by NASA and the European Space Agency, *HST* is designed to allow astronomers to probe the universe with at least 10 times finer resolution and with some 30 times greater sensitivity to light than existing Earth-based devices. At the heart of *HST* is a 2.4-m-diameter mirror designed to capture optical, ultraviolet, and infrared radiation before it reaches Earth's murky atmosphere. The accompanying figure shows the telescope being lifted out of the cargo bay of the space shuttle *Discovery* in the spring of 1990.

The telescope reflects light from its primary mirror back to a smaller, 0.3-m secondary mirror, which sends the light through a hole in the primary mirror and into the aft bay of the spacecraft (see the chapter opening illustration). There, any of five major scientific instruments, each about the size of a telephone booth, wait to analyze the incoming radiation. They are designed to be maintained by NASA astronauts, and indeed most of the telescope's original instruments have been upgraded or replaced since *HST* was launched.

(NASA)

Soon after launch, astronomers discovered that the telescope's primary mirror had been polished to the wrong shape. The mirror is too flat by 2 μm, about 1/50 the width of a human hair, making it impossible to focus light as well as expected. In 1993, astronauts aboard the space shuttle *Endeavour* visited *HST*. They replaced *Hubble*'s gyroscopes to help the telescope point more accurately, installed sturdier versions of the solar panels that power the telescope's electronics, and—most importantly—inserted an intricate set of small mirrors to compensate for the faulty primary mirror. *Hubble*'s resolution is now close to the original design specifications, and the telescope has regained much of its lost sensitivity. Additional service missions were performed in 1997 and 1999 to upgrade instruments and repair damaged systems.

A good example of *Hubble*'s current scientific capabilities can be seen by comparing the two images of the spiral galaxy M100, shown here. On the left is perhaps the best ground-based photograph of this beautiful galaxy, showing rich detail and color in its spiral arms. On the right, to the same scale and orientation, is an *HST* image showing improvements in both resolution and sensitivity.

computerized image-processing techniques were used to correct for known instrumental problems in the *Hubble Space Telescope*, allowing much of the planned resolution of the telescope to be recovered even before its repair in 1993.

New Telescope Design

By analyzing the image while the light is still being collected, it is possible to adjust the telescope from moment to moment to reduce the effects of mirror distortion, temperature changes, and bad seeing. Some of these techniques, collectively known as **active optics**, were first combined in the New Technology Telescope (NTT) at the European Southern Observatory in Chile (Figure 3.12). This 3.5-m instrument achieves resolution of about 0.5″ by making minute modifications to the tilt of the mirror as its temperature and orientation change, thus maintaining the best possible focus at all times and greatly improving image quality (Figure 3.15). The Keck and VLT instruments employ similar methods and may ultimately achieve resolution as fine as 0.25″ by these means.

An even more ambitious undertaking is known as **adaptive optics**. This technique deforms the shape of a mirror's surface, under computer control while the image is being exposed, in order to undo the effects of

(The chevron-shaped field of view is caused by the corrective optics inserted into the telescope in 1993; an additional trade-off is that *Hubble*'s field of view is smaller than those of ground-based telescopes.) The inset shows *Hubble*'s exquisite resolution of small fields of view. Many spectacular examples of the telescope's remarkable capabilities appear throughout this book.

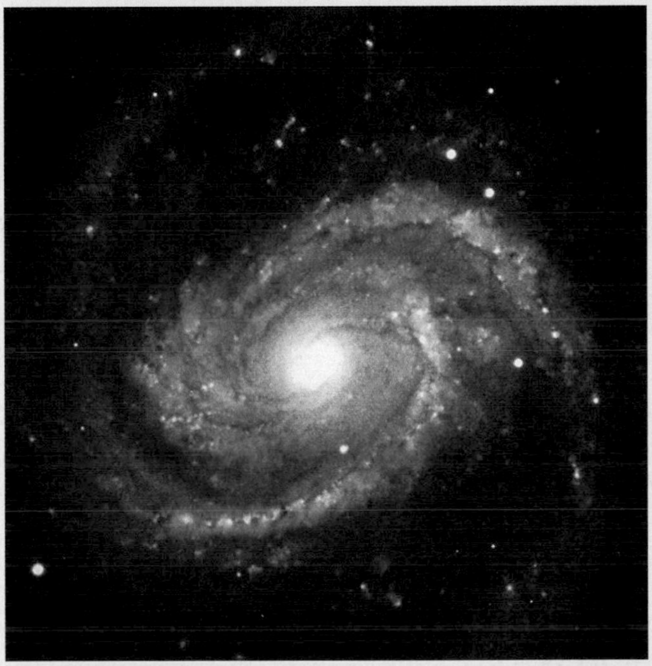

(Anglo-Australian Observatory/Royal Observatory Edinburgh. Photograph made from VK Schmidt Plates by David Malin)

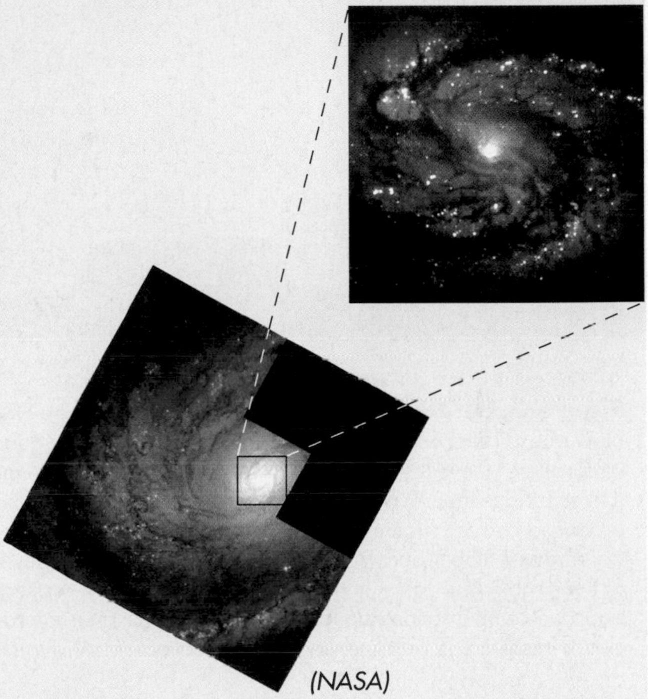

(NASA)

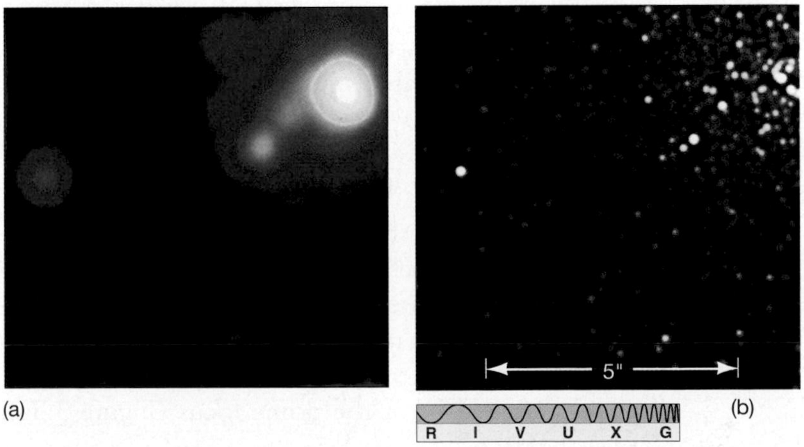

Figure 3.15 Active Optics Infrared images of part of the star cluster R136 contrast the resolution obtained (a) without and (b) with an active optics system. *(Courtesy of Bernhard Brandl (Max Planck Institute) and Sky & Telescope Magazine)*

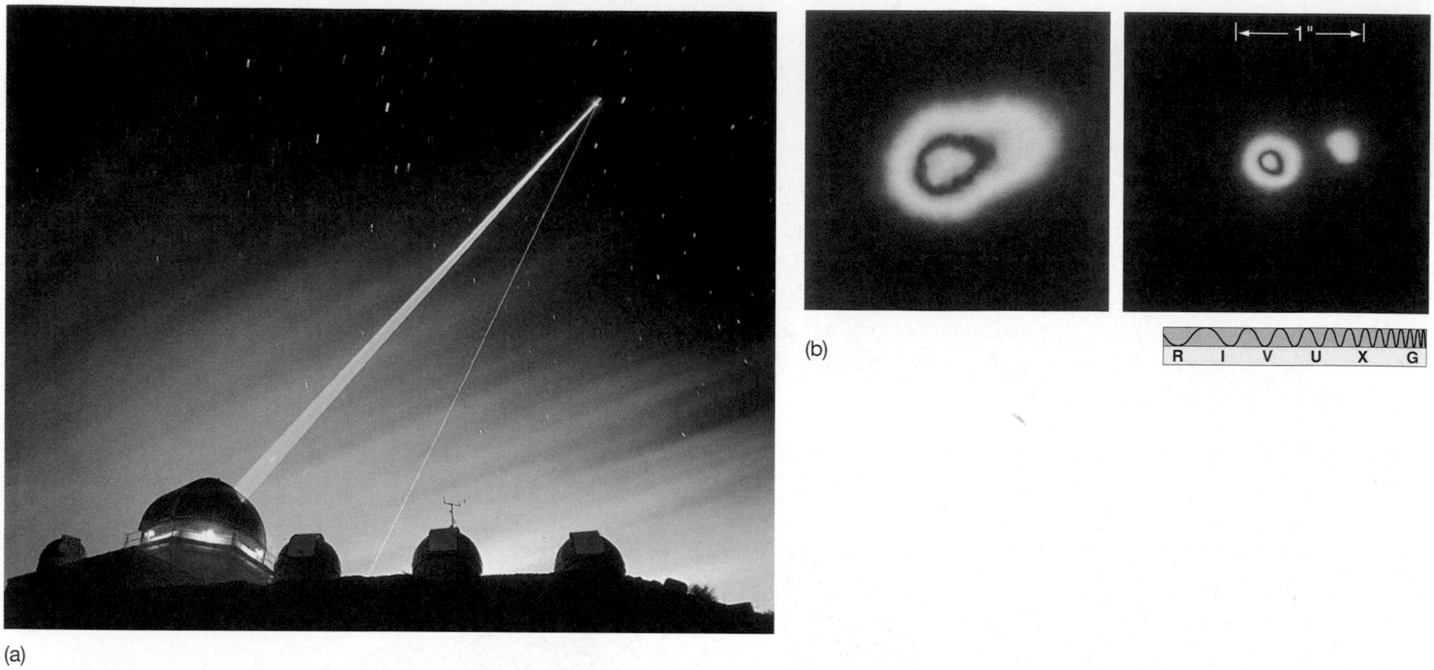

Figure 3.16 Adaptive Optics (a) Until mid-1991, the Starfire Optical Range at Kirtland Air Force Base in New Mexico was one of the U.S. Air Force's most closely guarded secrets. Here, beams of laser light probe the atmosphere above the system's 1.5-m telescope, allowing minute computer-controlled changes to be made to the mirror surface thousands of times each second. *(Roger Ressmeyer © 1994 Corbis)* (b) The improvement in seeing produced by such systems can be dramatic, as can be seen in these images acquired at another military observatory atop Mount Haleakala in Maui, Hawaii, employing similar technology. The uncorrected image (left) of the double star Castor is a blur spread over several arc seconds, with little hint of its binary nature. With adaptive compensation applied (right), the resolution is improved to a mere 0.1″ and the two components are clearly resolved. *(R. Ressmeyer/Starlight Collection, A Division of Corbis © 1994; MIT Lincoln Laboratory)*

atmospheric turbulence. In the experimental system shown in Figure 3.16(a), lasers probe the atmosphere above the telescope, returning information about the air's swirling motion to a computer that modifies the mirror thousands of times per second to compensate for poor seeing. Already, impressive improvements in image quality have been obtained (Figure 3.16b). These developing techniques are also being incorporated into Keck and VLT. In the next decade, it may well be possible to achieve with large ground-based telescopes the kind of resolution presently attainable only from space.

Concept Check

- What steps do optical astronomers take to overcome the obscuring and blurring effects of Earth's atmosphere?

3.4 Radio Astronomy

4 In addition to the visible radiation that penetrates Earth's atmosphere on a clear day, radio radiation also reaches the ground, and astronomers have built many ground-based **radio telescopes** capable of detecting cosmic radio waves. ∞ (Sec. 2.3) These devices have all been constructed since the 1950s—radio astronomy is a much younger field than optical astronomy.

Essentials of Radio Telescopes

Figure 3.17 shows a typical radio telescope, the large 43-m-diameter telescope located at the National Radio Astronomy Observatory in West Virginia. Conceptually, the operation of a radio telescope is similar to the operation of an optical reflector with the detecting instruments placed at the prime focus (Figure 3.5a).

Figure 3.17 **Radio Telescope** The 43-m-diameter radio telescope at the National Radio Astronomy Observatory in Green Bank, West Virginia. *(NRAO)*

They have a large, horseshoe-shaped mount supporting a huge, curved metal dish that serves as the collecting area. The dish captures cosmic radio waves and reflects them to the focus, where a receiver detects the signals and channels them to a computer.

Radio telescopes are built large in part because cosmic radio sources are extremely faint. In fact, the total amount of radio energy received by Earth's entire surface is less than a trillionth of a watt. Compare this with the roughly 10 *million* watts our planet's surface receives in the form of infrared and visible light from any of the bright stars visible in the night sky. In order to capture enough radio energy to allow detailed measurements to be made, a large collecting area is essential.

Because of diffraction, the angular resolution of radio telescopes is generally quite poor compared with that of their optical counterparts. Typical wavelengths of radio waves are about a million times longer than those of visible light, and these longer wavelengths impose a corresponding crudeness in angular resolution. Even the enormous sizes of radio dishes only partly offset this effect. The 43-m telescope shown in Figure 3.17 can achieve resolution of about 1′ at a wavelength of 1 cm, although the instrument was designed to be most sensitive to wavelengths around 5 cm, where the resolution is only 6′. The best angular resolution obtainable with a single radio telescope is about 10″ (for the largest instruments operating at millimeter wavelengths), at least 10 times coarser than the capabilities of the largest optical mirrors.

Figure 3.18 shows the world's largest radio telescope, built in 1963 in Arecibo, Puerto Rico. Approximately 300 m in diameter, the telescope's reflecting surface lies in a natural depression in a hillside and spans nearly 20 acres. The receiver is strung among several limestone hills. Its enormous collecting area makes this telescope the most sensitive on Earth. The huge size of the dish creates one distinct disadvantage, however: The Arecibo telescope cannot be pointed very well to follow cosmic objects across the sky, limiting its observations to those objects that happen to pass within about 20° of overhead as Earth rotates.

The Value of Radio Astronomy

Despite the inherent disadvantage of relatively poor angular resolution, radio astronomy enjoys many advantages. Radio telescopes can observe 24 hours a day. Darkness is not needed for receiving radio signals because the Sun is a relatively weak source of radio energy, so its emission does not swamp radio signals arriving at Earth from elsewhere in the sky. In addition, radio observations can often be made through cloudy skies, and radio telescopes can detect the longest-wavelength radio waves even during rain or snowstorms.

However, perhaps the greatest value of radio astronomy (and, in fact, of all invisible astronomies) is that it opens up a whole new window on the universe. There are two main reasons for this. First, just as objects that are bright in the visible part of the spectrum (the Sun, for example) are not necessarily strong radio emitters, many of the strongest radio sources in the universe emit little or no visible light. Second, visible light may be strongly absorbed by interstellar dust along the line of sight to a source. Radio waves, on the other hand, are generally unaffected by intervening matter. Many parts of the universe cannot be seen at all by optical means but are easily detectable at radio wavelengths. Radio observations therefore allow us to see whole new classes of objects that would otherwise be completely unknown.

Figure 3.19 shows a visible-light photograph of a distant galaxy called Centaurus A. Superimposed on the optical image is a radio map of the same region. Notice just how different the galaxy looks at radio and optical different wavelengths. The visible galaxy emits little or

Figure 3.18 Arecibo Observatory The 300-m-diameter dish at the National Astronomy and Ionospheric Center near Arecibo, Puerto Rico. The receivers that detect the focused radiation are suspended nearly 150 m (about 45 stories) above the center of the dish. The two insets show close-ups of the receivers high above the dish (left) and technicians tweaking the dish surface to make it smoother (right). *(D. Parker/Science Photo Library; T. Acevedo/NAIC; Cornell University Laboratory of Planetary Studies)*

no radio energy, while the large "blobs" (called *radio lobes*, and thought to have been ejected from the center of the galaxy by explosive events long ago) are completely invisible. Centaurus A (an example of a *radio galaxy*, and studied in more detail in Chapter 16) actually emits more energy in the form of radio waves than it does as visible light, but without radio astronomy we would be entirely ignorant of this fact, and of the galaxy's violent past.

Interferometry

5 Radio astronomers can sometimes overcome the problem of poor angular resolution by using a technique known as **interferometry**. This technique makes it possible to produce radio images of much higher angular resolution than can be achieved with even the best optical telescopes, on Earth or in space. In interferometry, two or more radio telescopes are used in tandem to observe the same object at the same wavelength and at the same time. The combined instruments together make up an **interferometer** (Figure 3.20). By means of electronic cables or radio links, the signals received by each antenna in the array making up the interferometer are sent to a central computer that combines and stores the data as the antennas track their target. After extensive computer processing, a high-resolution image of the target results.

As far as resolving power is concerned, the effective diameter of an interferometer is the distance between its outermost dishes. In other words, two small dishes can act as opposite ends of an imaginary but huge single radio telescope, dramatically improving the an-

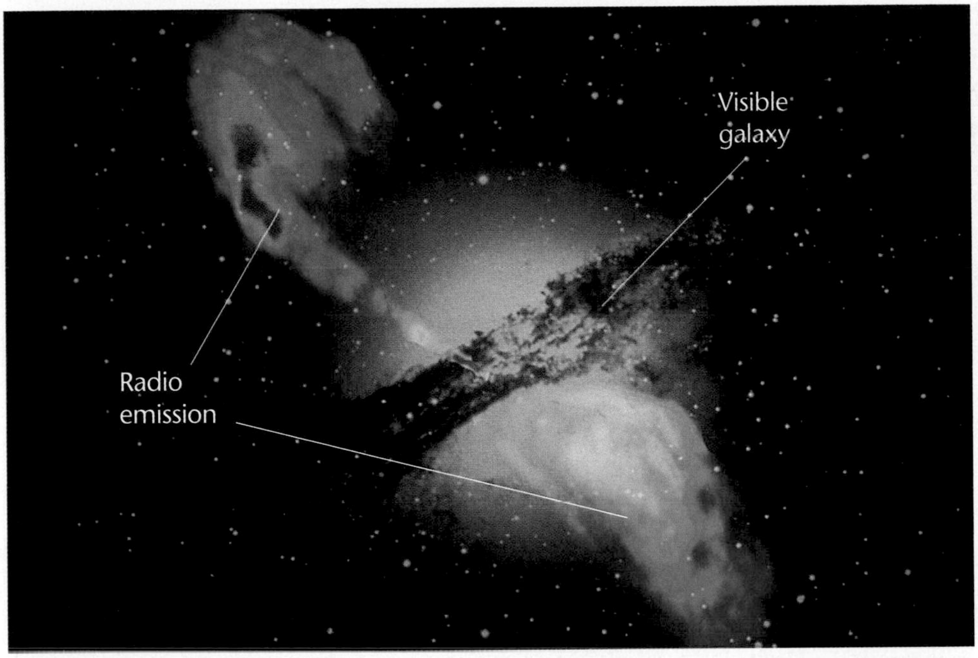

Figure 3.19 Radio Galaxy Visible image of the radio galaxy Centaurus A, with the radio emission from the region superimposed (in false color, with red indicating greatest radio intensity, blue the least). The resolution of the optical image is about 1″; that of the radio map is 12″. *(J. Burns)*

R I V U X G

gular resolution. Large interferometers made up of many dishes, like the Very Large Array (VLA) shown in Figure 3.20, now routinely attain radio resolution comparable to that of optical images. Figure 3.21 compares an interferometric radio map of a nearby galaxy with a photograph of that same galaxy made using a large optical telescope. (The radio image in Figure 3.19 was also obtained by interferometric means.) The larger the distance (or *baseline*) separating the telescopes, the better the resolution attainable. Very-long-baseline interferometry (VLBI), using instruments separated by thousands of kilometers, can achieve resolution in the milliarcsecond (0.001″) range.

Although the technique was originally developed by radio astronomers, interferometry is no longer restricted to the radio domain. Radio interferometry became feasible when electronic equipment and computers achieved speeds great enough to combine and analyze radio signals from separate radio detectors without loss of data. As the technology has improved, it has become possible to apply the same methods to higher-frequency radiation. Millimeter-wavelength interferometry has already become an established and important observational technique, and infrared interferometry will become commonplace in the next few years. Optical interferometry is the subject of intensive research. The

Figure 3.20 VLA Interferometer This large interferometer, located on the Plain of San Augustin in New Mexico, comprises 27 dishes spread along a Y-shaped pattern about 30 km across. The most sensitive radio device in the world, it is called the Very Large Array, or VLA. The dishes are mounted on railroad tracks so that they can be repositioned easily. *(NRAO)*

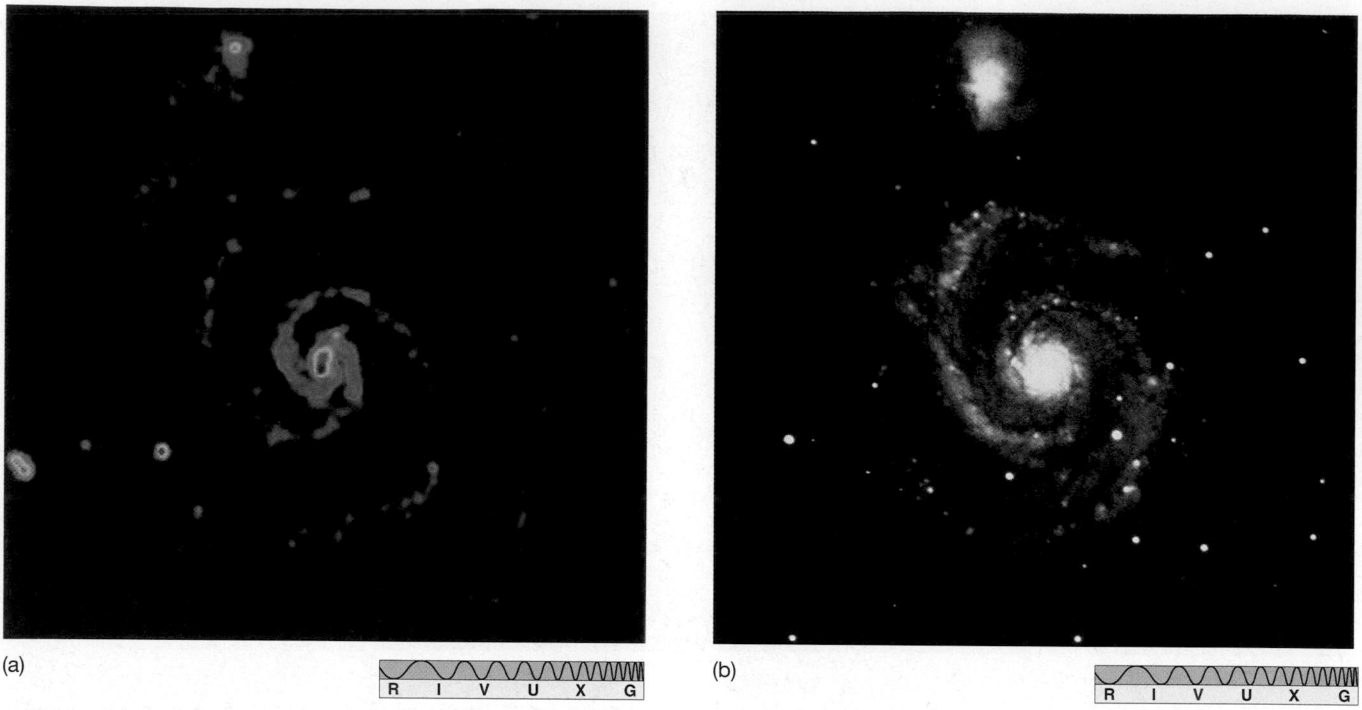

Figure 3.21 **Radio-optical Comparison** (a) A VLA radio image of the spiral galaxy M51, observed at radio frequencies having an angular resolution of a few arc seconds. (b) A visible-light image of M51, made with the 4-m Kitt Peak optical telescope. *(NRAO; AURA)*

Keck and VLT telescopes are designed to be used for infrared—and perhaps someday for optical—interferometric work.

Concept Check

- What is the main reason for the poor angular resolution of radio telescopes? How do radio astronomers overcome this problem?

3.5 Other Astronomies

6 As just noted, the *types* of astronomical objects that can be observed may differ markedly from one wavelength range to another. Full-spectrum coverage is essential not only to see things more clearly, but even to see some things at all. Because of the transmission characteristics of Earth's atmosphere, astronomers must study most wavelengths other than optical and radio from space. The rise of these "other astronomies" has therefore been closely tied to the development of the space program.

Infrared and Ultraviolet Astronomy

Infrared studies are a very important component of modern observational astronomy. Generally, **infrared telescopes** resemble optical telescopes (indeed, many optical telescopes are also used for infrared work), but the infrared detectors are sensitive to longer-wavelength radiation.

Although most infrared radiation is absorbed by the atmosphere (primarily by water vapor), there are a few windows in the high-frequency part of the infrared spectrum where the opacity is low enough to permit ground-based observations. ∞ (Sec. 2.3) Indeed, some of the most useful infrared observing is done from the ground, even though the radiation is somewhat diminished in intensity by our atmosphere. Because of its 4-km altitude, Mauna Kea is one of the finest locations on Earth for both optical and infrared ground-based astronomy.

Astronomers can make still better infrared observations if they can place their instruments above most or all of Earth's atmosphere. Improvements in balloon-, aircraft-, rocket-, and satellite-based telescope technologies have made infrared research a powerful tool

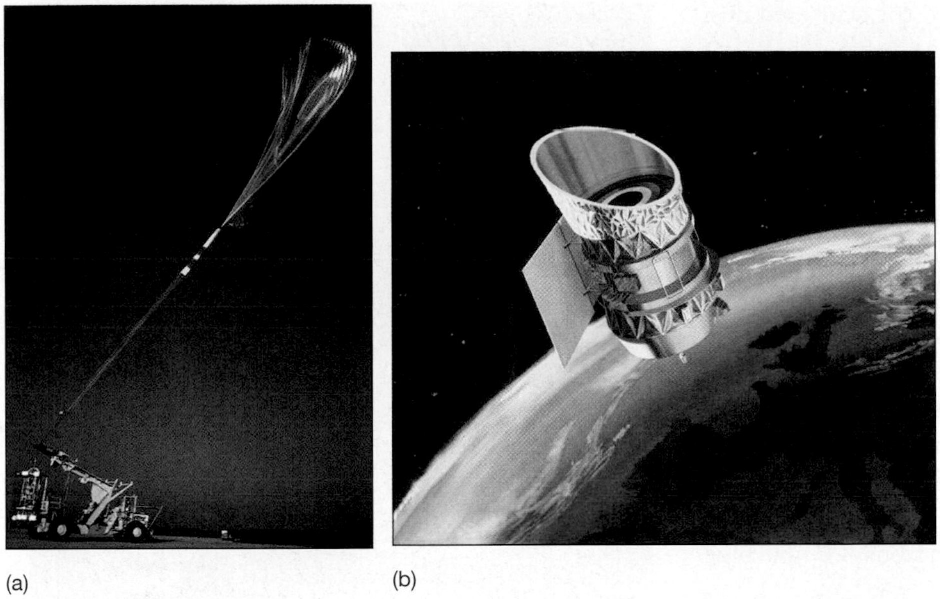

Figure 3.22 Infrared Telescopes (a) A gondola containing a 1-m infrared telescope (lower left) is readied for its balloonborne ascent to an altitude of about 30 km, where it will capture infrared radiation that cannot penetrate the atmosphere. (b) An artist's conception of the *Infrared Astronomy Satellite*, placed in orbit in 1983. This 0.6-m telescope surveyed the infrared sky at wavelengths ranging from 10 to 100 μm. During its 10 months of operation, it greatly increased astronomers' understanding of many aspects of the universe, from the formation of stars and planets to the evolution of galaxies. *(Smithsonian Astrophysical Observatory; NASA)*

for studying the universe (Figure 3.22). However, as might be expected, the infrared telescopes that can be carried above the atmosphere are considerably smaller than the massive instruments found in ground-based observatories.

Figure 3.23(a) shows an infrared image of the Orion region as seen by the 0.6-m British–Dutch–U.S. *Infrared Astronomy Satellite* (*IRAS*, shown in Figure 3.22b) in 1983. The image shows clouds of warm dust and gas, believed to play a critical role in star formation, and groups of bright young stars that are completely obscured at visible wavelengths. During its 10-month lifetime (and long afterwards—the data archives are still heavily used today), *IRAS* contributed greatly to our knowledge of

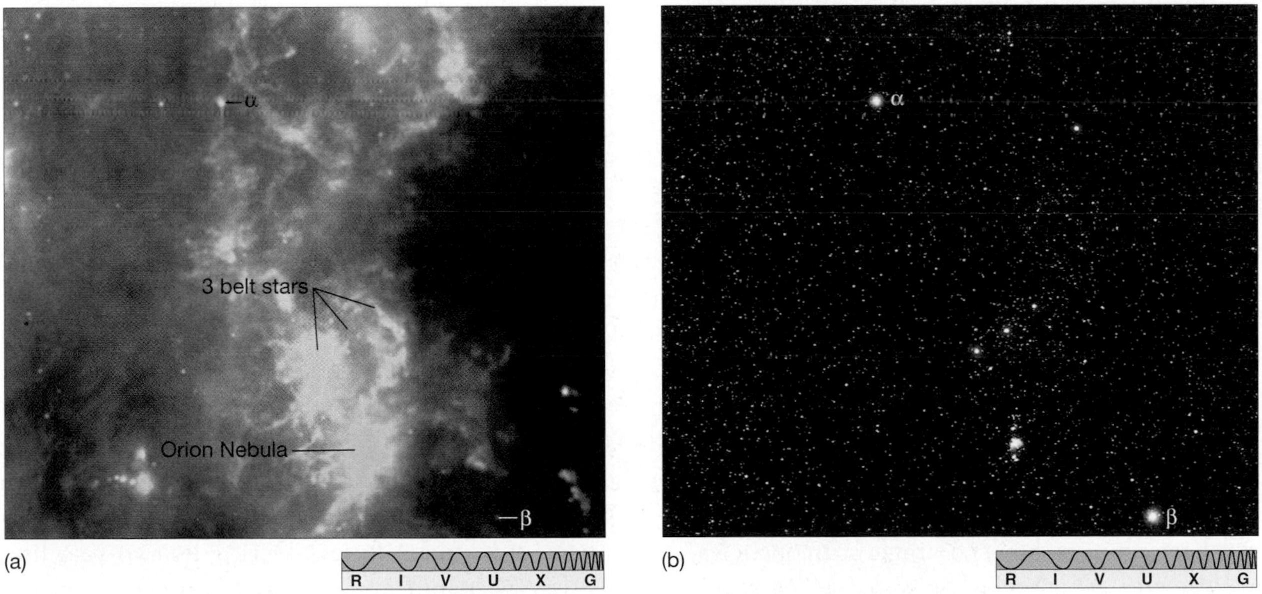

Figure 3.23 Infrared Image (a) This infrared image of Orion was made by the *Infrared Astronomy Satellite*. The whiter regions denote greater strength of infrared radiation; the false colors denote different temperatures, descending from white to red to black. Resolution is about 1′. (b) The same region photographed in visible light. The labels α and β refer, respectively, to Betelgeuse and Rigel, the two brightest stars in the constellation. Note how the red star Betelgeuse can be seen in the infrared (part a), but the blue star Rigel cannot. *(NASA; J. Sanford/Astrostock-Sanford)*

comets, planets, stars, galaxies, and the scattered dust and rocky debris found in interstellar space.

Unfortunately, by Wien's law, telescopes radiate strongly in the infrared unless they are cooled to extremely low temperatures. ∞ (Sec. 2.4) The end of the *IRAS* mission came not because of any equipment malfunction or unexpected mishap but simply because its supply of liquid helium coolant ran out. *IRAS*'s own thermal emission then overwhelmed the radiation the instrument was built to detect. In 1995 the European Space Agency launched the 0.6-m *Infrared Space Observatory* (*ISO*). It is now in orbit, refining and extending the groundbreaking work begun by *IRAS*.

To the short-wavelength side of the visible spectrum lies the ultraviolet domain. This region of the spectrum, extending in wavelength from 400 nm (blue light) down to a few nanometers ("soft" X rays), has only recently begun to be explored. Because Earth's atmosphere is partially opaque to radiation below 400 nm and is totally opaque to radiation below about 300 nm, astronomers cannot conduct any useful ultraviolet observations from the ground, not even from the highest mountaintop. Rockets, balloons, or satellites are therefore essential to any **ultraviolet telescope**—a device designed to capture and analyze this high-frequency radiation.

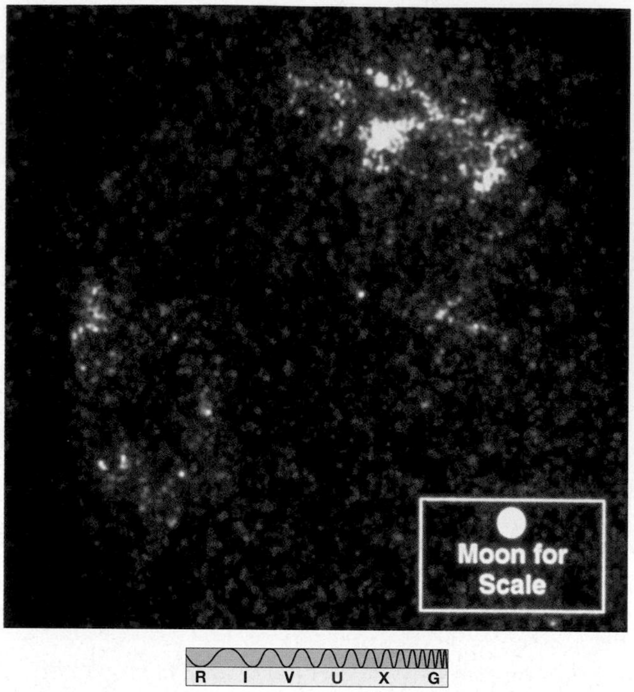

Figure 3.24 Ultraviolet Image A camera on board the *Extreme Ultraviolet Explorer* satellite captured this image of the Vela supernova remnant, the result of a massive star blowing itself virtually to smithereens. The release of energy was prodigious, and the afterglow has lingered for millennia. The glowing field of debris shown here lies some 1500 light-years from Earth. Based on the velocity of the outflowing debris, astronomers estimate that the explosion itself must have occurred about 12,000 years ago. For scale, the inset shows the angular size of the Moon. *(NASA)*

Figure 3.24 shows an image of a supernova remnant—the remains of a violent stellar explosion that occurred some 12,000 years ago (see Chapter 12)—obtained by the *Extreme Ultraviolet Explorer* (*EUVE*) satellite, launched in 1992. Since its launch *EUVE* has mapped out our local cosmic neighborhood as it presents itself in the far ultraviolet, and has radically changed astronomers' conception of interstellar space in the vicinity of the Sun. The *Hubble Space Telescope* (*Interlude 3-1*), best known as an optical telescope, is also a superb ultraviolet instrument.

High-Energy Astronomy

High-energy astronomy studies the universe as it presents itself to us in X rays and gamma rays—the types of radiation whose photons have the highest frequencies and hence the greatest energies. How do we detect radiation of such short wavelengths? First, it must be captured high above Earth's atmosphere because none of it reaches the ground. Second, its detection requires the use of equipment basically different from that used to capture the relatively low-energy radiation discussed up to this point.

The difference in the design of **high-energy telescopes** comes about because X-rays and gamma rays cannot be reflected easily by any kind of surface. Rather, these rays tend to either pass straight through or else be absorbed by any material they strike. When X rays barely graze a surface, however, they can be reflected from it in a way that yields an image, although the mirror design is fairly complex (Figure 3.25).

The *Einstein Observatory*, launched by NASA in 1978, was the first X-ray telescope capable of forming an image of its field of view. During its two-year lifetime, this spacecraft made major advances in our understanding of high-energy phenomena throughout the universe; its observational database is still heavily used. More recently, the German *ROSAT* (short for Röntgen Satellite, after Wilhelm Röntgen, the discoverer of X rays) was launched in 1991 and has generated a wealth of high-quality observational data. It was turned off in

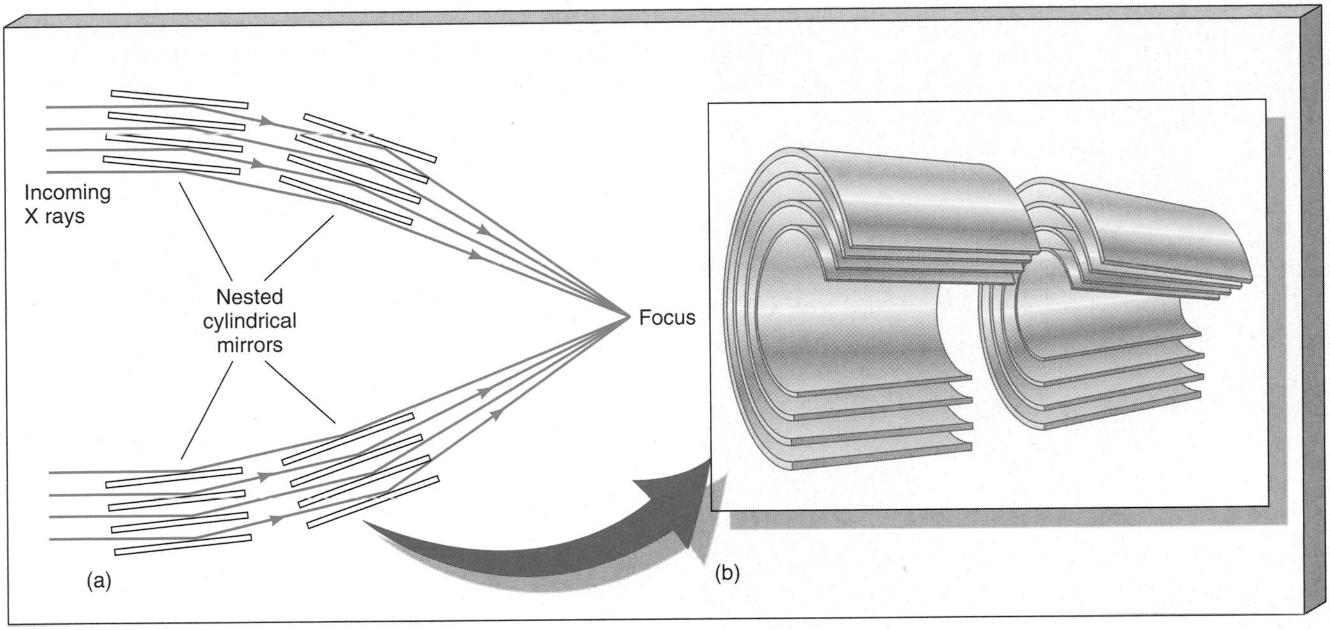

Figure 3.25 **X-ray Telescope** The arrangement of mirrors in an X-ray telescope allows the rays to be reflected at grazing angles and focused into an image.

1999, a few months after its electronics were irreversibly damaged when the telescope was accidentally pointed too close to the Sun. In 1999 NASA launched the *Advanced X-Ray Astrophysics Facility* (*AXAF*, renamed *Chandra* after launch, in honor of the Indian astrophysicist Subramanyan Chandrasekhar, and shown in Figure 3.26). With greater sensitivity, a wider field of view, and better resolution than either *Einstein* or *ROSAT*, *Chandra* is providing high-energy astronomers with new levels of observational detail (Figure 3.27). Several examples of *Chandra* imagery are found throughout our text.

Gamma-ray astronomy is the youngest entrant into the observational arena. For gamma rays, no method of producing an image has yet been devised—present-day gamma-ray telescopes simply point in a specified direction and count photons received. As a result, only fairly coarse (1° resolution) observations can be made. Nevertheless, even at this resolution, there is much to be learned. Cosmic gamma rays were originally detected in the 1960s by the U.S. *Vela* series of satellites, whose primary mission was to monitor illegal nuclear detonations on Earth. Since then, several X-ray

Figure 3.26 **Chandra Observatory** The Chandra X-ray telescope is the most recent "great observatory" to be deployed in space. The left end of the mirror arrangement depicted in Figure 3.25 is at the bottom of the satellite as oriented here. Chandra's effective angular resolution is 1″, allowing this spacecraft to produce images of quality comparable to that of optical photographs. *(NASA)*

Figure 3.27 X-ray Image An X-ray image of the Orion region, taken by the *ROSAT* X-ray satellite. (Compare with Figures P.6, 3.23, and 5.12.) Objects that are most intensely emitting X-rays are colored blue, including the three stars of Orion's belt and the glowing nebula below them at bottom center in the photograph. *(Courtesy of Konrad Dennuerl and Wolfgang Voges, Max–Planck Institut fur Extraterrestrische Physik, Garching Germany)*

telescopes have also been equipped with gamma-ray detectors. By far the most advanced instrument is the *Gamma Ray Observatory* (GRO), launched by space shuttle in 1991, and now nearing the end of its operational lifetime. This satellite can scan the sky and study individual objects in much greater detail than previously attempted. Figure 3.28 shows a false-color *GRO* image of a highly energetic outburst in the nucleus of a distant galaxy.

Concept Check

- List some scientific benefits of placing telescopes in space. What are the drawbacks of space-based astronomy?

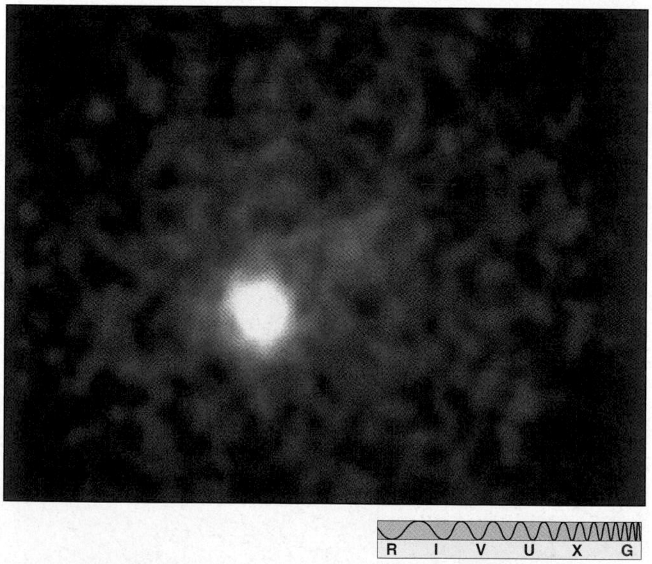

Figure 3.28 Gamma-ray Image A false-color gamma-ray image from the *Gamma Ray Observatory*, showing a violent event known as a *gamma-ray blazar* in the distant galaxy 3C279. *(Compton Gro/NASA)*

Full-Spectrum Coverage

7 Table 3.1 lists the basic regions of the electromagnetic spectrum and describes objects typically studied in each frequency range. Bear in mind that the list is far from exhaustive and that many astronomical objects are now routinely observed at many different electromagnetic wavelengths. As we proceed through the text, we will discuss more fully the wealth of information that high-precision astronomical instruments can provide us.

As an illustration of the sort of comparison that full-spectrum coverage allows, Figure 3.29 shows a series of images of the Milky Way Galaxy. They were made by several instruments, at wavelengths ranging from radio to gamma ray, over a period of about five years. By comparing the features visible in each, we immediately see how multiwavelength observations can complement each other, greatly extending our perception of the universe around us.

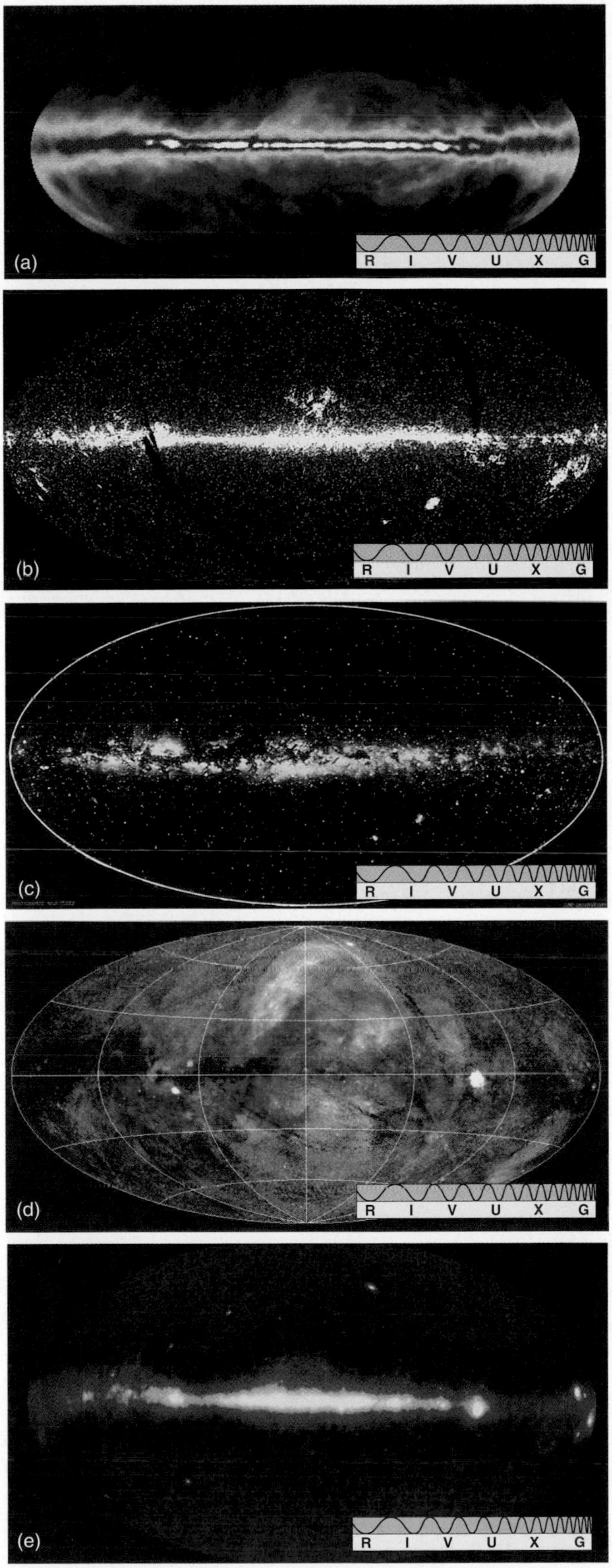

Figure 3.29 Multiple Wavelengths The Milky Way Galaxy as it appears at (a) radio, (b) infrared, (c) visible, (d) X-ray, and (e) gamma-ray wavelengths. *(NRAO; NASA; Lund Observatory; Courtesy of Konrad Dennuerl and Wolfgang Voges, Max–Planck Institut fur Extraterrestrische Physik, Garching Germany; NASA)*

TABLE 3.1 Astronomy at Many Wavelengths

RADIATION	GENERAL CONSIDERATIONS	COMMON APPLICATIONS (Chapter Reference)
Radio	Can penetrate dusty regions of interstellar space. Earth's atmosphere largely transparent to radio wavelengths. Can be detected in daytime as well as at night. High resolution at long wavelengths requires very large telescopes.	Radar studies of planets (1, 6) Planetary magnetic fields (7) Interstellar gas clouds (11) Center of Milky Way Galaxy (14) Galactic structure (14, 15) Active galaxies (16) Cosmic background radiation (17)
Infrared	Can penetrate dusty regions of interstellar space. Earth's atmosphere only partially transparent to IR radiation, so some observations must be made from space.	Star formation (11) Cool stars (11, 12) Center of Milky Way Galaxy (14) Active galaxies (16) Large-scale structure of the universe (16, 17)
Visible	Earth's atmosphere transparent to visible light.	Planets (6, 7) Stars and stellar evolution (9, 10, 12) Galactic structure (14, 15) Large-scale structure of the universe (16, 17)
Ultraviolet	Earth's atmosphere opaque to UV radiation, so observations must be made from space.	Interstellar medium (11) Hot stars (12)
X ray	Earth's atmosphere opaque to X rays, so observations must be made from space. Special mirror configurations needed to form images.	Stellar atmospheres (9) Neutron stars and black holes (13) Hot gas in galaxy clusters (15) Active galactic nuclei (16)
Gamma ray	Earth's atmosphere opaque to gamma rays, so observations must be made from space. Cannot form images.	Neutron stars (13) Active galactic nuclei (16)

Chapter Review www

Summary

A **telescope** (p. 72) is a device designed to collect as much light as possible from some distant source and deliver it to a detector for detailed study. **Reflecting telescopes** (p. 72) use a mirror to concentrate and focus the light. **Refracting telescopes** (p. 72) use a lens; **refraction** (p. 72) is the bending of light as it passes from one medium to another. The **prime focus** (p. 72) of a telescope is the point where the incoming beam is focused to form an **image** (p. 72), and where analysis instruments may be placed. The **Newtonian** (p. 75) and **Cassegrain telescope** (p. 75) designs employ secondary mirrors to avoid placing heavy equipment at the prime focus. All astronomical telescopes larger than about 1 m in diameter use mirrors in their design.

The light-gathering power of a telescope depends on its **collecting area** (p. 76), which is proportional to the square of the mirror diameter. To study the faintest sources of radiation, astronomers must use large telescopes.

Angular resolution (p. 77) is a telescope's ability to distinguish between light sources lying close together on the sky. One limitation on resolution is **diffraction** (p. 77), which makes it impossible to focus a beam perfectly. The amount of diffraction is proportional to the wavelength of the radiation under study and inversely proportional to the size of the mirror. Thus, at any given wavelength, larger telescopes suffer least from the effects of diffraction.

The resolution of most ground-based optical telescopes is limited by **seeing** (p. 79)—the blurring effect of Earth's turbulent atmosphere, which smears the point-like images of a star into a **seeing disk** (p. 79) a few arc seconds in diameter. Because radio and space-based telescopes do not suffer from atmospheric effects, their resolution is determined mainly by the effects of diffraction.

Most modern telescopes use **charge-coupled devices** (p. 81) instead of photographic plates to collect data. CCDs are many times more sensitive than photographic plates, and the resultant data are easily saved directly on disk or tape for later image processing.

Using **active optics** (p. 82), in which a telescope's environment and focus are carefully monitored and controlled, and **adaptive optics** (p. 82), in which the blurring effects of atmospheric turbulence are corrected for in real time, it may soon be possible to achieve diffraction-limited resolution in ground-based optical instruments.

Radio telescopes (p. 84) are conceptually similar in construction to optical reflecting telescopes. However, radio telescopes are generally much larger than optical instruments, for two reasons. First, the amount of radio radiation reaching Earth from space is much less than the amount of visible radiation, so a large collecting area is essential. Second, the long wavelengths of radio waves mean that diffraction severely limits its resolution unless large instruments are used.

In order to increase the effective area of a telescope, and hence improve its resolution, several instruments may be combined into an **interferometer** (p. 86). Using **interferometry** (p. 86), radio telescopes can produce images much sharper than those from the best optical equipment.

Infrared (p. 88) and **ultraviolet telescopes** (p. 90) are similar in basic design to optical systems. Infrared studies in some parts of the infrared range can be carried out using large ground-based systems. Ultraviolet astronomy must be carried out from space.

High-energy telescopes (p. 90) study the X-ray and gamma-ray regions of the electromagnetic spectrum. X-ray telescopes can form images of their field of view, although the mirror design is more complex than for lower-energy instruments. Gamma-ray telescopes simply point in a certain direction and count photons received. Because the atmosphere is opaque at these short wavelengths, both types of telescope must be placed in space.

Review and Discussion

1. List three advantages of reflecting telescopes over refracting telescopes.
2. What are the largest optical telescopes in use today?
3. How does Earth's atmosphere affect what is seen through an optical telescope?
4. What advantages does the *Hubble Space Telescope* have over ground-based telescopes? List some disadvantages.
5. What are the advantages of a CCD over a photographic plate?
6. Is the resolution of a 2-m telescope on Earth's surface limited more by atmospheric turbulence or by the effects of diffraction?
7. Why do radio telescopes have to be very large?
8. What kind of astronomical objects can we best study with radio techniques?
9. What is interferometry, and what problem in radio astronomy does it address?
10. Compare the highest resolution attainable with optical telescopes to the highest resolution attainable with radio telescopes (including interferometers).
11. What special conditions are required to conduct observations in the infrared?
12. In what ways do the mirrors in X-ray telescopes differ from those found in optical instruments?
13. Why can't gamma-ray astronomy be done from the ground?
14. What are the main advantages of studying objects at many different wavelengths of radiation?
15. Our eyes can see light with an angular resolution of about 1′—equivalent to about a third of a millimeter at arm's length. Suppose our eyes detected only infrared radiation, with 1° angular resolution. Would we be able to make our way around on Earth's surface? To read? To sculpt? To create technology?

True or False?

____ **1.** A Newtonian telescope has no secondary mirror.

____ **2.** A Cassegrain telescope has a hole in the middle of the primary mirror to allow light reflected from its secondary mirror to reach a focus behind the primary mirror.

____ **3.** The term "seeing" is used to describe how faint an object can be detected by a telescope.

____ **4.** Because of diffraction, the image made by a telescope becomes sharper as the telescope diameter decreases.

____ **5.** The main advantage to using the *Hubble Space Telescope* is the increased amount of "night" time available to it.

____ **6.** One of the primary advantages of CCDs over photograph plates is the former's high efficiency in detecting light.

____ **7.** The *Hubble Space Telescope* can observe objects in the optical, infrared, and ultraviolet parts of the spectrum.

____ **8.** The Keck telescopes contain the largest single mirrors ever produced.

____ **9.** Radio telescopes are large in part to improve their angular resolution, which is poor because of the long wavelengths at which they observe.

____ **10.** Radio telescopes are large in part because the sources of radio radiation they observe are very faint.

____ **11.** For technological reasons, interferometry is limited to radio astronomy.

____ **12.** Infrared astronomy can only be done from space.

____ **13.** X-ray astronomy can only be done from space.

____ **14.** X-ray and gamma-ray telescopes employ the same basic design as optical instruments.

____ **15.** Because gamma rays have very short wavelengths, gamma-ray telescopes can achieve extremely high angular resolution.

Fill in the Blank

1. A telescope that uses a lens to focus light is called a ________ telescope.

2. A telescope that uses a mirror to focus light is called a ________ telescope.

3. All large modern telescopes are of the ________ type.

4. The light-gathering power of a telescope is determined by the ________ of its mirror or lens.

5. The angular resolution of a telescope is determined by the ________ of the telescope and the ________ of the radiation being observed.

6. The angular resolution of ground-based optical telescopes is more seriously limited by Earth's ________ than by diffraction.

7. Optical telescopes on Earth can see angular detail down to about ________ arc second.

8. CCDs produce images in ________ form that can be easily transmitted, stored, and processed by computers.

9. Active optics and adaptive optics are both being used to improve the ________ of ground-based optical telescopes.

10. All radio telescopes are of the ________ design.

11. An ________ is two or more telescopes used in tandem to observe the same object, in order to improve angular resolution.

12. *Chandra* is a telescope designed to make observations in the ________ part of the spectrum.

13. An object having a temperature of 300 K would be best observed with an ________ telescope.

14. Orbiting ________ telescopes must be cooled to very low temperatures.

15. Gamma-ray telescopes are unable to form ________ of their fields of view.

Problems

1. A certain telescope has a $10' \times 10'$ field of view that is recorded using a CCD chip having 1024×1024 pixels. What angle on the sky corresponds to one pixel? What would be the diameter of a typical seeing disk ($2''$ radius), in pixels?
2. Estimate the fraction of incoming starlight blocked by the observer and observing cage at the prime focus of the Hale 5-m telescope (see Figure 3.6).
3. A 2-m telescope can collect a given amount of light in one hour. Under the same observing conditions, how much time would be required for a 6-m telescope to perform the same task? A 12-m telescope?
4. A certain space-based telescope can achieve (diffraction-limited) angular resolution of $0.05''$ for red light (wavelength 700 nm). What would its resolution be (a) in the infrared, at wavelength 3.5μm, and (b) in the ultraviolet, at wavelength 140 nm?
5. The photographic equipment on a telescope is replaced by a CCD. If the photographic plate records five percent of the light reaching it but the CCD records 75 percent, how much time will the new system take to collect as much information as the old detector recorded in a one-hour exposure?
6. Based on collecting areas, how much more sensitive would you expect the Arecibo telescope (Figure 3.18) to be, compared with the 43-m Green Bank instrument (Figure 3.17)?
7. The Andromeda Galaxy lies about 2.9 million light-years away. To what distances do the angular resolutions of *HST* ($0.05''$) and a radio interferometer ($0.001''$) correspond at that distance?
8. What would be the equivalent single-mirror diameter of a telescope constructed from two separate 10-m mirrors? Four separate 8-m mirrors?
9. Estimate the angular resolutions of (a) a radio interferometer with a 5000-km baseline, operating at a frequency of 5 GHz, and (b) an infrared interferometer with a baseline of 50 m, operating at a wavelength of 1 μm.
10. An instrument aboard the *Gamma Ray Observatory* detects a burst of energy coming from some distant source. Simultaneously, an X-ray telescope with an effective diameter of 0.5 m observes the same event at a wavelength of 1 nm. Both detectors record the location of the event on the sky, to within the accuracy of their respective measurements. You are about to make a follow-up observation using a large optical telescope with a $5'$ field of view. Which position—the X-ray or the gamma-ray measurement—should you use, and why?

Projects

1. Here's how to take some easy pictures of the night sky. You will need a location with a clear, dark sky; a 35-mm camera with a standard 50-mm lens, tripod, and cable release; a watch with a seconds display visible in the dark; and a roll of high-speed color slide film. Set your camera to the "bulb" setting for the exposure and attach the cable release so that you can take a long exposure. Set the focus on infinity. Point the camera to a favored constellation, seen through your viewfinder, and take a 20-s to 30-s exposure. In order to minimize vibration, don't hold on to the cable release during the exposure. Keep a log of your shots. When finished, have the film developed in the standard way.
2. For some variations, vary your exposure times, use different films, take hours-long exposures for star trails, use a wide-angle or telephoto lens, place the camera piggyback on a telescope that is tracking and take exposures that are a few minutes long. Experiment and have fun!

4 THE SOLAR SYSTEM

Interplanetary Matter and the Birth of the Planets

LEARNING GOALS

Studying this chapter will enable you to:

1 Describe the scale and structure of the solar system and list the basic differences between the terrestrial and the jovian planets.

2 Summarize the orbital and physical properties of the major groups of asteroids.

3 Describe the composition and structure of a typical comet and explain how a cometary tail forms.

4 Explain what cometary orbits tell us about the probable origin of comets.

5 Summarize the orbital and physical properties of meteoroids and explain how these bodies are related to asteroids and comets.

6 List the major facts that any theory of solar system formation must explain and indicate how the leading theory accounts for them.

7 Outline the process by which planets form as natural by-products of star formation.

(Opposite page) A truly great comet—one of the brightest of the twentieth century—illuminated the skies of the Northern Hemisphere in early 1997. Comet Hale–Bopp is shown here in this wide-angle photograph, taken at the moment of perihelion, or closest approach to the Sun, yet still some 0.9 A.U. away. Note how the gas (glowing bluish light) boiling off the comet extends straight away from the Sun, while the comet's heavier dust particles (reflected whitish light) lag behind, forming a more gently curved tail. The full extent of the tail measured nearly 40 degrees, sweeping a huge arc across the nighttime sky. Almost surely, Hale–Bopp was the most photographed comet in history. *(© Copyright 1997 Jerry Lodriguss)*

In less than a generation we have learned more about our solar system—the Sun and everything that orbits it—than in all the centuries that went before. By studying the planets, their moons, and the countless fragments of material that orbit in interplanetary space, astronomers have gained a richer outlook on our own home in space. The discoveries of the past few decades have revolutionized our understanding not only of the present state of our cosmic neighborhood but also of its history, for our solar system is filled with clues to its own origin and evolution. Paradoxically, the richest sources of information about the earliest days of the solar system are the not the large bodies orbiting the Sun but rather the asteroids, comets, and meteoroids that pepper interplanetary space. These objects may seem to be only rocky and icy "debris," but more than the planets themselves, they hold a record of the formative stages of our planetary system and have much to teach us about the origin of our world.

4.1 An Inventory of the Solar System

1 Our **solar system** contains 1 star (the Sun), 9 planets orbiting that star, 63 moons (at last count) orbiting those planets, 6 *asteroids* larger than 300 km in diameter, more than 7000 smaller (but well-studied) asteroids, myriad *comets* a few kilometers in diameter, and countless *meteoroids* less than 100 meters across. This list will likely grow as we continue to explore our cosmic neighborhood.

The arrangement of the major bodies in the solar system is shown in Figure 4.1. The planet closest to the Sun is Mercury. Moving outward, we encounter in turn Venus, Earth, Mars, Jupiter, Saturn, Uranus, Neptune, and Pluto. The asteroids lie mainly in a broad belt between the orbits of Mars and Jupiter. Figure 4.2 shows the sizes of the planets relative to the Sun. Table 4.1 lists some basic orbital and physical properties of the nine planets, with a few other solar-system objects (the Sun, the Moon, an asteroid, and a comet) included for comparison.

Planetary Properties

The planets' distances from the Sun are known from Kepler's laws once the scale of the solar system is set by radar ranging on Venus. ∞ (Sec. 1.3) Earth's radius has long been known through conventional surveying techniques and, more recently, via satellite measurements. Other planetary radii are found by measuring the planets' angular sizes and then employing elementary geometry. ∞ (*More Precisely P-2*) Mass is determined by

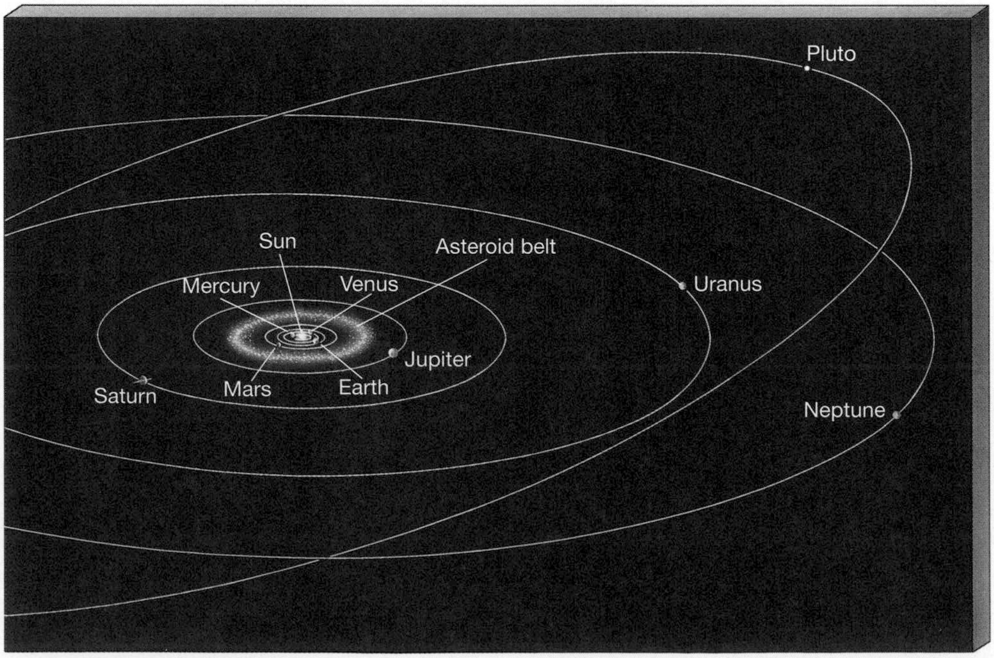

Figure 4.1 Animation **Solar System** Major bodies of the solar system: Sun, planets, asteroids. Except for Mercury and Pluto, the orbits of the planets all lie nearly in the same plane. As we move outward from the Sun, the distance between adjacent orbits increases. The entire solar system spans nearly 80 A.U. The solar system is seen obliquely here, from a vantage point about 30° above the ecliptic plane. The planets' orbits are, for the most part, almost circular.

Animation Animation

Figure 4.2 **Sun and Planets** Relative sizes of the planets and our Sun. Notice that Jupiter, Saturn, Uranus, and Neptune are much larger than Pluto and much larger than Earth and the other inner planets. However, even these large planets are dwarfed by the still larger Sun.

observing a planet's gravitational influence on some nearby object and applying Newton's laws of motion and gravity. (*More Precisely 1-2*) Before the Space Age, astronomers calculated planetary masses either by tracking the orbits of the planets' moons (if any) or by measuring the small (but detectable) distortions the planets produce in each other's orbits. Today, the masses of most of the objects in Table 4.1 are accurately known through their gravitational effects on artificial satellites and space probes launched from Earth. Measurements of the orbits, masses, and radii of asteroids and comets are discussed in Section 4.2.

Note in Figure 4.1 that the planetary orbits are not evenly spaced. Roughly speaking, the spacing between adjacent orbits doubles as we move outward from the Sun. The distance from the Sun to Pluto is about

TABLE 4.1 Properties of Some Solar System Objects

OBJECT	ORBITAL SEMI-MAJOR AXIS (A.U.)	ORBITAL PERIOD (EARTH YEARS)	MASS (EARTH MASSES)	RADIUS (EARTH RADII)	NUMBER OF KNOWN MOONS	AVERAGE DENSITY (kg/m^3)
Mercury	0.39	0.24	0.055	0.38	0	5400
Venus	0.72	0.62	0.82	0.95	0	5200
Earth	1.0	1.0	1.0	1.0	1	5500
Moon	1.0	—	0.012	0.27	—	3300
Mars	1.5	1.9	0.11	0.53	2	3900
Ceres (asteroid)	2.8	4.7	0.00015	0.073	0	2700
Jupiter	5.2	11.9	318	11.2	16	1300
Saturn	9.5	29.4	95	9.5	18	700
Uranus	19.2	84	15	4.0	17	1300
Neptune	30.1	164	17	3.9	8	1600
Pluto	39.5	249	0.002	0.2	1	2100
Comet Hale–Bopp	180	2400	1.0×10^{-9}	0.004	—	100
Sun	—	—	332,000	109	—	1400

40 A.U., almost a million times Earth's radius and roughly 15,000 times the distance from Earth to the Moon. Yet despite the solar system's vast extent, the planets all lie very close to the Sun, astronomically speaking. The diameter of Pluto's orbit is less than 1/1000 of a light year, while the star nearest the solar system is several light years distant.

All the planets orbit the Sun counterclockwise as seen from above Earth's North Pole and in nearly the same plane as Earth (the ecliptic plane). ∞ (Sec. 1.2) Mercury and Pluto deviate from this rule—their orbital planes lie at 7° and 17° to the ecliptic, respectively. Still, we can think of the solar system as being quite flat. If we were to view the planets' orbits from a vantage point in the ecliptic plane about 50 A.U. from the Sun, only Pluto's orbit would be noticeably tilted.

The final column in Table 4.1 lists a quantity called **density**, a measure of the "compactness" of an object, computed by dividing the object's mass (in kilograms) by its volume (in cubic meters). Earth's average density is 5500 kg/m^3. For comparison, the density of water is 1000 kg/m^3, rocks on Earth's surface have densities in the range 2000 to 3000 kg/m^3, and iron has a density of about 8000 kg/m^3. Earth's atmosphere at sea level has a density of only a few kilograms per cubic meter.

Terrestrial and Jovian Planets

As discussed further in Chapters 5 through 8, astronomers can infer a planet's overall composition from a combination of density measurements, spacecraft data, and theoretical calculations. Based on this, we can draw a clear distinction between the inner and outer members of our planetary system. Simply put, the inner planets—Mercury, Venus, Earth, and Mars—are small, dense, and *rocky* in composition. The outer worlds—Jupiter, Saturn, Uranus, and Neptune (but not Pluto)—are large, of low density, and *gaseous*.

Because the physical and chemical properties of Mercury, Venus, and Mars are somewhat similar to Earth's, the four innermost planets are called the **terrestrial planets**. (The word *terrestrial* derives from the Latin word *terra*, meaning "land" or "earth.") The larger outer planets—Jupiter, Saturn, Uranus, and Neptune—are all similar to one another chemically and physically (and very different from the terrestrial worlds). They are labeled the **jovian planets**, after Jupiter, the largest member of the group. (The word *jovian* comes from *Jove*, another name for the Roman god Jupiter.) The jovian worlds are all much larger than the terrestrials and quite different from them in both composition and structure. Table 4.2 compares and contrasts some key properties of these two distinct planetary classes.

There are important differences among planets within each category. For example, when we take into account how the weight of overlying layers squeezes the interiors of the terrestrial planets to different extents (greatest for Earth, least for Mercury), we find that the average *uncompressed densities* of the terrestrial worlds—that is, the densities they would have in the absence of any gravitational compression—decrease steadily as we move farther from the Sun. This decrease in density indicates that, despite their other similarities, the overall compositions of these planets differ significantly one from the other. A similar trend is found among the jovian worlds. All four are thought to have large, dense "terrestrial" cores some ten to fifteen times the mass of Earth, but these cores account for an increasing fraction of the planet's total mass as we move outward from the Sun.

Finally, beyond the outermost jovian planet, Neptune, lies one more small world, frozen and mysterious. Pluto doesn't fit well into either planetary category. Indeed, there is ongoing debate among planetary scientists as to whether it should be classified as a planet at all. In both mass and composition, it has much more in common with the icy jovian moons than with any terrestrial or jovian planet. Many astronomers suspect that it may in fact be the largest member of a newly recognized class of solar system objects that reside beyond the jovian worlds.

TABLE 4.2 Comparison Between the Terrestrial and Jovian Planets

TERRESTRIAL	JOVIAN
close to the Sun	far from the Sun
closely spaced orbits	widely spaced orbits
small masses	large masses
small radii	large radii
predominantly rocky	predominantly gaseous
solid surface	no solid surface
high density	low density
slower rotation	faster rotation
weak magnetic fields	strong magnetic fields
no rings	many rings
few moons	many moons

Interplanetary Matter

In the vast space among the nine known planets move countless small chunks of matter ranging in size from a few hundred kilometers in diameter down to tiny grains of interplanetary dust. The three major constituents of this cosmic "debris" are asteroids, comets, and meteoroids. **Asteroids** and **meteoroids** are fragments of rocky material, somewhat similar in composition to the outer layers of the terrestrial planets. The distinction between the two is simply a matter of size—anything larger than 100 m in diameter (corresponding to a mass exceeding 10,000 tons) is conventionally termed an asteroid; anything smaller is a meteoroid. **Comets** are predominantly icy, rather than rocky in composition (although they do contain some rocky material), and have typical diameters in the 1–10 km range.

Taken together, these small bodies account for a negligible fraction of the total mass of the solar system. They play no important role in the present-day workings of the planets or their moons. Why then do we study them in any detail? The answer is that they are of crucial importance to our studies, for they are the keys to answering some very fundamental questions about our planetary environment.

One of the goals of planetary science is to understand how the solar system formed and to explain the physical conditions now found on Earth and elsewhere in our planetary system. Ironically, studies of the most accessible planet—Earth itself—do not help us much in our quest, because information about our planet's early stages was obliterated long ago by atmospheric erosion and geological activity. Much the same holds true for the other planets (with the possible exception of Mercury, where a near-airless and geologically inactive surface still retains some imprint of the distant past). The problem is that the large bodies of the solar system have all *evolved* significantly since they formed, making it very difficult to decipher the circumstances of their births.

A much better place to look for hints of conditions in the early solar system is on its *small* bodies—the planetary moons and the asteroids, meteoroids, and comets that make up interplanetary debris—for nearly all such fragments contain traces of solid and gaseous matter from the earliest times. Many have scarcely changed since the solar system formed billions of years ago. Therefore, as a prologue to our study of the formation of the solar system, we will examine in a little more detail the contents of interplanetary space.

✓ Concept Check

- Why do astronomers draw such a clear distinction between the inner and the outer planets?

4.2 Solar System Debris

Asteroids

2 The name asteroid literally means "starlike body," but asteroids are definitely not stars. They are too small even to be classified as planets. Astronomers often refer to them as "minor planets." Researchers have so far cataloged over 7000 asteroids with well-determined orbits. As sketched in Figure 4.3, the vast majority are found in a region of the solar system known as the **asteroid belt**, located between 2.1 and 3.3 A.U. from the Sun—roughly midway between the orbits of Mars (at 1.5 A.U.) and Jupiter (at 5.2 A.U.). All but one of the known asteroids revolve about the Sun in prograde orbits (that is, in the same sense as Earth and the other planets; see Section 1.4), and, like the planets, most asteroids have orbits fairly close to the ecliptic plane (with inclinations less than 10°–20°). Unlike the almost circular paths of the major planets, however, asteroid orbits are generally noticeably elliptical in shape.

In addition to the main-belt asteroids, a few hundred **Trojan asteroids** share an orbit with Jupiter, remaining a constant 60° ahead of or behind that planet as it circles the Sun. This peculiar orbital behavior is not a matter of chance—the Trojan asteroids are held in place by a delicate but stable balance between the gravitational fields of Jupiter and the Sun. Calculations first performed by the eighteenth-century French mathematician Joseph Louis Lagrange show that interplanetary matter that happens to stray into one of the two regions of space now occupied by the Trojan asteroids can remain there indefinitely, perfectly synchronized with Jupiter's motion.

The orbits of most asteroids have eccentricities lying in the range 0.05 to 0.3, ensuring that they always remain between the orbits of Mars and Jupiter. However, a few **Earth-crossing asteroids** move on orbits that intersect the orbit of Earth, most likely because the gravitational field of Mars or Jupiter has deflected these small bodies into the inner solar system. From a human perspective, an important consequence of this fact is the very real possibility of a collision with Earth. For

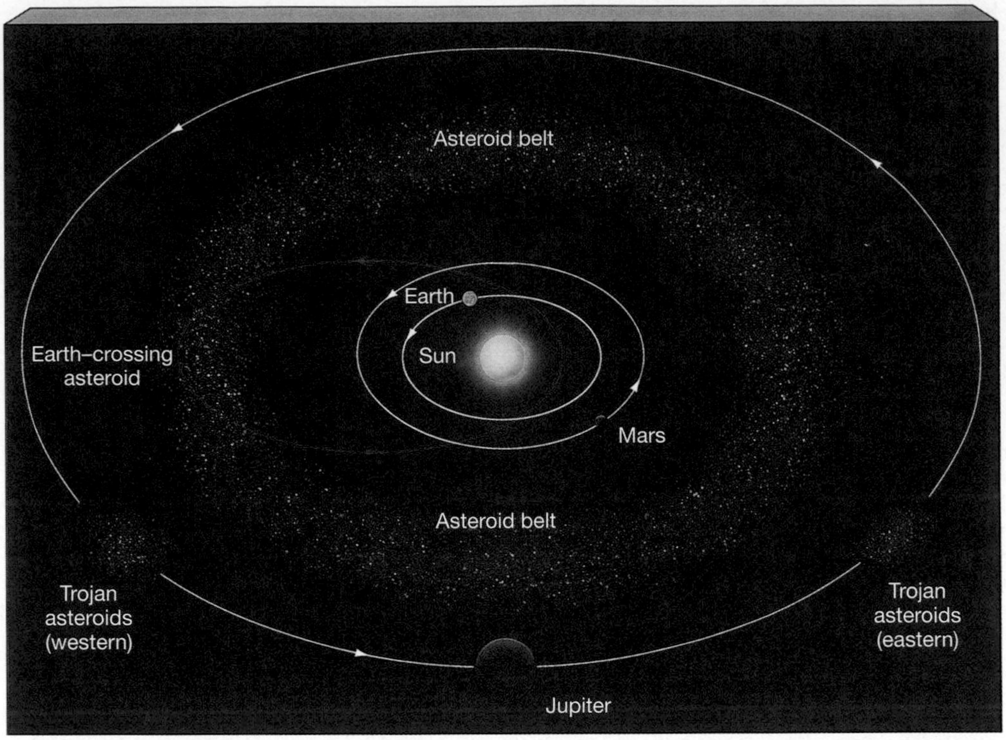

Figure 4.3 Inner Solar System The main asteroid belt, along with the orbits of Earth, Mars, and Jupiter. Note the Trojan asteroids at two locations in Jupiter's orbit. The orbits of the main-belt asteroids are typically somewhat eccentric, but generally lie within the "Belt" region of the figure. An Earth-crossing asteroid orbit is also shown.

example, the Earth-crossing asteroid Icarus (Figure 4.4) missed our planet by "only" six million km in 1968—a close call by cosmic standards. In 1994, asteroid 1994 XM1 came much closer, missing us by a mere 105,000 km (just over a quarter of the distance to the Moon). Encounters within one million km are relatively commonplace, happening perhaps a few times per decade.

In fact, calculations indicate that most Earth-crossing asteroids will eventually collide with Earth and that, during any given million-year period, roughly three asteroids strike our planet. Several dozen large basins and eroded craters on Earth are suspected to be sites of ancient asteroid collisions. The many large impact craters on the Moon, Venus, and Mars are direct evidence of similar events on other worlds. Such an impact could be catastrophic by human standards. Even a 1-km object carries more than 100 times more energy than all the nuclear weapons currently in existence on Earth and would devastate an area hundreds of kilometers in diameter. Should a sufficiently large asteroid hit our planet, it might even cause the extinction of entire species. Indeed, many scientists think that the extinction of the dinosaurs was the result of just such an impact (see *Interlude 18-1*).

Asteroids are generally too small to be resolved by Earth-based telescopes. However, astronomers have estimated the sizes of around 1000 of these tiny worlds by measuring the amount of sunlight they reflect and the

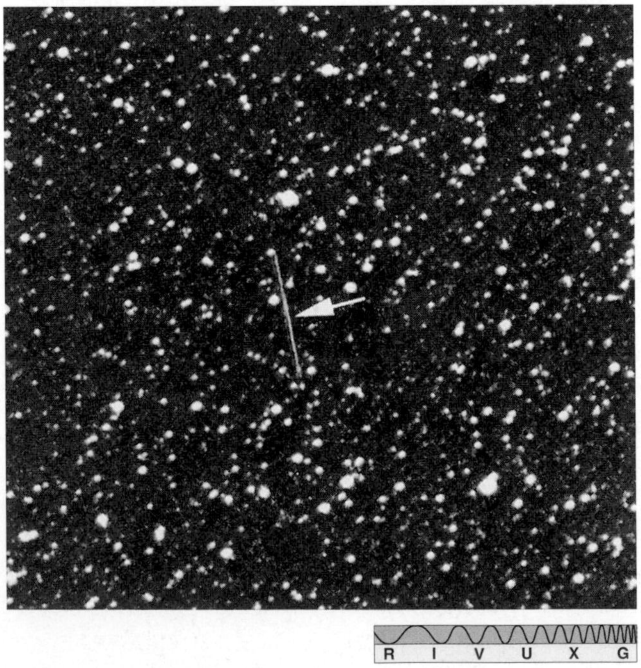

Figure 4.4 Asteroid Icarus The Earth-crossing asteroid Icarus has an orbit that passes within 0.2 A.U. of the Sun, well within Earth's orbit. Icarus occasionally comes close to Earth, making it one of the best-studied asteroids in the solar system. Its motion relative to the stars makes it appear as a streak (marked) in this long-exposure photograph. *(Palomar Observatory/California Institute of Technology)*

heat they radiate. In addition, from time to time an asteroid happens to pass directly in front of a star, as seen from Earth, allowing astronomers to determine the asteroid's size and shape with great accuracy. The largest asteroid, Ceres, has a diameter of 940 km. Only about two dozen are more than 200 km across, and most are far smaller. Large asteroids are roughly spherical because, as with planets, gravity is the dominant force determining their shape, but smaller bodies can be highly irregular.

Masses are known only for the half-dozen or so largest asteroids, mainly because the gravitational effect of an asteroid on its neighbors is very small and hard to measure accurately. The mass of Ceres is just 1/10,000 the mass of Earth. The total mass of all the known asteroids probably amounts to less than one-tenth the mass of the Moon.

Asteroid compositions are determined from spectroscopic and other measurements in the infrared, visible, and ultraviolet parts of the spectrum. The darkest (least reflective) asteroids contain a large fraction of carbon and are known as *carbonaceous* asteroids. The more reflective *silicate* asteroids are composed of rocky material. The few densities that are known are generally consistent with compositions inferred by spectroscopic means. Generally speaking, silicate asteroids predominate in the inner portions of the asteroid belt and the fraction of carbonaceous bodies steadily increases as we move outward. Overall, about 15 percent of all asteroids are silicate, 75 percent are carbonaceous, and 10 percent are other types (such as those containing large fractions of iron). Most planetary scientists believe that the carbonaceous asteroids consist of material representative of the earliest stages of the solar system and have not experienced significant heating or chemical evolution since they formed.

The first close-up views of asteroids were provided by the Jupiter probe *Galileo*, which, on its path to the giant planet, passed twice through the asteroid belt, making close encounters with asteroid Gaspra in October 1991 and asteroid Ida in August 1993 (Figure 4.5).

(a)

(b)

(c)

Figure 4.5 Asteroid Images (a) The asteroid Gaspra as seen from a distance of 1600 km by the space probe *Galileo*. (b) The asteroid Ida photographed by *Galileo* from a distance of 3400 km. (Ida's moon, Dactyl, is visible at the right.) The resolution in these photographs is on the order of 100 m. True-color images showed the surfaces of both bodies to be a fairly uniform gray. Spacecraft sensors indicated that the amount of infrared radiation absorbed by these surfaces varies from place to place, probably as a result of variations in the thickness of the dust layer blanketing them. (c) Asteroid Mathilde, imaged by the *NEAR* spacecraft on its way to the asteroid Eros. The largest craters in this image are about 20 km across—much larger than those seen on Ida or Gaspra. The reason may be Mathilde's low density and rather soft composition. *(NASA)*

Both Gaspra and Ida are irregularly shaped bodies having maximum diameters of about 20 and 60 km, respectively. They are pitted with craters ranging in size from a few hundred meters to two km across and are covered with layers of dust of variable thickness. Both asteroids are thought to be fragments of much larger objects that broke up following violent collisions long ago. Based on the extent of cratering on the asteroids' surfaces, astronomers estimate that the collision responsible for Ida occurred nearly a billion years ago, compared to just 200 million years for the impact that created Gaspra.

To the surprise of most astronomers, close inspection of the Ida image (Figure 4.5b) revealed the presence of a tiny moon (now named Dactyl), just 1.5 km across, orbiting the asteroid. By studying the *Galileo* images, astronomers were able to obtain some information on Dactyl's orbit around Ida and hence (using Newton's law of gravity) to estimate Ida's mass at about $5–10 \times 10^{16}$ kg, and its density as 2200–2900 kg/m^3. ∞ *(More Precisely 1-2)*

In June 1997 the Near Earth Asteroid Rendezvous (NEAR) spacecraft, en route to the mission's main target, the asteroid Eros, visited the asteroid Mathilde. Shown in Figure 4.5(c), Mathilde is some 60 km across. By sensing its gravitational pull, NEAR measured Mathilde's mass to be about 10^{17} kg, implying a density of just 1400 kg/m^3. To account for this low density, scientists speculate that the asteroid's interior must be quite porous. Indeed, many smaller asteroids seem to be more like loosely bound "rubble piles" than pieces of solid rock. On arrival at Eros in February 2000, *NEAR* (now renamed *NEAR-Shoemaker*) went into orbit around it, making detailed measurements of its size, shape, gravitational and magnetic fields, composition, and structure.

The concentration of most asteroids in a well-defined belt suggests one of two possible origins: either they are the fragments of a planet broken up long ago, or they are primal rocks that somehow never managed to form a genuine planet. Researchers strongly favor the latter view. There is far too little mass in the belt to constitute a planet, and the marked chemical differences between individual asteroids indicate that they could not all have originated in a single body. Instead, as discussed in more detail below, astronomers believe that the strong gravitational field of Jupiter continuously disturbs the motions of these chunks of primitive matter, nudging and pulling at them, preventing them from aggregating into a planet.

✓ Concept Check

- What are some similarities and differences between asteroids and the inner planets?

Comets

3 4 Comets are usually discovered as faint, fuzzy patches of light on the sky while still several astronomical units away from the Sun. Traveling in a highly elliptical orbit with the Sun at one focus, a comet brightens and develops an extended **tail** as it nears the Sun. (The name comet derives from the Greek *kome*, meaning "hair.") As the comet departs from the Sun's vicinity, its brightness and its tail diminish until it once again becomes a faint point of light receding into the distance. The various parts of a typical comet are shown in Figure 4.6(a). Like the planets, comets emit no visible light of their own—they shine by reflected (or reemitted) sunlight.

The **nucleus**, or main solid body, of a comet is only a few kilometers in diameter. During most of the comet's orbit, far from the Sun, only this frozen nucleus exists. When a comet comes within a few astronomical units of the Sun, however, its icy surface becomes too warm to remain stable. Part of it becomes gaseous and expands into space, forming a diffuse **coma** ("halo") of dust and evaporated gas around the nucleus. The coma gets larger and brighter as the comet nears the Sun. At maximum size, it can measure 100,000 km in diameter—almost as large as Saturn or Jupiter. Engulfing the coma, an invisible **hydrogen envelope** stretches across millions of kilometers of space. The comet's tail, most pronounced when the comet is closest to the Sun, is larger still, sometimes spanning as much as 1 A.U. From Earth, only the coma and tail of a comet are visible to the naked eye. Despite the size of the tail, most of the comet's light comes from the coma. Most of the mass resides in the nucleus.

Two types of comet tail may be distinguished. An **ion tail** is approximately straight, often made of glowing, linear streamers like those shown in Figure 4.7(a). Its emission spectrum indicates numerous ionized atoms and molecules that have lost some of their normal complement of electrons. ∞ (Sec. 2.6) A **dust tail** is usually broad, diffuse, and gently curved (Figure 4.7b). It is rich in microscopic dust particles that reflect sunlight, making the tail visible from afar. Both

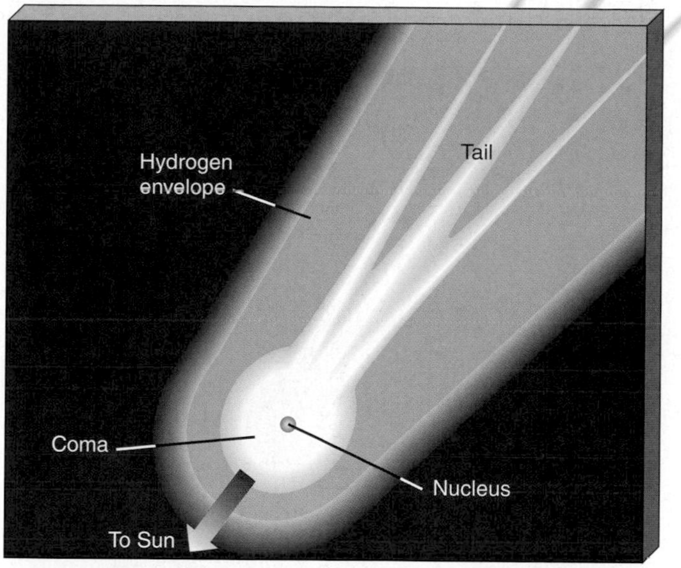

(a)

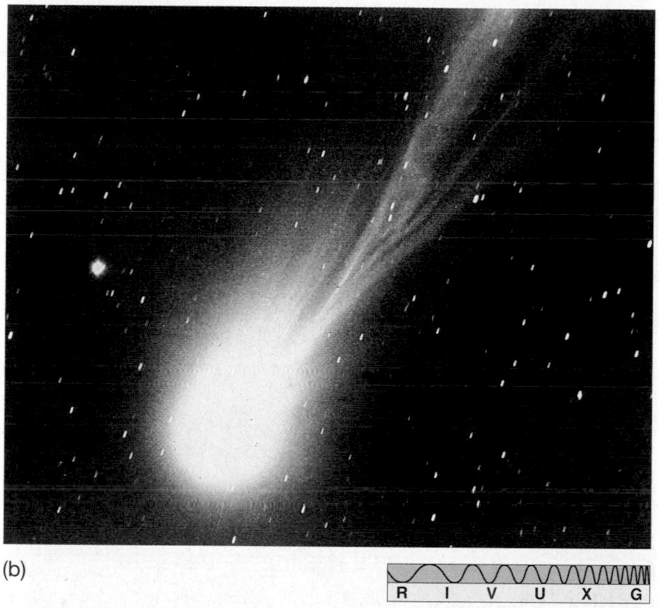

(b)

Figure 4.6 Halley's Comet (a) Diagram of a typical comet, showing the nucleus, coma, hydrogen envelope, and tail. The tail is not a sudden streak in time across the sky, as in the case of meteors or fireworks. Instead, it travels along with the rest of the comet as long as the comet is close enough to the Sun. (b) Halley's Comet in 1986, about one month before it rounded the Sun. *(NOAO)*

Animation

types of tails are in all cases directed away from the Sun by the **solar wind**, an invisible stream of matter and radiation escaping from the Sun. (In fact, astronomers first inferred the existence of the solar wind from observations of comet tails.) Consequently, as depicted in Figure 4.8, the tail, be it ion or dust, always lies outside a comet's orbit and actually *leads* the comet during the portion of the orbit that is outbound from the Sun.

Probably the most famous comet of all is Halley's Comet (Figure 4.6b), whose appearance at 76-year intervals has been documented at every passage since

Figure 4.7 Comet Tails (a) A comet having a primarily ion tail—Comet Giacobini–Zinner, seen here in 1959. The coma of this comet measured 70,000 km across, and its tail was well over 500,000 km long. (b) A comet having (mostly) a dust tail. This photograph, of Comet West in 1976, shows both the gentle curvature of the dust tail and its inherent fuzziness. This tail stretched 13° across the sky. Being composed largely of ice, comets tend to be fragile. Shortly after this photograph was taken, Comet West split into several fragments. *(U.S. Naval Observatory; NASA)*

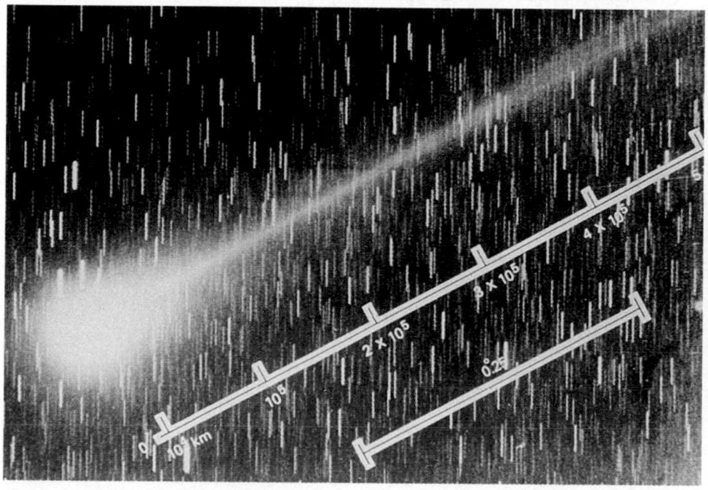

(a)

(b)

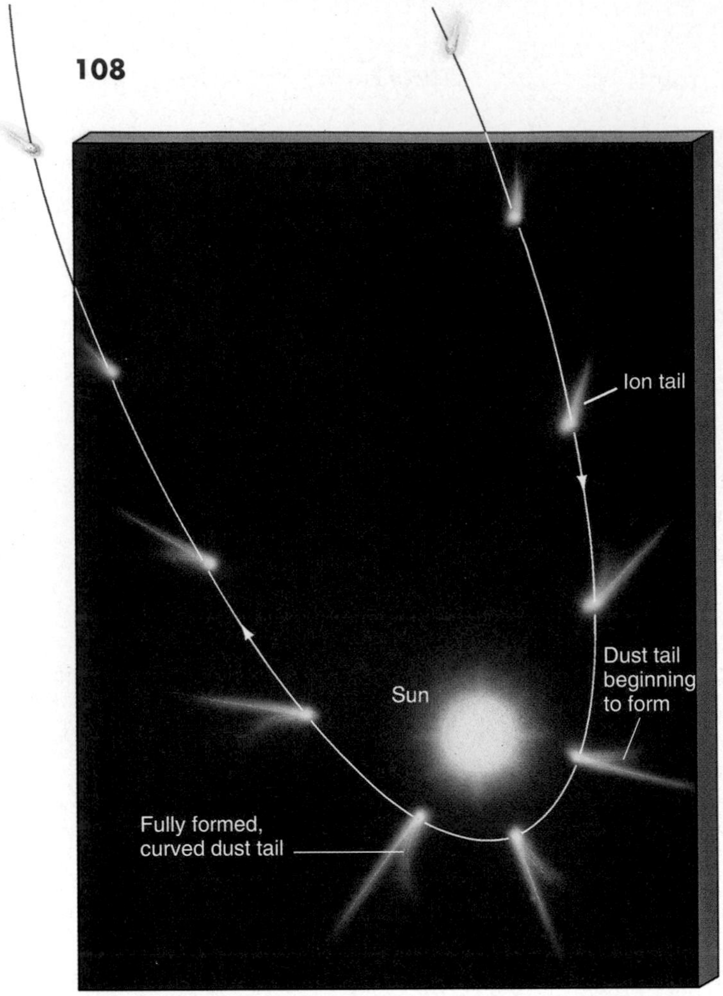

Figure 4.8 Comet Trajectory As it approaches the Sun, a comet develops an ion tail, which is always directed away from the Sun. Closer in, a curved dust tail, also directed generally away from the Sun, may appear. Notice that the ion tail always points directly away from the Sun on both the inbound and the outgoing portion of the orbit. The dust tail has a marked curvature and tends to lag behind the ion tail. Compare this figure with the comet Hale-Bopp, shown in the chapter opening photograph.

Animation

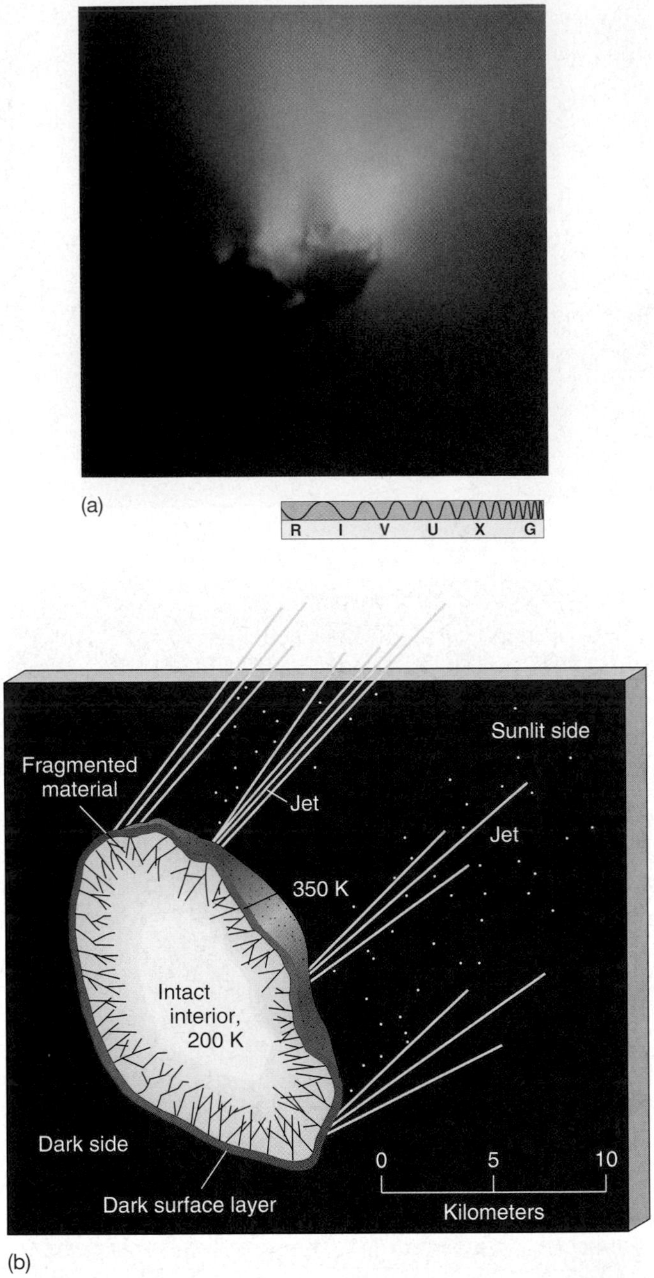

Figure 4.9 Halley's Comet Closeup (a) The *Giotto* spacecraft resolved the nucleus of Halley's Comet, showing it to be very dark, although heavy dust in the area obscured any surface features. Resolution here is about 50 m. The Sun is toward the right in this image. The brightest areas are jets of evaporated gas and dust spewing from the comet's nucleus. (b) A diagram of the nucleus of Halley's Comet. *(European Space Agency/Max Planck Institute/ H.U. Keller)*

Animation

240 B.C. A spectacular show, the tail of Halley's Comet can reach almost a full astronomical unit in length, stretching many tens of degrees across the sky.

When Halley's Comet rounded the Sun in 1986, a small armada of spacecraft launched by the USSR, Japan, and a group of western European countries went to meet it. The Soviet craft *Vega 2* traveled through the comet's coma, approaching to within 8000 km of the nucleus. Using data gained from the *Vega* encounter, the European spacecraft *Giotto* was navigated to within 600 km of the nucleus. Figure 4.9 shows *Giotto*'s view of Halley's nucleus, along with a sketch of its structure. The comet's next visit to the inner solar system is expected in 2061.

Based on the best available observations, experts now consider the nucleus of a comet to be composed of dust particles plus some small rocky fragments all trapped within a loosely packed mixture of methane, ammonia, and ordinary water ice, the density of the mixture being about 100 kg/m^3. Comets are often described as "dirty snowballs." Even as atoms, molecules, and dust particles boil off into space, creating a comet's coma and tail, the nucleus remains frozen at a temperature of only a few tens of kelvins. Estimates of typical cometary masses range from 10^{12} to 10^{16} kg, comparable to the masses of small asteroids.

Their highly elliptical orbits take most comets far beyond Pluto, where, in accordance with Kepler's second law, they spend most of their time. ∞ (Sec. 1.3) The majority take hundreds of thousands, even millions, of years to complete a single orbit around the Sun. However, a few *short-period* comets (conventionally defined as those having a period of less than 200 years) return for another encounter within a relatively short period of time. According to Kepler's third law, short-period comets never venture far beyond the orbit of Pluto.

Unlike the orbits of other solar system objects, the orbits of comets are *not* confined to within a few degrees of the ecliptic plane. Short-period comets do tend to have prograde orbits lying close to the ecliptic, but long-period comets exhibit all inclinations and all orientations, both prograde and retrograde, roughly uniformly distributed in all directions from the Sun.

Astronomers believe that short-period comets originate beyond the orbit of Neptune, in a region of the outer solar system called the **Kuiper Belt** (after Gerard Kuiper, a pioneer in infrared and planetary astronomy). A little like the asteroids in the inner solar system, most Kuiper Belt comets move in roughly circular orbits between about 30 and 100 A.U. from the Sun, always remaining outside the orbits of the jovian planets. Occasionally, however, either a chance encounter between two comets or (more likely) the gravitational influence of Neptune "kicks" a Kuiper Belt comet into an eccentric orbit that brings it into the inner solar system and into our view. The observed orbits of these comets reflect the flattened structure of the Kuiper Belt.

What of the long-period comets? How do we account for their apparently random orbital orientations? Only a tiny portion of a typical long-period cometary orbit lies within the inner solar system, so it follows that, for every comet we see, there must be many more similar objects far from the Sun. On these general grounds, astronomers reason that there must be a huge "cloud" of comets lying far beyond the orbit of Pluto, completely surrounding the Sun. It is named the **Oort Cloud**, after the Dutch astronomer Jan Oort, who first wrote in the 1950s of the possibility of such a vast and distant reservoir of inactive, frozen comets. The Kuiper Belt and the orbits of some typical Oort Cloud comets are sketched in Figure 4.10.

Based on the observed orbital properties of long-period comets, researchers believe that the Oort Cloud

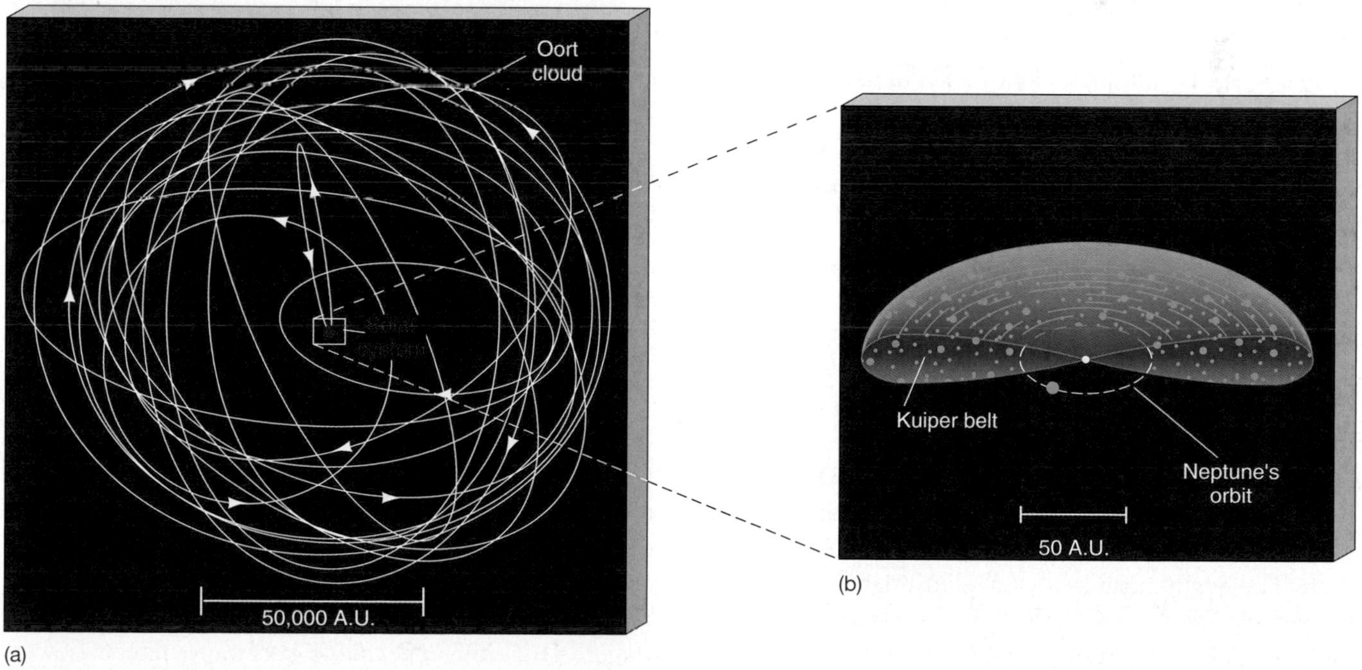

Figure 4.10 Comet Reservoirs (a) Diagram of the Oort Cloud, showing a few cometary orbits. Of all the orbits shown, only the most elongated ellipse represents a comet that will enter the solar system (which, on the scale of this drawing, is much smaller than the red dot at the center of the figure) and possibly become visible from Earth. (b) The Kuiper Belt, believed to be the source of short-period comets, whose orbits predominantly hug the ecliptic plane.

may be up to 100,000 A.U. (0.5 light-years) in diameter. Like their Kuiper Belt counterparts, however, most Oort Cloud comets never come anywhere near the Sun. Indeed, Oort Cloud comets rarely approach even the orbit of Pluto, let alone that of Earth. Only when the gravitational field of a passing star happens to deflect a comet into an extremely eccentric orbit that passes through the inner solar system do we get to see it at all. Because the Oort Cloud surrounds the Sun in all directions instead of being confined to the ecliptic plane like the Kuiper Belt, the long-period comets we see can come from any direction in the sky.

✓ Concept Check

- In what sense are the comets we see very *un*representative of comets in general?

Meteoroids

5 On a clear night, it is possible to see a few *meteors*—"shooting stars"—every hour. A **meteor** is a sudden streak of light in the night sky caused by friction between air molecules in Earth's atmosphere and an incoming piece of asteroid, meteoroid, or comet. This friction heats and excites the air molecules, which then emit light as they return to their ground states, producing the characteristic bright streak shown in Figure 4.11. Note that the brief flash that is a meteor is in no way similar to the broad, steady swath of light associated with a comet's tail. A meteor is a fleeting event in Earth's atmosphere whereas a comet tail exists in deep space, and can remain visible in the sky for weeks or even months.

Before encountering the atmosphere, the chunk of debris causing a meteor was almost certainly a meteoroid, simply because these small interplanetary fragments are far more common than either asteroids or comets. Any piece of interplanetary debris that survives its fiery passage through our atmosphere and finds its way to the ground is called a **meteorite**.

The smallest meteoroids are mainly the rocky remains of broken-up comets. Each time a comet passes near the Sun, some fragments dislodge from the main body. The fragments initially travel in a tightly knit group of dust or pebble-sized objects called a **meteoroid swarm**, moving in nearly the same orbit as the parent comet. Over the course of time, the swarm gradually disperses along the orbit, so that eventually the **micrometeoroids**, as these small meteoroids are known, become more or less smoothly spread around the parent comet's orbit. If Earth's orbit happens to intersect the orbit of such a young cluster of meteoroids, a spectacular *meteor shower* can result. Earth's motion takes it across a given comet's orbit at most twice a year (depending on the precise orbit of each body). Because the intersection occurs at the same time each year (Figure 4.12), the appearance of certain meteor showers is a regular and (fairly) predictable event (Table 4.3).

Meteor showers are usually named for their *radiant*, the constellation from whose direction they appear

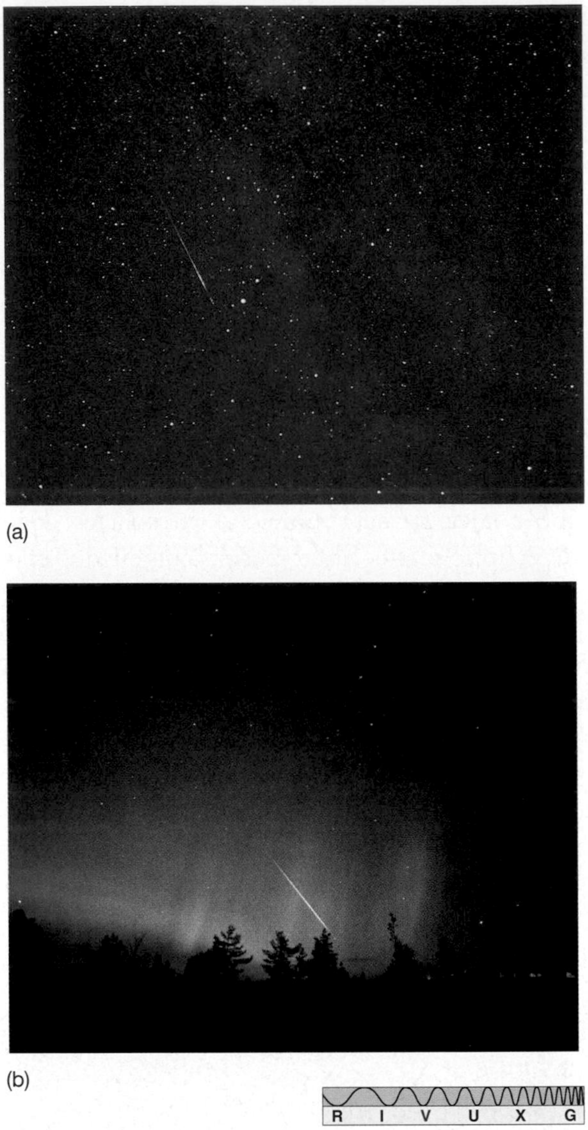

Figure 4.11 **Meteor Trails** A bright streak of light is produced when a fragment of interplanetary debris plunges into the atmosphere, heating the air to incandescence. (a) A small meteor photographed against a backdrop of stars. (b) The Northern Lights provide the background for a brighter meteor trail. *(P. Parviainen/Science Photo Library/Photo Researchers, Inc.)*

to come. For example, the Perseid shower is seen to emanate from the constellation Perseus. It can last for several days, but reaches maximum every year on the morning of August 12, when upward of 50 meteors per hour can be observed. Astronomers use the speed and direction of a meteor's flight to compute its interplanetary trajectory. This is how certain meteoroid swarms have come to be identified with well-known comet orbits.

Larger meteoroids—those more than a few centimeters in diameter—are usually *not* associated with comets. They are more likely small bodies that have strayed from the asteroid belt, possibly as the result of asteroid collisions. Their orbits can sometimes be reconstructed in a manner similar to that used to determine the orbits of meteor showers. In most cases, their computed orbits do indeed intersect the asteroid belt, providing the strongest evidence we have that this is where they originated. These objects are responsible for most of the cratering on the surfaces of the Moon, Mercury, Venus, Mars, and some of the moons of the jovian planets.

TABLE 4.3 Some Prominent Meteor Showers

MORNING OF MAXIMUM ACTIVITY	SHOWER NAME	ROUGH HOURLY COUNT	PARENT COMET
Jan. 3	Quadrantid	40	—
Apr. 21	Lyrid	10	1861I (Thatcher)
May 4	Eta Aquarid	20	Halley
June 30	Beta Taurid	25	Encke
July 30	Delta Aquarid	20	—
Aug. 12	Perseid	50	1862III (Swift–Tuttle)
Oct. 9	Draconid	up to 500	Giacobini–Zimmer
Oct. 20	Orionid	30	Halley
Nov. 7	Taurid	10	Encke
Nov. 16	Leonid	12[1]	1866I (Tuttle)
Dec. 13	Geminid	50	3200 Phaeton[2]

[1]Every 33 years, as Earth passes through the densest region of this meteoroid swarm, we see intense showers that can reach 1000 meteors per minute for brief periods of time. This occurred most recently in 1999 and 2000.

[2]Phaeton is actually an asteroid and shows no signs of cometary activity, but its orbit matches the meteoroid paths very well.

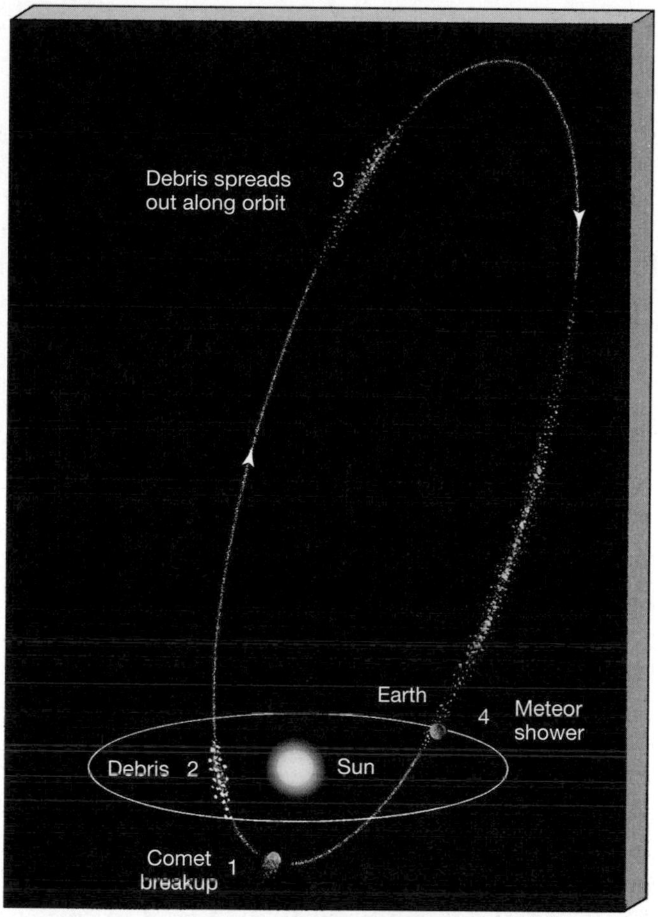

Figure 4.12 **Meteor Showers** A meteoroid swarm associated with a given comet intersects Earth's orbit at specific locations, giving rise to meteor showers at specific times of the year. A portion of the comet breaks up as it rounds the Sun, at the point marked 1. The fragments continue along the comet orbit, gradually spreading out (points 2 and 3). The rate at which the debris disperses around the orbit is much slower than depicted here. It takes many orbits for the material to spread out as shown, but eventually the fragments extend all around the orbit, more or less uniformly. If the orbit happens to intersect Earth's, the result is a meteor shower each time Earth passes through the intersection (point 4).

Meteoroids less than about a meter across (roughly a ton in mass) generally burn up in Earth's atmosphere. Larger bodies reach the surface, where they can cause significant damage, such as the kilometer-wide Barringer Crater shown in Figure 4.13. From the size of this crater, we can estimate that the meteoroid responsible must have had a mass of about 200,000 tons. Only 25 tons of iron meteorite fragments have been found at the crash site. The remaining mass must have been scattered by the explosion at impact, broken down by subsequent erosion, or buried in the ground. Currently, Earth is scarred with nearly 100 craters larger than 0.1 km in diameter.

Figure 4.13 **Barringer Crater** The Barringer Meteor Crater, near Winslow, Arizona, is 1.2 km in diameter and 0.2 km deep, the result of a meteorite impact about 25,000 years ago. The meteoroid was probably about 50 m across and likely weighed around 200,000 tons. *(U.S. Geological Survey, Denver)*

Most of these are so heavily eroded by weather and geological activity that they can be identified only in satellite photography, as in Figure 4.14. Fortunately, such major collisions between Earth and large meteoroids are thought to be rare events now. Researchers believe that, on average, they occur only once every few hundred thousand years (see *Interlude 18-1*).

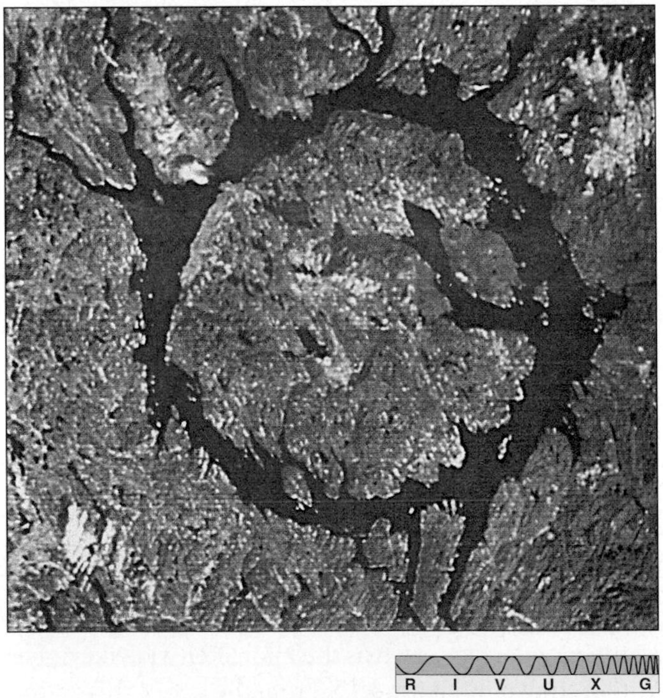

Figure 4.14 **Manicouagan Reservoir** This photograph, taken from orbit by the U.S. *Skylab* space station, shows the ancient impact basin that forms Quebec's Manicouagan Reservoir. A large meteorite landed there about 200 million years ago. The central floor of the crater rebounded after the impact, forming an elevated central peak. The lake, 70 km in diameter, now fills the resulting ring-shaped depression. *(NASA)*

One of the most recent documented meteoritic events occurred in central Siberia on June 30, 1908 (Figure 4.15). The presence of only a shallow depression as well as a complete lack of fragments implies that this Siberian intruder exploded several kilometers above the ground, leaving a blasted depression at ground level but no well-formed crater. Recent calculations suggest that the object in question was a rocky meteoroid about 30 m across. The explosion, estimated to have been equal in energy to a 10-megaton nuclear detonation, was heard hundreds of kilometers away and produced measurable increases in atmospheric dust levels all across the Northern Hemisphere.

Figure 4.15 **Tunguska Debris** The Tunguska event of 1908 leveled trees over a vast area. Although the impact of the blast was tremendous and its sound audible for hundreds of kilometers, the Siberian site is so remote that little was known about the event until scientific expeditions arrived to study it many years later. *(Sovfoto/Eastfoto)*

(a) (b)

Figure 4.16 Meteorite Samples (a) A stony meteorite often has a dark crust, created when its surface is melted by the tremendous heat generated during passage through the atmosphere. (b) Iron meteorites, usually contain some nickel as well. Most show characteristic crystalline patterns when their surfaces are cut, polished, and etched with acid. *(Science Graphics)*

Most meteorites are rocky (Figure 4.16a), although a few are composed mainly of iron and nickel (Figure 4.16b). Their basic composition is much like that of the rocky inner planets or the Moon, except that some of their lighter elements—such as hydrogen and oxygen—are depleted. It is thought that these light elements boiled away long ago when the bodies from which the meteorites originated were wholly or partially molten. Some meteorites show clear evidence of strong heating in the past. One possible explanation is that their parent asteroids were partially melted during the collisions that liberated the meteoritic fragments. Alternatively, those asteroids may have experienced geological activity of some sort before the collisions occurred. Others show no evidence of past heating, and probably date back to the formation of the solar system.

Most primitive of all are carbonaceous meteorites, black or dark gray and most likely related to carbonaceous asteroids. (Similarly, the silicate-rich rocky meteorites are probably associated with silicate asteroids.) Many carbonaceous meteorites contain significant amounts of ice and other volatile substances, and they are usually rich in organic (carbon-based) molecules.

Finally, almost all meteorites are *old*. Radioactive dating shows most of them to be between 4.4 and 4.6 billion years old—roughly the age of the oldest Moon rocks brought back to Earth by *Apollo* astronauts. Meteorites, along with asteroids, comets, and some lunar rocks, provide essential clues to the original state of matter in the solar neighborhood and to the birth of our planetary system.

✓ Concept Check

- Why are astronomers so interested in interplanetary matter?

4.3 The Formation of the Solar System

You might be struck by the vast range of physical and chemical properties found in the solar system. Indeed, our astronomical neighborhood may seem more like a great junkyard than a smoothly running planetary system. Can we really make any sense of solar system matter? Is there some underlying principle that unifies the facts we have outlined in this chapter? Remarkably, the answer is "yes." The origin of the solar system is a complex and as yet incompletely solved puzzle, but the basic outlines are now quite well understood.

Despite the recent exciting discoveries of planets orbiting other stars (see *Interlude 4-1*), astronomers still have little detailed information on the properties of those planets, and so far there is no firm evidence that planets like our own exist anywhere beyond our solar system. For that reason, our theories of planet formation concentrate on the planetary system in which we live. Bear in mind, however, that no part of the scenario we describe here is in any way unique to our own system. The same basic processes could have occurred—and, many astronomers believe, probably did occur—during the formative stages of many of the stars in our Galaxy (but see *Interlude 4-1* for hints that the layout of our solar system may *not* be the only way a planetary system can form).

Model Requirements

6 Based on the measured ages of the oldest meteorites, as well as the ages of Earth and lunar rocks, planetary scientists believe that the age of the solar system is 4.6 billion years. What happened 4.6 billion years ago to create the planetary system we see today? Any theory of the origin and architecture of our planetary system must adhere to these nine known facts:

1. *Each planet is relatively isolated in space*. The planets exist as independent bodies at progressively larger distances from the central Sun; they are not bunched together.
2. *The orbits of the planets are nearly circular*. In fact, with the exceptions of Mercury and Pluto, which we will argue are special cases, each planetary orbit closely describes a perfect circle.
3. *The orbits of the planets all lie in nearly the same plane*. The planes swept out by the planets' orbits are accurately aligned to within a few degrees. Again, Mercury and Pluto are slight exceptions.
4. *The direction in which the planets orbit the Sun (counterclockwise as viewed from above Earth's north pole) is the same as the direction in which the Sun rotates on its axis*. Virtually all the large-scale motions in the solar system (other than comet orbits) are in the same plane and in the same sense. The plane is that of the Sun's equator, and the sense is that of the Sun's rotation.
5. *The direction in which most planets rotate on their axis is roughly the same as the direction in which the Sun rotates on its axis*. This property is considerably less general than the one just described for revolution, as three planets—Venus, Uranus, and Pluto—do not share it.
6. *The direction in which most of the known moons revolve about their parent planet is the same as the direction in which the planet rotates on its axis*. Of the large moons in the solar system, only Neptune's moon Triton is an exception.
7. *Our planetary system is highly differentiated*. The terrestrial planets are characterized by high densities,

INTERLUDE 4-1
Planets Beyond the Solar System

The past few years have seen enormous strides in the search for extrasolar planets. These advances have come not through any dramatic scientific or technical breakthrough, but rather through steady improvements in both telescope and detector technology and computerized data analysis. The count of potential extrasolar planets currently stands at 28. However, it is not yet possible to image any of these newly discovered worlds. The techniques used to find them are indirect, based on analysis of light from the parent star, not from the unseen planet.

The accompanying figure shows two sets of data that betray the presence of planets. As a planet orbits a star, gravitationally pulling one way and then the other, the star wobbles slightly. The more massive the planet or the closer its orbit to the star, the greater its gravitational pull and hence the star's movement.

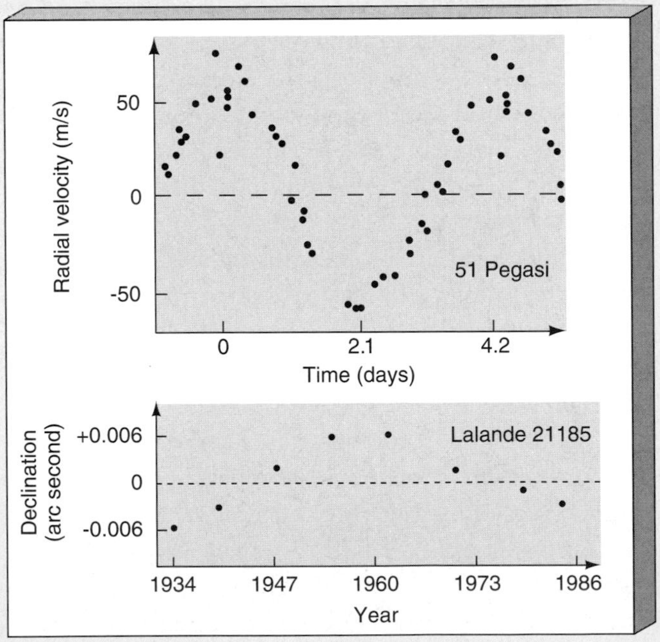

If the wobble happens to occur along our line of sight to the star (as illustrated in the upper graph, which shows the line-of-sight velocity of the star 51 Pegasi, a near-twin of our Sun lying some 40 light-years away), then we see small fluctuations in the star's radial velocity, which can be measured using the Doppler effect. ∞ (*More Precisely* 2-3) These data were acquired in 1994 by Swiss astronomers using the 1.9-m telescope at Haute-Provence Observatory in France, and imply that a planet at least half the mass of Jupiter orbits 51 Peg with a period of just 4.2 days. Alternatively, if the wobble is predominantly perpendicular to our line of sight (as indicated in the lower graph, which plots the declination of the star Lalande 21185 over the last half century), then the star's position in the sky changes slightly from night to night. ∞ (Sec. P.2) The 30-year-moderate atmospheres, slow rotation rates, and few or no moons. The jovian planets (Pluto excepted, as usual) have low densities, thick atmospheres, rapid rotation rates, and many moons.

8. *Asteroids are very old and exhibit a range of properties not characteristic of either the terrestrial or the jovian planets or their moons.* Asteroids share, in rough terms, the bulk orbital properties of the planets. However, they appear to be made of primitive, unevolved material, and the meteorites that strike Earth are the oldest rocks known.
9. *Comets are primitive, icy fragments that do not necessarily orbit in the ecliptic plane and reside primarily at large distances from the Sun, in the Kuiper Belt and the Oort Cloud.*

All these observed facts taken together strongly suggest a high degree of order in our solar system. The large-scale architecture is too neat, and the ages of the components too uniform, to be the result of random chaotic events. The overall organization points toward a single formation, an ancient but one-time event 4.6 billion years ago.

In the next few chapters, we will see that some planetary properties (such as atmospheric composition and interior structure) have gradually *evolved* into their present states during the billions of years since the planets formed. However *no* such evolutionary explanation exists for the items in the preceding list. For example, Newton's laws imply that the planets must move in elliptical orbits with the Sun at one focus, but they offer no explanation of why the observed orbits should be roughly circular, coplanar, and prograde. We know of no way in which the planets could have started off in random paths, then later evolved into the orbits we see today. Their basic orbital properties must have been established at the outset.

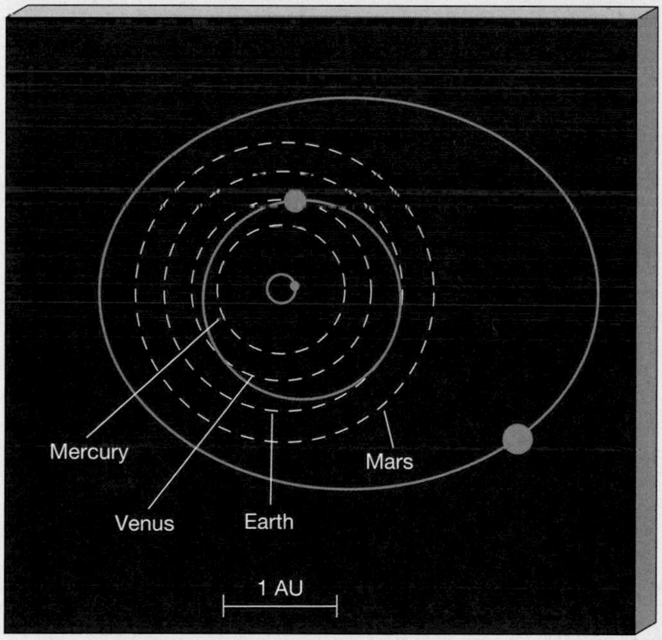

period wobble at the level of about 0.01″ is roughly what would be expected for our Sun due to a Jupiter-sized object orbiting at a distance of a few A.U. In at least one case, the existence of a planet has been inferred from observations of variations in its parent star's light as the planet periodically passes between the star and Earth.

These techniques are strongly biased toward finding massive planets, since these bodies induce the largest wobbles. At present, they can be used only to detect planets comparable in mass to Jupiter (in fact, one of the most recently discovered planets appears to have a mass similar to that of Saturn), so we should not expect any reports of the detection of Earth-sized planets in the immediate future. Even so, as the numbers mount, comparative studies are beginning to suggest that the layout of our own planetary system might be the exception rather than the rule. Most of the findings to date indicate "hot Jupiters"—giant planets surprisingly close (well within one A.U.) to their parent star—and the remainder suggest planets on wider but still very eccentric orbits. Furthermore, all but one system has been found to have just a single planet. The one exception is the Sun-like star Upsilon Andromedae, whose system of three planets resides some 44 light-years away; its orbital configuration is shown in the second figure (in orange), compared to the inner planet orbits of our Solar System (in dashed white).

So do not think that the newly discovered planetary systems closely resemble our own solar system. At least so far, they don't—and that has theorists puzzled. If current trends continue, our solar system may turn out to represent a minority architecture among planetary system designs—an initial surprise and an ongoing puzzle.

In addition to its many regularities, our solar system also has many notable *irregularities*, some of which we have already mentioned. Far from threatening our theory, however, these irregularities are important facts for us to consider in shaping our explanations. For example, any theory explaining solar system formation must not insist that *all* planets rotate in the same sense or have only prograde moons because that is not what we observe. Instead, the theory should provide strong reasons for the observed planetary characteristics, yet be flexible enough to allow for and explain the deviations, too. And, of course, the existence of the asteroids and comets that tell us so much about our past must be an integral part of the picture. That's a tall order, yet many researchers now believe we are close to that level of understanding.

Concept Check

- Why is it important that the theory of planet formation not be *too* rigid in its predictions?

Nebular Contraction

7 Modern theory holds that planets are in effect by-products of the process of star formation (Chapter 11). Imagine a large cloud of interstellar dust and gas (called a **nebula**) a light-year or so across. Now suppose that, due to some external influence, such as a collision with another interstellar cloud or perhaps the explosion of a nearby star, the nebula starts to contract under the influence of its own gravity. As it contracts, it becomes denser and hotter, eventually forming a star—the Sun—at its center.

In 1796 the French mathematician-astronomer Pierre Simon de Laplace showed mathematically that conservation of angular momentum (see *More Precisely 4-1*) demands that our hypothetical nebula must spin faster as it contracts. The increase in rotation speed, in turn, causes the nebula's *shape* to change as it shrinks. Centrifugal forces (due to rotation) tend to oppose the contraction in directions perpendicular to the rotation axis, with the result that the nebula collapses most rapidly along the rotation axis. As shown in Figure 4.17, by the time it has shrunk to about 100 A.U., the cloud has flattened into a pancake-shaped disk. This swirling mass destined to become our solar system is usually referred to as the **solar nebula**.

If we now suppose that planets form out of this spinning disk, we can begin to understand the origin of much of the architecture observed in our planetary system today, such as the near-circularity of the planets' orbits and the fact that they move in prograde orbits in almost the same plane. The idea that planets form from such a disk is called the **nebular theory**.

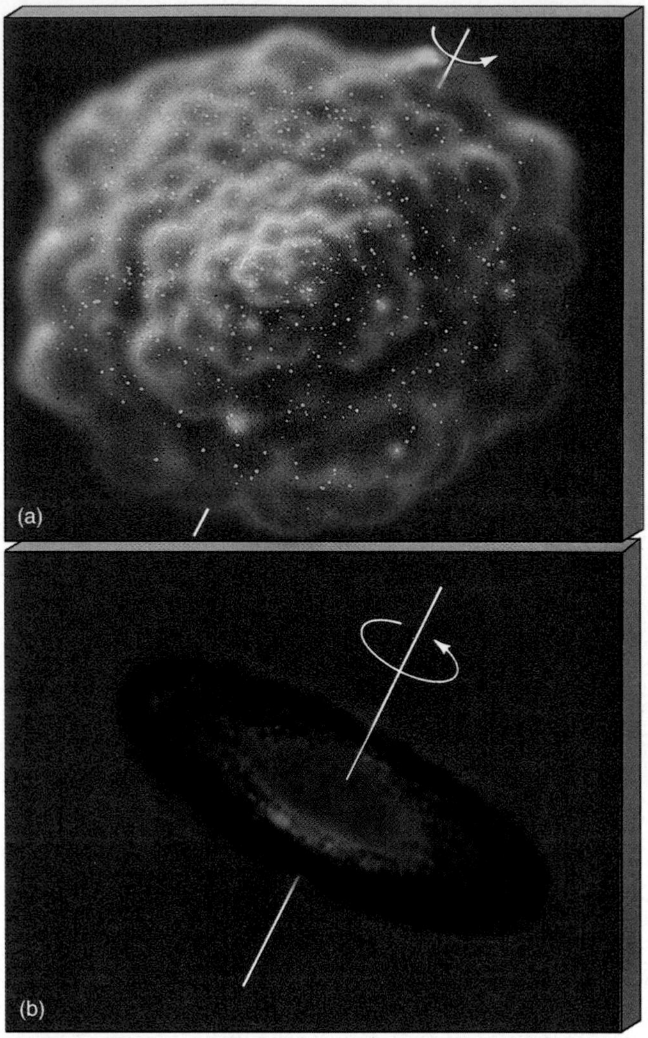

Figure 4.17 **Angular Momentum** (a) Conservation of angular momentum demands that a contracting, rotating cloud must spin faster as its size decreases. (b) Eventually, the solar nebula, which is that particular nebula destined to become the solar system, came to resemble a gigantic pancake. The large blob at the center ultimately became the Sun.

Astronomers are fairly confident that the solar nebula formed such a disk, because similar disks have been observed (or inferred) around other stars. Figure 4.18 shows a visible-light image of the region around the star Beta Pictoris, which lies about 50 light-years from the Sun. When the light from Beta Pictoris is suppressed and the resulting image enhanced by a computer, a faint disk of warm matter (viewed almost edge-on here) can be seen. This particular disk is roughly 1000 A.U. across—about 10 times the diameter of Pluto's

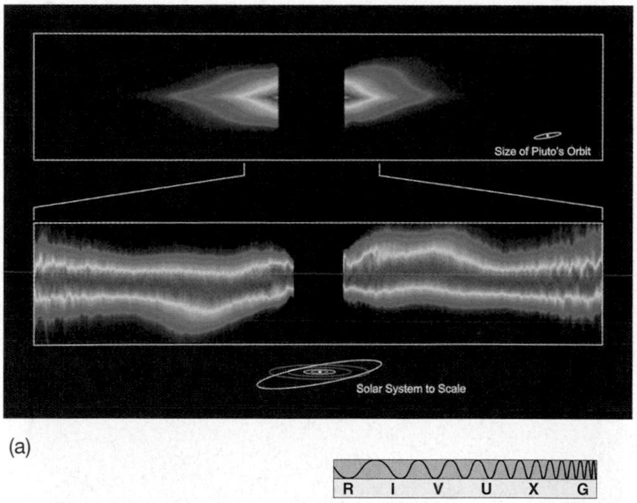

(a)

(b)

Figure 4.18 **Beta Pictoris** (a) A computer-enhanced view of a disk of warm matter surrounding the star Beta Pictoris. Both images show actual data taken at visible wavelengths, but are presented here in false color to accentuate the details; the bottom image is a closeup of the inner regions of the disk, implying a warp in the disk possibly caused by the gravitational pull of unseen companions. In both images the overwhelmingly bright central star has been removed in order to let us see the much fainter disk surrounding it. The disk is nearly edge-on to our line of sight. For scale, the dimension of Pluto's orbit (78 A.U.) has been drawn adjacent to the images. (b) An artist's conception of the disk of clumped matter, showing the warm disk with a young star at the center and several large bodies already forming. The colors are thought to be accurate. At the outer edges of the disk the temperature is low, and the color is a dull red. Progressing inward, the colors brighten and shift to a more yellowish tint as the temperature increases. *(NASA; D. Berry)*

Animation

orbit. Astronomers believe that Beta Pictoris is a very young star, perhaps only 100 million years old, and that we are witnessing its passage through an evolutionary stage similar to that experienced by our own Sun 4.6 billion years ago.

Unfortunately for the original theory, while Laplace's description of the collapse and flattening of the solar nebula was essentially correct, a disk of gas would *not* form clumps of matter that would subsequently evolve into planets. In fact, computer calculations predict just the opposite: any clumps in the gas would tend to disperse, not contract further. However, the model currently favored by most astronomers, known as the **condensation theory**, rests squarely on the old nebular theory, combining its basic physical reasoning with new information about interstellar chemistry to avoid most of the original theory's problems.

The key new ingredient is *interstellar dust* in the solar nebula. Astronomers now recognize that the space between the stars is strewn with microscopic dust grains, an accumulation of the ejected matter of many long-dead stars. These dust particles probably formed in the cool atmospheres of old stars, then grew by accumulating more atoms and molecules from the interstellar gas. The end result is that interstellar space is littered with tiny chunks of icy and rocky matter having typical diameters of about 10^{-5} m. Figure 4.19 shows one of many such dusty regions in the vicinity of the Sun.

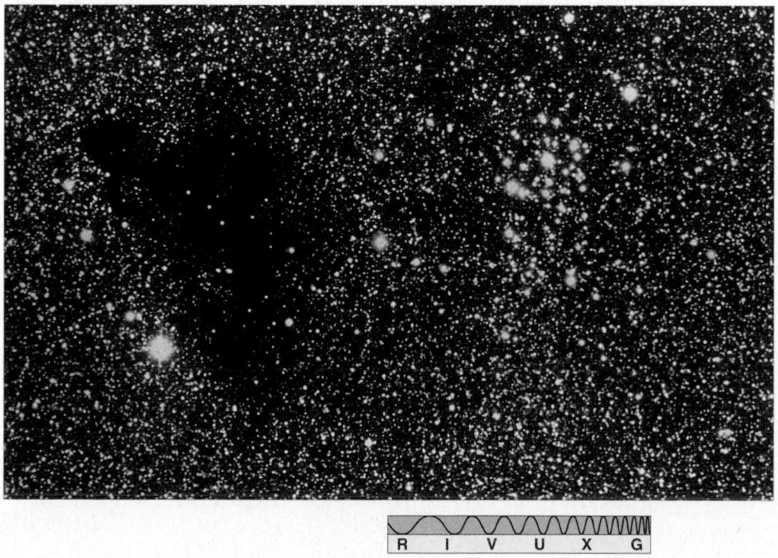

Figure 4.19 **Dark Cloud** Interstellar gas and dark dust lanes mark this region of star formation. The dark cloud known as Barnard 86 (left) flanks a cluster of young blue stars called NGC 6520 (right). Barnard 86 may be part of a larger interstellar cloud that gave rise to these stars. *(D. Malin/Anglo-Australian Observatory)*

Dust grains play an important role in the evolution of any gas cloud. Dust helps to cool warm matter by efficiently radiating heat away in the form of infrared radiation, allowing the nebula to collapse more easily. Furthermore, the dust grains greatly speed up the process of collecting enough atoms to form a planet. They act as **condensation nuclei**—microscopic platforms to which other atoms can attach, forming larger and larger balls of matter. This is similar to the way that raindrops form in Earth's atmosphere; dust and soot in the air act as condensation nuclei around which water molecules cluster.

Planet Formation

According to the condensation theory, the planets formed in three stages (Figure 4.20). Early on, dust grains in the solar nebula formed condensation nuclei around which matter began to accumulate. This vital step greatly hastened the critical process of forming the first small clumps of matter. Once these clumps formed, they grew rapidly by sticking to other clumps. As the clumps grew larger, their surface areas increased and consequently the rate at which they swept up new material accelerated. They gradually grew into objects of pebble size, baseball size, basketball size, and larger.

Eventually, this process of **accretion**—the gradual growth of small objects by collision and sticking—created objects a few hundred kilometers across. At the end of this first stage of planet formation, the solar system was made up of hydrogen and helium gas and millions of **planetesimals**—objects the size of small moons, having gravitational fields just strong enough to affect their neighbors. By then, their gravity was strong enough to sweep up material that would otherwise not have collided with them, so their rate of growth became faster still. During this second phase of planet formation, gravitational forces between planetesimals caused them to collide and merge, forming larger and larger objects. Because larger bodies exert stronger gravitational pulls, eventually almost all the planetesimal material was swept up into a few large **protoplanets**—accumulations of matter that would eventually evolve into the planets we know today.

As the protoplanets grew, another process became important. Their strong gravitational fields produced many high-speed collisions between planetesimals and protoplanets. These collisions led to **fragmentation** as small objects broke into still smaller chunks that were then swept up by the protoplanets. Only a relatively small number of 10- to 100-km fragments escaped capture by a planet or a moon and became the asteroids and comets. After about 100 million years, the primitive solar system had evolved into nine protoplanets, dozens of protomoons, and a glowing **protosun** at the center. Computer simulations generally reproduce the increasing spacing between the planets, although the reasons for the regularity seen in the actual planetary spacing remain unclear. Roughly a billion more years were required to sweep the system clear of interplanetary trash.

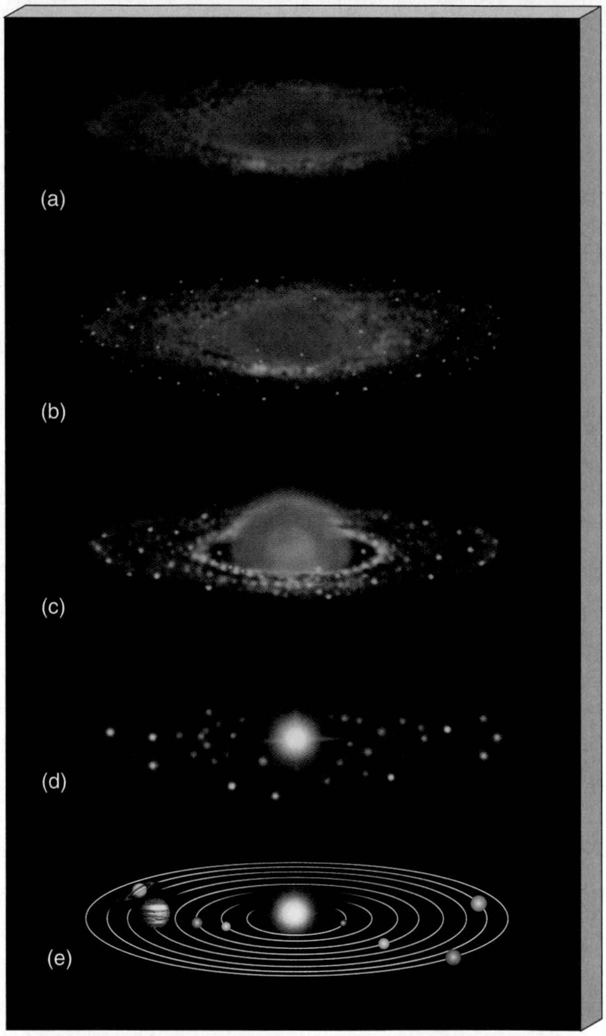

Figure 4.20 **Solar System Formation** The condensation theory of planet formation (not drawn to scale; Pluto not shown in part e). (a) The solar nebula after it has contracted and flattened to form a spinning disk (Figure 4.17b). The large blob in the center will become the Sun. Smaller blobs in the outer regions may become jovian planets. (b) Dust grains act as condensation nuclei, forming clumps of matter that collide, stick together, and grow into moon-sized planetesimals. (c) Strong winds from the still-forming Sun expel the nebular gas. (d) Planetesimals continue to collide and grow. (e) Over the course of a hundred million years or so, planetesimals form a few large planets that travel in roughly circular orbits.

MORE PRECISELY 4-1

Conservation of Angular Momentum

Angular momentum is a fundamental property of any rotating object. It is a measure of the object's tendency to keep spinning, or, equivalently, of how much effort must be expended to stop it. Intuitively, we know that the more massive an object, or the larger it is, or the faster it spins, the harder it is to stop. In fact, angular momentum depends on the object's *mass*, *rotation rate* (measured in, say, revolutions per second), and *radius*, in a very specific way:

$$\text{angular momentum} \propto \text{mass} \times \text{rotation rate} \times \text{radius}^2$$

(the constant of proportionality depends on the details of how the object's mass is distributed).

According to Newton's laws of motion, angular momentum must be *conserved* at all times—that is, it must remain constant before, during, and after a physical change in any object (so long as no external forces act). This allows us to relate changes in size to changes in rotation rate. Figure skaters use this principle to spin faster by drawing in their arms (as shown in the accompanying figure) and slow down by extending them. The mass of the human body remains the same, but its overall radius changes, causing the body's rotation rate to increase in order to keep the angular momentum unchanged.

To take an example from the text, suppose that the interstellar gas cloud from which our solar system formed was initially one light-year across and rotating very slowly—once every 10 million years, say. The cloud's mass doesn't change (we assume) during the contraction, but its radius decreases. The rotation rate must therefore *increase* in order to keep the total angular momentum unchanged. By the time it has collapsed to a radius of 100 A.U., its radius has shrunk by a factor of (one light-year / 100 A.U.) $\approx$ 630. Conservation of angular momentum then implies that its (average) spin rate increases by a factor of $630^2 \approx 400{,}000$, to roughly one revolution every 25 years, close to the orbital period of Saturn. Incidentally, the law of conservation of angular momentum also applies to planetary orbits (where now "radius" is the distance from the planet to the Sun)—in fact, Kepler's second law *is* just conservation of angular momentum expressed another way. ∞ (Sec. 1.3)

(AP/Wide World Photo)

The four largest protoplanets became large enough to enter a third phase of planetary development, sweeping up large amounts of gas from the solar nebula to form what would ultimately become the jovian planets. The smaller, inner protoplanets never reached that point, and as a result their masses remained relatively low. The outer protoplanets may have grown faster simply because there was more raw material available to them (see below). Alternatively, some or all of them may have formed directly, through instabilities in the cool gas of the outer solar nebula, mimicking on small scales the collapse of the initial interstellar cloud, and skipping the initial accretion stage. In either case, the large size and substantial cores of the jovian planets reflects the "head start" they had in the formation process.

What of the gas that made up most of the original cloud? Why don't we see it today throughout the planetary system? All young stars apparently experience a highly active evolutionary stage known as the *T-Tauri phase* (Chapter 11), during which their radiation emission and stellar winds are very intense. When our Sun entered this phase, just before nuclear burning started at its center, any gas remaining between the planets was blown away into interstellar space by the solar wind and the Sun's radiation pressure. Afterwards, all that remained were protoplanets and planetesimal fragments, ready to continue their long evolution into the solar system we know today.

The Differentiation of the Solar System

6 We can use the condensation theory to understand the basic composition differences in the terrestrial planets, jovian planets, and smaller bodies that constitute the solar system. Indeed, it is in this context that the term *condensation* derives its true meaning.

As the solar nebula contracted under the influence of gravity, it heated up as it flattened into a disk. The density and temperature were greatest near the central protosun and much lower in the outlying regions. In the hot inner regions, dust grains broke apart into molecules, which in turn split into atoms. Most of the original dust in the inner solar system disappeared at this stage, although the grains in the outermost parts probably remained largely intact.

The destruction of the dust in the inner solar nebula introduced an important new ingredient into the theoretical mix, one that we omitted from our earlier discussion. With the passage of time, the gas radiated away its heat, and the temperature decreased at all locations except in the very core, where the Sun was forming. Everywhere beyond the protosun, new dust grains began to condense out, much as raindrops, snowflakes, and hailstones condense from moist, cooling air here on Earth. It may seem strange that although there was plenty of interstellar dust early on, it was mostly destroyed, only to form again later. However, a critical change had occurred. Initially, the nebular gas was uniformly peppered with dust grains. When the dust reformed later, the distribution of grains was very different.

Figure 4.21 plots the temperature in various parts of the primitive solar system just before the onset of accretion. At any given location, the only materials to condense out were those able to survive the temperature there. As marked on the figure, in the innermost regions, around Mercury's present orbit, only metallic grains could form—it was simply too hot for anything else to exist. A little farther out, at about 1 A.U., it was possible for rocky, silicate grains to form, too. Beyond about 3 or 4 A.U., water ice could exist, and so on, with the condensation of more and more material possible at greater and greater distances from the Sun. The composition of the material that could condense out at any given distance from the Sun ultimately determined the type of planet that formed there.

Beyond about 5 A.U. from the center, the temperature was low enough to allow several abundant gases—water vapor, ammonia, and methane—to condense into solid form. (These compounds are still the primary constituents of jovian atmospheres.) Consequently, the planetesimals destined to become the cores of the jovian planets were formed under cold conditions out of low-density, icy material. Because more material could condense out of the solar nebula at these radii than in the inner regions near the protosun, accretion began sooner, with more resources to draw on. The outer planets grew rapidly to the point where they could accrete not just grains but nebular gas also, and the eventual result was the hydrogen-rich jovian worlds we see today.

In the inner regions of the primitive solar system, the environment was too hot for ices to survive. Many of the abundant heavier elements, such as silicon, iron, magnesium, and aluminum, combined with oxygen to produce a variety of rocky materials. Planetesimals in the inner solar system were therefore rocky or metallic, as were the protoplanets and planets they ultimately formed. Here is an additional reason why the jovian planets grew so much bigger than the terrestrial worlds. The inner regions of the nebula had to wait for the temperature to drop so that a few rocky grains could appear

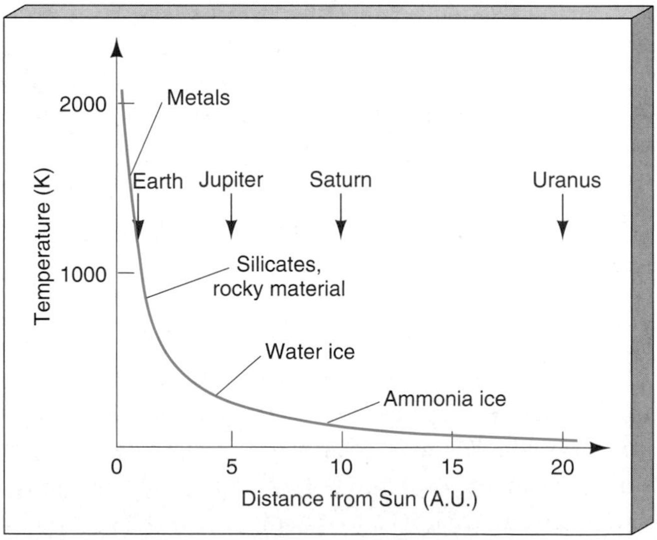

Figure 4.21 Temperature in Early Solar Nebula Theoretically computed variation of temperature across the primitive solar nebula. In the hot central regions, only metals could condense out of the gaseous state to form grains. At greater distances from the central protosun, the temperature was lower, so rocky and icy grains could also form. The labels indicate the minimum radii at which grains of various types could condense out of the nebula.

and begin the accretion process, whereas accretion in the outer solar system began almost as soon as the solar nebula collapsed to a disk.

The heavier materials condensed into grains in the outer solar system, too, of course. However, there they would have been vastly outnumbered by the far more abundant light elements. The outer solar system is not deficient in heavy elements—rather, the inner solar system is *underrepresented in light material.*

Comets and Asteroids

With the formation of the four jovian planets, the remaining planetesimals were subject to those planets' strong gravitational fields. Over a period of hundreds of millions of years, and after repeated gravitational "kicks" from the giant planets, most of the interplanetary fragments in the outer solar system were flung into orbits taking them far from the Sun (Figure 4.22). Astronomers believe that those fragments now make up the Oort Cloud.

A key prediction of this theory is that some of the original planetesimals should have remained behind beyond the orbit of Neptune, in fact forming the Kuiper Belt, whose existence has been inferred from observations of short-period comets. In 1993, several asteroid-sized objects were discovered between 30 and 35 A.U. from the Sun, lending strong support to the condensation theory. Over 60 Kuiper-belt objects, having diameters ranging from 100 to 400 km, are now known. Estimates of the total number of such objects larger than 100 km run into the tens of thousands, so the combined mass of all the debris in the Kuiper Belt could be hundreds of times larger than the asteroid belt (although still less than the mass of Earth).

During this time many icy planetesimals were also deflected into the inner solar system, where they played an important role in the evolution of the terrestrial planets. A long-standing puzzle in the condensation theory's account of the formation of the inner planets was where the water and other volatile gases on Earth and elsewhere originated. At formation, the inner planets' surface temperatures were far too high, and their gravity too low, to capture and retain those gases. The answer seems to be that icy fragments—comets—from the outer solar system bombarded the newly born inner planets, supplying them with water after their formation.

The myriad rocks of the asteroid belt failed to accumulate into a protoplanet, probably because nearby Jupiter's huge gravitational field prevented them from doing so. Many of the Trojan asteroids have likely been locked in their odd orbits by the combined gravitational pulls of Jupiter and the Sun since these earliest times. The result is a band of rocky planetesimals, still colliding and occasionally fragmenting, but never coalescing into a larger body—surviving witnesses to the birth of the planets.

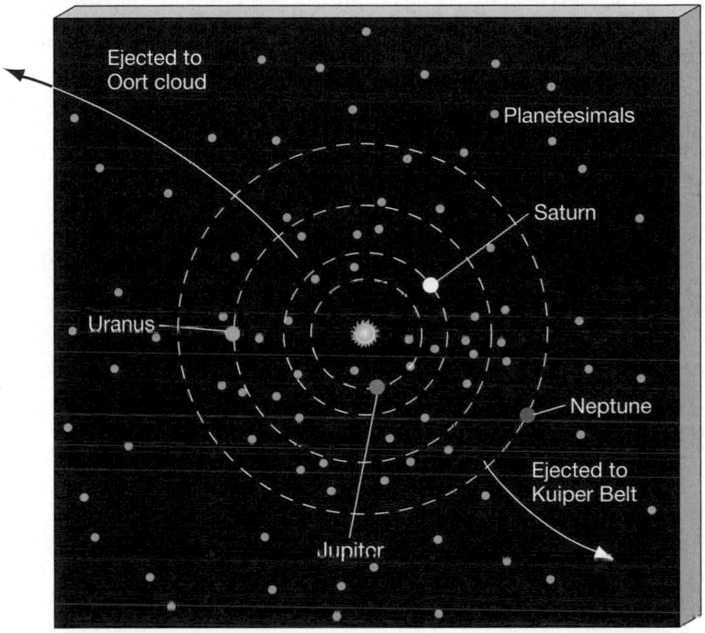

(a)

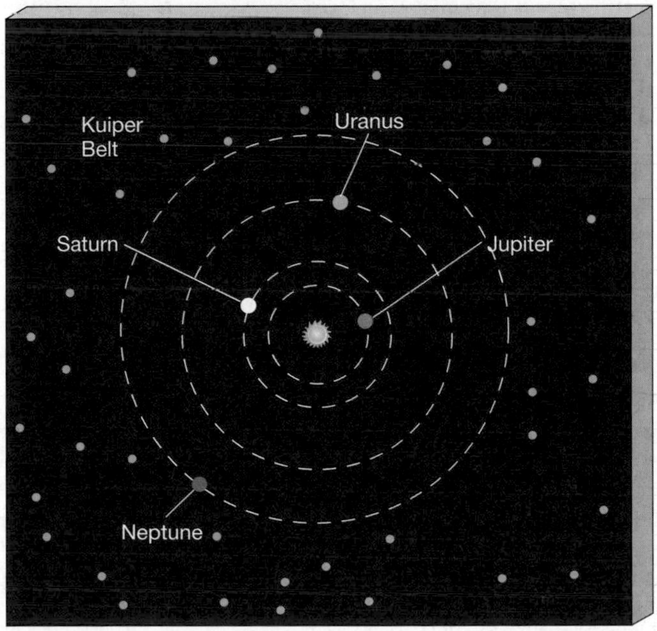

(b)

Figure 4.22 Planetesimal Ejection Ejection of planetesimals to form the Oort Cloud and Kuiper Belt.

Random Encounters in the Solar Nebula

The condensation theory accounts for the nine facts listed at the start of this section. The growth of planetesimals throughout the solar nebula, with each protoplanet ultimately sweeping up the material near it, accounts for the planets' wide spacing (point 1, although the theory does not adequately explain the regularity of the spacing). That the planets' orbits are circular (2), in the same plane (3), and in the same direction as the Sun's rotation on its axis (4) is a direct consequence of the solar nebula's shape and rotation. The rotation of the planets (5) and the orbits of the moon systems (6) are due to the tendency of smaller structures to inherit the overall sense of rotation of their parent. The heating of the nebula and the Sun's ignition resulted in the observed differentiation (7), while the debris from the accretion–fragmentation stage naturally accounts for the asteroids (8) and comets (9).

We mentioned earlier that an important aspect of any theory of solar system formation is its ability to accommodate deviations. In the condensation theory, that capacity is provided by the randomness inherent in the encounters that combined planetesimals into protoplanets. As the numbers of large bodies decreased and their masses increased, individual collisions acquired greater and greater importance. The effects of these collisions can still be seen today in many parts of the solar system. A case in point is the anomalously slow and retrograde rotation of Venus (Chapter 6), which can be explained if we assume that the last major collision in the formation history of that planet just happened to involve a near-head-on encounter between two protoplanets of comparable mass. Scientists usually do not like to invoke random events to explain observations. However, there seem to be many instances where pure chance has played an important role in determining the present state of the universe.

✓ Concept Check

- Would you expect to find asteroids and comets orbiting other stars?

Chapter Review

Summary

The planets that make up our **solar system** (p. 100) all orbit the Sun counterclockwise, as viewed from above Earth's North Pole, on roughly circular orbits that lie close to the ecliptic plane. The orbits of the innermost planet, Mercury, and the outermost, Pluto, are the most eccentric and also have the greatest orbital inclination. The spacing between planetary orbits increases as we move outward from the Sun. The four planets closest to the Sun—Mercury, Venus, Earth, Mars—are the **terrestrial planets** (p. 102). Jupiter, Saturn, Uranus, and Neptune are the **jovian planets** (p. 102). **Density** (p. 102) is a measure of the compactness of any object. The terrestrial planets are all of comparable density and generally rocky, whereas all the jovian planets have much lower densities and are made up mostly of gaseous or liquid hydrogen and helium.

Most **asteroids** (p. 103) orbit in a broad band called the **asteroid belt** (p. 103) lying between the orbits of Mars and Jupiter. The **Trojan asteroids** (p. 103) share Jupiter's orbit, remaining 60° ahead or behind that planet as it moves around the Sun. A few **Earth-crossing asteroids** (p. 103) have orbits that intersect Earth's orbit and will probably collide with our planet one day. The largest asteroids are a few hundred kilometers across. Most are much smaller.

Comets (p. 103) are fragments of icy and rocky material that normally orbit far from the Sun. We see a comet by the sunlight reflected from the dust and vapor released if its orbit happens to bring it near the Sun. The kilometer-sized **nucleus** (p. 107) of a comet is surrounded by a **coma** (p. 107) of dust and gas. Surrounding this is an extensive invisible **hydrogen envelope** (p. 107). Stretching behind the comet is a long **tail** (p. 107), formed by the interaction between the cometary material and the **solar wind** (p. 107). The **ion tail** (p. 107) consists of ionized gas particles and always points directly away from the Sun. The **dust tail** (p. 107) is less affected by the solar wind and has a curved shape. Comet orbits are generally highly elongated and not confined to the ecliptic plane. Most comets are thought to reside in the **Oort Cloud** (p. 109), a vast reservoir of cometary material tens of thousands of astronomical units across and completely surrounding the Sun. Short-period comets having an orbital period of less than about 200 years are thought to originate in the **Kuiper Belt** (p. 109), a broad band of material beyond the orbit of Neptune.

A **meteoroid** (p. 110) is a piece of rocky interplanetary debris smaller than 100 m across. Any meteoroid, asteroid, or comet fragment that enters Earth's atmosphere produces a **meteor** (p. 110), a bright streak of light across the sky. If part of the body causing the meteor reaches the ground, the remnant is called a **meteorite** (p. 110). Each time a comet rounds the Sun, some cometary material becomes dislodged, forming a **meteoroid swarm** (p. 110)—a group of small **micrometeoroids** (p. 110) that travel in the comet's orbit. Larger meteoroids are pieces of material chipped off asteroids following collisions in the asteroid belt. Most meteorites are between 4.4 and 4.6 billion years old.

The organization of the solar system points to formation as a one-time event that occurred 4.6 billion years ago. An ideal theory of the solar system must provide strong reasons for the observed characteristics of our planetary system, yet at the same time be flexible enough to allow for deviations.

A **nebula** (p. 116) is any large cloud of interstellar dust and gas. In the **nebular theory** (p. 116) of solar system formation, the **solar nebula** (p. 116) began to collapse under its own gravity. As it did so, it began to spin faster, eventually forming a disk. **Protoplanets** (p. 116) formed in the disk and became planets, while the central **protosun** (p. 118) eventually evolved into the Sun. The **condensation theory** (p. 117) incorporates into the nebular theory the effects of particles of interstellar dust, which help to cool the solar nebula and act as **condensation nuclei** (p. 118), allowing the planet-building process to begin.

Small clumps of matter grew by **accretion** (p. 118), gradually sticking together and growing into moon-sized **planetesimals** (p. 118) whose gravitational fields were strong enough to accelerate accretion. Competing with accretion in the solar nebula was **fragmentation** (p. 118), whereby small bodies were broken up following collisions with larger ones. Eventually, only a few planet-sized objects remained. The planets in the outer solar system became so large that they could capture the hydrogen and helium gas in the solar nebula, forming the jovian worlds. When the Sun became a star, its strong winds blew away any remaining gas in the solar nebula.

The terrestrial planets are rocky because they formed in the hot inner regions of the solar nebula, near the Sun. Farther out, the nebula was cooler, so water and ammonia ice could also form. Many leftover planetesimals were ejected into the Oort Cloud by the gravitational fields of the outer planets, leaving the Kuiper belt behind. The asteroid belt is a collection of planetesimals that never formed a planet, probably because of Jupiter's gravitational influence.

Review and Discussion

1. Name three differences between terrestrial and jovian planets.
2. Why are asteroids and meteoroids important to planetary scientists?
3. Do all asteroid orbits lie between Mars and Jupiter?
4. What might be the consequences of a 10-km asteroid striking Earth today?
5. What are comets like when they are far from the Sun? What happens when they enter the inner solar system?
6. What are the typical ingredients of a comet nucleus?
7. Why can comets approach the Sun from any direction but asteroids generally orbit close to the ecliptic plane?
8. Explain the difference between a meteor, a meteoroid, and a meteorite.
9. What causes a meteor shower?
10. What do meteorites reveal about the age of the solar system?
11. Describe the basic features of the nebular theory of solar system formation, and give three examples of how this theory explains some observed features of the present-day solar system.
12. What key ingredient in the modern condensation theory of solar system formation was missing from the original nebular theory?
13. Why are the jovian planets so much larger than the terrestrial planets?
14. How did the temperature at various locations in the solar nebula determine planetary composition?
15. Where did Earth's water come from?

True or False?

____ 1. Most planets orbit the Sun in nearly the same plane as Earth.

____ 2. The largest planets also have the highest densities.

____ 3. All planets have moons.

____ 4. Asteroids generally move on almost circular orbits.

____ 5. Asteroids were recently formed by the breakup of a planet orbiting in the asteroid belt.

____ 6. The Kuiper Belt is another name for the asteroid belt.

____ 7. Some comets orbit as much as 50,000 A.U. from the Sun.

____ **8.** Most comets have short periods and orbit close to the ecliptic plane.

____ **9.** Comet tails always lie along the comet's orbit.

____ **10.** The light emitted from a meteor is due to a meteoroid burning up in Earth's atmosphere.

____ **11.** Comets are the sources of meteor showers.

____ **12.** The direction of planetary revolution is the same as the direction of the Sun's rotation.

____ **13.** The solar system is highly differentiated.

____ **14.** The terrestrial planets formed out of the original dust that made up the solar nebula.

____ **15.** Accretion occurred faster in the inner part of the solar system than in the outer regions.

Fill in the Blank

1. The two planets with the highest eccentricities and orbital tilts are _______ and _______.
2. Asteroids and meteoroids are often _______ in composition, in contrast to comets, which are generally _______ in composition.
3. Most asteroids are found between the orbits of _______ and _______.
4. The largest asteroids are _______ kilometers in diameter; the smallest are only _______ meters across.
5. The Trojan asteroids share an orbit with _______.
6. The nucleus of a comet is typically _______ kilometers across, and its tail may be up to _______ long.
7. Passage of a comet near the Sun may leave a _______ moving in the comet's orbit.
8. When a meteoroid, asteroid, or comet fragment enters Earth's atmosphere, you see a _______.
9. The oldest meteorites are _______ years old.
10. By the time planetesimals had formed, the accretion process was accelerated by the effect of _______.
11. In the final stage of accretion, the largest protoplanets were able to attract large quantities of _______ from the solar nebula.
12. Unlike the planetesimals that formed the terrestrial planets, those that formed the jovian planets were made up of _______ material.
13. The large number of left-over planetesimals formed beyond 5 A.U. were destined to become _______.
14. The reason the planetesimals of the asteroid belt did not form a larger object was probably the gravitational influence of _______.
15. Angular momentum depends on the mass, radius, and _______ of an object.

Problems

1. Only Mercury, Mars, and Pluto have orbits that deviate significantly from circles. Calculate the perihelion and aphelion distances from the Sun (see *More Precisely 1-1* and Table 1.1) of these planets.
2. Suppose the average mass of each of 10,000 asteroids in the solar system is 10^{17} kg. Compare the total mass of these asteroids to the mass of Earth. Assuming a spherical shape and a density of 3000 kg/m^3, estimate the diameter of an asteroid having this average mass.
3. The asteroid Icarus (Figure 4.4) has a perihelion distance of 0.2 A.U. and an orbital eccentricity of 0.7. What is its semi-major axis and aphelion distance from the Sun?
4. The largest asteroid, Ceres, has a radius 0.073 times the radius of Earth and a mass of 0.0002 Earth masses. How much would a 80-kg astronaut weigh on Ceres?
5. The number of asteroids (or meteoroids) of a given diameter is roughly inversely proportional to the square of the diameter. Assuming a density of 3000 kg/m^3, and approximating the actual distribution of asteroids and meteoroids as a single 1000-km body (Ceres), one hundred 100-km bodies, ten thousand 10-km bodies, and so on, calculate the total mass in the form of 1000-km bodies, 100-km bodies, 10-km bodies, and 1-km bodies.
6. (a) Using the version of Kepler's laws of planetary motion from Section 1.4, calculate the orbital period of an Oort-Cloud comet if the semi-major axis of the comet's orbit is 50,000 A.U. (b) What is the maximum possible aphelion distance for a short-period comet with an orbital period of 200 years?

7. The orbital angular momentum of a planet in a circular orbit is the product of mass times orbital speed times distance from the Sun. (a) Compare the orbital angular momenta of Jupiter, Saturn, and Earth. (b) What is the orbital angular momentum of an Oort-Cloud comet of mass 10^{13} kg, moving in a circular orbit 50,000 A.U. from the Sun?

8. Consider a planet growing by accretion of material from the solar nebula. As it grows, its density remains roughly constant. By what factor will the planet's surface gravity change as the radius doubles?

9. How many 100-km-diameter rocky (3000 kg/m^3) planetesimals would have been needed to form Earth?

10. A typical comet contains some 10^{13} kg of water ice. How many comets would have to strike Earth in order to account for the roughly 2×10^{21} kg of water presently found on our planet? If this amount of water accumulated over a period of 0.5 billion years, how frequently would Earth be hit by a comet during that time?

Projects

1. The only way to tell an asteroid from a star is to watch over several nights so that you can detect the asteroid's movement in front of the star background. The astronomy magazines *Sky & Telescope* and *Astronomy* often publish charts for prominent asteroids, the three brightest of which are Ceres, Pallas, and Vesta. Aiming binoculars at that star field, you should be able to pick out the asteroid from its location in the chart. Come back a night or two later, and look again. The "star" that has moved is the asteroid.

2. Although a spectacular naked-eye comet comes along only about once a decade, fainter comets can be seen with binoculars and telescopes in the course of every year. *Sky & Telescope* often runs a "Comet Digest" column announcing the whereabouts of comets. In addition, a comprehensive list of periodic comets expected to return in a given year can be found in Guy Ottewell's *Astronomical Calendar*. This calendar contains a wealth of other sky information as well, including monthly star charts. At the time of this writing, it costs $15 a year and can be purchased from Astronomical Workshop, Furman University, Greenville, South Carolina 29613 (803-294-2208).

3. There are a number of major meteor showers every year, but if you plan to watch one, be sure to notice the phase of the Moon because bright moonlight or city lights can obliterate a meteor shower. Most meteors don't become visible until they are 20 or 30 degrees from the radiant. They can appear in all parts of the sky! Just relax and let your eyes rove among the stars. You will generally see many more meteors in the hours before dawn than in the hours after sunset. Why do you suppose meteors have different brightnesses? Can you detect their variety of colors? Watch for meteors that appear to "explode" as they fall, and for vapor trails that linger after the meteor has disappeared.

A
B
C
D
C
B
A

5 EARTH AND ITS MOON

Our Cosmic Backyard

LEARNING GOALS

Studying this chapter will enable you to:

1 Summarize and compare the basic properties of Earth and the Moon.

2 Describe the consequences of gravitational interactions between Earth and the Moon.

3 Discuss how Earth's atmosphere helps heat us as well as protect us.

4 Outline our current model of Earth's interior structure and describe some experimental techniques used to establish this model.

5 Summarize the evidence for continental drift and discuss the physical processes that drive it.

6 Explain how dynamic events early in the Moon's history formed its major surface features.

7 Describe the nature and origin of Earth's magnetosphere.

8 Discuss the formation and evolution of Earth and the Moon.

(Opposite page, background) The *Apollo 15* mission to the Moon explored a geological fault, called Hadley Rille, where molten lava once flowed. In the large photograph printed here, the rille—a system of valleys—runs along the base of the Apennine Mountains (lower right) at the edge of Mare Imbrium (to the left). For scale, the lower of the two large craters at top left, Autolycus, spans 40 km. The shadow-sided, most prominent peak at lower right, Mount Hadley, rises almost 5 km high.

(Inset A) An astronaut embarks from his lunar rover, preparing to explore Hadley Rille in the distance. The width of the rille is about 1.5 km and its depth averages 300 m.

(Inset B) Photograph of an astronaut and a lunar rover in front of Mount Hadley.

(Inset C) An astronaut collects samples at the edge of Hadley Rille.

(Inset D) This photographic mosaic of a small part of an interior wall of Hadley Rille shows evidence for subsurface horizontal layering. Each distinct layer (denoted A, B, C) is a few meters deep and presumably represents successive lava flows that helped form the extensive lava plain called Mare Imbrium. *(NASA)*

If we are to appreciate the universe, we must first come to know our own home. By cataloging Earth's properties and attempting to explain them, we set the stage for a comparative study of all the other planets. From an astronomical perspective, this is a compelling reason to study the structure and history of our own world.

The Moon is by far our closest neighbor in space. Yet, despite its nearness, it is a world very different from our own. It has no air, no sound, no water. Weather is nonexistent; clouds, rainfall, and blue sky are all absent, boulders and pulverized dust litter the landscape. Why then do we study it? In part, simply because it is our nearest neighbor and dominates our night sky. Beyond that, however, we study this body because it holds important clues to our own past. The very fact that it hasn't changed much since its formation means the Moon is a vital key to unlocking the secrets of the solar system.

5.1 Earth and the Moon in Bulk

Physical Properties

1 Table 5.1 presents some basic properties of Earth and the Moon. The first five columns list mass, radius, and average density for Earth and the Moon, determined as described in Chapter 4. ∞ (Sec. 4.1) The next two columns present important measures of a body's gravitational field. **Surface gravity** is the strength of the gravitational force at the body's surface. ∞ (*More Precisely 1-2*) **Escape speed** is the speed required for any object—an atom, a baseball, or a spaceship—to escape forever from the body's gravitational pull (see *More Precisely 5-1*). By either measure, the Moon's gravitational pull is much weaker than Earth's. As we will see, this fact has played an important role in determining the very different evolutionary paths followed by the two worlds. The final column in Table 5.1 lists (sidereal) rotation periods, which have long been accurately known from Earth-based observations.

These data will form the basis for our comparative study of our home planet and its nearest neighbor. We will expand our catalog of "planetary" characteristics and peculiarities in the next three chapters as we study in turn the other members of the solar system.

One important quantity not listed in Table 5.1 is the *distance* from Earth to the Moon. Parallax, with Earth's diameter as a baseline, has long provided astronomers with reasonably accurate measurements of the Earth–Moon distance. ∞ (Sec. P.5) However, radar ranging and laser ranging now yield far more accurate results (in fact, the Moon's precise distance from Earth at any instant is known to within a few centimeters). The semi-major axis of the Moon's orbit around Earth is 384,000 km.

Overall Structure

Figure 5.1 compares and contrasts the main regions of these two very dissimilar worlds. As indicated in Figure 5.1(a), our planet may be divided into six main regions. In Earth's interior, a thick **mantle** surrounds a smaller, two-part **core**. At the surface we have a relatively thin **crust**, comprising the solid continents and the seafloor, and the **hydrosphere**, comprising rivers, lakes and the liquid oceans. An **atmosphere** of air lies just above the surface. At much greater altitudes, a zone of charged particles trapped by our planet's magnetic field forms Earth's **magnetosphere**.

The Moon lacks a hydrosphere, an atmosphere, and a magnetosphere. Its internal structure (Figure 5.1b) is not as well studied as that of Earth, for the very good reason that the Moon is much less accessible. Nevertheless, as we will see, the same basic interior regions as found on Earth—*crust*, *mantle*, and *core*—may be dis-

TABLE 5.1 Some Properties of Earth and the Moon

	MASS (kg)	MASS (Earth = 1)	RADIUS (km)	RADIUS (Earth = 1)	AVERAGE DENSITY (kg/m^3)	SURFACE GRAVITY (Earth = 1)	ESCAPE SPEED (km/s)	ROTATION PERIOD
Earth	6.0×10^{24}	1.00	6400	1.00	5500	1.00	11.2	$23^h 56^m$
Moon	7.3×10^{22}	0.012	1700	0.27	3300	0.17	2.4	27.3 days

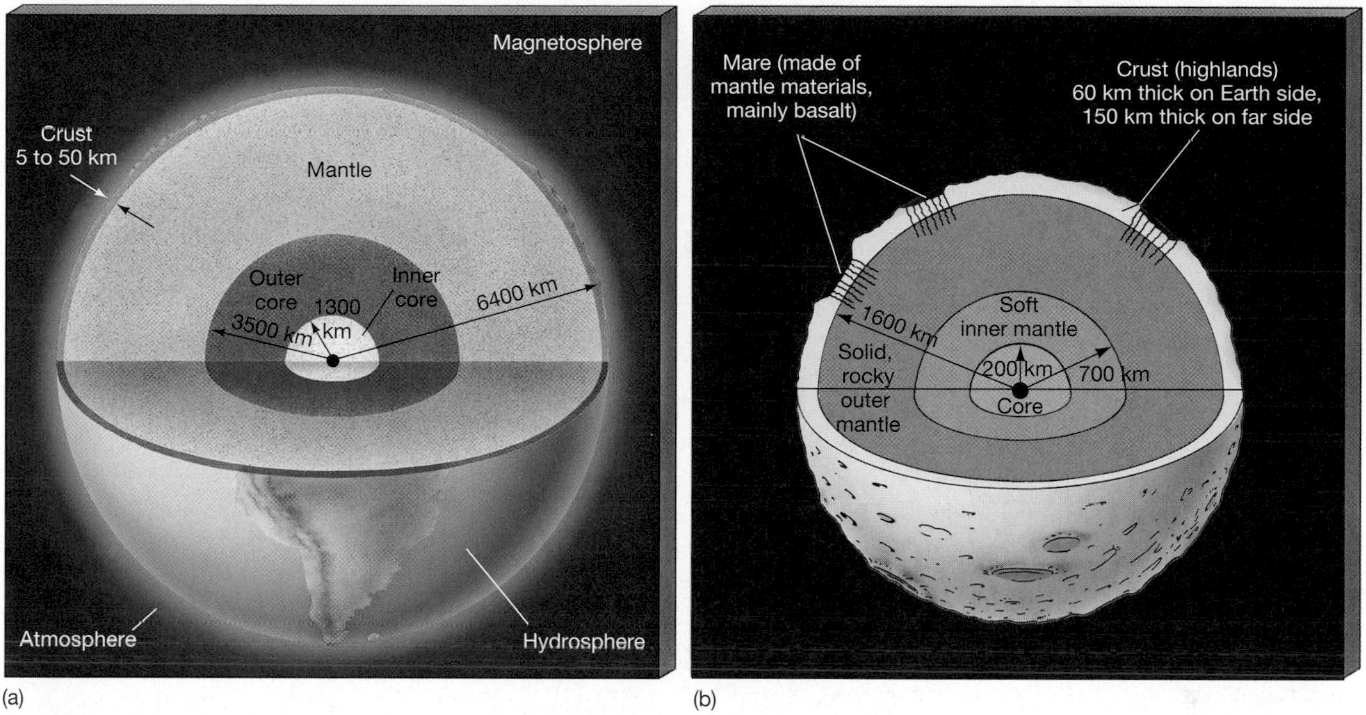

Figure 5.1 Earth and Moon The main regions of Earth and the Moon (not drawn to scale). (a) At Earth's center is an inner core, with a radius of about 1300 km. Surrounding the inner core is a liquid outer core 3500 km in radius. Most of the rest of Earth's interior is taken up by the mantle (radius 6400 km), which is topped by a thin crust only a few tens of kilometers thick. The liquid portions of Earth's surface make up the hydrosphere. Above the hydrosphere and solid crust lies the atmosphere, most of it within 50 km of the surface. Earth's outermost region is the magnetosphere, extending thousands of kilometers into space. (b) The Moon's rocky outer mantle is about 900 km thick. The inner mantle is a semisolid layer similar to the upper regions of Earth's mantle. At the center lies the core, which may be partly molten. The surface mare and highland regions are discussed in Section 5.6.

cerned for the Moon, although their properties differ somewhat from those of their Earthly counterparts.

5.2 The Tides

Earth is unique among the planets in that it has large quantities of liquid water on its surface. Approximately three-quarters of Earth's surface is covered by water, to an average depth of some 3.6 km. A familiar hydrospheric phenomenon is the daily fluctuation in ocean level known as the **tides**. At most coastal locations on Earth, there are two low tides and two high tides each day. The "height" of the tides—the magnitude of the variation in sea level—can range from a few centimeters to many meters, depending on the location and time of year, averaging about a meter on the open sea. An enormous amount of energy is contained in the daily tidal motion of the oceans.

Gravitational Deformation

2 What causes the tides? A clue comes from the fact that they exhibit daily, monthly, and yearly cycles. In fact, the tides are a direct result of the gravitational influence of the Moon and the Sun on Earth. We have already seen how gravity keeps Earth and the Moon in orbit about one another, and both in orbit around the Sun. For simplicity, let's first consider just the interaction between Earth and the Moon.

Recall that the strength of the gravitational force depends on the inverse square of the distance separating any two objects. ∞ (Sec. 1.4) Thus, the Moon's gravitational attraction is greater on the side of Earth that faces the Moon than on the opposite side, some 12,800 km (Earth's diameter) farther away. This difference in the gravitational force is small, only about three percent, but it produces a noticeable effect—a **tidal bulge**. Earth's oceans undergo the greatest deformation because

MORE PRECISELY 5-1

Tidal Forces

Consider the gravitational force exerted on Earth by the Moon. As discussed in the text, the *tidal* force is the name we give to the *variation* of the Moon's gravitational pull from one part of Earth to another. Specifically, as illustrated in Figure 5.2(b), it is the *difference* between the actual force exerted on a chunk of Earth matter and the force exerted on the same chunk if it were placed at Earth's center.

With the help of a little calculus, we can write down an equation for the tidal force exerted by the Moon (of mass M and distance D) on a body (of mass m) placed at various locations on Earth's surface (of radius R). When the body is placed along the line joining Earth to the Moon, the tidal force is directed *away* from Earth's center (that is, vertically up), and has magnitude

$$\text{tidal force} = 2\,\frac{G\,M\,m\,R}{D^3}.$$

When the body is placed on a line through Earth's center perpendicular to the Earth–Moon line, the force has a magnitude one-half of the above value, and is directed *toward* Earth's center (that is, vertically down). Tidal forces tend to stretch bodies along the line joining them and compress them perpendicular to that line. The result is the characteristic tidal bulge shown in Figure 5.2.

Note that, unlike the familiar inverse-square law of gravity, the tidal force falls off as the inverse *cube* of the distance between the bodies involved. This rapid decline with increasing distance means that an object has to be very close, or very massive, in order to have a significant tidal effect on another. As an example, let's calculate the tidal forces exerted by first the Moon and then the Sun on a 1-kg mass placed on Earth's surface directly between the centers of Earth and the other body (i.e., at the peak of the bulge). For comparison, Earth's gravitational pull on the body (the body's *weight*) is $GM_{Earth} \times 1\text{ kg} / R_{Earth}{}^2 \approx 9.8$ newton (using the numbers in *More Precisely 1-2* and Table 5.1; recall that the newton is the SI unit of force).

The above equation tells us that the tidal force due to the Moon is $2GM_{Moon}R_{Earth} / D^3{}_{Moon} \approx 1.1 \times 10^{-6}$ newton—tiny compared to Earth's pull, but nevertheless important, as we have seen. For the Sun, the result is $2GM_{Sun} R_{Earth} / D^3{}_{Sun} \approx 5.1 \times 10^{-7}$ newton. Even though the Sun is around 375 times farther away from Earth than is the Moon, the Sun's mass is so much greater (by about a factor of 27 million) that its tidal influence is still significant—about half that of the Moon.

---

liquid can most easily move around on our planet's surface. As illustrated in Figure 5.2, the ocean becomes a little deeper in some places (along the line joining Earth to the Moon) and shallower in others (along the line perpendicular to the Earth–Moon line). The daily tides we see result as Earth rotates beneath this deformation.

The variation of the Moon's (or the Sun's) gravity across Earth is an example of a *differential force*, or **tidal force**. The *average* gravitational interaction between two bodies determines their orbit around one another. However, the tidal force, superimposed on that average, tends to deform the bodies. Tidal forces are discussed in more detail in *More Precisely 5-1*. We will see many situations in this book where such forces are critically important in understanding astronomical phenomena. Notice that we still use the word *tidal* in these other contexts, even though we are not discussing oceanic tides, and possibly not even planets at all.

Notice in Figure 5.2 that the side of Earth opposite the Moon also experiences a tidal bulge. The different gravitational pulls—greatest on that part of Earth closest to the Moon, weaker at Earth's center, and weakest of all on Earth's opposite side—cause average tides on opposite sides of our planet to be approximately equal in height. On the side nearer the Moon, the ocean water is pulled slightly toward the Moon. On the opposite side, the ocean water is "left behind" as Earth is pulled closer to the Moon. Thus, high tide occurs *twice*, not once, every day.

Both the Moon and the Sun exert tidal forces on our planet. Thus, instead of one tidal bulge being created on Earth, there are two—one pointing toward the Moon, the other toward the Sun—and the interaction between them accounts for the changes in the height of the tides over the course of a month or a year. When Earth, Moon, and Sun are roughly lined up—at new or full Moon—the gravitational effects reinforce one another, and the highest tides occur (Figure 5.3a). These tides are known as *spring tides*. When the Earth–Moon line is perpendicular to the Earth–Sun line (at the first and third quarters), the daily tides are smallest (Figure 5.3b). These are termed *neap tides*.

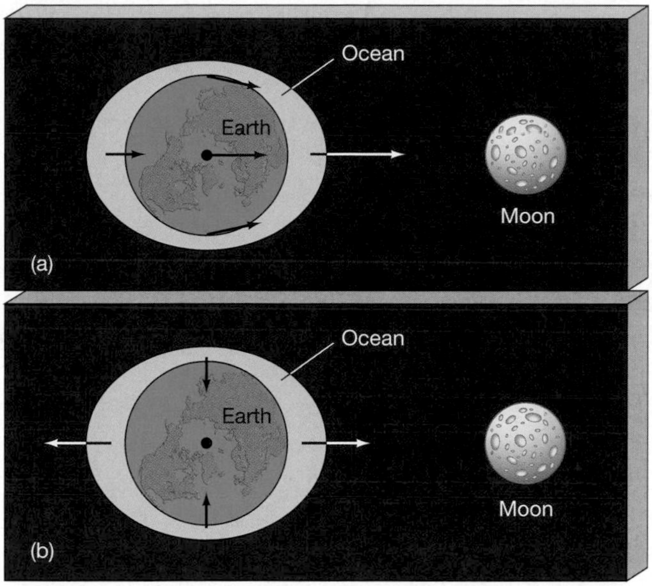

Figure 5.2 Lunar Tides This exaggerated illustration shows how the Moon induces tides on both the near and far sides of Earth. The lengths of the arrows indicate the relative strengths of the Moon's gravitational pull on various parts of Earth. (a) The lunar gravitational forces acting on several locations on and inside Earth. The force is greatest on the side nearest the Moon and smallest on the opposite side. (b) The difference between the lunar forces experienced at the locations shown in part (a) and the force acting on Earth's center. The arrows represent the force with which the Moon tends to either pull matter away from or squeeze it toward Earth's center.

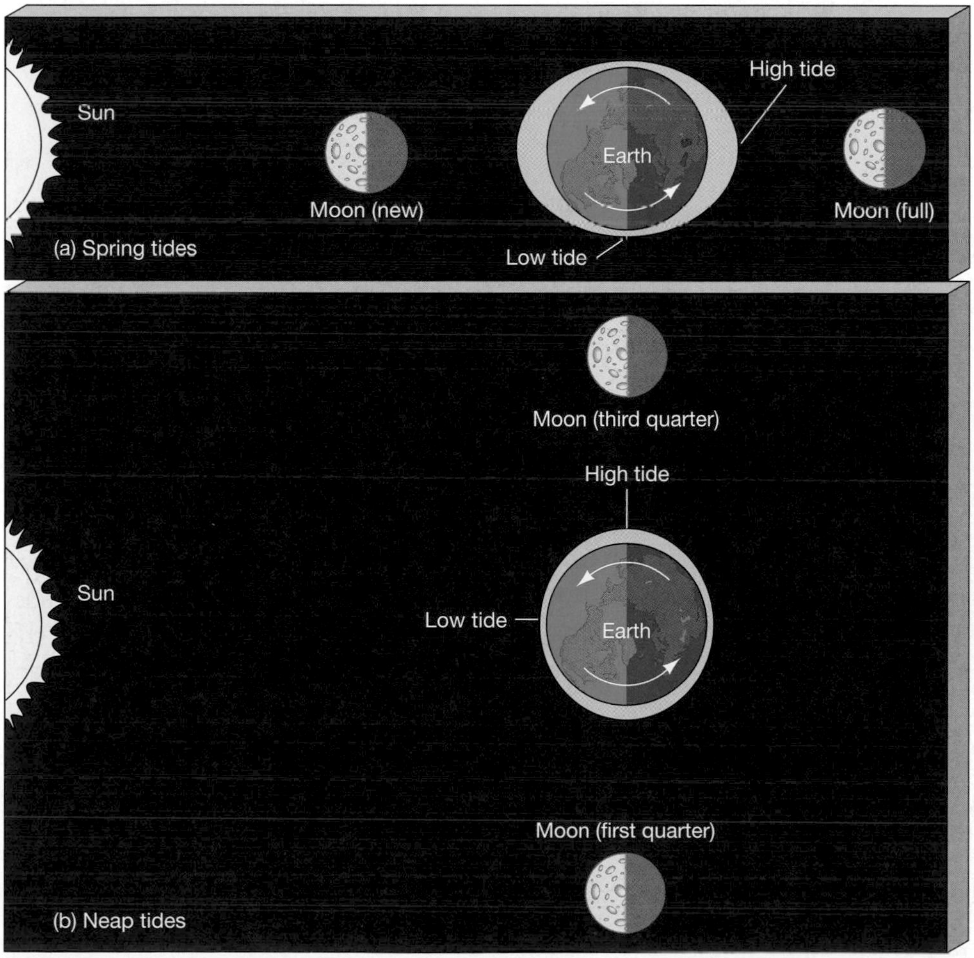

Figure 5.3 Solar and Lunar Tides The combined effects of the Sun and the Moon produce variations in high and low tides. (a) When the Moon is either full or new, Earth, Moon, and Sun are approximately aligned, and the tidal bulges raised in Earth's oceans by the Moon and the Sun reinforce one another. (b) When the Moon is in either its first or its third quarter, the tidal effects of the Moon and the Sun partially cancel each other, and the tides are smallest. Because the Moon's tidal effect is greater than that of the Sun, the net bulge points toward the Moon.

Tidal Locking

The Moon rotates once on its axis in 27.3 days—exactly the same time as it takes to complete one revolution around Earth. (Sec. P.4) As a result, the Moon presents the *same* face toward Earth at all times; that is, the Moon has a "near" side, which is always visible from Earth, and a "far" side, which never is. This condition, in which the rotation period of a body is precisely equal to (or *synchronized* with) its orbital period around another body, is known as a **synchronous orbit**.

The fact that the Moon is in a synchronous orbit around Earth is no accident. It is an inevitable consequence of the tidal gravitational interaction between those two bodies. To see how this can occur, let's continue our discussion of tides on Earth. Because of Earth's rotation, the tidal bulge raised by the Moon does not point directly at the Moon, as indicated in Figure 5.2. Instead, through the effects of friction, both between the crust and the oceans and within Earth's interior, Earth's rotation tends to drag the tidal bulge around with it, causing the bulge to be displaced slightly ahead of the Earth–Moon line (Figure 5.4). The Moon's asymmetrical gravitational pull on this slightly offset bulge acts to slow our planet's rotation.

According to fossil measurements, the rate of decrease is just two milliseconds every century— not much on the scale of a human lifetime, but it means that half a billion years ago, the day was just over 21 hours long and the year contained 410 days. At the same time, the Moon spirals slowly away from Earth, increasing its average distance from our planet by about 4 cm per century. This process will continue until (billions of years from now) Earth rotates on its axis at exactly the same rate as the Moon orbits Earth—that is, Earth's rotation will become synchronized with the Moon's motion. At that time the Moon will always be above the same point on Earth and will no longer lag behind the bulge it raises.

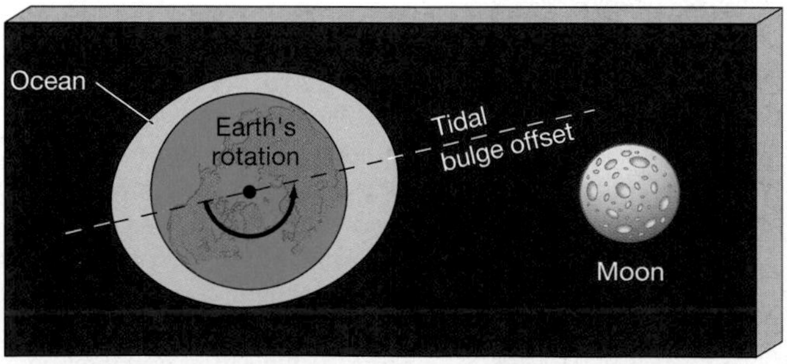

Figure 5.4 Tidal Bulge The tidal bulge raised in Earth by the Moon does not point directly at the Moon. Instead, because of the effects of friction, the bulge points slightly "ahead" of the Moon, in the direction of Earth's rotation. (The magnitude of the effect is greatly exaggerated in this diagram.) Because the Moon's gravitational pull on the near-side part of the bulge is greater than the pull on the far side, the overall effect is to decrease Earth's rotation rate.

Basically the same process is responsible for the Moon's synchronous orbit. Just as the Moon raises tides on Earth, Earth also produces a tidal bulge in the Moon. Indeed, because Earth is so much more massive than the Moon, the lunar tidal bulge is considerably larger (*More Precisely 5-1*), and the synchronization process correspondingly faster. The Moon's orbit became synchronous long ago—the Moon is said to have become *tidally locked* to Earth. Most of the moons in the solar system are similarly locked by the tidal fields of their parent planets.

Concept Check

- In what ways do tidal forces differ from the familiar inverse-square force of gravity?

5.3 Atmospheres

Earth's Atmosphere

Our planet's atmosphere is a mixture of gases, the most abundant of which are nitrogen (78 percent by volume), oxygen (21 percent), argon (0.9 percent), and carbon dioxide (0.03 percent). Water vapor is a variable constituent of the atmosphere, making up anywhere from 0.1 to three percent, depending on location and climate. The presence of a large amount of free oxygen makes our atmosphere unique in the solar system—Earth's oxygen is a direct consequence of the emergence of life on our planet. We will discuss the formation and evolution of planetary atmospheres in more detail in Chapter 6, when we consider the terrestrial planets as a whole.

Figure 5.5 shows a cross section of Earth's atmosphere. Compared with Earth's overall dimensions, the extent of the atmosphere is not great. Half of it lies within 5 km of the surface, and all but 1 percent of it is found below 30 km. The region below about 12 km is called the **troposphere**. Above it, extending up to an altitude of 40 to 50 km, lies the **stratosphere**. Between 50 and 80 km from the surface lies the **mesosphere**. Above about 80 km, in the **ionosphere**, the atmosphere is kept partly ionized by solar ultraviolet radiation. These various

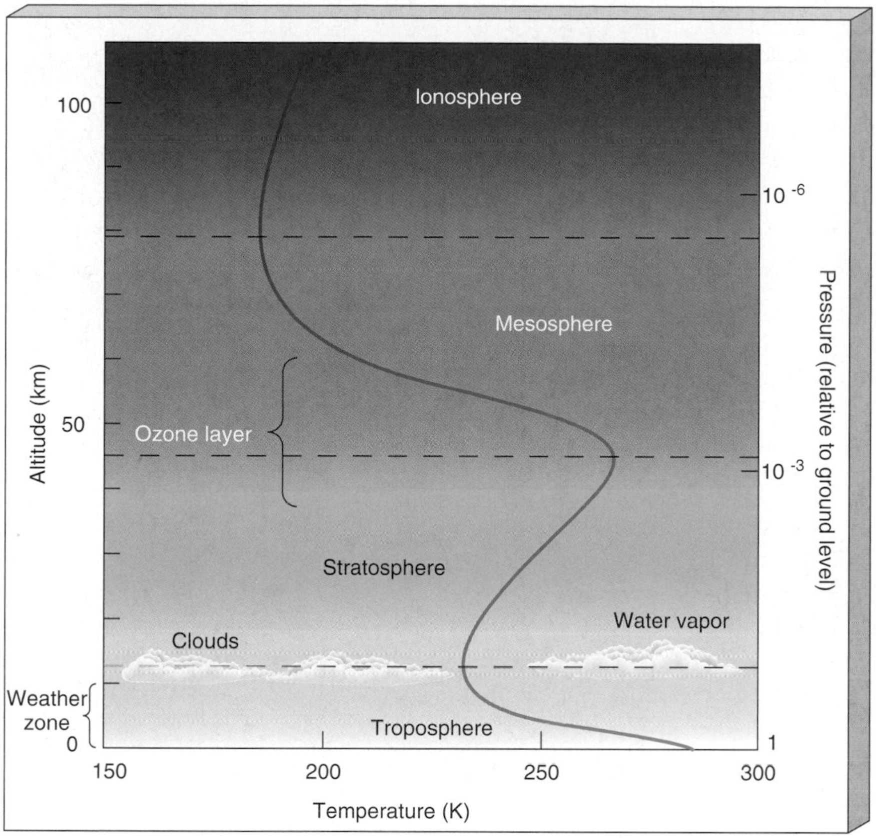

Figure 5.5 **Earth's Atmosphere** Diagram of Earth's atmosphere, showing the changes in temperature (blue line, bottom axis) and pressure (right-hand axis) from the surface to the lower part of the ionosphere. Pressure decreases steadily with increasing altitude, but the temperature may fall or rise, depending on height above the ground.

atmospheric regions are distinguished from one another by the behavior of the temperature (decreasing or increasing with altitude) in each. Atmospheric pressure decreases steadily with increasing altitude.

The troposphere is the region of Earth's (or any other planet's) atmosphere where **convection** occurs. Convection is the constant upwelling of warm air and the concurrent downward flow of cooler air to take its place. In Figure 5.6, part of Earth's surface is heated by the Sun. The air immediately above the warmed surface is heated, expands a little, and becomes less dense. As a result, it becomes buoyant and starts to rise. At higher

Figure 5.6 **Convection** Convection occurs whenever cool matter overlies warm matter. The resulting circulation currents are familiar to us as the winds in Earth's atmosphere, caused by the solar-heated ground. Over and over, hot air rises, cools, and falls back to Earth. Eventually, steady circulation patterns are established and maintained, provided the source of heat remains intact.

altitudes, the opposite effect occurs: The air gradually cools, grows denser, and sinks back to the ground. The cool air at the surface rushes in to replace the hot buoyant air that has risen. In this way, a circulation pattern is established. These *convection cells* of rising and falling air contribute to atmospheric heating and are responsible for surface winds and all the weather we experience. Above the troposphere, the atmosphere is stable and the air is calm.

Straddling the boundary between the stratosphere and the mesosphere is the **ozone layer**, where incoming solar ultraviolet radiation is absorbed by atmospheric oxygen, ozone, and nitrogen. (Ozone is a form of oxygen, consisting of three oxygen atoms combined into a single molecule.) The ozone layer is one of the insulating layers that protect life on Earth from the harsh realities of outer space. By absorbing potentially dangerous high-frequency radiation, it acts as a planetary umbrella. Without it, advanced life (at least on Earth's surface) would be at best unlikely and at worst impossible.

The Greenhouse Effect

3 Most of the Sun's energy is emitted in the visible and near-infrared (that is, wavelengths only slightly longer than red light) regions of the electromagnetic spectrum. Because Earth's atmosphere is largely transparent to radiation of this type, almost all of the solar radiation not absorbed by or reflected from clouds in the upper atmosphere shines directly onto Earth's surface, heating it up. ∞ (Sec. 2.3)

Earth's surface reradiates the absorbed energy. As the temperature rises, the amount of radiated energy increases rapidly, according to Stefan's law, and eventually our planet radiates as much energy back into space as it receives from the Sun. In the absence of any complicating effects, this balance would be achieved at an average surface temperature of about 250 K (−23°C). At that temperature, Wien's law tells us that most of the reemitted energy is in the form of far-infrared (roughly 10 μm) radiation. ∞ (*More Precisely 2-2*)

However, there is a complication. Long-wavelength infrared radiation is partially blocked by Earth's atmosphere, mainly because of carbon dioxide and water vapor, both of which absorb very efficiently in the infrared portion of the spectrum. Even though these two gases are only trace constituents of our atmosphere, they manage to absorb a large fraction of all the infrared radiation emitted from the surface. Consequently, only some of that radiation escapes back into space. The rest is radiated back to the surface, causing the temperature to rise.

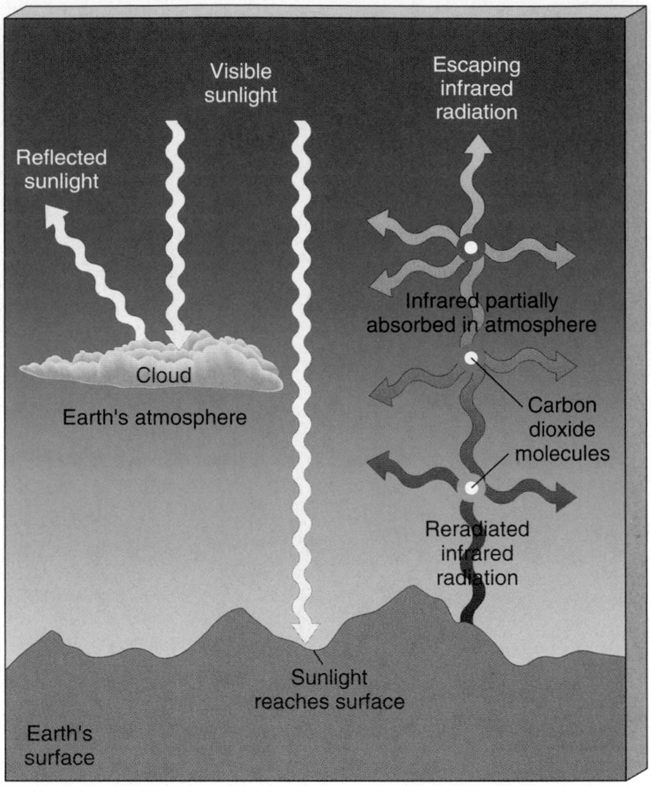

Figure 5.7 Greenhouse Effect Sunlight that is not reflected by clouds reaches Earth's surface, warming it up. Infrared radiation reradiated from the surface is partially absorbed by carbon dioxide (and also water vapor, not shown here) in the atmosphere, causing the overall surface temperature to rise.

This partial trapping of solar radiation is known as the **greenhouse effect**. The name comes about because a similar process operates in a greenhouse—sunlight passes relatively unhindered through glass panes, but much of the infrared radiation reemitted by the plants is blocked by the glass and cannot get out.[1] Consequently, the interior of the greenhouse heats up, and flowers, fruits, and vegetables can grow even on cold winter days. The radiative processes that determine the temperature of Earth's atmosphere are illustrated in Figure 5.7. Earth's greenhouse effect makes our planet about 40 K hotter than would otherwise be the case—a very important difference, as it raises the average temperature *above* the freezing point of water.

[1]Although this process does contribute to warming the interior of a greenhouse, it is not the most important effect. A greenhouse works mainly because its glass panes prevent convection from carrying heat up and away from the interior. Nevertheless, the name "greenhouse effect" to describe the heating effect due to Earth's atmosphere has stuck.

The magnitude of the greenhouse effect is very sensitive to the concentration of *greenhouse gases* (that is, gases that absorb infrared radiation efficiently) in the atmosphere. Of greatest importance among these is carbon dioxide, although water vapor also plays a significant role. The amount of carbon dioxide in Earth's atmosphere is increasing, largely as a result of the burning of fossil fuels (principally oil and coal) in the industrialized world. Carbon dioxide levels have increased by more than 20 percent in the last century, and they continue to rise at a present rate of four percent per decade. Many scientists believe that this increase, if left unchecked, may result in global temperature increases of several kelvins over the next half-century, enough to melt much of the polar ice caps and cause dramatic, perhaps even catastrophic, changes in Earth's climate.

Lunar Air?

What about the Moon's atmosphere? That's easy—for all practical purposes, there is none! All of it escaped long ago. More massive objects have a better chance of retaining their atmospheres because the more massive an object, the greater the speed needed for atoms and molecules to escape (see *More Precisely* 5-2). The Moon's escape speed is only 2.4 km/s, compared with 11.2 km/s for Earth (Table 5.1). Simply put, the Moon has a lot less pulling power—any atmosphere it might once have had is gone forever.

Lacking the moderating influence of an atmosphere, the Moon experiences wide variations in surface temperature. Noontime temperatures can reach 400 K, well above the boiling point of water (373 K). At night (which lasts nearly 14 Earth days) or in the shade, temperatures fall to about 100 K, well below water's freezing point (273 K). With such high daytime temperatures and no atmosphere to help retain it, any water that ever existed on the Moon's surface has most likely evaporated and escaped. Not only is there no lunar hydrosphere, but all the lunar samples returned by the American and Soviet Moon programs were absolutely bone dry. Lunar rock doesn't even contain minerals containing water molecules locked within their crystal structure. Terrestrial rocks, conversely, are almost always one or two percent water.

The Moon is not entirely devoid of water, however. In 1994 radar data from the U.S. *Clementine* spacecraft suggested the presence of water ice at the lunar poles. Since the Sun never rises more than a few degrees above the horizon, as seen from these regions, temperatures on the permanently shaded floors of craters near the poles never exceed about 50 K. Consequently, scientists had theorized, ice there could have remained permanently frozen since the very early days of the solar system, never melting or vaporizing and hence never escaping into space. In March, 1998, sensitive equipment on board NASA's *Lunar Prospector* spacecraft detected large amounts of water ice—perhaps hundreds of millions of tons—in the form of tiny crystals mixed with the lunar dust.

✓ Concept Check

- Why is the greenhouse effect important for life on Earth?

5.4 Interiors

Seismology

4 Although we reside on Earth, we cannot easily explore our planet's interior because drilling gear can penetrate rock only so far before breaking. No substance—not even diamond, the hardest known material—can withstand the conditions below a depth of about 10 km. That's rather shallow compared with Earth's 6400-km radius. Fortunately, geologists have developed techniques that can indirectly probe the deep recesses of our planet.

A sudden dislocation of rocky material near Earth's surface—an **earthquake**—causes the entire planet to vibrate a little. It literally rings like a bell (but one tuned so low that human ears cannot detect the sound). These vibrations are not random. They are systematic waves, called **seismic waves** (after the Greek word for "earthquake"), that move outward from the site of the quake. Like all waves, they carry information. This information can be detected and recorded using sensitive equipment—a *seismograph*—designed to monitor Earth tremors.

Decades of earthquake research have demonstrated several types of seismic waves. The speed of each type depends on the density and physical state of the matter through which it is traveling. Consequently, if we can measure the time taken for waves to move from the site of an earthquake to one or more monitoring stations on Earth's surface, we can infer the density of matter in the interior.

MORE PRECISELY 5-2

Why Air Sticks Around

Why does Earth have an atmosphere but the Moon does not? Why doesn't our atmosphere disperse into space? The answer is that *gravity* holds it down. However, gravity is not the only influence acting, for if it were, all of Earth's air would have fallen to the surface long ago. *Heat* competes with gravity to keep the atmosphere buoyant. Let's explore this competition between gravity and heat in a little more detail.

All gas molecules are in constant random motion. The temperature of any gas is a direct measure of this motion—the hotter the gas, the faster the molecules are moving. (*More Precisely 2-1*) The rapid movement of heated molecules creates *pressure* which tends to oppose the force of gravity, preventing our atmosphere from collapsing under its own weight.

An important measure of the strength of a body's gravity is its *escape speed*—the speed needed for any object to escape forever from its surface. This speed increases with increased mass or decreased radius of the parent body. In fact, the escape speed is proportional to the speed of a circular orbit at the body's surface. (*More Precisely 1-2*) In convenient units, it can be expressed as

$$\text{escape speed (km/s)} = 11.2\sqrt{\frac{\text{mass of body (Earth masses)}}{\text{radius of body (Earth radii)}}}.$$

For example, if the *mass* of the parent body is quadrupled, the escape speed doubles. If the parent body's *radius* quadruples, then the escape speed is halved.

To determine whether or not a planet will retain an atmosphere, we must compare the planet's escape speed with the average speed of the gas particles making up the atmosphere. This speed depends not only on the temperature of the gas but also on the mass of the individual molecules—the hotter the gas or the smaller the molecular mass, the higher the average speed of the molecules:

$$\text{average molecular speed (km/s)} = 0.157\sqrt{\frac{\text{gas temperature (K)}}{\text{molecular mass (hydrogen mass)}}}.$$

Thus, increasing the absolute temperature of a sample of gas by a factor of four—from, for example, 100 K to 400 K—doubles the average speed of its constituent molecules. And, at a given temperature, molecules of hydrogen in air move, on average, four times faster than molecules of oxygen, which are 16 times heavier.

At any instant, a tiny fraction of the molecules in any gas have speeds much greater than average, and some molecules are always moving fast enough to escape, even when the average molecular speed is much less than the escape speed. As a result all planetary atmospheres slowly leak away into space. Don't be alarmed, however—the leakage is usually very gradual. As a rule of thumb, if the escape speed from a planet or moon exceeds the average speed of a given type of molecule by a factor of six or more, then molecules of that type will not have escaped from the planet's atmosphere in significant quantities since the solar system formed.

For air on Earth (with a temperature of about 300 K), the mean molecular speed of oxygen (mass = 32 hydrogen masses) and nitrogen (mass = 28) is about 0.6 km/s, comfortably below one-sixth of the escape speed (11.2 km/s). As a result, Earth is able to retain its atmosphere. However, if the Moon originally had an Earthlike atmosphere, the lunar escape speed of just 2.4 km/s means that any original atmosphere long ago dispersed into interplanetary space. The same reasoning can be used to understand atmospheric composition. For example, hydrogen molecules (mass = 2) move, on average, at about 2 km/s in Earth's atmosphere at sea level. Consequently, they have had plenty of time to escape since our planet formed, which is why we find very little hydrogen in Earth's atmosphere today.

Modeling Earth's Interior

Because earthquakes occur often and at widespread places across the globe, geologists have accumulated a large amount of data about seismic-wave properties. They have used these data, along with direct knowledge of surface rocks, to model Earth's interior.

Figure 5.8 illustrates some paths followed by seismic waves from the site of an earthquake. Some of the waves (called shear, or S-waves, and colored red in the figure) are blocked by Earth's **outer core**. The result is the *shadow zone* shown in Figure 5.8. The explanation for this behavior is that these particular waves cannot pass through liquid. The observation that every earth-

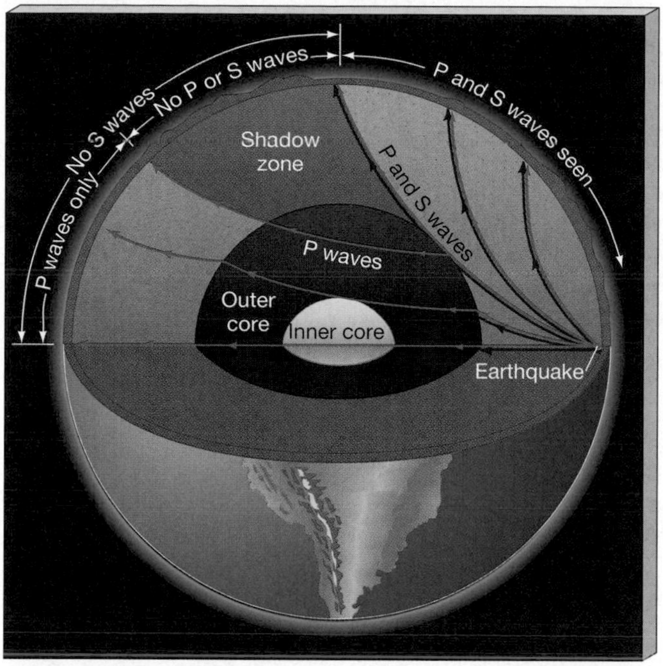

Figure 5.8 Seismic Waves Earthquakes generate seismic waves that can be detected at seismographic stations around the world. (The waves bend as they move through Earth's interior because of the variation in density and temperature within our planet.) Some waves are not detected by stations "shadowed" by the outer core of Earth, indicating that the outer core is liquid.

quake exhibits these shadow zones is the best evidence we have that the outer core of our planet is liquid. The radius of the outer core, as determined from seismic data, is about 3500 km. There is also evidence that some waves (pressure, or P-waves, shown in green), which can pass through the liquid outer core, are reflected off the surface of a solid **inner core**, of radius 1300 km.

Figure 5.9 presents a model that most scientists accept. Earth's outer core is surrounded by a thick mantle and topped with a thin crust. The mantle is about 3000 km thick and accounts for the bulk (80 percent) of our planet's volume. The crust has an average thickness of only 15 km—a little less (around 8 km) under the oceans and somewhat more (20–50 km) under the continents. The average density of crust material is around 3000 kg/m^3. Density and temperature both increase with depth. From Earth's surface to its very center, the density increases from roughly 3000 kg/m^3 to a little more than 12,000 kg/m^3, while the temperature rises from just under 300 K to well over 5000 K. Much of the mantle has a density midway between the densities of the core and crust: about 5000 kg/m^3.

The high central density suggests to geologists that the inner parts of Earth must be rich in nickel and iron. Under the heavy pressure of the overlying layers, these metals (whose densities under surface conditions are around 8000 kg/m^3) can be compressed to the high densities predicted by the model. The sharp density increase at the mantle–core boundary results from the difference in composition between the two regions. The mantle is composed of dense but rocky material—compounds of silicon and oxygen. The core consists primarily of even denser metallic elements. There is no similar jump in density or temperature at the inner core boundary—the material there simply changes from the liquid to the solid state.

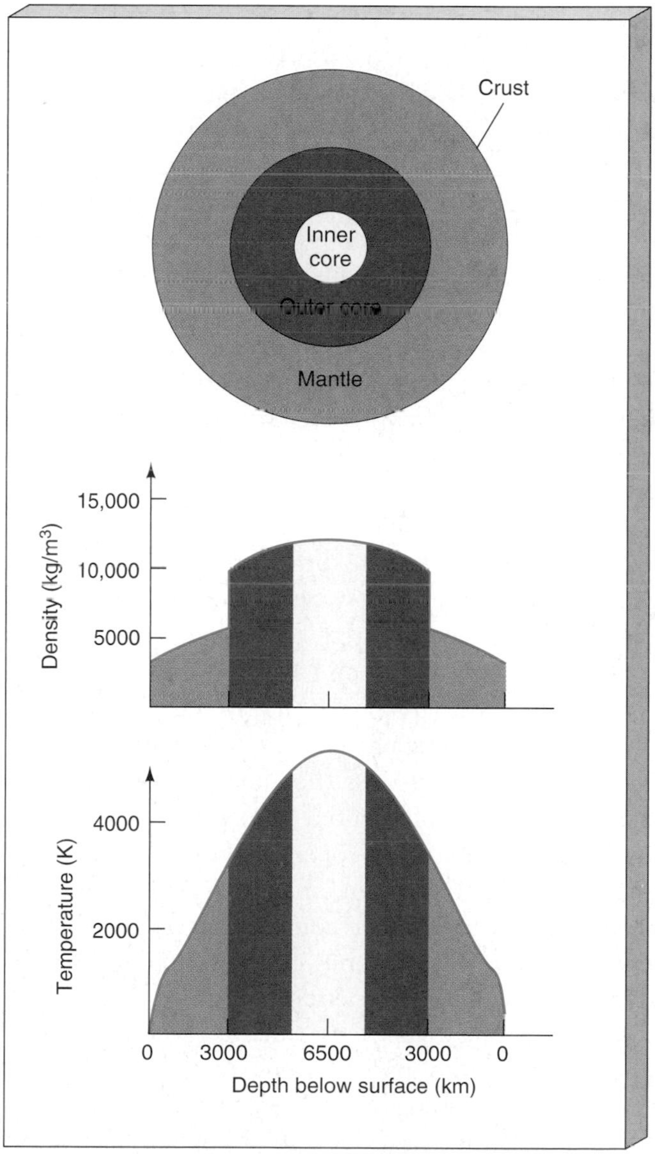

Figure 5.9 Earth's Interior Computer models of Earth's interior imply that the density and temperature vary considerably through the mantle and the core. Note the sharp density change at the mantle–outer core boundary.

Because geologists have been unable to drill deeper than about 10 km, no experiment has yet recovered a sample of the mantle. However, we are not entirely ignorant of the mantle's properties. In a *volcano*, molten rock upwells from below the crust, bringing a little of the mantle to us in the form of lava and providing some inkling of the composition of Earth's interior. The composition of the upper mantle is probably quite similar to the dark gray material known as *basalt* often found near volcanoes. With a density between 3000 and 3300 kg/m^3, it contrasts with the lighter *granite* (density 2700–3000 kg/m^3) that constitutes much of Earth's crust.

Differentiation

Earth, then, is not a homogeneous ball of rock. Instead, it has a layered structure, with a low-density rocky crust at the surface, intermediate-density rocky material in the mantle, and a high-density metallic core. This variation in density and composition is known as **differentiation**.

Why isn't our planet just one big, rocky ball of uniform density? The answer is that, at some time in the distant past, much of Earth was *molten*, allowing the higher-density matter to sink to the core, displacing lower-density material toward the surface. A remnant of this ancient heating exists today: Earth's central temperature is nearly equal to that of the Sun's surface.

Two processes played important roles in heating Earth to the point where differentiation could occur. First, very early in its history, our planet experienced a violent bombardment by interplanetary debris, an essential part of the process by which Earth and the other planets formed and grew. ∞ (Sec. 4.3) This bombardment probably generated enough heat to melt much of our planet.

The second process that heated Earth after its formation and contributed to differentiation was **radioactivity**—the release of energy by certain elements, such as uranium, thorium, and plutonium. These elements emit energy as their complex, heavy nuclei break up into simpler, lighter ones. Rock is such a poor conductor of heat that the energy released through radioactivity took a very long time to reach the surface and leak away into space. As a result, the heat built up in the interior, adding to the energy left there by Earth's formation. Geologists believe that enough radioactive elements were originally spread throughout the primitive planet, like raisins in a cake, that the entire planet—from crust to core—could have melted and remained molten, or at least semisolid, for about a billion years.

The Lunar Interior

The Moon's average density, about 3300 kg/m^3, is quite similar to the density of the lunar surface rock obtained by U.S. and Soviet missions. This similarity all but eliminates any chance that the Moon has a large, massive, dense nickel–iron core like that within Earth. The low average lunar density suggests that the Moon contains substantially fewer heavy elements (such as iron) than does Earth.

Most of our detailed information on the Moon's interior comes from seismic data obtained from equipment left on the lunar surface by astronauts. These measurements indicate only very weak *moonquakes* deep within the lunar interior. Even if you stood directly above one of these quakes, you would not feel the vibrations. The average moonquake releases about as much energy as a firecracker, and no large quakes have ever been detected. This barely perceptible seismic activity confirms the idea that the Moon is geologically dead. Nevertheless, researchers can use these weak lunar vibrations to obtain information about the Moon's interior.

Modeling all available data indicates that the Moon's interior is of almost uniform density but chemically differentiated—that is, the chemical properties change from core to surface. As noted in Figure 5.1(b), the models suggest a central core perhaps 200 km in radius, surrounded by a 500-km-thick inner mantle of semisolid rock having properties similar to those of Earth's upper mantle. Above these regions lies a 900-km-thick outer mantle of solid rock, topped by a 60- to 150-km-thick crust.

The core is probably somewhat more iron-rich than the rest of the Moon, although it is still iron-poor relative to Earth's core. The models imply that, near the center, the current temperature may be as low as 1500 K, too cool to melt rock. However, some of the seismic data suggest that the inner core may be at least partially molten, implying a higher temperature. Our knowledge of the Moon's deep interior is still very limited.

The crust on the Moon's far side is considerably thicker (150 km) than the crust on the side nearer Earth (60 km). Why is this? The answer is probably connected to Earth's gravitational pull. Just as heavier material tries to sink to the center of Earth, the denser far-side lunar mantle tended to sink below the lighter far-side crust in the presence of Earth's gravitational field. In other words, while the Moon was cooling and solidifying, the far-side lunar mantle was pulled a little closer to Earth than the far-side crust. In this way the crust and mantle became slightly off-center with respect to one another. The result was the thicker crust on the Moon's far side.

✓ Concept Check

- How would our knowledge of Earth's interior be changed if our planet were geologically dead, like the Moon?

5.5 Surface Activity on Earth

Earth is geologically alive today. Its interior seethes, and its surface constantly changes. Many clear indicators of geological activity, in the form of earthquakes and volcanic eruptions, are scattered across our globe. Erosion by wind and water has obliterated much of the evidence from ancient times, but the sites of more recent activity are well documented.

Continental Drift

5 Figure 5.10 is a map of the currently active areas of our planet. Nearly all these sites have experienced surface activity during this century. The intriguing aspect of the figure is that the active sites are not spread evenly across our planet. Instead, they trace well-defined lines of activity, where crustal rocks shift (resulting in earthquakes) or mantle material upwells (in volcanoes).

In the mid-1960s, scientists realized that these lines are the outlines of gigantic *plates*, or slabs of Earth's surface, and that these plates are slowly drifting around the surface of our planet. These plate motions, popularly known as *continental drift*, have created the mountains, oceanic trenches, and many other large-scale features across the face of planet Earth. The technical term for the study of plate movement and the reasons for it is **plate tectonics**. Some plates are made mostly of continental landmasses, some are made of a continent plus a large part of an ocean floor, and some contain no continental land at all and are made solely of ocean floor. For the most part, the continents are just passengers riding atop much larger plates.

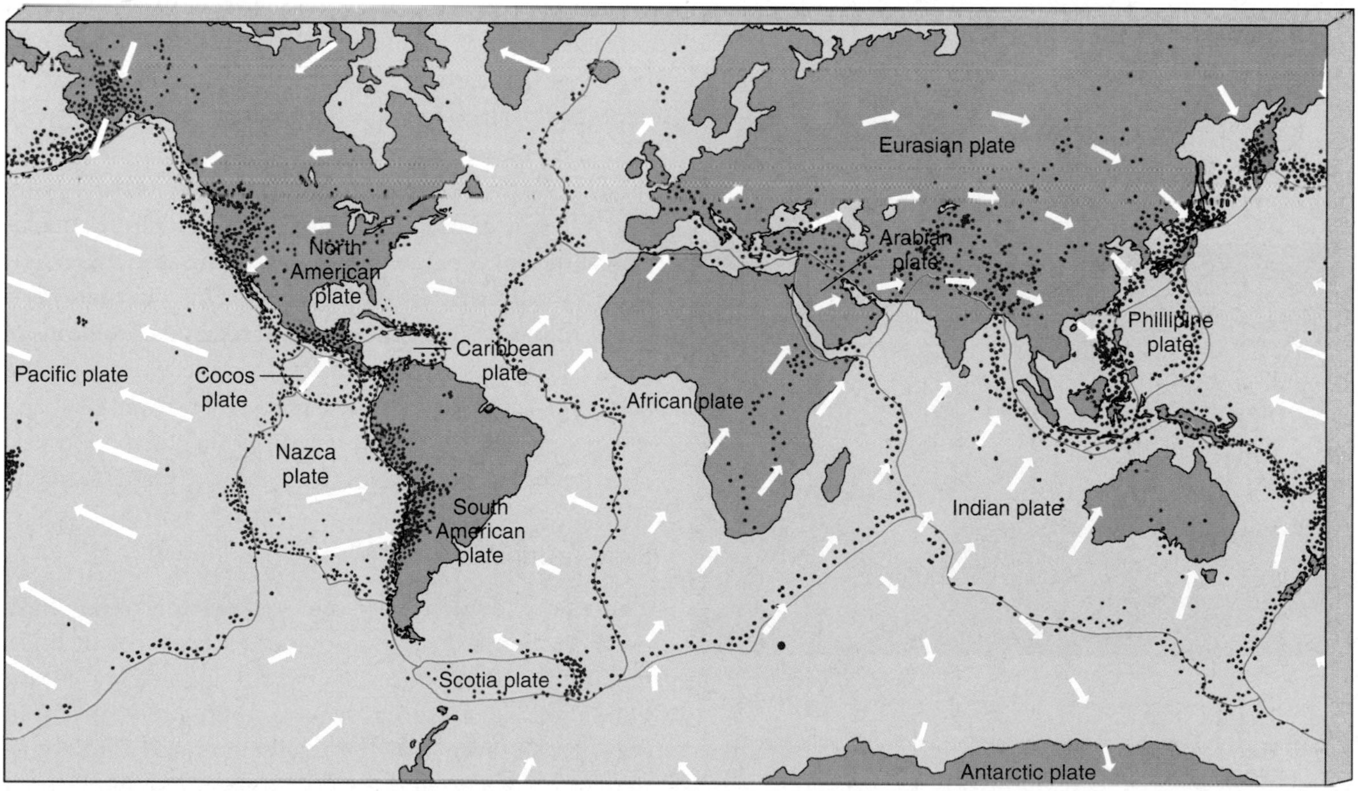

Figure 5.10 Global Plates The red dots represent active sites where major volcanoes or earthquakes have occurred in the twentieth century. Taken together, the sites delineate vast plates, outlined in blue, that drift around on the surface of our planet. The white arrows show the general directions and speeds of plate motions.

The plates move at an extremely slow rate. Typical speeds amount to only a few centimeters per year—about the same speed as your fingernails grow. Nevertheless, over the course of Earth history, the plates have had plenty of time to move large distances. For example, a drift rate of 2 cm per year can cause two continents (for example, Europe and North America) to separate by 4000 km—the width of the Atlantic Ocean—in 200 million years. That's a long time by human standards, but it represents only about five percent of Earth's age.

As the plates drift around, we might expect collisions to be routine. Indeed, plates do collide, but they are driven by enormous forces and do not stop easily. Instead, they just keep crunching into one another. Figure 5.11 shows the result of an ongoing collision as the Indian Plate thrusts northward into the Eurasian Plate. As Earth's rocky crust crumples and folds, mountains are formed—in this case, Mount Everest in the Himalayan range.

Not all plates collide head-on. As indicated by the arrows on Figure 5.10, sometimes plates slide or shear past one another. A good example is the most famous active region in North America—the San Andreas Fault in California (Figure 5.12), which forms part of the boundary between the Pacific and North American plates. Along the fault, these two plates are not moving in quite the same direction, nor at quite the same speed. Like parts in a poorly oiled machine, their motion is neither steady nor smooth. Instead, they tend to stick, then suddenly lurch forward as surface rock gives way. The resulting violent, jerky movements have been the cause of many major earthquakes along the fault line.

Figure 5.11 Himalayas Mountain building results largely from plate collisions. Here, the folding of rock is visible (in the foreground) near Mount Everest, which is part of the Himalayan mountain range at the northern end of the Indian Plate. *(L. Day/Black Star)*

At still other locations, such as the seafloor under the Atlantic Ocean, the plates are moving apart. As they recede, new mantle material wells up between them, forming mid-ocean ridges. Today, hot mantle material is rising through a crack all along the Mid-Atlantic Ridge, which extends, like a seam on an enormous baseball, all the way from the North Atlantic to the southern tip of South America. Radioactive dating indicates that material has been upwelling along the ridge more or less steadily for the past 200 million years. The Atlantic seafloor is slowly growing, as the North and South American plates move away from the Eurasian and African plates.

What Drives the Plates?

What is responsible for the enormous forces that drag plates apart in some locations and ram them together in others? The answer (Figure 5.13) is *convection*—the same process we encountered earlier in our study of Earth's atmosphere. Each plate is made up of crust plus a small portion of upper mantle. Below the plates, at a depth of perhaps 50 km, the temperature is sufficiently high that the mantle at that depth is soft enough to flow, very slowly, although it is not molten.

This is a perfect setting for convection—warm matter underlying cool matter. The warm mantle rock rises, just as hot air rises in our atmosphere, and large circulation patterns become established. Riding atop these convection patterns are the plates. The circulation is extraordinarily sluggish. Semisolid rock takes millions of years to complete one convection cycle. Although the details are far from certain and remain controversial, many researchers believe that it is large-scale convection patterns near plate boundaries that cause the plates to move.

Plate Tectonics on the Moon

There is no evidence for tectonic motion of any kind on the Moon today—no obvious extensive fault lines, no significant seismic activity, no ongoing mountain building. Plate tectonics requires both a relatively thin outer rocky layer, which is easily fractured into continent-sized pieces, and a soft, convective region under it, to make the pieces move. On the Moon, neither of these ingredients exists. The Moon's thick crust and

Figure 5.12 Californian Faults The San Andreas and associated faults in California. This fault system is the result of the North American and Pacific plates sliding past one another. The percentages are estimates of the probability of a major earthquake occurring at various locations along the faults. The inset at the top right shows a small part of the fault line separating the two plates. *(Inset: P. Menzel/P. Menzel Photography)*

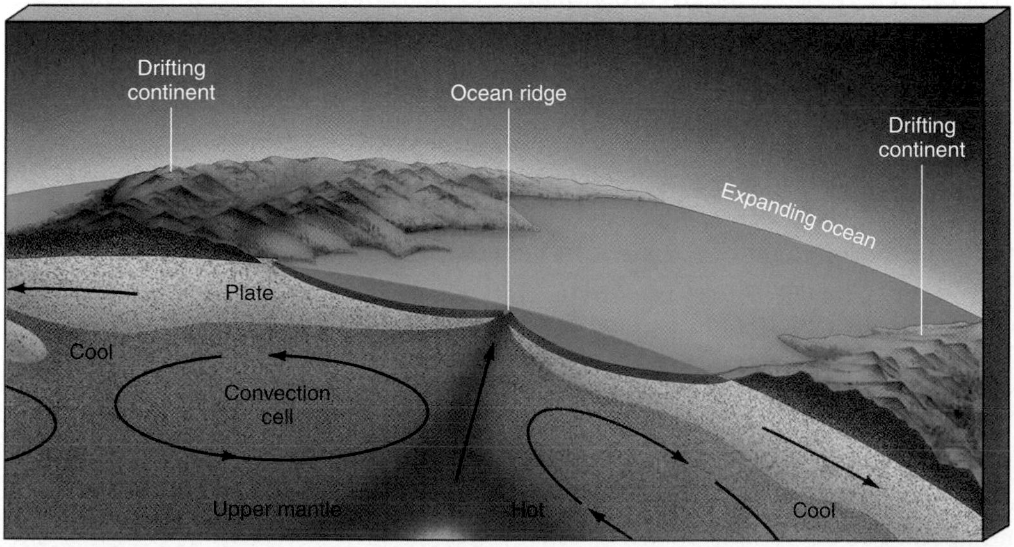

Figure 5.13 Plate Drift Plate drift is caused by convection patterns in the upper mantle that drag the plates across Earth's surface.

solid upper mantle make it impossible for pieces of the surface to move relative to one another. There simply isn't enough energy left in the lunar interior for plate tectonics to work.

Concept Check

- Describe the causes and some consequences of plate tectonics on Earth.

Figure 5.14 Full Moon, Near Side A photographic mosaic of the full Moon, north pole at the top. Some prominent maria are labeled. *(UC/Lick Observatory)*

5.6 The Surface of the Moon

On Earth, the combined actions of air, water, and geological activity erode our planet's surface and reshape its appearance almost daily. As a result, most of the ancient history of our planet's surface is lost to us. The Moon, though, has no air, no water, no ongoing volcanic or other geological activity. Consequently, features dating back almost to the formation of the Moon are still visible today. For this reason, studies of the lunar surface are of great importance to Earth geologists, and have played a major role in shaping theories of the early development of our planet.

Large-Scale Features

The first observers to point their telescopes at the Moon noted large, roughly circular, dark areas that resemble (they thought) Earth's oceans. They called these regions **maria**, a Latin word meaning "seas" (singular: *mare*). The largest of them (Mare Imbrium, the "Sea of Showers") is about 1100 km in diameter. Today we know that the maria are actually extensive flat plains that resulted from the spread of lava during an earlier, volcanic period of lunar evolution. In a sense, then, the maria *are* oceans—ancient seas of molten lava, now solidified.

Early observers also saw light-colored areas that resembled Earth's continents. Originally dubbed *terrae*, from the Latin word for "land," these regions are now known to be elevated several kilometers above the maria. Accordingly, they are usually called the lunar **highlands**. Both types of region are visible in Figure 5.14, a photographic mosaic of the full Moon. These light and dark surface features are also evident to the naked eye, creating the face of the familiar "Man-in-the-Moon."

Based on studies of lunar rock brought back to Earth by *Apollo* astronauts and unmanned Soviet landers, geologists have found important differences in both *composition* and *age* between the highlands and the maria. The highlands are made largely of rocks rich in aluminum, making them lighter in color and lower in density (2900 kg/m^3). The maria's basaltic matter contains more iron, giving it a darker color and greater density (3300 kg/m^3). Loosely speaking, the highlands represent the Moon's crust, while the maria are made of mantle material. Maria rock is quite similar to terrestrial basalt, and geologists believe that it arose on the Moon much as basalt did on Earth, through the upwelling of molten material through the crust. Radioactive dating indicates ages of more than four billion years for highland rocks, and from 3.2 to 3.9 billion years for those from the maria.[2]

Until spacecraft flew around the Moon, no one on Earth had any idea what the Moon's hidden half looked like. To the surprise of most astronomers, when the far side of the Moon was mapped first by Soviet and later by American spacecraft, no major maria were found. The lunar far side (Figure 5.15) is composed almost entirely of highlands.

[2]Radioactive dating compares the rates at which different radioactive elements in a sample of rock decay into lighter elements. The "age" returned by this technique is the time since the rock solidified.

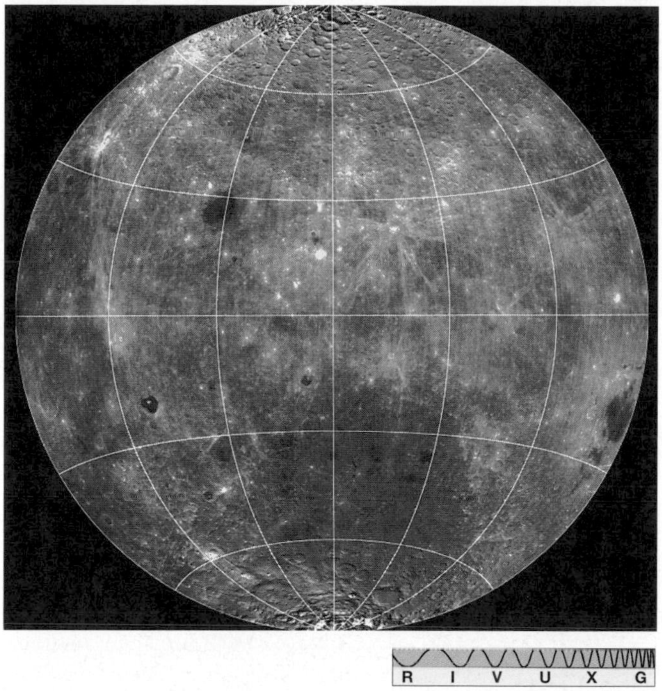

Figure 5.15 Full Moon, Far Side The far side of the Moon, as seen by the *Clementine* military spacecraft. The large, dark region at center bottom outlines the South Pole-Aitken Basin, the largest and deepest impact basin known in the Solar System. This image shows only a few small maria on the far side. *(Naval Research Laboratory)*

Cratering

6 Because the smallest lunar features we can distinguish with the naked eye are about 200 km across, we see little more than the maria and highlands when we gaze at the Moon. Through a telescope, however (Figure 5.16), we find that the lunar surface is scarred by numerous bowl-shaped **craters** (after the Greek word for "bowl").

Most craters formed eons ago as the result of meteoritic impact. ∞ (Sec. 4.2) Meteoroids generally strike the Moon at speeds of several kilometers per second. At these speeds, even a small piece of matter carries an enormous amount of energy—for example, a 1-kg object hitting the Moon's surface at 10 km/s would release as much energy as the detonation of 10 kg of TNT.

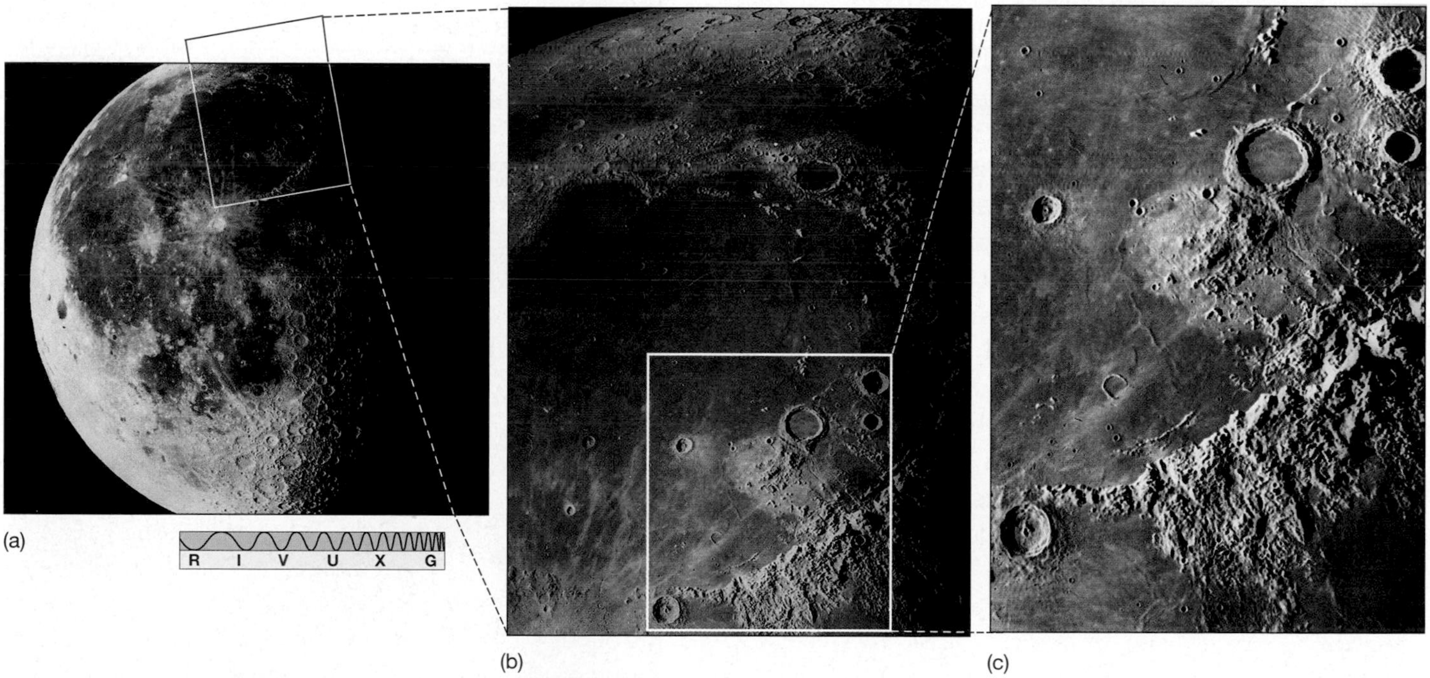

Figure 5.16 Moon, Close Up (a) The Moon near third quarter. Surface features are most visible near the *terminator*, the line separating light from dark, where sunlight strikes at a sharp angle and shadows highlight the topography.
(b) Magnified view of a region near the terminator, as seen from Earth through a large telescope. Crater Copernicus is at bottom left, and the central dark area is Mare Imbrium, ringed at the bottom right by the Apennine mountains.
(c) Enlargement of the lower right portion of (b). The smallest craters visible here have diameters of about 2 km, about twice the size of the Barringer crater shown in Figure 4.13. *(UC/Lick Observatory; California Institute of Technology)*

As illustrated in Figure 5.17, impact by a meteoroid causes sudden and tremendous pressures to build up on the lunar surface, heating the normally brittle rock and deforming the ground. The ensuing explosion pushes previously flat layers of rock up and out, forming a crater.

The material thrown out by the explosion surrounds the crater in a layer called an *ejecta blanket*, the ejected debris ranging in size from fine dust to large boulders. The larger pieces of ejecta may themselves form secondary craters. Many of the rock samples brought back by the *Apollo* astronauts show patterns of repeated shattering and melting—direct evidence of the violent shock waves and high temperatures produced in meteoritic impacts.

Lunar craters come in all sizes, reflecting the range in sizes of the impactors that create them. The largest craters are hundreds of kilometers in diameter, the smallest microscopic. Because the Moon has no protective atmosphere, even tiny interplanetary fragments can reach the lunar surface unimpeded. Figure 5.18(a) shows the result of a large meteoritic impact on the Moon, Figure 5.18(b) a crater formed by a micrometeoroid.

Craters are found everywhere on the Moon's surface, but the older highlands are much more heavily cratered than the younger maria. Knowing the ages of the highlands and maria, researchers can estimate the rate of cratering in the past. They conclude that the Moon, and presumably the entire inner solar system, experienced a sudden sharp drop in meteoritic bombardment rate about 3.9 billion years ago. The rate of cratering has been roughly constant since that time.

This time—3.9 billion years in the past—is taken to represent the end of the accretion process through which planetesimals became planets. ∞ (Sec. 4.3) The lunar highlands solidified and received most of their craters before that time. The great basins that formed the maria are thought to have been created during the final stages of heavy meteoritic bombardment between about 4.1 and 3.9 billion years ago. Subsequent volcanic activity filled the craters with lava, creating the formations we see today.

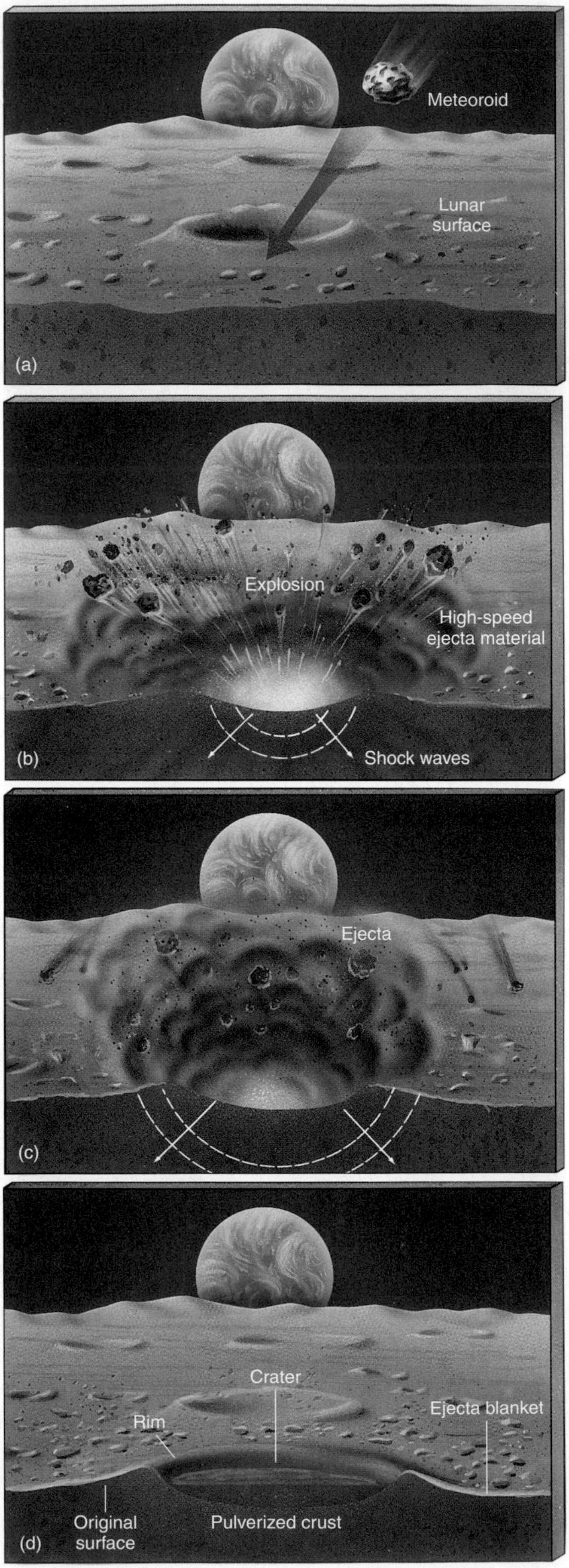

Figure 5.17 Meteoroid Impact Stages in the formation of a crater by meteoritic impact. (a) The meteoroid strikes the surface, releasing a large amount of energy. (b, c) The resulting explosion ejects material from the impact site and sends shock waves through the underlying surface. (d) Eventually, a characteristic crater surrounded by a blanket of ejected material results.

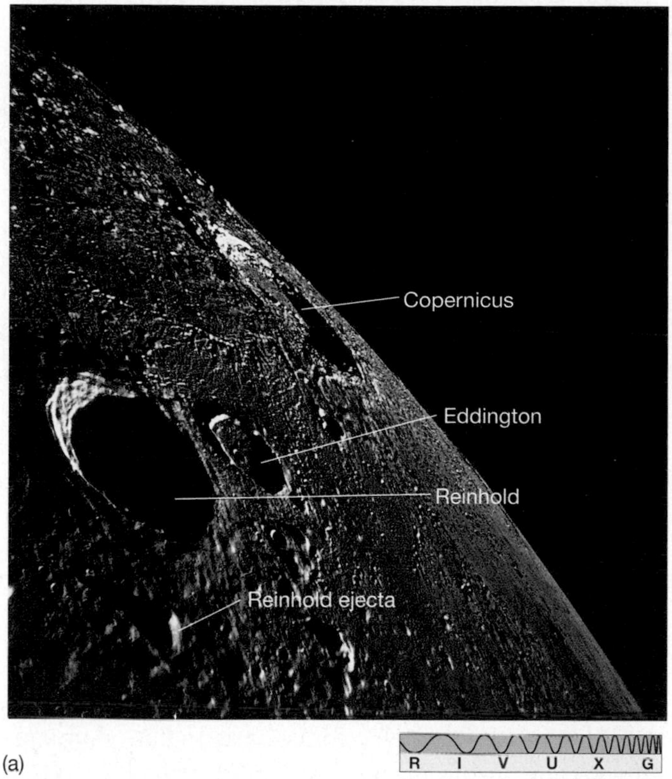

(a)

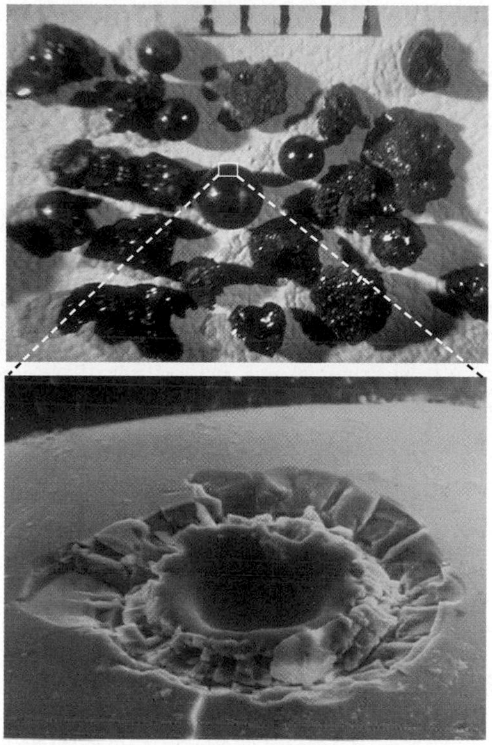

(b)

Figure 5.18 Lunar Craters (a) Two smaller craters called Reinhold and Eddington sit amid the secondary cratering resulting from the impact that created the 90-km-wide Copernicus Crater (near the horizon) about a billion years ago. The ejecta blanket from crater Reinhold, 40 km across, in the foreground, can be seen clearly. (b) Craters of all sizes litter the lunar landscape. Some shown here, embedded in glassy beads retrieved by *Apollo* astronauts, measure only 0.01 mm across. (The scale at the top is in millimeters.) The beads themselves were formed during the explosion following a meteoroid impact, when surface rock was melted, ejected, and rapidly cooled. *(NASA)*

Lunar Erosion

Meteoritic impact is the only important source of erosion on the Moon. Over billions of years, collisions with meteoroids, large and small, have scarred, cratered, and sculpted the lunar landscape. At the present average rates, one new 10-km-diameter lunar crater is formed every 10 million years, one new 1-m-diameter crater is created about once a month, and 1-cm-diameter craters are formed every few minutes. In addition, a steady "rain" of micrometeoroids also eats away at the lunar surface (Figure 5.19). The accumulated dust from countless impacts (called the lunar *regolith*) covers the lunar surface to an average depth of about 20 m, thinnest on the maria (about 10 m) and thickest on the highlands (more than 100 m in places).

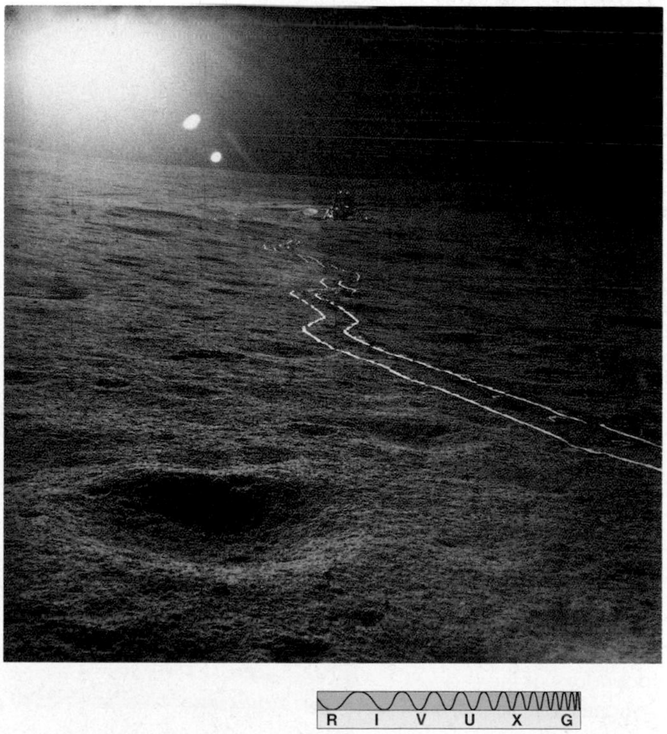

Figure 5.19 Lunar Surface The lunar surface is not entirely changeless. Despite the complete lack of wind and water on the airless Moon, the surface has still eroded a little under the constant "rain" of impacting meteoroids, especially micrometeoroids. Note the soft edges of the craters visible in the foreground of this image. In the absence of erosion, these features would be as jagged and angular today as they were when they formed. (The twin tracks were made by the *Apollo* lunar rover.) *(NASA)*

Despite this barrage from space, the Moon's present-day erosion rate is still very low—about 10,000 times less than on Earth. For example, the Barringer Meteor Crater (Figure 4.13) in the Arizona desert, one of the largest meteor craters on Earth, is only 25,000 years old, but it is already decaying. It will probably disappear completely in a mere million years, quite a short time geologically. If a crater that size had formed on the Moon even a billion years ago, it would still be plainly visible today.

✓ Concept Check

- Describe two important ways in which the lunar maria differ from the highlands.

5.7 Magnetospheres

Simply put, the magnetosphere is the region around a planet that is influenced by that planet's magnetic field. It forms a buffer zone between the planet and the high-energy particles of the solar wind. It can also provide important insights into the planet's interior structure.

Earth's Magnetosphere

7 Shown in Figure 5.20, Earth's magnetic field extends far above the atmosphere, completely surrounding our planet. The *magnetic field lines*, which indicate the strength and direction of the field at any point in space, run from south to north, as indicated by the white arrowheads in the figure. The north and south *magnetic poles*, where the axis of the imaginary bar magnet intersects Earth's surface, are very roughly aligned with Earth's spin axis.

Earth's inner magnetosphere contains two doughnut-shaped zones of high-energy charged particles, one located about 3000 km and the other 20,000 km above Earth's surface. These zones are named the **Van Allen belts**, after the American physicist whose instruments on board some early rocket flights during the late 1950s first detected them. We call them "belts" because they are most pronounced near Earth's equator and because they completely surround the planet. Figure 5.21 shows how these invisible regions envelop Earth except near the north and south poles.

The particles that make up the Van Allen belts originate in the solar wind. Traveling through space, neutral particles and electromagnetic radiation are unaffected by Earth's magnetism, but electrically charged particles are strongly influenced. As illustrated in Figure 5.22, a magnetic field exerts a force on a moving charged particle, causing the particle to spiral around the magnetic field lines. In this way, charged particles—mainly electrons and protons—from the solar wind can become trapped by Earth's magnetism. Earth's magnetic field exerts electromagnetic control over these particles, herding them into the Van Allen belts. The outer belt contains mostly electrons; the much heavier protons accumulate in the inner belt.

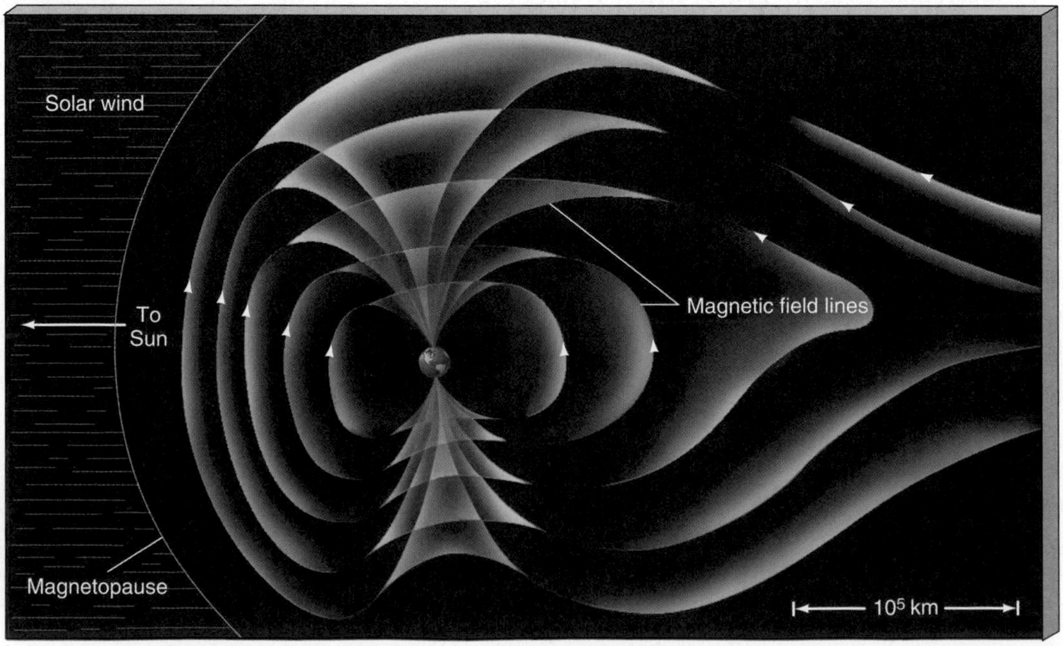

Figure 5.20 Earth's Magnetosphere Earth's magnetic field resembles somewhat the field of an enormous bar magnet situated inside our planet. The white arrowheads on the field lines indicate the direction in which a compass needle would point. Far from Earth, the magnetosphere is greatly distorted by the solar wind, with a long tail extending from the nighttime side of Earth well into space. The magnetopause is the boundary of the magnetosphere in the sunward direction.

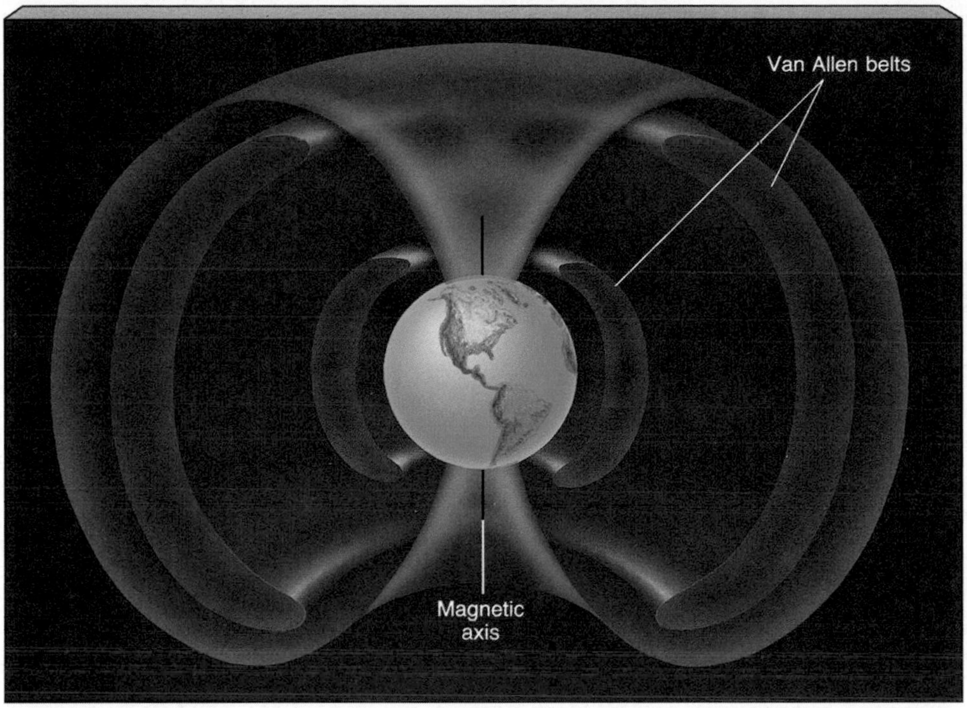

Figure 5.21 **Van Allen Belts** High above Earth's atmosphere, the magnetosphere (blue-green area) contains two doughnut-shaped regions (pink and purple areas) of magnetically trapped charged particles. These are the Van Allen belts.

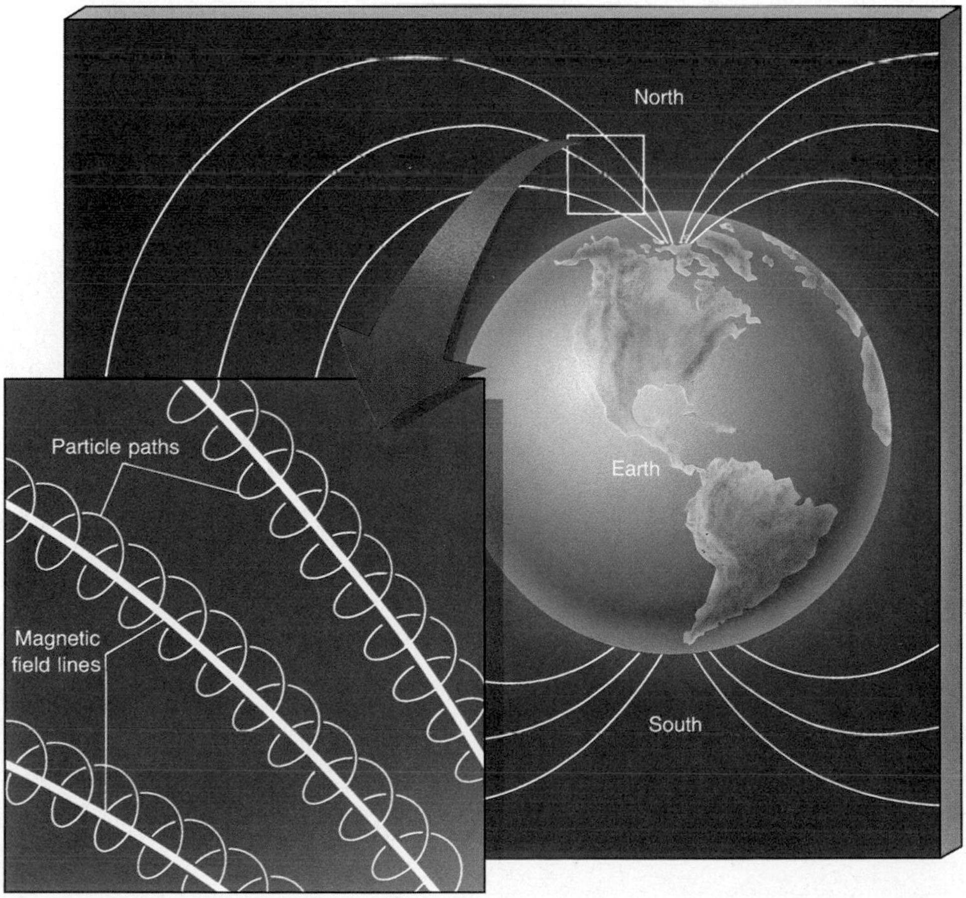

Figure 5.22 **Magnetic Field Effect** Charged particles in a magnetic field spiral around the field lines. Thus, the particles tend to become trapped by strong magnetic fields.

Particles from the Van Allen belts often escape from the magnetosphere near Earth's north and south magnetic poles, where the field lines intersect the atmosphere. Their collisions with air molecules create a spectacular light show called an **aurora** (Figure 5.23). This colorful display results when atmospheric molecules, excited upon collision with the charged particles, fall back to their ground states and emit visible light. ∞ (Sec. 2.6) Aurorae are most brilliant at high latitudes, especially inside the Arctic and Antarctic circles. In the north, the spectacle is called the *aurora borealis*, or *Northern Lights*. In the south, it is called the *aurora australis*, or *Southern Lights*.

Just as Earth's magnetosphere influences the charged particles of the solar wind, the stream of incoming solar-wind particles also affects our magnetosphere. As shown in Figure 5.20, the sunward (daytime) side is squeezed toward Earth's surface, while the opposite side has a long "tail" often extending hundreds of thousands of kilometers into space.

Earth's magnetosphere plays an important role in controlling many of the potentially destructive charged particles that venture near our planet. Without the magnetosphere, Earth's atmosphere—and perhaps the planet surface, too—would be bombarded by harmful particles, possibly damaging many forms of life. Some researchers have even suggested that had the magnetosphere not existed in the first place, life might never have arisen on planet Earth.

Earth's magnetic field is not an intrinsic part of our planet. Instead, it is thought to be continuously generated in Earth's core and exists only because the planet is rotating. As in the dynamos that run industrial machines, Earth's magnetism is produced by the spinning, electrically conducting, liquid metal core deep within our planet. Both rapid rotation *and* a conducting liquid core are needed for this mechanism to work. As we will see, this link between magnetism and internal structure is very important to studies of other planets, as we have few other probes of planetary interiors.

Lunar Magnetism

No Earth-based observation or spacecraft measurement has ever detected any lunar magnetic field. Based on our current understanding of how Earth's magnetic field is created, this is not surprising. As we have just seen, researchers believe that planetary magnetism requires a rapidly rotating liquid metal core. Because the Moon rotates slowly and because its core is probably neither molten nor particularly rich in metals, the absence of a lunar magnetic field is exactly what we expect.

✓ Concept Check

- What does the existence of a planetary magnetic field tell us about the planet's interior?

Figure 5.23 **Aurora Borealis** A colorful aurora results from the emission of light radiation after magnetospheric particles collide with atmospheric molecules. The aurora rapidly flashes across the sky, looking like huge wind-blown curtains glowing in the dark. *(NCAR)*

5.8 History of the Earth–Moon System

8 Given all the data, can we construct a reasonably consistent history of Earth and the Moon? The answer seems to be yes. Many specifics are still debated, but a consensus now exists.

Formation

Sometime around 4.6 billion years ago, Earth formed by accretion in the solar nebula. ∞ (Sec. 4.3) The formation of the Moon is somewhat less certain. The basic problem is that Earth and the Moon are too *dissimilar* in both density and composition for them simply to have formed together out of the same preplanetary matter (the *coformation* theory). However, there are enough *similarities*, particularly between their mantles, to make it unlikely that Earth and the Moon formed entirely independently of one another and subsequently became bound to one another, presumably after a close encounter (the *capture* theory). In addition, there are good theoretical arguments against each scenario—the former because it is hard to reconcile it with the theory of solar-system formation outlined in Chapter 4, the latter because of the extreme improbability of the event.

Today, many astronomers favor a scenario often called the *impact* theory, which postulates a glancing collision between a large, Mars-sized object and a youthful, molten Earth. Such collisions were probably quite frequent in the early solar system. ∞ (Sec. 4.3) Computer simulations of such a catastrophic event (Figure 5.24) show that most of the bits and pieces of splattered Earth could have coalesced into a stable orbit, forming the Moon. If Earth had already formed an iron core by the time the collision occurred, the Moon could indeed have ended up with a composition similar to Earth's mantle. During the collision, any iron core in the impacting object would have been left behind, eventually to become part of Earth's core. Thus, both the Moon's overall similarity to Earth's mantle and its lack of a dense central core are explained.

Evolution

The approximate age of the oldest rocks discovered in the lunar highlands is 4.4 billion years, so we know that at least part of the lunar crust must already have solidified by that time. The oldest known Earth rocks are substantially younger—"only" 3.9 billion years old. This

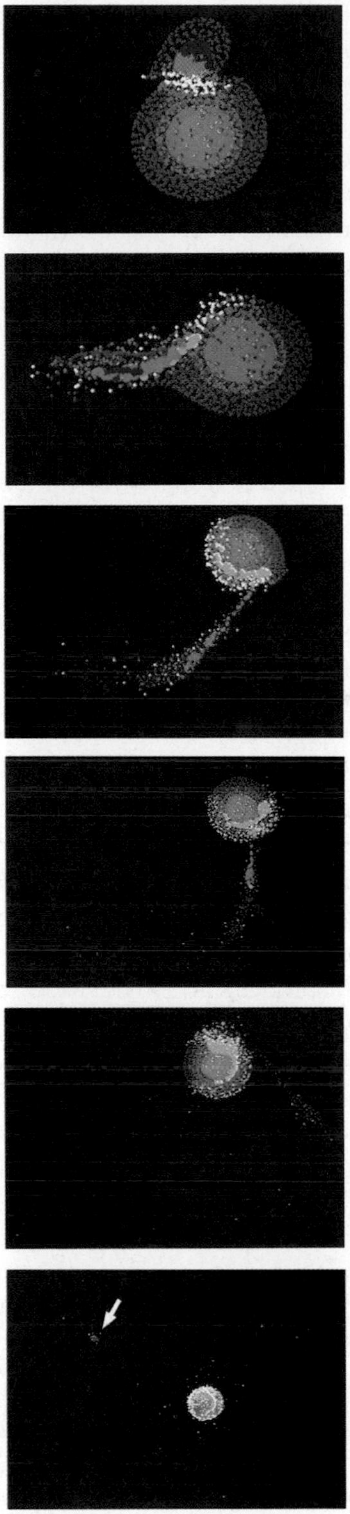

Figure 5.24 Moon Formation A simulated collision between Earth and an object the size of Mars. (The sequence proceeds from top to bottom and zooms out dramatically.) The arrow in the final frame shows the newly formed Moon. Red and blue colors represent rocky and metallic regions, respectively, and the direction of motion of the blue material in frames 2, 3, 4, and 5 is toward Earth. Note how most of the impactor's metallic core becomes part of Earth, leaving the Moon composed primarily of rocky material. *(W. Benz)*

implies that Earth's surface stayed molten a little longer, or possibly that erosion on Earth has been more effective at hiding the details of our planet's distant past. At formation, the Moon was already depleted in heavy metals relative to Earth.

Earth was at least partially molten during most of its first billion years of existence. Denser matter sank toward the core while lighter material rose to the surface—Earth became differentiated. The intense meteoritic bombardment that helped melt Earth at early times subsided about 3.9 billion years ago. Radioactive heating in the interior continued even after Earth's surface cooled and solidified, but it too diminished with time.

As our planet cooled, it did so from the outside in, because regions closest to the surface could most easily lose their excess heat to space. In this way, the surface developed a crust, and the differentiated interior attained the layered structure now implied by seismic studies. Today, radioactive heating continues throughout Earth, but there is probably not enough of it to melt any part of our planet. The high temperatures in the core are mainly the trapped remnant of a much hotter Earth that existed eons ago.

The lunar interior evolved quite differently from Earth's, in large part because of the Moon's smaller size. Small objects cool more quickly than large ones (basically because heat from the interior has less distance to go to reach the surface), and so the Moon rapidly lost its internal heat to space. During the earliest phases of the Moon's existence—roughly the first half-billion years—the meteoritic bombardment was violent enough to heat, and keep molten, most of the surface layers, perhaps to a depth of 400 km in places. However, the intense heat derived from these collisions probably did not penetrate into the deep lunar interior, because rock is a very poor conductor of heat. As on Earth, radioactivity probably heated the Moon, but (because the heat could escape more easily) not sufficiently to transform it from a warm, semisolid object to a completely liquid one. The Moon must have differentiated during this period. If it has a small iron core, that core also formed at this time.

About 3.9 billion years ago, when the heaviest bombardment ceased, the Moon was left with a solid crust dented with numerous large basins. The crust ultimately became the highlands, and the basins soon flooded with lava and became the maria. Between 3.9 and 3.2 billion years ago, lunar volcanism filled the basins with the basaltic material we see today. The age of the youngest maria—3.2 billion years—apparently indicates the time when this volcanic activity finally subsided. Not all these great craters became flooded with lava, however. One of the youngest is the Orientale Basin (Figure 5.25), which formed about 3.9 billion years ago. It did not undergo much subsequent volcanism, so we can recognize it as an impact crater rather than a mare. Similar "unflooded" basins can be seen on the lunar far side.

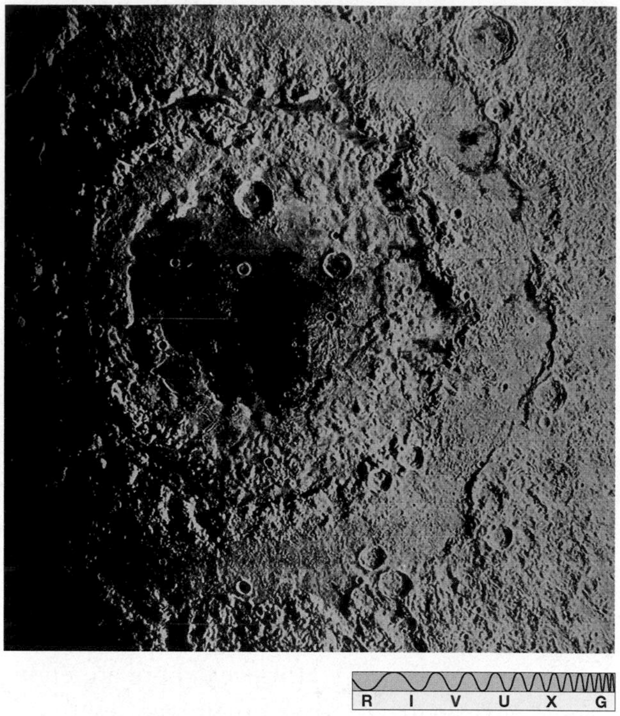

Figure 5.25 Large Lunar Crater A large lunar crater, called the Orientale Basin. The meteoroid that produced this crater upthrust much surrounding matter, which can be seen as concentric rings of cliffs called the Cordillera Mountains. The outermost ring is nearly 1000 km in diameter. *(NASA)*

Because of Earth's gravitational pull, the lunar crust became thicker on the far side than on the near side. Therefore, lava from the interior had a shorter route through the crust to the surface on the Moon's Earth-facing side. As a result, relatively little volcanic activity occurred on the far side, and no large maria were created—the crust was simply too thick to allow that to occur.

As the Moon continued to cool, volcanic activity ended as the thickness of the solid surface layer increased. The crust is now far too thick for volcanism or plate tectonics to occur. With the exception of a few meters of surface erosion from eons of meteoritic bombardment, the lunar landscape has remained more or less structurally frozen for the past three billion years. The Moon is dead now, and it has been dead for a long time.

Concept Check

- What single factor is most responsible for the evolutionary differences between Earth and the Moon?

Chapter Review

Summary

Surface gravity (p. 128) is the strength of the gravitational force at a body's surface. **Escape speed** (p. 128) is the speed required for any object to escape the body's gravitational pull.

The six main regions of Earth are (from inside to outside) a central metallic **core** (p. 128), which is surrounded by a thick rocky **mantle** (p. 128) and topped with a thin **crust** (p. 128). The liquid oceans on our planet's surface make up the **hydrosphere** (p. 128). Above the surface is the **atmosphere** (p. 128), which is composed primarily of nitrogen and oxygen. Higher still lies the **magnetosphere** (p. 128), where charged particles from the Sun are trapped by Earth's magnetic field.

The daily **tides** (p. 129) in Earth's oceans are caused by the gravitational effects of the Moon and the Sun, which raise **tidal bulges** (p. 129) in the oceans. Their size depends on the orientations of the Sun and the Moon relative to Earth. A differential gravitational force is called a **tidal force** (p. 130), even when no oceans or even planets are involved. The tidal interaction between Earth and the Moon is causing Earth's spin to slow and is responsible for the Moon's **synchronous orbit** (p. 132), in which the same side of the Moon always faces our planet.

Convection (p. 133) is the process by which heat is moved from one place to another by the upwelling or downwelling of a fluid, such as air or water. Convection occurs in the **troposphere** (p. 132), the lowest region of Earth's atmosphere. It is the cause of surface winds and weather. Above the troposphere, in the **stratosphere** and **mesosphere** (p. 132), the air is calm. At even higher altitudes, in the **ionosphere** (p. 132), the atmosphere is kept ionized by high-energy radiation and particles from the Sun. Straddling the stratosphere and mesosphere is the **ozone layer** (p. 134), where incoming solar ultraviolet radiation is absorbed. The Moon has no atmosphere because lunar gravity is too weak to retain any gases.

The **greenhouse effect** (p. 134) is the absorption and trapping of infrared radiation emitted by Earth's surface by atmospheric gases (primarily carbon dioxide and water vapor). It makes our planet's surface some 40 K warmer than would otherwise be the case.

We study Earth's interior by observing how **seismic waves** (p. 135) produced by **earthquakes** (p. 135) travel through the mantle. Earth's iron core consists of a solid **inner core** (p. 137) surrounded by a liquid **outer core** (p. 136).

The process by which dense material sinks to the center of a planet while lighter material rises to the surface is called **differentiation** (p. 138). The differentiation of Earth implies that our planet must have been at least partially molten in the past. This molten state was caused by a combination of heat resulting from bombardment by material from interplanetary space, and heat released by **radioactivity** (p. 138), the decay of unstable elements present in the material from which Earth formed.

Earth's surface is made up of enormous slabs, or plates. The slow movement of these plates across the surface is known as continental drift, or **plate tectonics** (p. 139). Earthquakes, volcanism, and mountain building are associated with plate boundaries, where plates may collide, move apart, or rub against one another. The motion of the plates is thought to be driven by convection in Earth's mantle. On the Moon, the crust is too thick and the mantle too cool for plate tectonics to occur.

The main surface features on the Moon are the dark **maria** (p. 142) and the lighter-colored **highlands** (p. 142). **Craters** (p. 143) of all sizes, caused by impacting meteoroids, are found everywhere on the lunar surface. The highlands are older than the maria and are much more heavily cratered. Meteoritic impacts are the main source of erosion on the lunar surface. There is no volcanic activity on the Moon because all volcanism was stifled by the Moon's cooling mantle shortly after extensive lava flows formed the maria more than 3 billion years ago.

Charged particles from the solar wind are trapped by Earth's magnetic field lines to form the **Van Allen belts** (p. 146). When particles from the Van Allen belts hit Earth's atmosphere, they heat and ionize the atoms there, causing the atoms to glow in an **aurora** (p. 148). Planetary magnetic fields are produced by the motion of rapidly rotating, electrically conducting fluid (such as molten iron) in a planet's core. The Moon rotates slowly, and lacks a conducting liquid core, which accounts for the absence of a lunar magnetic field.

The most likely explanation for the formation of the Moon is that the newly formed Earth was struck by a Mars-sized object. The core of the impacting body remained behind as part of the core of our planet, and debris splattered into space formed the Moon.

Review and Discussion

1. Explain how the Moon produces tides in Earth's oceans.
2. What is a synchronous orbit? How did the Moon's orbit become synchronous?
3. What is convection? What effect does it have on (a) Earth's atmosphere and (b) Earth's interior?
4. In contrast to Earth, the Moon undergoes extremes in temperature. Why?

5. Use the concept of escape speed to explain why the Moon has no atmosphere.
6. What is the greenhouse effect? Is the greenhouse effect operating in Earth's atmosphere helpful or harmful? What are the consequences of an enhanced greenhouse effect?
7. The density of water in Earth's hydrosphere and the density of rocks in the crust are both lower than the average density of the planet as a whole. What does this fact tell us about Earth's interior?
8. Give two reasons geologists believe that part of Earth's core is liquid.
9. What clue does Earth's differentiation provide to our planet's history?
10. What process is responsible for the surface mountains, oceanic trenches, and other large-scale features on Earth's surface?
11. In what sense were the lunar maria once "seas"?
12. What is the primary source of erosion on the Moon? Why is the average rate of lunar erosion so much less than on Earth?
13. Name two pieces of evidence indicating that the lunar highlands are older than the maria.
14. Give a brief description of Earth's magnetosphere. Why does the Moon have no magnetosphere?
15. Describe the theory of the Moon's origin currently favored by many astronomers.

True or False?

____ 1. The average density of Earth is less than the density of water.

____ 2. There is one high tide and one low tide per day at any given coastal location on Earth.

____ 3. Because of tidal forces, Earth's rotation rate is increasing.

____ 4. Because of the tides, the Moon is in a synchronous orbit around Earth.

____ 5. Lunar maria are extensive lava-flow regions.

____ 6. Except for the layer of air closest to Earth's surface, the ozone layer is the warmest part of the atmosphere.

____ 7. Earth's atmosphere is composed primarily of oxygen.

____ 8. Water vapor and nitrogen are the primary greenhouse gases in Earth's atmosphere.

____ 9. The Moon has no detectable atmosphere.

____ 10. Earth's magnetic field is the result of our planet's large, permanently magnetized iron core.

____ 11. The Moon has a weak magnetic field.

____ 12. Motion of the crustal plates is driven by convection in Earth's upper mantle.

____ 13. Samples of Earth's core are available from volcanos.

____ 14. Volcanic activity continues today on the surface of the Moon.

____ 15. Like Earth, the Moon has a liquid metal core.

Fill in the Blank

1. The radius of the Moon is about _______ Earth's radius. (Give your answer as a simple fraction, not as a decimal.)
2. Of Earth's crust, mantle, outer core, and inner core, which layer is the thinnest?
3. Earth is unique among the planets in that it has _______ on its surface.
4. The tidal force is due to the _______ in the gravitational force from one side of Earth to the other.
5. The _______ on the Moon are dark, flat, roughly circular regions hundreds of kilometers in diameter.
6. Craters on the Moon are primarily caused by _______.
7. The lunar maria's dark, dense rock originally was part of the _______ of the Moon.
8. Earth's atmosphere is 78 percent _______ and 21 percent _______.
9. The troposphere is where the process called _______ occurs.
10. Sunlight is absorbed by Earth's surface and is reemitted in the form of _______ radiation.
11. An increase in the level of carbon dioxide in Earth's atmosphere would _______ our planet's temperature.
12. When trapped electrons and protons from the magnetosphere eventually collide with the upper atmosphere, they produce an _______.
13. Observations of seismic waves imply that Earth's inner core is _______ and the outer core is _______.
14. For differentiation to have occurred, Earth's interior must, at some time in the past, have been largely _______.
15. _______ is the study of continental drift, and the resulting volcanism, earthquakes, faults, and mountain building.

Problems

1. What would Earth's surface gravity and escape speed be if the entire planet had a density equal to that of the crust (3000 kg/m^3, say)?
2. The Moon's mass is 1/80 that of Earth, and the lunar radius is 1/4 Earth's radius. Based on these figures, calculate the total weight on the Moon of a 100-kg astronaut with a 50-kg spacesuit and backpack, relative to his weight on Earth.
3. Most of Earth's ice is found in Antarctica, whose permanent ice caps cover approximately 0.5 percent of Earth's surface and are 3 km thick, on average. Earth's oceans cover roughly 71 percent of our planet, to an average depth of 3.6 km. Taking water and ice to have roughly the same density, estimate by how much sea level would rise if global warming were to cause the Antarctic ice caps to melt.
4. You are standing on Earth's surface, and the Moon is directly overhead. By what fraction is your weight decreased due to the Moon's tidal gravitational force?
5. Compare the magnitude of the tidal gravitational force on Earth due to Jupiter with that due to the Moon. Assume an Earth-Jupiter distance of 4.2 A.U. Based on your answer, do you think that the tidal stresses caused by a "cosmic convergence"—a chance alignment of the four jovian planets, so that they all appear from Earth to be in exactly the same direction in the sky—would have any noticeable effect on our planet?
6. Approximating Earth's atmosphere as a layer of gas 7.5 km thick, with a uniform density of 1.3 kg/m^3, calculate the atmosphere's total mass. Compare this with the mass of Earth.
7. As discussed in the text, without the greenhouse effect, Earth's average surface temperature would be about 250 K. With the greenhouse effect, it is some 40 K higher. Use this information and Stefan's Law to calculate the fraction of infrared radiation leaving Earth's surface that is absorbed by greenhouse gases in the atmosphere.
8. Following an earthquake, how long would it take a seismic wave moving in a straight line with speed 5 km/s to reach the opposite side of Earth?
9. Using the rate given in the text for the formation of 10-km craters on the Moon, estimate how long it would take to cover the Moon with new craters of this size.
10. The *Hubble Space Telescope* has a resolution of about 0.05 arc second. What is the smallest object it can see on the surface of the Moon? Give your answer in meters.

Projects

1. Write or draw an explanation of spring tides and neap tides. Look in an almanac to find a month during which the Moon's *perigee* (the point in its orbit where the Moon is closest to Earth) and the day of new or full Moon more or less coincide. Then watch the newspapers and television weather reports for news of especially high spring tides. Do especially high tides actually occur on the day of the new or full Moon?
2. Go to a sporting goods store and get a tide table; many stores near the ocean provide them free. Choose a month and plot the height of one high and one low tide versus the day of the month. Now mark the dates when the primary phases of the Moon occur. How well does the phase of the Moon predict the tides?
3. The best place to aim a telescope or binoculars on the Moon is along the terminator line, the line between the Moon's light and dark hemispheres. Why? If you were standing on the lunar terminator, where would the Sun be in your sky? What time of day is it when you're standing on Earth's terminator line?
4. Watch the Moon over a period of hours on a night when you can see one or more bright stars near it. Estimate how many Moon diameters the Moon moves per hour. Knowing the Moon is about 0.5° in diameter, how many degrees per hour does it move? What is your estimate of its orbital period?

6 THE TERRESTRIAL PLANETS

A Study in Contrasts

LEARNING GOALS

Studying this chapter will enable you to:

1. Explain how Mercury's rotation has been influenced by its orbit around the Sun.
2. Describe how the atmospheres of Venus and Mars differ from one another and from Earth's.
3. Compare the surface of Mercury with that of the Moon and the surfaces of Venus and Mars with that of Earth.
4. Describe how we know that Mars once had running water and a thick atmosphere.
5. Discuss the similarities and differences in the geological histories of the four terrestrial planets.
6. Explain why the atmospheres of Venus, Mars, and Earth are now so different from one another.

(Opposite page, background) The red planet, Mars, shown in true color here, displays an array of fascinating surface features. Most prominent in this *Viking*-mission mosaic is the Mariner Valley, a vast "canyon" named after an earlier spacecraft, *Mariner*, that discovered it. The valley extends for approximately 4000 km—roughly the width of the United States. *(NASA)*

With Earth and the Moon as our guides, we now expand our field of view to study the other planets of the inner solar system. As we explore these worlds and seek to understand the similarities and differences among them, we begin our comparative study of the only planetary system we know. Mercury, the smallest terrestrial planet, in many ways is kin to Earth's Moon, and much can be learned by comparing Mercury with our own satellite. Venus and Mars both have properties more like Earth's, and we learn about these two terrestrial worlds by drawing parallels with our own planet. It is possible that Venus, Earth, and Mars had many similarities when they formed, yet Earth today is vibrant, teeming with life, while Venus is an uninhabitable inferno and Mars is a dry, dead world. What were the factors leading to these present conditions? In answering this question, we will discover that a planet's environment, as well as its composition, can play a critical role in determining its future.

6.1 Orbital and Physical Properties

Mercury, the innermost planet, lies close to the Sun and is visible above the horizon for at most two hours before the Sun rises or after it sets. Being somewhat farther from the Sun, the next planet, Venus, is visible for a little longer—up to three hours, depending on the time of year.

Venus is the third brightest object in the entire sky (only the Sun and the Moon are brighter). Like all the planets, it shines by reflected sunlight. It is so bright because almost all the sunlight reaching it is reflected from thick clouds that envelop the planet. You can even see Venus in the daytime if you know where to look. The much fainter Mercury is visible to the naked eye only when the Sun's light is blotted out—just before dawn, just after sunset, and during a total solar eclipse. Orange-red Mars is also quite easy to spot in the night sky. Because of its less reflective surface, smaller size, and greater distance from the Sun, Mars does not appear as bright as Venus, as seen from Earth. However, at its brightest, Mars is still brighter than any star.

Table 6.1 expands Table 5.1 to include Mercury, Venus, and Mars, and adds two additional properties: surface temperature and surface atmospheric pressure. In seeking to understand the other terrestrial planets, astronomers are guided by our much more detailed knowledge of Earth and the Moon. Mercury's high average density, for instance, tells us that this planet must have a large iron core, consistent with the condensation theory's account of its formation. ∞ (Sec. 4.3) However, in many other respects Mercury is similar to Earth's Moon, leading us to use the Moon as a model for understanding Mercury's past. Venus and, to a lesser extent, Mars are more similar to Earth, so Earth provides the natural starting point for studies of those planets. For example, despite the lack of seismic data, we nevertheless assume that Venus has a metallic core and rocky mantle similar to those shown for Earth in Figure 5.1. Even the widely different atmospheres of these two worlds can be explained in familiar earthly terms.

TABLE 6.1 Some Properties of the Terrestrial Planets and Earth's Moon

	MASS		RADIUS		AVERAGE DENSITY	SURFACE GRAVITY	ESCAPE SPEED	SURFACE ROTATION PERIOD	SURFACE TEMPERATURE	ATMOSPHERIC PRESSURE
	(kg)	(Earth = 1)	(km)	(Earth = 1)	(kg/m^3)	(Earth = 1)	(km/s)	(solar days)	(K)	(Earth = 1)
Mercury	3.3×10^{23}	0.055	2400	0.38	5400	0.38	4.3	59	100–700	—
Venus	4.9×10^{24}	0.82	6100	0.95	5300	0.91	10.4	−243[1]	730	90
Earth	6.0×10^{24}	1.00	6400	1.00	5500	1.00	11.2	1.00	290	1.0
Mars	6.4×10^{23}	0.11	3400	0.53	3900	0.38	5.0	1.03	180–270	0.007
Moon	7.3×10^{22}	0.012	1700	0.27	3300	0.17	2.4	27.3	100–400	—

[11]The minus sign indicates retrograde rotation.

6.2 Rotation Rates

Mercury's Curious Spin

1 From Earth, even through a large telescope, we see Mercury only as a slightly pinkish, almost featureless disk. The largest ground-based telescopes can resolve features on the surface of Mercury about as well as we can perceive features on our Moon with our unaided eyes. Figure 6.1 is one of the few photographs of Mercury taken from Earth that show any indication of surface features. In the days before close-up images were obtainable from space, astronomers could only speculate about the faint, dark markings this photograph reveals.

In the mid-nineteenth century, an Italian astronomer named Giovanni Schiaparelli attempted to measure Mercury's rotation rate by watching surface features move around the planet. He concluded that Mercury always keeps one side facing the Sun, much as our Moon always presents only one face to Earth. The explanation suggested for this synchronous rotation was the same as for the Moon: The tidal bulge raised in Mercury by the Sun had modified the planet's rotation rate until the bulge always points directly at the Sun. ∞ (Sec. 5.2) Although the surface features could not be seen clearly, the combination of Schiaparelli's observations and a plausible physical explanation was enough to convince most astronomers. The belief that Mercury rotated synchronously with its revolution about the Sun (once every 88 Earth days) persisted for almost a century.

In 1965 astronomers making radar observations of Mercury from the Arecibo radio telescope in Puerto Rico discovered that this long-held view was in error. ∞ (Sec. 3.4) The technique they used is illustrated in Figure 6.2, which shows a radar signal reflecting from the surface of a hypothetical planet. Because the planet is rotating, a pulse of outgoing radiation of a single frequency is broadened—"smeared out"—by the Doppler effect, by an amount that depends on the planet's rotation speed. ∞ (*More Precisely 2-3*) The radiation reflected from the side moving toward us returns at a slightly higher frequency than does the radiation reflected from the receding

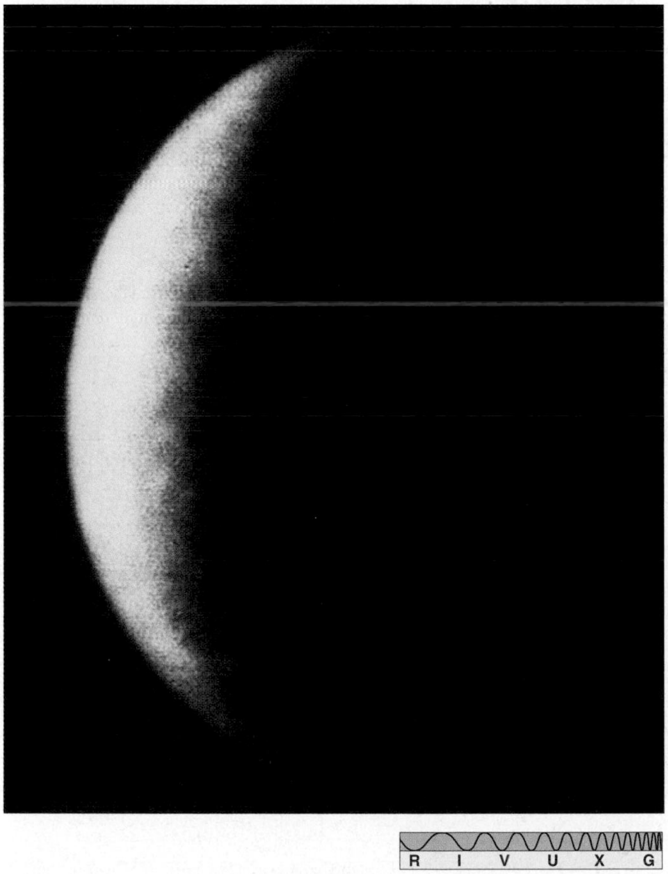

Figure 6.1 Mercury Photograph of Mercury taken from Earth with one of the largest ground-based optical telescopes. Only a few faint surface features are discernible. *(Palomar Observatory/California Institute of Technology)*

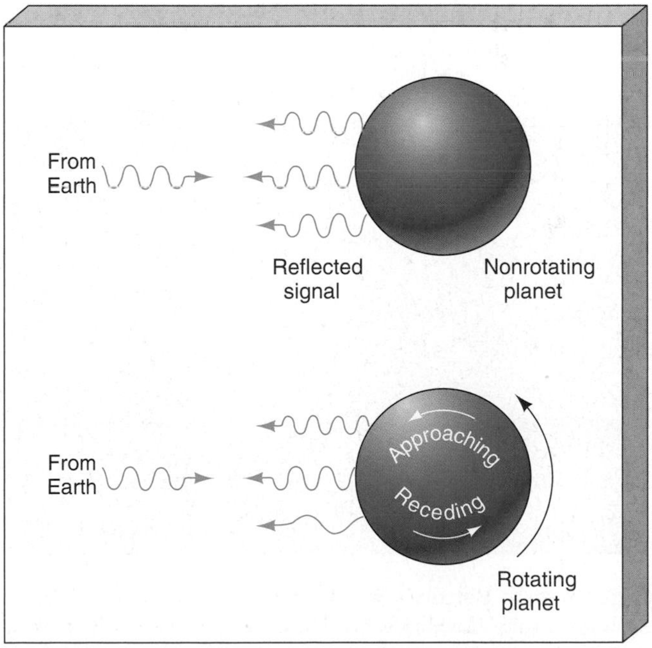

Figure 6.2 Planetary Radar A radar beam reflected from a rotating planet yields information both about the planet's line-of-sight motion and about its rotation rate.

side. (Think of the two hemispheres as separate sources of radiation moving at slightly different velocities, one toward us and one away from us.) By measuring the extent of this broadening, we can determine the rate at which the planet rotates.

In this way, astronomers found that the rotation period of Mercury is not 88 days, as had previously been believed, but just under 59 days—in fact, exactly two-thirds of Mercury's year. This odd state of affairs surely did not occur by chance. In fact, nineteenth-century astronomers were correct in thinking that Mercury's rotation was governed by the tidal effect of the Sun. However, the combination of the Sun's gravity and Mercury's eccentric orbit has caused the planet's rotation to be more complicated than that of the Moon. Unable to come into a state of precisely synchronous rotation (because Mercury's orbital speed changes significantly from place to place in its orbit), Mercury did the next best thing. It presents the same face to the Sun not every time around, but every other time. Figure 6.3 illustrates the implications of this odd rotation for a hypothetical inhabitant of Mercury. The planet's solar day—the time from one noon to the next—is two Mercury years long!

The Sun also influences the tilt of Mercury's spin axis. Because of the Sun's tides, Mercury's rotation axis is almost exactly perpendicular to its orbit plane. Thus, the noontime Sun is always directly overhead for someone standing on the equator and always on the horizon for someone standing at either pole.

Venus and Mars

The same clouds whose reflectivity make Venus so easy to see in the night sky also make it impossible for us to discern any surface features, at least in visible light. Figure 6.4, one of the best photographs of Venus taken with an Earth-based telescope, shows an almost uniform yellow-white disk, with rare hints of clouds. Because of the cloud cover, astronomers did not know Venus's rotation period until the development of radar techniques in the 1960s, when Doppler broadening of returning radar echoes indicated an unexpectedly sluggish 243-day rotation period. Furthermore, Venus's spin was found to be *retrograde*—that is, opposite that of Earth and most other solar system objects and in the direction opposite Venus's orbital motion. The planet's rotation axis is almost exactly perpendicular to its orbit plane, just as Mercury's is.

We have no "evolutionary" explanation for Venus's anomalous rotation. It is not the result of any known interaction with the Sun, Earth, or any other solar-system body. At present, the best explanation planetary scientists can offer is that, during the final stages of Venus's formation in the early solar system, the planet was struck by a large body, much like the one that may have hit Earth and formed the Moon. ∞ (Sec. 4.3, 5.8) That impact was sufficient to reduce the planet's spin almost to zero, leaving Venus rotating in the manner we now observe.

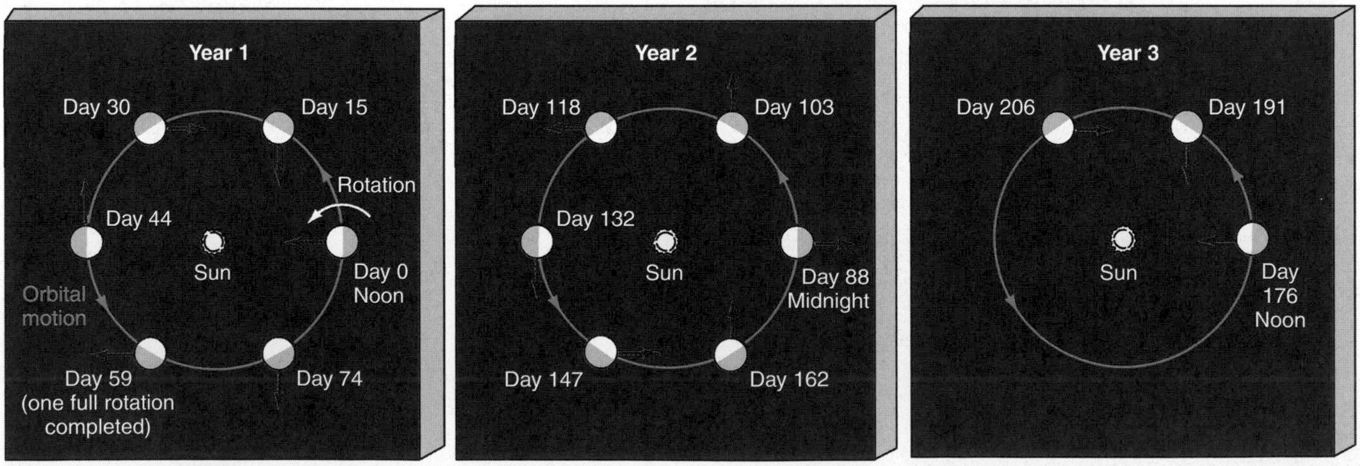

Figure 6.3 **Mercury's Rotation** Mercury's orbital and rotational motions combine to produce a solar day that is two Mercury years long. The pink arrows represent an observer standing on the surface of the planet. At day 0 (center right in Year 1 drawing), it is noon for our observer and the Sun is directly overhead. By the time Mercury has completed one full orbit around the Sun and moved from day 0 to day 88 (center right in Year 2 drawing), it has rotated on its axis exactly 1.5 times, so that it is now midnight at the observer's location. After another complete orbit, it is noon once again on day 176 (center right in Year 3 drawing).

Animation

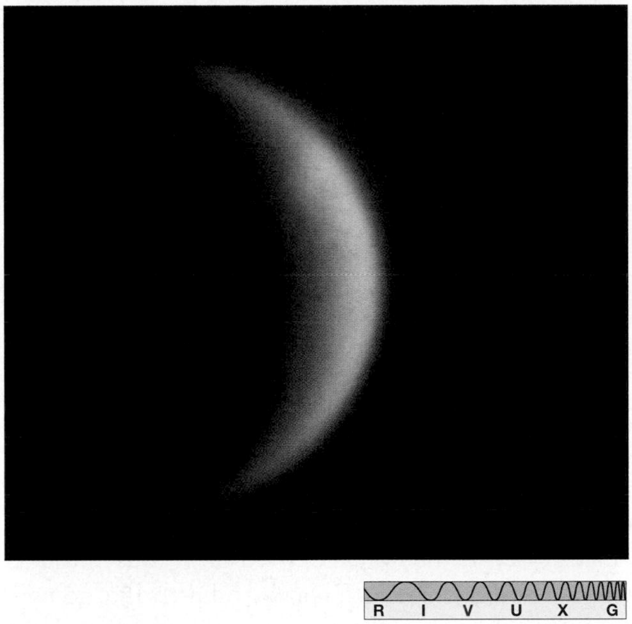

Figure 6.4 Venus This photograph taken from Earth shows Venus with its creamy yellow mask of clouds. *(NOAO)*

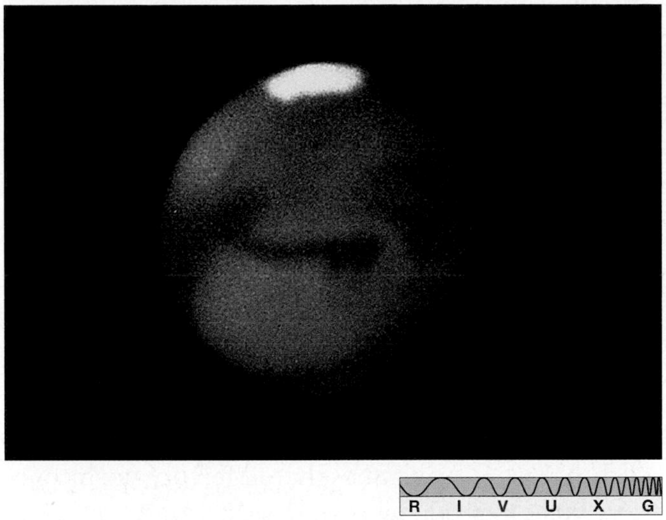

Figure 6.5 Mars A deep-red (800-nm) image of Mars taken in 1991 at Pic du Midi, an exceptionally clear site in the French Alps. One of the planet's polar caps appears at the top, and numerous other surface markings are visible. *(Copyright Observatoire du Pic du Midi (France, CNRS and Université Paul Sabatier)*

In contrast to Mercury and Venus, surface markings are easily seen on Mars (Figure 6.5), allowing astronomers to track the planet's rotation. Mars rotates once on its axis every 24.6 hours—close to one Earth day. The planet's equator is inclined to the orbit plane at an angle of 24.0°, very similar to Earth's inclination of 23.5°. Thus, as Mars orbits the Sun, we find both daily and seasonal cycles, just as on Earth. In the case of Mars, however, the seasons are complicated somewhat by variations in solar heating due to the planet's eccentric orbit. Figure 6.6 summarizes the rotations and revolutions of the four terrestrial planets.

✓ Concept Check

- How has the Sun's gravity influenced Mercury's rotation?

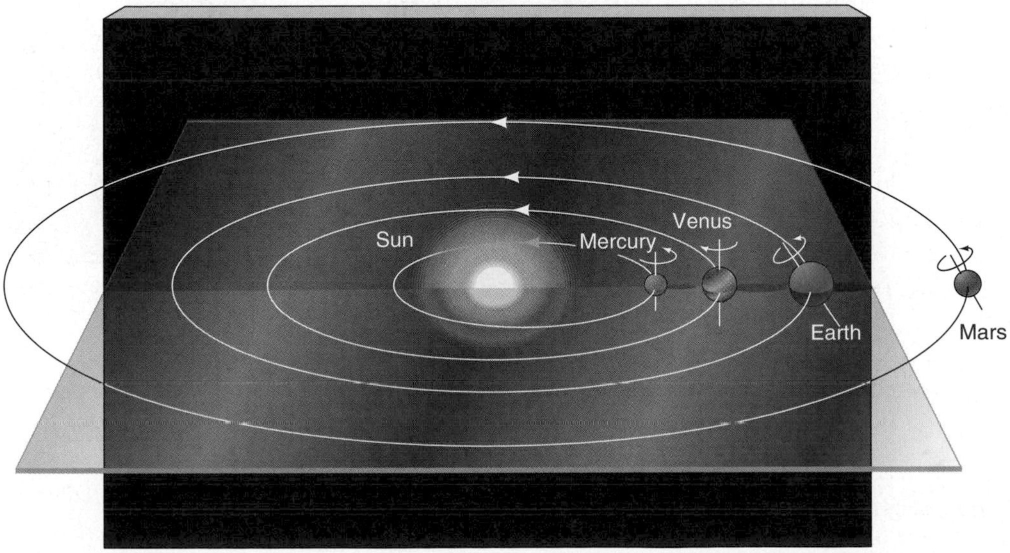

Figure 6.6 Terrestrial Planets' Spin The inner planets of the solar system—Mercury, Venus, Earth, and Mars—display widely different rotational properties. Although all orbit the Sun in the same direction and in nearly the same plane, Mercury's rotation is slow and prograde, that of Venus is slow and retrograde, Earth's is fast and prograde, and that of Mars is fast and prograde. Venus rotates clockwise as seen from above the plane of the ecliptic, but Mercury, Earth, and Mars all rotate counterclockwise.

6.3 Atmospheres

Mercury

Mercury has no appreciable atmosphere. Although the U.S. space probe *Mariner 10*[1] did find a trace of what was at first thought to be an atmosphere, the gas is now known to be temporarily trapped hydrogen and helium from the solar wind. Mercury holds this gas for just a few weeks before it leaks away into space. The absence of an atmosphere on Mercury is readily explained by the planet's high surface temperature (up to 700 K at noon on the equator) and low mass (only 4.5 times the mass of the Moon). Any atmosphere Mercury might once have had escaped long ago. ∞ (*More Precisely 5-2*)

With no atmosphere to retain heat, Mercury's surface temperature falls to about 100 K during the planet's long night. Mercury's 600 K temperature range is the largest of any planet or moon in the solar system. Near the poles, where the Sun's light arrives almost parallel to the surface, the temperature remains low at all times. Recent Earth-based radar studies suggest that Mercury's polar temperature could be as low as 125 K and that the poles may be permanently covered with extensive thin sheets of water ice.

Venus

2 In the 1930s, scientists used spectroscopy to measure the temperature of Venus's upper atmosphere and found it to be about 240 K, not much different from that of Earth's stratosphere. ∞ (Sec. 5.3) Taking into account the cloud cover and Venus's nearness to the Sun, and assuming that Venus had an atmosphere much like our own, researchers concluded that Venus might have an average surface temperature only a few degrees higher than Earth's. In the 1950s, however, when radio observations of the planet penetrated the cloud layer and gave the first indication of conditions near the surface, they revealed a temperature exceeding 600 K! Almost overnight, the popular conception of Venus changed from lush tropical jungle to arid, uninhabitable desert.

Since then, spacecraft data have revealed the full extent of the differences between the atmospheres of Venus and Earth. Venus's atmosphere is much more massive than our own, and it extends to a much greater height above the planet's surface. The surface pressure on Venus is about 90 times the pressure at sea level on Earth, equivalent to an (Earth) underwater depth of about 1 km. (Unprotected humans cannot dive much below 100 m.) The surface temperature is a sizzling 730 K.

The dominant constituent (96.5 percent) of Venus's atmosphere is carbon dioxide. Almost all of the remaining 3.5 percent is nitrogen. Given Venus's similarity to Earth in mass, radius, and location in the solar system, it is commonly assumed that Venus and Earth must have started off looking somewhat alike. However, there is no sign of the large amount of water vapor that would be present if a volume of water equivalent to Earth's oceans had once existed on Venus and later evaporated. If Venus started off with Earth-like composition, something happened to its water, for the planet is now an exceedingly dry place. Even the highly re-

Figure 6.7 Venus, Up Close Venus as photographed by the *Pioneer* spacecraft's cameras 200,000 km away from the planet. This image was made by capturing solar ultraviolet radiation reflected from the planet's clouds, which are composed mostly of sulfuric acid droplets, much like the corrosive acid in a car battery. *(Edmund Scientific Company)* Animation

[1]*Mariner 10* is one of the few spacecraft to have visited Mercury. In 1973 it was placed in an eccentric orbit around the Sun that brings it close to Mercury every 176 days (two Mercury years). The three close encounters it made with Mercury in 1974 and 1975 (after which it ran out of fuel) are the source of virtually all the detailed information we have on that planet.

flective clouds are composed not of water vapor, as on Earth, but of sulfuric acid droplets.

Venus's atmospheric patterns are much more evident when examined with equipment capable of detecting ultraviolet radiation. Some of Venus's upper-level clouds absorb this high-frequency radiation, thereby increasing the contrast. Figure 6.7 is an ultraviolet image taken in 1979 by the U.S. *Pioneer Venus* spacecraft from a distance of 200,000 km (compare the optical image shown in Figure 6.4). The large, fast-moving cloud patterns lie between 50 and 70 km above the surface. Upper-level winds reach speeds of 400 km/h relative to the planet. Below the clouds, extending down to an altitude of 30 km, is a layer of haze. Below 30 km, the air is clear.

Mars

2 Well before the arrival of spacecraft, astronomers knew from Earth-based spectroscopy that the Martian atmosphere is quite thin and composed primarily of carbon dioxide. Spacecraft measurements confirmed these results, indicating that the atmospheric pressure is only about 1/150 the pressure of Earth's atmosphere at sea level. The Martian atmosphere is 95.3 percent carbon dioxide, 2.7 percent nitrogen, and 1.6 percent argon, plus small amounts of oxygen, carbon monoxide, and water vapor. While there is some superficial similarity in composition between the atmospheres of Mars and Venus, the two planets clearly must have had very different atmospheric histories. Average surface temperatures on Mars are about 50 K cooler than on Earth.

Concept Check

- Describe some important differences between the atmospheres of Venus and Earth.

6.4 The Surface of Mercury

3 Figure 6.8 shows a picture of Mercury taken when the *Mariner 10* spacecraft was 200,000 km from the planet, and Figure 6.9 shows a higher-resolution photograph of the planet from a distance of 20,000 km. The similarities to our Moon are striking. Indeed, much of the cratered surface bears a strong resemblance to the Moon's highlands. The crater walls are generally not as high as on the Moon, and the ejected material landed closer to the impact site, exactly as we would expect on the basis of Mercury's greater surface gravity (which is a little more than twice that of the Moon). Mercury, however, shows no extensive lava flow regions akin to the lunar maria.

As on the Moon, Mercury's craters are the result of meteoritic bombardment. The craters are not so densely packed as their lunar counterparts, however, and there are extensive **intercrater plains**. One likely explanation for Mercury's relative lack of craters is that the older impact craters have been filled in by volcanic activity, in much the same way as the Moon's maria filled in craters as they formed. However, Mercury's intercrater plains do not look much like mare material. They are much lighter in color and not as flat.

Figure 6.8 Mercury, Up Close Mercury is imaged here as a mosaic of photographs taken by the *Mariner 10* spacecraft in the mid-1970s during its approach to the planet. At the time, the spacecraft was 200,000 km away from Mercury. *(NASA)*

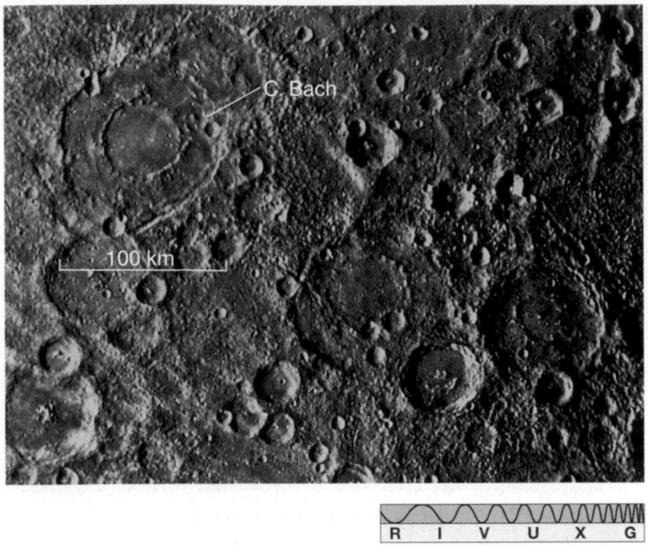

Figure 6.9 Mercury, Very Close Another photograph of Mercury by *Mariner 10*, this time from a distance of about 20,000 km. The double-ringed crater at the upper left, named C. Bach and about 100 km across, exemplifies how many of the large craters on Mercury tend to form double, rather than single, rings. The reason is not yet understood. *(NASA)*

Mercury has at least one type of surface feature not found on the Moon. Figure 6.10 shows a **scarp**, or cliff, that does not appear to be the result of volcanic or other familiar geological activity. The scarp cuts across several craters, which indicates that whatever produced it occurred after most of the meteoritic bombardment was over. Numerous scarps have been found in the *Mariner* images. Mercury shows no evidence of crustal motions, so the scarps could not have been formed by tectonic processes such as those responsible for fault lines on Earth. Instead, the scarps probably formed when the crust cooled, shrank, and split long ago. After the splitting, one side of the crack moved upward relative to the other side, forming the scarp cliff face. If we can apply to Mercury the cratering age estimates we use for the Moon, the scarps appeared about four billion years ago.

Much of the discussion in Chapter 5 about the surface of Earth's Moon applies equally well to Mercury. (Sec. 5.6) Figure 6.11 shows the result of what was probably the last great event in the geological history of Mercury—an immense bull's-eye crater called the Caloris

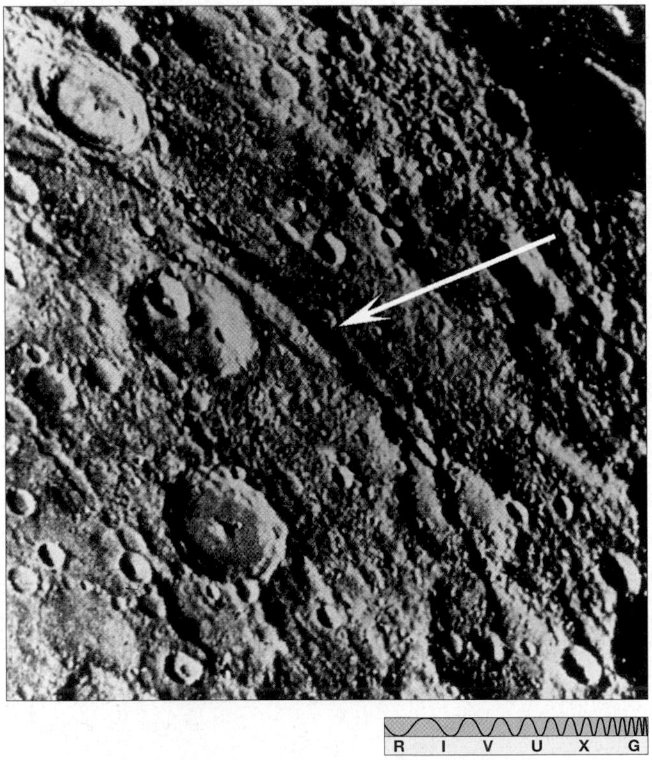

Figure 6.10 Mercury's Surface Discovery Scarp on Mercury's surface. This cliff appears to have formed when the planet's crust cooled and shrank early in its history. The shrinking caused the crust to split. Subsequently the crust on one side of the split moved upward relative to the crust on the other side, forming the scarp. Several hundred kilometers long and up to 3 km high in places, it runs diagonally across the center of the frame in this photograph. *(NASA)*

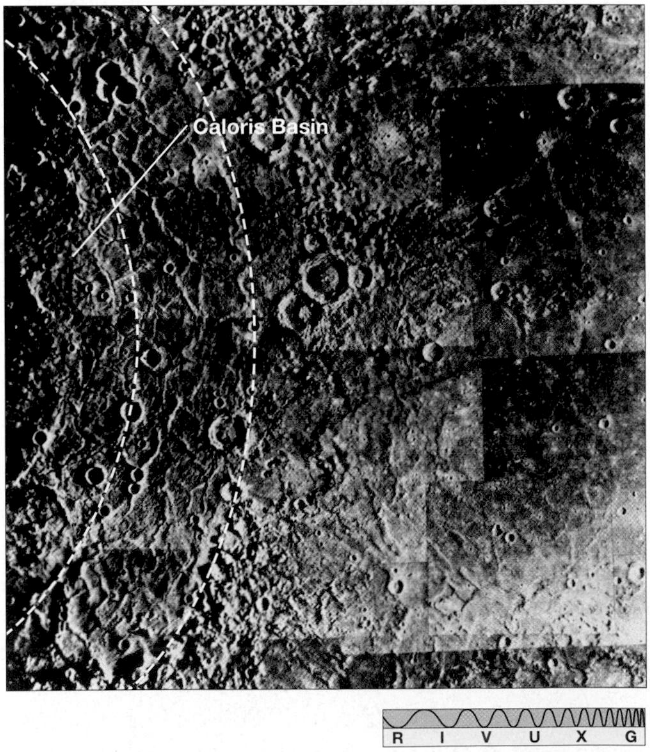

Figure 6.11 Mercury's Basin Mercury's most prominent geological feature—the Caloris Basin—measures about 1400 km across and is ringed by concentric mountain ranges (dashed lines) that reach more than 3 km high in places. This huge circular basin, only half of which shows in this *Mariner 10* photograph, is similar in size to the Moon's Mare Imbrium and spans more than half of Mercury's radius. *(NASA)*

Basin, formed eons ago by the impact of a large asteroid. (Because of the orientation of the planet during *Mariner 10*'s flybys, only half the basin is visible. The center is off the left-hand side of the photograph.) Compare this basin with the Orientale Basin on the Moon (Figure 5.25).

✓ Concept Check

- How do scarps on Mercury differ from geological faults on Earth?

6.5 The Surface of Venus

3 Although the clouds are thick and the surface of Venus totally shrouded, we are by no means ignorant of the planet's surface. Radar astronomers have bombarded Venus with radio signals, and analysis of the radar echoes yields a map of surface features. Except for Figure 6.16, all the views of Venus in this section are "radargraphs" (not photographs) created in this way. Most recently, the U.S. *Magellan* spacecraft has provided very-high-resolution radar images of the planet.

Large-Scale Topography

Figure 6.12(a) shows a relatively low-resolution map of Venus made by *Pioneer Venus* in 1979. Surface elevation above the average radius of the planet's surface is indicated by color, with white representing the highest elevations, blue the lowest. (Note that the blue has nothing to do with oceans!) For comparison, Figure 6.12(b) shows a map of Earth to the same scale and at the same spatial resolution. Figure 6.13 is a 1995 mosaic of *Magellan* images of Venus. The orange color is based on optical data returned from spacecraft that landed on the planet.

Venus's surface appears to be mostly smooth, resembling rolling plains with modest highlands and lowlands. Only two continent-sized features, called Ishtar Terra and Aphrodite Terra, adorn the landscape, and these contain mountains comparable in height to those on Earth. The highest peaks rise some 14 km above the level of the deepest surface depressions. (The highest point on Earth, the summit of Mount Everest, lies about 20 km above the deepest section of the ocean floor.) The elevated "continents" occupy only eight percent of Venus's total surface area whereas continents on Earth make up about 25 percent of our planet's surface.

The larger continent-sized formation, Aphrodite Terra, is located on Venus's equator. It is comparable in size to Africa. Before *Magellan*'s arrival, some researchers had speculated that Aphrodite Terra might have been the site of something equivalent to seafloor spreading on Earth, where two tectonic plates moved apart and molten rock rose to the surface in the gap between them, forming an extended ridge. This is just what is happening today at Earth's Mid-Atlantic Ridge, which is clearly visible in Figure 6.12(b). However, the Magellan images now seem to rule out any tectonic activity on Venus, and the Aphrodite region shows no signs of spreading. The crust appears buckled and fractured, suggesting large compressive forces, and there seem to have been numerous periods when extensive lava flows occurred.

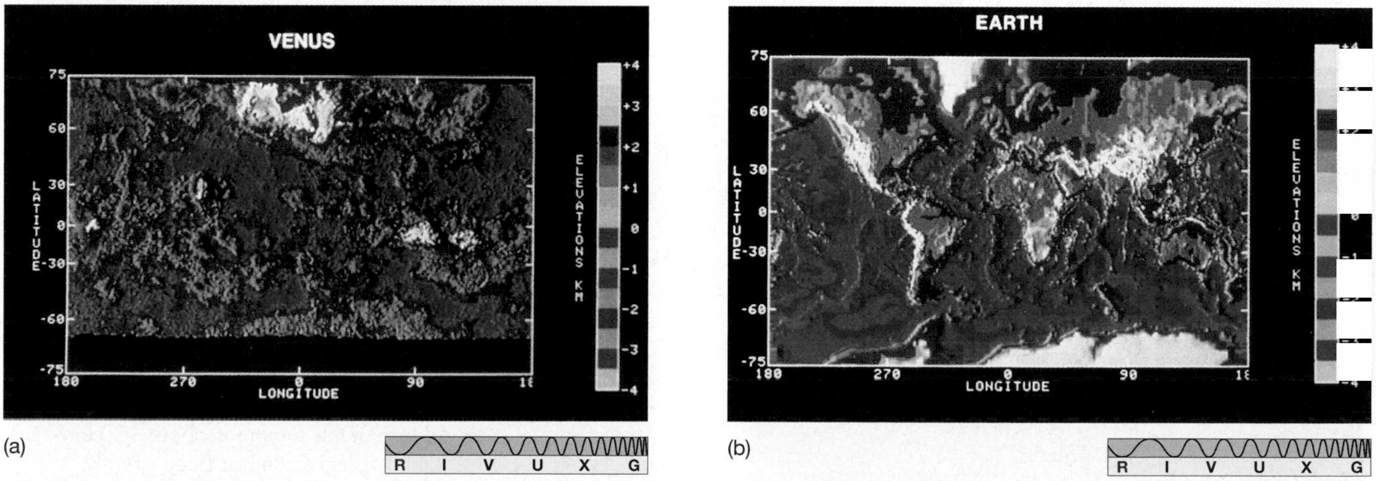

Figure 6.12 Venus Radar Map (a) Radar map of the surface of Venus, based on *Pioneer Venus* data. Color represents elevation, with white the highest areas and blue the lowest. (b) A similar map of Earth, at the same spatial resolution. *(NASA)*

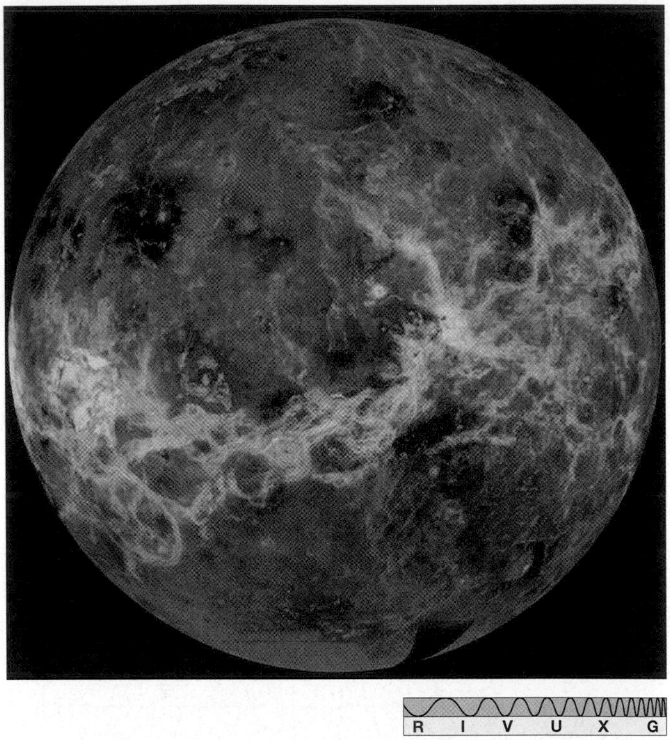

Figure 6.13 **Venus Magellan Map** A planetwide mosaic of Venus made from *Magellan* images. The largest "continent" on Venus, Aphrodite Terra, is the yellow "dragon-shaped" area across the center of the image. *(Russian Space Agency)*

Volcanism and Cratering

Although erosion by the planet's atmosphere may play some part in obliterating surface features, the most important factor is volcanism, which appears to "resurface" the planet every few hundred million years. Many areas of Venus have volcanic features. Figure 6.14(a) shows a *Magellan* image of seven pancake-shaped lava domes, each about 25 km across. They probably formed when lava oozed out of the surface, formed the dome, then withdrew, leaving the crust to crack and subside. Lava domes such as these are found in several locations on Venus. Figure 6.14b shows a computer-generated three-dimensional view of the domes.

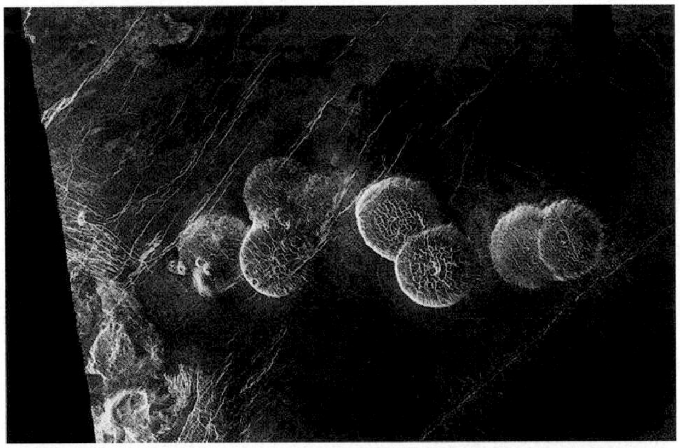

(a)

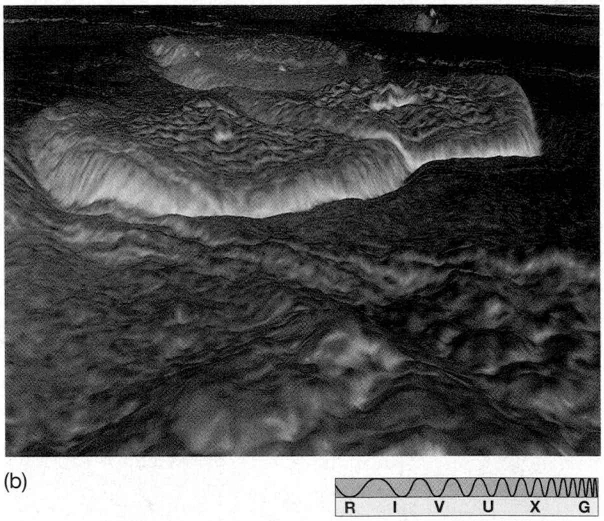

(b)

(c)

Figure 6.14 **Venus's Surface Features** (a) A series of dome-shaped structures on Venus. They are the result of molten rock having bulged out of the ground and then retreated, leaving behind a thin, solid crust that subsequently cracked and subsided. *Magellan* found features like this in several locations on Venus. (b) A computer-generated three-dimensional representation of four of the domes. (c) A three-dimensional *Magellan* view of the large shield volcano known as Gula Mons. The volcanic caldera at the summit is about 100 km across. As in part (b), the vertical scale has been greatly exaggerated. The volcano is about 4 km high. Color in (b) and (c) is based on data returned by Soviet landers. *(NASA)*

The most common volcanoes on the planet are of the type known as **shield volcanoes**. Those on Earth are associated with lava welling up through a "hot spot" in the crust (like the Hawaiian Islands). They are built up over long periods of time by successive eruptions and lava flows. A characteristic of shield volcanoes is the formation of a *caldera*, or crater, at the summit when the underlying lava withdraws and the surface collapses. A large shield volcano, called Gula Mons, is shown (again as a computer-generated view) in Figure 6.14(c).

The largest volcanic structures on Venus are huge, roughly circular regions known as **coronae**. A large corona, called Aine, can be seen in Figure 6.15. Coronae are unique to Venus. They appear to have been caused by upwelling motions in the mantle that caused the surface to bulge outward, but never developed into full-fledged convection as on Earth. Coronae generally have volcanoes both in and around them, and their rims usually show evidence of extensive lava flows into the plains below.

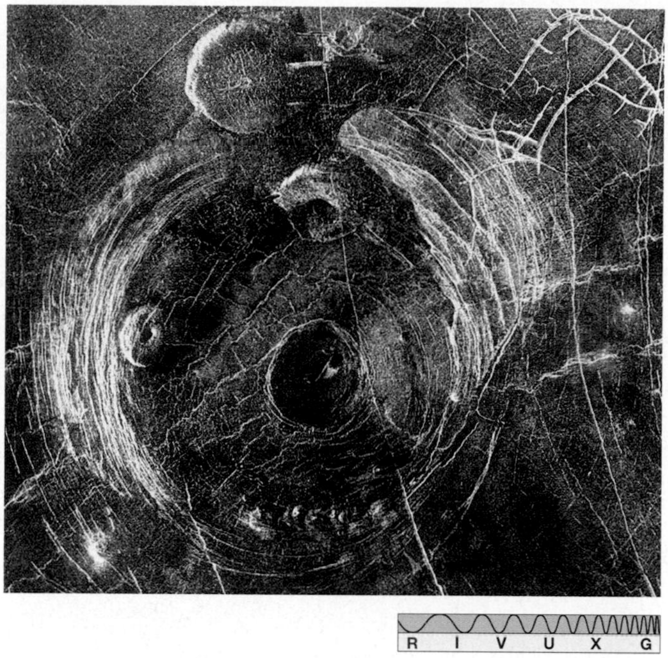

Figure 6.15 **Venus Corona** This corona, called Aine, lies in the plains south of Aphrodite Terra. It is about 300 km across. Notice the cracked and fissured surface surrounding it. Coronae are probably the result of upwelling mantle material causing the surface to bulge outward. *(NASA)*

Two pieces of indirect evidence suggest that volcanism on Venus continues today. First, the level of sulfur dioxide above Venus's clouds shows large and fairly frequent fluctuations, possibly as the result of volcanic eruptions on the surface. Second, orbiting spacecraft have observed bursts of radio energy from the planet's surface similar to those produced by the lightning discharges that often occur in the plumes of erupting volcanoes on Earth. However, while quite persuasive, these pieces of evidence are still only circumstantial. No "smoking gun" (or erupting volcano) has yet been seen, so the case for active volcanism is not yet complete.

A few Soviet *Venera* spacecraft have landed on Venus's surface. Each survived for about an hour before being destroyed by the intense heat, its electronic circuitry melting in this planetary oven. Figure 6.16 shows one of the first photographs of the surface of Venus radioed back to Earth. The flat rocks visible in this image show little evidence of erosion and are apparently quite young, supporting the idea of ongoing surface activity. Later Soviet landers performed simple chemical analyses of the surface. Some of the samples studied were found to be predominantly basaltic, again implying a volcanic past. Others resembled terrestrial granite.

Figure 6.16 **Venus in Situ** One of the first true-color views of the surface of Venus, radioed back to Earth from the Russian *Venera 14* spacecraft, which made a soft landing in 1975. The amount of sunlight penetrating Venus's cloud cover is about the same as the amount reaching Earth's surface on a heavily overcast day. *(Russian Space Agency)*

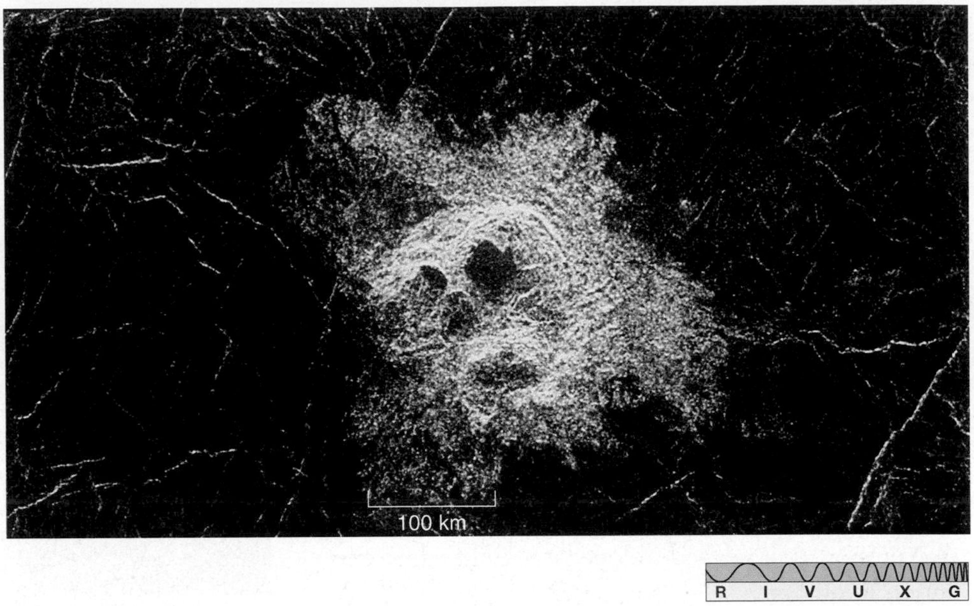

Figure 6.17 Venus's Craters A *Magellan* image of an apparent multiple-impact crater in Venus's southern hemisphere. The irregular shape of the light-colored ejecta blanket seems to be the result of a meteorite that fragmented just prior to impact. The dark regions in the crater may be pools of solidified lava associated with the individual fragments. *(NASA)*

Not all the craters on Venus are volcanic in origin. Some were formed by meteoritic impact. The largest impact craters on Venus are generally circular, but those less than about 15 km in diameter can be quite asymmetric. Figure 6.17 shows a *Magellan* image of a relatively small impact crater, about 10 km across, in Venus's southern hemisphere. Geologists believe that the light-colored region is an ejecta blanket—material thrown from the crater following the impact. Its irregular shape may be the result of a large meteorite that broke up just before impact, with the separate pieces hitting the surface near one another. Numerous impact craters, again identifiable by their ejecta blankets, can also be discerned in Figure 6.15.

Concept Check

- Are the volcanoes on Venus mainly associated with the movement of tectonic plates, as on Earth?

6.6 The Surface of Mars

The View From Earth

4 Earth-based observations of Mars at closest approach can distinguish surface features as small as 100 km across—about the same resolution as the unaided human eye can achieve when viewing the Moon. However, when Mars is closest to us and most easily observed, it is also full, so the angle of the Sun's rays does not permit us to see any topographical detail, such as craters or mountains. Even through a large telescope Mars appears only as a reddish disk with some light and dark patches and prominent polar caps (Figure 6.5). Figure 6.18 shows a *Hubble Space Telescope* image of Mars, along with a photograph taken by one of the two U.S. *Viking* spacecraft en route to the planet.

Mars's surface features undergo slow seasonal changes over the course of a Martian year—a consequence of Mars's axial tilt and somewhat eccentric orbit. The polar caps grow or shrink according to the seasons, almost disappearing during the Martian summer. The dark features also vary in size and shape. To fanciful observers around the start of the twentieth century, these changes suggested the annual growth of vegetation but, as with Venus, however, these speculations were not confirmed. The changing polar caps are mostly frozen carbon dioxide, not water ice as at Earth's north and south poles (although smaller "residual" caps of water ice also exist), and the dark regions are just highly cratered and eroded areas on the surface.

During summer in the Martian southern hemisphere, planetwide dust storms sweep up the dry dust and carry it aloft, sometimes for months at a time, eventually depositing it elsewhere on the planet. Repeated covering and uncovering of the Martian landscape gives the impression from a distance of surface variability, but it is only the thin dust cover that changes.

Large-Scale Topography

A striking feature of the terrain of Mars is the marked difference between the northern and southern hemispheres (Figure 6.19). The northern hemisphere is made

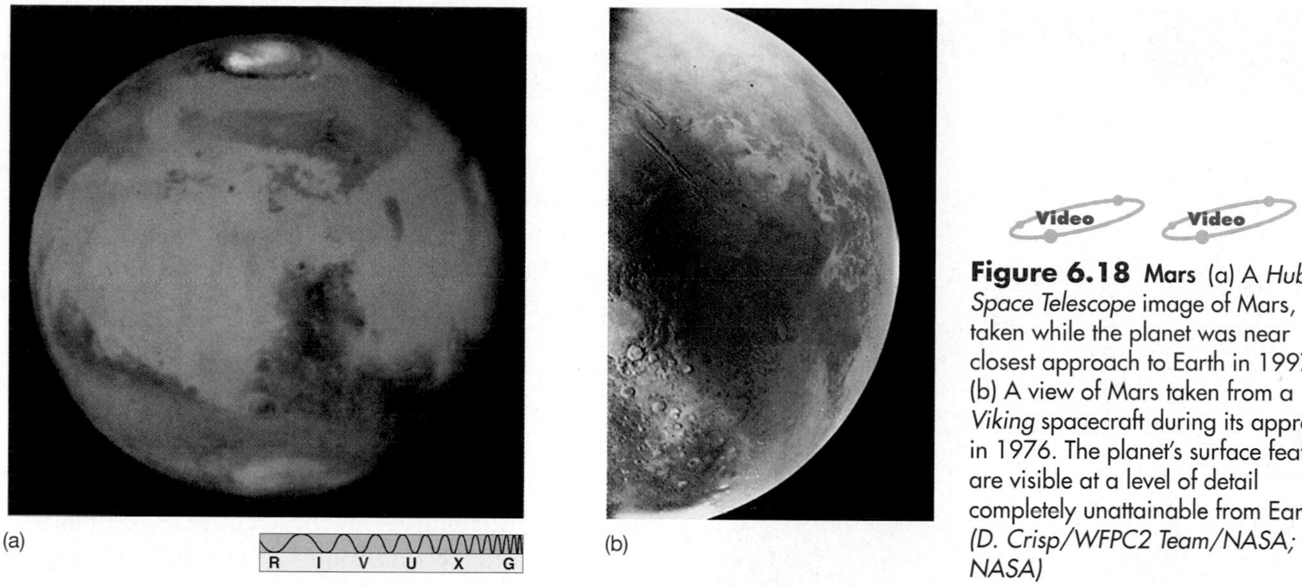

Figure 6.18 **Mars** (a) A *Hubble Space Telescope* image of Mars, taken while the planet was near closest approach to Earth in 1997. (b) A view of Mars taken from a *Viking* spacecraft during its approach in 1976. The planet's surface features are visible at a level of detail completely unattainable from Earth. *(D. Crisp/WFPC2 Team/NASA; NASA)*

up largely of rolling volcanic plains somewhat like the lunar maria but much larger than any plains found on Earth or the Moon. They were apparently formed by eruptions involving enormous volumes of lava. The plains are strewn with blocks of volcanic rock, as well as with boulders blasted out of impact areas by infalling meteoroids. The southern hemisphere consists of heavily cratered highlands lying several kilometers above the level of the lowland north.

The northern plains are cratered much less than the southern highlands, suggesting that the northern surface is younger—perhaps three billion years old, compared with four billion in the south. In places, the boundary between the southern highlands and the northern plains is quite sharp. The surface level can drop by as much as 4 km in a distance of 100 km or so. Most scientists assume that the southern terrain is the original crust of the planet. How most of the northern hemisphere could have been lowered in elevation and flooded with lava remains a mystery.

Figure 6.20 is a large-scale view showing Mars's major geological feature, the *Tharsis bulge*. This region, roughly the size of North America, lies on the Martian equator and rises 10 km higher than the rest of the

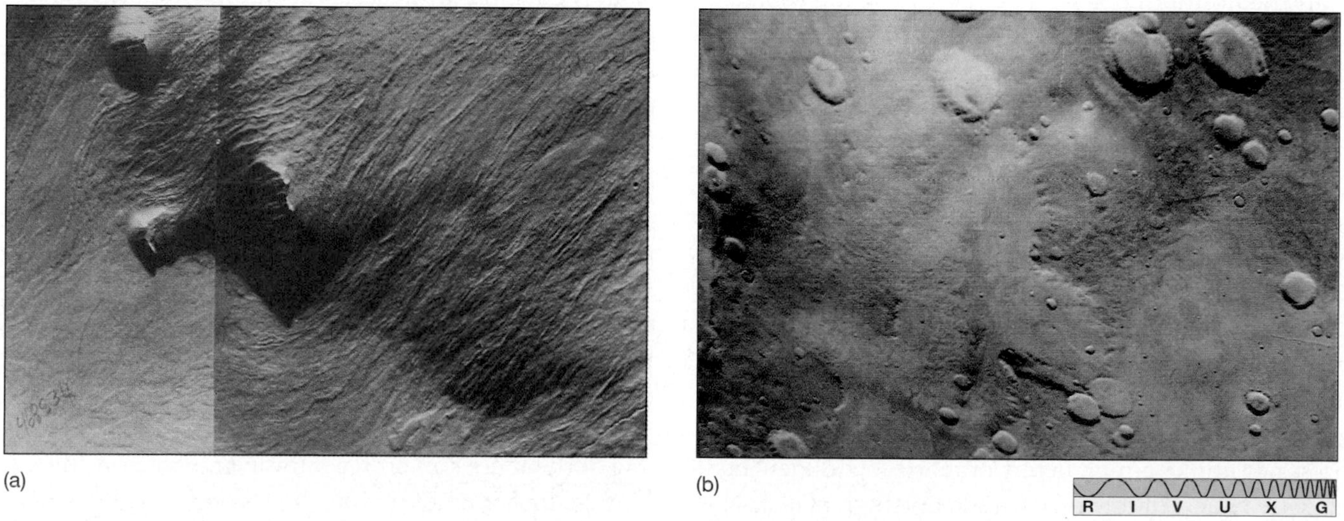

Figure 6.19 **Mars, Up Close** (a) Mars's northern hemisphere consists of rolling volcanic plains (false color image.) (b) The southern Martian highlands are heavily cratered (true color). Both photographs show roughly the same scale, nearly 1000 km across. *(NASA)*

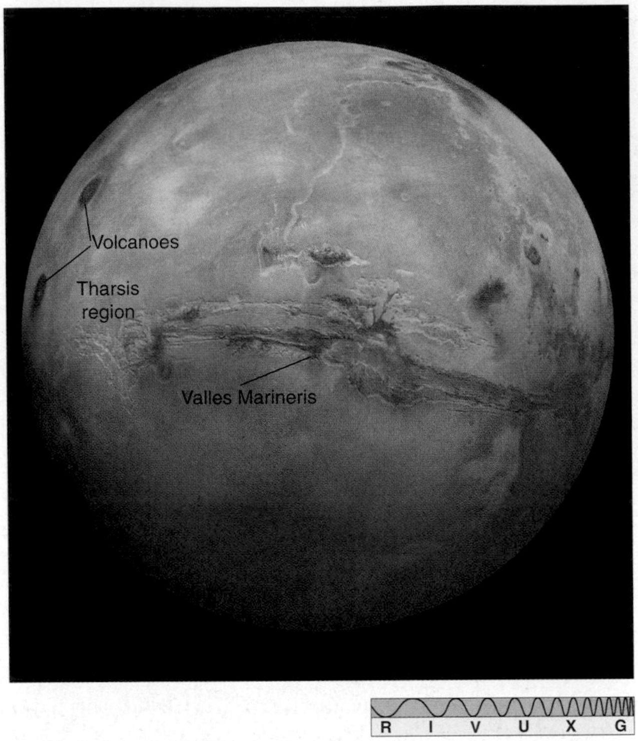

Figure 6.20 Mars Globe The Tharsis region of Mars, 5000 km across, bulges out from the planet's equatorial region, rising to a height of about 10 km. The two large volcanoes on the left mark the approximate peak of the bulge. Dominating the center of the field of view is a vast "canyon" known as Valles Marineris—the Mariner Valley. *(NASA)*

Martian surface. To the east and west of Tharsis lie wide depressions, hundreds of kilometers across and up to 3 km deep. Tharsis appears to be even less heavily cratered than the northern hemisphere, making it the youngest region on the planet. It is estimated to be only two to three billion years old. If we wished to extend the idea of "continents" from Earth and Venus to Mars, we would say that Tharsis is the only continent on the Martian surface. However, as on Venus, there is no sign of plate tectonics—the continent of Tharsis is not drifting as its Earth counterparts are.

Both U.S. *Viking* missions dispatched landers to the Martian surface. Figure 6.21(a) is the view from *Viking 1*, which touched down near the planet's equator just east of Tharsis. The photograph shows a windswept, gently rolling, desolate plain littered with rocks of all sizes, not unlike a high desert on Earth. This view may be quite typical of the low-latitude northern plains. The *Viking* landers performed numerous chemical analyses of the rock on Mars's surface. One important finding of these studies was the high iron content of the planet's surface. Chemical reactions between the iron-rich surface soil and trace amounts of oxygen in the atmosphere are responsible for the iron oxide ("rust") that gives Mars its characteristic red color.

The most recent successful mission to the Martian surface was *Mars Pathfinder*. During the unexpectedly long lifetime of its mission (it lasted almost three months instead of the anticipated one), the lander performed measurements of the Martian atmosphere and atmospheric dust while its robot rover *Sojourner* (Figure 6.21b) carried out chemical analyses of the soil and rocks within about 50 m of the parent craft. In addition, over 16,000 images of the region were returned to Earth. Figure 6.21(c) shows a panorama of the *Pathfinder* landing site. The soil chemistry in the vicinity of the landing site was similar to that found by the Viking landers. Analyses of the nearby rocks revealed a chemical makeup different from that of the Martian meteorites found on Earth (see *Interlude 6-1*).

In December 1999, *Mars Polar Lander* failed to make contact with mission scientists on Earth after parachuting to the surface near the Martian south pole on a mission to study the polar ice and the planet's climatic history. The cause of the failure is still uncertain, and NASA's planned future missions to Mars are currently under review.

Volcanism

Mars contains the largest known volcanoes in the solar system. Four particularly large ones are found on the Tharsis bulge, two of them visible in Figure 6.20. Biggest of all is Olympus Mons (Figure 6.22), which lies on the northwestern slope of Tharsis, just over the left horizon of Figure 6.20. Olympus Mons is 700 km in diameter at its base—only slightly smaller than the state of Texas—and rises to a height of 25 km above the surrounding plains.

Like the volcanoes on Venus, those on Mars are not associated with plate motion but instead are shield volcanoes, sitting atop hot spots in the Martian mantle. Spacecraft images of the Martian surface reveal many hundreds of volcanoes. Most of the largest are associated with the Tharsis bulge, but many smaller ones are also found in the northern plains. It is not known whether any of them are still active. However, based on the extent of impact cratering on their slopes, some of them apparently erupted as recently as 100 million years ago.

The great height of Martian volcanoes is a direct consequence of the planet's low surface gravity. As a

Figure 6.21 **Mars Panorama** (a) Panoramic view from the perspective of the *Viking 1* spacecraft now parked on the surface of Mars. The fine-grained soil and the rock-strewn terrain stretching toward the horizon are reddish. Containing substantial amounts of iron ore, the surface of Mars is literally rusting away. The sky is a pale pink, the result of airborne dust. (b) The Mars rover *Sojourner*, seen here using a device called an Alpha Proton X-Ray Spectrometer to determine the chemical composition of a large Martian rock, nicknamed Yogi by mission scientists. For scale, the rock is about 1 m tall, and lies some 10 m from Sojourner's mother ship, *Pathfinder*. (c) This 360° panorama of Mars was taken by the *Pathfinder* lander in 1997. *Sojourner* can be seen up against a rock at center. The "Twin Peaks" on the left-center horizon are 1 to 2 km away. *(NASA)*

shield volcano forms and lava flows and spreads, the new mountain's height depends on its ability to support its own weight. The lower the surface gravity, the less the lava weighs, and the higher the mountain can be. Mars has a surface gravity only 40 percent that of Earth, and its volcanoes rise roughly 2.5 times as high.

The Martian "Grand Canyon"

Associated with the Tharsis bulge is a great "canyon" known as Valles Marineris (the Mariner Valley). Shown in its entirety cutting across the center of Figure 6.20, this enormous crack in the Martian surface is not really a canyon in the terrestrial sense, because running water played no part in its formation. Astronomers believe it was formed by the same crustal forces that pushed the Tharsis region upward, causing the surface to split and crack. Cratering studies suggest that Valles Marineris is at least 2 billion years old. Similar (but smaller) cracks, formed in a similar way, have been found in the Aphrodite Terra region of Venus.

Valles Marineris runs for almost 4000 km along the Martian equator and extends about one-fifth of the way

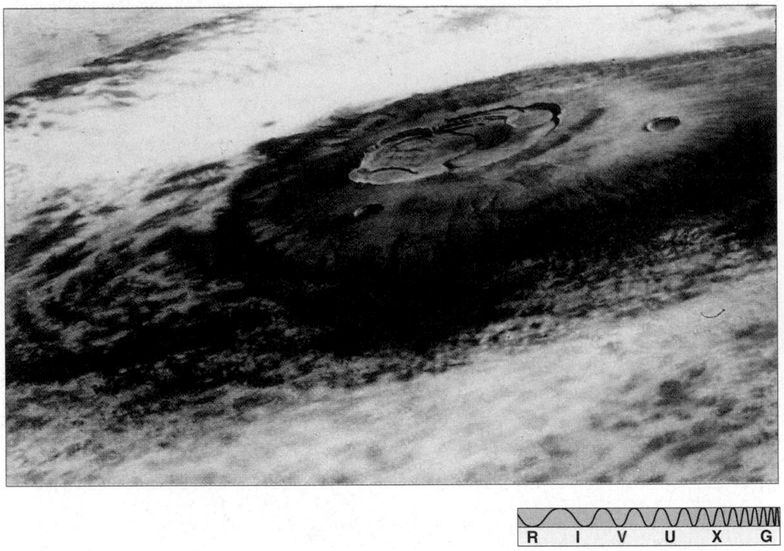

Figure 6.22 Martian Volcano Olympus Mons, the largest volcano known on Mars or anywhere else in the solar system. Nearly three times taller than Mount Everest on Earth, this mountain measures about 700 km across the base, and its peak rises 25 km above the Martian surface. It seems currently inactive and may have been extinct for at least several hundred million years. By comparison, the largest volcano on Earth, Hawaii's Mauna Loa, measures a mere 120 km across and peaks about 9 km above the Pacific Ocean floor. *(NASA)*

around the planet. At its widest, it is 120 km across, and it is as deep as 7 km in places. Earth's Grand Canyon in Arizona would easily fit into one of its side "tributary" cracks. It is so large that it can even be seen from Earth. This Martian geological feature is not a result of Martian plate tectonics, however. For some reason, the crustal forces that formed it never developed into full-fledged plate motion as on Earth.

Evidence for Water on Mars

4 Although the Valles Marineris was not formed by running water, photographic evidence reveals that liquid water once existed in great quantity on the surface of Mars. Two types of flow features are seen: **runoff channels** and **outflow channels**.

The runoff channels (Figure 6.23) are found in the southern highlands. They are extensive systems—some-

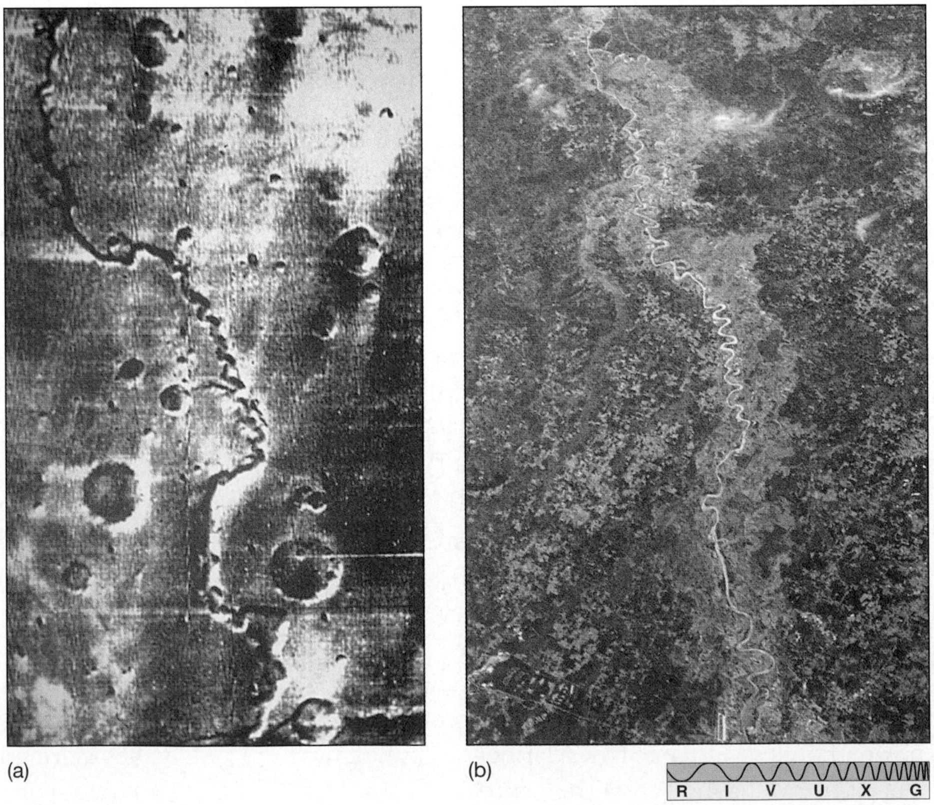

Figure 6.23 Martian Channel (a) This runoff channel on Mars is about 400 km long and 5 km wide. (b) The Red River running from Shreveport, Louisiana, to the Mississippi. Martian runoff channels and rivers on Earth differ mainly in that there is currently no liquid water in this, or any other, Martian channel. *(NASA)*

times hundreds of kilometers in total length—of interconnecting, twisting channels that merge into larger, wider channels. They bear a strong resemblance to river systems on Earth, and geologists believe that that is just what they are: the dried-up beds of long-gone rivers that once carried rainfall on Mars from the mountains down into the valleys. These runoff channels speak of a time four billion years ago (the age of the Martian highlands) when the atmosphere was thicker, the surface warmer, and liquid water widespread.

The outflow channels (Figure 6.24) are probably relics of catastrophic flooding on Mars long ago. They appear only in equatorial regions and generally do not form the extensive interconnected networks that characterize the runoff channels. Instead, they are probably the paths taken by huge volumes of water draining from the southern highlands into the northern plains. Judging from the width and depth of the channels, the flow rates must have been truly enormous—perhaps as much as a hundred times greater than the 10^5 tons per second carried by the Amazon River, the largest river system on Earth. Flooding shaped the outflow channels about three billion years ago, about the same time as the northern volcanic plains formed.

The landing site for the *Pathfinder* mission had been carefully chosen to lie near the mouth of an outflow channel, and the size distribution and composition of the many rocks and boulders surrounding the lander were consistent with their having been deposited there by flood waters. In addition, the presence of numerous rounded pebbles strongly suggested the erosive action of running water at some time in the past.

There is no evidence for liquid water anywhere on Mars today, and the amount of water vapor in the Martian atmosphere is tiny. Yet the extent of the outflow channels indicates that a huge total volume of water existed on Mars in the past. Where did all that water go? One possible answer is that much of Mars's original water is now locked in a layer of **permafrost**, which is water ice lying just below the planet's surface, with some more water contained in the polar caps. Figure 6.25 shows evidence for the permafrost layer in the form of a fairly typical Martian impact crater named Yuty. Unlike the lunar craters discussed in Chapter 5, ∞ (Sec. 5.4) Yuty's ejecta blanket gives the distinct impression of a liquid that has splashed or flowed out of the crater. Most likely, the explosive impact heated and liquefied the permafrost, resulting in the fluid appearance of the ejecta.

Scientists believe that four billion years ago, as the Martian climate changed, the running water that formed the runoff channels began to freeze, forming the permafrost and drying out the river beds. Mars remained frozen for about a billion years, until volcanic (or some other) activity heated large regions of the surface, melting the permafrost and causing the flash

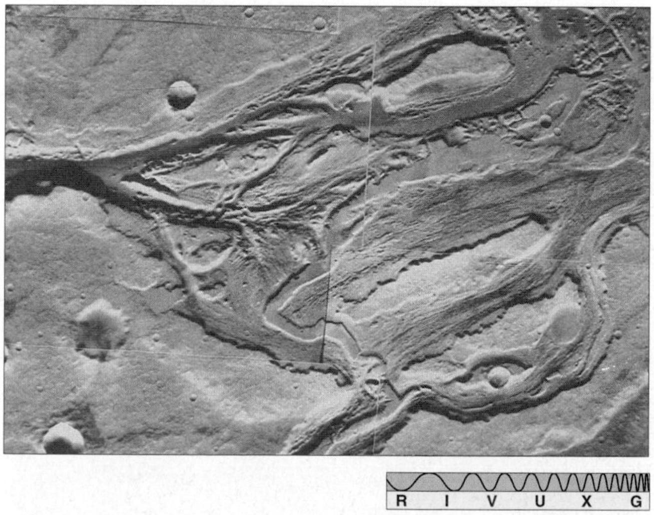

Figure 6.24 Martian Outflow An outflow channel near the Martian equator bears witness to a catastrophic flood that occurred about three billion years ago. *(NASA)*

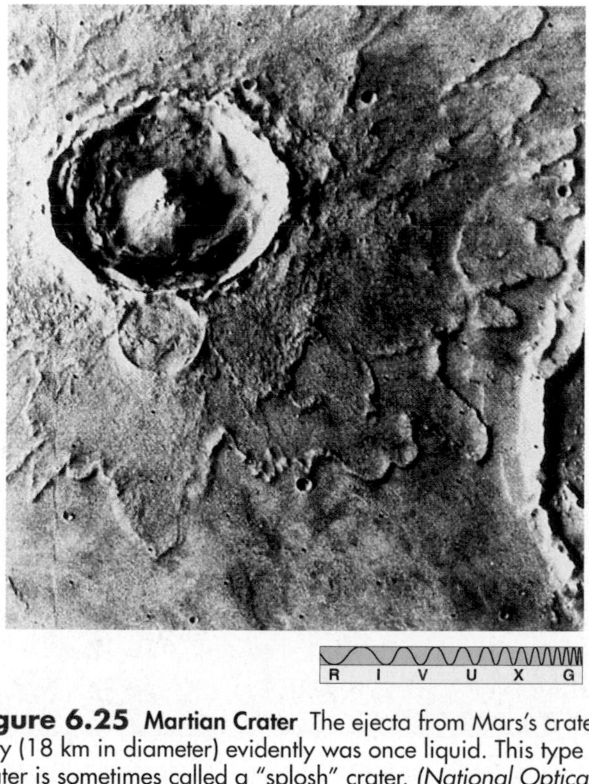

Figure 6.25 Martian Crater The ejecta from Mars's crater Yuty (18 km in diameter) evidently was once liquid. This type of crater is sometimes called a "splosh" crater. *(National Optical Astronomy Observatories)*

floods that created the outflow channels. Subsequently, volcanic activity subsided, the water refroze, and Mars once again became a dry world.

Concept Check

- Is Valles Marineris a true canyon? How did it form?

6.7 Internal Structure and Geological History

Mercury

5 Mercury's magnetic field, discovered by *Mariner 10*, is about 1/100 of Earth's field. The discovery that Mercury has a magnetic field came as a surprise to planetary scientists, who, having detected no magnetic field in the Moon, expected Mercury to have none either. In Chapter 5, we saw how a combination of liquid metal core and rapid rotation is necessary for the production of a planetary magnetic field. ∞ (Sec. 5.7) Mercury certainly does not rotate rapidly and it may also lack a liquid metal core, yet a magnetic field undeniably surrounds it. Although weak, the field is strong enough to deflect the solar wind and create a small magnetosphere around the planet. Scientists have no clear understanding of its origin. Possibly it is simply a "fossil remnant" dating back to the distant past when the planet's core solidified.

Mercury's magnetic field and high average density (roughly 5400 kg/m^3) together imply that most of the planet's interior is dominated by a large, iron-rich core having a radius of perhaps 1800 km. The ratio of core volume to total planet volume is greater for Mercury than for any other object in the solar system. Figure 6.26 illustrates the relative sizes and internal structures of Earth, the Moon, Mercury, and Mars.

Like the Moon, Mercury seems to have been geologically dead for roughly the past four billion years. Again as on the Moon, the lack of present-day geological activity on Mercury results from the mantle's being solid, preventing volcanism or tectonic motion. Largely on the basis of studies of the Moon, scientists have pieced together the following outline of Mercury's early history.

When Mercury formed 4.6 billion years ago, its location in the hot inner region of the early solar system ensured a dense, largely metallic overall composition. ∞ (Sec. 4.3) During the next half-billion years, Mercury melted and differentiated, just as the other terrestrial worlds did. It suffered the same intense meteoritic bombardment as the Moon. Being more massive than the Moon, Mercury cooled more slowly, so its crust was thinner than the Moon's and volcanic activity more common. More craters were erased by lava, leading to the intercrater plains found by *Mariner 10*.

As the planet's large iron core formed and then cooled, the planet began to shrink, causing the surface to contract. This compression produced the scarps seen

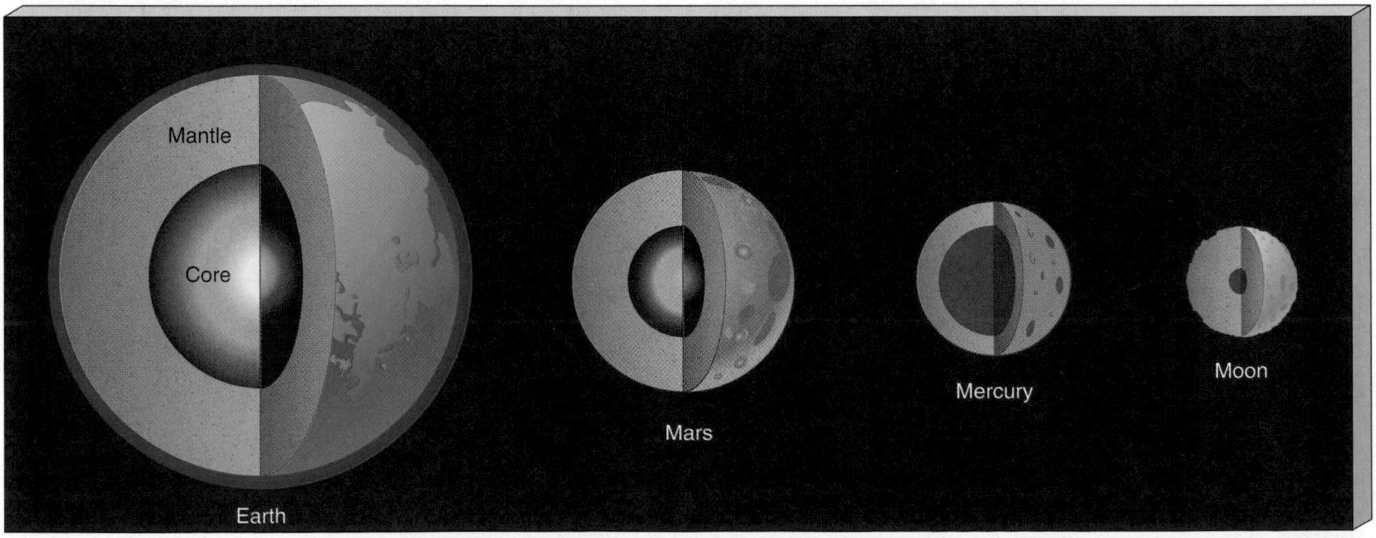

Figure 6.26 Terrestrial Planet Interiors The internal structure of Earth, the Moon, Mercury, and Mars, drawn to the same scale. Note how large a fraction of Mercury's interior is core. Animation

on Mercury's surface and may have prematurely terminated volcanic activity by squeezing shut the cracks and fissures on the surface. Thus, Mercury did not experience the subsequent extensive volcanic outflows that formed the lunar maria. Despite its larger mass and greater internal temperature, Mercury has probably been geologically inactive for even longer than the Moon.

Venus

Both U.S. and Soviet spacecraft have failed to detect any magnetosphere around Venus. Given that Venus's average density is similar to Earth's, it seems likely that Venus has an Earth-like overall composition and a partially molten iron-rich core. The lack of any detectable magnetic field, then, is almost surely the result of the planet's extremely slow rotation.

Because none of the *Venera* landers carried seismic equipment, no direct measurements of the planet's interior have ever been made, and theoretical models of the interior have very little hard data to constrain them. However, to many geologists the surface of Venus resembles that of the young Earth, at an age of perhaps a billion years. At that time, volcanic activity had already begun on Earth, but the crust was still relatively thin, and the convective processes in the mantle that drive plate-tectonic motion were not yet established.

Why has Venus remained in that immature state and not developed plate tectonics as Earth did? That question remains to be answered. Some planetary geologists have speculated that the high surface temperature on Venus has inhibited evolution by slowing the planet's cooling rate. Possibly the high surface temperature has made the crust too soft for Earth-style plates to develop. Or perhaps the high temperature and soft crust led to more volcanism, tapping the energy that might otherwise have gone into convective motion.

Mars

In September 1997, the orbiting *Mars Global Surveyor* succeeded in detecting a very weak Martian field, about 1/800 that of Earth. However, this is most likely a local anomaly, not a global field. Because Mars rotates rapidly, the weakness of the magnetic field is taken to mean that the planet's core is nonmetallic, or nonliquid, or both. ∞ (Sec. 7.4)

Mars's small size means that any internal heat would have been able to escape more easily than in a larger planet like Earth or Venus. The evidence of ancient surface activity, especially volcanism, suggests that at least parts of the Martian interior must have melted at some time in the past, but the lack of current activity and the absence of any significant magnetic field indicate that the melting was never as extensive as on Earth. Scientists now believe that Mars's core has a diameter of about 2500 km and is composed largely of iron sulfide (a compound about twice as dense as surface rock).

The history of Mars appears to be that of a planet where large-scale tectonic activity almost started but was stifled by the planet's rapidly cooling outer layers. On a larger, warmer planet, the large upwelling of material that formed the Tharsis bulge might have developed into full-fledged plate tectonic motion, but the Martian mantle became too rigid and the crust too thick for that to occur. Instead, the upwelling continued to fire volcanic activity, perhaps even up to the present day, but geologically much of the planet died two billion years ago.

6.8 Atmospheric Evolution on Earth, Venus, and Mars

6 Now that we have studied the structure and history of the three terrestrial worlds that still have atmospheres today— Venus, Earth, and Mars—we can return to the question of why their atmospheres are so different from one another. Let's begin our study with Earth. The chain of events that occurred on Earth serves as a useful baseline for the histories of the other two planets.

The Development of Earth's Atmosphere

When Earth formed, any atmosphere it might have had—sometimes called the **primary atmosphere**—would have consisted of the gases most common in the early solar system: light gases such as hydrogen, helium, methane, ammonia, and water vapor, a far cry from the atmosphere we enjoy today. Almost all this light material, and especially any hydrogen or helium, escaped into space during the first half-billion or so years after Earth was formed. As discussed in Chapter 5, Earth's gravity is simply too weak to retain these light gases. ∞ (*More Precisely 5-2*)

Subsequently, Earth developed a **secondary atmosphere**, which was released from the planet's

INTERLUDE 6-1

Life on Mars?

Even before the *Viking* missions reached Mars in 1976, astronomers had abandoned hope of finding life on the planet. Scientists knew there were no large-scale canal systems, no surface water, almost no oxygen in the atmosphere, and no seasonal vegetation changes. The present lack of liquid water on Mars especially dims the chances for life there now. However, running water and possibly a dense atmosphere in the past may have created conditions suitable for the emergence of life long ago. In the hope that some form of microbial life might have survived to the present day, the *Viking* landers carried out experiments designed to detect biological activity. The accompanying pair of photographs show the robot arm of one of the landers digging a shallow trench.

(Before) (After)

(NASA)

All three *Viking* biological experiments assumed some basic similarity between hypothetical Martian bacteria and those found on Earth. A *gas-exchange* experiment offered a nutrient broth to any residents of a sample of Martian soil and looked for gases that would signal metabolic activity. A *labeled-release* experiment added compounds containing radioactive carbon to the soil, then waited for results signaling that Martian organisms had either eaten or inhaled this carbon. Finally, a *pyrolitic-release* experiment added radioactively tagged carbon dioxide to a sample of Martian soil and atmosphere, waited awhile, then removed the gas and tested the soil (by heating it) for signs that something had absorbed the tagged gas. In all cases, contamination by terrestrial bacteria was a major concern. Indeed, any release of Earth organisms would have invalidated these and all future such experiments on Martian soil. Both *Viking* landers were carefully sterilized prior to launch.

Initially, all three experiments appeared to be giving positive signals. However, subsequent careful studies showed that the results could all be explained by inorganic (that is, nonliving) chemical reactions. Thus, we have no evidence for even microbial life on the Martian surface. The *Viking* robots detected peculiar reactions that mimic in some ways the basic chemistry of living organisms, but they did not detect life itself.

One criticism of the *Viking* experiments is that they searched only for life now living. Today, Mars seems locked in an ice age—the kind of numbing cold that would prohibit sustained life as we know it. If bacterial life did arise on an Earth-like early Mars, however, then we might be able to find its fossilized remains preserved on or near the Martian surface. This is one reason that scientists are so eager to land more spacecraft on the planet, particularly at the poles, whose ice caps may offer the best environment for finding Martian life (or its remains).

Surprisingly, one alternative place to look for evidence of life on Mars is right here on Earth. The accom-

interior as a result of volcanic activity—a process called *outgassing*. Volcanic gases are rich in water vapor, carbon dioxide, sulfur dioxide, and compounds containing nitrogen. As Earth's surface temperature fell and the water vapor condensed, oceans formed. Much of the carbon dioxide and sulfur dioxide became dissolved in the oceans or combined with surface rocks. Oxygen is such a reactive gas that any free oxygen that appeared at early times was removed as quickly as it formed. As solar ultraviolet radiation liberated nitrogen from its chemical bonds with other elements, a nitrogen-rich atmosphere slowly appeared.

The final major development in the story of Earth's atmosphere was the appearance of *life* in the oceans panying figure shows ALH84001, a blackened, 2-kg meteorite about 17 cm across, found in 1984 in Antarctica. Chemical analysis strongly indicates that it originated on Mars. It was apparently blasted off that planet long ago by a meteoritic impact of some sort, thrown into space, and eventually captured by Earth's gravity. Based on the estimated cosmic-ray exposure it received before reaching Earth, it left Mars about 16 million years ago.

In 1996, a group of scientists argued, based on all the data accumulated from studies of ALH84001, that they had discovered fossilized evidence for life on Mars. They pointed to globular structures similar to those produced by bacteria on Earth (top left inset), the presence of chemical compounds sometimes associated with Earth biology, and curved, rodlike structures (right image) resembling Earth bacteria, which the researchers interpreted as fossils of primitive Martian organisms.

These claims are very controversial. Many experts categorically disagree that evidence of Martian life has been found—not even fossilized life. They maintain that, as in the case of the *Viking* experiments, the "evidence" could all be due to chemical reactions not requiring any kind of biology. The scale of the supposed fossil structures is also important. They are only about 0.5 μm across, 1/30 the size of ancient bacterial cells found fossilized on Earth. Furthermore, several key experiments have not yet been done, such as testing the suspected fossils for evidence of cell walls, or of any internal cavities where body fluids would have resided. Nor has anyone yet found in ALH84001 any amino acids, the basic building blocks of life as we know it.

At the frontiers of science, issues are not always as clear-cut as we might hope. Only additional analysis and new data will tell for sure if primitive Martian life existed long ago.

(NASA)

3.5 billion years ago. Living organisms began to produce atmospheric oxygen and eventually the ozone layer formed, shielding the surface from the Sun's harmful radiation. Eventually, life spread to the land and flourished. The fact that oxygen is a major constituent of our present-day atmosphere is a direct consequence of the evolution of life on Earth.

The Runaway Greenhouse Effect on Venus

Given the distance of Venus from the Sun, the planet was not expected to be such a pressure cooker. Why is Venus so hot? And if, as we believe, Venus started off like Earth, why is its atmosphere now so different from Earth's?

The answer to the first question is fairly easy. Given the present composition of its atmosphere, Venus is hot because of the greenhouse effect. Recall that *greenhouse gases* in Earth's atmosphere, particularly water vapor and carbon dioxide, tend to warm our planet. ∞ (Sec. 5.3) By stopping the escape of much of the infrared radiation emitted by Earth's surface, these gases increase the planet's equilibrium temperature much as an extra blanket keeps you warm on a cold night. Venus's dense atmosphere is made up almost entirely of a prime greenhouse gas, carbon dioxide (Figure 6.27). This thick blanket absorbs about 99 percent of all the infrared radiation released from the surface of Venus and is the immediate cause of the planet's sweltering 730 K surface temperature.

The answer to the second question—Why is Venus's atmosphere so different from Earth's?—is more complex. The initial stages of atmospheric development on Venus probably took place in more or less the same way as just described for our own planet. However, on Earth much of the secondary atmosphere became part of the planet surface, as carbon dioxide and sulfur dioxide dissolved in the oceans or combined with surface rocks. If all the dissolved or chemically combined carbon dioxide on Earth were released back into our present-day atmosphere, its new composition would be 98 percent carbon dioxide and two percent nitrogen, and it would have a pressure about 70 times its current value. In other words, apart from the presence of oxygen (which appeared on Earth only after the development of life) and water (whose absence on Venus will be explained shortly), Earth's atmosphere would look a lot like that of Venus. The real difference between Earth and Venus, then, is that Venus's greenhouse gases never left the atmosphere the way they did on Earth.

When Venus's secondary atmosphere appeared, the temperature was higher than on Earth because Venus is closer to the Sun, but the exact atmospheric temperature is uncertain. If it was so high that water vapor could not

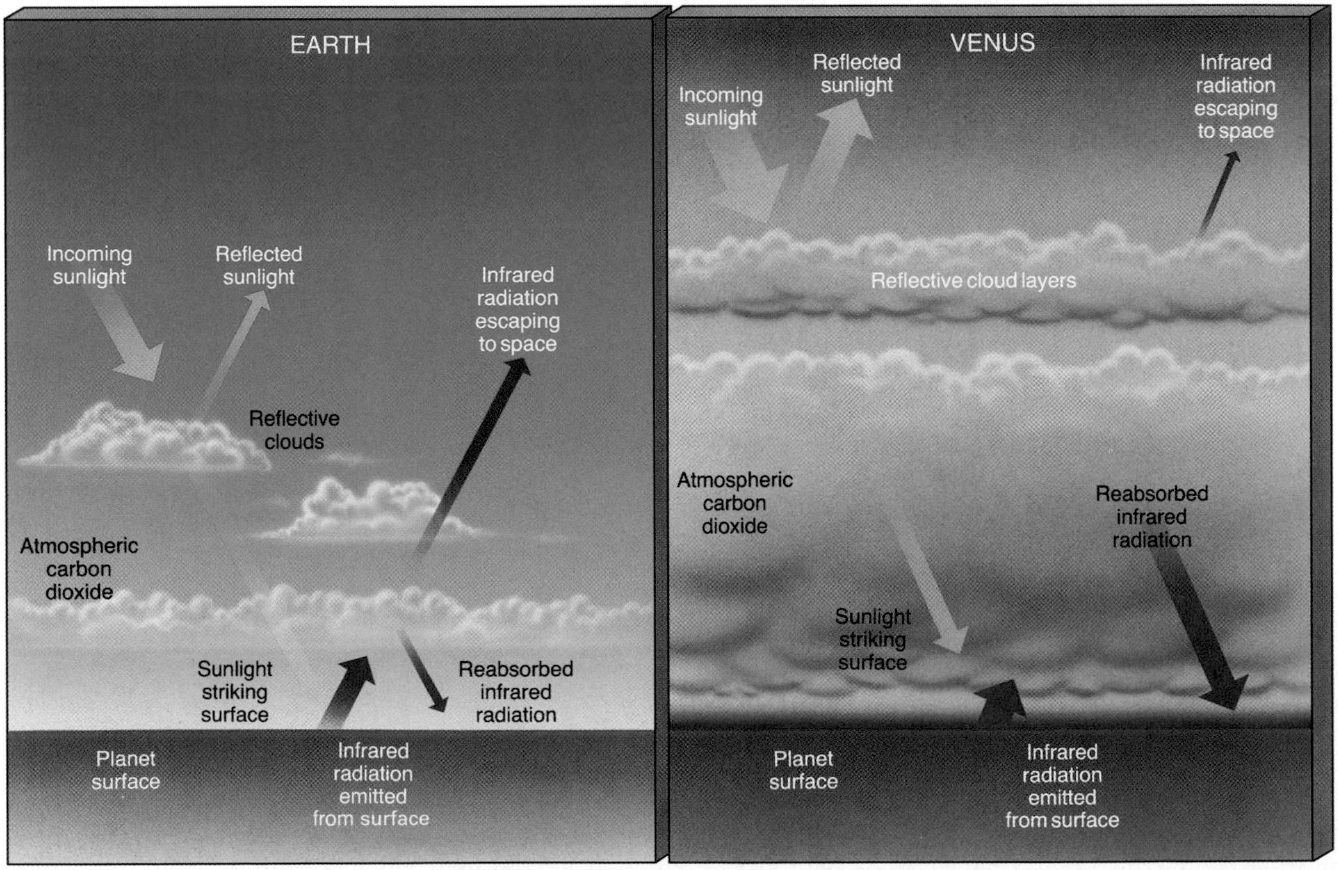

Figure 6.27 Venus's Atmosphere Because Venus's atmosphere is much deeper and denser than Earth's, a much smaller fraction of the infrared radiation leaving the planet's surface escapes into space. The result is a much stronger greenhouse effect than on Earth and a correspondingly hotter planet. The outgoing infrared radiation is not absorbed at a single point in the atmosphere. Instead, absorption occurs at all atmospheric levels. (The arrows indicate only that absorption occurs, not that it occurs at any specific level.)

become liquid, no oceans would have formed. Consequently, outgassed water vapor and carbon dioxide would have remained in the atmosphere, and the full greenhouse effect would have gone into operation immediately. If oceans did form and most of the greenhouse gases left the atmosphere to become dissolved in the water, the temperature must still have been sufficiently high to allow a process known as the **runaway greenhouse effect** to come into play.

To understand the runaway greenhouse effect, imagine that we took Earth from its present orbit and placed it in Venus's orbit. At its new distance from the Sun, the amount of sunlight hitting Earth's surface would be almost twice its present level, and so the planet would warm up. More water would evaporate from the oceans, leading to an increase in atmospheric water vapor. At the same time, the ability of both the oceans and surface rocks to hold carbon dioxide would diminish, allowing more carbon dioxide to enter the atmosphere. As a result, the greenhouse heating would increase, and the planet would warm still further, leading to a further increase in atmospheric greenhouse gases, and so on. Once started, the process would "run away," eventually leading to the complete evaporation of the oceans, restoring all the original greenhouse gases to the atmosphere. Basically the same thing would have happened on Venus long ago, leading to the planetary inferno we see today.

The greenhouse effect on Venus was even more extreme in the past, when the atmosphere also contained water vapor. By intensifying the blanketing effect of the carbon dioxide, the water vapor helped the surface of Venus reach temperatures perhaps twice as hot as at present. At those high temperatures, the water vapor was able to rise high into the planet's upper atmosphere—so high that it was broken up by solar ultraviolet radiation into its components, hydrogen and oxygen. The light hydrogen rapidly escaped, the reactive oxygen quickly combined with other atmospheric gases, and all water on Venus was lost forever.

Evolution of the Martian Atmosphere

Presumably, Mars also had first a primary and then a secondary (outgassed) atmosphere early in its history. Around 4 billion years ago, Mars may have had a fairly dense atmosphere, perhaps even with blue skies and rain. Despite Mars's distance from the Sun, the greenhouse effect would have kept conditions fairly comfortable, and an average surface temperature above 0°C seems quite possible.

Sometime during the next billion years, however, most of the Martian atmosphere disappeared. Possibly some of it was lost because of impacts with large bodies in the early solar system, and a large part may have leaked away into space because of the planet's weak gravity. ∞ (*More Precisely 5-2*) Most of the remainder probably became unstable in a kind of reverse runaway greenhouse effect. In this scenario, much of Mars's atmospheric carbon dioxide dissolved in the liquid water of the planet's rivers and lakes (and oceans, if any), ultimately to combine with Martian surface rocks. Calculations show that much of the Martian atmospheric carbon dioxide could have been depleted in this way in a relatively short period of time, perhaps as quickly as a few hundred million years. As the level of carbon dioxide declined and the greenhouse heating diminished, the planet cooled. The water froze out of the atmosphere, lowering still further the level of atmospheric greenhouse gases and accelerating the cooling.

The present level of water vapor in the Martian atmosphere is the maximum possible given the atmosphere's present density and temperature. Estimates of the total amount of water stored as permafrost and in the polar caps are quite uncertain, but it is likely that if all the water on Mars were to become liquid, it would cover the surface to a depth of several meters.

✓ Concept Check

- If Venus and Mars had formed at Earth's distance from the Sun, what might their climates be like today?

6.9 The Moons of Mars

The two Martian moons—named Phobos (Fear) and Deimos (Panic) for the horses that drew the Roman war god's chariot—were discovered in 1877 by American astronomer Asaph Hall. Both are small, irregularly shaped, and heavily cratered (Figure 6.28). The larger of the two, Phobos, is about 28 km long and 20 km wide. Deimos is just 16 km long by 10 km wide.

Based on measurements of the moons' gravitational effects on orbiting spacecraft, astronomers have estimated their densities to be around 2000 kg/m^3, far less than the density of Earth's Moon or any terrestrial world, implying that the moons' compositions are quite *dissimilar* from that of Mars. Most likely, Phobos

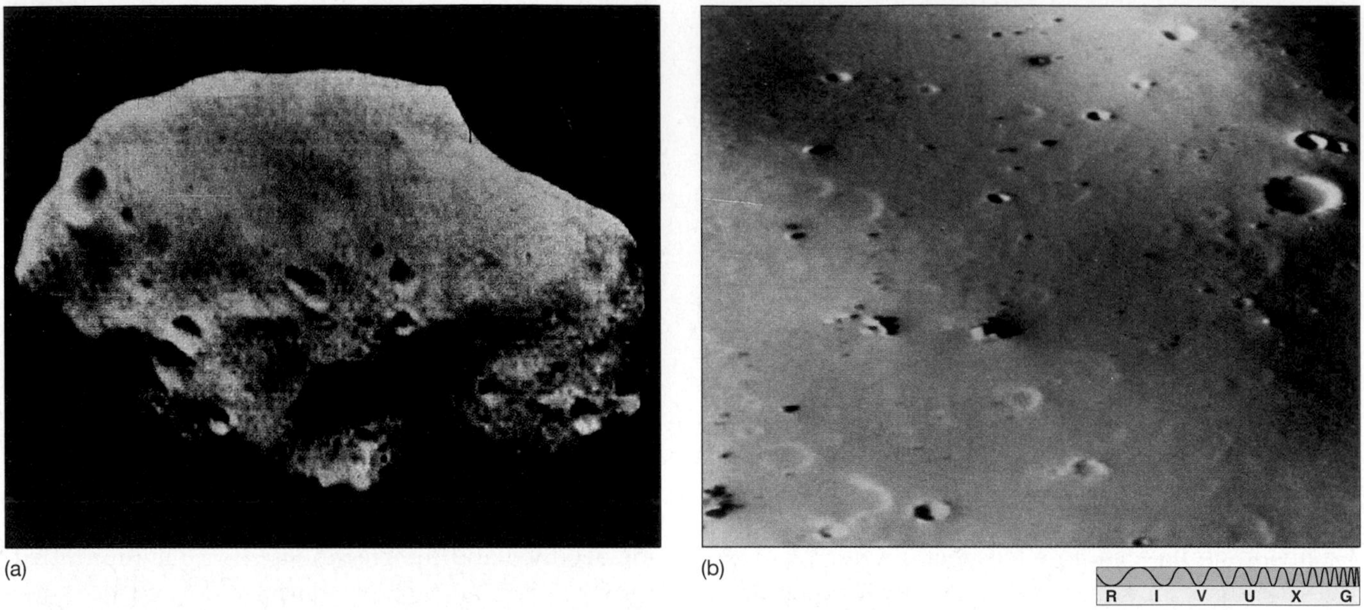

Figure 6.28 Martian Moons (a) A *Mariner 9* photograph of Phobos, not much larger than Manhattan Island. (b) A photograph of part of the smaller Martian moon Deimos, taken by a *Viking* orbiter. The field of view is only 2 km across. Most of the boulders shown are about the size of a house. *(NASA)*

and Deimos did not form along with Mars, but instead are asteroids that were slowed and captured by the outer fringes of the early Martian atmosphere (which, as we have just seen, was probably much denser than the atmosphere today). It is even possible that the two moons are remnants of a single object that broke up during capture.

Chapter Review

Summary

In size and appearance, Mercury is similar to the Moon. Venus is comparable in mass and radius to Earth. Mars is smaller than Earth, but has many Earth-like characteristics. Mercury's rotation rate is strongly influenced by the tidal effect of the Sun. The planet rotates exactly one and a half times for every one revolution around the Sun. Venus's rotation is slow and retrograde. Mars rotates at almost the same rate as Earth.

Mercury has no permanent atmosphere. The atmospheres of Venus and Mars are mainly carbon dioxide. Venus's atmosphere is extremely hot and 90 times denser than Earth's, while the density of the cool Martian atmosphere is only 0.7 percent that of Earth's.

Mercury's surface is heavily cratered, much like the lunar highlands. Mercury lacks lunar-like maria but has extensive **intercrater plains** (p. 161) and cliff-forming cracks, or **scarps** (p. 162), in its crust. The plains were caused by lava flows early in Mercury's history. The scarps were apparently formed when the planet's core cooled and shrank, causing the surface to crack. Mercury has a large impact crater called the Caloris Basin, the diameter of which is comparable to the radius of the planet.

Because of a thick cloud cover, Venus's surface cannot be seen in visible light from Earth, but it has been thoroughly mapped by radar. Many lava domes and **shield volcanoes** (p. 165) have been found on the surface. No eruptions have been observed, although there is indirect evidence that Venus is still volcanically active today. The planet's surface shows no sign of plate tectonics. Features called **coronae** (p. 165) are thought to have been caused by an upwelling of mantle material that never developed into full convective motion.

Mars's major surface feature is the Tharsis bulge. Associated with the bulge is the largest known volcano in the solar system and a huge crack, called the Valles Marineris, in the planet's surface. There is clear evidence that water once existed in great quantity on Mars. The **runoff channels** (p. 170) are the remains of ancient Martian rivers. The **outflow channels** (p. 170) are the paths taken by floods that cascaded from the southern highlands into the northern plains. Today, a large amount of that water may be locked up in the polar caps and in a layer of **permafrost** (p. 171) lying under the Martian surface.

Neither Venus nor Mars has any detectable magnetic field. Mercury's weak field seems to be a fossil remnant from long ago, dating to the time when the planet's iron core solidified.

Any **primary atmospheres** (p. 173) of light gases that the Earth, Venus, and Mars had at formation rapidly escaped. All the terrestrial planets probably developed **secondary atmospheres** (p. 173), outgassed from volcanoes, early in their lives. On Mercury, the atmosphere escaped, as on the Moon. On Earth, much of the outgassed material became absorbed in surface rocks or dissolved in the oceans. The development of life produced oxygen. On Venus, the **runaway greenhouse effect** (p. 177) has resulted in all the planet's carbon dioxide being left in the atmosphere, leading to the extreme conditions we observe today. On Mars, part of the secondary atmosphere escaped into space. Of the remainder, most of the carbon dioxide is now locked up in surface rock, and most of the water vapor is now stored in the permafrost and the polar caps.

Mars has two small moons, which are probably asteroids captured after encountering the planet's atmosphere long ago.

Review and Discussion

1. In contrast to Earth, Mercury undergoes extremes in temperature. Why?
2. What do Mercury's magnetic field and large average density imply about the planet's interior?
3. How is Mercury's evolutionary history like that of the Moon? How is it different?
4. Mercury used to be called "the Moon of the Sun." Why do you think it had this name? What piece of scientific evidence proved the name inappropriate?
5. Why does Venus appear so bright to the naked eye?
6. Venus probably has a molten iron-rich core like Earth. Why doesn't it also have a magnetic field?
7. How did radio observations of Venus made in the 1950s change our conception of that planet?
8. What are the main constituents of Venus's atmosphere? What are clouds in the upper atmosphere made of?
9. What is the runaway greenhouse effect, and how might it have altered the climate of Venus?
10. Why is Mars red?
11. What is the evidence that water once flowed on Mars? Is there water on Mars today?
12. Why were Martian volcanoes able to become so large?
13. Given that Mars has an atmosphere and its composition is mostly carbon dioxide, why isn't there a significant greenhouse effect to warm its surface?
14. Do you think that sending humans to Mars in the near future is a reasonable goal? Why or why not?
15. Compare and contrast the evolution of the atmospheres of Mars, Venus, and Earth. Include a discussion of the importance of volcanoes and water.

True or False?

____ 1. Mercury has no detectable atmosphere.

____ 2. Mercury has nighttime low temperatures of 100 K, well below the freezing point of water.

____ 3. Mercury's solar day is longer than its solar year.

____ 4. Some volcanic activity continues today on the surface of Mercury.

____ 5. The average surface temperature of Venus is about 250 K.

____ 6. Numerous surface features on Venus can be seen from Earth-based observations made in the ultraviolet part of the spectrum.

____ 7. Lava flows are common on the surface of Venus.

____ 8. There is strong circumstantial evidence that active volcanism continues on Venus.

____ 9. Venus has a magnetic field similar to that of Earth.

____ 10. Mars has the largest volcanoes in the solar system.

____ 11. The northern hemisphere of Mars is much older than the southern hemisphere.

____ 12. There are many indications of plate tectonics on Mars.

____ 13. Valles Marineris is similar in size to Earth's Grand Canyon.

____ 14. The orange-red color of the surface of Mars is primarily due to rust (iron oxide) in its soil.

____ 15. One of the two moons of Mars is larger than Earth's Moon.

Fill in the Blank

1. Mercury's iron core contains a _______ fraction of the planet's mass than does Earth's core.
2. Mercury's rate of rotation was first measured using _______.
3. Although Mercury's daytime temperatures are always very hot, it may still be possible for the planet to have sheets of water ice at its _______.
4. Mercury's _______ is about 1/100 that of Earth and was originally thought not to exist at all.
5. Venus's rotation is unusual because it is _______.
6. The most abundant gas in the atmosphere of Venus is _______.
7. The runaway greenhouse effect on Venus was a result of the planet's being _______ to the Sun than is Earth.
8. The surface of Venus has been mapped using _______.
9. The surface of Venus appears to have been resurfaced by _______ every few hundred million years.
10. The main difficulties in using landers to study Venus's surface are the planet's extremely high _______ and _______.
11. The southern hemisphere of Mars consists of heavily _______ highlands.
12. The Tharsis region of Mars is a large equatorial _______.
13. The great height of Martian volcanoes is a direct result of the planet's low _______ .
14. The flowing-liquid appearance of the ejecta surrounding Martian impact craters is evidence of a layer of _______ just under the surface.
15. Runoff channels on Mars carried _______ from the southern mountains into the valleys.

Problems

1. How long does it take a radar signal to travel from Earth to Mercury and back when Mercury is at its closest point to Earth?
2. Assume that a planet will have lost its initial atmosphere by the present time if the average (daytime) molecular speed exceeds one-sixth of the escape speed (see *More Precisely* 5-2). What would Mercury's mass have to be in order for it still to have a nitrogen (molecular weight: 28) atmosphere?
3. Given that the *Hubble Space Telescope* orbits 600 km above Earth's surface with a period of 95 minutes, what was the orbital period of the *Magellan* spacecraft as it orbited 500 km above the surface of Venus?
4. Approximating Venus's atmosphere as a layer of gas that is 50 km thick and has a uniform density of 21 kg/m^3, calculate the total mass of this atmosphere. Compare your answer with the mass of Earth's atmosphere (Chapter 5, problem 6) and with the mass of Venus.
5. *Pioneer Venus* observed high-level clouds moving around Venus's equator in four days. What was their speed in km/h? In mph?
6. What is the size of the smallest feature that can be distinguished on the surface of Venus by the Arecibo radio telescope, at an angular resolution of 0.5′?
7. Verify that the surface gravity on Mars is 40 percent that of Earth.
8. The outflow channel shown in Figure 6.24 is about 10 km across and 100 m deep. If it carried 10^7 tons (10^{10} kg) of water per second, as stated in the text, estimate the speed at which the water must have flowed.
9. Calculate the total mass of a uniform layer of water covering the entire Martian surface to a depth of 2 m (see ∞ Section 6.8). Compare it with the mass of Venus's atmosphere (problem 4).
10. Phobos orbits Mars at a distance of 9380 km from the planet's center and has an orbital period of 7.7 hours. Calculate the mass of Mars from these data.

Projects

1. Try to spot Mercury in morning or evening twilight. (In the Northern Hemisphere, the best evening sightings of the planet take place in the spring, the best morning sightings in the fall.) What color is the planet? Do you think this is the planet's actual color, or does Mercury appear this way because we generally see it in a twilight sky?
2. Consult an almanac to determine the next time Venus will pass between Earth and the Sun. How many days before and after this event can you glimpse the planet with your naked eyes? Use the almanac to find out the next time Venus will pass on the far side of the Sun from Earth. How many days before and after this event can you glimpse the planet with the unaided eye?
3. Using a powerful pair of binoculars or a small telescope, examine Venus as it goes through its phases. Note each phase and its relative size. (You can compare its size to the field of view in a telescope; always use the same eyepiece for this.) Look at the planet every few days or once a week. Make a table of the shape of the phase, the size, and the relative brightness to the naked eye. After you have observed through a significant change in phase, can you see the correlations in these three properties first recognized by Galileo?
4. Track the motion of Mars relative to background stars for several months following its appearance in the predawn sky. (Consult an almanac to determine where Mars is in the sky this year.) You will see that the planet moves rapidly eastward, crossing many constellation boundaries, while the background stars shift continuously toward the west. Can you explain why Mars's orbit appears as it does?

A
B
C

7 THE JOVIAN PLANETS

Giants of the Solar System

LEARNING GOALS

Studying this chapter will enable you to:

1. Explain how both chance and calculation played major roles in the discovery of Uranus and Neptune.
2. Describe the similarities and the differences among the four jovian worlds.
3. Discuss some of the processes responsible for the properties of the jovian atmospheres.
4. Describe how the internal structure and composition of the jovian planets are inferred from external measurements.
5. Explain why three of the four jovian worlds radiate more energy into space than they receive from the Sun.

(Opposite page, background) Through the cameras aboard the *Hubble Space Telescope*, the planet Saturn appears much as our naked eyes would see it if it were only six times as far away as the Moon. Resolution here is 670 km, good enough to see clearly the band structure on the planet, as well as several gaps in the rings. True-color images like this one are made by combining separate images taken in red, green, and blue light, as shown in the inserts.

(Inset A) Saturn as seen through a red filter, centered at a wavelength of 718 nm.

(Inset B) Saturn as seen through a green filter, centered at a wavelength of 547 nm.

(Inset C) Saturn as seen through a blue filter, centered at a wavelength of 439 nm. *(NASA)*

Beyond the orbit of Mars, the solar system is a very different place. In sharp contrast to the small, rocky, terrestrial bodies found near the Sun, the outer solar system presents us with a totally unfamiliar environment—huge gas balls, peculiar moons, planet-circling rings, and a wide variety of physical and chemical properties, many of which are only poorly understood even today. U.S. spacecraft visiting the outer planets over the past two decades have revealed these worlds in detail that was only dreamed of by centuries of Earth-bound astronomers. The jovian planets—Jupiter, Saturn, Uranus, and Neptune—differ from each other in many ways, but they have much in common, too. As with the terrestrial planets, we learn from their differences as well as from their similarities.

7.1 Observations of Jupiter and Saturn

The View from Earth

Named after the most powerful god of the Roman pantheon, Jupiter is the third-brightest object in the night sky (after the Moon and Venus), making it easy to locate and study from Earth. Ancient astronomers could not have known the planet's true size, but their name choice was very apt—Jupiter is by far the largest planet in the solar system.

Figure 7.1(a) is a *Hubble Space Telescope* image of Jupiter. Notice the alternating light and dark bands that cross the planet parallel to its equator, and also the large oval at lower right. These atmospheric features are quite unlike anything found on the inner planets. Also in contrast to the terrestrial worlds, Jupiter has many moons that differ greatly in size and other properties (Figure 7.1b). The four largest are visible from Earth with a small telescope (and, for a few people, with the naked eye). They are known as the *Galilean moons* after their discoverer, Galileo Galilei.

Saturn (Figure 7.2), the next major body we encounter as we continue outward beyond Jupiter, was the most distant planet known to Greek astronomers. Named after the father of Jupiter in Greco–Roman mythology, Saturn orbits at almost twice Jupiter's distance from the Sun, making it considerably fainter (as seen from Earth) than either Jupiter or Mars. The planet's banded atmosphere is somewhat similar to Jupiter's, but Saturn's atmospheric bands are much less distinct. Overall, the planet has a rather uniform butterscotch hue. Again like Jupiter and unlike the inner planets,

(a)

R I V U X G

(b)

Animation

Figure 7.1 Jupiter (a) A *Hubble Space Telescope* image of Jupiter, in true color. Features as small as a few hundred kilometers across are resolved. (b) Photograph of Jupiter made with an Earth-based telescope, showing the planet and its four Galilean moons. *(NASA/AURA)*

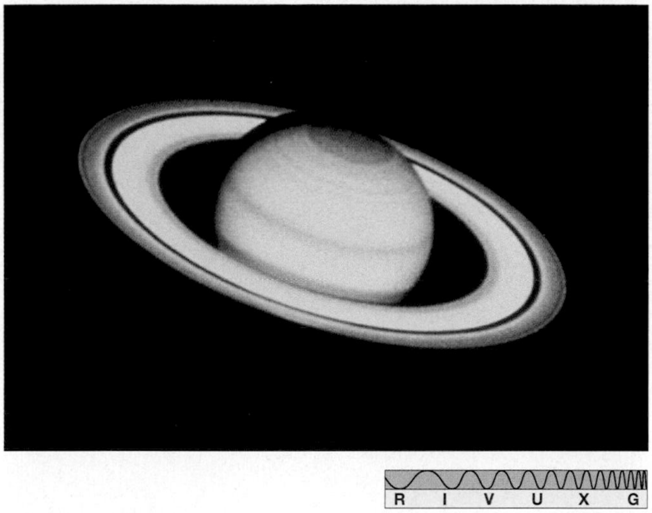

Figure 7.2 Saturn Saturn as seen by the *Hubble Space Telescope.* Resolution is about 700 km. *(NASA)*

Saturn has many moons orbiting it. Saturn's best-known feature, its spectacular *ring system*, is clearly visible in Figure 7.2. The moons and rings of the jovian planets are the subject of Chapter 8.

Spacecraft Exploration

Most of our detailed knowledge of the jovian planets has come from NASA spacecraft visiting those worlds. Of particular importance are the two *Voyager* probes and the recent *Galileo* mission.

The *Voyager* spacecraft left Earth in 1977, reaching Jupiter in March (*Voyager 1*) and July (*Voyager 2*) of 1979. Both craft subsequently used Jupiter's strong gravity to send them on to Saturn in a maneuver called a *gravity assist* (see *More Precisely 7-1*). *Voyager 2* took advantage of a rare planetary configuration that enabled it to use Saturn's gravity to propel it to Uranus and then on to Neptune in a spectacularly successful "grand tour" of the outer planets. Each craft carried equipment to study planetary magnetic fields and magnetospheres, as well as radio, visible-light, and infrared sensors to analyze reflected and emitted radiation from the jovian planets and their moons. The data they returned revolutionized our knowledge of all the jovian worlds. Much of the information presented in this and the next chapter came from *Voyager* sensors. The two *Voyager* craft are now headed out of the solar system, racing toward interstellar space.

Galileo was launched in 1989 and reached Jupiter in December 1995, after a circuitous route that took it three times through the inner solar system, receiving gravity assists from both Venus and Earth before finally reaching its target. The mission had two components: an atmospheric probe and an orbiter. The probe, slowed by a heat shield and a parachute, descended into Jupiter's atmosphere, making measurements and chemical analyses as it went. The orbiter executed a complex series of orbits through Jupiter's moon system, returning to some moons studied by *Voyager* and visiting others for the first time. The mission was scheduled to end in December 1997, but was so successful that NASA extended its lifetime for two more years to obtain even more detailed data on Jupiter's inner moons. As of mid-2000, the spacecraft was still operational.

In October 1997, NASA launched the *Cassini* mission to Saturn. When the craft reaches its destination in 2004, it will dispatch a probe built by the European Space Agency into the atmosphere of Titan, Saturn's largest moon, and orbit among the planet's moons for four years, much like *Galileo* at Jupiter. If experience with *Galileo* is any guide, *Cassini* will probably resolve many outstanding questions about the Saturn system, but it is sure to pose many new ones too.

7.2 The Discoveries of Uranus and Neptune

1 The British astronomer William Herschel discovered the planet Uranus in 1781. Because this was the first new planet discovered in recorded history, the event caused quite a stir. The story goes that Herschel's first instinct was to name the new planet "Georgium Sidus" (Latin for "George's star") after his king, George III of England. The world was saved from a planet named George by the wise advice of another astronomer, Johann Bode, who suggested instead that the tradition of using names from ancient mythology be continued and that the planet be named after Uranus, the father of Saturn.

Uranus is just barely visible to the naked eye. It looks like a faint, undistinguished star. Even through a large optical telescope, Uranus appears hardly more than a tiny, pale, slightly greenish disk. Figure 7.3 shows a close-up visible-light image taken by *Voyager 2* in 1986. Uranus's apparently featureless atmosphere contrasts sharply with the bands and spots visible in the atmospheres of the other jovian planets.

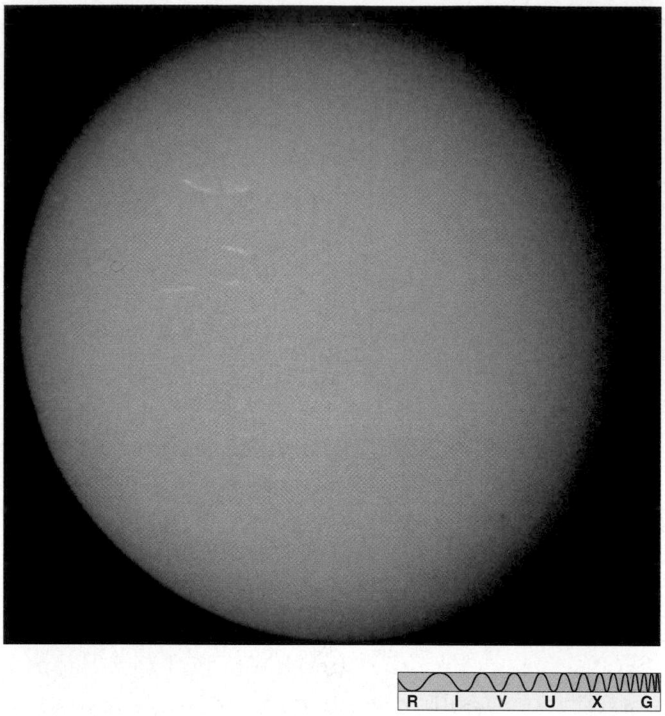

Figure 7.3 **Uranus** A close-up view of Uranus sent back to Earth by *Voyager 2* while the spacecraft was whizzing past this gigantic planet at 10 times the speed of a rifle bullet. *(NASA)*

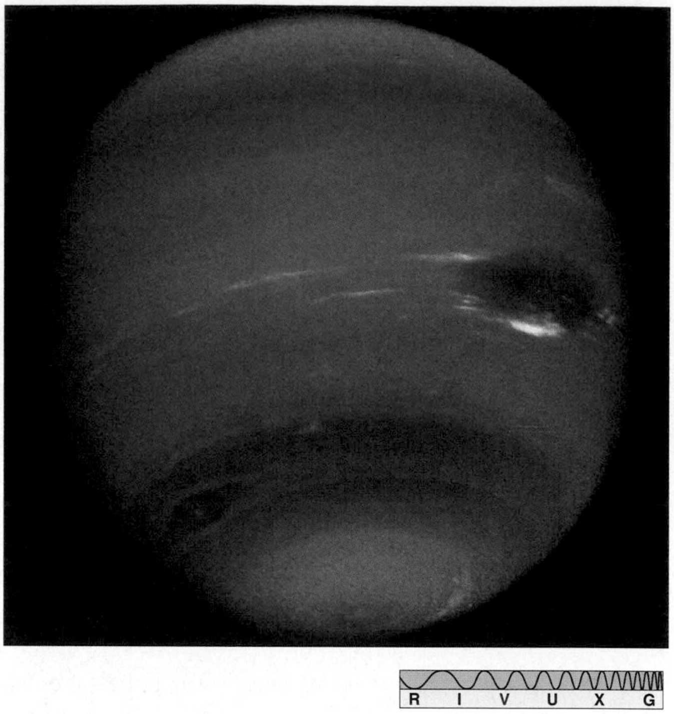

Figure 7.4 **Neptune** Neptune as seen by *Voyager 2* from a distance of 1,000,000 km. *(NASA)*

Once Uranus was discovered, astronomers set about charting its orbit. They quickly discovered a discrepancy between the planet's predicted and observed positions. Try as they might, they could not find an elliptical orbit that fit the planet's path. By the early nineteenth century, the discrepancy had grown to a quarter of an arc minute, far too large to be explained away as observational error. Uranus seemed to be violating—slightly—Kepler's laws of planetary motion. ∞ (Sec. 1.3)

Astronomers realized that, although the Sun's gravitational pull dominates the planet's orbital motion, the small deviation meant that some unknown body was exerting a much weaker but still measurable gravitational force on Uranus. There had to be *another* planet in the solar system influencing Uranus. In September 1845, after almost two years of work, an English mathematician named John Adams solved the problem of determining the new planet's mass and orbit. In June 1846 a French mathematician, Urbain Leverrier, independently came up with essentially the same answer. Later that year, a German astronomer named Johann Galle found the new planet within one or two degrees of the predicted position. The new planet was named Neptune, and Adams and Leverrier are now jointly credited with its discovery.

Unlike Uranus, distant Neptune cannot be seen with the naked eye, although it can be seen through binoculars or a small telescope. In fact, according to his notes, Galileo may actually have seen Neptune, although he surely had no idea what it really was at the time. Through a large telescope on Earth, the planet appears only as a small, bluish disk, with few surface features visible. Even under the best observing conditions, only a few markings, suggestive of multicolored cloud bands, can be seen. With *Voyager 2*'s arrival, much more detail emerged (Figure 7.4). Superficially, at least, Neptune resembles a blue-tinted Jupiter, with atmospheric bands and spots clearly evident.

Concept Check

- How did observations of Uranus's orbit lead to the discovery of Neptune?

MORE PRECISELY 7-1

Gravitational "Slingshots"

Sending a spacecraft to another planet requires a lot of energy—often more than can be conveniently provided by a rocket launched from Earth or safely transported in a shuttle for launch from orbit. Faced with these limitations, mission scientists make use of a "slingshot" maneuver to boost an interplanetary probe to a more energetic orbit and also aid navigation toward the target, all at no additional cost!

The accompanying figure illustrates a gravitational slingshot, or *gravity assist*, in action. A spacecraft approaches a planet, passes close by, then escapes along a new trajectory. Obviously, the spacecraft's *direction* of motion is changed by the encounter. Less obvious, the spacecraft's *speed* is also altered as the planet's gravity propels the spacecraft in the direction of the planet's motion. By careful choice of incoming trajectory, the craft can either speed up (by passing "behind" the planet, as shown) *or* slow down (by passing in front), by as much as twice the planet's orbital speed. Of course, there is no "free lunch"—the spacecraft gains energy from, or loses it to, the planet's motion, causing its orbit to change ever so slightly. However, since planets are so much more massive than spacecraft, the effect is tiny.

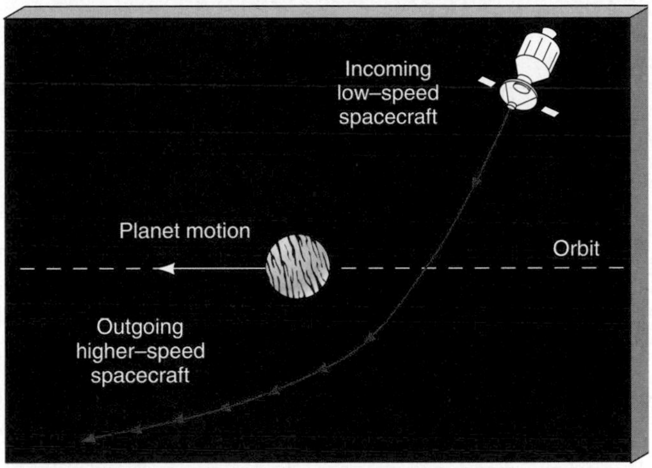

Such a slingshot maneuver has been used many times in missions to both the inner and the outer planets, as illustrated in the second figure, which shows *Voyager 2*'s "grand tour" through the jovian planets. The gravitational pulls of these giant worlds whipped the craft around at each visitation, enabling flight controllers to get considerable extra "mileage" out of the probe. More recently, as described in the text, the *Galileo* mission used multiple gravity assists, both to reach Jupiter and subsequently to navigate among that planet's moons. Every encounter with a moon had a slingshot effect—sometimes accelerating and sometimes slowing the probe, but each time moving it onto a different orbit—and every one of these effects was carefully calculated long before *Galileo* ever left Earth.

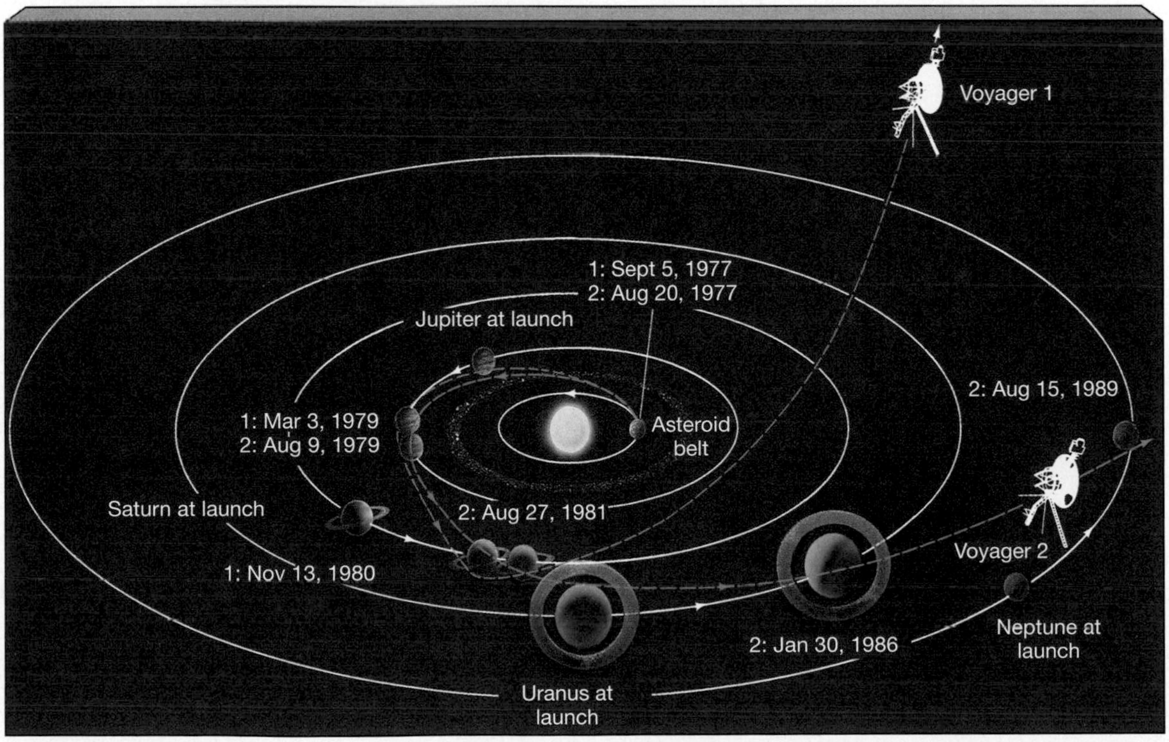

7.3 The Jovian Planets in Bulk

Overall Properties

2 Table 7.1 extends Table 6.1 to include the four jovian planets. Figure 7.5 shows them to scale, along with Earth for comparison. The large masses and radii and relatively low average densities of these gigantic worlds imply that they differ radically in both composition and structure from the terrestrial planets. (Sec. 4.1)

Most of the masses of Jupiter and Saturn, and about half those of Uranus and Neptune, are made up of hydrogen and helium—light gases on Earth (having densities 0.08 and 0.16 kg/m^3, respectively, at room temperature and sea level), but compressed enormously by these planets' strong gravitational fields. The abundance of hydrogen and helium on these worlds is itself a consequence of the strong jovian gravity. The jovian planets are massive enough to have retained even the lightest gas, hydrogen, and very little of their original atmospheres have escaped since the birth of the solar system 4.6 billion years ago. (*More Precisely 5-2*) At the heart of each jovian planet lies a dense, compact core many times more massive than Earth.

None of the jovian planets has a solid surface of any kind. Their gaseous atmospheres just become denser with depth because of the pressure of the overlying layers, eventually becoming liquid in the interior. The surface we see from Earth is simply the top of the outermost layer of atmospheric clouds. Where relevant, the term *surface* in Table 7.1 and elsewhere in this chapter refers to this level in the atmosphere.

Differential Rotation

Because the jovian planets have no solid surfaces, different parts of the atmosphere can and do move at different rates. This state of affairs—when the rotation rate is not constant from one location to another—is called **differential rotation**. It is not possible in solid objects like the terrestrial worlds, but is normal for fluid bodies such as the jovian planets.

The amount of differential rotation on Jupiter is small—the equatorial regions rotate once every 9^h 50^m, while the higher latitudes take about 6 minutes longer. On Saturn, the difference between the equatorial and polar rotation rates is a little greater (26 minutes, again slower at the poles). On Uranus and Neptune, the differences are greater still—more than

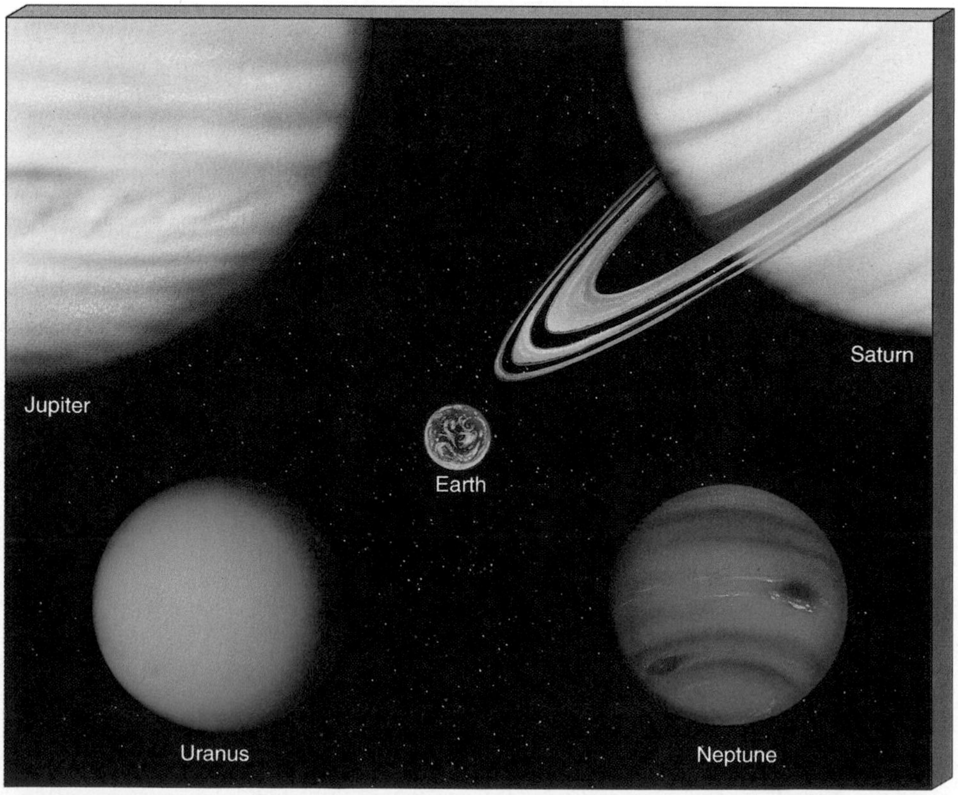

Figure 7.5 Jovian Planets Jupiter, Saturn, Uranus, and Neptune, drawn to scale and compared to Earth.

two hours in the case of Uranus—this time with the poles rotating more rapidly. In each case, differential rotation is the result of large-scale wind flows in the planet's atmosphere.

More meaningful measurements of overall planetary rotation rates are provided by measurements of their magnetospheres. All four jovian worlds have strong magnetic fields and emit radiation at radio wavelengths. The strength of this radio emission varies with time and repeats itself periodically. Scientists assume that this period matches the rotation of the planet's deep interior, where (as on Earth) the magnetic field arises. ∞ (Sec. 5.7) The rotation periods listed in Table 7.1 were obtained in this way. There is no clear relationship between the interior and atmospheric rotation rates of the jovian planets. In Jupiter and Saturn, the interior rotation rate matches the polar rotation. In Uranus, the interior rotates more slowly than any part of the atmosphere. In Neptune, the reverse is true.

The observations used to measure a planet's rotation period also let us determine the orientation of its rotation axis. Jupiter's rotation axis is almost perpendicular to the orbit plane, the axial tilt being only 3°, quite small compared with Earth's tilt of 23°. The tilts of Saturn and Neptune (Table 7.1) are similar to those of Earth and Mars.

Each planet in our solar system seems to have some outstanding peculiarity. In the case of Uranus, it is its rotation axis. Unlike the other major planets, which have their spin axes (very) roughly perpendicular to the ecliptic plane, Uranus's rotation axis lies almost within that plane. We might say that Uranus is "tipped over" onto its side, for its axial tilt is 98°. As a result, the "north" pole[1] of Uranus, at some point in the planet's orbit, points almost directly toward the Sun. Half a Uranian year later, the "south" pole faces the Sun, as illustrated in Figure 7.6. When *Voyager 2* encountered the planet in 1986, the north pole happened to be pointing nearly directly at the Sun, so it was midsummer in the northern hemisphere.

No one knows why Uranus is tilted in this way. Astronomers speculate that a catastrophic event, such as a grazing collision between the planet and another

[1]We adopt here the convention that a planet's rotation is counterclockwise as seen from above the north pole (that is, planets always rotate from west to east).

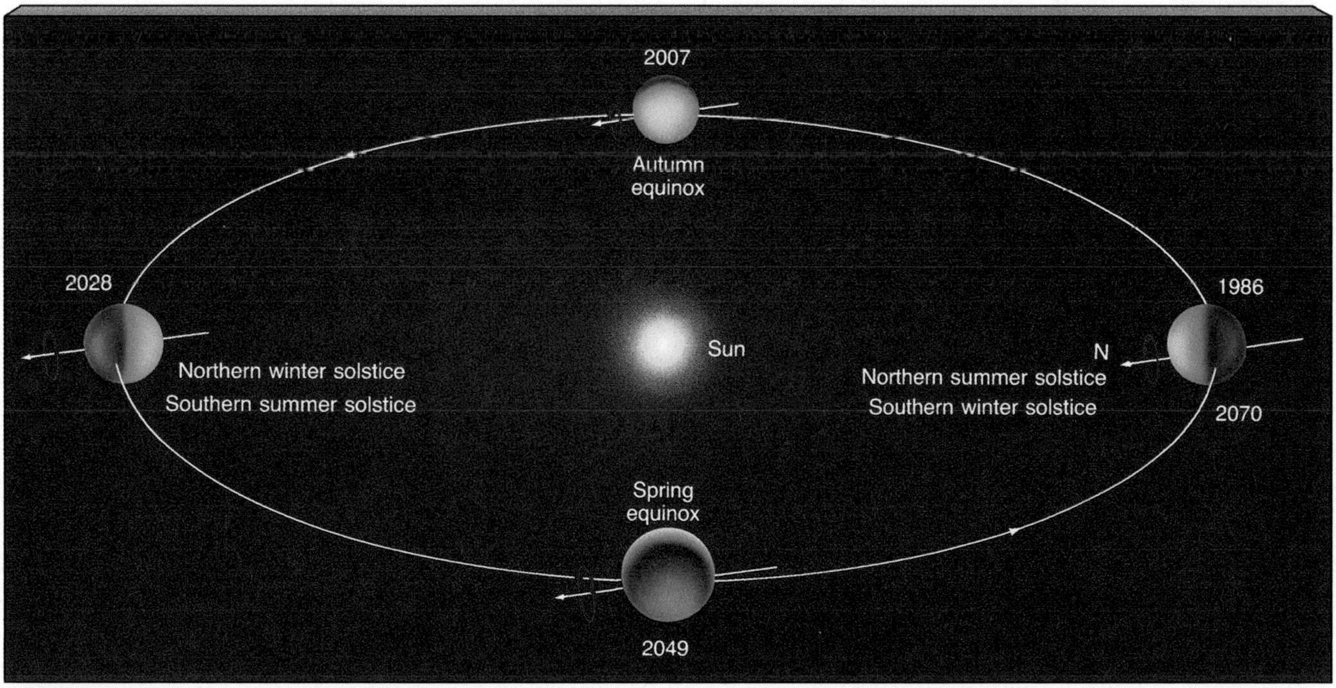

Figure 7.6 Uranus's Seasons Because of Uranus's axial tilt of 98°, the planet experiences the most extreme seasons known in the solar system. The equatorial regions experience two "summers" (warm seasons, at the times of the two equinoxes) and two "winters" (cool seasons, at the times of the solstices) each year, and the poles are alternately plunged into darkness for 42 years at a time.

TABLE 7.1 Planetary Properties

	MASS (kg)	MASS (Earth = 1)	RADIUS (km)	RADIUS (Earth = 1)	AVERAGE DENSITY (kg/m^3)	SURFACE GRAVITY (Earth = 1)	ESCAPE SPEED (km/s)	ROTATION PERIOD (solar days)	AXIAL TILT (degrees)	SURFACE TEMPERATURE (K)	MAGNETIC FIELD (Earth = 1)
Mercury	3.3×10^{23}	0.055	2,400	0.38	5400	0.38	4.3	59	0	100–700	0.01
Venus	4.9×10^{24}	0.82	6,100	0.95	5300	0.91	10.4	−243[1]	179	730	0.0
Earth	6.0×10^{24}	1.00	6,400	1.00	5500	1.00	11.2	1.00	23	290	1.0
Mars	6.4×10^{23}	0.11	3,400	0.53	3900	0.38	5.0	1.03	24	180–270	0.0
Jupiter	1.9×10^{27}	318	71,000	11	1330	2.5	60	0.41	3	124	14
Saturn	5.7×10^{26}	95	60,000	9.5	710	1.1	36	0.43	27	97	0.7
Uranus	8.7×10^{25}	15	26,000	4.0	1240	0.91	21	−0.69[1]	98	58	0.7
Neptune	1.0×10^{26}	17	25,000	3.9	1670	1.1	24	0.72	30	59	0.4

[1]The minus sign indicates retrograde rotation.

planet-sized body, might have altered the planet's spin axis. However, there is no direct evidence for such an occurrence and no theory to tell us how we should seek to confirm it.

✓ Concept Check

- Why do observations of a planet's magnetosphere allow astronomers to measure the rotation rate of the interior?

7.4 Jupiter's Atmosphere

3 Just as Earth has been our yardstick for the terrestrial worlds, Jupiter will be our guide to the outer planets. We therefore begin our study of jovian atmospheres with Jupiter itself.

Overall Appearance and Composition

Visually, Jupiter is dominated by two atmospheric features (both clearly visible in Figure 7.1); a series of ever-changing atmospheric cloud bands arranged parallel to the equator and an oval atmospheric blob called the **Great Red Spot**. The cloud bands display many colors—pale yellows, light blues, deep browns, drab tans, and vivid reds, among others. The Red Spot is one of many features associated with Jupiter's weather. This egg-shaped region, the long diameter of which is approximately twice Earth's diameter, seems to be a hurricane that has persisted for hundreds of years.

The most abundant gas in Jupiter's atmosphere is molecular hydrogen (roughly 86 percent of all molecules), followed by helium (nearly 14 percent). Small amounts of atmospheric methane, ammonia, and water vapor are also found. None of these gases can, by itself, account for Jupiter's observed coloration. For example, frozen ammonia and water vapor would produce white clouds, not the many colors we see. Scientists believe that complex chemical processes occurring in Jupiter's turbulent atmosphere are responsible for these colors, although the details of the chemistry are not fully understood. The trace elements sulfur and phosphorus may play important roles in influencing the cloud colors—particularly the reds, browns, and yellows. The energy that powers the reactions comes in many forms: the planet's own internal heat, solar ultraviolet radiation, aurorae in the planet's magnetosphere, and lightning discharges within the clouds.

Astronomers describe the banded structure of Jupiter's atmosphere as consisting of a series of lighter-colored **zones** and darker **belts** crossing the planet. The zones and belts vary in both latitude and intensity during the year, but the general pattern is always present. These variations appear to be the result of convective motion in the planet's atmosphere. The zones lie above upward-moving convective currents in Jupiter's atmosphere. The belts are the downward part of the cycle, where material is generally sinking, as illustrated in Figure 7.7.

Because of the upwelling material below them, the zones are regions of high pressure. The belts, conversely, are low-pressure regions. Thus, the belts and zones are the planet's equivalents of the familiar high- and low-

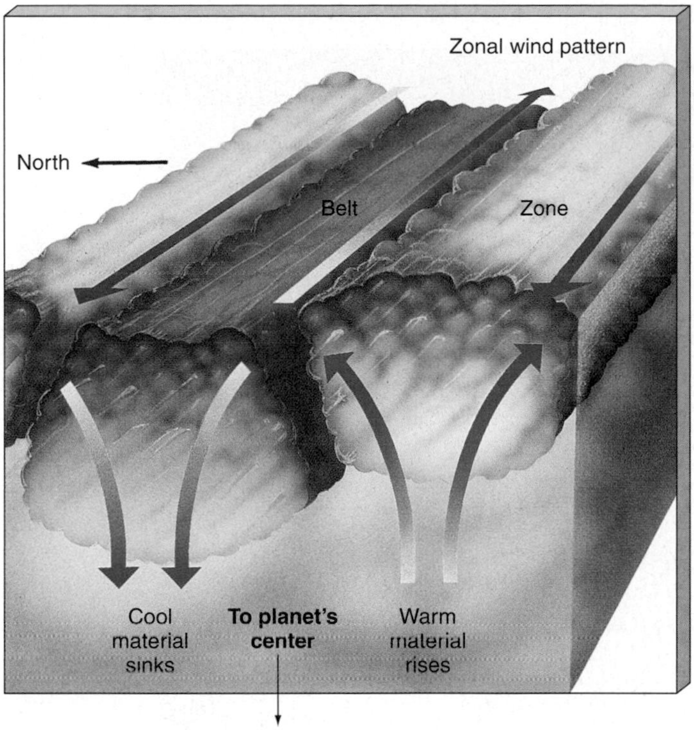

Figure 7.7 Jupiter's Convection The colored bands in Jupiter's atmosphere are associated with vertical convective motion. Upwelling warm gas results in the lighter-colored zones. The darker belts lie atop lower-pressure regions, where cooler gas is sinking back down into the atmosphere. As on Earth, surface winds tend to blow in the direction from high-pressure regions to low-pressure regions. Jupiter's rotation channels these winds into an east–west flow pattern, as indicated by the three arrows drawn atop the belts and zones.

pressure systems that cause weather on Earth. A major difference is that Jupiter's rapid rotation causes these systems to wrap all the way around the planet, instead of forming localized circulating storms as on our own world. Because of the pressure difference, the zones lie slightly higher in the atmosphere than the belts. Cloud chemistry is very sensitive to temperature, and the temperature difference between the two levels is the main reason for the different colors of the belts and zones.

Underlying the bands is an apparently very stable pattern of eastward and westward wind flow called Jupiter's *zonal flow*. The equatorial regions of Jupiter's atmosphere rotate faster than the planet as a whole, with an average flow speed of about 300 km/h in the easterly direction, somewhat similar to the jet stream on Earth. At higher latitudes, there are alternating regions of westward and eastward flow, the alternations corresponding to the pattern of belts and zones. The flow speed diminishes toward the poles. Near the poles, where the zonal flow disappears, the band structure vanishes also.

Atmospheric Structure

Planetary scientists believe that Jupiter's clouds are arranged in several layers, with white ammonia clouds generally overlying the colored layers, whose composition we will discuss in a moment. Above the ammonia clouds lies a thin, faint layer of haze created by chemical reactions similar to those that cause smog on Earth. When we observe Jupiter's colors, we are looking down to many depths in the planet's atmosphere.

Figure 7.8 is a diagram of Jupiter's atmosphere, based on observations and computer models. Because the planet lacks a solid surface to use as a reference level for measuring altitude, scientists conventionally take the top of the troposphere (the turbulent region containing the clouds we see) to lie at 0 km. With this level as the zero point, the troposphere's colored cloud layers all lie at negative altitudes in the diagram. As on other planets, weather on Jupiter is the result of convection in the troposphere. The haze layer lies at the upper edge of Jupiter's troposphere. The temperature at this level is about 110 K. Above the troposphere, as on Earth, the temperature rises as the atmosphere absorbs solar ultraviolet light.

Below the haze layer, at an altitude of −30 km, lie white, wispy clouds of ammonia ice. The temperature at this level is 125–150 K. A few tens of kilometers below the ammonia clouds, the temperature is a little warmer—above 200 K—and the clouds are made up mostly of droplets or crystals of ammonium hydrosulfide, produced by reactions between ammonia and hydrogen sulfide. At deeper levels, the ammonium hydrosulfide clouds give way to clouds of water ice or water vapor. The top of this lowest cloud layer, which is not seen in visible-light images of Jupiter, lies some 80 km below the top of the troposphere.

In December 1995, the *Galileo* atmospheric probe arrived at Jupiter. The probe survived for about an hour before being crushed by atmospheric pressure at an altitude of −150 km (that is, just at the bottom of Figure 7.8). While initially the data from the probe appeared not to support the picture just presented, most of the discrepancies were the result of improperly calibrated instruments. In addition, the probe's entry location (Figure 7.9) was in Jupiter's equatorial zone and, as luck would have it, coincided with an atypical hole almost devoid of upper-level clouds, and with abnormally low water content. *Galileo*'s revised findings on wind speed, temperature, and composition were in good agreement with the description presented above.

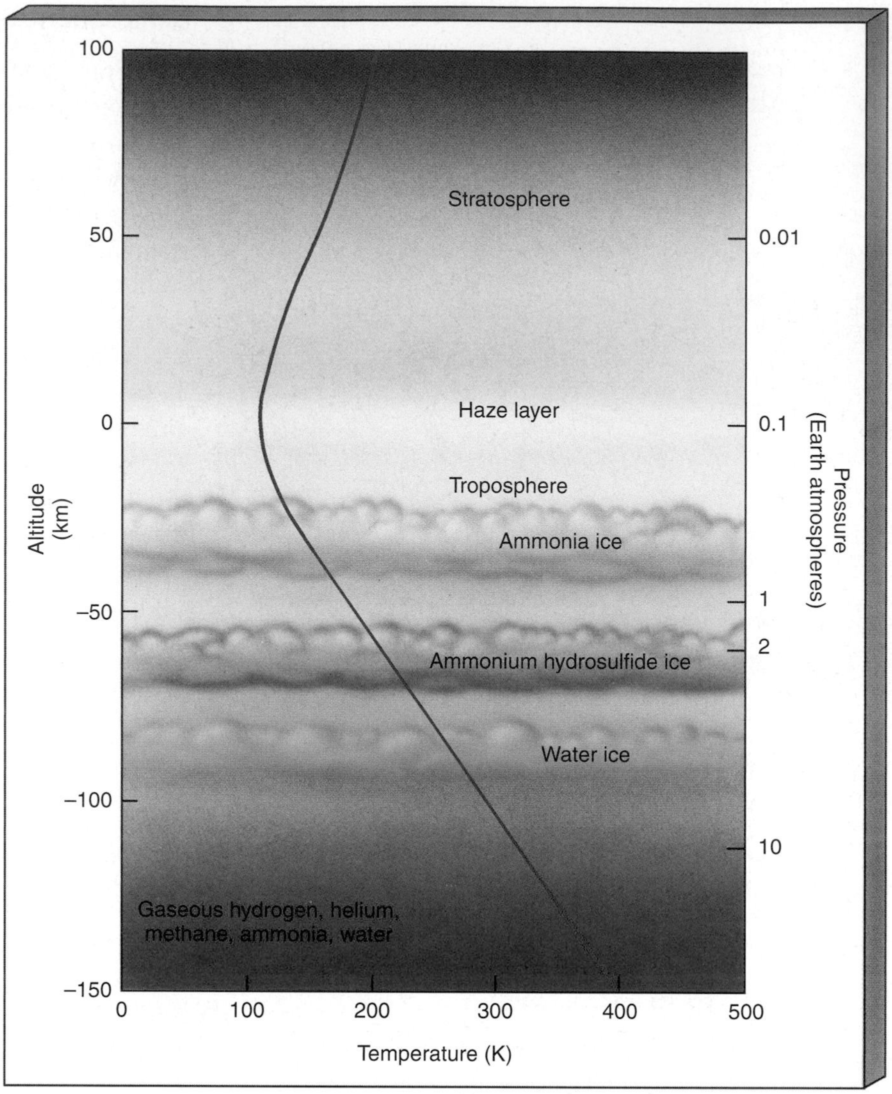

Figure 7.8 Jupiter's Atmosphere The vertical structure of Jupiter's atmosphere contains clouds that are arranged in three main layers, each with quite different colors and chemistry. The colors we see in photographs of the planet depend on the cloud cover. The white regions are the tops of the upper ammonia clouds. The yellows, reds, and browns are associated with the second cloud layer, which is composed of ammonium hydrosulfide ice. The lowest (bluish) cloud layer is water ice, however, the overlying layers are sufficiently thick that this level is not seen in visible light.

Weather on Jupiter

In addition to the large-scale zonal flow, Jupiter has many smaller-scale weather patterns. The Great Red Spot (Figure 7.10) is a prime example. Observational records indicate that it has existed continuously in one form or another for more than 300 years, and it may well be much older. *Voyager* observations showed the Spot to be a region of swirling, circulating winds, like a whirlpool or a terrestrial hurricane, a persistent and vast atmospheric storm. The size of the Spot varies its length averaging about 25,000 km. It rotates around Jupiter at a rate similar to that of the planet's interior, suggesting that its roots lie far below the atmosphere.

Astronomers think that the Spot is somehow sustained by Jupiter's large-scale atmospheric motion. The zonal motion north of the Spot is westward, while that to the south is eastward (Figure 7.10), supporting the idea that the Spot is confined and powered by the zonal flow. Turbulent eddies form and drift away from the spot's edge. The center, however, remains tranquil in appearance, like the eye of a hurricane on Earth.

The *Voyager* mission discovered many smaller spots that are also apparently circulating storm systems. Examples of these smaller systems are the **white ovals** seen in many images of Jupiter (see, for example, the one just south of the Great Red Spot in Figure 7.10). Their high cloud tops give these regions their white color. The white oval in Figure 7.10 is known to be at least 40 years old.

Figure 7.11 shows a **brown oval**—a hole in the overlying clouds that allows us to look down into

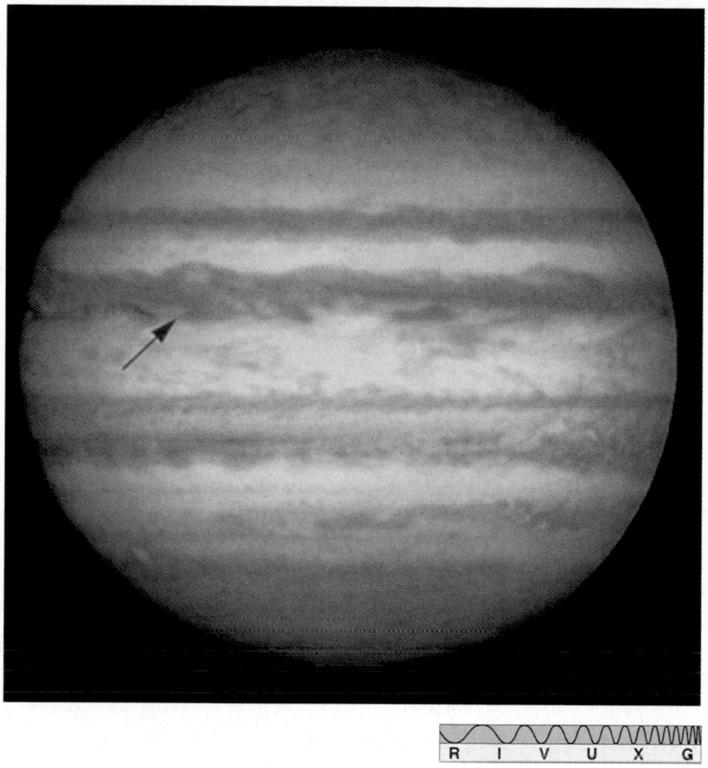

Figure 7.9 ***Galileo's* Entry Site** The arrow on this *Hubble Space Telescope* image shows where the *Galileo* atmospheric probe plunged into Jupiter's cloud deck on December 7, 1995. The entry location was in Jupiter's equatorial zone and apparently almost devoid of upper-level clouds. Until its demise, the probe took numerous meteorological measurements, transmitting those signals to the orbiting mother ship, which then relayed them to Earth. *(NASA)*

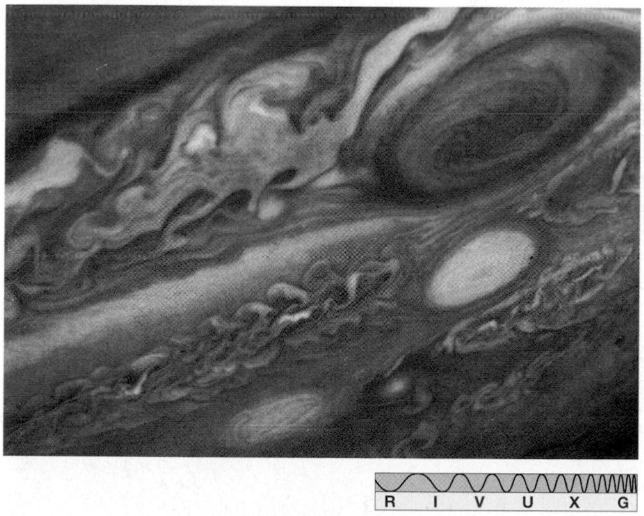

Figure 7.10 Jupiter's Red Spot *Voyager 1* took this photograph of Jupiter's Great Red Spot from a distance of about 100,000 km. Resolution is about 100 km. Note the complex flow patterns to the left of both the Red Spot and the smaller white oval below it. For scale, Earth is about the size of the white oval. The arrows indicate direction of gas flow above, below, and inside the Great Red Spot. *(NASA)*

the lower atmosphere. For unknown reasons, brown ovals appear only in latitudes around 20° north. Although not as long-lived as the Red Spot, they can persist for many years.

Although we cannot explain their formation, we can at least offer a partial explanation for the longevity of storm systems on Jupiter. On Earth a large storm, such as a hurricane, forms over the ocean and may survive for many days, but it dies quickly once it encounters land because the land disrupts the energy supplied and flow patterns that sustain the storm. Jupiter has no continents, so once a storm is established and has reached a size at which other storm systems cannot destroy it, apparently little affects it. The larger the system, the longer its lifetime.

✓ Concept Check

- List some similarities and differences between Jupiter's belts and zones and weather systems on Earth.

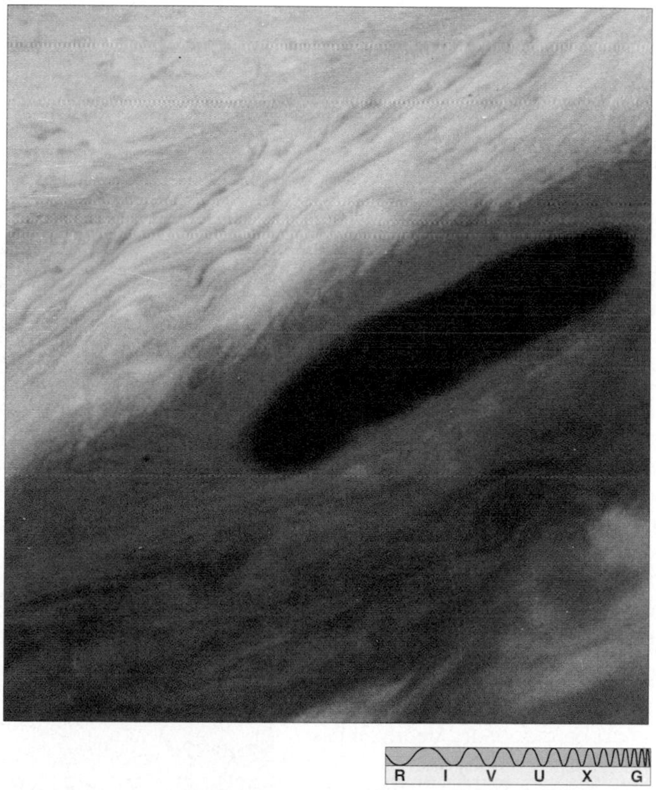

Figure 7.11 Jupiter's Brown Oval A brown oval in Jupiter's northern hemisphere is a break in the upper cloud layer, allowing us to see deeper in, to where the clouds are brown. The oval's long diameter is approximately equal to Earth's diameter. *(NASA)*

7.5 The Atmospheres of the Outer Jovian Worlds

Saturn's Atmospheric Composition

2 3 Saturn is not as colorful as Jupiter. Figure 7.12 shows yellowish and tan cloud bands that parallel the equator, but these regions display less atmospheric structure than do the bands on Jupiter. No obvious large spots or ovals adorn Saturn's cloud decks. Storms do exist, but the color changes that distinguish them on Jupiter are largely absent on Saturn.

Spectroscopic studies show that Saturn's atmosphere consists of molecular hydrogen (92.4 percent) and helium (7.4 percent), with traces of methane and ammonia. As on Jupiter, hydrogen and helium dominate. However, the fraction of helium on Saturn is substantially less than on Jupiter. It is unlikely that the processes that created the outer planets preferentially stripped Saturn of nearly half its helium or that the missing helium somehow escaped from the planet while the hydrogen remained behind. Instead, astronomers believe that at some time in Saturn's past the helium liquefied and then sank toward the center of the planet, reducing the abundance of that gas in the outer layers. The reason for this gas-to-liquid transformation is discussed in Section 7.7.

Figure 7.13 illustrates the vertical structure of Saturn's atmosphere. In many respects, it is similar to Jupiter's (Figure 7.8), except that Saturn's overall atmospheric temperature is a little lower because of its greater distance from the Sun. As on Jupiter, the troposphere contains clouds arranged in three distinct layers, composed (in order of increasing depth) of ammonia ice, ammonium hydrosulfide ice, and water ice. Above the clouds lies a layer of haze.

The total thickness of the three cloud layers in Saturn's atmosphere is roughly 250 km—about three times the cloud thickness on Jupiter—and each layer is thicker than its counterpart on Jupiter. The reason for this difference is Saturn's weaker gravity. At the haze level, Jupiter's gravitational field is nearly 2.5 times stronger than Saturn's, so Jupiter's atmosphere is pulled much more powerfully toward the center of the planet and the cloud layers are squeezed more closely together.

The colors of Saturn's cloud layers, as well as the planet's overall yellowish coloration, are likely due to the same basic cloud chemistry as on Jupiter. However, because Saturn's clouds are thicker, there are fewer holes and gaps in the top layer, with the result that we rarely glimpse the more colorful levels below. Instead, we see only the topmost layer, which accounts for Saturn's less varied appearance.

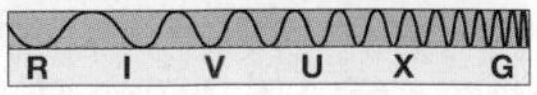

Figure 7.12 Saturn, Up Close The banded structure of Saturn's atmosphere appears clearly in this image, taken as *Voyager 2* approached the planet. (Three of Saturn's moons appear at bottom; a fourth casts a black shadow on the cloud tops.) This image is approximately true color. *(NASA)*

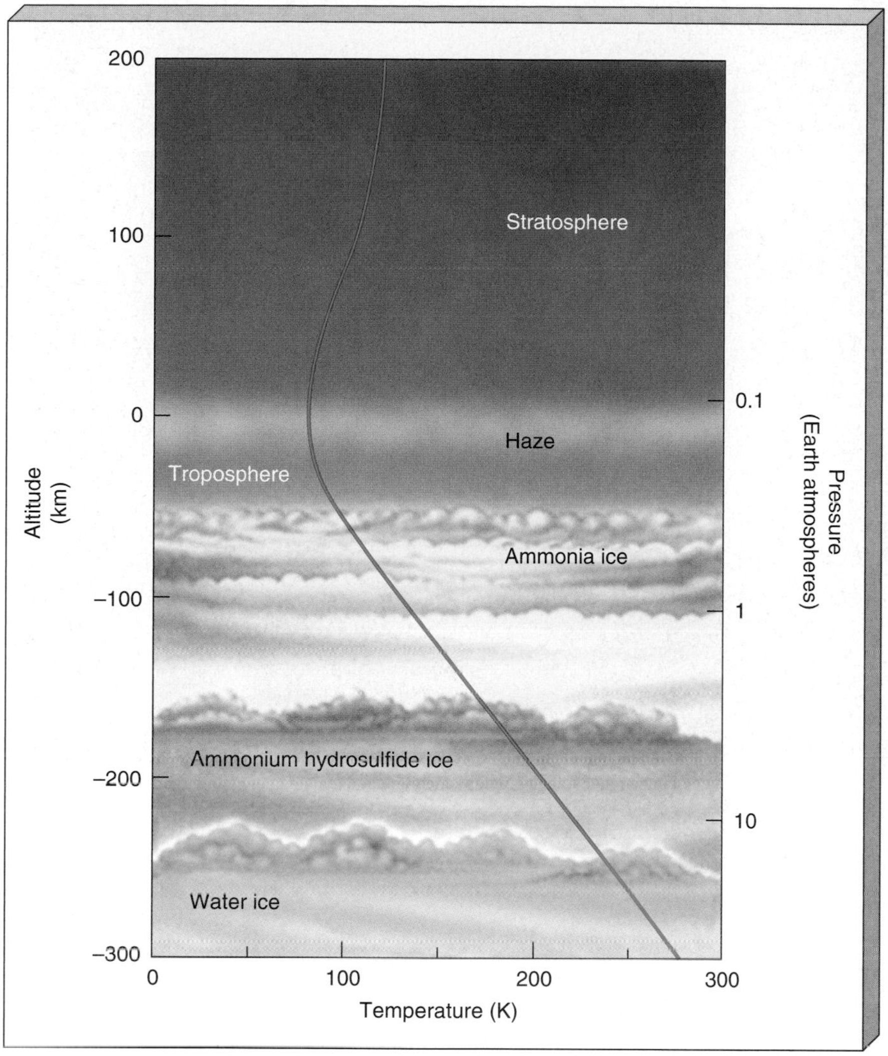

Figure 7.13 Saturn's Atmosphere The vertical structure of Saturn's atmosphere contains several cloud layers, like Jupiter, but Saturn's weaker gravity results in thicker layers and a more uniform appearance.

Weather on Saturn

Computer-enhanced images of Saturn (Figure 7.14) clearly show the existence of bands, oval storm systems, and turbulent flow patterns, all looking very much like those on Jupiter. In addition, there is an apparently quite stable east–west zonal flow, although the wind speed is considerably greater than on Jupiter and shows fewer east–west alternations. The equatorial eastward jet stream reaches a brisk 1500 km/h. Astronomers do not fully understand the reasons for the differences between Jupiter's and Saturn's flow patterns.

Figure 7.15 shows a rare storm system on Saturn. First detected in September 1990 as a large white spot just south of Saturn's equator, by November of that year the spot had developed into a band of clouds completely encircling the planet's equator. Astronomers believe that the white coloration arose from crystals of ammonia ice that formed when an upwelling plume of warm gas penetrated the cool upper cloud layers. Because the crystals were freshly formed, they had not yet been affected by the chemical reactions that color the planet's other clouds. The turbulent flow patterns seen in and around the spot had many similarities to the flow around Jupiter's Great Red Spot. Scientists hope that routine observations of such temporary atmospheric phenomena on the outer worlds will give them greater insight into the dynamics of planetary atmospheres.

Atmospheric Composition of Uranus and Neptune

The atmospheres of Uranus and Neptune are very similar in composition to that of Jupiter—the most abundant gas is molecular hydrogen (84 percent), followed by

Figure 7.14 Saturn's Cloud Structure More structure is seen in Saturn's cloud cover when computer processing and artificial color are used to enhance the image contrast, as in these *Voyager* images of the total planet and of a smaller, magnified piece of it. *(NASA)*

helium (about 14 percent) and methane, which is more abundant on Neptune (about 3 percent) than on Uranus (2 percent). Ammonia is not observed in any significant quantity in the atmosphere of either planet, most likely because, at the low atmospheric temperatures of those planets, the ammonia exists in the form of ice crystals, which are much harder to detect spectroscopically than gaseous ammonia.

The blue color of the outer jovian planets is the result of their relatively high percentages of methane. Because methane absorbs long-wavelength red light quite efficiently, sunlight reflected from these planets' atmospheres is deficient in red and yellow photons and appears blue-green or blue. The greater the concentration of methane, the bluer the reflected light. Therefore Uranus, having less methane, looks blue-green, while Neptune, having more methane, looks distinctly blue.

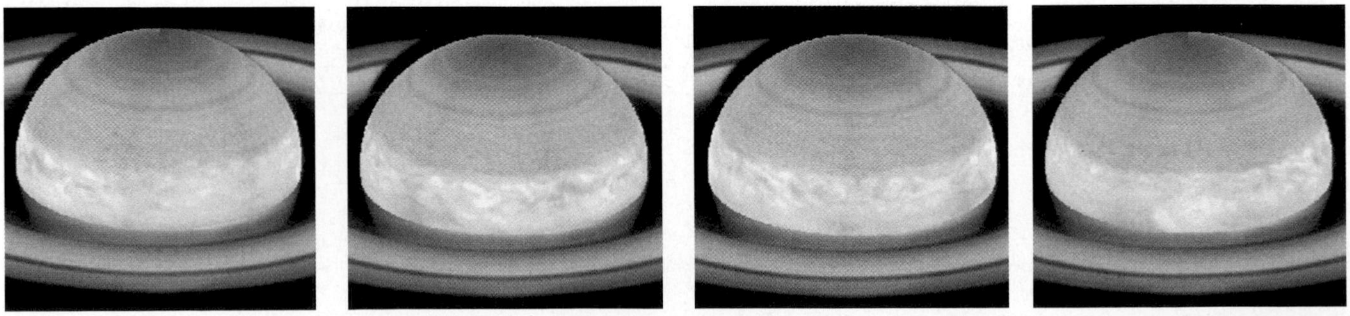

Figure 7.15 Saturn's Rotation Circulating and evolving cloud systems on Saturn, imaged by the *Hubble Space Telescope* at approximately two-hour intervals. *(NASA)*

Video

R I V U X G

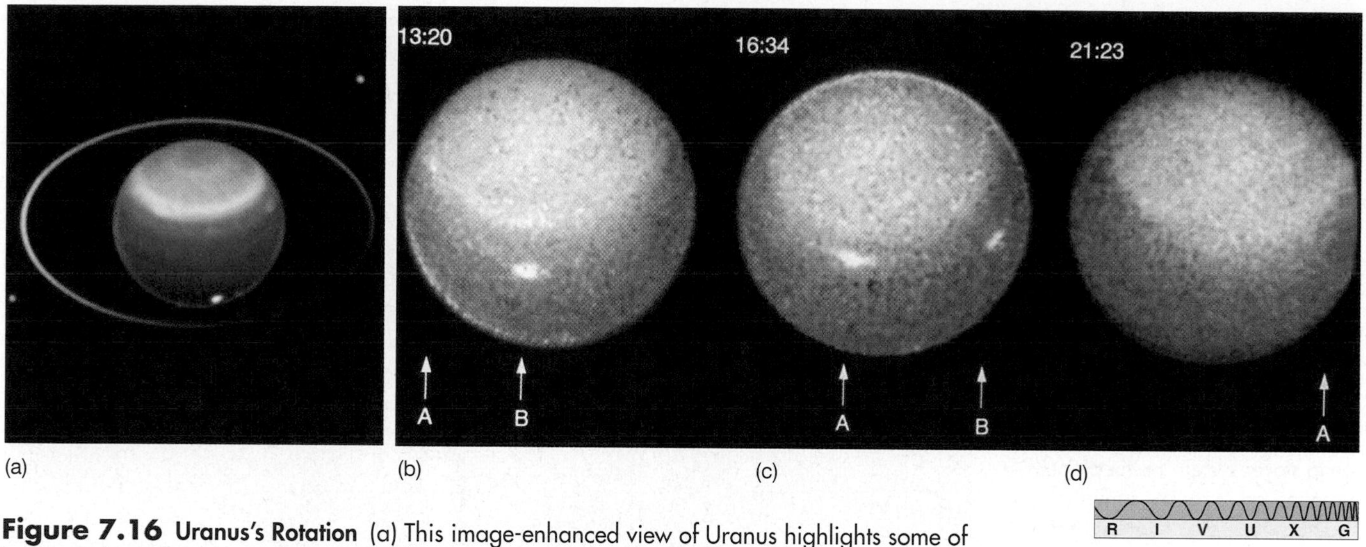

Figure 7.16 Uranus's Rotation (a) This image-enhanced view of Uranus highlights some of its atmospheric features and its surrounding ring, yet approximates the planet's true color. (b), (c), and (d) Image-processed *Hubble Space Telescope* photographs made at roughly four-hour intervals, showing the motion of a pair of bright clouds (labeled A and B) in the planet's southern hemisphere. (The numbers at the top give the time of each photograph.) *(NASA)*

Weather on Uranus and Neptune

Voyager 2 detected only a few atmospheric features on Uranus (Figure 7.16a), and even they became visible only after extensive computer enhancement. Few clouds exist in Uranus's cold upper atmosphere. Clouds of the kind found on Jupiter and Saturn are found only at lower, warmer atmospheric levels. The absence of high-level clouds means we must look deep into the planet's atmosphere to see any structure, and the bands and spots that characterize flow patterns on the other jovian worlds are largely washed out on Uranus by intervening stratospheric haze. Series of computer-processed images such as Figures 7.16(b–d) have shown astronomers that Uranus's atmospheric clouds and flow patterns move around the planet in the same direction as the planet's rotation, with wind speeds ranging from 200 to 500 km/h.

Although it lies farther from the Sun, Neptune's upper atmosphere is slightly warmer than that of Uranus, and this is probably why Neptune's atmospheric features are more easily visible. The extra warmth causes Neptune's stratospheric haze to be thinner than the haze on Uranus, and also makes the cloud layers less dense than they would otherwise be, with the result that they lie at higher levels in the atmosphere. *Voyager 2* detected numerous white methane clouds (some of them visible in Figure 7.4) lying about 50 km above the main cloud tops. Neptune's equatorial winds blow from east to west with speeds of over 2000 km/h relative to the planet's interior. Why such a cold planet should have such rapid winds, and why they should be retrograde relative to the west-to-east rotation of the interior, remains a mystery.

Neptune sports several storm systems similar in appearance to those seen on Jupiter, and probably produced and sustained by the same basic processes. The largest such storm, the **Great Dark Spot** (Figure 7.17a), was discovered by *Voyager 2* in 1989. Comparable in size to Earth, the Spot was located near the planet's equator, and exhibited many of the same general characteristics as the Great Red Spot on Jupiter. The flow around it was counterclockwise, as with the Red Spot, and there appeared to be turbulence where the winds associated with the Great Dark Spot interacted with the zonal flow to its north and south. Astronomers did not have long to study the Dark Spot's properties, however. As shown in Figure 7.17(b), when the *Hubble Space Telescope* viewed Neptune in 1994, the Spot had disappeared.

✓ Concept Check

- Why are atmospheric features much less evident on Uranus than on the other jovian planets?

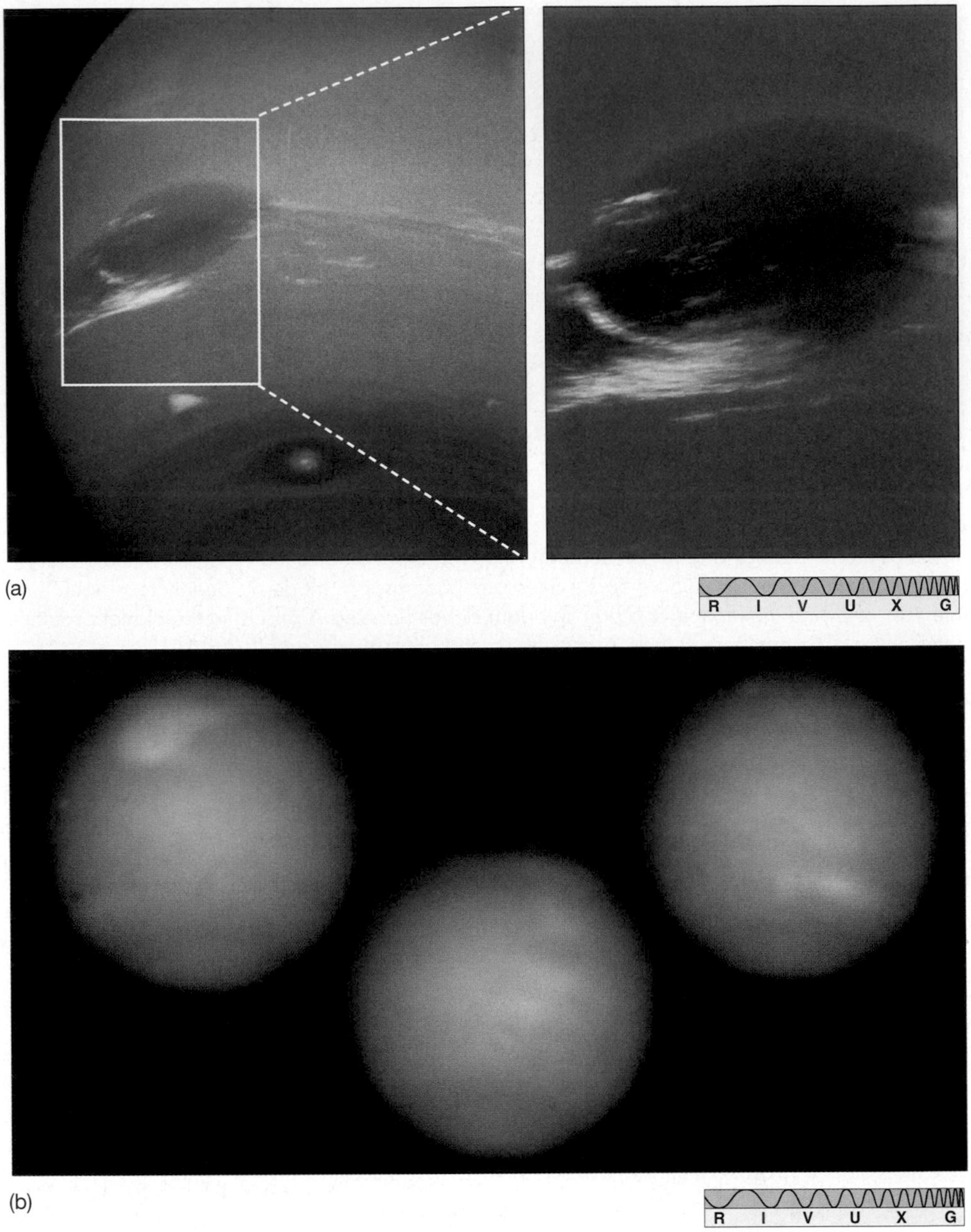

Figure 7.17 Neptune's Dark Spot (a) Close-up views of the Great Dark Spot of Neptune, which astronomers believe to be a large storm system in the planet's atmosphere, possibly similar in structure to Jupiter's Great Red Spot. Resolution in the image on the right is about 50 km. The dark spot is roughly the size of Earth. (b) These three *Hubble Space Telescope* views of Neptune were taken about 10 days apart in 1994, when the planet was 4.5 billion km from Earth. The planet appears aqua because methane in its clouds has absorbed red light from the sunlight striking the planet, with the result that the reflected light we see is blue-green in color. The cloud features (mostly methane ice crystals) are tinted pink here because they were imaged in the infrared, but they are white in visible light. Neptune apparently has a remarkably dynamic atmosphere that changes every few days. Notice that the Great Dark Spot has now disappeared. *(NASA)*

7.6 Jovian Interiors

2 4 The visible cloud layers of the jovian planets are all less than a few hundred kilometers thick. How then do we determine the conditions beneath the clouds, in the unseen interiors of these distant worlds? We have no seismographic information on these regions (but see *Interlude 7-1*), nor do we happen to live on a more or less similar planet that we can use for comparison. Instead, we must use all available bulk data on each planet, along with our knowledge of physics and chemistry, to construct a *model* of its interior that agrees with observations. Our statements about the interiors of the jovian worlds, then, are really statements about the models that best fit the facts.

Internal Structure

As illustrated in Figure 7.18, both the temperature and the pressure of Jupiter's atmosphere increase with depth. At a depth of a few thousand kilometers, the gas makes a gradual transition into the liquid state. By a depth of about 20,000 km, the pressure is about three million times greater than atmospheric pressure on Earth. Under these conditions, the hot liquid hydrogen is compressed so much that it undergoes another transition, this time to a "metallic" state, with properties in many ways similar to a liquid metal.

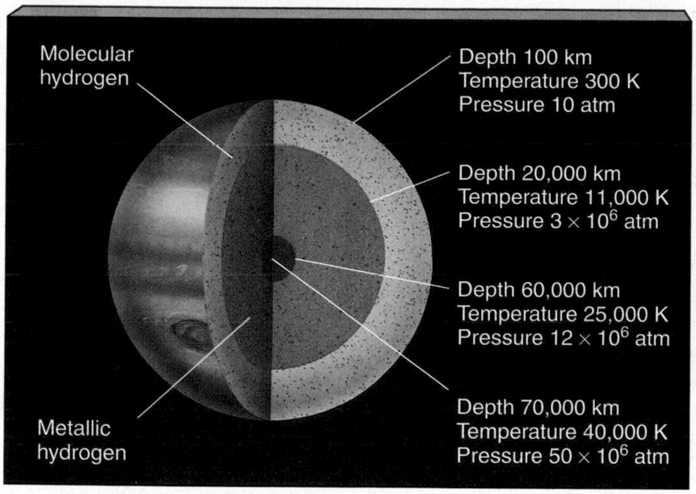

Figure 7.18 **Jupiter's Interior** Jupiter's internal structure, as deduced from spacecraft measurements and theoretical modeling. The outer radius shown in the diagram represents the bottom of the cloud layers, 70,000 km from the planet's center and about 100 km below the visible cloud top. Pressure and temperature increase with depth, and the atmosphere gradually liquefies over the outermost few thousand kilometers. Below 20,000 km, the hydrogen behaves like a liquid metal. At the center of the planet lies a rocky core, somewhat terrestrial in composition but much larger than any of the inner planets. The temperature at the center is probably about 40,000 K, and the pressure there is thought to be 50 million Earth atmospheres.

Of particular importance for Jupiter's magnetic field, this metallic hydrogen is an excellent conductor of electricity.

The strong outward push arising from Jupiter's rotation causes a pronounced bulge at the planet's equator—Jupiter's equatorial radius (Table 7.1) exceeds its polar radius by nearly seven percent. However, careful calculations indicate that, if Jupiter were composed of hydrogen and helium alone, it should be *more* flattened than is observed. The actual degree of flattening implies a dense *core* having a mass perhaps 10 times the mass of Earth. The core's precise composition is unknown, but many scientists think it consists of materials similar to those found in the terrestrial worlds: molten, perhaps semisolid, rock. Because of Jupiter's enormous central pressure—approximately 50 million times that on Earth's surface, or 10 times that at Earth's center—the core must be compressed to very high densities (perhaps twice the core density of Earth). It is probably not much more than 20,000 km in diameter.

Saturn has the same basic internal parts as Jupiter, but their relative proportions are different: Its metallic hydrogen layer is thinner, and the central dense core (again inferred from studies of the planet's polar flattening) is larger—around 15 Earth masses. Because of its lower mass, Saturn has a less extreme core temperature, density, and pressure than Jupiter. The central pressure is around a tenth of Jupiter's—not too different from the pressure at the center of Earth.

The interior pressures of Uranus and Neptune are sufficiently low that hydrogen stays in its molecular form all the way into the planets' cores. Currently, astronomers theorize that, deep below the cloud layers, Uranus and Neptune may have high-density, "slushy" interiors containing thick layers of highly compressed water clouds. It is also possible that much of the planets' ammonia could be dissolved in the water, producing a thick, electrically conducting layer that could conceivably explain the planets' magnetic fields. At the present time, however, we don't know enough about the interiors to assess the correctness of this picture.

The cores of Uranus and Neptune are inferred mainly from considerations of those planets' densities. Because of their lower masses, we would not expect the hydrogen and helium in the interiors of Uranus and Neptune to be compressed as much as in the two larger jovian worlds, yet the average densities of Uranus and Neptune are actually *greater* than the density of Saturn, and similar to that of Jupiter. Astronomers interpret this to mean that Uranus and Neptune each have Earth-sized cores some 10 to 15 times more massive than our planet, having compositions similar to the cores of Jupiter and Saturn.

Our current state of knowledge is summarized in Figure 7.19, which compares the internal structures of the four jovian worlds.

Magnetospheres

The combination of rapid overall rotation and an extensive region of highly conductive fluid in its interior gives Jupiter by far the strongest planetary magnetic field in the solar system. ∞ (Sec. 5.7) The planet's intrinsic field strength is some 20,000 times that of Earth. Jupiter is surrounded by a vast sea of energetic charged particles (mostly electrons and protons), somewhat similar to Earth's Van Allen belts but very much larger. Direct spacecraft measurements show

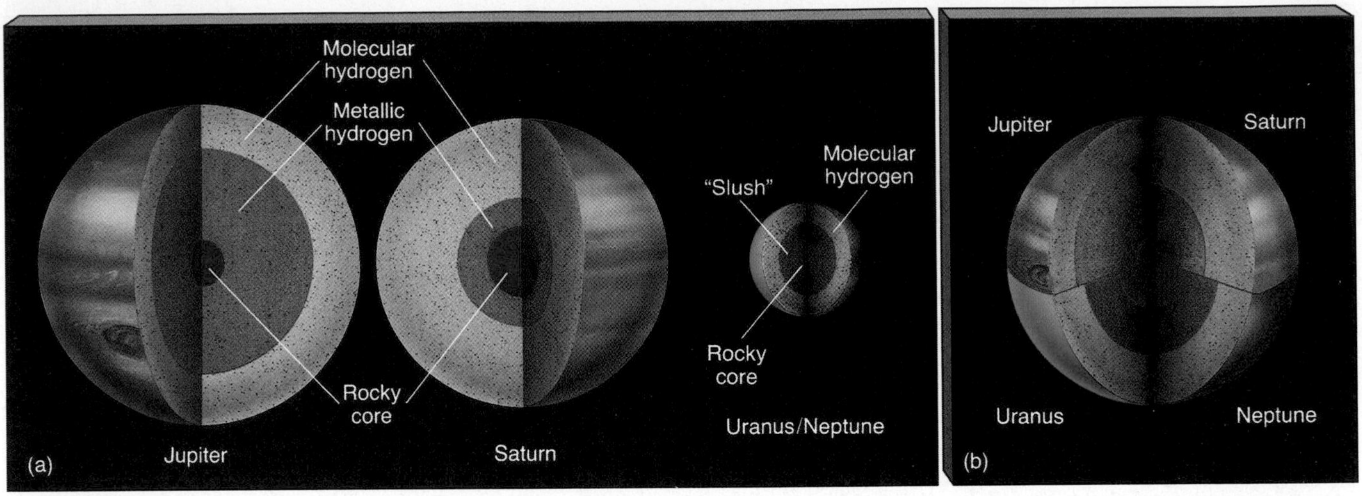

Figure 7.19 **Jovian's Interiors** A comparison of the interior structures of the four jovian planets. (a) The planets drawn to scale. (b) The relative proportions of the various internal zones. Animation

Jupiter's magnetosphere to be almost 30 million km across (north to south), roughly a million times more voluminous than Earth's magnetosphere and far larger than the entire Sun. All four Galilean moons lie within it. On the sunward side, the boundary of Jupiter's magnetic influence on the solar wind lies about three million km from the planet. On the opposite side, however, the magnetosphere has a long "tail" extending away from the Sun at least as far as Saturn's orbit (some 4 A.U., or 600 *million* km), as sketched in Figure 7.20.

Saturn's electrically conducting interior and rapid rotation also produce a strong magnetic field and an extensive magnetosphere. However, because of the considerably smaller mass of Saturn's metallic hydrogen zone, Saturn's basic magnetic field strength is only about 1/20 that of Jupiter. Saturn's magnetosphere extends about one million km toward the Sun and is large enough to contain the planet's ring system and most of its moons.

Voyager 2 found that both Uranus and Neptune have fairly strong magnetic fields, about 1/10 that of Saturn, or some 100 times stronger than Earth's field. Both planets have substantial magnetospheres, populated largely by electrons and protons either captured from the solar wind or created from hydrogen gas escaping from the planets below.

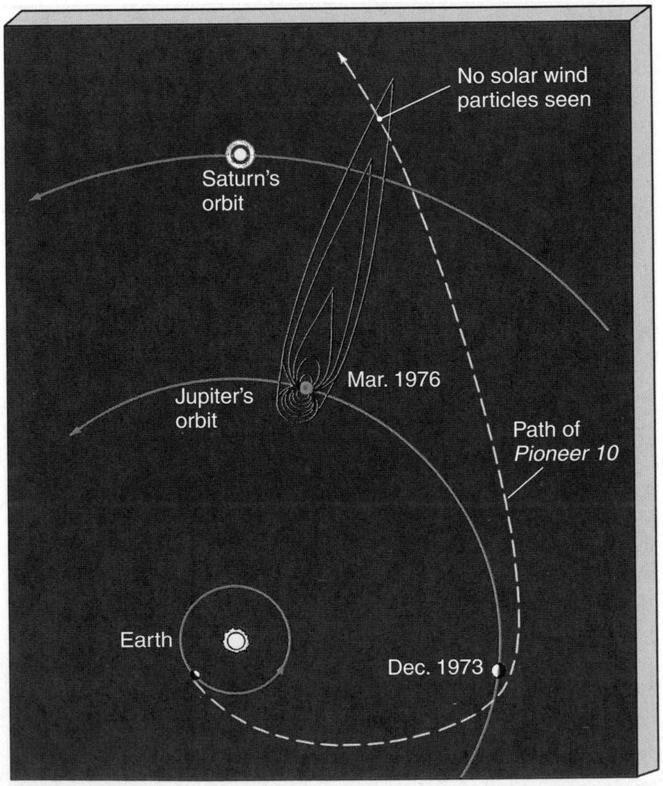

Figure 7.20 ***Pioneer 10* Mission** The *Pioneer 10* spacecraft (a forerunner to the *Voyager* missions) did not detect any solar-wind particles while moving far behind Jupiter in March 1976, indicating that the tail of Jupiter's magnetosphere extends beyond the orbit of Saturn.

Internal Heating

5 On the basis of Jupiter's distance from the Sun, astronomers expected to find the temperature of the cloud tops to be around 105 K. At that temperature, they reasoned, Jupiter would radiate back into space the same amount of energy as it receives from the Sun. Radio and infrared observations of the planet, however, revealed a black-body spectrum corresponding to a temperature of

125 K, implying (from Stefan's law) that Jupiter emits about twice—$(125/105)^4$—as much energy as reaches the planet in the form of sunlight. ∞ (Sec. 2.4) Astronomers theorize that the source of Jupiter's excess energy is heat left over from the planet's formation. That heat is slowly leaking out through the planet's heavy atmospheric blanket, resulting in the excess emission we observe.

Infrared measurements indicate that Saturn too has an internal energy source. Saturn's cloud-top temperature is 97 K, substantially higher than the temperature at which Saturn would reradiate all the energy it receives from the Sun. In fact, Saturn radiates away almost *three times* more energy than it absorbs. However, the explanation for Jupiter's excess emission doesn't work for Saturn. Saturn is smaller than Jupiter and so must have cooled more rapidly—rapidly enough, in fact, that its original supply of heat should have been used up long ago. What then is happening inside Saturn to produce this extra heat?

The explanation for this strange state of affairs also explains of Saturn's atmospheric helium deficit. At the temperatures and high pressures found in Jupiter's interior, liquid helium *dissolves* in liquid hydrogen. In Saturn, where the internal temperature is lower, the helium doesn't dissolve so easily and tends to form droplets instead. (The basic phenomenon is familiar to cooks who know that it is generally much easier to dissolve ingredients in hot liquids than in cold ones.) Saturn probably started out with a fairly uniform mix of hydrogen and helium, but the helium tended to condense out of the surrounding hydrogen, much as water vapor condenses out of Earth's atmosphere to form a mist. The amount of helium condensation was greatest in the planet's cool outer layers, where the mist turned to rain about two billion years ago. A light shower of liquid helium has been falling through Saturn's interior ever since. This helium precipitation is responsible for depleting the outer layers of their helium content.

As the helium sinks toward the center, the planet's gravitational field compresses it and heats it up. The energy thus released is the source of Saturn's internal heating. In the distant future, the helium rain will stop and Saturn will cool until its outermost layers radiate only as much energy as they receive from the Sun. When that happens, the temperature at Saturn's cloud tops will be 74 K.

Uranus apparently has no internal energy source, radiating into space the same amount of energy as it receives from the Sun. It never experienced helium precipitation, and its initial supply of heat was lost to space long ago. What is surprising, though, given its other similarities to Uranus, is that Neptune *does* have an internal source of heat—the planet emits 2.7 times more energy than it receives from the Sun. The cause of this heating is still uncertain. Some scientists have suggested that Neptune's relatively high concentration of methane insulates the planet, helping to maintain its initially high internal temperature.

✓ Concept Check

- In general terms, how are the massive, dense cores and extended gaseous envelopes of the jovian planets accounted for by the picture of planet formation presented in Section 4.3?

INTERLUDE 7-1

A Cometary Impact

In 1994, astronomers were granted a rare alternative means of studying Jupiter's atmosphere and interior when comet Shoemaker–Levy 9 collided with the planet. Discovered in 1993, the comet had an unusual flattened nucleus made up of several large pieces (shown below), the largest no more than 1 km across. Tracing the orbit backward in time, researchers found that the comet had narrowly missed Jupiter in 1992. They realized that the objects shown in the figure were fragments produced when a previously "normal" comet was captured by Jupiter and torn apart by the planet's strong tidal forces. (*More Precisely 5-1*)

On their next approach to Jupiter, the comet fragments struck Jupiter's upper atmosphere at speeds of over 60 km/s, causing a series of enormous explosions. Each impact created, for a period of a few minutes, a brilliant fireball hundreds of kilometers across and having a temperature of many thousands of kelvins. The energy released in each explosion was comparable to a billion terrestrial nuclear detonations. One of the largest pieces of the comet, fragment G, produced the spectacular fireball shown in the second image. The effects on the planet's atmosphere and the vibrations produced throughout Jupiter's interior were observable for days after the impact.

As far as we can determine, none of the cometary fragments breached the jovian clouds. Only *Galileo* had a direct view of the impacts on the back side of Jupiter, and in every case the explosions seemed to occur high in the atmosphere, above the uppermost cloud layer. Spectral lines from water vapor, silicon, magnesium, and iron were observed, but all apparently came from the comet itself—which really did resemble a loosely packed snowball. Debris from the impacts spread slowly around Jupiter's bands, and after five months reached completely around the planet. It took years for all the cometary matter to settle into Jupiter's interior.

(NASA)

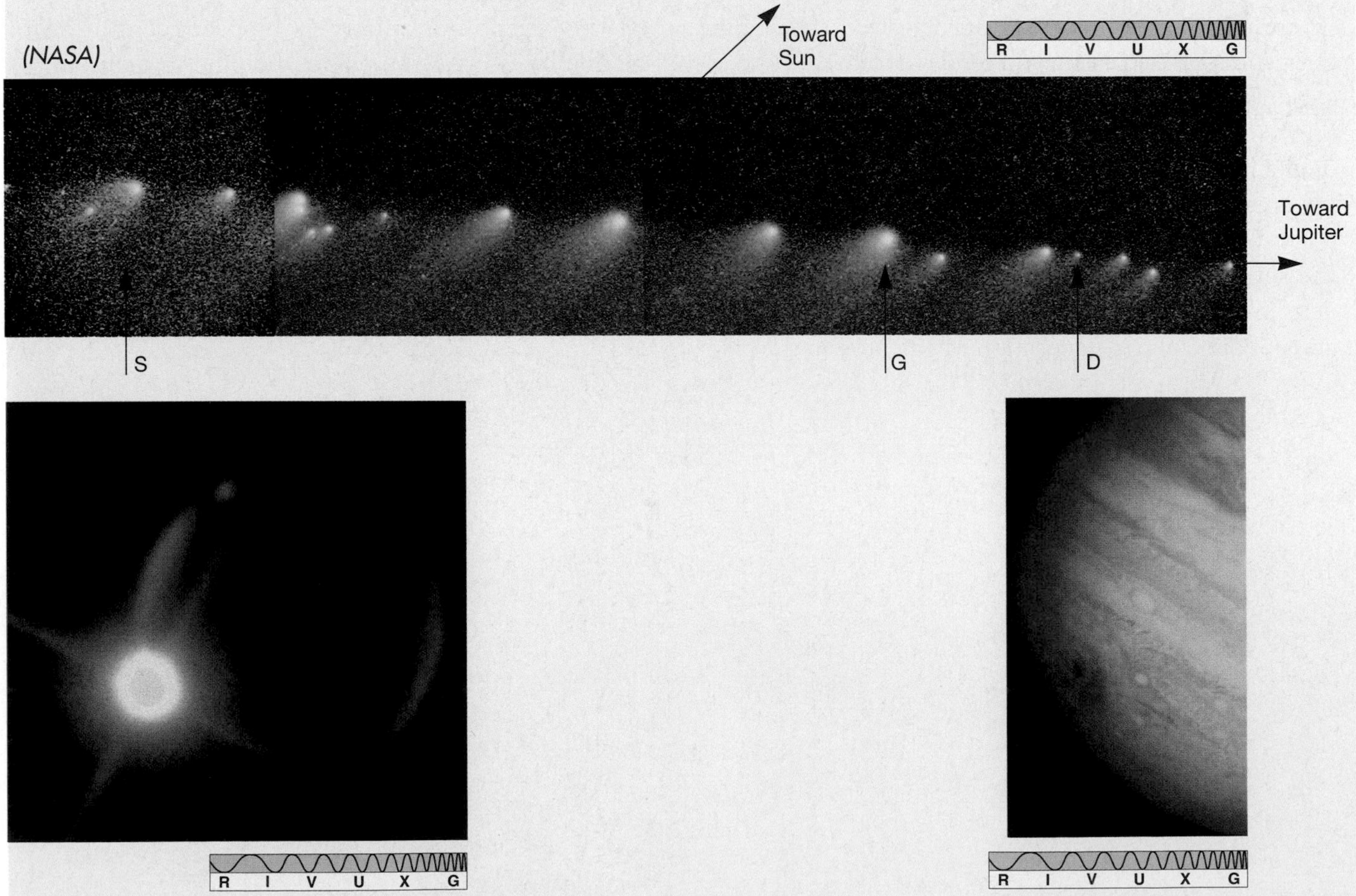

Chapter Review

Summary

Jupiter and Saturn were the outermost planets known to ancient astronomers. Uranus was discovered in the eighteenth century, by chance. Neptune was discovered after mathematical calculations of Uranus's slightly non-Keplerian orbit revealed the presence of an eighth planet.

Having no solid surface, the jovian planets display **differential rotation** (p. 188)—the rotation rate varies from place to place in the atmosphere. Radio waves emitted from the planets' magnetospheres provide a measure of their interior rotation rates. For unknown reasons, Uranus's spin axis lies nearly in the ecliptic plane, leading to extreme seasonal variations in solar heating on the planet as it orbits the Sun.

The cloud layers on all the jovian worlds are arranged into bands of light-colored **zones** (p. 190) and darker **belts** (p. 190) crossing the planets parallel to the equator. The bands are the result of (1) convection in the planets' interiors and (2) the planets' rapid rotation. Underlying the bands is a stable pattern of eastward or westward wind flow. The wind direction alternates as we move north or south away from the equator.

Jupiter's atmosphere consists of three main cloud layers. The colors we see are the result of chemical reactions at varying depths below the cloud tops. A major weather pattern on Jupiter is the **Great Red Spot** (p. 190), a huge hurricane that apparently has been raging for at least three centuries. Other, smaller, atmospheric features—**white ovals** (p. 192) and **brown ovals** (p. 192)—are also observed. Similar systems are found on the other jovian planets, although they are less distinct on Saturn and Uranus. Short-lived storms on Saturn are occasionally seen. The **Great Dark Spot** (p. 197) on Neptune had many similarities to Jupiter's Red Spot.

The atmospheres of the jovian planets become hotter and denser with depth, eventually becoming liquid. In Jupiter and Saturn, the interior pressures are so high that the hydrogen near the center exists in a liquid-metallic form rather than as molecular hydrogen. In Uranus and Neptune, this change from molecular hydrogen to metallic hydrogen apparently does not occur. All four planets have large, dense cores, of chemical composition similar to that of the terrestrial planets. The combination of rapid rotation and conductive interiors means that all four jovian planets have strong magnetic fields and extensive magnetospheres.

Three jovian planets radiate more energy than they receive from the Sun. On Jupiter and Neptune, the source of this energy is most likely heat left over from the planet's formation. On Saturn, the heating is the result of helium precipitation in the interior, where helium liquefies and forms droplets, which then fall toward the center of the planet. This process is also responsible for reducing the amount of helium observed in Saturn's surface layers.

Review and Discussion

1. How was Uranus discovered?
2. Why did astronomers suspect an eighth planet beyond Uranus?
3. Why have the jovian planets retained most of their original atmospheres?
4. What is differential rotation, and what evidence do we have that it occurs on Jupiter?
5. Why does Saturn have a less varied appearance than Jupiter?
6. Compare the thicknesses of Saturn's various layers (clouds, molecular hydrogen, metallic hydrogen, and core) to the equivalent layers in Jupiter.
7. What is the Great Red Spot? What is known about the source of its energy?
8. Describe the cause of the colors in the jovian planets' atmospheres.
9. Compare and contrast the atmospheres of the jovian planets, describing how the differences affect the appearance of each.
10. Explain the theoretical model that accounts for Jupiter's internal heat source.
11. Describe the theoretical model that accounts for Saturn's missing helium and the surprising amount of heat the planet radiates.
12. What is unusual about the rotation of Uranus?
13. What is responsible for Jupiter's enormous magnetic field?
14. How are the interiors of Uranus and Neptune thought to differ from those of Jupiter and Saturn?
15. How much information do you think we might have gathered about the outer solar system without the *Voyager* and *Galileo* missions?

True or False?

____ **1.** Uranus was discovered by Galileo.

____ **2.** After the discovery of Uranus, astronomers started looking for other planets and discovered Neptune less than a century later.

____ **3.** There is no evidence to suggest that either Jupiter or Saturn has a rocky core.

____ **4.** There is no evidence to suggest that either Uranus or Neptune has a rocky core.

____ **5.** Jupiter's solid surface lies just below the cloud layers visible from Earth.

____ **6.** The thickness of Jupiter's cloud layer is less than one percent of the planet's radius.

____ **7.** Jupiter has only one large storm system.

____ **8.** In general, a storm system in Jupiter's atmosphere is much longer-lived than a storm system in Earth's atmosphere.

____ **9.** The element helium plays an important role in producing the colors in jovian atmospheres.

____ **10.** Like that of Jupiter, the rotation of Saturn is rapid and shows differential rotation.

____ **11.** The atmosphere of Saturn contains only half the helium of Jupiter's atmosphere, but overall Saturn probably has a normal helium abundance.

____ **12.** Like the other jovian worlds, Uranus has prominent belts and zones in its atmosphere.

____ **13.** Jupiter's magnetic field is much stronger than Earth's.

____ **14.** Both Uranus and Neptune have layers of metallic hydrogen surrounding their central cores.

____ **15.** The fraction of Uranus or Neptune that is core is larger than the fraction of Jupiter or Saturn that is core.

Fill in the Blank

1. The _______ of Jupiter indicates that the planet's overall composition differs greatly from that of the terrestrial planets.

2. The planet _______ is blue-green and virtually featureless.

3. The planet _______ is dark blue, with white clouds and visible storm systems.

4. The overall bluish coloration of Uranus and Neptune is caused by the gas _______.

5. The main constituents of Jupiter and Saturn are _______ and _______.

6. Jupiter's clouds consist of a series of light-colored _______ and darker _______.

7. Jupiter's Great Red Spot has similarities to _______ on Earth.

8. Features in the atmospheres of Saturn and Jupiter are mostly hidden by an upper layer of frozen _______ clouds.

9. Saturn's cloud layers are thicker than those of Jupiter because of Saturn's weaker _______.

10. The Great Dark Spot was a storm system on _______.

11. Jupiter's rapid _______ produces a significant equatorial bulge.

12. Uranus's rotation axis is almost _______ to the ecliptic plane.

13. Jupiter emits about _______ times more radiation than it receives from the Sun.

14. Jupiter's magnetic field is generated by its rapid rotation and by the element _______, which becomes metallic in the interior of the planet.

15. Saturn's excess energy emission is caused by _______.

Problems

1. The *Hubble Space Telescope* has an angular resolution of 0.05″. What is the size of the smallest feature visible on (a) Jupiter, at 4.2 A.U., and (b) Neptune, at 29 A.U.?
2. What is the gravitational force exerted on Uranus by Neptune, at closest approach? Compare your answer with the Sun's gravitational force on Uranus.
3. What is the round-trip travel time of light from Earth to Jupiter (at the minimum Earth–Jupiter distance of 4.2 A.U.)? How far would a spacecraft orbiting Jupiter at a speed of 20 km/s travel in that time?
4. What would be the mass of Jupiter if it were composed entirely of hydrogen at a density of 0.08 kg/m^3 (the density of hydrogen at sea level on Earth)?
5. Verify the statement made in the text that the force of gravity at Jupiter's cloud tops is nearly 2.5 times that at Saturn's cloud tops.
6. Given Jupiter's current atmospheric temperature, what is the smallest possible mass the planet could have and still have retained its hydrogen atmosphere? (See *More Precisely 5-2*.)
7. How long does it take for Saturn's equatorial flow, moving at 1500 km/h, to encircle the planet?
8. If Saturn's surface temperature is 97 K and the planet radiates three times more energy than it receives from the Sun, use Stefan's law to calculate what the surface temperature would be in the absence of any internal heat source. (See *More Precisely 2-2*.)
9. If the core of Uranus has a radius twice that of planet Earth and an average density of 8,000 kg/m^3, calculate the mass of Uranus outside the core. What fraction of the planet's total mass is core?
10. From Wien's law, at what wavelength does Neptune's thermal emission peak? In what part of the electromagnetic spectrum does this wavelength lie? (See *More Precisely 2-2* and Figure 2.8.)

Projects

1. Look in an almanac to find out where Jupiter is now. What constellation is it in? Do any stars in the sky look as bright as Jupiter?
2. Saturn moves more slowly among the stars than any other visible planet. It crosses a constellation boundary only once every few years. Look in an almanac to see where the planet is now, and find Saturn in the sky. What constellation is it in? Can you explain why it moves so slowly, in contrast to Mars or Jupiter? How many stars in the night sky are brighter than Saturn at its brightest? What do you notice about the difference in appearance between Saturn and a star of about equal brightness?
3. The major astronomy magazines *Sky & Telescope* and *Astronomy* generally print, in their January issues, charts showing the whereabouts of the planets. Consult the charts in one of these magazines and locate Neptune and Uranus in the sky. Uranus may be visible to the naked eye, but binoculars make the search much easier. (*Hint*: Uranus shines more steadily than do the background stars.) With the eye alone, can you detect a color to Uranus? Does color become more intense through binoculars?
4. The search for Neptune requires a much more determined effort. A telescope is best, but high-powered binoculars mounted on a steady support also reveal the planet. If you have an opportunity to see Neptune and Uranus through a telescope (they are currently quite close together on the sky), contrast their colors. Which planet appears bluer? Through a telescope, can you see that Uranus appears as a disk rather than a point? Does Neptune also look like a disk, or does it look more like a pinpoint of light?

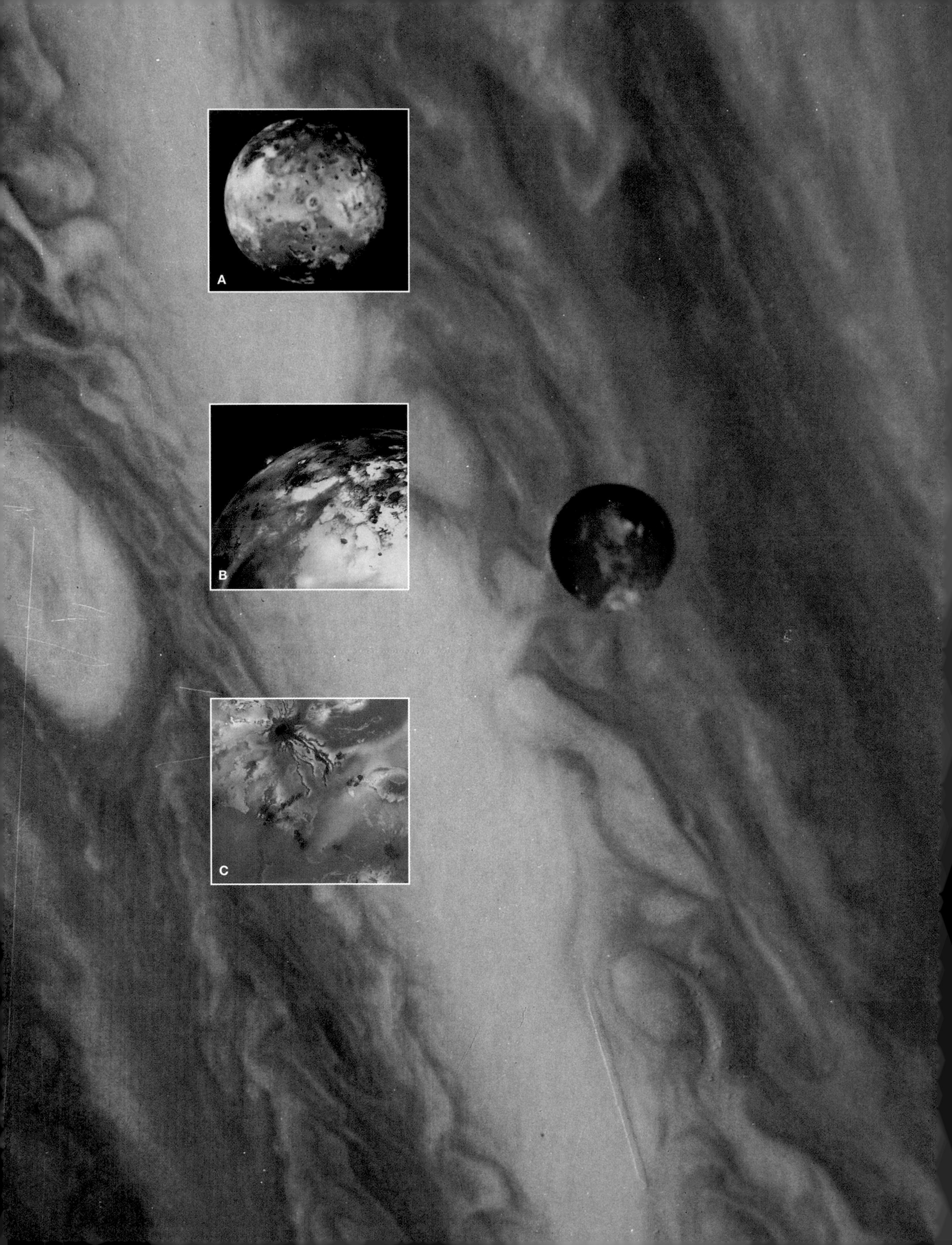
A
B
C

8 MOONS, RINGS, AND PLUTO

Small Worlds among Giants

LEARNING GOALS

Studying this chapter will enable you to:

1. Describe how the Galilean moons form a miniature solar system around Jupiter and exhibit a wide range of properties.
2. Discuss the composition and origin of the atmosphere on Titan, Saturn's largest moon.
3. Explain why astronomers think Neptune's moon Triton was captured by that planet.
4. Describe the nature and detailed structure of Saturn's rings.
5. Discuss the differences between the rings of Saturn and those of the other jovian planets.
6. Explain how the Pluto–Charon system is fundamentally different from all other planet–moon systems.

(Opposite page, background) In this close-up image of Jupiter, taken by the *Voyager 2* spacecraft, the moon Io is clearly visible. One of the most peculiar moons in the solar system, and certainly the most active, Io's multi-hued surface is caused by various chemical compounds deposited by currently active volcanoes.

(Inset A) Photo of Io taken by *Voyager 1* with a remarkable resolution of 7 km. The red color results largely from sulfur compounds, the white areas probably covered with sulfur dioxide frost.

(Inset B) At top left, a plume from an erupting volcano, known as Prometheus, is clearly seen at the limb of Io against the blackness of space.

(Inset C) Many lava flows can be seen on Io, like this one stretching for several hundred kilometers. *(NASA)*

All four jovian planets have systems of moons and rings that exhibit fascinating variety and complexity. The six largest jovian moons have many planetary features, providing us with insight into the terrestrial worlds. Two of these moons, Jupiter's Europa and Saturn's Titan, have become prime candidates for study in the search for life elsewhere in the solar system. Like the moons, the rings of the jovian planets differ greatly from one to the next. The rings of Saturn—the best-known ring system among the jovian planets—are one of the most spectacular displays in the sky. Pluto is a planet, not a moon, but it is very much smaller than even the terrestrial worlds and in many ways is much more moonlike than planetlike. For that reason, we study it here along with the moons and rings of its giant neighbors.

8.1 The Galilean Moons of Jupiter

All four jovian planets have extensive moon systems—Jupiter has at least 16 moons, Saturn 18, Uranus 17, and Neptune 8. However, these moons range greatly in size and properties, from planet-sized worlds with their own distinctive geologies and histories, to irregularly shaped chunks of ice just a few tens of kilometers across. Six "large" jovian moons, each comparable in size to Earth's Moon, are listed in Table 8.1. We begin our study of the large moons with the Galilean satellites of Jupiter.

A "Miniature Solar System"

1 Jupiter's four large moons travel on nearly circular, prograde orbits in Jupiter's equatorial plane. Moving outward from the planet, they are named Io, Europa, Ganymede, and Callisto, after the mythical attendants of the Roman god Jupiter. The two *Voyager* spacecraft and, more recently, the *Galileo* mission returned many remarkable images of these small worlds, allowing scientists to discern much fine surface detail on each. Figure 8.1 is a *Voyager 1* image of Io and Europa, with Jupiter providing a spectacular backdrop. Figure 8.2 shows the four Galilean moons to scale.

The Galilean moons range in size from slightly smaller than Earth's Moon (Europa) to slightly larger than Mercury (Ganymede). Their densities decrease with increasing distance from Jupiter in a manner very reminiscent of the falloff in density of the terrestrial planets with increasing distance from the Sun. ∞ (Sec. 4.1) Indeed, many astronomers think that the formation of Jupiter and its Galilean satellites may have mimicked on a small scale the formation of the Sun and the inner planets, with generally denser moons forming closer to the intense heat of the young Jupiter. ∞ (Sec. 4.3)

Detailed gravity measurements made by the *Galileo* orbiter support this view. They suggest that Io and Europa have large, iron-rich cores surrounded by thick mantles of generally rocky composition, perhaps similar to the crusts of the terrestrial planets. Europa has a thin water and ice outer shell about 150 km thick. Ganymede and Callisto are of more lightweight overall composition. Low-density materials, such as water ice, may account for as much as half their total mass. Ganymede has a relatively small metallic core topped by a rocky mantle and a thick icy outer shell. Callisto seems to be a largely undifferentiated mixture of rock and ice.

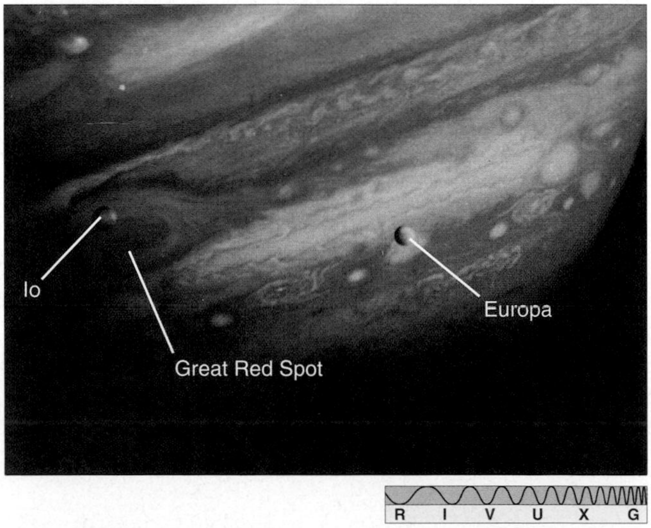

Figure 8.1 **Jupiter, Up Close** *Voyager 1* took this photograph of Jupiter with ruddy Io on the left and pearllike Europa toward the right. Note the scale of objects here: Io and Europa are each comparable in size to our Moon; the Great Red Spot is roughly twice as large as Earth. *(NASA)*

TABLE 8.1 The Large Moons of the Solar System

NAME	PARENT PLANET	DISTANCE FROM PARENT PLANET (km)	(planet radii)	ORBIT PERIOD (Earth days)	RADIUS (km)	MASS (Earth's Moon = 1)	DENSITY (kg/m^3)
Moon	Earth	384,000	60.2	27.3	1740	1.00	3300
Io	Jupiter	422,000	5.90	1.77	1820	1.22	3500
Europa	Jupiter	671,000	9.38	3.55	1570	0.65	3000
Ganymede	Jupiter	1,070,000	15.0	7.15	2630	2.02	1900
Callisto	Jupiter	1,880,000	26.3	16.7	2400	1.46	1800
Titan	Saturn	1,220,000	20.3	16.0	2580	1.83	1900
Triton	Neptune	355,000	14.3	−5.88[1]	1350	0.29	2100

[1]The minus sign indicates a retrograde orbit.

Io: The Most Active Moon

Io is the most geologically active object in the entire solar system. In mass and radius it is fairly similar to Earth's Moon, but there the resemblance ends. Shown in Figure 8.3, Io's uncratered surface is a collage of oranges, yellows, and blackish browns.

As *Voyager 1* sped past Io, it made an outstanding discovery: Io has active volcanoes! The spacecraft photographed eight erupting volcanoes. Six of them were still erupting when *Voyager 2* passed by four months later. The inset in Figure 8.3 shows one volcano ejecting matter at speeds of up to 2 km/s to an altitude of about 150 km. The orange color immediately surrounding the volcanoes most likely results from sulfur compounds in the ejected material. Io's exceptionally smooth surface is apparently the result of molten

Figure 8.2 Galilean Moons *Voyager 1* spacecraft images of the four Galilean moons of Jupiter. Shown here to scale as they would appear from a distance of about one million km are (clockwise from upper left) Io, Europa, Callisto, and Ganymede. For scale, Earth's Moon is slightly smaller than Io. *(NASA)*

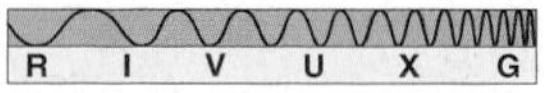

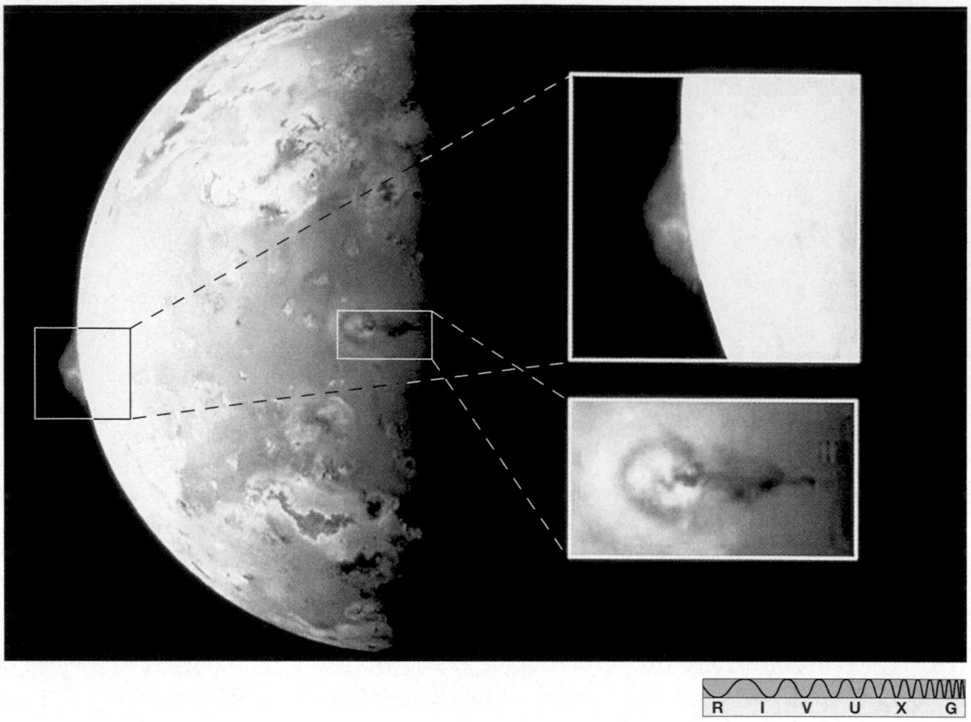

Animation Animation

Figure 8.3 Io Jupiter's innermost moon, Io, is quite different from the other three Galilean satellites. Its surface is kept smooth and brightly colored by constant volcanism. The resolution of this *Galileo* spacecraft image is about 6 km. The dark, circular features are volcanoes. The top inset shows one of Io's umbrellalike volcanoes erupting as *Galileo* flew past this fascinating moon in 1997, the bottom inset shows another volcano face-on. Surface features here are resolved to a few kilometers. The plume measures about 150 km high and 300 km across. (See also the Chapter Opener.) *(NASA)*

matter constantly filling in any dents and cracks. Io has a thin, temporary atmosphere made up mainly of sulfur dioxide, presumably produced by volcanic activity. By the time *Galileo* arrived, some of the volcanoes observed by *Voyager 2* had subsided, but many new ones were seen. In all, more than 80 active volcanoes have been identified.

Io is far too small to have geological activity similar to that on Earth. It should be long dead, like our own Moon, its internal heat lost to space billions of years ago. Instead, the source of Io's energy is *external*—Jupiter's gravity. Because Io orbits very close to Jupiter, the planet's huge gravitational field produces strong tidal forces on the moon. If Io were Jupiter's only satellite, it would long ago have come into a state of synchronous rotation with the planet. ∞ (Sec. 5.2) In that case, Io would move in a perfectly circular orbit with one face permanently turned toward Jupiter. The tidal bulge would be stationary with respect to the moon, and there would be no internal stresses and no volcanism.

But Io is not alone. As it orbits, it is constantly tugged by the gravity of its nearest large neighbor, Europa. These tugs are small and not enough to cause any great tidal effect in and of themselves, but they are sufficient to make Io's orbit slightly noncircular, preventing the moon from settling into a precisely synchronous state. As a result, seen from Jupiter, Io wobbles slightly from side to side as it moves. The conflicting forces acting on Io result in enormous tidal stresses that continually flex and squeeze its interior. Just as repeated back-and-forth bending of a wire can produce heat through friction, Io is constantly energized by this ever-changing distortion. The large amount of heat generated within Io causes huge jets of gas and molten rock to squirt out of the surface. It is likely that much of Io's interior is soft or molten, with only a relatively thin solid crust overlying it.

Europa: Liquid Water Locked in Ice

Europa (Figure 8.4) is a world very different from Io. Europa has relatively few craters on its surface, again suggesting that recent activity has erased the scars of ancient meteoric impacts. However, the surface displays a vast network of lines criss-crossing bright, clear fields of water ice. Many of these linear bands extend halfway around the satellite.

Based on the *Voyager* images, planetary scientists theorized that Europa is covered by an ocean of liquid water whose surface layers are frozen. More recent *Galileo* images seem to support this idea. Figure 8.4(b) is a high-resolution image of this weird moon, showing ice-filled surface cracks similar to those seen in Earth's polar ice floes. Figure 8.4(c) shows what appear to be "icebergs"—flat chunks of ice that have been broken apart and reassembled, perhaps by the action of water below. Mission scientists speculate that Europa's ice may be several kilometers thick, and that the ocean below it may be 100 km deep. As on Io, the ultimate source of the energy that keeps the ocean liquid and also drives the internal motion leading to the cracks in the icy surface is Jupiter's gravity. The heating mechanism is essentially similar to that on Io, but the effects are less extreme on Europa because of its greater distance from the planet.

If Europa does have an ocean of liquid water below its surface ice, it opens up many interesting avenues of speculation about the possible development of life there. In the rest of the solar system, only Earth has liquid water on or near its surface, and most scientists agree that water played a key role in the appearance of life here (see Chapter 18). However, bear in mind that the existence of water does not *necessarily* imply the emergence of life. Europa is a very hostile environment compared with Earth. The surface temperature on Europa is just 130 K, and the atmospheric pressure is only a billionth the pressure on our planet.

Ganymede and Callisto: Fraternal Twins

Ganymede, shown in Figure 8.5, is the largest moon in the solar system, exceeding not only Earth's Moon but also the planets Mercury and Pluto in size. It has many impact craters on its surface, along with patterns of dark and light markings reminiscent of the highlands and maria on Earth's Moon. In fact, Ganymede's history has many parallels with that of the Moon (with water ice replacing lunar rock).

As on the inner planets, we can estimate ages on Ganymede by counting craters. The darker regions are the oldest parts of Ganymede's surface. They are the original icy surface, just as the ancient highlands on our own Moon are its original crust. Ganymede's surface has darkened with age as micrometeorite dust has slowly covered it. Being much less heavily cratered, the light-colored parts of Ganymede must be younger. They probably formed in a manner similar to the maria on the

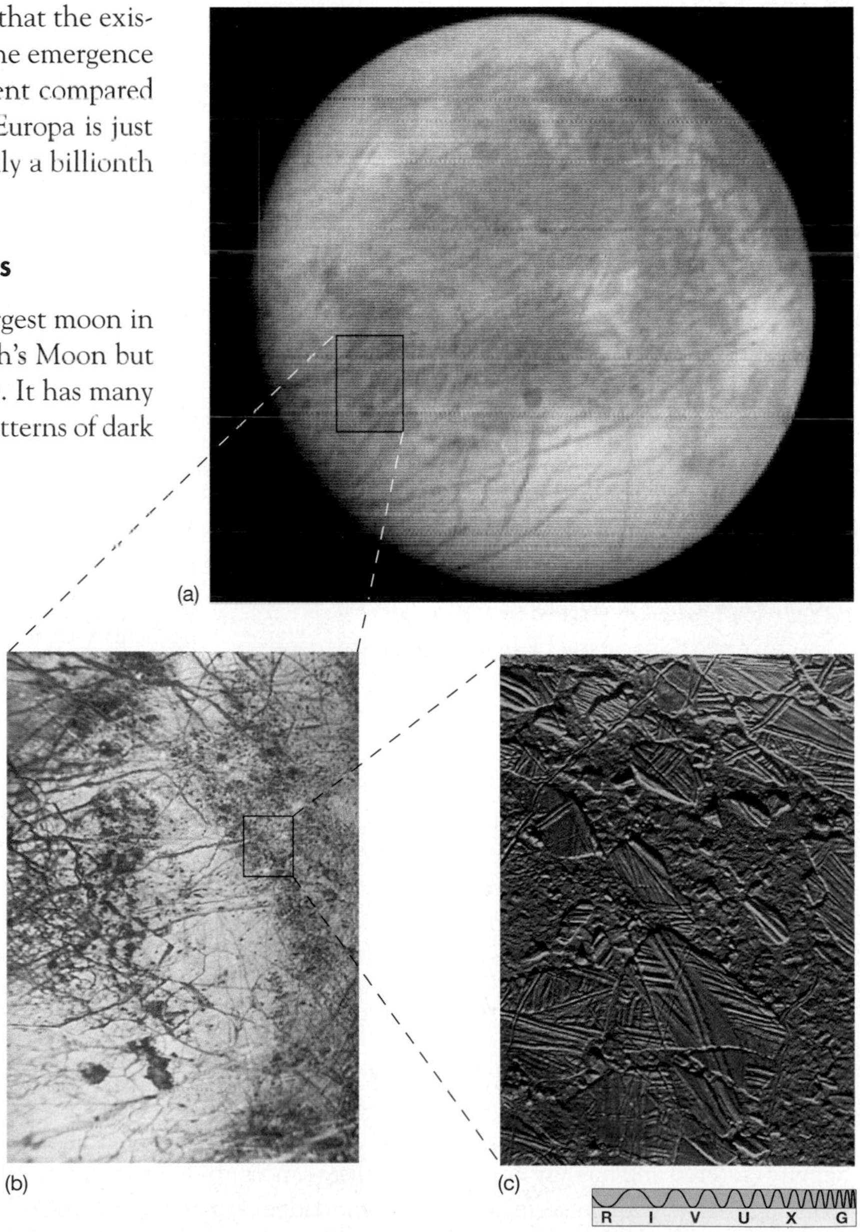

Video

Figure 8.4 Europa (a) A *Voyager 1* mosaic of Europa. Resolution is about 5 km. (b) Europa's icy surface is only lightly cratered, indicating that some ongoing process must be obliterating impact craters soon after they form. The origin of the cracks criss-crossing the surface is uncertain. (c) At 20-m resolution—the width of a typical house—this image from the *Galileo* spacecraft shows a smooth yet tangled surface resembling the huge ice floes that reside in Earth's polar regions. *(NASA)*

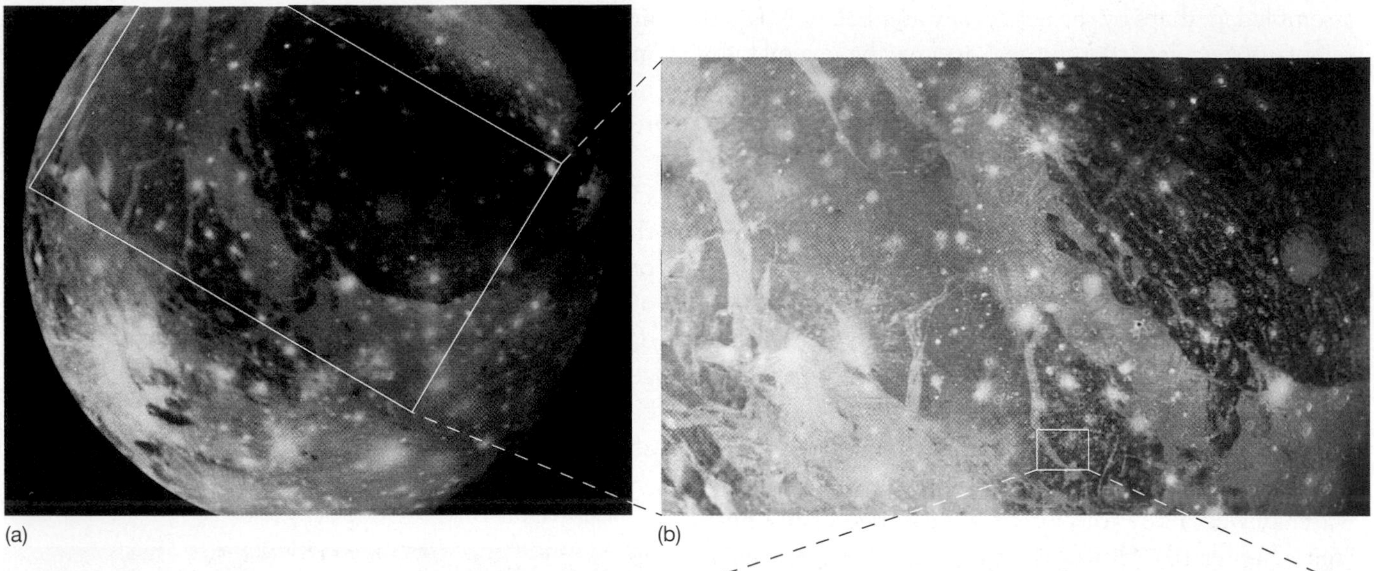

(a) (b)

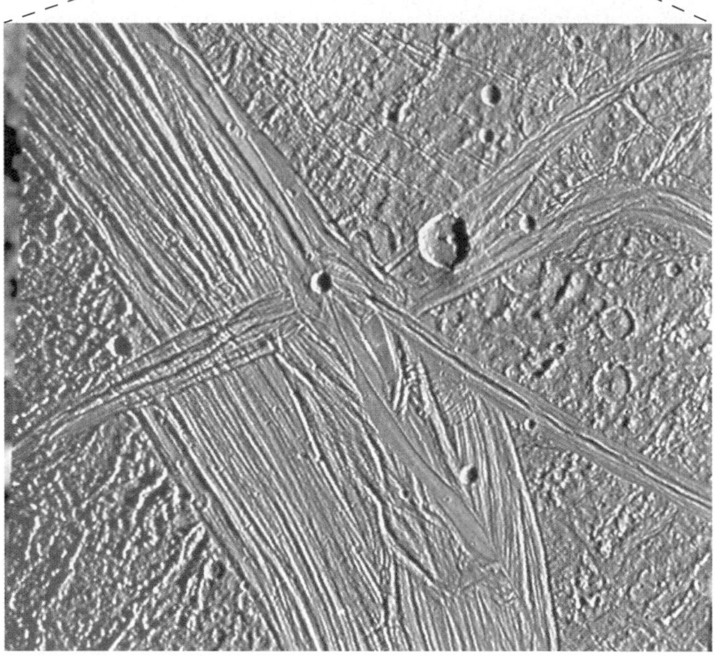

(c)

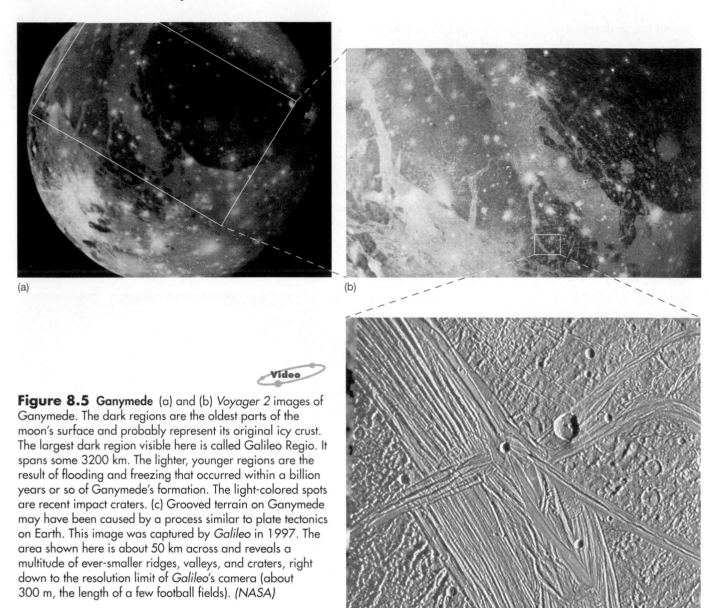

Figure 8.5 **Ganymede** (a) and (b) *Voyager 2* images of Ganymede. The dark regions are the oldest parts of the moon's surface and probably represent its original icy crust. The largest dark region visible here is called Galileo Regio. It spans some 3200 km. The lighter, younger regions are the result of flooding and freezing that occurred within a billion years or so of Ganymede's formation. The light-colored spots are recent impact craters. (c) Grooved terrain on Ganymede may have been caused by a process similar to plate tectonics on Earth. This image was captured by *Galileo* in 1997. The area shown here is about 50 km across and reveals a multitude of ever-smaller ridges, valleys, and craters, right down to the resolution limit of *Galileo*'s camera (about 300 m, the length of a few football fields). *(NASA)*

R I V U X G

Moon. Intense meteoritic bombardment caused liquid water—Ganymede's counterpart to our own Moon's molten lava—to upwell from the interior, flooding the impacting regions before solidifying. Not all of Ganymede's surface features follow the lunar analogy, however. Ganymede has a system of grooves and ridges (Figures 8.5b,c) that may have resulted from crustal tectonic motion, much as Earth's surface undergoes mountain building and faulting at plate boundaries.

Callisto, shown in Figure 8.6, is in many ways similar in appearance to Ganymede, although Callisto is more heavily cratered and has few fault lines. Callisto's most obvious feature is a huge series of concentric ridges surrounding two large basins, one clearly visible in Figure 8.6. The ridges resemble the ripples made as a stone hits water, but on Callisto they probably resulted from a cataclysmic impact with a meteorite. The upthrust ice was partially melted, but it resolidified quickly, before the ripples had a chance to subside. Today, both the ridges and the rest of the crust are frigid ice and show no obvious signs of geological activity. Apparently, Callisto froze before plate tectonic or other activity could start. The density of impact craters on the large impact basins indicates that they formed long ago, perhaps four billion years in the past.

Ganymede's internal differentiation indicates that the moon was largely molten at some time in the past, yet Callisto is undifferentiated, and hence apparently never melted. Researchers are uncertain why two such

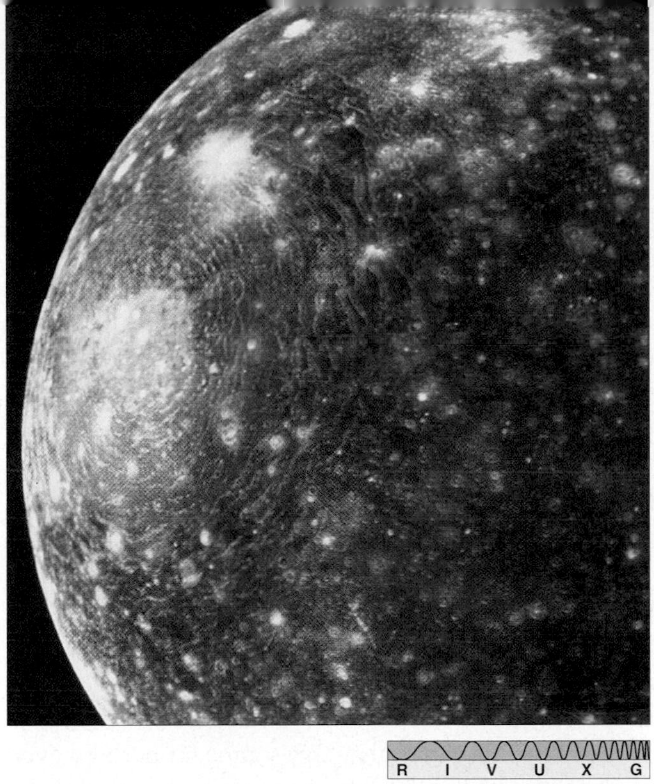

Figure 8.6 Callisto Callisto is similar to Ganymede in overall composition, but is more heavily cratered. The large series of concentric ridges visible at left is known as Valhalla. Extending nearly 1500 km from the basin center, the ridges formed when "ripples" from a large meteoritic impact froze before they could disperse completely. Resolution here is around 10 km. *(NASA)*

similar bodies should have evolved so differently. Complicating things further is the discovery by the *Galileo* orbiter that Ganymede has a weak magnetic field, about one percent that of Earth, suggesting that the moon's core may still be partly molten. ∞ (Sec. 5.7) If that is so, then Ganymede's heating and differentiation must have happened relatively recently—less than a billion years ago, based on recent estimates of how rapidly the moon's heat escapes into space. Scientists have no clear explanation for how this could have occurred, although some speculate that interactions among the inner moons may have caused Ganymede's orbit to have changed significantly about one billion years ago, and that prior tidal heating by Jupiter helped melt the moon's interior.

✓ Concept Check

- What is the source of energy for all the activity observed on Jupiter's Galilean moons?

8.2 The Large Moons of Saturn and Neptune

Titan: A Moon with an Atmosphere

2 The largest and most intriguing of Saturn's moons is Titan (Figure 8.7). Scientists believe that Titan's overall composition must be similar to those of Ganymede and Callisto, because these three moons have such similar masses and radii, and hence average densities. However, at least until the arrival of *Cassini* in 2004, the moon's internal structure is unknown.

From Earth, Titan is visible only as a barely resolved reddish disk. Long before the *Voyager* missions, astronomers knew from spectroscopic observations that the moon's reddish coloration is caused by something quite special—an atmosphere. So anxious were mission planners to obtain a closer look that they programmed *Voyager 1* to pass very close to Titan, even though that meant the spacecraft could not then use Saturn's gravity to continue on to Uranus and Neptune.

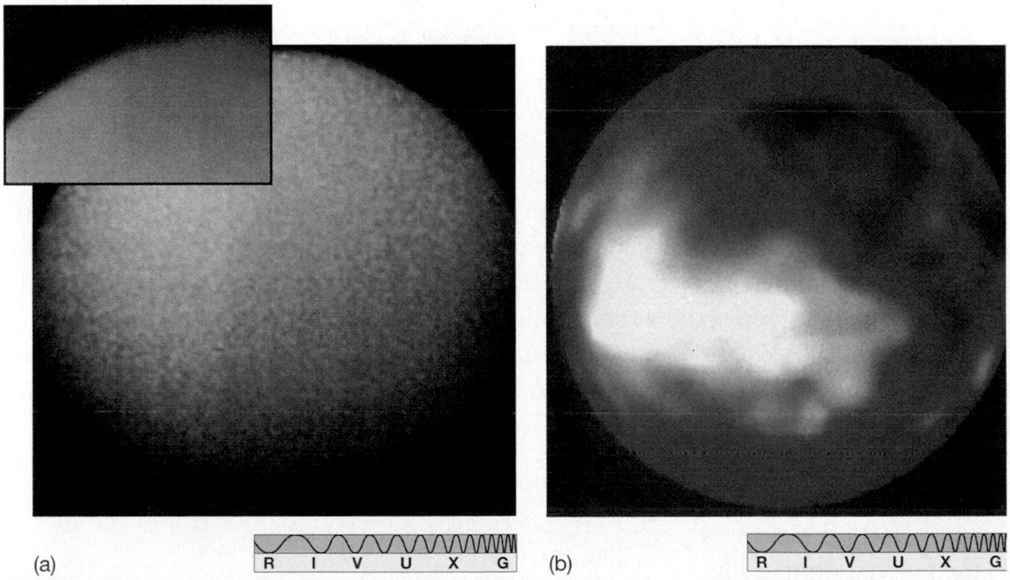

Figure 8.7 Titan (a) Titan, larger than Mercury and roughly half the size of Earth, was photographed from only 4000 km away as the *Voyager 1* spacecraft passed by in 1980. All we can see here is Titan's upper cloud deck. The inset shows a contrast-enhanced image of the haze layer (falsely colored blue). (b) An infrared image from the *Hubble Space Telescope* shows some surface features. The bright area is nearly 400 km across—about the size of Australia. *(NASA)*

Unfortunately, despite *Voyager 1*'s close pass, the moon's surface remains a mystery. A thick, uniform haze layer, in some ways similar to the smog found over many cities on Earth, envelops the moon and completely obscured the spacecraft's view. Titan's atmosphere is thicker and denser even than Earth's, and it is certainly more substantial than that of any other moon. Radio and infrared observations made by *Voyager 1* showed that the atmosphere is made up mostly of nitrogen (roughly 90 percent) and possibly argon (up to 10 percent), with a few percent of methane and trace amounts of other gases. Titan's surface temperature is a frigid 94 K, roughly what we would expect on the basis of its distance from the Sun. At the temperatures typical of the lower atmosphere, methane and ethane may behave rather like water on Earth, raising the possibility of methane rain, snow, and fog and even ethane oceans. Scientists eagerly await the arrival of *Cassini*'s atmospheric probe to explore further this remarkable world.

Why does Titan have such a thick atmosphere when Ganymede and Callisto do not? The answer lies largely in Titan's 94 K surface temperature, which is considerably lower than that of Jupiter's moons, making it easier for the moon to retain an atmosphere. Also, gases may have become trapped in the surface ice when the moon formed. As Titan's internal radioactivity warmed the moon, the ice released the trapped gases, forming a thick methane–ammonia atmosphere. Sunlight split the ammonia into hydrogen, which escaped into space, and nitrogen, which remained in the atmosphere. The methane, which was less easily broken apart, survived intact. Together with argon outgassed from Titan's interior, these gases form the basis of the atmosphere we see today.

Triton: Captured from the Kuiper Belt?

Neptune's large moon, Triton, is the smallest of the six large moons in the solar system, with about half the mass of the next smallest, Europa. Lying 4.5 billion km from the Sun, and possessing an icy surface that reflects much of the solar radiation reaching it, Triton has a temperature of just 37 K. *Voyager 2* found that this moon has an extremely thin nitrogen atmosphere and a solid, frozen surface probably composed primarily of water ice.

A *Voyager 2* mosaic of Triton's south polar region is shown in Figure 8.8. The moon's low temperatures produce a layer of nitrogen frost that forms, evaporates, and reforms seasonally over the polar caps. The frost is visible as the pinkish region at the bottom right of Figure 8.8. Overall, there is a marked lack of cratering on Triton, presumably indicating that surface activity has obliterated the evidence of most impacts. There are many other signs of an active past. Triton's face is scarred by large fissures similar to those seen on Ganymede, and the moon's odd cantaloupelike terrain may indicate repeated faulting and deformation over the moon's lifetime. In addition, Triton has numerous frozen lakes of water ice, which are believed to be volcanic in origin.

Triton's surface activity is not just a thing of the past. As *Voyager 2* passed the moon, its cameras detected great jets of nitrogen gas erupting several kilometers into the sky. These geysers may result when liquid nitrogen below Triton's surface is heated and vaporized by some internal energy source, or perhaps even by the Sun's feeble light. Scientists conjecture that nitrogen geysers may be very common on Triton and are perhaps responsible for much of the moon's thin atmosphere.

Triton is unique among the large moons of the solar system in that its orbit around Neptune is *retrograde*. Also, with an orbital inclination of about 20°, Triton is the only large jovian moon not to orbit in its parent planet's equatorial plane. The event or events that placed Triton on a tilted, retrograde orbit are the subject of considerable speculation. Triton's peculiar orbit and surface features suggest to many astronomers that the moon did not form as part of the Neptune system but instead was captured by the planet, probably not too long ago. The surface deformations on Triton certainly suggest violent and relatively recent events in the moon's past, but the details remain uncertain. The absence of a "normal" mid-sized moon system around Neptune (see Section 8.3) is also often cited as evidence for some catastrophic event in the planet's history.

Whatever Triton's past, its future seems quite clear. Because of its retrograde orbit, the tidal bulge that Triton raises on Neptune tends to make the moon spiral *toward* the planet rather than away from it (as our Moon moves away from Earth). ∞ (Sec. 5.2) Careful calculations indicate that Triton is doomed to be torn apart by Neptune's tidal gravitational field, probably in no more than about 100 million years.

Concept Check

- Why was *Voyager 1* unable to photograph the surface of Titan?

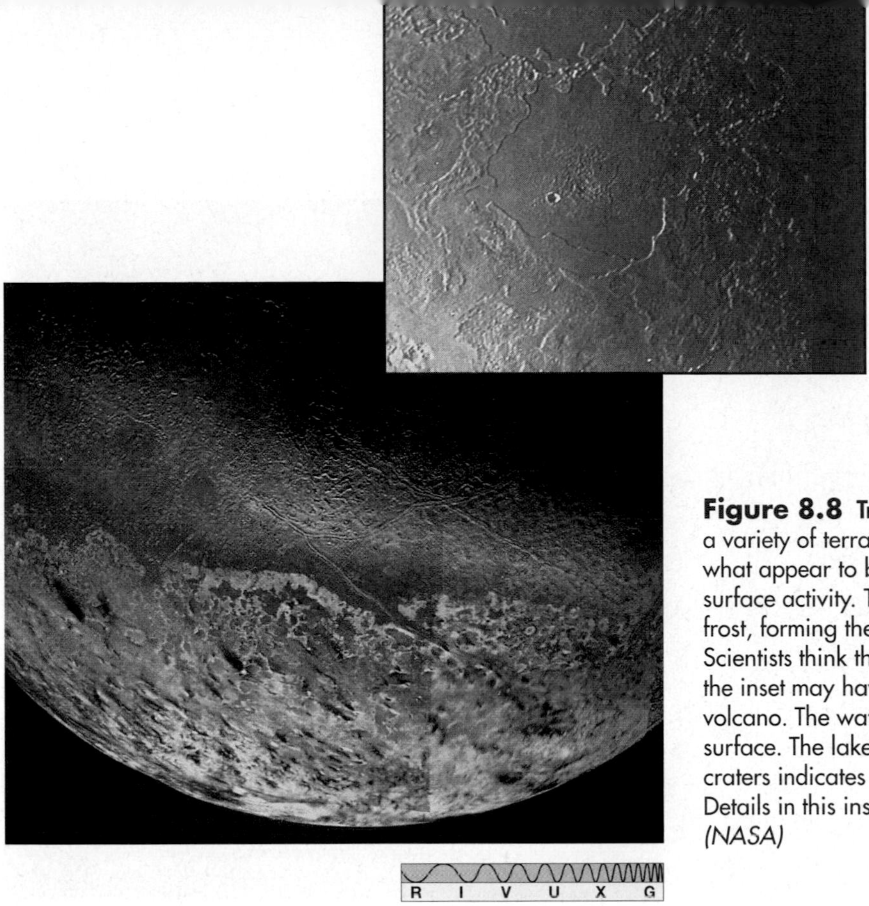

Figure 8.8 **Triton** The south polar region of Triton, showing a variety of terrains ranging from deep ridges and gashes to what appear to be frozen water lakes, all indicative of past surface activity. The pinkish region at the lower right is nitrogen frost, forming the moon's polar cap. Resolution is about 4 km. Scientists think that the roughly circular lakelike feature shown in the inset may have been caused by the eruption of an ice volcano. The water "lava" has since solidified, leaving a smooth surface. The lake is about 200 km in diameter. The absence of craters indicates that the eruption was a relatively recent event. Details in this inset image are resolved to a remarkable 1 km. *(NASA)*

8.3 The Medium-Sized Jovian Moons

3 Table 8.2 lists the 12 medium-sized satellites of the solar system, conventionally defined as those bodies having radii between about 200 and 800 km. The six that orbit Saturn are shown in Figure 8.9. Figure 8.10 shows the remainder (five orbiting Uranus, the other orbiting Neptune), to the same scale as Figure 8.9. Based on their densities, scientists believe that all are composed largely of rock and water ice. All move on nearly circular paths and are tidally locked by their parent planet's gravity into synchronous rotation.

Most of these moons show no evidence for significant geological activity. The light-colored *wispy terrain* visible on Saturn's Rhea in Figure 8.9 (and similar bright streaks on Saturn's Dione) are thought to have been caused by events in the distant past during which water

TABLE 8.2 The Medium-Sized Moons of the Solar System

NAME	PARENT PLANET	DISTANCE FROM PARENT PLANET (km)	DISTANCE FROM PARENT PLANET (planet radii)	ORBIT PERIOD (Earth days)	RADIUS (km)	MASS (Moon = 1)	DENSITY (kg/m^3)
Mimas	Saturn	186,000	3.10	0.94	200	0.00051	1100
Enceladus	Saturn	238,000	3.97	1.37	250	0.00099	1000
Tethys	Saturn	295,000	4.92	1.89	530	0.0085	1000
Dione	Saturn	377,000	6.28	2.74	560	0.014	1400
Rhea	Saturn	527,000	8.78	4.52	760	0.032	1200
Iapetus	Saturn	3,560,000	59.3	79.3	720	0.022	1000
Miranda	Uranus	130,000	5.09	1.41	240	0.00090	1200
Ariel	Uranus	191,000	7.48	2.52	580	0.018	1700
Umbriel	Uranus	266,000	10.4	4.14	590	0.016	1400
Titania	Uranus	436,000	17.1	8.71	790	0.048	1700
Oberon	Uranus	583,000	22.8	13.5	760	0.041	1600
Proteus	Neptune	118,000	4.76	1.12	210	—	—

Figure 8.9 Saturnian Moons Saturn's six medium-sized satellites, to scale. All are heavily cratered. Lapetus shows a clear contrast between its light-colored (icy) trailing surface (top and center in this image) and its dark leading hemisphere (at bottom). The light-colored wisps on Rhea are thought to be water and ice released from the moon's interior during some long-ago period of activity. Dione and Tethys arguably show evidence of ancient geological activity. Enceladus appears to still be volcanically active, but the cause of the volcanism is unexplained. Mimas's main surface feature is the large crater Herschel. For scale, the diameter of Earth's Moon is about the height of the figure. *(NASA)*

was released from the moons' interiors and condensed on their surfaces. Dione also has "maria" of sorts, where flooding appears to have obliterated the older craters. Both Tethys in the Saturn system and Ariel orbiting Uranus have extensive cracks on their surfaces, which may have been caused by cooling and shrinking of the moons' outer layers, but are more likely the result of meteoritic bombardment.

All the moons show heavy cratering, indicating the cluttered and violent planetary environment that existed just after the outer planets formed, as the remaining planetesimals were accreted by larger bodies or ejected from the vicinity of the jovian planets to form the Oort cloud. ∞ (Sec. 4.3) The innermost mid-sized moons Mimas (Saturn) and Miranda (Uranus) show perhaps the clearest evidence of violent meteoritic impacts. The impact that caused the large crater visible on the image of Mimas in Figure 8.9 must have come very close to shattering the moon. Miranda (Figure 8.11) displays a wide range of surface terrains, including ridges, valleys, faults, and many other geological features. To explain why Miranda seems to combine so many types of terrain, some researchers have hypothesized that it has been catastrophically disrupted several times, with the pieces falling back together in a chaotic, jumbled way. However, it will be a long time before we can obtain more detailed information to test this theory.

An exception to most of the above statements is Saturn's moon Enceladus, which is so bright and shiny—reflecting virtually 100 percent of the sunlight falling on it—that astronomers believe its surface must be completely coated with fine crystals of pure ice, possibly the icy "ash" of water volcanoes. Much of the moon's surface is devoid of impact craters, which seem to have been erased by "lava flows" composed of water temporarily liquefied during internal upheavals and now frozen again. It is not known why there should be such activity on so small a moon.

✓ Concept Check

- What surface property of the medium-sized moons tells us that the outer solar system was once a very violent place?

Figure 8.11 Miranda Miranda, the asteroid-sized innermost moon of Uranus, photographed by *Voyager 2*. The fractured surface suggests a violent past, but the cause of the grooves and cracks is unknown. The resolution in the inset is about 2 km. *(NASA)*

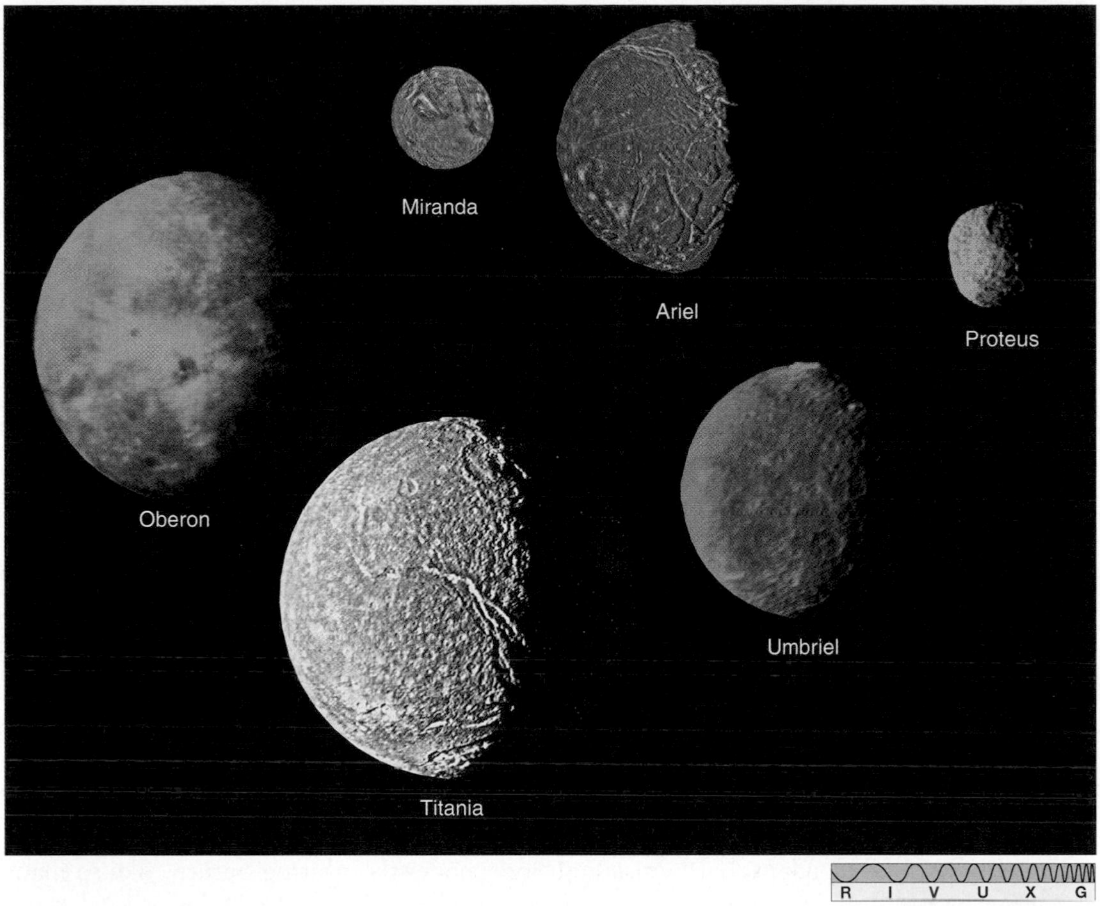

Figure 8.10 Moons of Uranus and Neptune The five medium-sized moons of Uranus and the single medium-sized moon of Neptune, Proteus, drawn to the same scale as Figure 8.9. In order of increasing distance from the planet, the moons of Uranus are Miranda, Ariel, Umbriel, Titania, and Oberon. Like their counterparts among Saturn's moons, all six moons are heavily cratered, and most show no clear evidence of any present or past geological activity. *(NASA)*

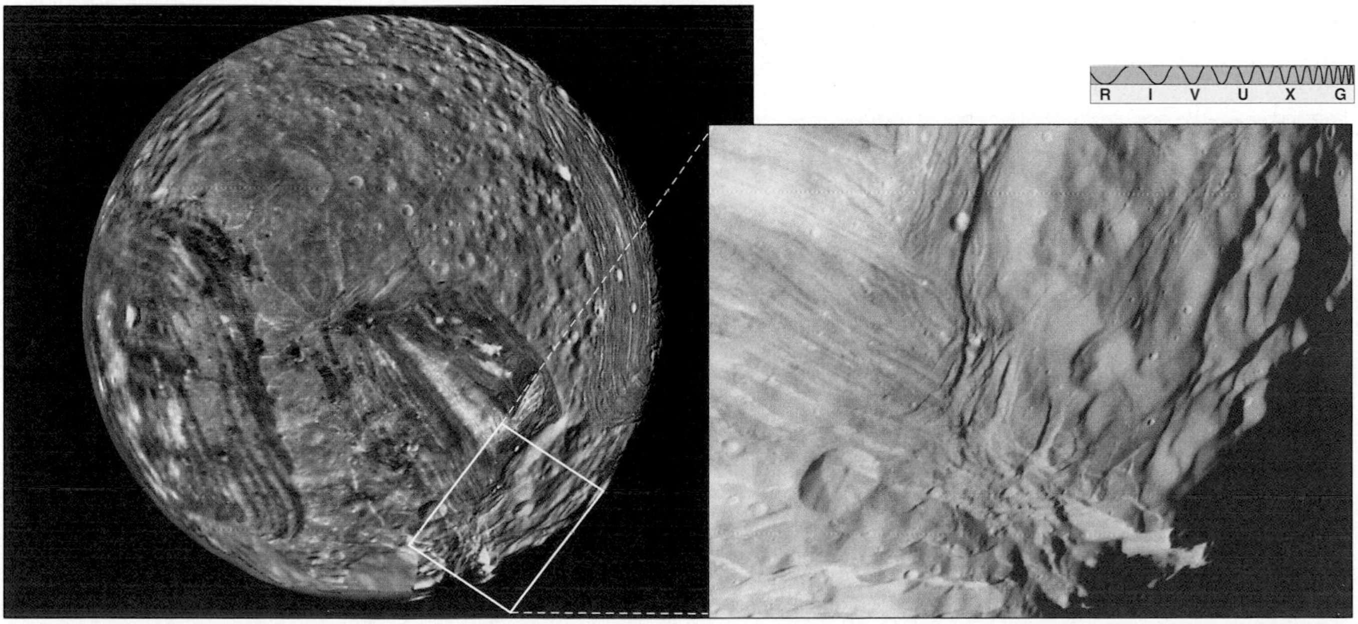

8.4 Planetary Rings

All four jovian worlds have planetary ring systems girdling their equators. The properties and appearance of each planet's rings are intimately connected with the planet's small and medium-sized moons, with many of the inner jovian moons orbiting close to (or even within) the parent planet's rings. Saturn's rings are by far the best known and best observed, and we study them here in greatest detail. However, the rings of all the jovian planets have much to tell us about the environment in the outer solar system.

Saturn's Spectacular Ring System

4 5 Figure 8.12 shows the main features of Saturn's rings. The dark gap about two-thirds of the way out from the inner edge is named the **Cassini Division**. Outside the Cassini Division lies the **A ring**. The inner ring, between the Cassini Division and the planet, is in reality two rings, the **B** and **C rings**. A smaller gap, known as the **Encke Division**, is found in the outer part of the A ring. Its width is 270 km. No finer ring details are visible from Earth. Of the three main rings, the B ring is brightest, followed by the somewhat fainter A ring, and then by the almost translucent C ring. The additional rings listed in Table 8.3 cannot be seen in Figure 8.12.

Because they lie in the planet's equatorial plane and Saturn's rotation axis is tilted with respect to the ecliptic, the appearance of Saturn's rings (as seen from Earth) changes seasonally, as shown in Figure 8.13. As Saturn orbits the Sun, the angle at which the rings are illuminated varies. When the planet's north or south pole is tipped toward the Sun (which means it is either summer or winter on Saturn), the highly reflective rings are at their brightest. In Saturn's spring and fall, the rings are close to being edge-on, both to the Sun and to observers on Earth, so they seem to disappear. One important deduction we can make from this simple observation is that the rings are very *thin*. In fact, we now know that their thickness is less than a few hundred meters, even though they are more than 200,000 km in diameter.

TABLE 8.3 The Rings of Saturn

RING	INNER RADIUS (km)	INNER RADIUS (planet radii)	OUTER RADIUS (km)	OUTER RADIUS (planet radii)	WIDTH (km)
D	60,000	1.00	74,000	1.23	14,000
C	74,000	1.23	92,000	1.53	18,000
B	92,000	1.53	117,600	1.96	25,600
A	122,200	2.04	136,800	2.28	14,600
F	140,500	2.34	140,600	2.34	100
E	210,000	3.50	300,000	5.00	90,000

Saturn's rings are not solid objects. In 1857, Scottish physicist James Clerk Maxwell realized that they must be composed of a great number of small solid *particles*, all independently orbiting Saturn, like so many tiny moons. Measurements of the Doppler shift of sunlight reflected from the rings reveal that the orbital speeds decrease with increasing distance from the planet in precise accordance with Kepler's laws and Newton's law of gravity. (Sec. 1.4) The highly reflective nature of the rings suggests that the particles are made of ice, and infrared observations in the 1970s confirmed that water ice is in fact a prime ring constituent. Radar

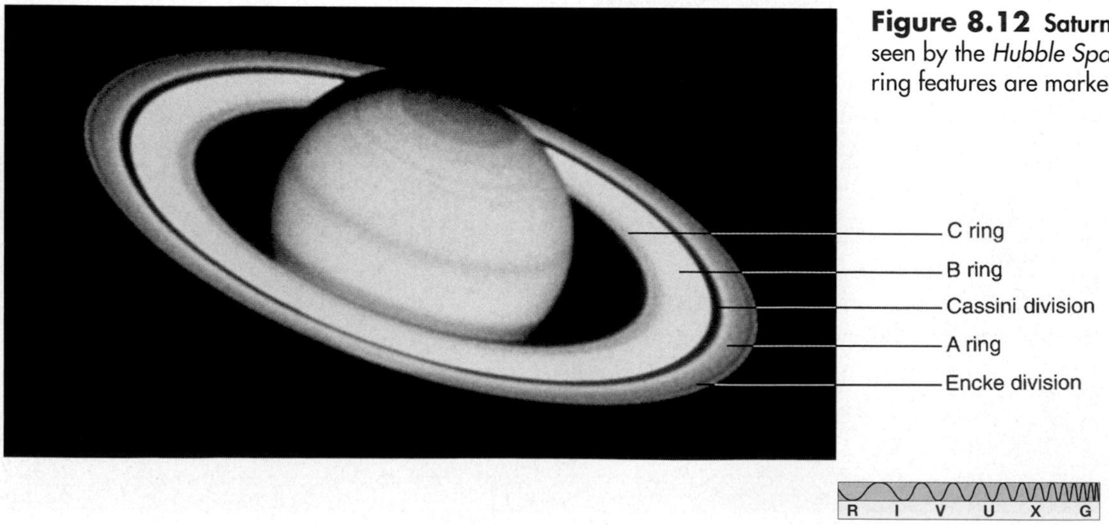

Figure 8.12 **Saturn's Rings** Saturn as seen by the *Hubble Space Telescope*. The main ring features are marked. *(NASA)*

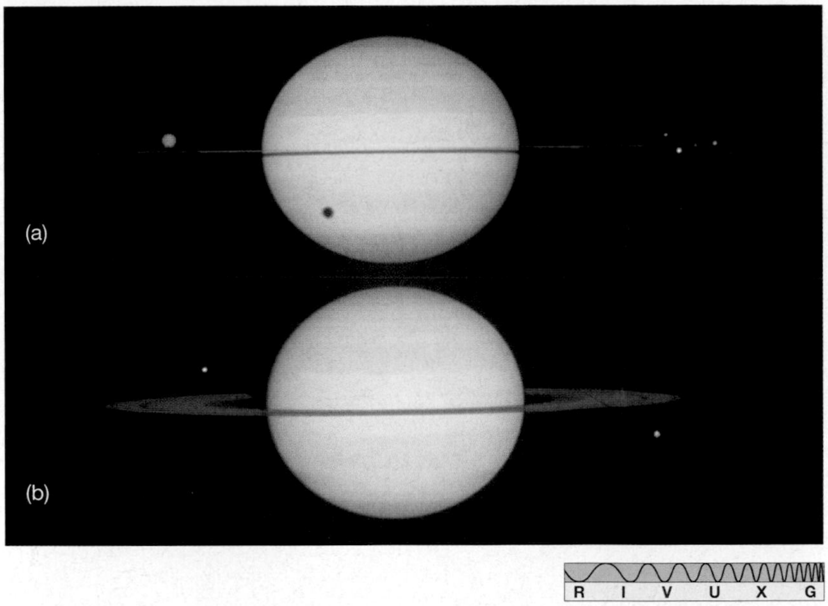

Figure 8.13 **Saturn's Tilt** Over time, Saturn's rings change their appearance to terrestrial observers as the tilted ring plane orbits the Sun. These true-color images are separated by a few months, having been taken by the *Hubble Space Telescope* in the summer and fall of 1995. (a) The ring plane is precisely edge-on as viewed from Earth. Several moons are visible, including Titan's shadow (black dot). (b) The rings are more apparent as our vantage point changes. *(NASA)*

observations and later *Voyager* studies of scattered sunlight showed that the sizes of the particles range from fractions of a millimeter to tens of meters in diameter, with most particles being about the size (and composition) of a large snowball.

The Roche Limit

To understand why rings form, consider the fate of a small moon orbiting close to a massive planet such as Saturn (Figure 8.14). As we bring our hypothetical moon closer to the planet, the tidal force exerted on it by the planet increases, stretching it along the direction to the planet. ∞ (*More Precisely 5-1*) The tidal force increases rapidly as the distance to the planet decreases. Eventually a point is reached where the planet's tidal force becomes greater than the internal forces holding the moon together, and the moon is torn apart by the planet's gravity. The pieces of the satellite move on their own individual orbits around that planet, eventually spreading all the way around it in the form of a ring.

For any given planet and any given moon, this critical distance inside which the moon is destroyed is known as the **Roche limit**, after the nineteenth-century French mathematician Edouard Roche, who first calculated it. If our hypothetical moon is held together by its own gravity, and if its average density is similar to that of the parent planet (both reasonably good assumptions for Saturn's larger moons), then the Roche limit is approximately 2.5 times the radius of the planet. For example, in the case of Saturn (of radius 60,000 km), the Roche limit lies roughly 150,000 km from the planet's center—just outside the outer edge of the A ring. The rings of Saturn occupy the region inside Saturn's Roche limit. These considerations apply equally well to the other jovian worlds. As indicated in Figure 8.15, all four ring systems in our solar system are found within (or close to) the Roche limit of their parent planet.

Note that the calculation of the Roche limit applies only to moons massive enough for their own gravity to be the dominant binding force. Sufficiently small moons can survive within the Roche limit because they are held together not by gravity but rather by interatomic (electromagnetic) forces.

Fine Structure

In 1979, the *Voyager* probes found that Saturn's main rings are actually composed of tens of thousands of narrow **ringlets** (Figure 8.16). Furthermore, this fine ring structure is not fixed, but instead varies both with time and with position in the rings. The ringlets are formed when the mutual gravitational attraction of the ring particles, possibly in combination with the effects of Saturn's inner moons, creates waves of matter—regions of high and low density—that move in the plane of the rings, like ripples on the surface of a pond.

The narrower gaps in the rings—about 20 in all—are most likely swept clean by the action of small *moonlets* embedded in them. Because these moonlets are much

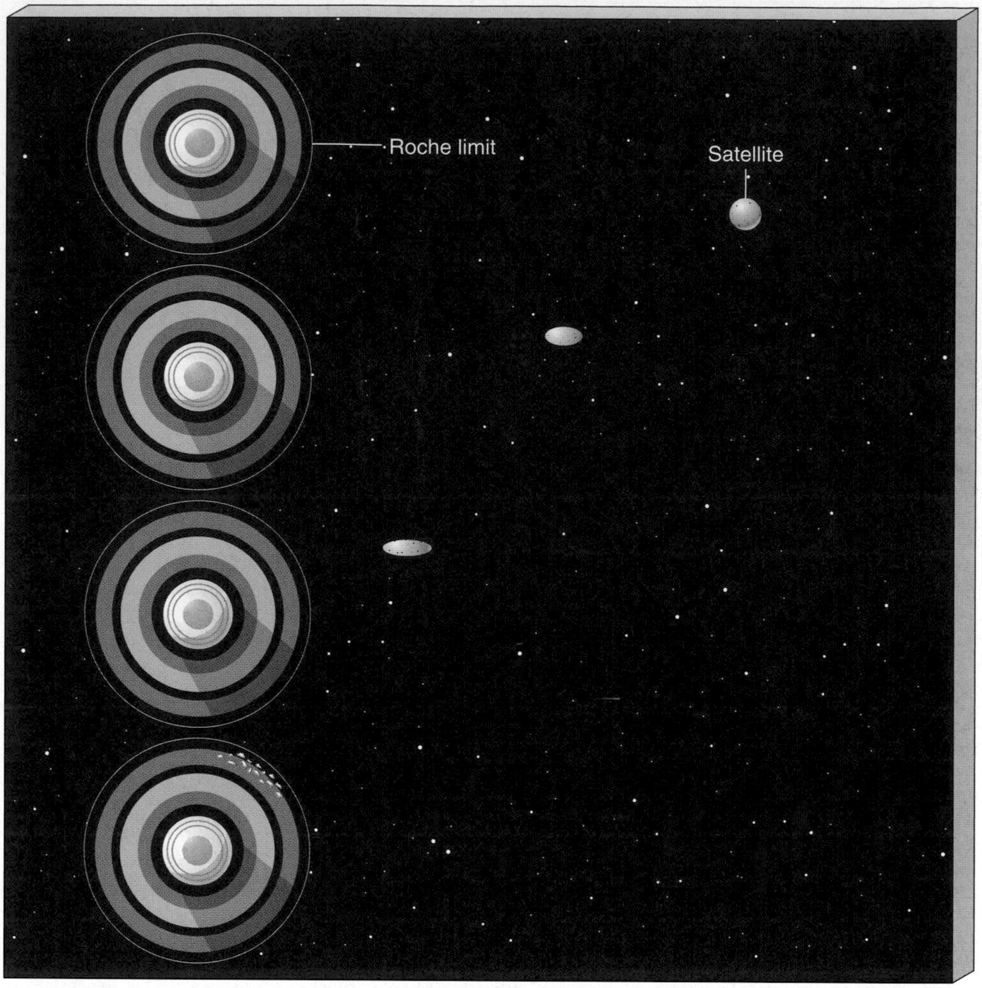

Figure 8.14 **Roche Limit** The tidal field of a planet first distorts and then destroys a moon that strays too close.

larger (perhaps 10 or 20 km across) than the largest ring particles, they simply scoop up ring material as they go. Only one of these moonlets has so far been found, despite many careful searches of the *Voyager* images. Nevertheless, moonlets are still regarded as the best explanation for the small gaps in Saturn's ring system.

The largest gap—the Cassini Division—has a different origin. It owes its existence to the gravitational influence of Saturn's innermost medium-sized moon, Mimas. Over time, particles orbiting within the Division have been deflected by Mimas's gravity into eccentric orbits that cause them to collide with other ring particles, effectively moving them into new orbits at different radii. The net result is that the number of ring particles in the Cassini Division is greatly reduced.

Outside the A ring lies perhaps the strangest ring of all: the faint, narrow **F ring** (Figure 8.17). It was discovered by *Pioneer 11* in 1979, but its full complexity became evident only when the *Voyager* spacecraft took a closer look. Its oddest feature is that it looks as though it is made up of several strands braided together. It seems as though the ring's intricate structure, as well as its thinness, arises from the influence of two small moons, known as **shepherd satellites**, that orbit about 1000 km on either side of it. Their gravitational effect gently guides any particle straying too far out of the F ring back into the fold. However, the details of how the shepherd moons produce braids, and why the moons are there at all, remain unclear.

Voyager 2 also found a faint series of rings, now known collectively as the **D ring**, inside the inner edge of the C ring, stretching down almost to Saturn's cloud tops. The D ring contains relatively few particles and is so dark that it is completely invisible from Earth. The faint **E ring**, also a *Voyager 2* discovery, lies well outside the main ring structure. It is thought to be the result of volcanism on the moon Enceladus.

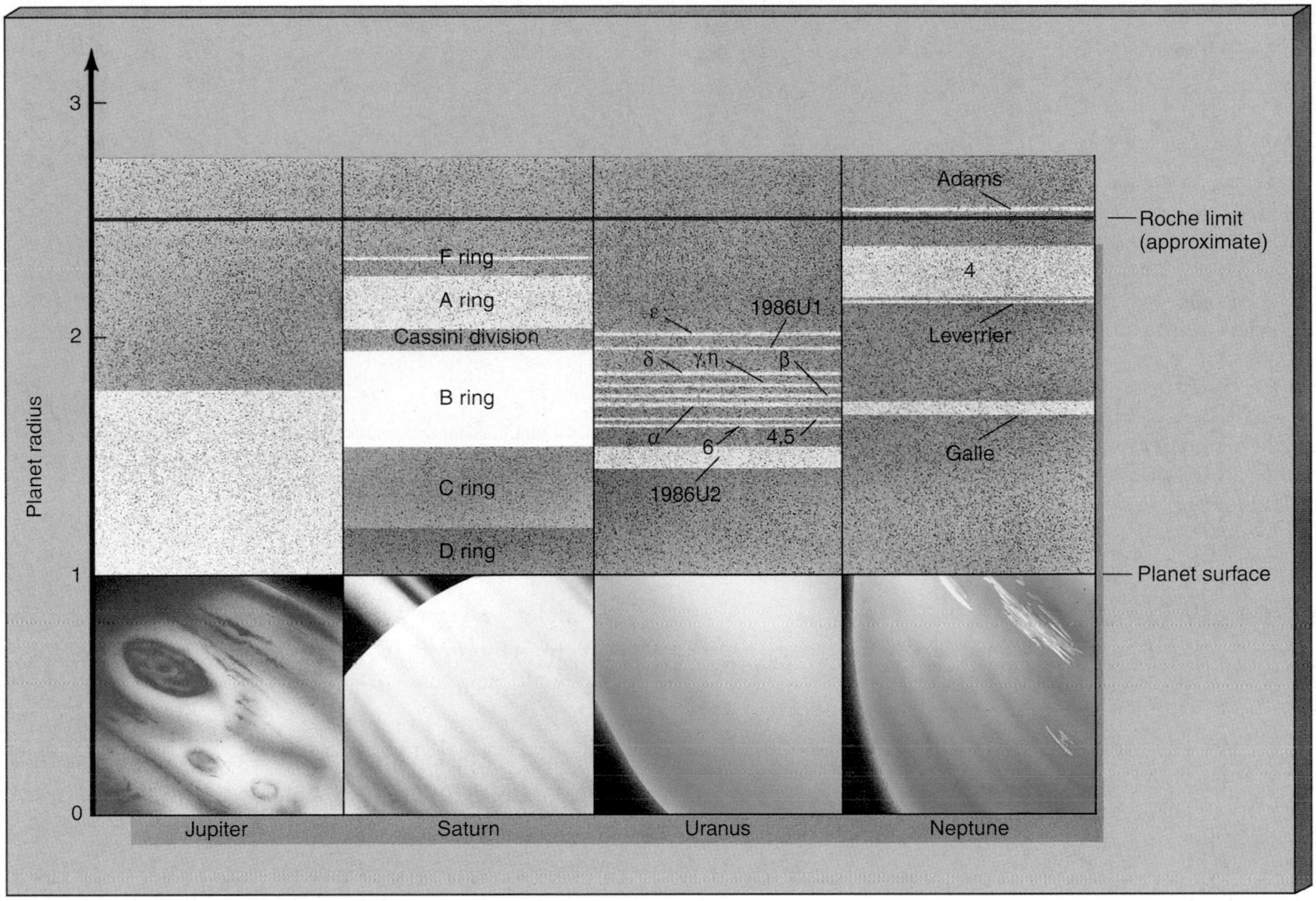

Figure 8.15 Jovian Ring Systems The rings of Jupiter, Saturn, Uranus, and Neptune. All distances are expressed in planetary radii. The red line represents the Roche limit. All the rings lie within (or very close to) the Roche limit of the parent planet.

Figure 8.16 Saturn's Rings, Up Close *Voyager 2* took this false-color close-up of the ring structure just before flying through the tenuous outer rings of Saturn. Earth is superposed, to scale, for a size comparison. *(NASA)*

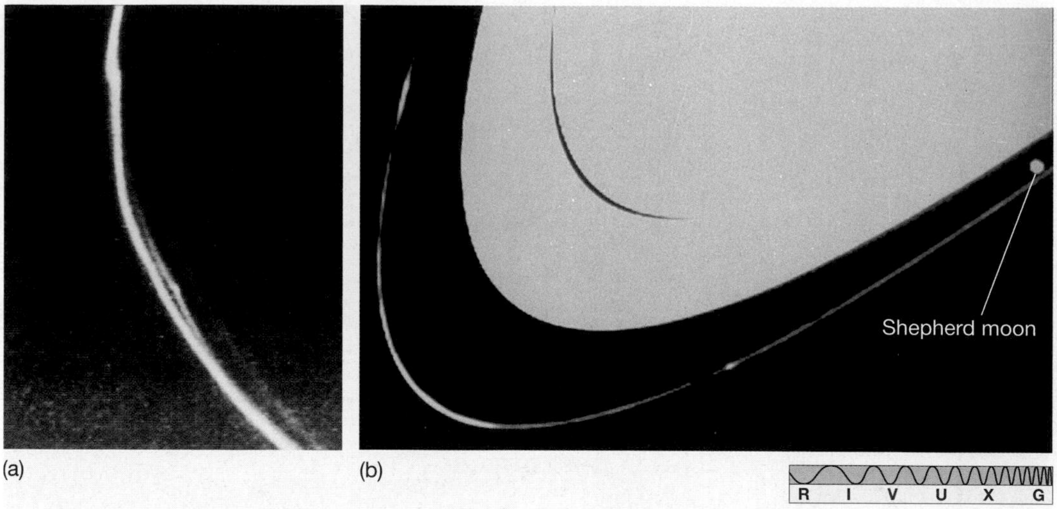

Figure 8.17
Shepherd Moon
(a) Saturn's narrow F ring appears to contain kinks and braids, making it unlike any of Saturn's other rings. Its thinness can be explained by the effects of two shepherd satellites that orbit the ring, one a few hundred kilometers inside the ring and the other a few hundred kilometers outside it. (b) One of the shepherding satellites, roughly 100 km across, can be seen at the right. *(Morrison/NASA)*

The Rings of Jupiter, Uranus, and Neptune

5 One of the many remarkable findings of the 1979 *Voyager* missions was the discovery of a faint ring of matter encircling Jupiter in the plane of the planet's equator (Figure 8.18). This ring lies roughly 50,000 km above the top cloud layer, inside the orbit of the innermost moon. The outer edge of the ring is quite sharply defined, although a thin sheet of material may extend all the way down to Jupiter's cloud tops. In the direction perpendicular to the equatorial plane, the ring is only a few tens of kilometers thick.

The ring system surrounding Uranus was discovered in 1977, when astronomers observed the planet passing in front of a bright star, momentarily dimming the star's light. Such a **stellar occultation** happens a few times per decade and allows astronomers to measure planetary structures that are too small and faint to be detected directly. The ground-based observations revealed nine thin rings encircling Uranus. Two more were discovered by *Voyager 2*. The locations of all 11 rings are indicated in Figure 8.15.

The nine rings known from Earth-based observation are shown in Figure 8.19, a *Voyager 2* image which makes it readily apparent that Uranus's rings are quite different from those of Saturn. While Saturn's rings are bright and wide with relatively narrow, dark gaps between them, those of Uranus are dark, narrow, and widely spaced. However, like Saturn's rings, the rings of Uranus are all less than a few tens of meters thick (measured in the direction perpendicular to the ring plane). Like the F ring of Saturn, Uranus's narrow rings require shepherd satellites to keep them in place. During its 1986 flyby, *Voyager 2* detected the shepherds of the Epsilon ring (Figure 8.20). Many other, undetected shepherd satellites must also exist.

As shown in Figure 8.21, Neptune is surrounded by four dark rings. Three are quite narrow, like the rings of Uranus, and one is quite broad and diffuse, more like Jupiter's ring. The outermost ring is noticeably clumped in places. From Earth we see not a complete ring, but only partial arcs; the unseen parts of the ring are simply too thin (unclumped) to be detected. The connection between the rings and the planet's small inner satellites has not yet been firmly established, although many astronomers now believe the clumping is caused by shepherd satellites.

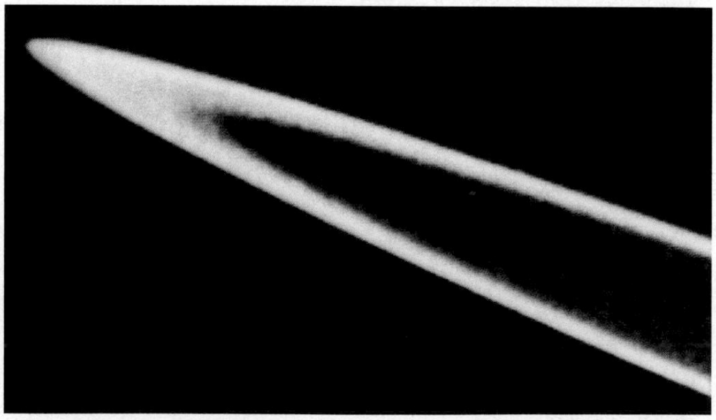

Figure 8.18 Jupiter's Faint Ring Jupiter's faint ring as photographed by *Voyager 2*. Made up of dark fragments of rock and dust possibly chipped off the innermost moons by meteorites, the ring was unknown before the two *Voyager* spacecraft arrived at the planet. *(NASA)*

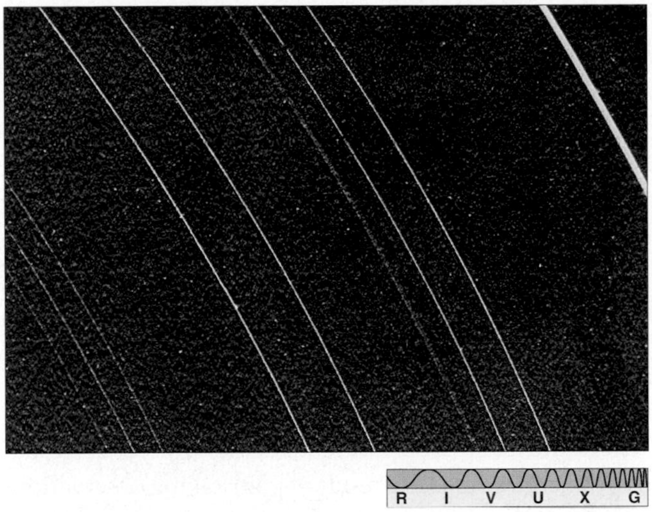

Figure 8.19 Uranus's Rings The main rings of Uranus, as imaged by *Voyager 2*. All nine of the rings known before *Voyager 2*'s arrival can be seen in this photo. Resolution is about 10 km, which is just about the width of most of these rings. The two rings discovered by *Voyager 2* are too faint to be seen here. *(NASA)*

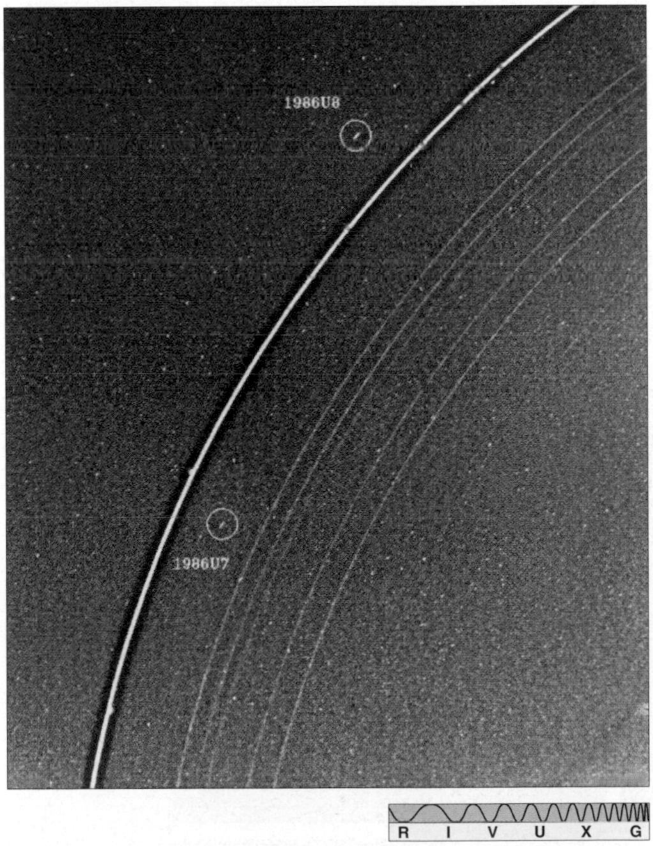

Figure 8.20 More Shepherd Moons These two small moons, named Cordelia and Ophelia, were discovered by *Voyager 2* in 1986. They shepherd Uranus's Epsilon ring, keeping it from diffusing away. *(NASA)*

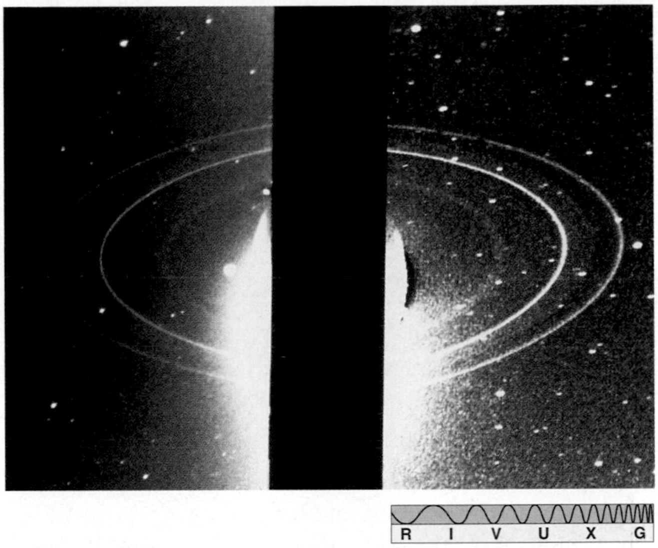

Figure 8.21 Neptune's Faint Rings In this long-exposure image, the planet (center) is heavily overexposed and has been artificially blotted out to make the rings easier to see. One of the two fainter rings lies between the inner bright ring and the planet. The other lies between the two bright rings. *(NASA)*

The Formation of Planetary Rings

The dynamic behavior observed in the rings of Saturn and the other jovian planets implies to many researchers that the rings are quite young—perhaps no more than 50 million years old, just one percent of the age of the solar system. If the rings are indeed so young, then either they are replenished from time to time, perhaps by fragments of moons chipped off by meteoritic impact, or they are the result of a relatively recent, possibly catastrophic event in the planet's system—a moon torn apart by tidal forces, or destroyed in an impact with a large comet or even with another moon.

Astronomers estimate that the total mass of Saturn's rings is enough to make a satellite about 250 km in diameter. If such a satellite strayed inside Saturn's Roche limit or was destroyed (perhaps by a collision) near that radius, a ring system could have resulted. While we do not know if Saturn's rings originated in this way, this is the likely fate of Neptune's large moon Triton. As mentioned earlier, Triton will be torn apart by Neptune's tidal gravitational field within 100 million years or so. By that time, it is quite possible that Saturn's ring system will have disappeared, and Neptune will then be the planet in the solar system with spectacular rings!

✓ Concept Check

- What is the Roche limit, and what is its relevance to planetary rings?

8.5 Pluto and Its Moon

The Discovery of Pluto

6 In Chapter 7 we saw how irregularities in Uranus's orbit led to the discovery of Neptune in the mid-nineteenth century. (Sec. 7.2) By the end of that century, observations of the orbits of Uranus and Neptune suggested that Neptune's influence was not sufficient to account for all of the observed irregularities in Uranus's motion. Furthermore, it seemed that Neptune itself was being affected by some other unknown body. Astronomers hoped to pinpoint the location of this new planet using techniques similar to those that had guided them to the discovery of Neptune.

One of the most ardent searchers was Percival Lowell, a capable and persistent observer and one of the best-known astronomers of his day. Basing his investigation primarily on the motion of Uranus (Neptune's orbit was still relatively poorly determined at the time), Lowell set about calculating where the supposed ninth planet should be. He searched for it, without success, during the decade preceding his death in 1916. In 1930, the American astronomer Clyde Tombaugh, working with improved equipment and photographic techniques at the Lowell Observatory, finally succeeded in finding Lowell's ninth planet, only 6° away from Lowell's predicted position. The new planet was named Pluto for the Roman god of the dead, who reigns over eternal darkness (and also because its first two letters and its astrological symbol ♇ are Lowell's initials).

Unfortunately, it now appears that the supposed irregularities in Neptune's orbit and the extra irregularities in the motion of Uranus do not really exist. Furthermore, the mass of Pluto, measured accurately only in the 1980s, is far too small to have caused them anyway. In the end, the discovery of Pluto owed much more to simple luck than to complex mathematics.

At nearly 40 A.U. from the Sun, Pluto is often hard to distinguish from the background stars. Like Neptune, it is never visible to the naked eye. Pluto is the only planet in the solar system not studied at close range by unmanned spacecraft, and there is little prospect of such a visit in the foreseeable future. Every new discovery about this remote world is the result of painstaking observations either from Earth or from Earth-orbiting instruments. Figure 8.22 is a composite map of Pluto's surface obtained by combining many *Hubble Space Telescope* images of the planet. It shows surface detail at about the same level as can be seen on Mars with a small telescope. Other than the bright polar caps, none of the surface features has been conclusively identified. Some of the dark regions may be craters or impact basins, as on Earth's Moon.

The Pluto–Charon System

In 1978, astronomers at the U.S. Naval Observatory discovered that Pluto has a satellite. It is now named Charon, after the mythical boatman who ferried the dead across the river Styx into Hades, Pluto's domain. The discovery photograph of Charon is shown in Figure 8.23(a). Charon is the large white bump at the top right of the image. Figure 8.23(b) shows a 1990 *Hubble Space Telescope* image

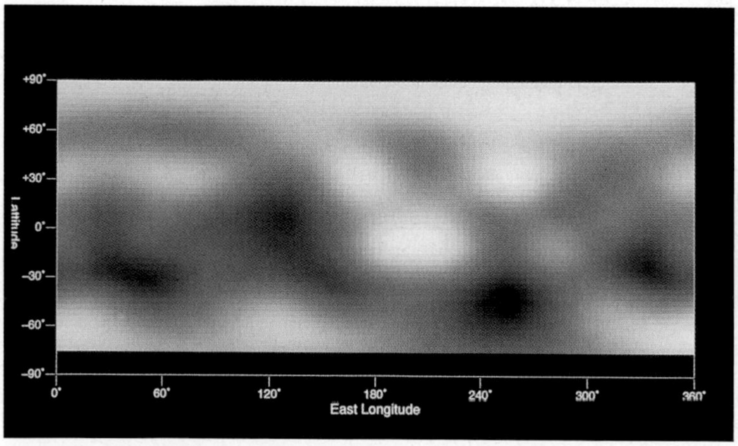

(a)

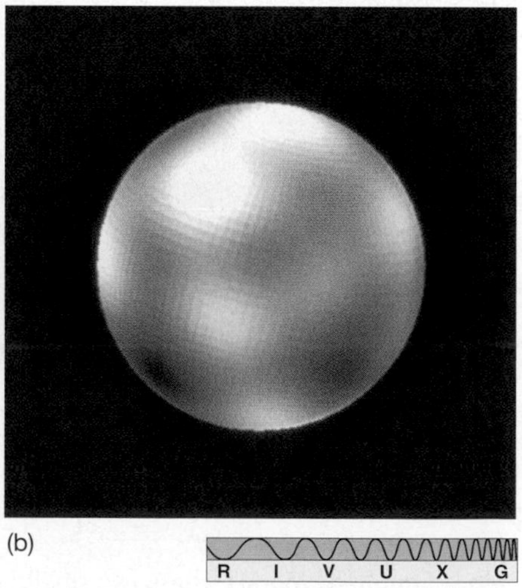

(b)

Figure 8.22 Pluto (a) A surface map of Pluto—not a photograph, but rather a modeled view created by carefully combining 24 *Hubble Space Telescope* images. (b) A computer-generated three-dimensional representation of the same data. *(NASA)*

Animation

that clearly resolves the two bodies. Based on studies of Charon's orbit, astronomers have measured the mass of Pluto to great accuracy. It is just 0.0021 Earth masses (1.3×10^{22} kg)—0.17 times the mass of Earth's Moon.

By pure chance, Charon's orbit over the six-year period from 1985 to 1991 produced a series of *eclipses* as Pluto and Charon repeatedly passed in front of one other, as seen from Earth. With more good fortune, these eclipses took place while Pluto was closest to the Sun, making for the best possible Earth-based observations. Basing their calculations on the variations in light as Pluto and Charon periodically hid each other, astronomers computed the masses and radii of both bodies, and determined their orbit plane. The mass of Charon is about 0.12 that of Pluto, giving the Pluto–Charon system by far the largest satellite-to-planet mass ratio in the solar system. Pluto's radius is 1150 km, about one-fifth that of Earth. Charon is about 600 km in radius, and orbits 19,600 km from Pluto.

By studying sunlight reflected from their surfaces, astronomers also found that Pluto and Charon are tidally locked as they orbit each other. Their orbital period, and the rotation period of each, is 6.4 days. As shown in Figure 8.24, Charon's orbit and the spin axes of both planet and moon are inclined at an angle of 118° to the plane of the ecliptic. Thus, Pluto is the third planet in the solar system found to have retrograde rotation (Venus and Uranus being the other two).

Pluto's Origin

Pluto's average density of 2000 kg/m^3 is too low for a terrestrial planet but far too high for a jovian one. Instead, the mass, radius, and density of Pluto are just what we expect for one of the icy moons of a jovian planet. Indeed, Pluto is comparable in both mass and radius to Neptune's large moon, Triton.

This similarity has fueled much speculation, both about Pluto's origin and about the strange state of Neptune's moon system. However, despite considerable effort, theorists have never been able to construct a plausible sequence of events that simultaneously accounts for Pluto's eccentric, inclined orbit and Triton's unusual retrograde path around Neptune. Instead, it now seems much more likely that the similarities between the two bodies stem from the fact that both formed and grew by accretion in the same environment—the Kuiper Belt. ∞ (Sec. 4.3)

Many researchers believe that there may have been thousands of Pluto-sized objects initially present in the outer solar system. Most have long since been ejected from the Kuiper Belt—into the inner solar system or out to the Oort Cloud—or, like Triton, have been captured by or collided with one of the outer planets. In this view, like Ceres in the inner asteroid belt, Pluto is simply the largest surviving member of the class. Charon was captured by Pluto late in the formation process, following a collision (or near-miss) between the two, in a manner reminiscent of the formation of Earth's own Moon. ∞ (Sec. 5.8)

✓ Concept Check

- What do Pluto and Triton have in common?

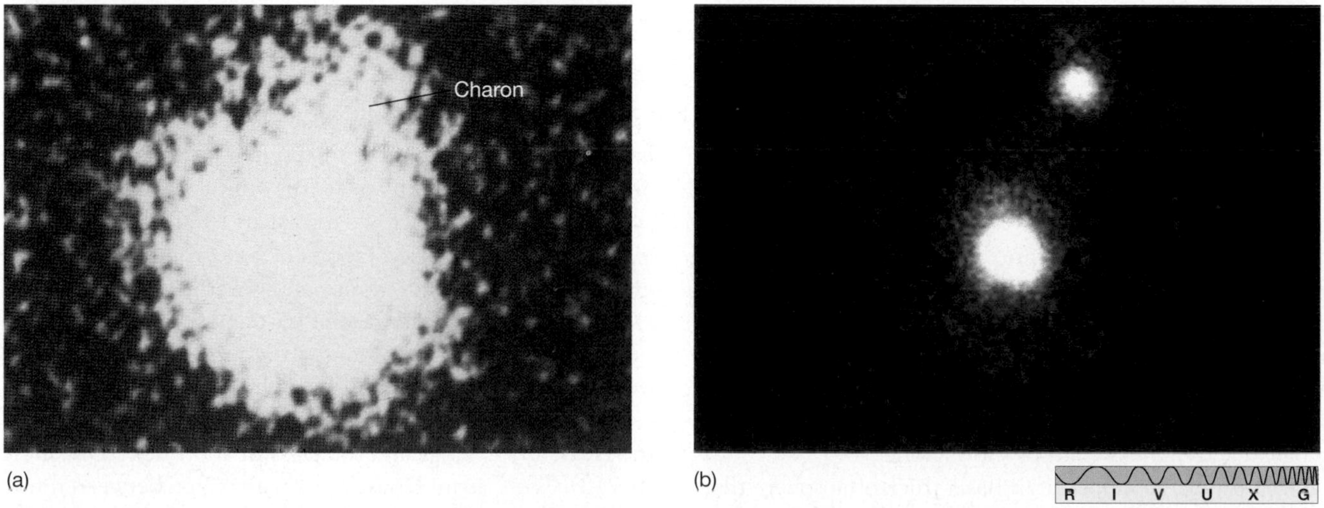

Figure 8.23 Pluto and Charon (a) The discovery photograph of Pluto's moon Charon. The moon is the small bump on the top right portion of the image. (b) The Pluto–Charon system, as seen by the *Hubble Space Telescope.* The angular separation of the planet and its moon is about 0.9″. *(U.S. Naval Observatory; NASA)*

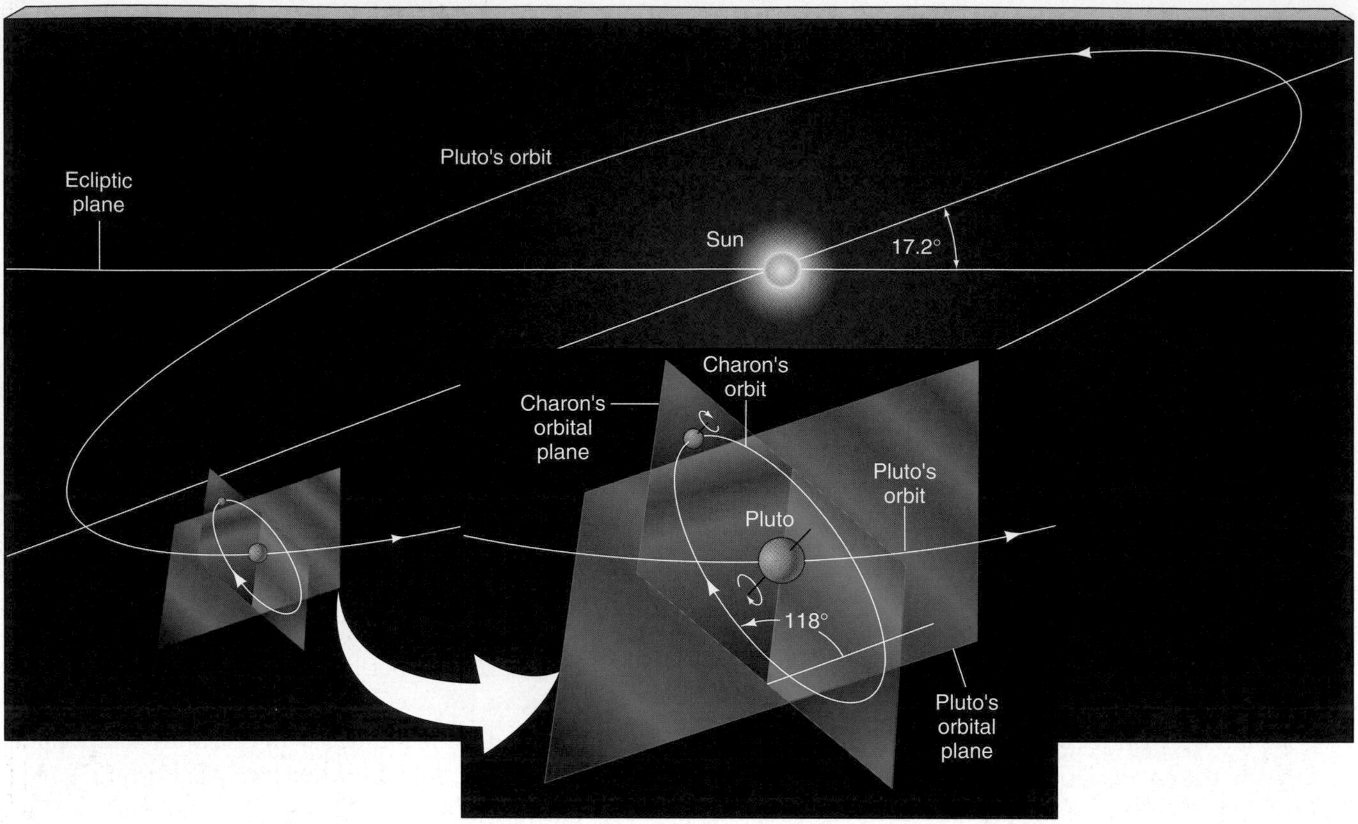

Figure 8.24 Pluto–Charon Orbit Charon's path around Pluto is circular, synchronous, and inclined at 118° to the orbit plane of the Pluto–Charon system about the Sun. The Pluto–Charon orbit plane is itself inclined at 17.2° to the plane of the ecliptic.

Animation

Chapter Review

www

Summary

There are 6 large moons in the outer solar system, each comparable to or larger than Earth's Moon in size and mass. In addition, there are 12 medium-sized moons and many smaller ones.

The four Galilean moons of Jupiter have densities that decrease with increasing distance from the planet. The innermost, Io, has active volcanoes powered by the constant flexing of the moon by Jupiter's tidal forces. Europa has a cracked, icy surface that may possibly conceal an ocean of liquid water. Ganymede and Callisto have heavily cratered surfaces. Ganymede's magnetic field implies relatively recent geological activity, but Callisto shows none.

Saturn's large moon Titan has a thick atmosphere that obscures the moon's surface and may be the site of complex cloud and surface chemistry. Neptune's moon Triton has a fractured surface of water ice and a thin atmosphere of nitrogen, probably produced by nitrogen "geysers" on its surface. Triton is the only large moon in the solar system to have a retrograde orbit around its parent planet. This orbit is unstable and will eventually cause Triton to be torn apart by Neptune's gravity.

The medium-sized moons of Saturn and Uranus are made up predominantly of rock and water ice. Many of them are heavily cratered and in some cases must have come close to being destroyed by the meteoritic impacts whose craters we now see.

From Earth, the main visible features of Saturn's rings are the **A**, **B**, and **C rings** and the **Cassini** and **Encke Divisions** (p. 218). The Cassini Division is a dark region between the A and B rings. The Encke Division lies near the outer edge of the A ring. The rings are made up of trillions of individual particles, ranging in size from dust grains to boulders. Their total mass is comparable to that of a small moon. Both divisions are dark because they are almost empty of ring particles.

The **Roche limit** (p. 219) of a planet is the distance within which the planet's tidal field overwhelms the internal gravity of an orbiting moon, tearing the moon apart and forming a ring. All the rings of the four jovian planets lie inside (or close to) their parent planets' Roche limits.

Saturn's rings are made up of tens of thousands of narrow **ringlets** (p. 219). Interactions between the ring particles and the planet's inner moons are responsible for much of the fine structure observed. Saturn's narrow **F ring** (p. 220) lies just outside the A ring. It has a kinked, braided structure, apparently caused by two small **shepherd satellites** (p. 220) that orbit close to it and prevent it from breaking up. The faint **D ring** (p. 220) lies between the C ring and Saturn's cloud layer. The **E ring** (p. 220) is apparently associated with volcanism on the moon Enceladus.

Jupiter has a faint, dark ring extending down to the planet's cloud tops. Uranus has a series of dark, narrow rings first detected from Earth by **stellar occultation** (p. 222), which occurs when a body passing between Earth and some distant star obscures the light received from that star. Shepherd satellites keep Uranus's rings from breaking apart. Neptune has three narrow rings like Uranus's and one broad ring, like Jupiter's.

Pluto was discovered in 1930 after a laborious search for a planet that was supposedly affecting Uranus's orbital motion. We now know that Pluto is far too small to have had any detectable influence on Uranus's path. Pluto has a moon, Charon, the mass of which is about one-eighth that of Pluto. Studies of Charon's orbit around Pluto have allowed the masses and radii of both bodies to be accurately determined. Pluto's properties are far more moonlike than planetlike. Triton, Pluto, and Charon are probably all examples of large Kuiper-Belt objects.

Review and Discussion

1. How does the density of the Galilean moons vary with increasing distance from Jupiter? Is there a trend to this variation? If so, explain the reason for it.
2. What is unusual about Jupiter's moon Io?
3. Why is there speculation that the Galilean moon Europa might be an abode for life?
4. Why do astronomers think that Ganymede may have experienced a relatively recent episode of internal heating?
5. What property of Saturn's largest moon, Titan, makes it of particular interest to astronomers?
6. What is the predicted fate of Triton?
7. What is unique about Miranda? Give a possible explanation for this uniqueness.
8. Seen from Earth, Saturn's rings sometimes appear broad and bright but at other times seem to disappear. Why?
9. Why do many astronomers think Saturn's rings formed quite recently?
10. What effect does Mimas have on the rings of Saturn?
11. Describe the behavior of shepherd satellites.
12. How do the rings of Neptune differ from those of Uranus and Saturn?
13. How was Pluto discovered?
14. In what respect is Pluto more like a moon than a jovian or terrestrial planet?
15. How was the mass of Pluto determined?

True or False?

____ **1.** The densities of the Galilean moons increase with increasing distance from Jupiter.

____ **2.** Io has a noticeable lack of impact craters on its surface.

____ **3.** The surface of Europa is completely covered by water ice.

____ **4.** Most of the medium-sized jovian moons rotate synchronously with their orbits around their parent planet.

____ **5.** Saturn is unique among the planets in having a ring system.

____ **6.** A typical particle in Saturn's rings is more than 100 m across.

____ **7.** Although Saturn's ring system is tens of thousands of kilometers wide, it is only a few tens of meters thick.

____ **8.** Saturn has small and medium-sized moons, but no large ones.

____ **9.** Water ice predominates in Saturn's moons.

____ **10.** Titan's atmosphere is denser than Earth's.

____ **11.** Titan's surface is obscured by thick cloud layers.

____ **12.** Uranus has no large moons.

____ **13.** Triton's surface has a marked lack of cratering, indicating significant amounts of surface activity.

____ **14.** Pluto's moon, Charon, is very small compared with Pluto.

____ **15.** Pluto is larger than Earth's Moon.

Fill in the Blank

1. The Galilean moon _______ is larger than Mercury.
2. In contrast to the inner two Galilean moons, the outer two have compositions that include significant amounts of _______.
3. Io is the only moon in the solar system having active _______.
4. As viewed from Earth, Saturn's ring system is conventionally divided into _______ broad rings.
5. The Cassini Division lies between the _______ and _______ rings.
6. Saturn's rings exist because they lie within Saturn's _______.
7. Two small moons, known as _______ satellites, are responsible for the unusually complex form of the F ring.
8. The composition of Titan's atmosphere is 90 percent _______.
9. There may be large amounts of liquid _______ under the frozen surface of Europa.
10. The Uranian moon that shows the greatest amount of geological activity and disruption over time is _______.
11. Triton's orbit is unusual because it is _______.
12. Neptune's moon Triton has a thin atmosphere made of _______.
13. Nitrogen geysers are found on _______.
14. Overall, Pluto is most similar to which object in the solar system? _______
15. Astronomers measured the radii of Pluto and Charon by observing a fortuitous series of _______during the late 1980s.

Problems

1. Io orbits Jupiter at a distance of six planetary radii once every 42 hours. If Jupiter completes one rotation every 10 hours, use Kepler's third law to calculate how far from the planet's center a satellite must orbit in order to appear "stationary" above the planet. ∞ (Sec. 1.3)
2. Estimate the strength of Jupiter's gravitational tidal force on Io, relative to the moon's own surface gravity. ∞ (*More Precisely 5-1*) Repeat the calculation for Saturn and Mimas.
3. Compare the apparent sizes of the Galilean moons, as seen from Jupiter's cloudtops, with the angular diameter of the Sun at Jupiter's distance. Would you expect ever to see a total solar eclipse from Jupiter's cloudtops? ∞ (*More Precisely P-2*)
4. Assuming a spherical shape and a uniform density of 2000 kg/m^3, calculate how small an icy moon of one of the outer planets would have to be before a fastball pitched at 40 m/s (about 90 mph) could escape.
5. Show that Titan's surface gravity is about one-seventh that of Earth. What is Titan's escape speed? ∞ (*More Precisely 5-2*)
6. What is the orbital speed, in kilometers per second, of ring particles at the inner edge of Saturn's B ring? Compare with the speed of a satellite in low Earth orbit (500 km altitude, say). ∞ (*More Precisely 1-2*)
7. The total mass of material in Saturn's rings is about 10^{15} tons (10^{18} kg). Suppose the average ring particle is 6 cm in radius (a large snowball) and has a density of 1000 kg/m^3. How many ring particles are there?
8. How close is Charon to Pluto's Roche limit?
9. What would be your weight on Pluto? On Charon?
10. What is the round-trip travel time of light from Earth to Pluto (at a distance of 40 A.U.)? How far would a spacecraft orbiting Pluto at a speed of 0.5 km/s travel during that time?

Projects

1. Use binoculars to look at Jupiter. Be sure to hold them steady (try propping your arms up on the hood of a car or sitting down and bracing them against your knees). Can you see any of Jupiter's four largest moons? If you come back the following evening, the moons' relative positions will have changed. Have some changed more than others? Before observing, look up the positions of the Galilean moons in a current magazine such as *Astronomy* or *Sky & Telescope*. Identify each moon. Watch Io over a period of at least an hour. Can you see its motion? Do the same for Europa.
2. Binoculars may not reveal the rings of Saturn, but most small telescopes will. Use a telescope to look at Saturn. Does Saturn appear to be spherical, or is it slightly flattened at its poles? Examine the rings. Are they tilted toward Earth, or do you seem to be looking at them edge-on? Can you see a dark line in the rings? This is the Cassini Division. Can you see the shadow of the rings on the surface of Saturn?
3. While looking at Saturn through a telescope, can you see any of its moons? The moons line up with the rings; Titan is often the farthest out, and always the brightest. How many moons can you see? Use an almanac to identify each one you find.
4. Pluto does not look like a circular disk, even through a powerful telescope. It always looks like a star. To see Pluto at all, you must have at least an 8-inch telescope. Then you must locate the field of stars in which it currently resides (check the January issue of *Astronomy* for this year's chart, or *Sky & Telescope* for the months when the planets are visible). Draw a picture of all the stars you see in that field of view. Come back a few nights later, and draw a picture of the field again. Pluto is the "star" that has moved.

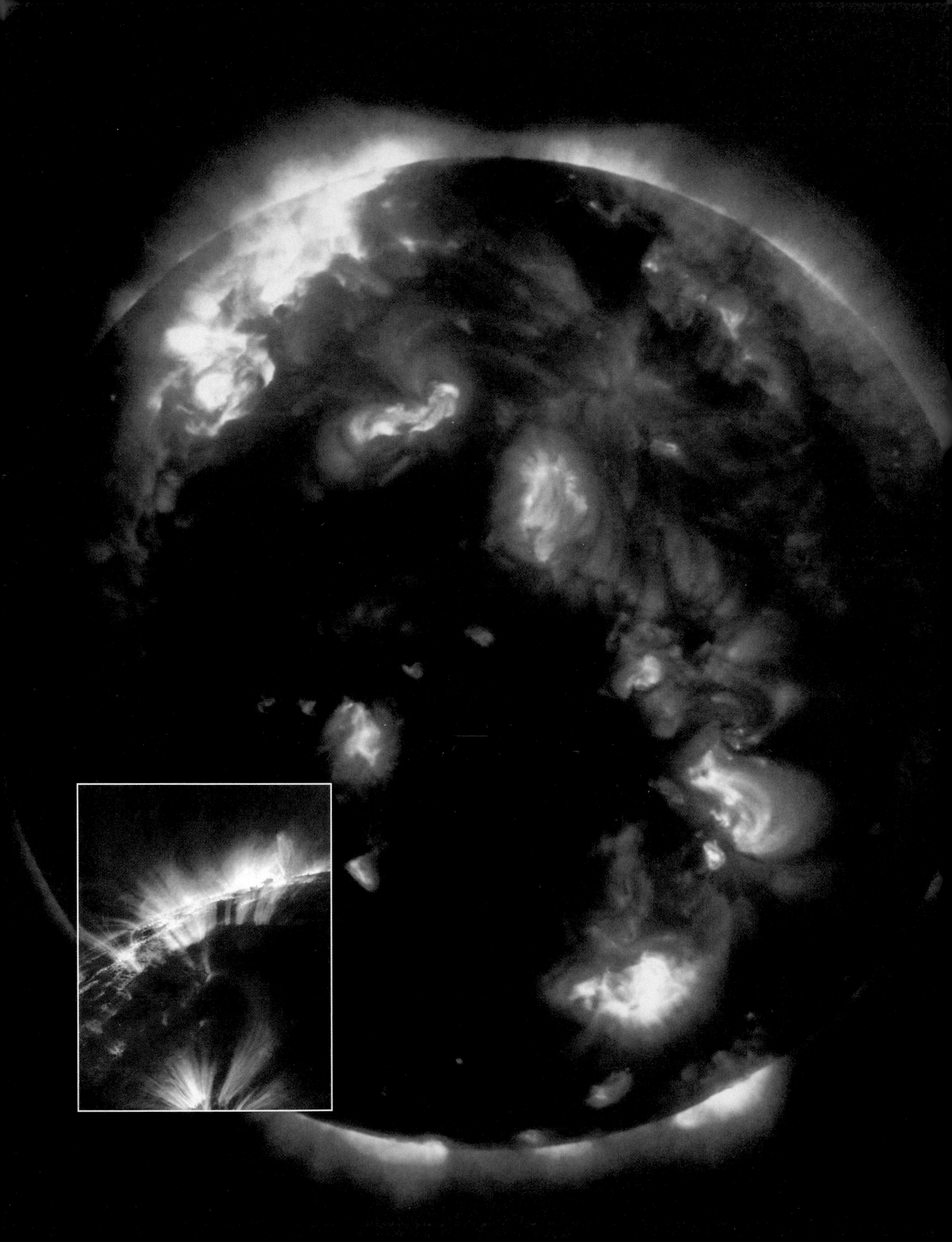

9 THE SUN

Our Parent Star

LEARNING GOALS

Studying this chapter will enable you to:

1. Summarize the overall properties of the Sun.
2. Explain how energy travels from the solar core, through the interior, and out into space.
3. Name the Sun's outer layers and describe what those layers tell us about the Sun's surface composition and temperature.
4. Discuss the nature of the Sun's magnetic field and its relationship to the various types of solar activity.
5. Outline the process by which energy is produced in the Sun's interior.
6. Explain how observations of the Sun's core challenge our present understanding of solar physics.

(Opposite page) This spectacular image of the Sun was made by capturing X rays emitted by our star's most active regions. It was taken by a camera on a rocket lofted shortly before the total solar eclipse of July 1991. (Note the shadow of the Moon approaching from the west, at top.) The brightest regions have temperatures of about three million kelvins.

(Inset) Up close and in the ultraviolet, as seen by the *Transition Region and Coronal Explorer* (TRACE) satellite in 1998, the limb of the Sun reveals a wealth of fine structure in and near the corona, including the "coronal mass ejections" shown here, each one of which releases billions of tons of matter into space, with an energy equivalent of about a hundred million atomic bombs. *(NASA)*

Living in the solar system, we have the chance to study, at close range, perhaps the most common type of cosmic object—a star. Our Sun is a star, and a fairly average star at that, but it has one unique feature: it is very close to us—300,000 times closer than our next nearest neighbor, Alpha Centauri. Whereas Alpha Centauri is 4.3 light-years distant, the Sun is only eight light-minutes away from us. Consequently, we know far more about the Sun than about any of the other distant points of light in the universe. A sizable fraction of all our astronomical knowledge is based on modern studies of the Sun. Just as we studied our parent planet, Earth, to set the stage for our exploration of the solar system, we now study our parent star, the Sun, as the next step in our exploration of the universe.

9.1 The Sun in Bulk

1 The Sun is the sole source of light and heat for the maintenance of life on Earth. It is a **star**, a glowing ball of gas held together by its own gravity and powered by nuclear fusion at its center. In its physical and chemical properties, the Sun is very similar to most other stars, regardless of when and where they formed. Far from detracting from our interest in the Sun, this very mediocrity is one of the main reasons we study it—most of what we know about the Sun also applies to many other stars in the universe.

Overall Structure

Table 9.1 lists some basic solar data. Having a radius of more than 100 Earth radii, a mass of more than 300,000 Earth masses, and a surface temperature well above the melting point of any known material, the Sun is clearly a body very different from any other we have encountered so far.

The Sun has a surface of sorts—not a solid surface (the Sun contains no solid material), but rather that part of the brilliant gas ball we perceive with our eyes or view through a heavily filtered telescope. The part of the Sun that emits the radiation we see is

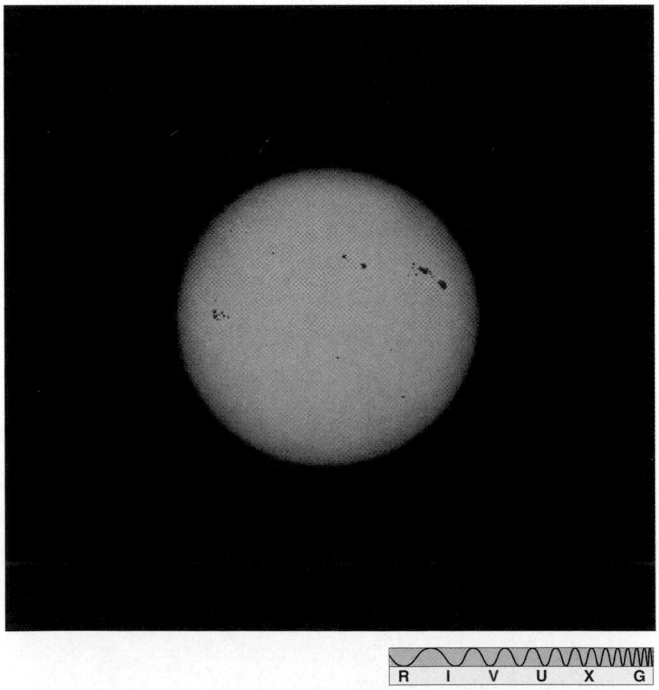

Figure 9.1 The Sun This image of the Sun, made using a heavily filtered telescope, has a sharp edge, although the Sun, like all stars, is made of gradually thinning gas. The edge appears sharp because the solar photosphere is very thin. *(NOAO)*

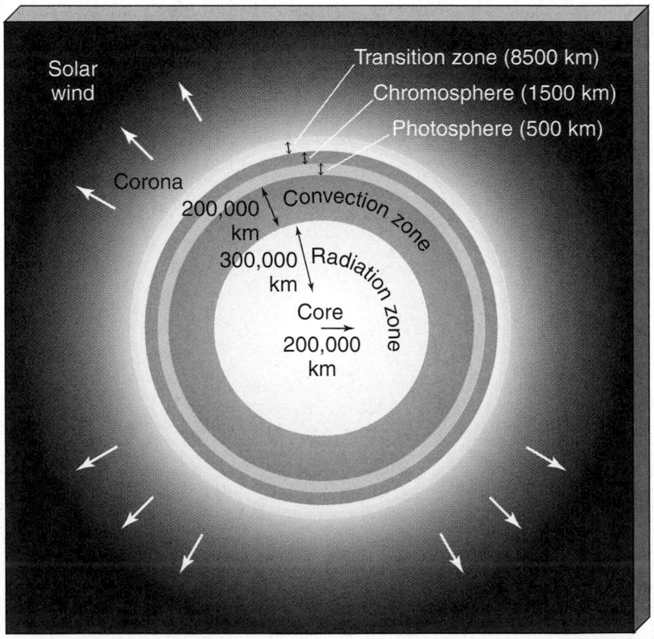

Figure 9.2 Solar Structure The main regions of the Sun, not drawn to scale, with some physical dimensions labeled. The photosphere is the visible "surface" of the Sun. Below it lie the convection zone, the radiation zone, and the core. Above the photosphere, the solar atmosphere consists of the chromosphere, the transition zone, and the corona.

TABLE 9.1 Some Solar Properties

Radius	696,000 km
Mass	1.99×10^{30} kg
Average density	1410 kg/m^3
Rotation period	24.9 days (equator); 29.8 days (poles)
Surface temperature	5780 K
Luminosity	3.86×10^{26} W

called the **photosphere**. Its radius (listed as the radius of the Sun in Table 9.1) is about 700,000 km. However, its thickness is probably no more than 500 km—less than 0.1 percent of the radius—which is why we perceive the Sun as having a well-defined, sharp edge (Figure 9.1).

The main regions of the Sun are illustrated in Figure 9.2. Just above the photosphere is the Sun's lower atmosphere, called the **chromosphere**, about 1500 km thick. Above it lies a region called the **transition zone**, where the temperature rises dramatically. Above 10,000 km, and stretching far beyond, is a thin, hot upper atmosphere, the solar **corona**. At still greater distances, the corona turns into the *solar wind*, which flows away from the Sun and permeates the entire solar system. ∞ (Sec. 4.2)

Extending down some 200,000 km below the photosphere is the **convection zone**, a region where the material of the Sun is in constant convective motion. Below the convection zone lies the **radiation zone**, where solar energy is transported toward the surface by radiation rather than by convection. The term *solar interior* is often used to mean both the radiation and convection zones. The central **core**, roughly 200,000 km in radius, is the site of powerful nuclear reactions that generate the Sun's enormous energy output.

Luminosity

The Sun radiates an enormous amount of energy into space. We can measure the Sun's total energy output in two stages, as follows. First, by holding a light-sensitive device—a solar cell, perhaps—perpendicular to the Sun's rays above Earth's atmosphere, we can measure how much solar energy is received per square meter of area every second. This quantity, known as the **solar constant**, is approximately 1400 watts per square meter (W/m^2).[1] About 70 percent of this energy reaches Earth's surface. Thus, for example, a sunbather's body having a total surface area of about 0.5 m^2 receives solar energy at a rate of roughly 500 watts, approximately equivalent to the output of a small electric room heater or five 100-W lightbulbs.

Knowing the solar constant, we can calculate the *total* amount of energy radiated in all directions from the Sun. Imagine a three-dimensional sphere that is centered on the Sun and is just large enough that its surface intersects Earth's center (Figure 9.3). The sphere's radius is one A.U., and its surface area is therefore $4\pi \times (1\ \text{A.U.})^2$, or approximately 2.8×10^{23} m^2. Assuming that the Sun radiates energy equally in all directions, we can determine the total rate at which energy leaves the Sun's surface simply by multiplying the solar constant by the total surface area of our imaginary sphere. The result is just under 4×10^{26} W. This quantity is called the **luminosity** of the Sun.

[1]The SI unit of energy is the *joule* (J). Probably more familiar to most readers is the closely related unit *watt* (W), which measures *power*, defined as the rate at which energy is emitted or expended by an object. One watt is equal to one joule per second; for instance, a light bulb with a power rating of 100 W radiates 100 J of energy each second.

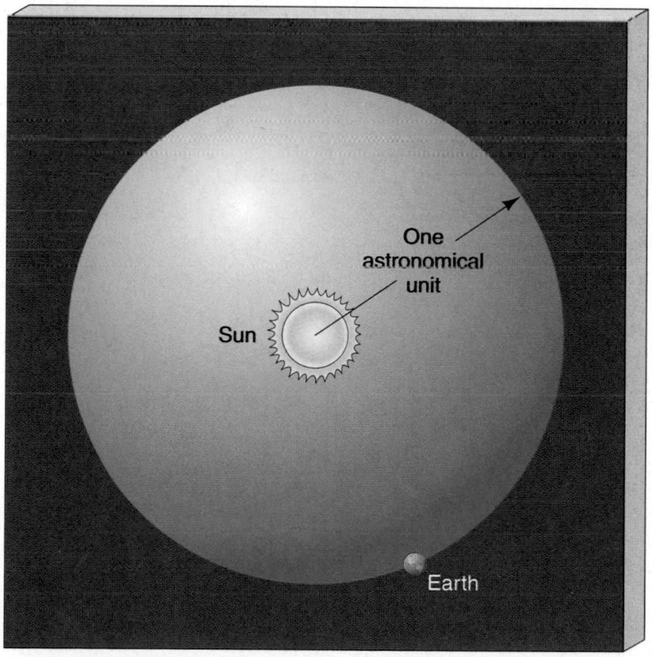

Figure 9.3 Astronomical Unit We can draw an imaginary sphere around the Sun so that the sphere's surface passes through Earth's center. The radius of this imaginary sphere is one A.U. By multiplying the sphere's surface area by the solar constant, we can measure the Sun's luminosity, the amount of energy it emits each second.

Take a moment to consider the magnitude of the solar luminosity. The Sun is an enormously powerful source of energy. *Every second*, it produces an amount of energy equivalent to the detonation of about 100 billion one-megaton nuclear bombs. There is simply nothing on Earth comparable to a star!

Concept Check

- Why must we assume that the Sun radiates equally in all directions when computing the solar luminosity from the solar constant?

9.2 The Solar Interior

Lacking any direct measurements of the Sun's interior, astronomers must use indirect means to probe the inner workings of our parent star. To accomplish this, they construct mathematical models of the Sun, combining all available observations with theoretical insight into solar physics. The result is the **Standard Solar Model**, which has gained widespread acceptance among astronomers.

Modeling the Structure of the Sun

An important technique for probing the Sun beneath the photosphere emerged in the 1960s, when it was discovered that the surface of the Sun vibrates like a complex set of bells. These vibrations, illustrated in Figure 9.4, are the result of internal pressure ("sound") waves that reflect off the photosphere and repeatedly cross the solar interior. These waves penetrate deep inside the Sun, and analysis of their surface patterns allows scientists to study conditions far below the Sun's surface. This process is similar to the way in which seismologists study Earth's interior by observing seismic waves produced by earthquakes. ∞ (Sec. 5.4) For this reason, study of solar surface patterns is usually called **helioseismology**, even though solar pressure waves have nothing whatever to do with solar seismic activity (which doesn't exist).

The most extensive study of solar vibrations is the ongoing GONG (short for Global Oscillations Network Group) project. By making continuous observations of the Sun from many clear sites around Earth, solar astronomers have obtained uninterrupted high-quality solar data spanning weeks at a time. The *Solar and Heliospheric Observatory* (SOHO), launched by the European Space Agency in 1995 and now permanently stationed between Earth and the Sun some 1.5 million km from our planet, provides continuous monitoring of the Sun's surface and atmosphere. Analysis of the data from all these sources provides important detailed information about the temperature, density, rotation, and convective state of the solar interior.

Figure 9.5 shows the solar density and temperature according to the Standard Solar Model, plotted as functions of distance from the Sun's center. Notice how the density drops sharply at first, then more slowly near the photosphere. The variation in density is large, ranging from a core value of about 150,000 kg/m^3, 20 times the density of iron, to an extremely small photospheric value of 2×10^{-4} kg/m^3, about 10,000 times less dense than air at Earth's surface. The *average* density of the Sun (Table 9.1) is 1400 kg/m^3, about the same as the density of Jupiter.

As shown in Figure 9.5, the solar temperature also decreases with increasing radius, but not as rapidly as the density. Computer models indicate a central temperature of about 15 million K. The temperature decreases steadily, reaching the observed value of 5800 K at the photosphere.

Energy Transport

2 The very hot solar interior ensures violent and frequent collisions among gas particles. In and near the core, the extremely high temperatures guarantee that

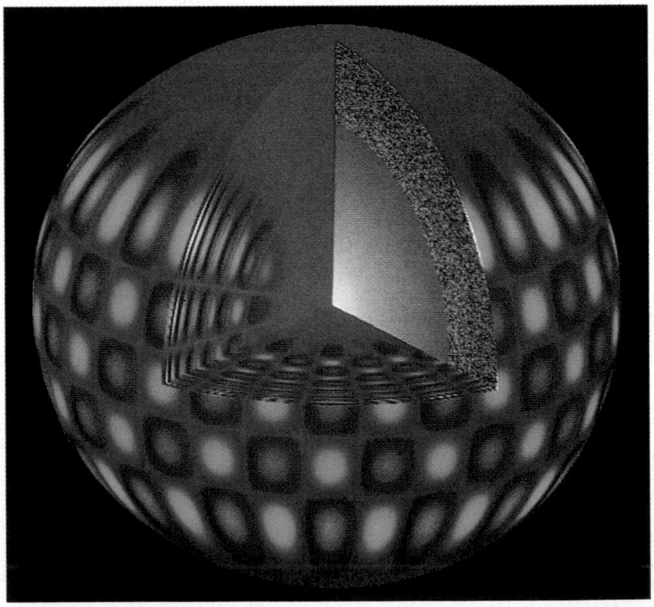

Figure 9.4 Solar Vibration The Sun vibrates in a very complex way as sound waves of many frequencies move through its interior. By observing the motion of the solar surface, scientists can determine the wavelengths and frequencies of the individual waves and thus deduce information about the solar interior not obtainable by other means. The alternating patches of color in this rendering of one particular wave pattern represent gas moving inward (red) and outward (blue). *(National Solar Observatory)*

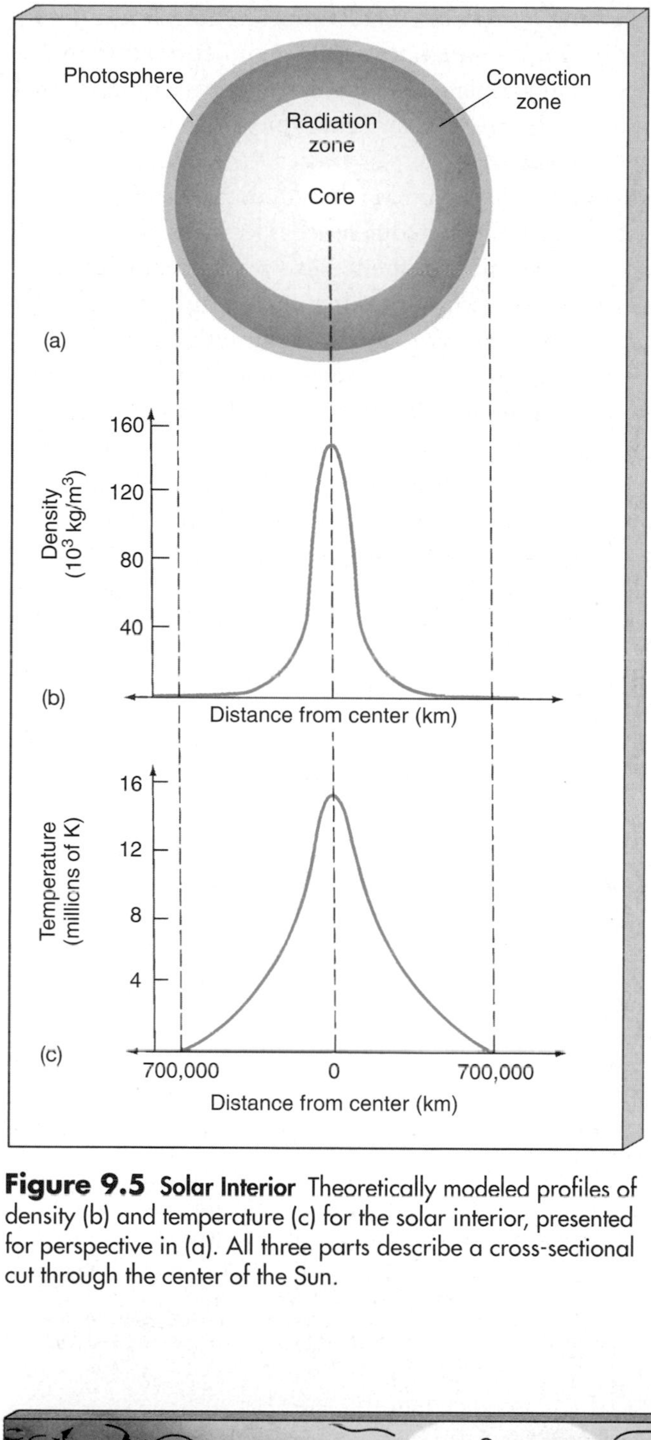

Figure 9.5 Solar Interior Theoretically modeled profiles of density (b) and temperature (c) for the solar interior, presented for perspective in (a). All three parts describe a cross-sectional cut through the center of the Sun.

the gas is completely ionized. With no electrons left on atoms to capture photons and move into more excited states, the deep solar interior is quite transparent to radiation. ∞ (Sec. 2.6) Only occasionally does a photon scatter off a free electron. The energy produced by nuclear reactions in the core travels outward toward the surface in the form of radiation with relative ease.

As we move outward from the core the temperature falls, and eventually some electrons can remain bound to nuclei. With more and more atoms retaining electrons that can absorb the outgoing radiation, the gas in the interior changes from being relatively transparent to being almost totally opaque. By the outer edge of the radiation zone, 500,000 km from the center, *all* of the photons produced in the Sun's core have been absorbed. Not one of them reaches the surface. But what happens to the energy they carry?

That energy is carried to the solar surface by *convection*—the same basic physical process we saw in our study of Earth's atmosphere. ∞ (Sec. 5.3) Convection can occur whenever cooler material overlies warmer material, and this is just what happens in the outer part of the Sun's interior. Hot solar gas physically moves outward, while cooler gas above it sinks, creating a characteristic pattern of convection cells. All through the convection zone, energy is transported to the surface by physical motion of the solar gas. Remember that there is no physical movement of material when radiation is the energy-transport mechanism; convection and radiation are *fundamentally different* ways in which energy can be transported from one place to another.

Figure 9.6 is a schematic diagram of the solar convection zone. There is a hierarchy of convection cells, organized in tiers at different depths. The deepest tier, about 200,000 km below the photosphere, is thought to contain cells tens of thousands of kilometers in

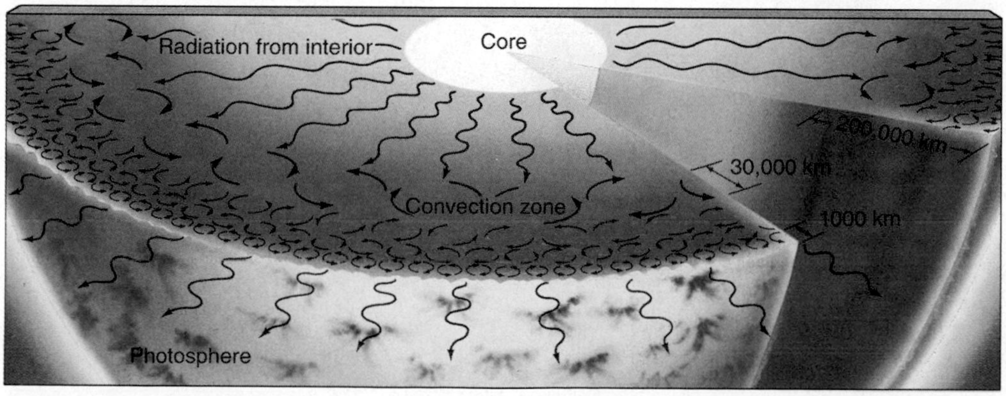

Figure 9.6 Solar Convection Transport of energy in the Sun's convection zone. We can visualize this region as a boiling, seething sea of gas. Each convective cell at the top of the convection zone is about 1000 km across. The cells are arranged in tiers, with cells of progressively smaller size at increasing distance from the center. (This is a highly simplified diagram.)

diameter. Energy is carried upward through a series of progressively smaller cells, stacked one upon another until, at a depth of about 1000 km below the photosphere, the individual cells are about 1000 km across. The top of this uppermost tier of convection is the solar photosphere.

Convection does not proceed into the solar atmosphere. In and above the photosphere, the density is so low that the gas is transparent and radiation once again becomes the mechanism of energy transport. Photons reaching the photosphere escape freely into space.

Evidence for Solar Convection

In part, our knowledge of solar convection is derived indirectly, from computer models of the solar interior. However, astronomers also have some direct evidence of conditions in the convection zone. Figure 9.7 is a high-resolution photograph of the solar surface. The visible surface is highly mottled with regions of bright and dark gas known as *granules*. This **granulation** of the solar surface is a direct reflection of motion in the convection zone. Each bright granule measures about 1000 km across and has a lifetime of between 5 and 10 minutes.

Each granule forms the topmost part of a solar convection cell. Doppler measurements indicate that the bright granules are moving outward with speeds of about 1 km/s, while the dark granules are moving down into the solar interior, exactly as we would expect for the topmost tier of convection in Figure 9.6. ∞ (*More Precisely 2-3*) The brightness variations of the granules result from differences in temperature. The upwelling gas is hotter and therefore (by Stefan's law) emits more radiation than the cooler downwelling gas. ∞ (Sec. 2.4) The adjacent bright and dark gases appear to contrast considerably, but in reality their temperature difference is less than about 500 K.

Careful measurements also reveal a much larger-scale flow beneath the solar surface. **Supergranulation** is a flow pattern quite similar to granulation except that supergranulation cells measure some 30,000 km across. As with granulation, material upwells at the center of the cells, flows across the surface, then sinks down again at the edges. Scientists believe that supergranules are the imprint on the photosphere of the deepest tier of large convective cells depicted in Figure 9.6.

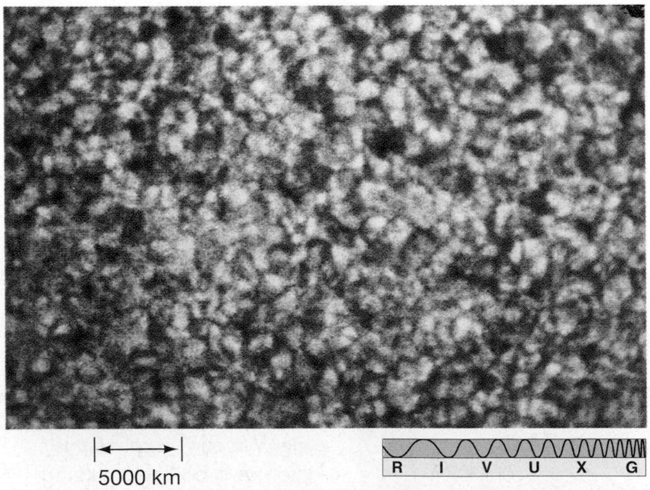

Figure 9.7 Solar Granulation A photograph of the granulated solar photosphere, taken from the *Skylab* space station. Typical solar granules are comparable in size to Earth's continents. The bright portions of the image are regions where hot material is upwelling from below, as shown in Figure 9.6. The dark regions correspond to cooler gas that is sinking back down into the interior. *(California Institute of Technology/Palomar/Hale Observatory)*

Concept Check

- Describe the two distinct ways in which energy moves outward from the solar core to the photosphere.

9.3 The Solar Atmosphere

3 Astronomers can glean an enormous amount of information about the Sun from analysis of absorption lines in the solar spectrum. ∞ (Sec. 2.5) Figure 9.8 (see also Figure 2.14) is a detailed visible spectrum of the Sun, spanning the range of wavelengths from 360 to 690 nm. Based on spectral analysis, Table 9.2 lists the 10 most

TABLE 9.2 The Composition of the Sun

ELEMENT	PERCENTAGE OF TOTAL NUMBER OF ATOMS	PECENTAGE OF TOTAL MASS
Hydrogen	91.2	71.0
Helium	8.7	27.1
Oxygen	0.078	0.97
Carbon	0.043	0.40
Nitrogen	0.0088	0.096
Silicon	0.0045	0.099
Magnesium	0.0038	0.076
Neon	0.0035	0.058
Iron	0.0030	0.14
Sulfur	0.0015	0.040

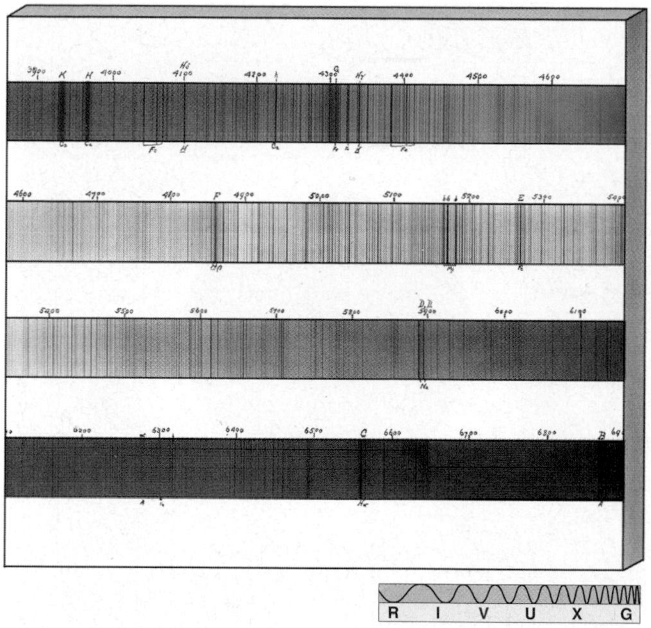

Figure 9.8 **Solar Spectrum** A detailed spectrum of our Sun in a portion of the visible domain shows thousands of dark absorption lines, indicating the presence of 67 elements in various stages of excitation and ionization in the lower solar atmosphere. *(California Institute of Technology/Palomar/Hale Observatory)*

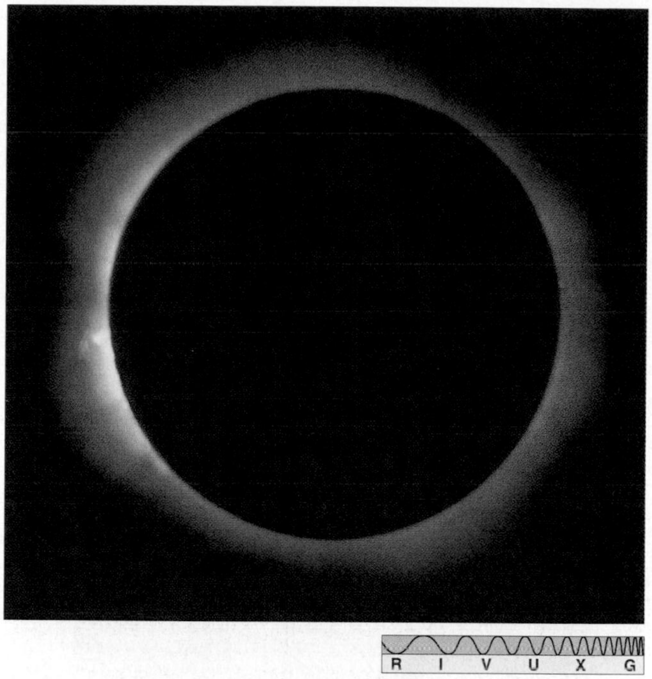

Figure 9.9 **Solar Chromosphere** This photograph of a total solar eclipse shows the solar chromosphere a few thousand kilometers above the Sun's surface. (G. Schneider)

common elements in the Sun. This distribution is very similar to what we saw on the jovian planets, and it is what we will find for the universe as a whole. Hydrogen is by far the most abundant element, followed by helium.

Strictly speaking, analysis of spectral lines allows us to draw conclusions only about the part of the Sun where the lines form—the photosphere and chromosphere. However, the data in Table 9.2 are thought to be representative of the entire Sun (with the exception of the solar core, where nuclear reactions are steadily changing the composition—see Section 9.5).

The Chromosphere

Density continues to decrease rapidly as we move outward through the solar atmosphere. The low density of the chromosphere means that it emits very little light of its own and cannot be observed visually under normal conditions. The photosphere is just too bright, dominating the chromosphere's radiation. Still, although it is not normally seen, astronomers have long been aware of the chromosphere's existence. Figure 9.9 shows the Sun during an eclipse in which the photosphere was obscured by the Moon but the chromosphere was not. The chromosphere's characteristic pinkish hue, the result of the red Hα emission line of hydrogen, is plainly visible. ∞ (Sec. 2.6)

The chromosphere is far from tranquil. Every few minutes, small solar storms erupt there, expelling jets of hot matter known as *spicules* into the Sun's upper atmosphere (Figure 9.10). These long, thin spikes of matter leave the Sun's surface at typical speeds of about 100 km/s, reaching several thousand kilometers above the photosphere. Spicules tend to accumulate around the edges of supergranules. The Sun's magnetic field is also known to be somewhat stronger than average in those regions. Scientists speculate that the downwelling material there tends to strengthen the solar magnetic field and that spicules are the result of magnetic disturbances in the Sun's churning outer layers.

The Transition Zone and Corona

During the brief moments of a solar eclipse, if the Moon's angular size is large enough that both the photosphere and the chromosphere are blocked, the ghostly solar corona can be seen, as in Figure 9.11. With the photospheric light removed, the pattern of spectral lines changes dramatically. The spectrum shifts from absorption to emission, and an entirely new set of spectral lines suddenly appears.

The shift from absorption to emission is entirely in accordance with Kirchhoff's laws, because we see the corona against the blackness of space, not against

Figure 9.10 Solar Spicules Solar spicules, short-lived, narrow jets of gas that typically last mere minutes, can be seen sprouting up from the chromosphere in this Hα image of the Sun. The spicules are the thin, dark, spikelike regions. They appear dark against the face of the Sun because they are cooler than the photosphere. *(NOAO)*

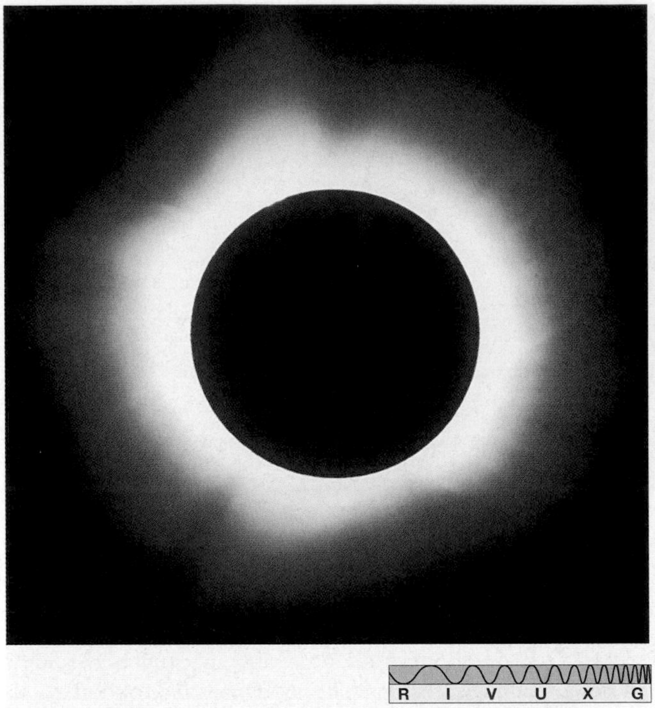

Figure 9.11 Solar Corona When both the photosphere and the chromosphere are obscured by the Moon during a solar eclipse, the faint solar corona becomes visible. This photograph shows clearly the emission of radiation from the corona. *(Sky Publishing Corporation)*

the bright continuous spectrum from the photosphere below. ∞ (Sec. 2.5) The new lines arise because atoms in the corona are much more highly *ionized* than atoms in the photosphere or chromosphere. Their internal electronic structure and hence their spectra are therefore quite different from those of atoms at lower atmospheric levels.

The cause of this extensive electron stripping is the high coronal temperature. Based on measurements made during total solar eclipses of the degree of ionization observed at different levels in the solar atmosphere, Figure 9.12 shows how the temperature varies with height above the photosphere. The temperature decreases to a minimum of about 4500 K some 500 km above the photosphere, after which it rises steadily. About 1500 km above the photosphere, in the transition zone, the gas temperature increases sharply, reaching more than one million K at an altitude of 10,000 km. Thereafter, in the corona, the temperature remains roughly constant, although *SOHO* and other orbiting instruments have detected coronal "hotspots" having

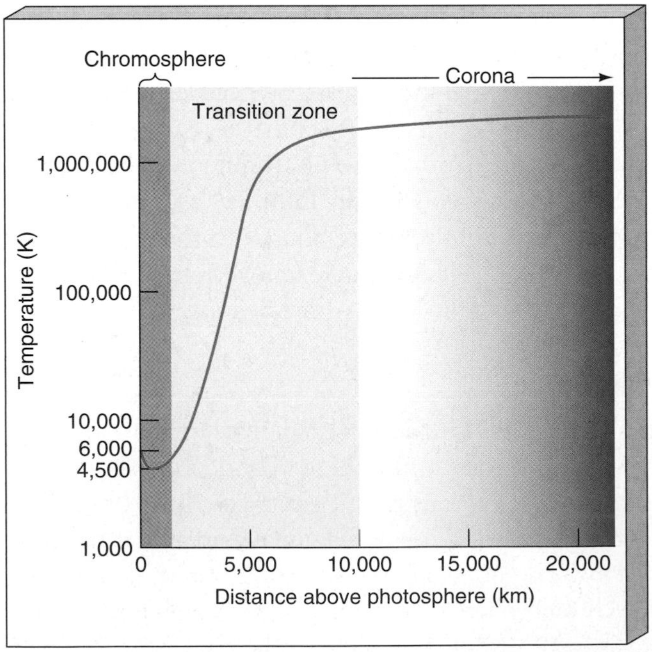

Figure 9.12 Solar Atmospheric Temperature The change in gas temperature in the solar atmosphere is dramatic. The temperature reaches a minimum of 4500 K in the chromosphere and then rises sharply in the transition zone, finally leveling off at more than one million K in the corona.

temperatures many times higher than this average value. The cause of the rapid temperature rise in the transition zone is thought to be magnetic activity on the solar surface, such as spicules and the more energetic phenomena described in the next section.

The solar wind is a direct consequence of the high coronal temperature. About 10 million km above the photosphere, the coronal gas is hot enough to escape the Sun's gravity, and it begins to flow outward into space. The Sun is, in effect, "evaporating"—constantly shedding mass through the solar wind. The wind is an extremely thin medium, however. Although it carries away about a million tons of solar matter each second, less than 0.1 percent of the Sun's mass has been lost since the solar system formed 4.6 billion years ago.

The Sun in X Rays

Unlike the 5800 K photosphere, which emits most strongly in the visible part of the electromagnetic spectrum, the million-kelvin coronal gas radiates at much higher frequencies—primarily in X rays. ∞ (Sec. 2.4) For this reason, X-ray telescopes have become important tools in studying the solar corona. Figure 9.13 shows a sequence of X-ray images of the Sun. The full corona extends well beyond the regions shown, but the density of coronal particles emitting the radiation diminishes rapidly with distance from the Sun. The intensity of X-ray radiation farther out is too dim to be seen here.

In the mid-1970s, instruments aboard NASA's *Skylab* space station revealed that the solar wind escapes mostly through solar "windows" called **coronal holes**—vast regions of the Sun's atmosphere where the density is about one-tenth its value in the rest of the corona. The dark area moving from left to right in Figure 9.13 represents such a coronal hole. The largest coronal holes can be hundreds of thousands of kilometers across. Structures of this size are seen only a few times each decade. Smaller holes—perhaps only a few tens of thousand kilometers in size—are much more common, appearing every few hours.

Like most aspects of the solar atmosphere, coronal holes are closely related to the Sun's magnetic field. In a coronal hole, the solar magnetic field lines extend from the photosphere far out into interplanetary space. Charged particles tend to follow the field lines, so they can escape, particularly from the Sun's polar regions, according to recent *SOHO* findings. In other parts of the corona, the solar magnetic field lines loop close to the Sun, keeping charged particles near the surface and inhibiting the outward flow of the solar wind. As a result, the coronal density remains (relatively) high.

✓ Concept Check

- List two ways in which the spectrum of the solar corona differs from that of the photosphere.

9.4 The Active Sun

4 Most of the Sun's luminosity results from continuous emission from the photosphere. However, superimposed on this steady, predictable aspect of our star's energy output is a much more irregular component, characterized by explosive and unpredictable surface activity. Solar activity contributes little to the Sun's total luminosity and has little effect on the evolution of the Sun as a star, but it does affect us here on Earth. The

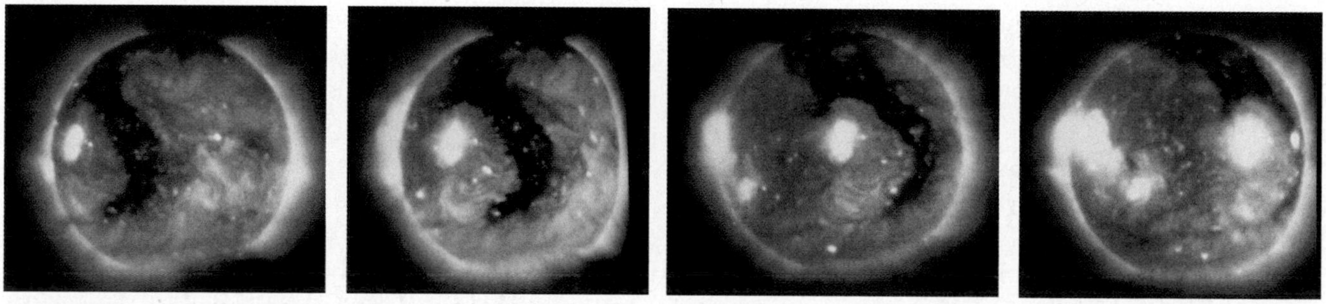

Figure 9.13 Coronal Holes Images of X-ray emission from the Sun observed by the *Skylab* space station. These frames were taken at one-day intervals. Note the dark, boot-shaped coronal hole traveling from left to right, where the X-ray observations outline in dramatic detail the abnormally thin regions through which the high-speed solar wind streams forth. *(NASA)*

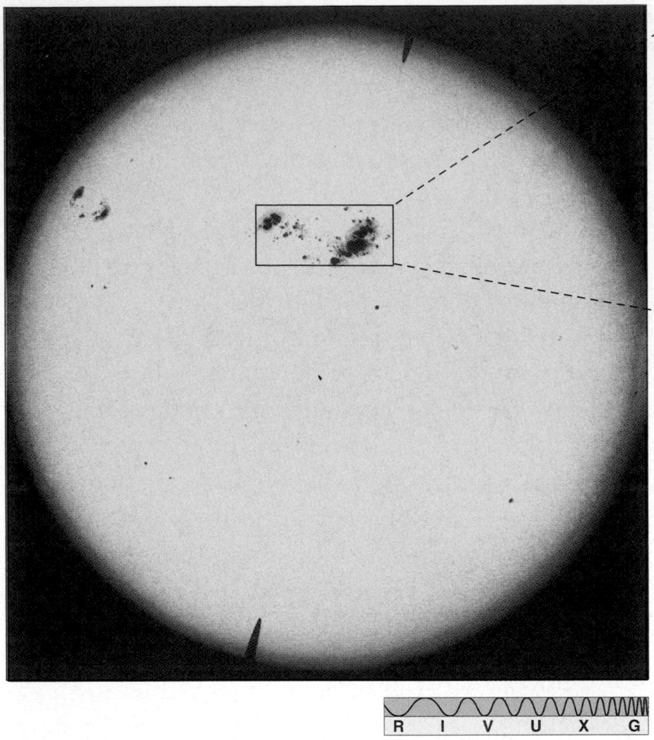

Figure 9.14 Sunspots This photograph of the Sun, taken during a period of maximum solar activity, shows several groups of sunspots. The largest spots in this image are more than 20,000 km across, nearly twice the diameter of Earth. Typical sunspots are only about half this size. *(California Institute of Technology/Palomar/Hale Observatory)*

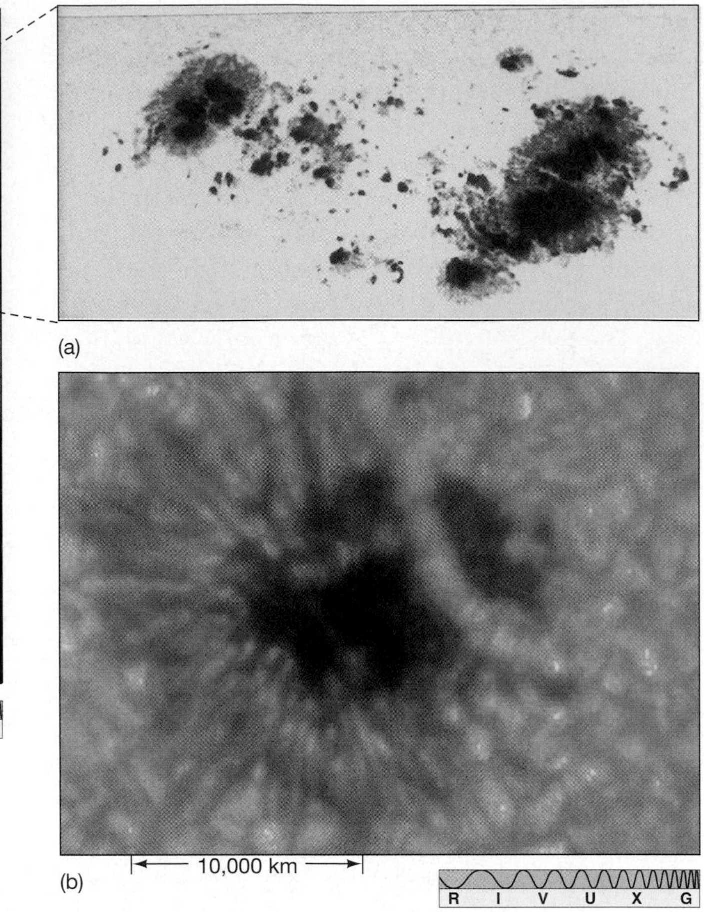

Figure 9.15 Sunspots, Up Close (a) The largest pair of sunspots in Figure 9.14. Each spot consists of a cool, dark inner region called the umbra surrounded by a warmer, less dark region called the penumbra. The spots appear dark because they are slightly cooler than the surrounding photosphere. (b) A high-resolution, true-color image of a single sunspot shows details of its structure as well as the surface granularity surrounding it. This spot is about the size of Earth. *(California Institute of Technology/Palomar/Hale Observatory; National Solar Observatory)*

size and duration of coronal holes are strongly influenced by the level of solar activity, and hence so is the strength of the solar wind. This, in turn, directly affects Earth's magnetosphere.

Sunspots

Figure 9.14 is an optical photograph of the entire Sun showing numerous dark **sunspots** on the surface. (Sec. 1.2) They typically measure about 10,000 km across—about the size of Earth. At any given instant, the Sun may have hundreds of sunspots, or it may have none at all.

Studies of sunspots show an *umbra*, or dark center, surrounded by a grayish *penumbra*. The close-up view of a pair of sunspots in Figure 9.15(a) shows both of these dark areas against the bright background of the undisturbed photosphere. This gradation in darkness indicates a gradual change in photospheric temperature. In other words, sunspots are simply relatively *cooler* regions of the photospheric gas. The temperature of the umbra is about 4500 K, that of the penumbra 5500 K. The spots, then, are certainly composed of hot gases. They seem dark only because they appear against an even brighter background (the 5800 K photosphere).

Observations of sunspots indicate that the Sun does not rotate as a solid body. Instead, it spins *differentially*—faster at the equator and more slowly at the poles, like Jupiter and Saturn. The solar photosphere rotates once every 25 days at the equator, but only once in 30 days at the poles. This combination of differential rotation and convection radically affects the character of the Sun's magnetic field.

Solar Magnetism

Spectroscopic studies indicate that the magnetic field in a typical sunspot is about 1000 times greater than the field in the surrounding, undisturbed photosphere (which is itself several times stronger than Earth's magnetic field). Furthermore, the field lines are not randomly oriented, but instead are directed roughly perpendicular to (out of or into) the Sun's surface. Scientists believe that sunspots are cooler than their surroundings because these abnormally strong fields interfere with the normal convective flow of hot gas toward the solar surface.

The **polarity** of a sunspot simply indicates which way its magnetic field is directed. We will conventionally label spots where the field lines emerge from the interior as "S" and those where the lines dive below the photosphere as "N" (so that field lines above the surface run from S to N, as on Earth). Sunspots almost always come in pairs whose members lie at roughly the same latitude and have opposite magnetic polarities. As illustrated in Figure 9.16, magnetic field lines emerge from the solar interior through one member (S) of a sunspot pair, loop through the solar atmosphere, then reenter the photosphere through the other member (N).

Despite the irregular appearance of the sunspots themselves, there is a great deal of order in the underlying solar field. *All* the sunspot pairs in the same solar hemisphere (north or south) at any instant have the *same* magnetic configuration. That is, if the leading spot (measured in the direction of the Sun's rotation) has N polarity, as shown in the figure, then all leading spots in that hemisphere have the same polarity. What's more, in the other hemisphere at the same time, all sunspot pairs have the *opposite* magnetic configuration (S polarity leading). To understand these regularities in sunspot polarities, we must look in more detail at the Sun's magnetic field.

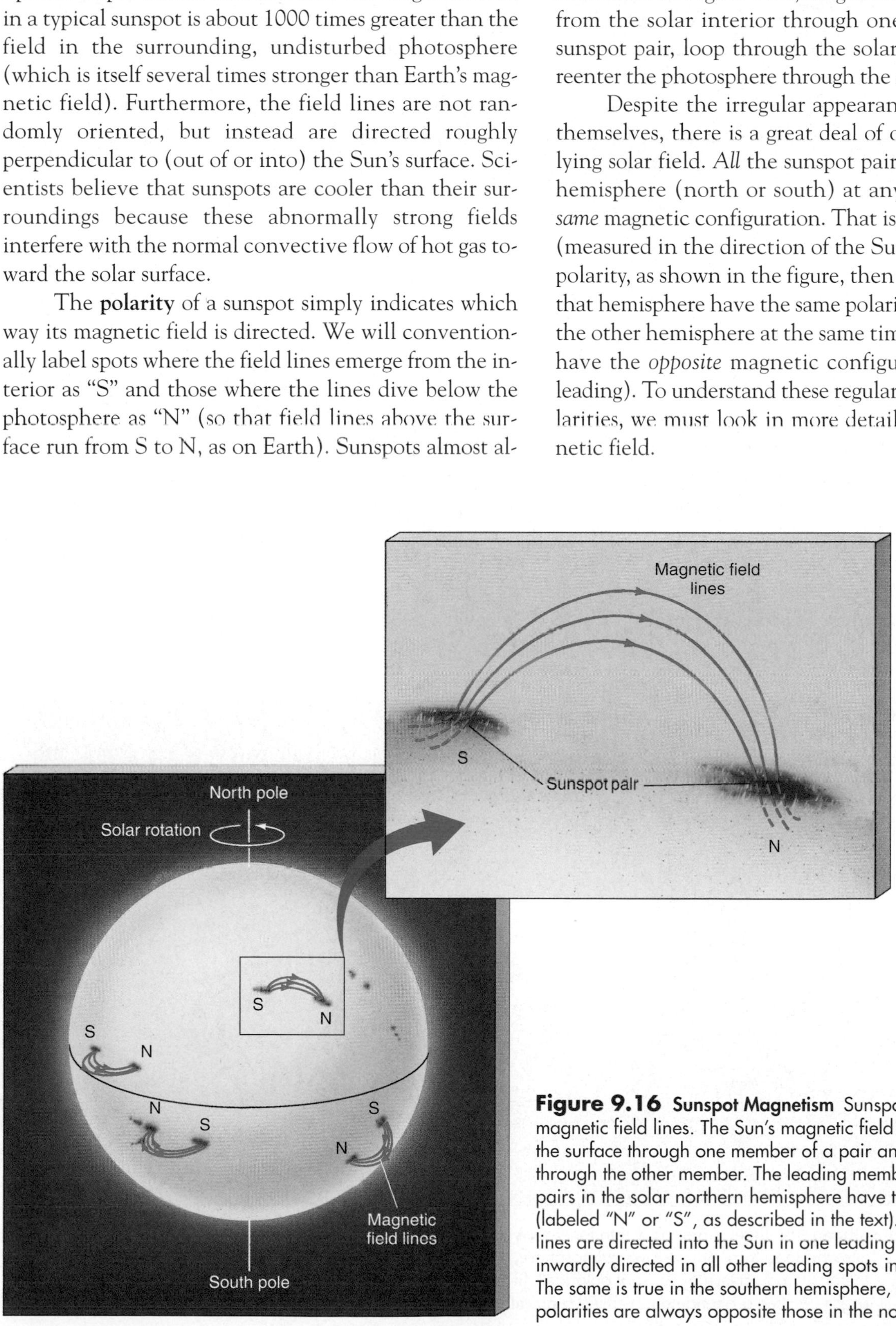

Figure 9.16 Sunspot Magnetism Sunspot pairs are linked by magnetic field lines. The Sun's magnetic field lines emerge from the surface through one member of a pair and reenter the Sun through the other member. The leading members of all sunspot pairs in the solar northern hemisphere have the same polarity (labeled "N" or "S", as described in the text). If the magnetic field lines are directed into the Sun in one leading spot, they are inwardly directed in all other leading spots in that hemisphere. The same is true in the southern hemisphere, except that the polarities are always opposite those in the north.

As illustrated in Figure 9.17, the Sun's differential rotation greatly distorts the solar magnetic field, "wrapping" it around the solar equator and eventually causing any originally north–south magnetic field to reorient itself in an east–west direction. At the same time, convection causes the magnetized gas to upwell toward the surface, twisting and tangling the magnetic field pattern. In some places, the field lines becomes kinked like a knot in a garden hose, causing the field strength to increase. Occasionally, the field becomes so strong that it overwhelms the Sun's gravity, and a "tube" of field lines bursts out of the surface and loops through the lower atmosphere, forming a sunspot pair. The general east–west organization of the underlying solar field accounts for the observed polarities of the resulting sunspot pairs in each hemisphere.

The Solar Cycle

Sunspots are not steady. Most change their size and shape, and all come and go. Individual spots may last anywhere from 1 to 100 days. A large group of spots typically lasts 50 days. However, centuries of observations have established a clear **sunspot cycle**. Figure 9.18(a) shows the number of sunspots observed each year during the twentieth century. The average number of spots reaches a maximum every 11 or so years, then falls off almost to zero before the cycle begins afresh. The latitudes at which sunspots appear vary as the sunspot cycle progresses. Individual sunspots do not move up or down in latitude, but new spots appear closer to the equator as older ones at higher latitudes fade away. Figure 9.18(b) is a plot of observed sunspot latitudes as a function of time.

In fact, the 11-year sunspot cycle is only half of a 22-year **solar cycle**. For the first 11 years of the solar cycle, the leading spots of all sunspot pairs in the same solar hemisphere have one polarity, and spots in the other hemisphere have the opposite polarity (Figure 9.16). These polarities then reverse their signs for the next 11 years.

Astronomers believe that the Sun's magnetic field is both generated and amplified by the constant stretching, twisting, and folding of magnetic field lines that results from the combined effects of differential rotation and convection. The theory is similar to the "dynamo" theory that accounts for the magnetic fields of Earth and the jovian planets, except that the solar dynamo operates much faster and on a much larger scale. One important prediction of this theory is that the Sun's magnetic field should rise to a maximum, then fall to

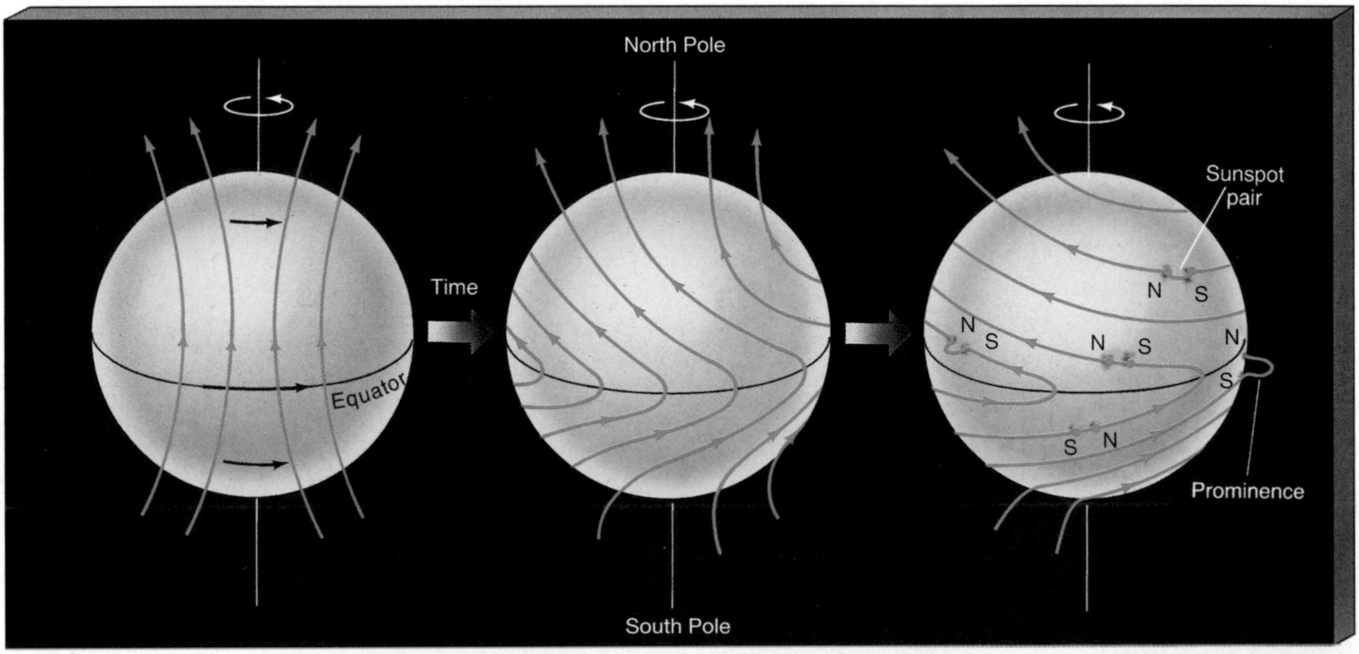

Figure 9.17 Solar Rotation The Sun's differential rotation wraps and distorts the solar magnetic field. Occasionally, the field lines burst out of the surface and loop through the lower atmosphere, thereby creating a sunspot pair. The underlying pattern of the solar field lines explains the observed pattern of sunspot polarities. (If the loop happens to occur near the edge of the Sun and is seen against the blackness of space, we see a phenomenon called a prominence, see Figure 9.20.)

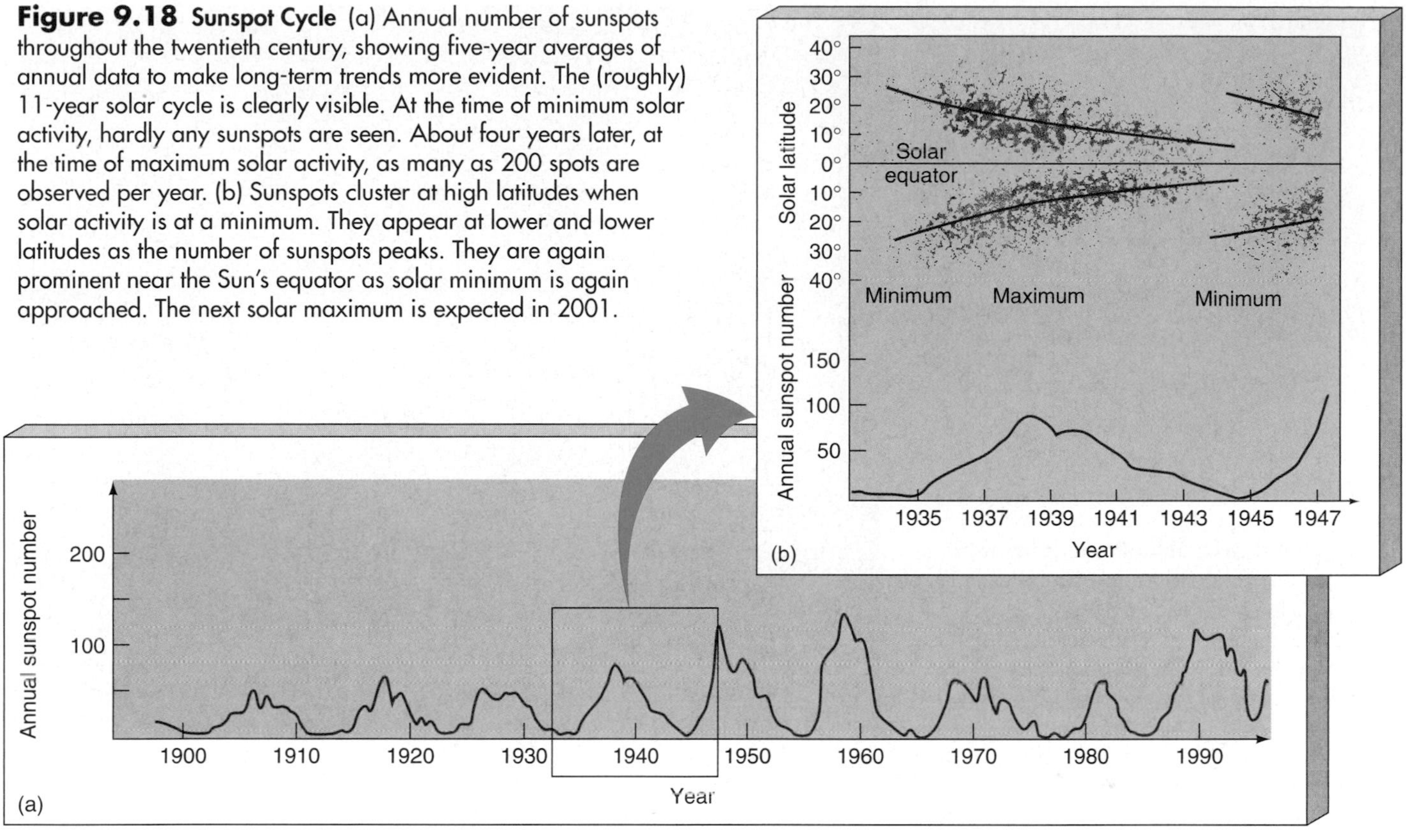

Figure 9.18 **Sunspot Cycle** (a) Annual number of sunspots throughout the twentieth century, showing five-year averages of annual data to make long-term trends more evident. The (roughly) 11-year solar cycle is clearly visible. At the time of minimum solar activity, hardly any sunspots are seen. About four years later, at the time of maximum solar activity, as many as 200 spots are observed per year. (b) Sunspots cluster at high latitudes when solar activity is at a minimum. They appear at lower and lower latitudes as the number of sunspots peaks. They are again prominent near the Sun's equator as solar minimum is again approached. The next solar maximum is expected in 2001.

zero and reverse itself in a roughly periodic way, just as observed. Solar surface activity, such as the sunspot cycle, simply follows the changes in the underlying field.

Figure 9.19 plots all sunspot data recorded since the invention of the telescope. As can be seen, the 11-year "periodicity" of the solar sunspot cycle is far from perfect. Not only does the period vary from 7 to 15 years, but the sunspot cycle has disappeared entirely in the relatively recent past. The lengthy period of solar inactivity that extended from 1645 to 1715 is called the *Maunder minimum*, after the British astronomer who drew attention to these historical records. (Interestingly, the period also seems to correspond fairly well to the coldest years of the so-called "Little Ice Age" that chilled northern Europe during the late seventeenth century, suggesting a link between solar activity and climate on Earth.) The cause of the Sun's century-long variations remains a mystery. Lacking a complete understanding of the solar cycle, we cannot easily explain how it could virtually shut down for such an extended period of time.

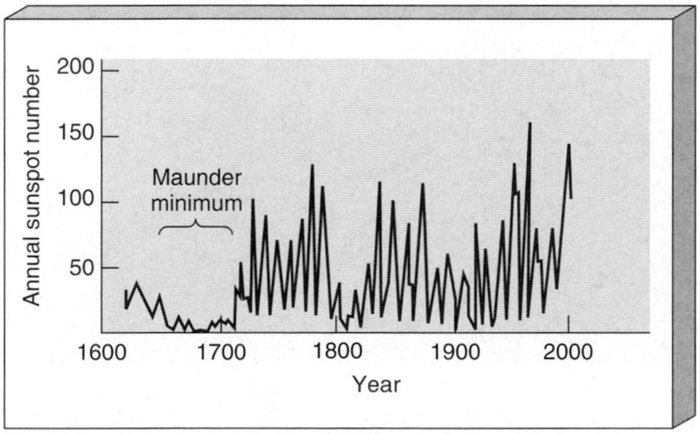

Figure 9.19 **11-year Sunspot Cycle** This graph depicts the number of sunspots occurring each year. Note the approximate 11-year "periodicity" and the absence of spots during the late seventeenth century.

Active Regions

Sunspots are relatively gentle aspects of solar activity. However, the photosphere surrounding them occasionally erupts violently, spewing forth into the corona large quantities of energetic particles. The sites of these explosive events are known simply as **active regions**. Most pairs or groups of sunspots have active regions associated with them. Like all other aspects of solar activity, these phenomena tend to follow the solar cycle and are most frequent and violent around the time of sunspot maximum.

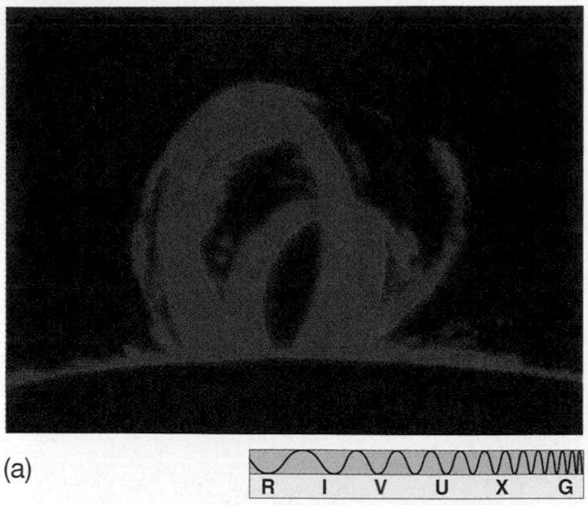

(a)

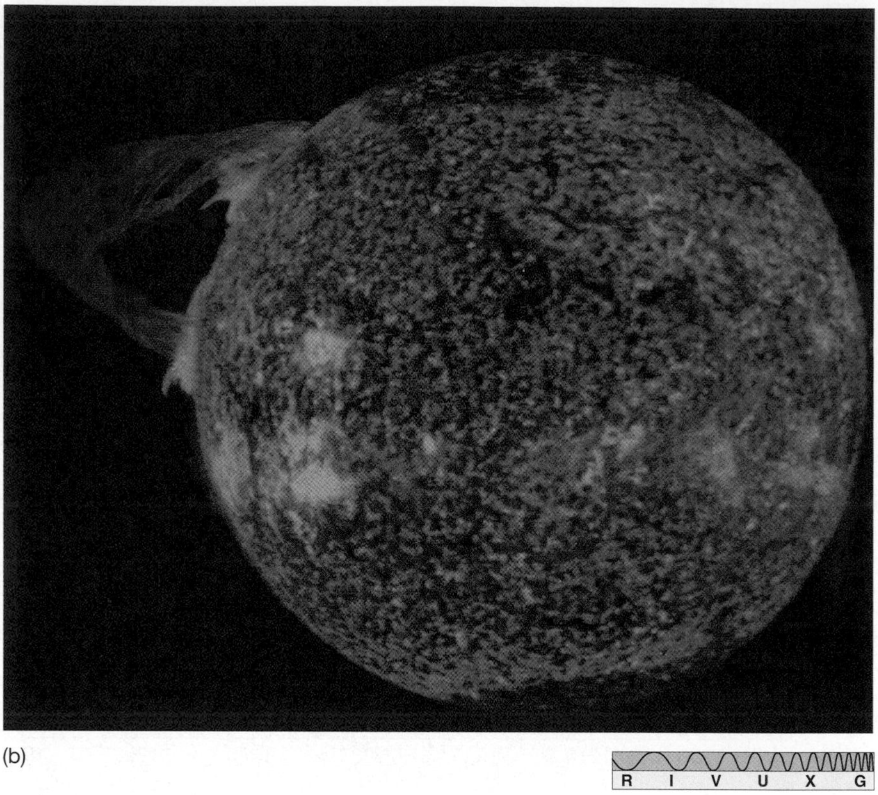

(b)

Video

Figure 9.20 **Solar Prominence** (a) The looplike structure of this prominence clearly reveals the magnetic field lines connecting the two members of a sunspot pair. (b) This image of a particularly large solar prominence was observed by ultraviolet detectors aboard the *Skylab* space station in 1979. *(National Solar Observatory; NASA)*

Figure 9.20 shows two solar **prominences**—loops or sheets of glowing gas ejected from an active region on the solar surface. Prominences move through the inner parts of the corona under the influence of the Sun's magnetic field. Magnetic instabilities in the strong fields found in and near sunspot groups may cause the prominences, although the details are still not completely understood. A typical solar prominence measures some 100,000 km in extent, nearly 10 times the diameter of planet Earth. Some may persist for days or even weeks. Prominences as large as that shown in Figure 9.20(b) (which traversed almost half a million kilometers of the solar surface) are less common and usually appear only at times of greatest solar activity.

Flares are another type of solar activity observed near active regions (Figure 9.21). Also the result of magnetic instabilities, flares are much more violent (and even less well understood) than prominences. They flash across a region of the Sun in minutes, releasing enormous amounts of energy as they go. Temperatures in the extremely compact hearts of flares can reach 100 million K. So energetic are these cataclysmic explosions that some researchers have likened them to bombs exploding in the lower regions of the Sun's atmosphere. Unlike the trapped gas that makes up the characteristic loop of a prominence, the particles produced by a flare are so energetic that the Sun's magnetic field is unable to hold them and shepherd them back to the surface. Instead, the particles are simply blasted into space by the violence of the explosion. Flares may also be responsible for many of the internal pressure waves that give rise to solar surface oscillations.

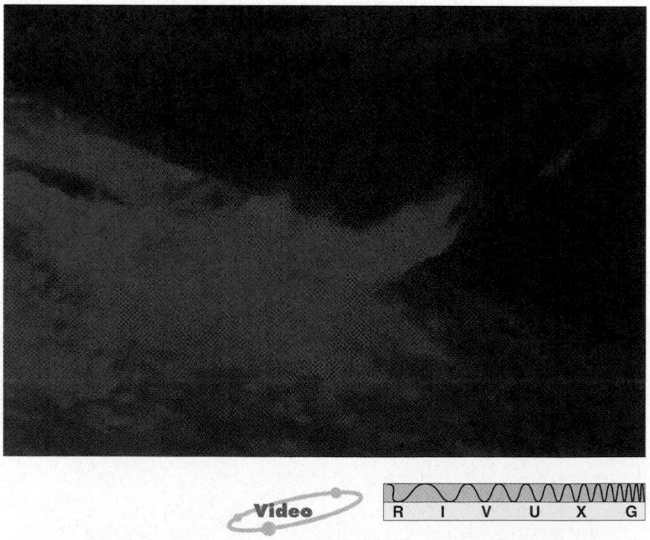

Video

Figure 9.21 **Solar Flare** Much more violent than a prominence, a solar flare is an explosion on the Sun's surface that sweeps across an active region in a matter of minutes, accelerating solar material to high speeds and blasting it into space. *(National Solar Observatory)*

The Changing Solar Corona

The solar corona also varies in step with the sunspot cycle. The photograph of the corona in Figure 9.11 shows the Sun at sunspot minimum. The corona is fairly regular in appearance and surrounds the Sun more or less uniformly. Compare this image with Figure 9.22, which was taken in 1991, close to a peak of the sunspot cycle. The active corona here is much more irregular than the one in Figure 9.11 and extends farther from the solar surface. The streamers of coronal material pointing away from the Sun are characteristic of this phase.

Astronomers think that the corona is heated primarily by magnetic activity on the solar surface, particularly prominences and flares, which can inject large amounts of energy into the corona, greatly distorting its shape. Extensive disturbances often move through the corona above an active site in the photosphere, distributing the energy throughout the coronal gas and often hurling large amounts (billions of tons) of matter into space. Both the appearance of the corona and the strength of the solar wind are closely correlated with the solar cycle.

✓ Concept Check

- What do observations of sunspot polarities tell us about the solar magnetic field?

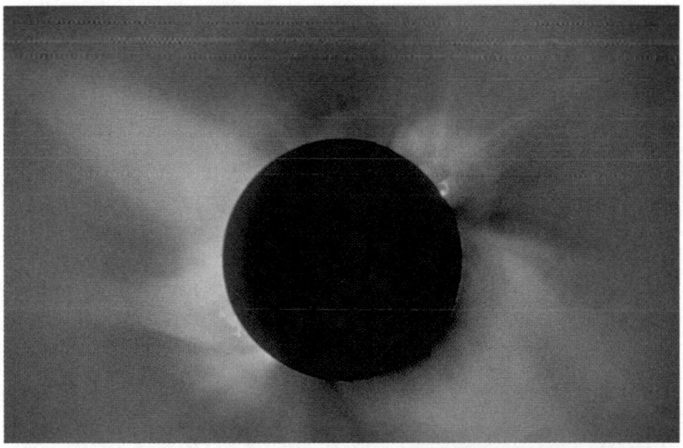

Figure 9.22 Solar Eclipse Photograph of the solar corona during the eclipse that occurred in July 1991 near the peak of the sunspot cycle. At these times, the corona is much less regular and much more extended than at sunspot minimum (compare Figure 9.11). Astronomers believe that coronal heating is caused by surface activity on the Sun. The changing shape and size of the corona are the direct result of variations in prominence and flare activity over the course of the solar cycle. *(National Solar Observatory)*

9.5 The Heart of the Sun

By what process does the Sun shine, day after day, year after year, eon after eon? The answer to this question is central to all of astronomy. Without it, we can understand neither the evolution of stars and galaxies in the universe nor the existence of life on Earth.

Nuclear Fusion

5 Only one known energy-generation mechanism can conceivably account for the Sun's enormous energy output. That process is **nuclear fusion**—the combining of light nuclei into heavier ones. We can represent a typical fusion reaction symbolically as

$$\text{nucleus 1} + \text{nucleus 2} \rightarrow \text{nucleus 3} + \text{energy.}$$

For powering the Sun, the most important piece of this equation is the energy produced.

The essential point here is that, during a fusion reaction, the total mass *decreases*—the mass of nucleus 3 is less than the combined masses of nuclei 1 and 2. Where does this mass go? It is converted to energy in accordance with Einstein's famous equation $E = mc^2$. In words, this equation says that, to determine the amount of energy corresponding to a given mass, simply multiply the mass by the square of the speed of light (c in the equation). The speed of light is so large that even a small amount of mass translates into an enormous amount of energy.

The production of energy by a nuclear fusion reaction is an example of the **law of conservation of mass and energy**, which states that the sum of mass and energy (properly converted to the same units, using Einstein's equation) must always remain constant in any physical process. There are no known exceptions.

The Proton–Proton Chain

The lightest and most common element in the universe is hydrogen, and it is the fusion of hydrogen nuclei (protons) into nuclei of the next lightest element (helium) that powers the Sun. Recall from Chapter 2 that protons, having the same (positive) electrical charge, repel one another, and this repulsion increases rapidly as the distance between the protons decreases. ∞ (Sec. 2.2) As a result, in order to get close enough together to come within the range of the strong nuclear force (see *More Precisely 9-1*) and fuse, two protons must be slammed together

MORE PRECISELY 9-1

The Strong and Weak Nuclear Forces

As far as we can tell, the behavior of all matter in the universe, from elementary particles to clusters of galaxies, is ruled by just four (or fewer) basic forces, which are fundamental to everything in the universe. We have already encountered two of them in this text—the *gravitational force*, which binds planets, stars and solar systems together, and the much stronger *electromagnetic force*, which governs the properties of atoms and molecules, as well as magnetic phenomena on planets and stars. ∞ (Secs. 1.4, 2.2)

Both of these forces are governed by inverse-square laws and are long range. In contrast, the other two fundamental forces, known as the *weak nuclear force* and the *strong nuclear force* are of extremely short range, comparable to the size of an atomic nucleus—about 10^{-15} m and 10^{-14} m, respectively.

The weak force governs the emission of radiation during some radioactive decays, as well as interactions between neutrinos and other particles. In fact, the electromagnetic and weak force are not really distinct, but are actually just different aspects of a single "electroweak" force.

The strong force holds atomic nuclei together. It binds particles with enormous strength, but only when they approach one another very closely. Two protons must come within about 10^{-15} m of one another before the attractive strong force can overcome their electromagnetic repulsion.

Very loosely speaking, we can say that the strong force is about 100 times stronger than electromagnetism, 100,000 times stronger than the weak force, and 10^{39} times stronger than gravity. However, not all particles are subject to all types of force. All particles interact through gravity because all have mass (or equivalently, energy). However, only charged particles interact electromagnetically. Protons and neutrons are affected by the strong force, but electrons are not. Under the right circumstances, the weak force can affect any type of subatomic particle, regardless of its charge.

---

at speeds in excess of a few hundred kilometers per second. For this to occur, the temperature of the solar gas must be at least 10 million K. ∞ (*More Precisely 5-2*) The radius at which the temperature drops below this value defines the outer edge of the solar core (Figure 9.5).

The basic nuclear reaction that powers the Sun (and the vast majority of all stars) is sketched in Figure 9.23. In fact it is not a single reaction, but rather a sequence of reactions called the **proton–proton chain**. Setting aside the temporary intermediate nuclei produced, the net effect is that four hydrogen nuclei (protons) combine to create one nucleus of the next lightest element, helium-4, creating two new lightweight particles called *neutrinos*, and releasing energy in the form of gamma-ray radiation:

$$4 \text{ protons} \rightarrow \text{helium-4} + 2 \text{ neutrinos} + \text{energy}.$$

Several alternative reaction sequences exist that produce the same end result. However, the sequence shown in Figure 9.23 is the simplest, and is responsible for almost 90 percent of the Sun's luminosity. Notice how, at each stage of the chain, more complex nuclei are created from simpler ones and energy is released in the process.

The **neutrino** is a chargeless and virtually massless elementary particle. (The name derives from the Italian for "little neutral one.") Neutrinos move at (or nearly at) the speed of light and interact with hardly anything. They can penetrate, without stopping, several light-years of lead. Yet despite their elusiveness, they can be detected with carefully constructed instruments. In the final section we discuss some rudimentary neutrino "telescopes" and the important contribution they have made to solar astronomy.

Energy Generated by the Proton–Proton Chain

Gargantuan quantities of protons are fused into helium by the proton–proton chain in the core of the Sun every second. Let's calculate the energy produced in the fusion process and compare it with the energy needed to account for the Sun's luminosity. Careful laboratory experiments have determined the masses of all the particles involved in the conversion of four protons to a helium-4 nucleus: The total mass

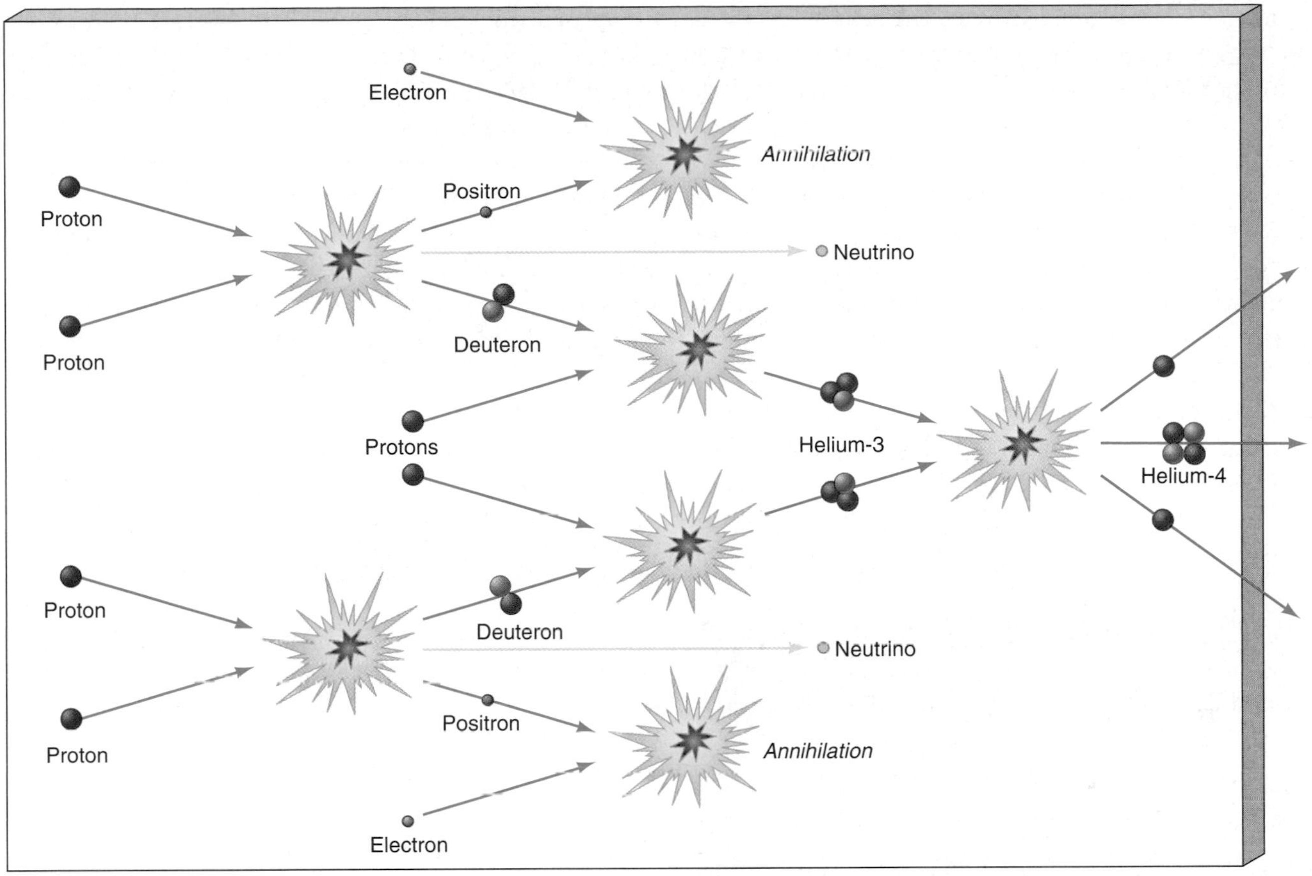

Figure 9.23 **Solar Fusion** In the proton–proton chain, a total of six protons (and two electrons) are converted to two protons, one helium-4 nucleus, and two neutrinos. The two leftover protons are available as fuel for new proton–proton reactions, so the net effect is that four protons are fused to form one helium-4 nucleus. Particles created in the intermediate reactions are deuterons (nuclei of "heavy hydrogen," having both a proton and a neutron in the nucleus), positrons (anti-electrons, which combine with free electrons to produce gamma rays), and helium-3 (a form of helium having only one neutron in the nucleus instead of the usual two). Energy, in the form of gamma rays, is produced at each stage.

of the protons is 6.6943×10^{-27} kg, the mass of the helium-4 nucleus is 6.6466×10^{-27} kg, and the neutrinos are virtually massless. The difference between the total mass of the protons and the mass of the helium nucleus, 0.0477×10^{-27} kg, is not great, but it is easily measurable.

Multiplying the vanished mass by the square of the speed of light yields 4.3×10^{-12} J. This is the energy produced in the form of radiation when 6.7×10^{-27} kg (the rounded-off mass of the four protons) of hydrogen fuses to helium. It follows that fusion of 1 kg of hydrogen generates $4.3 \times 10^{-12}/6.7 \times 10^{-27} = 6.4 \times 10^{14}$ J. Thus, to fuel the Sun's present energy output, hydrogen must be fused to helium in the core at a rate of 600 million tons per second—a lot of mass, but only a tiny fraction of the total amount available. The Sun will be able to sustain this rate of burning for about another five billion years.

The Sun's nuclear energy is produced in the core in the form of gamma rays. However, as it passes through the cooler layers of the solar interior and photons are absorbed and reemitted, the radiation's black-body spectrum steadily shifts toward lower and lower temperatures. The energy eventually leaves the photosphere mainly in the form of infrared and visible radiation. A comparable amount of energy is carried off by the neutrinos, which escape unhindered into space at almost the speed of light.

Observations of Solar Neutrinos

6 Because the gamma-ray energy created in the proton–proton chain is transformed into visible and infrared radiation by the time it emerges from the Sun, astronomers have no direct electromagnetic evidence of the core nuclear reactions. Instead, the *neutrinos* created in the proton–proton chain are our best bet for learning about conditions in the solar core. They travel cleanly out of the Sun, interacting with virtually nothing, and escape into space a few seconds after being created.

Of course, the fact that they can pass through the entire Sun without interacting also makes neutrinos difficult to detect on Earth! Nevertheless, with knowledge of neutrino physics it is possible to construct neutrino detectors. Over the past three decades, four experiments (one of them shown in Figure 9.24) have been designed to detect solar neutrinos reaching Earth's surface; the four are of widely different design and are sensitive to neutrinos of very different energies. They disagree somewhat on the details of their findings, but they seem to agree on one very important point: The number of solar neutrinos reaching Earth is substantially less (by 30 to 50 percent) than the prediction of the Standard Solar Model. This discrepancy is known as the **solar neutrino problem**.

Figure 9.24 Neutrino Telescope This swimming-pool-sized detector is a "neutrino telescope" of sorts, buried beneath a mountain near Tokyo, Japan. Called Super Kamiokande, it is filled (in operation) with 50,000 tons of purified water, and contains 13,000 light detectors (some shown here being inspected by technicians) to sense the telltale signature of a neutrino passing through the apparatus. *(ICRR Institute for Cosmic Ray Research, The University of Tokyo)*

Most scientists are confident that the experimental results are to be trusted, but they also regard it as very unlikely that conditions in the solar core could differ from the Standard Solar Model by a wide enough margin to account for the neutrino deficit. The properties of the Sun are just too well known. There is no plausible way to reduce the number of neutrinos produced by the core while still remaining consistent with the Sun's observed radius, temperature, and luminosity. In addition, interior measurements based on helioseismology (Section 9.2) seem to rule out a central temperature much different from the 15 million K of the Standard Model.

How then do we explain the clear disagreement between theory and observation? The most likely answer involves the properties of the neutrinos themselves. If neutrinos do have a minute amount of mass, it may be possible for them to change their properties, even to transform into other particles, during their eight-minute flight from the solar core to Earth, through a process known as **neutrino oscillations**. Neutrinos are produced in the Sun at the rate required by the Standard Solar Model, but some of them turn into something else (in particle-physics jargon, they are said to "oscillate" into other particles) on their way to Earth and so go undetected.

In June 1998, the Japanese group operating the detector shown in Figure 9.24 reported what many experts regard as compelling experimental evidence of neutrino oscillations (and hence of nonzero neutrino masses). However, the observed oscillations do *not* involve neutrinos of the type produced in the Sun, so they do not by themselves explain the solar neutrino deficit. Several groups have attempted to detect oscillations in the "solar-type" neutrinos produced by nuclear reactors here on Earth, so far without success.

✓ Concept Check

- Why does the fact that we see sunlight imply that the Sun's mass is slowly decreasing?

Chapter Review

Summary

Our Sun is a **star** (p. 232), a glowing ball of gas held together by its own gravity and powered by nuclear fusion at its center. The solar **photosphere** (p. 233) is the region at the Sun's surface from which virtually all the visible light is emitted. Above the photosphere lies the **chromosphere** (p. 233), which is the Sun's lower atmosphere. In the **transition zone** (p. 233) above the chromosphere, the temperature increases from a few thousand to around a million kelvins. Above the transition zone is the Sun's thin, hot upper atmosphere, the solar **corona** (p. 233). The main interior regions of the Sun are the **core** (p. 233), where nuclear reactions generate energy; the **radiation zone** (p. 233), where the energy travels outward in the form of electromagnetic radiation; and the **convection zone** (p. 233), where the Sun's matter is in constant convective motion.

The amount of solar energy reaching a one square meter area at the top of Earth's atmosphere each second is a quantity known as the **solar constant** (p. 233). The Sun's **luminosity** (p. 233) is the total amount of energy radiated from the solar surface per second.

Much of our knowledge of the solar interior comes from mathematical models. The model that best fits the observed properties of the Sun is the **Standard Solar Model** (p. 234). **Helioseismology** (p. 234)—the study of vibrations of the solar surface caused by pressure waves in the interior—provides further insight into the Sun's structure.

The effect of the solar convection zone can be seen on the surface in the form of **granulation** (p. 236) of the photosphere. As hotter (and therefore brighter) gas rises and cooler (dimmer) gas sinks, a characteristic mottled appearance results. Lower levels in the convection zone also leave their mark on the photosphere in the form of larger transient patterns called **supergranulation** (p. 236).

At a distance of about 15 solar radii, the gas in the corona is hot enough to escape the Sun's gravity, and the corona begins to flow outward as the solar wind. Most of the solar wind flows from low-density regions of the corona called **coronal holes** (p. 239).

Solar activity is generally associated with disturbances in the Sun's magnetic field. **Sunspots** (p. 240) are Earth-sized regions on the solar surface that are a little cooler than the surrounding photosphere. They are regions of intense magnetism. The **polarity** (p. 241) of a sunspot indicates whether its magnetic field is directed away from or into the solar interior. Both the numbers and locations of sunspots vary in a roughly 11-year **sunspot cycle** (p. 242). The overall direction of the solar magnetic field reverses from one sunspot cycle to the next. The 22-year cycle that results when the direction of the field is taken into account is called the **solar cycle** (p. 242).

Solar activity tends to be concentrated in **active regions** (p. 243) associated with sunspot groups. **Prominences** (p. 244) are loop- or sheetlike structures produced when hot gas ejected by activity on the solar surface interacts with the Sun's magnetic field. **Flares** (p. 244) are violent surface explosions that blast particles and radiation into interplanetary space.

The Sun generates energy by converting hydrogen to helium in its core by the process of **nuclear fusion** (p. 245). When four protons are converted to a helium nucleus in the **proton–proton chain** (p. 246), some mass is lost. The **law of conservation of mass and energy** (p. 245) requires that this mass appear as energy, eventually resulting in the light we see.

Neutrinos (p. 246) are massless (or perhaps only nearly massless) particles that are produced in the proton–proton chain and escape from the Sun. Despite their elusiveness, it is possible to detect a small fraction of the neutrinos streaming from the Sun. Observations lead to the **solar neutrino problem** (p. 248)—substantially fewer neutrinos are observed than are predicted by theory. A leading explanation is that **neutrino oscillations** (p. 248) convert some neutrinos to other (undetected) particles en route from the Sun to Earth.

Review and Discussion

1. Name and briefly describe the main regions of the Sun.
2. How massive is the Sun, compared with Earth?
3. How hot is the solar surface? The solar core?
4. How do scientists construct models of the Sun?
5. Describe how energy generated at the center of the Sun reaches Earth.
6. Why does the Sun appear to have a sharp edge?
7. What evidence do we have for solar convection?
8. What is the solar wind?
9. What is the cause of sunspots, flares, and prominences?
10. What is the Maunder minimum?
11. What fuels the Sun's enormous energy output?
12. What is the law of conservation of mass and energy? How is it relevant to nuclear fusion in the Sun?
13. What are the ingredients and the end result of the proton–proton chain? Why is energy released in the process?

14. Why are scientists trying so hard to detect solar neutrinos?

15. What would we observe on Earth if the Sun's internal energy source suddenly shut off? Would the Sun darken instantaneously? If not, how long do you think it might take—minutes, days, years, millions of years—for the Sun's light to begin to fade? Repeat the question for solar neutrinos.

True or False?

____ **1.** The Sun is a fairly typical star.

____ **2.** The average density of the Sun is significantly greater than the density of Earth.

____ **3.** The Sun's diameter is about 100 times that of Earth.

____ **4.** The Sun's mass is comparable to the mass of Jupiter.

____ **5.** Convection involves cool gas rising to the solar surface and hot gas sinking into the interior.

____ **6.** Absorption lines in the solar spectrum are produced mainly in the corona.

____ **7.** There are as many absorption lines in the solar spectrum as there are elements present in the Sun.

____ **8.** The faintness of the chromosphere is a result of its low density and temperature.

____ **9.** The temperature of the solar corona decreases with increasing radius.

____ **10.** Sunspots are regions of intense magnetic fields.

____ **11.** Coronal holes are low-density regions of the upper solar atmosphere.

____ **12.** Prominences are violent nuclear explosions on the surface of the Sun.

____ **13.** Neutrinos are neutrons traveling close to the speed of light.

____ **14.** Nuclear fusion releases energy because the total mass of the nuclei involved increases.

____ **15.** The nuclear reactions that power the Sun create energy in the form of gamma rays.

Fill in the Blank

1. The part of the Sun we see is called the ________.

2. The three main regions of the solar atmosphere are the ________, the ________, and the ________.

3. Below the solar surface, in order of increasing depth, lie the ________ zone, the ________ zone, and the ________.

4. The ________ seen on the surface of the Sun is evidence of convective cells.

5. The Sun appears to have a well-defined edge because the thickness of the ________ is only 0.1 percent of the solar radius.

6. ________ is the most abundant element in the Sun.

7. ________ is the second most abundant element in the Sun.

8. The two most abundant elements in the Sun make up about ________ percent of its composition.

9. The ________ is formed as the hot outer regions of the corona expand into interplanetary space.

10. Sunspots appear dark because they are ________ than the surrounding gas of the photosphere.

11. The sunspot cycle is ________ years long; the solar cycle is ________ as long.

12. A ________ is a violent explosive event in a solar active region.

13. The Sun's luminosity is produced in the solar ________ (give the region).

14. The **net** result of the proton–proton chain is that ________ protons are fused into a nucleus of ________, two ________ are emitted, and energy is released in the form of ________.

15. The solar neutrino problem is the fact that astronomers observe too ________ neutrinos coming from the Sun.

Problems

1. Use the reasoning presented in Section 9.1 to calculate the value of the "solar constant" on Jupiter.

2. The largest-amplitude solar pressure waves have periods of about five minutes and move at the speed of sound in the outer layers of the Sun, about 10 km/s. How far does the wave move during one wave period? Also compare the wave period with the orbital period of an object moving just above the solar photosphere. ∞ *(More Precisely 1-2)*

3. Given that the solar spectrum corresponds to a temperature of 5800 K and peaks at a wavelength of 500 nm, use Wien's law to determine the wavelength corresponding to the peak of the black-body curve (a) in the core of the Sun, where the temperature is 10^7 K, (b) in the solar convection zone (10^5 K), and (c) just below the solar photosphere (10^4 K). ∞ (More Precisely 2-2) What form (visible, infrared, X-ray, etc.) does the radiation take in each case?

4. If convected solar material moves at 1 km/s, how long does it take to flow across the 1000-km expanse of a typical granule? Compare this length of time with the roughly 10-minute lifetimes observed for most solar granules.

5. Use Stefan's law to calculate how much less energy is emitted per unit area of a 4500-K sunspot than is emitted per unit area of the surrounding 5800-K photosphere. ∞ (More Precisely 2-2) Give your answer as a percentage.

6. The solar wind carries mass away from the Sun at a rate of about 900,000 ton/s (where 1 ton = 1000 kg). Compare this rate with the rate at which the Sun loses mass in the form of radiation.

7. The Sun's differential rotation is responsible for wrapping the solar magnetic field around the Sun (Figure 9.17). Using the figures presented in the Table 9.1, estimate how long it takes for material at the solar equator to "lap" the material near the poles—that is, to complete one extra trip around the Sun's rotation axis.

8. How long does it take the Sun to convert one Earth mass of hydrogen to helium?

9. Assuming that (1) the solar luminosity has been constant since the Sun formed, and (2) the Sun was initially of uniform composition throughout, as described by Table 9.2, estimate how long it would take the Sun to convert all of its original hydrogen into helium.

10. The entire reaction sequence shown in Figure 9.23 generates 4.3×10^{-12} joules of electromagnetic energy and releases two neutrinos. Assume that all of the Sun's energy comes from this sequence and that neutrino oscillations transform half of the neutrinos produced into other particles by the time they travel one A.U. from the Sun. Estimate the total number of solar neutrinos that pass through Earth each second.

Projects

The projects given here all require a special solar filter. Such filters are easily purchased from various sources. **NEVER LOOK DIRECTLY AT THE SUN WITHOUT A FILTER!**

1. An appropriately filtered telescope will easily show you sunspots, although the numbers will vary considerably during the sunspot cycle. Count the number of sunspots you see on the Sun's surface. Notice that sunspots often come in pairs or groups. Come back and look again a few days later and you'll see that (a) the Sun's rotation has caused the spots to move and (b) the appearance of the spots has changed. If you see a sufficiently large sunspot (or, more likely, sunspot group), continue to watch it as the Sun rotates. It will be out of view for about two weeks. Can you determine the rotation period of the Sun from these observations?

2. Solar granulation is not too hard to see. The atmosphere of Earth is most stable in the morning hours. Observe the Sun on a cool morning, one or two hours after it has risen, using a properly filtered telescope for safety. Use high magnification and look initially at the middle of the Sun's disk. Can you see changes in the granulation pattern? They are there but are not always easy to see.

3. View some solar prominences and flares through a properly filtered telescope. Hydrogen-alpha (Hα) filters, which transmit only a narrow band of radiation around the characteristic red spectral line of hydrogen, are commercially available for small telescopes. Using such a filter, you can often see prominences and flares even during times of sunspot minimum. Because you are actually viewing the chromosphere rather than the photosphere, the Sun looks quite different and can be very impressive in appearance.

10 MEASURING THE STARS

Giants, Dwarfs, and the Main Sequence

LEARNING GOALS

Studying this chapter will enable you to:

1. Explain how stellar distances are determined.
2. Distinguish between luminosity and apparent brightness, and explain how stellar luminosity is determined.
3. Explain the usefulness of classifying stars according to their colors, surface temperatures, and spectral characteristics.
4. Discuss how physical laws are used to estimate stellar sizes.
5. Describe how an Hertzsprung–Russell diagram is constructed and used to identify stellar properties.
6. Explain how stellar masses are measured and how mass is related to other stellar properties.

(Opposite page) Even in our own local region of the cosmos, the number of stars is virtually beyond our ability to count. Relatively few of them have actually been studied in detail. Yet we have learned an enormous amount about stars in general and the range of properties they exhibit—their masses, their temperatures, their luminosities, even their ages and (as we shall see in subsequent chapters) their destinies. *(J. Sanford/Astrostock-Sanford)*

We have studied Earth, the Moon, the solar system, and the Sun. As we continue our inventory of the contents of the universe, we now move away from our local environment and out into the depths of space. In this chapter we combine our knowledge of geometry, gravity, electromagnetism, and atomic structure to uncover the nature of stars, both the familiar members of the constellations we see each night and the myriad more distant stars too faint to be detected by our unaided eyes. Rather than studying the individual peculiarities of these faraway bodies, however, we instead concentrate on determining the physical and chemical properties they share. There is order in the legions of stars scattered across the sky. Just as we compared planets in our study of the solar system, we now compare and catalog the stars to further our understanding of the Galaxy and the universe we inhabit.

10.1 The Solar Neighborhood

As with the planets, knowing the *distances* to the stars is essential to understanding many of their properties. We therefore begin our study of stellar astronomy by reviewing the use of simple geometry in determining the distances to our neighbors.

Stellar Parallax

1 The distances to the nearest stars can be measured using *parallax*—the apparent shift of an object relative to some distant background as the observer's point of view changes. ∞ (Sec. P.5) However, even the closest stars are so far away that no baseline on Earth is adequate to measure their parallaxes. But, by comparing observations made of a star *at different times of the year*, as shown in Figure 10.1 (compare Figure P.22), we can effectively extend the baseline to the diameter of Earth's orbit around the Sun, two A.U. The parallax is determined by comparing photographs made from the two ends of the baseline. As indicated on the figure, a star's parallactic angle—or, more commonly, just its "parallax"—is conventionally defined to be *half* of its apparent shift relative to the background as we move from one side of Earth's orbit to the other.

Because stellar parallaxes are so small, astronomers find it convenient to measure them in arc seconds rather than in degrees. If we ask at what distance from the Sun a star must lie in order for its observed parallax to be exactly 1″, we get an answer of 206,265 A.U., or 3.1×10^{16} m. ∞ (*More Precisely P-2*) Astronomers call this distance one **parsec** (1 pc), from "*par*allax in arc *sec*onds." Because parallax decreases as distance increases, we can relate a star's parallax to its distance from the Sun by the following simple formula:

$$\text{distance (in parsecs)} = \frac{1}{\text{parallax (in arc seconds)}}.$$

Thus, a star with a measured parallax of 1″ lies at a distance of 1 pc from the Sun. The parsec is defined so as to make the conversion between parallactic angle and distance easy. An object with a parallax of 0.5″ lies at a distance of $1/0.5 = 2$ pc, an object with a parallax of 0.1″ lies at $1/0.1 = 10$ pc, and so on. One parsec is approximately equal to 3.3 light-years.

Our Nearest Neighbors

The closest star to the Sun is called Proxima Centauri. It is a member of a triple-star system (three stars orbiting one another, bound together by gravity) known as the Alpha Centauri complex. Proxima Centauri displays the largest known stellar parallax, 0.76″, which means that it is about $1/0.76 = 1.3$ pc away—about 270,000 A.U., or 4.3 light years. That's the *nearest* star to Earth—at almost 300,000 times the Earth–Sun distance! This is a fairly typical interstellar distance in the Milky Way Galaxy.

Vast distances can sometimes be grasped by means of analogies. Imagine Earth as a grain of sand orbiting a golfball-sized Sun at a distance of one meter. The nearest star, also a golfball-sized object, is then 270 *kilo*meters away. Except for the planets in our solar system, ranging in size from grains of sand to small marbles within 50 m of the "Sun," nothing of consequence exists in the 270 km separating the Sun and the other star.

The next nearest neighbor to the Sun beyond the Alpha Centauri system is called Barnard's Star. Its parallax is 0.55″, so it lies at a distance of 1.8 pc, or 6.0 light-years—370 km in our model. All told, fewer

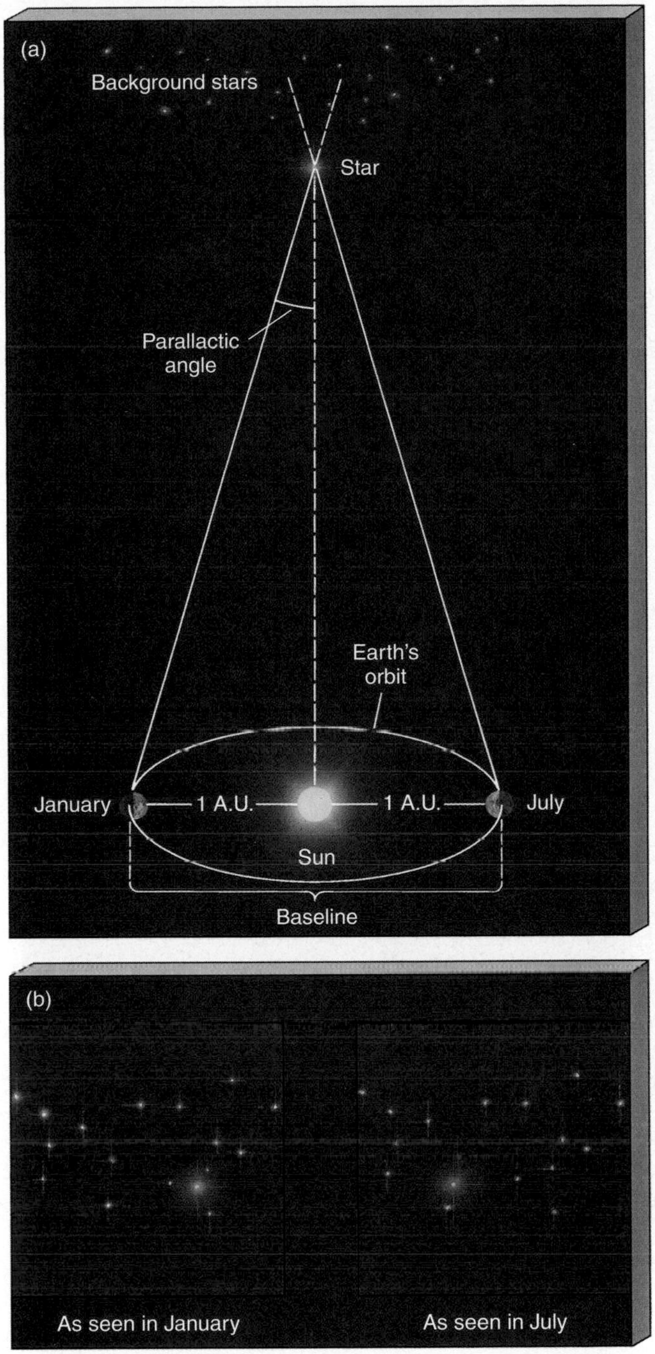

Figure 10.1 **Stellar Parallax** (a) The geometry of stellar parallax. For observations made six months apart, the baseline is twice the Earth–Sun distance, or two A.U. Compare Figure P.22, which shows exactly the same geometry, but on a much smaller scale. (b) The parallactic shift is usually measured photographically. Images of the same region of the sky made at different times of the year are used to determine a star's apparent movement relative to background stars.

than 100 stars lie within 5 pc (1000 km in our model) of the Sun. Such is the void of interstellar space. Figure 10.2 is a map of our nearest galactic neighbors—the 30 or so stars lying within 4 pc of Earth.

Ground-based images of stars are generally smeared out into a seeing disk of radius 1″ or so by turbulence in Earth's atmosphere. ∞ (Sec. 3.3) This blurring effectively limits ground-based measurements to stellar parallaxes exceeding about 0.03″, corresponding to stars within about 30 pc (100 light-years) of Earth. Several thousand stars lie within this range, most of them much dimmer than the Sun and invisible to the naked eye. In the 1990s, data from the European *Hipparcos* satellite extended this range to well over 100 pc, encompassing nearly a million stars. Even so, almost all of the stars in our Galaxy are far more distant.

Stellar Motion

In addition to the apparent motion caused by parallax, stars have real spacial motion, too. A star's *radial* velocity—along the line of sight—can be measured using the Doppler effect. ∞ (*More Precisely 2-3*) For many nearby stars, the *transverse* velocity—perpendicular to our line of sight—can also be determined by careful monitoring of the star's position on the sky.

Figure 10.3 shows two photographs of the sky around Barnard's Star, made on the same day of the year, but 22 years apart. Note that the star (marked by the arrow) has moved during the 22-year interval shown. Because Earth was at the same point in its orbit when the photographs were taken, the observed displacement is *not* the result of parallax. Instead, it indicates *real* space motion of Barnard's Star relative to the Sun. This annual movement of a star across the sky, as seen from Earth and corrected for parallax, is called **proper motion**. Barnard's Star moved 227″ in 22 years. Its proper motion is therefore 227″/22 years, or 10.3″/yr.

A star's transverse velocity is easily calculated once its proper motion and its distance are known. At the distance of Barnard's Star (1.8 pc), an angle of 10.3″ corresponds to a distance of 0.00009 pc, or about 2.8 billion km. Barnard's Star takes a year to travel this distance, so its transverse velocity is 2.8 billion km/3.2 × 10^7 s, or 88 km/s. Even though stars' transverse velocities are often quite large—tens or even hundreds of kilometers per second—their great distances from the Sun means that it usually takes many years for us to discern their movement across the sky. In fact, Barnard's Star has the largest known proper motion of any star. Only a few hundred stars have proper motions greater than 1″/yr.

✓ Concept Check

- Why can't astronomers use simultaneous observations from different parts of Earth's surface to determine stellar distances?

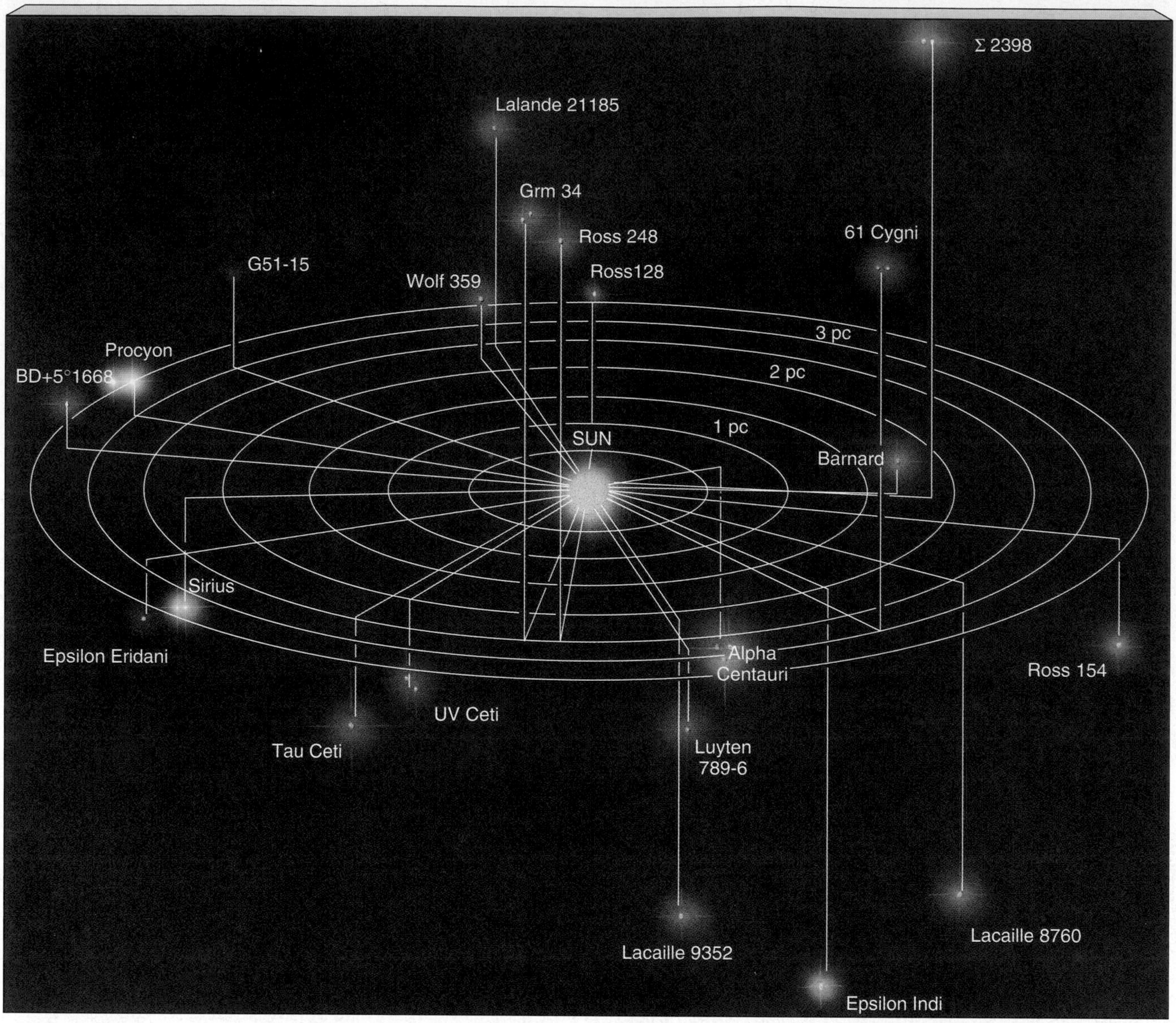

Figure 10.2 Sun's Neighborhood A plot of the 30 closest stars to the Sun, projected so as to reveal their three-dimensional relationships. Notice that many are members of multiple-star systems. All lie within 4 pc (about 13 light-years) of Earth. The gridlines represent distances in the Galactic plane.

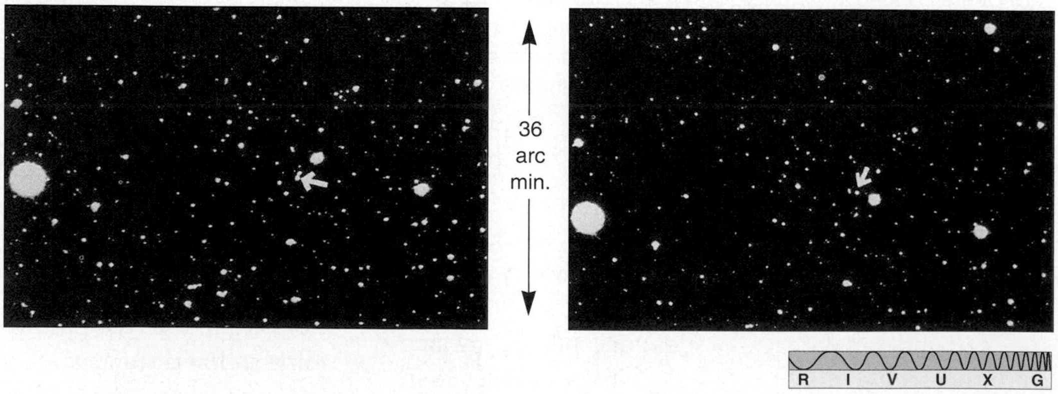

Figure 10.3 Real Space Motion Comparison of two photographs taken 22 years apart shows evidence of real space motion for Barnard's Star (denoted by an arrow). *(Harvard College Observatory)*

10.2 Luminosity and Apparent Brightness

2 Luminosity is an *intrinsic* property of a star—it does not depend in any way on the location or motion of the observer. It is sometimes referred to as the star's *absolute brightness*. However, when we look at a star, we see not its luminosity, but rather its **apparent brightness**—the amount of energy striking unit area of some light-sensitive surface or device (such as a human eye or a CCD chip) per unit time. In this section, we discuss how these important quantities are related to one another.

Another Inverse-Square Law

Figure 10.4 shows light leaving a star and traveling through space. Moving outward, the radiation passes through imaginary spheres of increasing radius surrounding the source. The amount of radiation leaving the star per unit time—the star's luminosity—is constant, so the farther the light travels from the source, the less energy passes through each unit of area. Think of the energy as being spread out over an ever-larger area, and therefore spread more thinly, or "diluted," as it expands into space. Because the area of a sphere grows as the square of the radius, the energy per unit area—the star's apparent brightness—is inversely proportional to the square of the distance from the star. Doubling the distance from a star makes it appear 2^2, or four, times dimmer. Tripling the distance reduces the apparent brightness by a factor of 3^2, or nine, and so on.

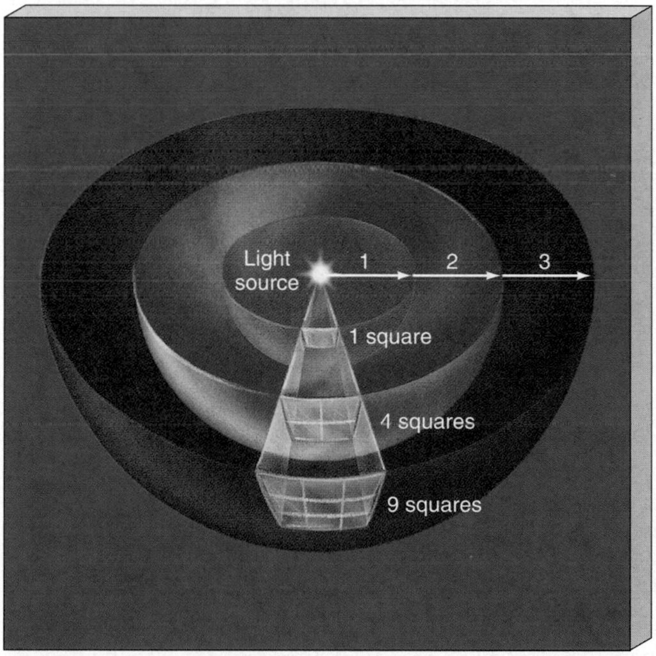

Figure 10.4 Inverse-square Law As it moves away from a source such as a star, radiation is steadily diluted while spreading over progressively larger surface areas (depicted here as sections of spherical shells). Thus, the amount of radiation received by a detector (the source's apparent brightness) varies inversely as the square of its distance from the source. Animation

Of course, the star's luminosity also affects its apparent brightness. Doubling the luminosity doubles the energy crossing any spherical shell surrounding the star and hence doubles the apparent brightness. The apparent brightness of a star is therefore directly proportional to the star's luminosity and inversely proportional to the square of its distance:

$$\text{apparent brightness} \propto \frac{\text{luminosity}}{\text{distance}^2}.$$

Thus, two identical stars can have the same apparent brightness if (and only if) they lie at the same distance from Earth. However, as illustrated in Figure 10.5, two non-identical stars can also have the same apparent brightness if the more luminous one lies farther away. A bright star (that is, one having large apparent brightness) is a powerful emitter of radiation (high luminosity), is near Earth, or both. A faint star (small apparent brightness) is a weak emitter (low luminosity), is far from Earth, or both.

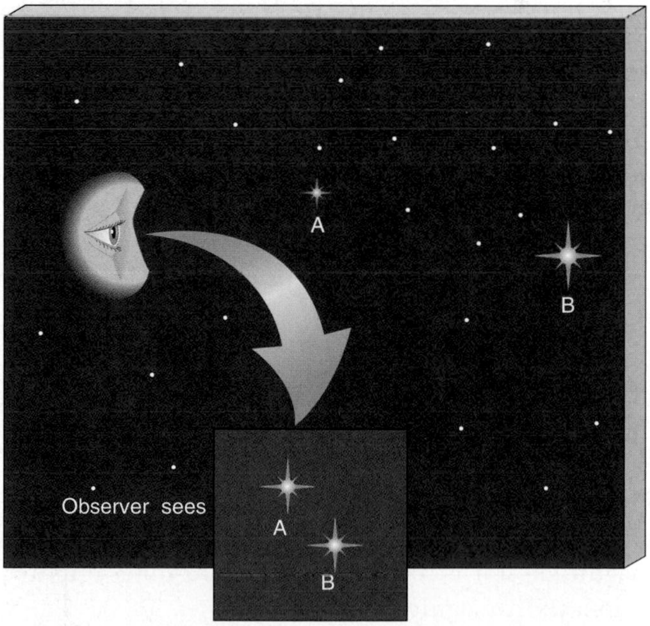

Figure 10.5 Luminosity Two stars A and B of different luminosities can appear equally bright to an observer on Earth if the brighter star B is more distant than the fainter star A.

Determining a star's luminosity is a twofold task. First, the astronomer must determine the star's apparent brightness by measuring the amount of energy detected through a telescope in a given amount of time. Second, the star's distance must be measured—by parallax for nearby stars and by other means (to be discussed later) for more distant stars. The luminosity can then be found using the inverse-square law. Note that this is basically the same reasoning we used earlier in our discussion of how astronomers measure the solar luminosity (in our new terminology, the solar constant is just the apparent brightness of the Sun). ∞ (Sec. 9.1)

The Magnitude Scale

Instead of measuring apparent brightness in SI units (for example, watts per square meter W/m^2, the unit in which we expressed the solar constant in Section 9.1), optical astronomers find it more convenient to work in terms of a construct called the **magnitude scale**. This scale dates back to the second century B.C., when the Greek astronomer Hipparchus ranked the naked-eye stars into six groups. The brightest stars were categorized as first magnitude. The next brightest stars were labeled second magnitude, and so on, down to the faintest stars visible to the naked eye, which were classified as sixth magnitude. The range one (brightest) through six (faintest) spanned all the stars known to the ancients. Notice that a *large* magnitude means a *faint* star.

When astronomers began using telescopes with sophisticated detectors to measure the light received from stars, they quickly discovered two important facts about the magnitude scale. First, the one through six magnitude range defined by Hipparchus spans about a factor of 100 in apparent brightness—a first-magnitude star is approximately 100 times brighter than a sixth-magnitude star. Second, the characteristics of the human eye are such that a change of one magnitude corresponds to a *factor* of about 2.5 in apparent brightness. In other words, to the human eye a first-magnitude star is roughly 2.5 times brighter than a second-magnitude star, which is roughly 2.5 times brighter than a third-magnitude star, and so on. (By combining factors of 2.5, we confirm that a first-magnitude star is indeed $(2.5)^5 \approx 100$ times brighter than a sixth-magnitude star.)

In the modern version of the magnitude scale, astronomers *define* a change of five in the magnitude of an object to correspond to *exactly* a factor of 100 in apparent brightness. Because we are really talking about apparent (rather than absolute) brightnesses, the numbers in Hipparchus's ranking system are now called **apparent magnitudes**. In addition, the scale is no longer limited to whole numbers, and magnitudes outside the original range 1–6 are allowed—very bright objects can have apparent magnitudes much less than 1, and very faint objects can have apparent magnitudes far greater than six. Figure 10.6 illustrates the apparent magnitudes of some astronomical objects, ranging from the Sun, at −26.8, to the faintest object detectable by the *Hubble* or *Keck* telescopes, at an apparent magnitude of +30—

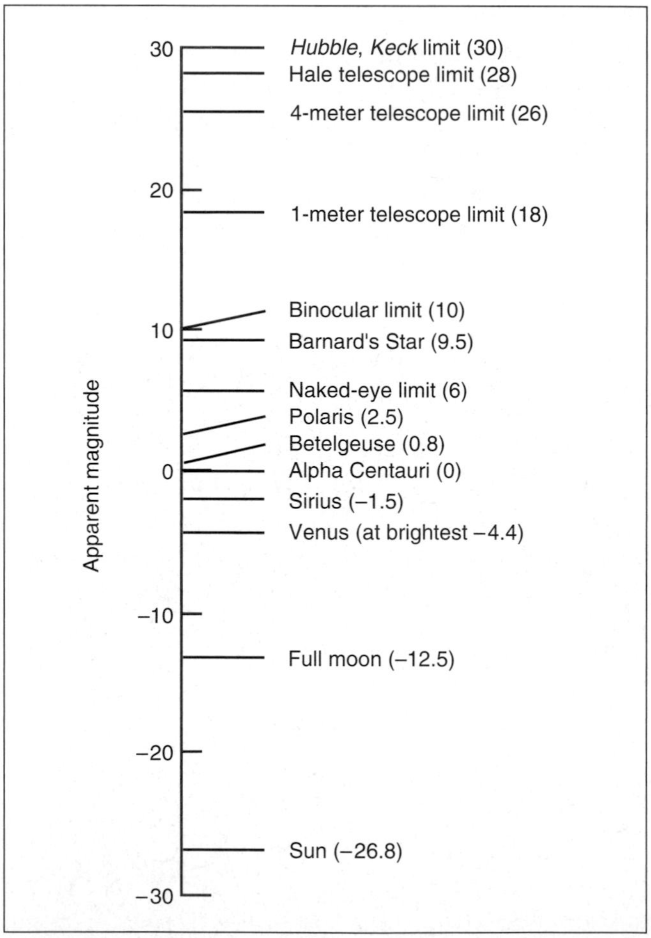

Figure 10.6 Apparent Magnitude This graph illustrates the apparent magnitudes of some astronomical objects. The original magnitude scale was defined so that the brightest stars in the night sky had magnitude one while the faintest stars visible to the naked eye had magnitude six. It has since been extended to cover much brighter and much fainter objects. An increase of one in apparent magnitude corresponds to a decrease in apparent brightness by a factor of approximately 2.5.

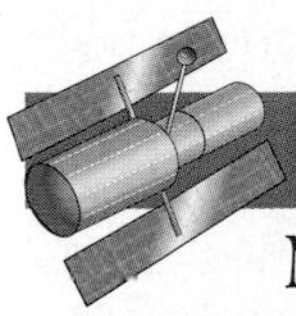

MORE PRECISELY 10-1

More on the Magnitude Scale

Let's restate our discussion of two important topics—stellar luminosity and the inverse-square law—in terms of magnitudes.

Absolute magnitude is equivalent to luminosity—an intrinsic property of a star. Given that the Sun's absolute magnitude is 4.8, we can construct a "conversion chart" (shown at right) relating these two quantities. Since an increase in brightness by a factor of 100 corresponds to a decrease in magnitude by five units, it follows that a star with luminosity 100 times that of the Sun has absolute magnitude $4.8 - 5 = -0.2$, while a 0.01 solar luminosity star has absolute magnitude $4.8 + 5 = 9.8$. We can fill in the gaps by noting that one magnitude corresponds to a factor of $100^{1/5} \approx 2.512$, two magnitudes to $100^{2/5} \approx 6.310$, and so on. A factor of 10 in brightness corresponds to 2.5 magnitudes. You can use this chart to convert between solar luminosities and absolute magnitudes in many of the figures in this and later chapters.

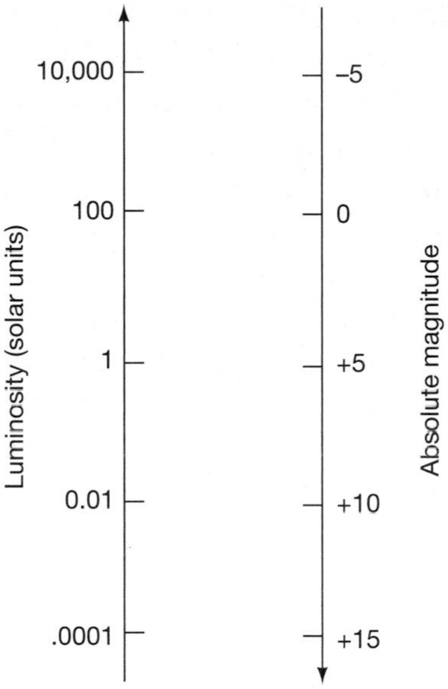

To cast the *inverse-square law* in these terms, recall that increasing the distance to a star by a factor of 10 decreases its apparent brightness by a factor of 100 (by the inverse-square law), and hence increases its apparent magnitude by five units. Increasing the distance by a factor of 100 increases the apparent magnitude by 10, and so on. Since absolute magnitude is simply apparent magnitude at a distance of 10 pc, we can write:

$$\text{apparent magnitude} - \text{absolute magnitude} = 5\log_{10}\left(\frac{\text{distance}}{10\text{ pc}}\right).$$

(The *logarithm* function is defined by the property that, if $a = \log_{10}(b)$, then $b = 10^a$.) For stars more than 10 pc from Earth, the apparent magnitude is greater than the absolute magnitude, while the reverse is true for stars closer than 10 pc. The numerical difference between the apparent and absolute magnitudes of an object is called its *distance modulus*. In terms of distance modulus (DM), we can rewrite the preceding equation as:

$$\text{distance} = 10\text{ pc} \times 10^{DM/5}.$$

This doesn't look much like the inverse-square law presented in the text! Nevertheless, it contains exactly the same information and is in widespread use in this form throughout astronomy.

about as faint as a firefly seen from a distance equal to Earth's diameter.

Apparent magnitude measures a star's apparent brightness when seen at the star's actual distance from the Sun. To compare intrinsic, or absolute, properties of stars, however, astronomers imagine looking at all stars from a standard distance of 10 pc. (There is no particular reason to use 10 pc—it is simply convenient.) A star's **absolute magnitude** is its apparent magnitude when viewed from a distance of 10 pc. Because distance is fixed in this definition, absolute magnitude is a measure of a star's absolute brightness, or luminosity. The Sun's absolute magnitude is 4.8. In other words, if the Sun were moved to a distance of 10 pc from Earth, it would be only a little brighter than the faintest stars visible in the night sky. As discussed further in *More Precisely 10-1*, the numerical difference between a star's absolute and apparent magnitudes is a measure of the distance to the star.

✓ Concept Check

- Two stars are observed to have the same apparent magnitude. Based on this information, what, if anything, can be said about their luminosities?

10.3 Stellar Temperatures

Color and the Blackbody Curve

3 Looking at the night sky, you can tell at a glance which stars are hot and which are cool. In Figure 10.7, which shows the constellation Orion as it appears through a small telescope, the colors of the cool red star Betelgeuse (α) and the hot blue star Rigel (β) are clearly evident. Astronomers can determine a star's surface temperature by measuring its apparent brightness (radiation intensity) at several frequencies, then matching the observations to the appropriate blackbody curve. ∞ (Sec. 2.4) In the case of the Sun, the theoretical curve that best fits the emission describes a 5800-K emitter. The same technique works for any star, regardless of its distance from Earth.

Because the basic shape of the blackbody curve is so well understood, astronomers can estimate a star's temperature using as few as *two* measurements at selected wavelengths. This is accomplished through the use of telescope filters that block out all radiation except that within specific wavelength ranges. For example, a B (blue) filter rejects all radiation except for a certain range of violet to blue light. Similarly, a V (visual) filter passes only radiation in the green to yellow range (the part of the spectrum to which human eyes happen to be particularly sensitive).

Figure 10.8 shows how these filters admit different amounts of light for objects of different temperatures. In curve (a), corresponding to a very hot 30,000-K emitter, considerably more radiation is received through the B filter than through the V filter. In curve (b) the temperature is 10,000 K and the B and V intensities are about the same. In the cool 3000-K curve (c) far more energy is received in the V range than in the B range. In each case, it is possible to reconstruct the entire blackbody curve on the basis of only those two measurements because no other blackbody curve can be drawn through both measured points.

To the extent that a star's spectrum is well approximated as a blackbody, measurements of the B and V intensities are enough to specify the star's blackbody curve and thus yield its surface temperature. Astronomers often refer to the ratio of a star's B to V in-

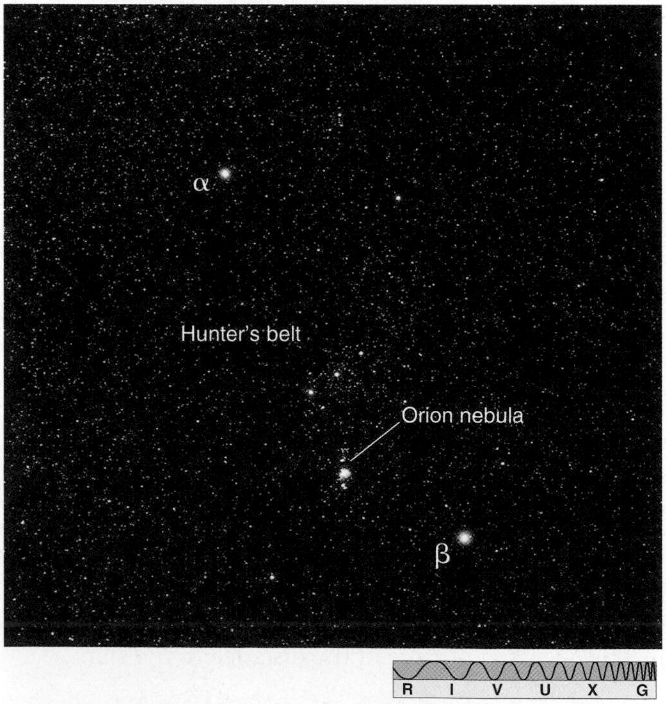

Figure 10.7 Orion A photograph of the constellation Orion as it appears through a small telescope or binoculars. The different colors of the member stars are easily distinguished. The bright red star at the upper left is Betelgeuse (α); the bright blue-white star at the lower right is Rigel (β). (Compare with Figure P.6.) The scale of this photograph is about 3° across. *(J Sanford/Astrostock-Sanford)*

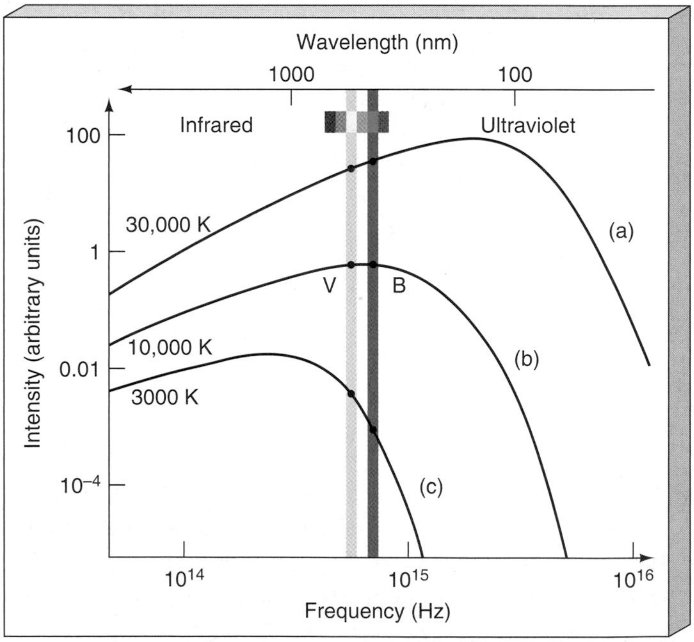

Figure 10.8 Blackbody Curves Blackbody curves for three temperatures, along with the locations of the B (blue) and V (visual) filters. Star (a) is very hot—30,000 K—so its B intensity is considerably greater than its V intensity. Star (b) has roughly equal B and V readings and so appears white. Its temperature is about 10,000 K. Star (c) is red. Its V intensity greatly exceeds the B value, and its temperature is 3000 K.

TABLE 10.1 Stellar Colors and Temperatures

SURFACE TEMPERATURE (K)	COLOR	FAMILIAR EXAMPLES
30,000	electric blue	Mintaka (δ Orionis)
20,000	blue	Rigel
10,000	white	Vega, Sirius
7000	yellow-white	Canopus
6000	yellow	Sun, Alpha Centauri
4000	orange	Arcturus, Aldebaran
3000	red	Betelgeuse, Barnard's Star

tensities (or, equivalently, the difference between the star's apparent magnitudes measured through the B and V filters) as the star's *color index* (or just its "color"). Table 10.1 lists surface temperatures and colors for a few well-known stars.

Stellar Spectra

3 Figure 10.9 compares the spectra of seven stars, arranged in order of decreasing surface temperature. All the spectra extend from 400 to 650 nm. Like the solar spectrum, each shows a series of dark absorption lines superimposed on a background of continuous color. ∞ (Sec. 9.3) However, the line patterns differ greatly from one star to another. Some stars display strong lines in the long-wavelength part of the spectrum, while others have their strongest lines at short wavelengths. Still others show strong absorption lines spread across the whole visible spectrum. What do these differences tell us?

Although spectral lines of many elements are present at widely varying strengths, the differences among the spectra in Figure 10.9 are not due to differences in chemical composition—all seven stars are, in fact, more or less solar in their makeup. Instead, the differences are due almost entirely to the stars' *temperatures*. The spectrum at the top of Figure 10.9 is exactly what we expect from a star having solar composition and a surface temperature of about 30,000 K, the second from a 20,000-K star, and so on, down to the 3000-K star at the bottom.

The spectra of stars with surface temperatures exceeding 25,000 K usually show *strong* absorption lines of singly ionized helium and multiply ionized heavier elements, such as oxygen, nitrogen, and silicon (these latter lines are not shown in the spectra of Figure 10.9). These strong lines are not seen in the spectra of cooler stars because only very hot stars can excite and ionize these tightly bound atoms. In contrast, the hydrogen absorption lines in the spectra of very hot stars are relatively *weak*. The reason is not a lack of hydrogen—it is by far the most abundant element in all stars. At these high temperatures, however, much of the hydrogen is ionized, so there are few intact hydrogen atoms to produce strong spectral lines.

Hydrogen lines are strongest in stars having intermediate surface temperatures of around 10,000 K. This temperature is just right for electrons to move frequently between hydrogen's second and higher orbitals, producing the characteristic visible hydrogen spectrum. ∞ (Sec. 2.6)

Hydrogen lines are again weak in stars with surface temperatures below about 4000 K. However, unlike

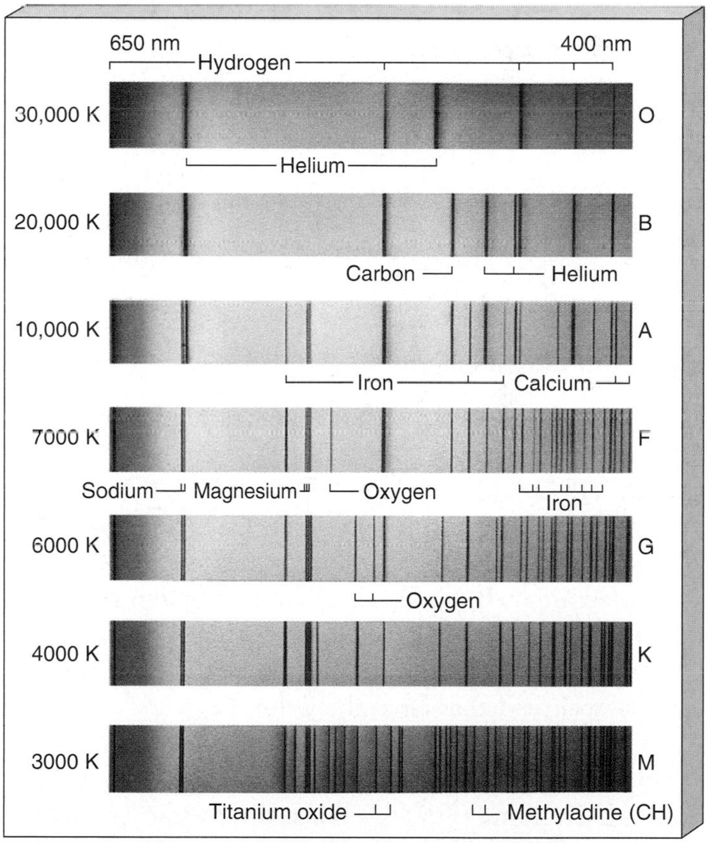

Figure 10.9 Stellar Spectra Comparison of spectra observed for seven stars having a range of surface temperatures. The spectra of the hottest stars, at the top, contain lines for helium and (although not indicated in the O spectrum shown here) lines for multiply ionized heavy elements. In the coolest stars, at the bottom, there are no lines for helium, but lines of neutral atoms and molecules are plentiful. At intermediate temperatures, hydrogen lines are strongest. All seven stars have about the same chemical composition.

TABLE 10.2 Spectral Classes

SPECTRAL CLASS	SURFACE TEMPERATURE (K)	PROMINENT ABSORPTION LINES	FAMILIAR EXAMPLES
O	30,000	Ionized helium strong; multiply ionized heavy elements; hydrogen faint	Mintaka (O9)
B	20,000	Neutral helium moderate; singly ionized heavy elements; hydrogen moderate	Rigel (B8)
A	10,000	Neutral helium very faint; singly ionized heavy elements; hydrogen strong	Vega (A0), Sirius (A1)
F	7000	Singly ionized heavy elements; neutral metals; hydrogen moderate	Canopus (F0)
G	6000	Singly ionized heavy elements; neutral metals; hydrogen relatively faint	Sun (G2), Alpha Centauri (G2)
K	4000	Singly ionized heavy elements; neutral metals strong; hydrogen faint	Arcturus (K2), Aldebaran (K5)
M	3000	Neutral atoms strong; molecules moderate; hydrogen very faint	Betelgeuse (M2) Barnard's Star (M5)

the hottest stars, in which most hydrogen atoms are highly excited or ionized, the hydrogen lines in these cool stars are faint because most of the atoms are in the ground state. The most intense lines are due to weakly excited heavy atoms and some molecules (note the lines of the molecules titanium oxide and methyladine at the bottom of Figure 10.9).

Spectral Classification

Stellar spectra like those shown in Figure 10.9 were obtained for numerous stars well before the start of the twentieth century. Lacking full understanding of how atoms produce spectra, researchers classified stars primarily according to their hydrogen-line intensities. They adopted an A, B, C, D, . . . scheme in which A stars, with the strongest hydrogen lines, were thought to have more hydrogen than did B stars, and so on. The classification extended as far as the letter P.

In the 1920s, scientists began to understand the intricacies of atomic structure and the causes of spectral lines. Astronomers quickly realized that stars could be more meaningfully classified according to surface temperature. Instead of adopting an entirely new scheme, however, they chose to shuffle the existing alphabetical categories—those based on the strength of hydrogen lines—into a new sequence based on temperature. In the modern scheme, in order of decreasing temperature, the letters now run O, B, A, F, G, K, M. (The other letter classes have been dropped.) These stellar designations are called **spectral classes** (or *spectral types*). Use the mnemonic "**O**h, **B**e **A** **F**ine **G**uy, **K**iss **M**e" to remember them in the correct order.

Astronomers further subdivide each lettered spectral class into 10 subdivisions, denoted by the numbers zero to nine. By convention, the lower the number, the hotter the star. Thus, for example, our Sun is classified as a G2 star (a little cooler than G1 and a little hotter than G3), Vega is A0, Barnard's Star is M5, Betelgeuse is M2, and so on. Table 10.2 lists the main properties of each stellar spectral class for the stars presented in Table 10.1.

✓ Concept Check

- Why does a star's spectrum depend on its temperature?

10.4 Stellar Sizes

Direct and Indirect Measurements

4 Virtually all stars are unresolvable points of light in the sky, even when viewed through the largest telescopes. Still, a few are big enough, bright enough, and close enough to allow us to measure their sizes *directly*. In some cases, the results are even detailed enough to allow a few surface features to be distinguished (Figure 10.10). By measuring a star's angular size and knowing its distance from the Sun, astronomers can determine its radius by simple geometry. ∞ (*More Precisely P-2*) The sizes of a few dozen stars have been measured in this way.

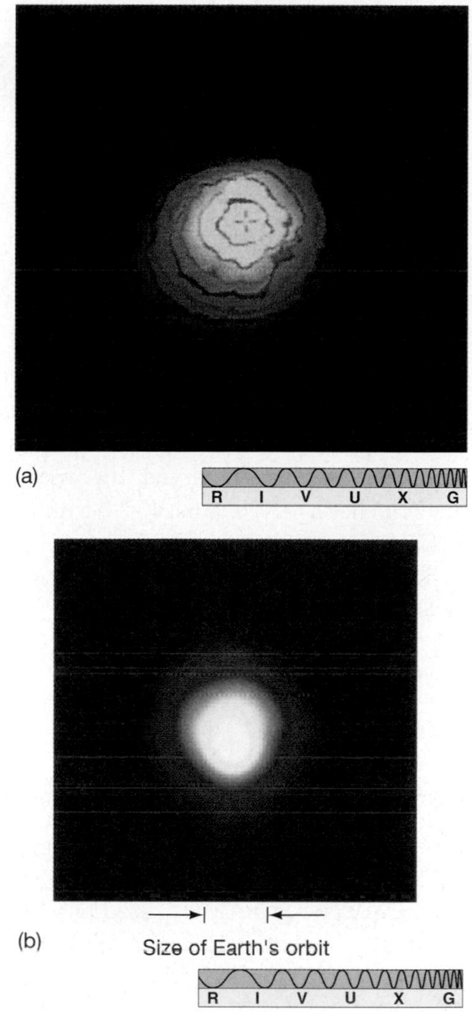

Figure 10.10 **Betelgeuse** (a) The swollen star Betelgeuse (shown here in false color) is close enough for us to directly resolve its size, along with some surface features thought to be storms similar to those that occur on the Sun. Betelgeuse is such a huge star (300 times the size of the Sun) that its photosphere spans roughly the size of Mars's orbit. Most of the surface features discernible here are larger than the entire Sun. (The dark lines are drawn contours, not part of the star itself.) (b) An ultraviolet view of Betelgeuse, in false color, as seen by a European camera onboard the *Hubble Space Telescope*. The bright spot at the bottom right of the star is more than 10 times bigger than Earth and at least 2000 K hotter than the surrounding 3000-K photosphere. *(NOAO; NASA)*

Most stars are too distant or too small for such direct measurements to be possible. Their sizes must be inferred by more indirect means, using the radiation laws. ∞ (Sec. 2.4) According to the Stefan-Boltzmann law, the rate at which a star emits energy into space—the star's *luminosity*—is proportional to the fourth power of the star's *surface temperature*. ∞ (*More Precisely* 2-2) However, the luminosity also depends on the star's *surface area*, because large bodies radiate more energy than do small bodies having the same surface temperature. Because the surface area of a star is proportional to the square of its radius, we can combine these proportionalities to say

$$\text{luminosity} \propto \text{radius}^2 \times \text{temperature}^4,$$

where $\propto$ is the symbol representing proportionality. This **radius–luminosity–temperature relationship** is important because it demonstrates that knowledge of a star's luminosity and temperature can yield an estimate of its radius—an *indirect* determination of stellar size.

Giants and Dwarfs

Let's consider some examples to clarify these ideas. The star Mira has a surface temperature of about 3000 K and a luminosity of 1.6×10^{29} W. Thus, its surface temperature is roughly 0.5, and its luminosity about 400, times the corresponding quantities for our Sun. The radius–luminosity–temperature relationship (see *More Precisely 10-2*) then implies that the star's radius is approximately 80 times that of our Sun. A star as large as Mira is known as a *giant*. More precisely, **giants** are stars having radii between 10 and 100 times that of the Sun. Even larger stars, ranging up to 1000 solar radii in size, are known as **supergiants**. Since the color of any 3000 K object is red, Mira is a **red giant**.

Now consider Sirius B—a faint companion to Sirius A, the brightest star in the night sky. Sirius B's surface temperature is roughly 24,000 K, four times that of the Sun. Its luminosity is 10^{25} W, about 0.04 times the solar value. Substituting these quantities into our radius–luminosity–temperature relationship we obtain a radius of 0.01 times the solar radius—roughly the size of Earth. Sirius B is much hotter but smaller and fainter than our Sun. Such a star is known as a *dwarf*. In astronomical parlance, the term **dwarf** refers to any star of radius comparable to or smaller than the radius of the Sun (including the Sun itself). Because any 24,000 K object glows bluish-white, Sirius B is an example of a **white dwarf**.

Stellar radii range from less than 0.01 to over 100 times the radius of the Sun. Figure 10.11 illustrates the sizes of a few well-known stars.

✓ Concept Check

- Can we measure the radius of a star without knowing its distance?

MORE PRECISELY 10-2

Estimating Stellar Radii

We can combine the Stefan-Boltzmann law $F = \sigma T^4$ (see *More Precisely 2-2*) with the formula for the area of a sphere, $A = 4\pi r^2$, to obtain the relationship between radius (r), luminosity (L), and temperature (T) described in the text:

$$L = 4\pi\sigma\, r^2\, T^4.$$

If we adopt convenient "solar" units, in which L is measured in solar luminosities (3.9×10^{26} W), r is measured in solar radii (696,000 km), and T in units of the solar temperature (5800 K), we can remove the constant $4\pi\sigma$ and write this equation as

$$L \text{ (in solar luminosities)} = r^2 \text{ (in solar radii)} \times T^4 \text{ (in units of 5800 K)}.$$

To compute the radius of a star from its luminosity and temperature, we rearrange this equation to read (in the same units)

$$R = \frac{\sqrt{L}}{T^2}.$$

This simple application of the radiation laws is the basis for almost every estimate of stellar size made in this text.

Let's take as examples the two stars discussed in the text. The star Mira has $L = 400$ units, $T = 0.5$ units, so its radius is $\sqrt{400/0.5^2} = 80$ solar radii. Sirius B has $L = 0.04$, $T = 4$, so its radius is $\sqrt{0.04/4^2} = 0.01$ times the radius of the Sun.

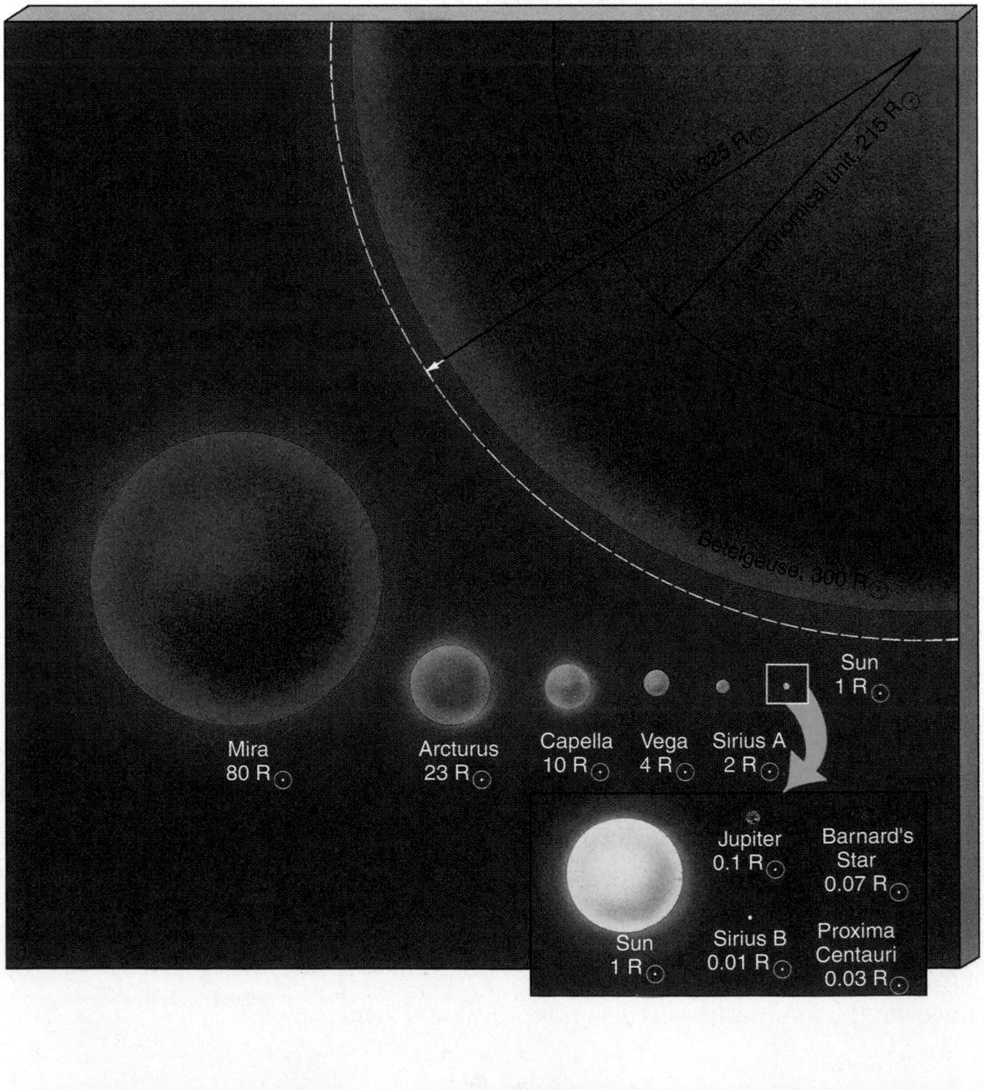

Figure 10.11 Stellar Sizes Star sizes vary greatly. Shown here are the estimated sizes of several well-known stars, including a few of those discussed in this chapter. The symbol "$R_\odot$" means "solar radius."

10.5 The Hertzsprung–Russell Diagram

5 Figure 10.12 plots luminosity versus temperature for a few well-known stars. A figure of this sort is called a *Hertzsprung–Russell diagram*, or **H–R diagram**, after Danish astronomer Ejnar Hertzsprung and U.S. astronomer Henry Norris Russell, who independently pioneered the use of such plots in the second decade of the twentieth century. The vertical luminosity scale, expressed in units of solar luminosity (3.9×10^{26} W), extends over a large range, from 10^{-4} to 10^4. The Sun appears right in the middle of the luminosity range, at a luminosity of one. Surface temperature is plotted on the horizontal axis, although in the unconventional sense of temperature increasing to the *left*, so that the spectral sequence O, B, A, . . . reads left to right.

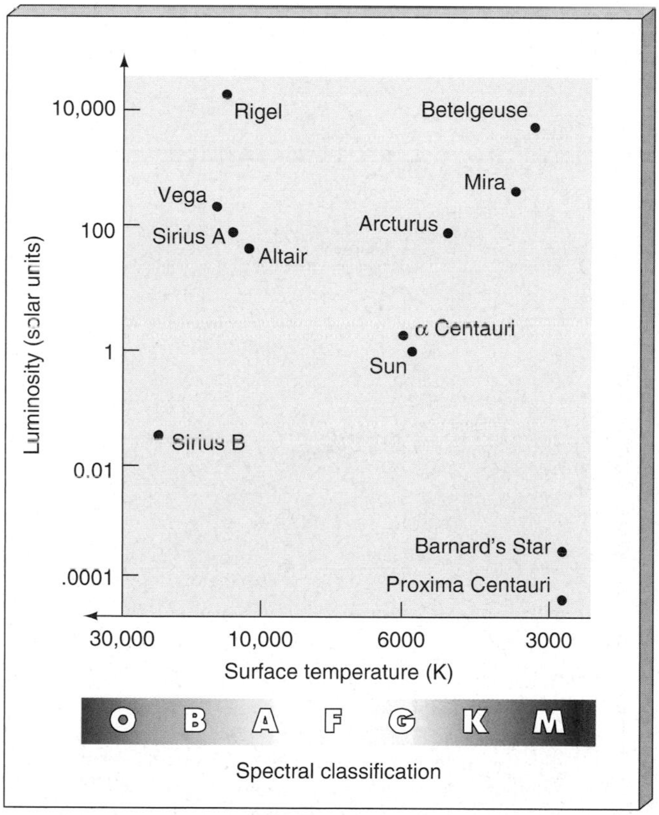

Figure 10.12 H–R Diagram of Prominent Stars A plot of luminosity against surface temperature (or spectral class), known as an H–R diagram, is a useful way to compare stars. Plotted here are the data for some stars mentioned earlier in the text. The Sun, of course, has a luminosity of one solar unit. Its temperature, read off the bottom scale, is 5800 K—a G-type star. Similarly, the B-type star Rigel, at top left, has a temperature of about 15,000 K and a luminosity more than 10,000 times that of the Sun. The M-type star Proxima Centauri, at bottom right, has a temperature of less than 3000 K and a luminosity less than 1/10,000 that of the Sun.

The Main Sequence

The few stars plotted in Figure 10.12 give little indication of any particular connection between stellar properties. However, as Hertzsprung and Russell plotted more and more stellar temperatures and luminosities, they found that a relationship does in fact exist. Stars are *not* uniformly scattered across the H–R diagram. Instead, most are confined to a fairly well-defined band stretching diagonally from top left (high-temperature, high-luminosity) to bottom right (low-temperature, low-luminosity). In other words, cool stars tend to be faint (less luminous) and hot stars tend to be bright (more luminous). This band of stars spanning the H–R diagram is known as the **main sequence**. Figure 10.13 is an H–R diagram for stars lying within 5 pc of the Sun. Note that

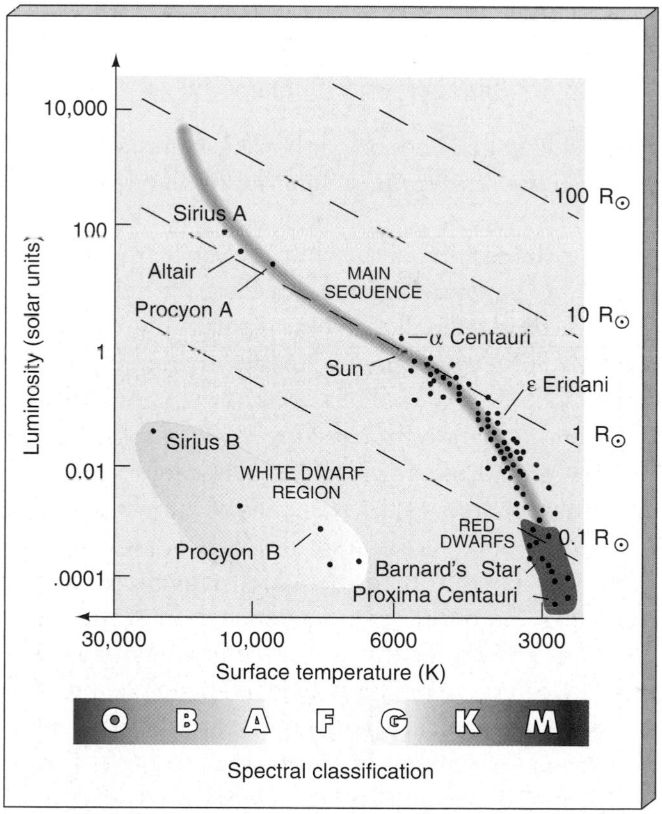

Figure 10.13 H–R Diagram of Nearby Stars Most stars have properties within the shaded region of the H–R diagram known as the main sequence. The points plotted here are for stars lying within about 5 pc of the Sun. Each dashed diagonal line corresponds to a constant stellar radius, allowing us to indicate stellar size on the same diagram as stellar luminosity and temperature. (Here and in other figures, the symbol "☉" stands for the Sun.)

most stars in the solar neighborhood lie on the main sequence.

The surface temperatures of main-sequence stars range from about 3000 K (spectral class M) to more than 30,000 K (spectral class O). This temperature range is relatively small—only a factor of 10. In contrast, the observed range in luminosities is very large, covering eight orders of magnitude (that is, a factor of 100 million), from 10^{-4} to 10^4 times the luminosity of the Sun.

Using the radius–luminosity–temperature relationship (Section 10.4), astronomers find that stellar radii also vary along the main sequence. The faint, red M-type stars in the bottom right of the H–R diagram are only about 1/10 the size of the Sun, whereas the bright, blue O-type stars in the upper left are about 10 times larger than the Sun. The dashed lines in Figure 10.13 represent constant stellar radius, meaning that any star lying on a given line has the same radius, regardless of its luminosity or temperature. Along a constant-radius line, the radius–luminosity–temperature relationship implies

$$\text{luminosity} \propto \text{temperature}^4.$$

By including such lines on our H–R diagrams, we can indicate stellar temperatures, luminosities, and radii on a single plot.

At the top end of the main sequence, the stars are large, hot, and very luminous. Because of their size and color, they are referred to as **blue giants**. The very largest are called **blue supergiants**. At the other end, stars are small, cool, and faint. They are known as **red dwarfs**. Our Sun lies right in the middle.

Figure 10.14 shows an H–R diagram for a different group of stars—the 100 stars of known distance having the greatest apparent brightness, as seen from Earth. Notice that here there are many more stars at the upper end of the main sequence than at the lower end. The reason for this excess of blue giants is simple—we can see very luminous stars a long way off. The stars shown in Figure 10.14 are scattered through a much greater volume of space than those in Figure 10.13, and the Figure 10.14 sample is heavily biased toward the most luminous objects. In fact, of the 20 brightest stars in the sky, only six lie within 10 pc of us. The rest are visible, despite their great distances, because of their high luminosities.

If very luminous blue giants are overrepresented in Figure 10.14, low-luminosity red dwarfs are surely underrepresented. In fact, no dwarfs appear on this diagram. This absence is not surprising, because low-luminosity stars are difficult to observe from Earth. In the 1970s, astronomers began to realize that they had greatly underestimated the number of red dwarfs in our galaxy. As hinted at by the H–R diagram in Figure 10.13, which shows an unbiased sample of stars in the solar neighborhood, red dwarfs are actually the most common type of star in the sky. They probably account for upward of 80 percent of all stars in the universe. In contrast, O- and B-type supergiants are extremely rare—only about one star in 10,000 falls into this category.

White Dwarfs and Red Giants

Some of the points plotted in Figures 10.12 to 10.14 clearly do not lie on the main sequence. One such point in Figure 10.12 represents Sirius B, the white dwarf we met in Section 10.4 (see *More Precisely 10-2*). Its surface temperature (24,000 K) is about four times that of the Sun, but its luminosity is only 0.04 times the solar

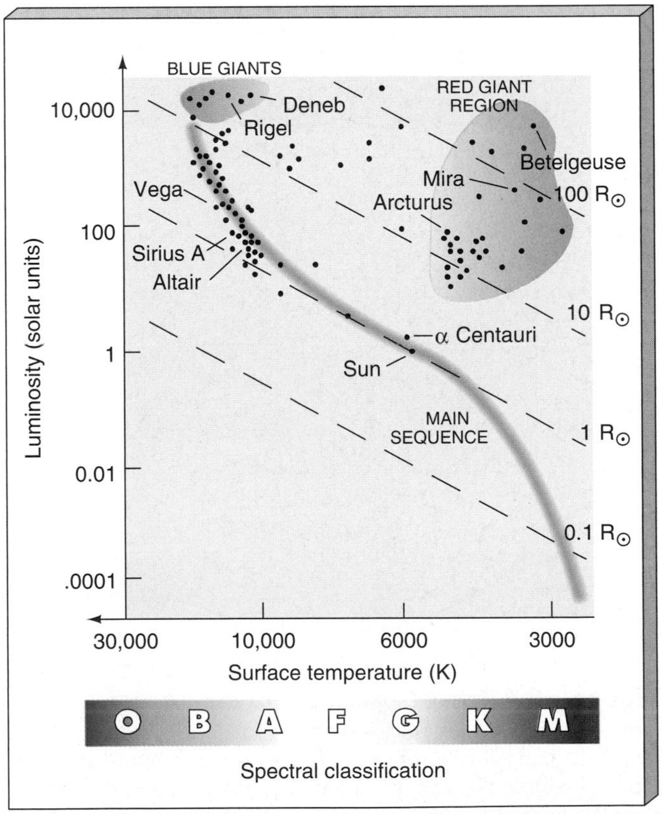

Figure 10.14 H–R Diagram of Brightest Stars An H–R diagram for the 100 brightest stars in the sky is biased in favor of the most luminous stars—which appear toward the upper left—because we can see them more easily than we can the faintest stars. (Compare with Figure 10.13, which shows only the closest stars.)

value. A few more such faint bluish-white stars can be seen in the bottom left-hand corner of Figure 10.13, in the **white dwarf region**.

Also shown in Figure 10.12 is Mira, with a surface temperature (3000 K) about half that of the Sun, but its luminosity is some 400 times greater than the Sun's. Another point represents Betelgeuse, the ninth brightest star in the sky, a little hotter than Mira, and more than 30 times more luminous. The upper right-hand corner of the H–R diagram (marked on Figure 10.14), where these stars are found, is called the **red giant region**. No red giants are found within 5 pc of the Sun (Figure 10.13), but many of the brightest stars seen in the sky are in fact red giants (Figure 10.14). Red giants are relatively rare, but they are so bright that they are visible to very great distances.

About 90 percent of all stars in our solar neighborhood, and presumably a similar percentage elsewhere in the universe, are main-sequence stars. About nine percent of stars are white dwarfs, and one percent are red giants.

Concept Check

- Only a tiny fraction of all stars are giants. Why then do giants account for so many of the brightest stars in the night sky?

10.6 Stellar Masses

6 As with all other objects, we measure a star's mass by observing its gravitational influence on some nearby body—another star, perhaps, or a planet. If we know the distance between the two bodies, then we can use Newton's laws to calculate their masses. *(More Precisely 1-2)*

Binary Stars

Most stars are members of *multiple-star systems*—groups of two or more stars in orbit around one another. The majority are found in **binary-star systems**, which consist of two stars in orbit about their common center of mass, held together by their mutual gravitational attraction. (The Sun is not part of a multiple-star system; if it has anything at all uncommon about it, it is this lack of stellar companions.)

Astronomers classify binary-star systems (or simply *binaries*) according to their appearance from Earth and the ease with which they can be observed. **Visual binaries** have widely separated members bright enough to be observed and monitored separately (Figure 10.15). The more common **spectroscopic binaries** are too distant from us to be resolved into separate stars, but they can be indirectly perceived by monitoring the back-and-forth Doppler shifts of their spectral lines as the stars orbit one another and their line-of-sight velocities vary periodically. *(More Precisely 2-3)*

In a *double-line* spectroscopic binary, two distinct sets of spectral lines—one for each component star—shift back and forth as the stars move. Because we see particular lines alternately approaching and receding, we know that the objects emitting the lines are in orbit. In the more common *single-line* systems (Figure 10.16), one star is too faint for its spectrum to be distinguished, so we see only one set of lines shifting back and forth. This shifting means that the detected star must be in orbit around another star, even though the companion cannot be observed directly.

In the much rarer **eclipsing binaries**, the orbital plane of the pair of stars is almost edge-on to our line of sight. In this situation, we observe a periodic decrease of starlight intensity as one member of the binary passes

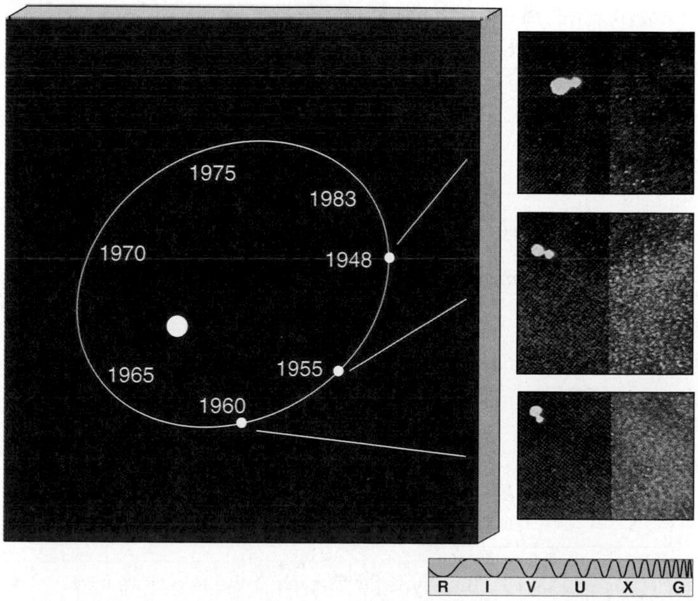

Figure 10.15 **Binary Stars** The periods and separations of binary stars can be observed directly if each star is clearly seen. *(Harvard College Observatory)*

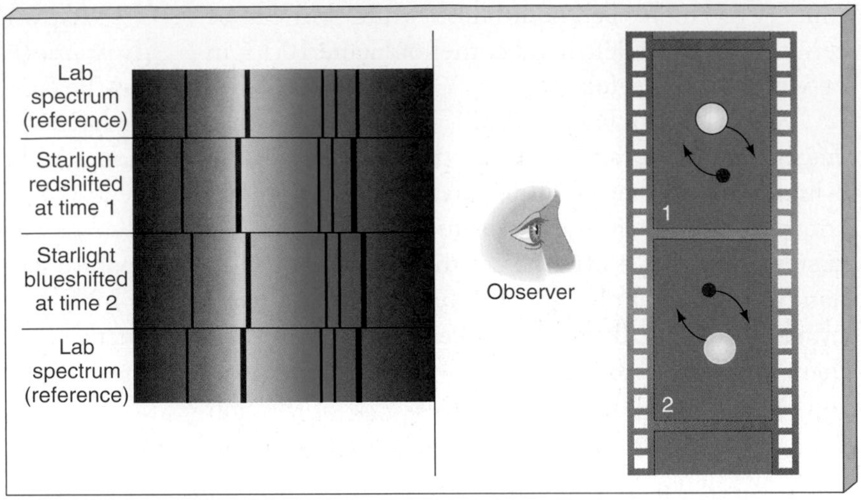

Figure 10.16 Binary Spectra Binary properties can be determined indirectly by measuring the periodic Doppler shift of one star relative to the other as they move in their orbits. The diagram shows a single-line system, in which only one spectrum (from the brighter component) is visible. (The observer is situated to the left of the diagram. Reference laboratory spectra are shown at top and bottom.)

in front of the other (Figure 10.17). By studying the variation of the light from an eclipsing binary system—called the binary's **light curve**—we can derive detailed information not only about the stars' orbits and masses but also about their radii.

Note that these categories are not mutually exclusive. For example, an eclipsing binary may also be (and, in fact, often is) a spectroscopic binary system.

Measuring Stellar Masses

The period of a binary can be measured by observing the orbits of the component stars, the back-and-forth motion of the stellar spectral lines, or the dips in the light curve—whatever information is available. Observed binary periods span a broad range—from hours to centuries. How much additional information can be extracted depends on the type of binary involved.

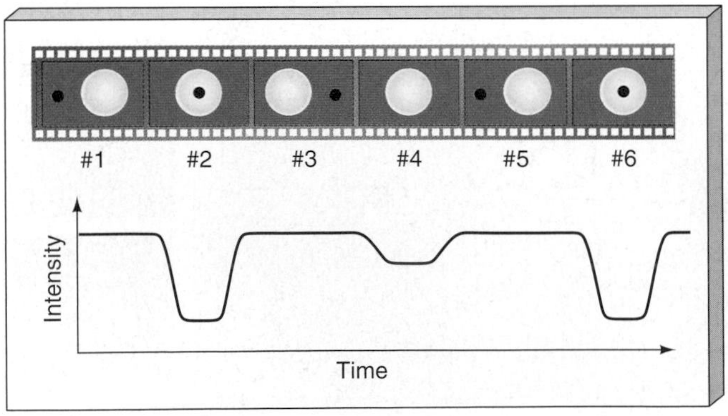

Figure 10.17 Binary Light Curves If the two stars in a binary happen to eclipse one another, additional information on their radii and masses can be obtained by observing the periodic decrease in starlight intensity as one star passes in front of the other.

If the distance to a *visual* binary is known, then the orbits of each component can be individually tracked, and the masses of the component can be determined.

For *spectroscopic* binaries, Doppler-shift measurements give us information only on the radial velocities of the component stars, and this limits the information we can obtain. Simply put, we cannot distinguish between a slow-moving binary seen edge-on and a fast-moving binary seen almost face-on (so that only a small component of the orbital motion is along the line of sight). For a double-line system, only lower limits on the individual masses can be obtained. For single-line systems, even less information is available and only a fairly complicated relation between the component masses (known as the mass function) can be derived.

If a spectroscopic binary happens also to be an *eclipsing* system, then the uncertainty in the orbital inclination is removed, as the binary is known to be edge-on (or very nearly so). In that case, both masses can be determined for a double-line binary. For a single-line system, the mass function is simplified to the point where the mass of the unseen component is known if the mass of the brighter component can be obtained by other means (for example, if it is recognized as a main sequence star of a certain spectral class—see Figure 10.18).

Individual component masses have been obtained for many nearby binary systems. Virtually all we know about the masses of stars is based on such observations. As a simple example, consider the nearby visual binary system made up of the bright star Sirius A and its faint companion Sirius B. Their orbital period is

50 years and their orbital semi-major axis is 20 A.U.—7.5″ at a distance of 2.7 pc—implying that the sum of their masses is 3.2 (= $20^3/50^2$) times the mass of the Sun. ∞ (Sec. 1.4) Further study of the orbit shows that Sirius A has roughly twice the mass of its companion. It follows that the masses of Sirius A and Sirius B are 2.1 and 1.1 solar masses, respectively.

Mass and Other Stellar Properties

Figure 10.18 is a schematic H–R diagram showing how stellar mass varies along the main sequence. There is a clear progression from low-mass red dwarfs to high-mass blue giants. With few exceptions, main-sequence stars range in mass from about 0.1 to 20 times the mass of the Sun. The hot O- and B-type stars are generally about 10 to 20 times more massive than our Sun. The coolest K- and M-type stars contain only a few tenths of a solar mass.

Figure 10.19 illustrates how the radii and luminosities of main-sequence stars depend on mass. The two plots are called the *mass–radius* (part a) and *mass–luminosity* (part b) relationships. Along the main sequence, both radius and luminosity increase with mass. As a (very rough) rule of thumb, radius rises in direct proportion to mass, whereas luminosity increases much faster—more like the *cube* of the mass. For example, a 2-solar-mass main-sequence star has a radius roughly twice that of the Sun and a luminosity of eight (2^3) solar

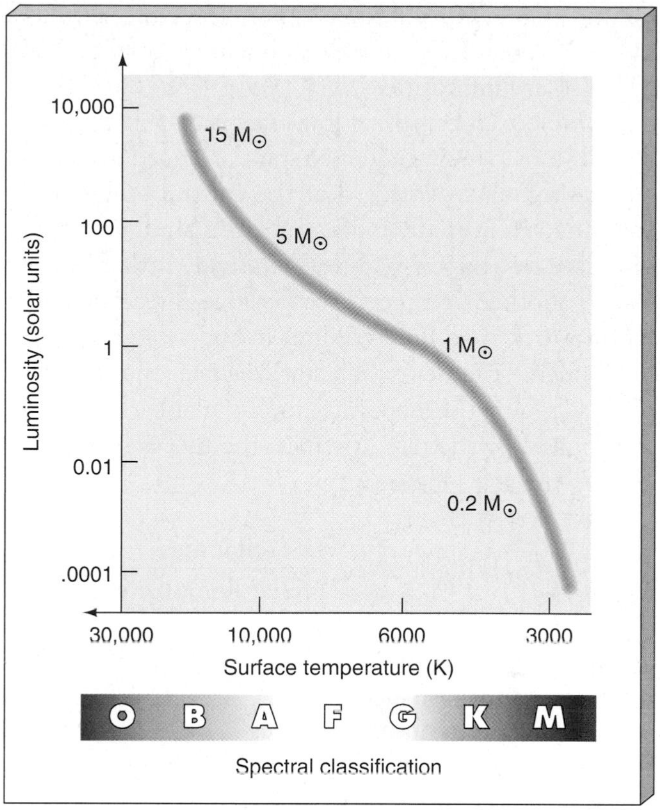

Figure 10.18 Stellar Masses More than any other stellar property, mass determines a star's position on the main sequence. Low-mass stars are cool and faint; they lie at the bottom of the main sequence. Very massive stars are hot and bright; they lie at the top of the main sequence. The symbol "$M_\odot$" means "solar mass."

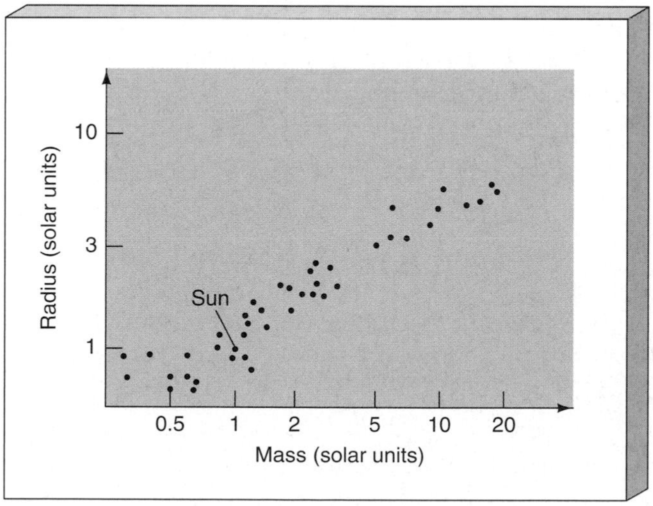

(a)

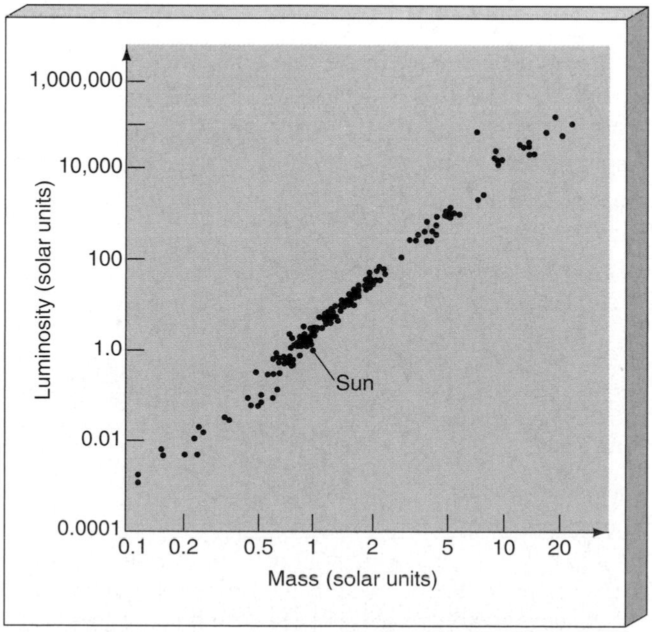

(b)

Figure 10.19 Stellar Radii and Luminosities (a) Dependence of stellar radius on mass for main-sequence stars. The radius increases roughly in proportion to the mass over much of the range. (b) Dependence of luminosity on mass. The luminosity increases roughly as the cube of the mass.

luminosities; a 0.2-solar-mass main-sequence star has a radius of about 0.2 solar radii and a luminosity of 0.008 (0.2^3) solar luminosities.

Table 10.3 compares some key properties of several well-known main-sequence stars, arranged in order of decreasing mass. Notice that the central temperature (obtained from mathematical models similar to those discussed in Chapter 9) differs relatively little from one star to another, compared to the large spread in stellar luminosities. The final column in the table presents a very rough estimate of each star's *lifetime*, obtained simply by dividing the amount of fuel available (that is, the star's mass) by the rate at which the fuel is being consumed (the star's luminosity):

$$\text{stellar lifetime} \propto \frac{\text{stellar mass}}{\text{stellar luminosity}},$$

and noting that the lifetime of the Sun (see Chapter 12) is about 10 billion years.

Because luminosity increases so rapidly with mass, the most massive stars are by far the shortest lived. For example, according to the mass–luminosity relationship, the lifetime of a 20-solar-mass O-type star is roughly $20/20^3 = 1/400$ that of the Sun, or about 25 million years. We can be sure that all the O- and B-type stars we now observe are quite young—less than a few tens of millions of years old. Their nuclear reactions proceed so rapidly that their fuel is quickly depleted despite their large masses. At the opposite end of the main sequence, the low core density and temperature of an 0.1-solar-mass M-type star mean that its proton–proton reactions churn away much more sluggishly than in the Sun's core, leading to a very low luminosity and a correspondingly long lifetime. ∞ (Sec. 9.5) Many of the K- and M-type stars now visible in the sky will shine on for at least another trillion years. The evolution of stars is the subject of the next two chapters.

Concept Check

- How do we know the masses of stars that aren't members of binaries?

10.7 Extending the Cosmic Distance Scale

1 We have already discussed the connections between luminosity, apparent brightness, and distance. Knowledge of a star's apparent brightness and its distance allows us to determine its luminosity using the inverse-square law. But we can also turn the problem around. If we somehow knew a star's luminosity and then measured its apparent brightness, the inverse-square law would tell us its distance from the Sun.

Spectroscopic Parallax

Most of us have a rough idea of the approximate intrinsic brightness (that is, the luminosity) of a typical traffic signal. Suppose you are driving down an unfamiliar street and see a red traffic light in the distance. Your knowledge of the light's luminosity enables you immediately to make a mental estimate of its distance. A light that appears relatively dim (low apparent brightness) must be quite distant (assuming it's not just dirty). A bright one must be relatively close. Thus, *measurement of the apparent brightness of a light source, combined with some knowledge of its luminosity, can yield an estimate of its distance.* For stars, the trick is to find an independent

TABLE 10.3 Key Properties of Some Well-Known Stars

STAR	SPECTRAL TYPE/ LUMINOSITY CLASS[1]	MASS (solar masses)	CENTRAL TEMPERATURE (10^6 K)	LUMINOSITY (solar luminosities)	ESTIMATED LIFETIME (millions of years)
Rigel	B8Ia	10	30	44,000	2
Sirius	A1V	2.3	20	23	1000
Alpha Centauri	G2V	1.1	17	1.4	8000
Sun	G2V	1.0	15	1.0	10,000
Proxima Centauri	M5V	0.1	0.6	0.00006	>10,000,000

[1]See Section 10.7.

measure of luminosity without knowing the distance. The H–R diagram can provide just that.

If we determine a star's spectral type and assume that the star lies on the main sequence, we can read the star's luminosity directly off a graph such as Figure 10.13. Then, knowing the star's luminosity, we can determine its distance by measuring its apparent brightness and using the inverse-square law. In practice, the "fuzziness" of the main sequence translates into a small (10–20 percent) uncertainty in the distance, but the basic idea remains valid. The existence of the main sequence allows us to make a connection between an easily measured quantity (spectral type) and the star's luminosity, which would otherwise be unknown. This process of using stellar spectra to infer distances is called **spectroscopic parallax**. (The name is very misleading, as the method has nothing in common with geometric parallax other than its use as a means of determining stellar distances.)

In Chapter 1 we introduced the first "rung" on a "ladder" of distance-measurement techniques that will ultimately carry us to the edge of the observable universe. That rung is radar ranging on the inner planets. ∞ (Sec. 1.3) It establishes the scale of the solar system and defines the astronomical unit. At the beginning of this chapter we discussed a second rung on the cosmic distance ladder—stellar parallax—which is based on the first, since Earth's orbit is the baseline. Now, having used the first two rungs to determine the distances and other physical properties of many nearby stars, we use that knowledge to construct the third rung—spectroscopic parallax. Figure 10.20 illustrates schematically how this third rung builds on the first two and expands our cosmic field of view still deeper into space.

Spectroscopic parallax can be used to determine stellar distances out to several thousand parsecs. Beyond that, spectra and colors of individual stars are difficult to obtain. Note that, in using this method, we are assuming (without proof) that distant stars are basically similar to those nearby—in particular, that main-sequence stars fall on the *same* main sequence. Only by making this assumption can we use spectroscopic parallax to expand the boundaries of our distance-measurement techniques.

Each rung in the distance ladder is calibrated using data from the lower rungs, so changes made at any level propagate to *all* larger scales. As an example, the recent *Hipparcos* mission (Section 10.1), in addition to determining hundreds of thousands of stellar parallaxes to

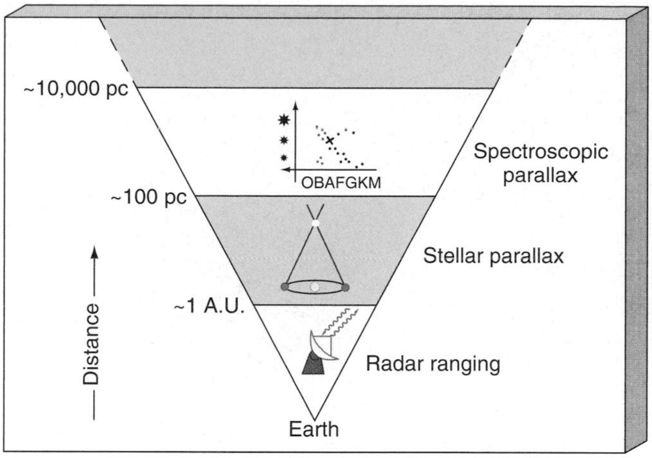

Figure 10.20 Stellar Distances Knowledge of a star's luminosity and apparent brightness can yield an estimate of its distance. Astronomers use this third rung on our distance ladder, called spectroscopic parallax, to measure distances as far out as individual stars can be clearly discerned—several thousand parsecs.

unprecedented precision, also measured the colors and luminosities of more than two million stars. Figure 10.21 shows an example of the richness of an H–R diagram based on *Hipparcos* data. The impact of this enormous dataset of high-quality observations extends far beyond the volume of space actually surveyed by the satellite. By recalibrating the local foundations of the cosmic distance scale, *Hipparcos* has caused astronomers to revise their estimates of distances on *all* scales—up to and including the scale of the universe itself. All distances quoted throughout this text reflect updated values based on *Hipparcos* data.

Luminosity Class

What if the star in question happens to be a red giant or a white dwarf, and does not lie on the main sequence? As mentioned in Chapter 2, detailed analysis of *line widths* can provide information on the pressure, and hence the density, of the gas where the line formed. ∞ (Sec. 2.6) The atmosphere of a red giant is much less dense than that of a main-sequence star, and this in turn is much less dense than the atmosphere of a white dwarf.

Over the years, astronomers have developed a system for classifying stars according to the widths of their spectral lines. Because line width depends on pressure in the stellar photosphere, and because this pressure in turn is well correlated with luminosity, this stellar property has come to be known as **luminosity class**. The

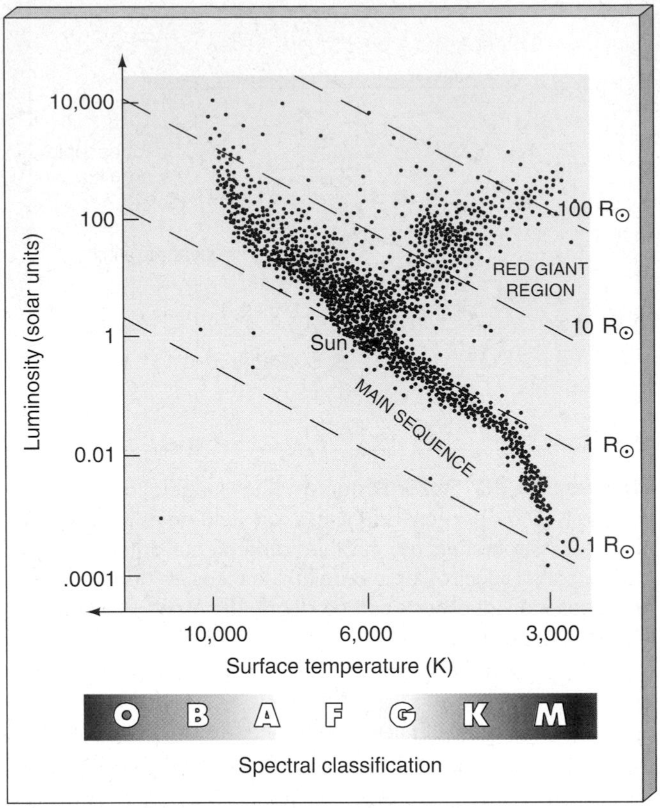

Figure 10.21 **Hipparcos H–R Diagram** This is a simplified version of the most complete H–R diagram ever compiled. It represents more than 20,000 data points, as measured by the European *Hipparcos* spacecraft for stars within about 1000 pc.

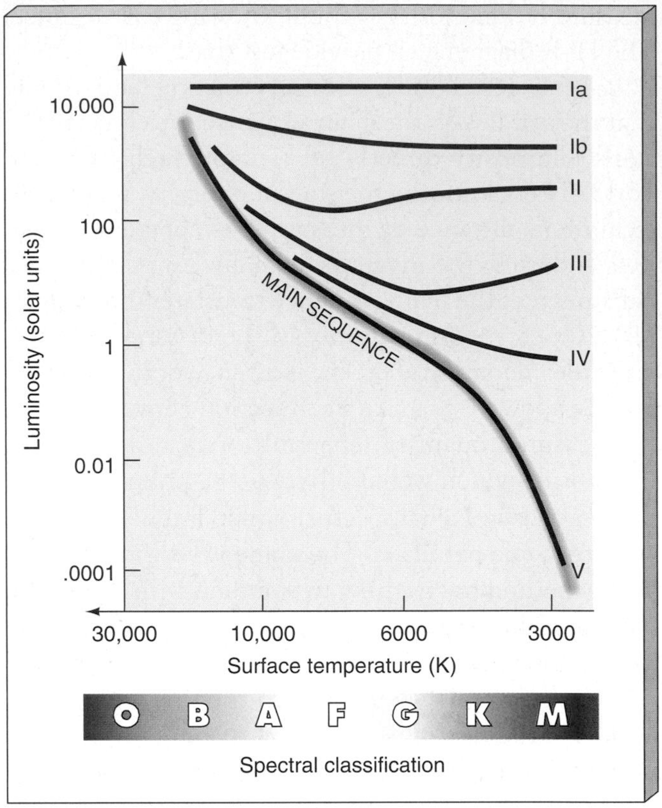

Figure 10.22 **Stellar Luminosities** Stellar luminosity classes in the H–R diagram. A star's location in the diagram could be specified by giving its spectral type and luminosity class instead of its temperature and luminosity.

standard luminosity classes are listed in Table 10.4 and shown on the H–R diagram in Figure 10.22. By determining a star's luminosity class, astronomers can usually tell with a high degree of confidence what sort of object it is, and the full specification of a star's spectral properties includes its luminosity class. For example, the Sun, on the main sequence, is of class G2V, the blue supergiant Rigel is B8Ia, the red dwarf Barnard's Star is M5V, the red supergiant Betelgeuse is M2Ia, and so on.

As an example, consider a K2-type star with a surface temperature of approximately 5000 K (Table 10.5).

TABLE 10.4 Stellar Luminosity Classes

CLASS	DESCRIPTION
Ia	Bright supergiants
Ib	Supergiants
II	Bright giants
III	Giants
IV	Subgiants
V	Main-sequence stars/dwarfs

TABLE 10.5 Variation in Stellar Properties within a Spectral Class

SURFACE TEMPERATURE (K)	LUMINOSITY (solar luminosities)	RADIUS (solar radii)	OBJECT	EXAMPLE
5000	0.3	0.8	K2V main-sequence star	ε Eridani
5000	100	14	K2III red giant	Arcturus
5000	5000	100	K2Ib red supergiant	ε Pegasi

If the widths of the star's spectral lines tell us that it lies on the main sequence (that is, it is a K2V star), then its luminosity is about 0.3 times the solar value. If its spectral lines are observed to be narrower than lines normally found in main-sequence stars, the star may be recognized as a K2III giant, with a luminosity of around 100 solar luminosities. If the lines are very narrow, the star might instead be classified as a K2Ib supergiant, brighter by a further factor of 50, at 5000 solar luminosities. In each case, knowledge of luminosity classes allows astronomers to identify the object and make an appropriate estimate of its luminosity, and hence its distance.

✓ Concept Check

- Suppose astronomers discover that, due to a calibration error, *all* distances measured by geometric parallax are 10 percent larger than currently thought. What effect would this have on the "standard" main sequence used in spectroscopic parallax?

Chapter Review

Summary

The distances to the nearest stars can be measured by stellar parallax. A star with a parallax of one arc second is one **parsec** (p. 254)—about 3.3 light-years—away from the Sun. A star's **proper motion** (p. 255), which is its true motion across the sky, is a measure of the star's velocity perpendicular to our line of sight.

The **apparent brightness** (p. 257) of a star is the rate at which energy from the star reaches a detector. Apparent brightness falls off as the inverse square of the distance. Optical astronomers use the **magnitude scale** (p. 258) to express and compare stellar brightnesses. The greater the magnitude, the fainter the star; a difference of five magnitudes corresponds to a factor of 100 in brightness. **Apparent magnitude** (p. 258) is a measure of apparent brightness. The **absolute magnitude** (p. 259) of a star is the apparent magnitude it would have if placed at a standard distance of 10 pc from the viewer. It is a measure of the star's luminosity.

Astronomers often measure the temperatures of stars by measuring their brightnesses through two or more optical filters, then fitting a blackbody curve to the results. Spectroscopic observations of stars provide an accurate means of determining both stellar temperatures and stellar composition. Astronomers classify stars according to the absorption lines in their spectra. The standard stellar **spectral classes** (p. 262), in order of decreasing temperature, are O B A F G K M.

Only a few stars are large enough and close enough to Earth that their radii can be measured directly. The sizes of most stars are estimated indirectly through the **radius–luminosity–temperature relationship** (p. 263). Stars are categorized as **dwarfs** (p. 263) comparable in size to, or smaller than, the Sun; **giants** (p. 263) up to 100 times larger than the Sun, and **supergiants** (p. 263) more than 100 times larger than the Sun. In addition to "normal" stars such as the Sun, two other important classes of star are **red giants** (p. 263), which are large, cool, and luminous, and **white dwarfs** (p. 263), which are small, hot, and faint.

A plot of stellar luminosity versus stellar spectral class (or temperature) is called an **H–R diagram** (p. 265). About 90 percent of all stars plotted on an H–R diagram lie on the **main sequence** (p. 265), which stretches from hot, bright **blue supergiants** (p. 266) and **blue giants** (p. 266), through intermediate stars such as the Sun, to cool, faint **red dwarfs** (p. 266). Most main-sequence stars are red dwarfs; blue giants are quite rare. About nine percent of stars lie in the **white dwarf region** (p. 267) and the remaining one percent lie in the **red giant region** (p. 267).

Most stars are not isolated in space but instead orbit other stars in **binary-star systems** (p. 267). In a **visual binary** (p. 267), both stars can be seen and their orbits charted. In a **spectroscopic binary** (p. 267), the stars cannot be resolved, but their orbital motion can be detected spectroscopically. In an **eclipsing binary** (p. 267), the orbits are oriented in such a way that one star periodically passes in front of the other as seen from Earth and dims the light we receive. The binary's **light curve** (p. 268) is a plot of its apparent brightness as a function of time. Studies of binary stars often allow stellar masses to be measured. The mass of a star determines its size, temperature, and brightness. Hot blue giants are much more massive than the Sun; cool red dwarfs are much less massive. High-mass stars burn their fuel rapidly and have much shorter lifetimes than the Sun. Low-mass stars consume their fuel slowly and may remain on the main sequence for trillions of years.

If a star is known to lie on the main sequence, measurement of its spectral type allows its luminosity to be estimated and its distance to be measured. This method of distance determination, which is valid for stars up to several thousand parsecs from Earth, is called **spectroscopic parallax** (p. 271). A star's **luminosity class** (p. 271) allows astronomers to distinguish main-sequence stars from giants and supergiants of the same spectral type.

Review and Discussion

1. How is stellar parallax used to measure the distances to stars?
2. What is a parsec? Compare it to the astronomical unit.
3. Explain how a star's real motion through space translates into motion observable from Earth.
4. Describe some characteristics of red giants and white dwarfs.
5. What is the difference between absolute and apparent magnitude?
6. How do astronomers measure star temperatures?
7. Briefly describe how stars are classified according to their spectral characteristics.
8. What information is needed to plot a star on the Hertzsprung–Russell diagram?
9. What is the main sequence? What basic property of a star determines where it lies on the main sequence?
10. How are distances determined using spectroscopic parallax?
11. Which stars are most common in the Milky Way Galaxy? Why don't we see many of them in H–R diagrams?
12. How can stellar masses be determined by observing binary star systems?
13. If a high-mass star starts off with much more fuel than a low-mass star, why doesn't the high-mass star live longer?
14. In general, is it possible to determine the age of an individual star simply by noting its position on an H–R diagram?
15. Visual binaries and eclipsing binaries are relatively rare compared to spectroscopic binaries. Why is this?

True or False?

____ 1. One parsec is a little more than 200,000 A.U.

____ 2. There are no other stars within 1 pc of the Sun.

____ 3. Parallax can be used to measure stellar distances out to about 10,000 pc.

____ 4. Most stars have radii between 0.1 and 10 times the radius of the Sun.

____ 5. Star A appears brighter than star B, as seen from Earth. Therefore, star A must be closer to Earth than star B.

____ 6. Star A and star B have the same luminosity, but star B is twice as distant as star A. Therefore, star A appears four times brighter than star B.

____ 7. A star of apparent magnitude five looks brighter than one of apparent magnitude two.

____ 8. Differences among stellar spectra are mainly due to differences in star composition.

____ 9. Stars with very weak hydrogen lines in their spectra contain no hydrogen.

____ 10. Red giants are very bright because they are extremely hot.

____ 11. Red dwarfs lie in the lower left part of the H–R diagram.

____ 12. The brightest stars visible in the night sky are all found in the upper part of the H–R diagram.

____ 13. In a spectroscopic binary, the orbital motion of the component stars appears as variations in the overall apparent brightness of the system.

____ 14. It is impossible to have a one-billion-year-old O- or B-type main-sequence star.

____ 15. Astronomers can distinguish between main-sequence and giant stars by purely spectroscopic means.

Fill in the Blank

1. Parallax measurements of the distances to the stars nearest Earth use ________ as a baseline.
2. To determine the transverse velocity of a star, both its ________ and its ________ must be known.
3. The radius of a star can be indirectly determined if the star's ________ and ________ are known.
4. The smallest stars normally plotted on the H–R diagram are ________.
5. The smallest main-sequence stars are ________ (larger/smaller) than the planet Jupiter.
6. Observations of stars through B and V filters are used to determine stellar ________.
7. The hottest stars show little evidence of hydrogen in their spectra because hydrogen is mostly ________ at these temperatures.
8. The coolest stars show little evidence of hydrogen in their spectra because hydrogen is mostly ________ at these temperatures.
9. The Sun has a spectral type of ________.
10. The H–R diagram is a plot of ________ on the horizontal scale versus ________ on the vertical scale.
11. The band of stars extending from the top left of the H–R diagram to its bottom right is known as the ________.

12. The large, cool stars found at the upper right of the H–R diagram are ________.

13. The small, hot stars found at the lower left of the H–R diagram are ________.

14. ________-star systems are important for providing measurements of stellar masses.

15. Going from spectral type O to M along the main sequence, stellar masses ________.

Problems

1. How far away is the star Spica, whose parallax is 0.013″? What would Spica's parallax be if it were measured from an observatory on Neptune's moon Triton as Neptune orbited the Sun?

2. A star lying 20 pc from the Sun has proper motion of 0.5″/yr. What is its transverse velocity?

3. (a) What is the luminosity of a star having three times the radius of the Sun and a surface temperature of 10,000 K? (b) A certain star has a temperature twice that of the Sun and a luminosity 64 times greater than the solar value. What is its radius, in solar units?

4. Two stars—A and B, of luminosities 0.5 and 4.5 times the luminosity of the Sun, respectively—are observed to have the same apparent brightness. Which one is more distant, and how much farther away is it than the other?

5. Two stars—A and B, of absolute magnitudes three and eight, respectively—are observed to have the same apparent magnitude. Which one is more distant, and how much farther away is it than the other?

6. Calculate the apparent brightness of the sun (amount of energy received per unit area per unit time) as seen from a distance of 10 pc. Compare this with the solar constant at Earth. ∞ (Sec. 9.1)

7. Astronomical objects visible to the naked eye range in apparent brightness from faint sixth-magnitude stars to the Sun, with magnitude −26.8. What is the range in apparent brightness corresponding to this magnitude range?

8. A star has apparent magnitude 7.5 and absolute magnitude 2.5. How far away is it?

9. Two stars in an eclipsing spectroscopic binary are observed to have an orbital period of 25 days. Further observations reveal that the orbit is circular, with a separation of 0.3 A.U., and that one star is 1.5 times the mass of the other. What are the masses of the stars? ∞ (*More Precisely 1-2*)

10. Given that the Sun's lifetime is about 10 billion years, estimate the life expectancy of (a) a 0.2-solar-mass, 0.01-solar-luminosity red dwarf, (b) a 3.0-solar-mass, 30-solar-luminosity star, (c) a 10-solar-mass, 1000-solar-luminosity blue giant.

Projects

1. Every winter, you can find an astronomy lesson in the evening sky. The Winter Circle is an *asterism*—or pattern of stars—made up of six bright stars in five constellations: Sirius, Rigel, Betelgeuse, Aldebaran, Capella, and Procyon. These stars span nearly the entire range of colors (and therefore temperatures) possible for normal stars. Rigel is a B-type star, Sirius an A-type, Procyon an F-type, Capella a G-type, Aldebaran a K-type, and Betelgeuse an M-type. The color differences of these stars are easy to see. Why do you suppose there is no O-type star in the Winter Circle?

2. In the winter sky, you'll find the red supergiant Betelgeuse in the constellation Orion. It's easy to see because it's one of the brightest stars visible in our night sky. Betelgeuse is a variable star with a period of about 6.5 years. Its brightness changes as it expands and contracts. At maximum size, Betelgeuse fills a volume of space that would extend from the Sun to beyond the orbit of Jupiter. Betelgeuse is thought to be about 10 to 15 times more massive than our Sun, and probably between four and 10 million years old. A similar star can be found shining prominently in midsummer. This is the red supergiant Antares in the constellation Scorpius. Depending on the time of year, can you find one of these stars? Why are they red?

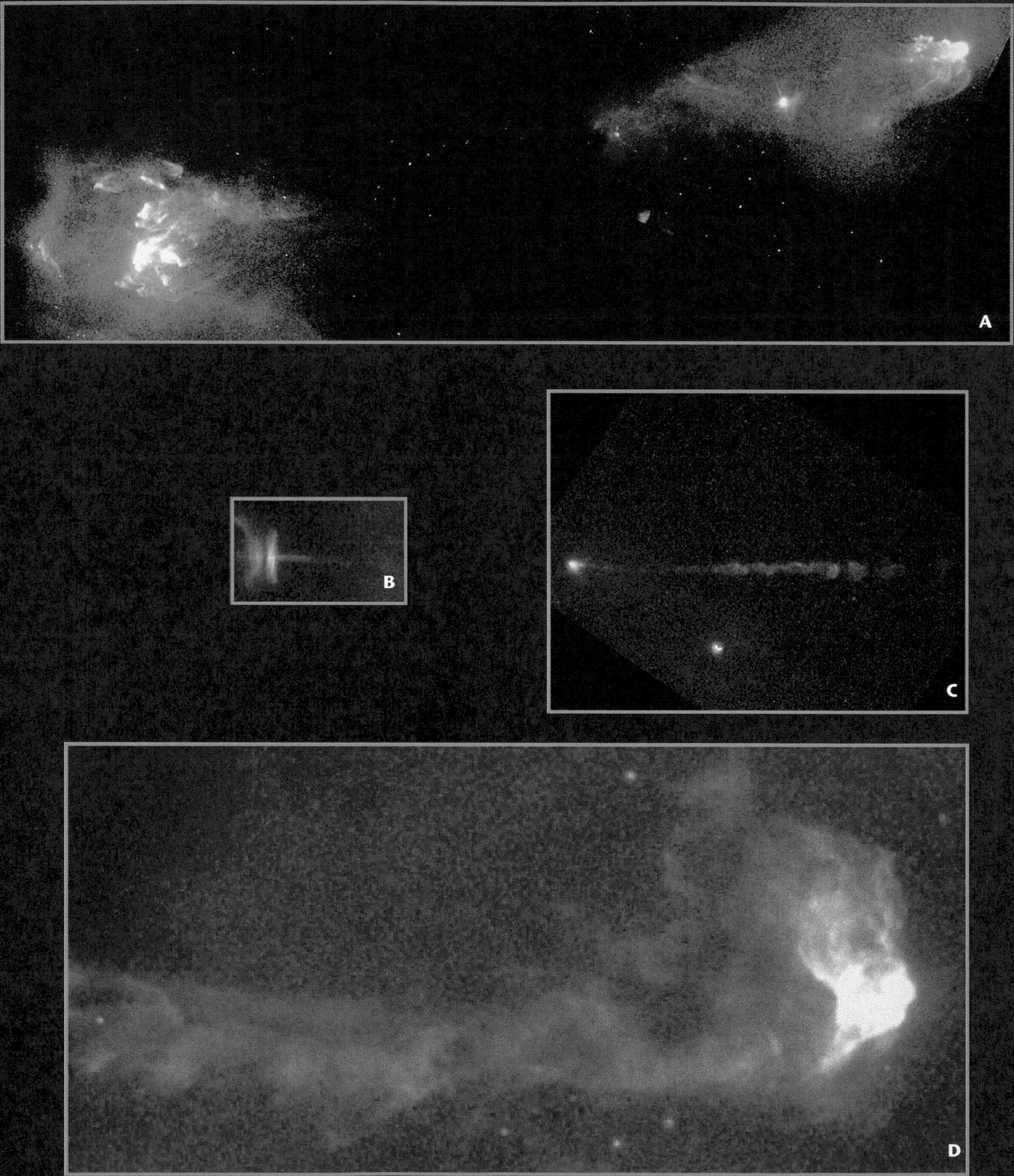
A
B
C
D

11 THE INTERSTELLAR MEDIUM

Birthplace of Stars

LEARNING GOALS

Studying this chapter will enable you to:

1. Summarize the composition and physical properties of the interstellar medium.
2. Describe the characteristics of emission nebulae, and explain their significance in the life cycle of stars.
3. Discuss the properties of dark interstellar dust clouds.
4. Specify some of the radio techniques used to probe the nature of interstellar matter.
5. Summarize the sequence of events leading to the formation of a star like our Sun.
6. Describe some of the observational evidence supporting the modern theory of star formation.
7. Explain how the formation of a star is affected by its mass.

Animation

(Photo A) Disks and jets pervade the universe on many scales. Here, they seem to be a natural result of a rotating cloud of gas contracting to form a star. Matter falling onto the embryonic star creates a pair of high-speed jets of gas perpendicular to the star's flattened disk, carrying away heat and angular momentum that might otherwise prevent the birth of the star. This image shows a small region near the Orion Nebula known as HH1/HH2, whose twin jets have blasted outward for several trillion km (nearly half a light-year) before colliding with interstellar matter. (HH stands for Herbig–Haro, after the investigators who first cataloged such objects.) The next three photos show stellar jets ejected from three different very young stars. Reproduced here to scale, these images collectively depict the propagation of a jet through space.

(Photo B) This image of HH30, spanning approximately 250 billion km, or about 0.01 pc, shows a thin jet (in red) emanating from a circumstellar disk (at left in grey) encircling a nascent star.

(Photo C) One of HH34's jets is longer, reaching some 600 billion km, yet remains narrow, with a beaded structure.

(Photo D) HH47 is more than a trillion km in length, or nearly 0.1 pc. This photo shows one of its jets plowing through interstellar space, creating bow shocks in the process. *(NASA)*

Stars and planets are not the only inhabitants of our Galaxy. The space around us harbors invisible matter throughout the dark voids between the stars. The density of this interstellar matter is extremely low—approximately a trillion trillion times less dense than matter in either stars or planets, far more tenuous than the best vacuum attainable on Earth. Only because the volume of interstellar space is so vast does the mass of interstellar matter amount to anything at all. Why bother to study this near-perfect vacuum? We do so for three important reasons. First, there is as much mass in the "voids" between the stars as there is in the stars themselves. Second, interstellar space is the region out of which new stars are born. Third, it is the region into which some old stars explode at death. It is one of the most significant crossroads through which matter passes in the history of our universe.

11.1 Interstellar Matter

1 Figure 11.1 shows a much greater expanse of universal "real estate" than anything we have studied thus far. The bright regions are congregations of innumerable stars, merging together into a continuous blur at the resolution of the telescope. However, the dark areas are *not* simply "holes" in the stellar distribution. They are regions of space where *interstellar matter* obscures the light from stars beyond.

Gas and Dust

The matter between the stars is collectively termed the **interstellar medium**. It is made up of two components—*gas* and *dust*—intermixed throughout all of space. The gas is made up mainly of individual atoms and small molecules. The dust consists of clumps of atoms and molecules—not unlike the microscopic particles that make up smoke or soot.

Apart from numerous narrow atomic and molecular absorption lines, the gas alone does not block electromagnetic radiation to any great extent. The obscuration evident in Figure 11.1 is caused by dust. Light from distant stars cannot penetrate the densest accumulations of interstellar dust any more than a car's headlights can illuminate the road ahead in a thick fog. As a rule of thumb, a beam of light can be absorbed or scattered only by particles having diameters comparable to or larger than the wavelength of the radiation involved, and the amount of obscuration (absorption or scattering) produced by particles of a given size increases with decreasing wavelength. The typical diameter of an interstellar dust particle—or **dust grain**—is about 10^{-7} m, comparable in size to the wavelength of visible light. Consequently, dusty regions of interstellar space are transparent to long-wavelength

Figure 11.1 Milky Way Photo A wide-angle photograph of a great swath of space, showing regions of brightness (vast fields of stars) as well as regions of darkness (where interstellar matter obscures the light from more distant stars). The field of view is roughly 30° across. *(California Institute of Technology/Palomar/Hale Observatory)*

radio and infrared radiation but opaque to shorter-wavelength optical, ultraviolet, and X-ray radiation.

Because the opacity of the interstellar medium increases with decreasing wavelength, light from distant stars is preferentially robbed of its higher-frequency ("blue") components. Hence, in addition to being generally diminished in overall brightness, stars also appear redder than they really are (Figure 11.2). This effect, known as **reddening**, is similar to the process that produces spectacular red sunsets here on Earth. As indicated in Figure 11.2, absorption lines in a star's spectrum are still recognizable in the radiation reaching Earth, allowing the star's spectral class, and hence its luminosity and color, to be determined. ∞ (Sec. 10.7) Astronomers can then measure the degree to which the star's light has been diminished and reddened en route to Earth.

Density and Composition of the Interstellar Medium

By measuring its effect on the light from many different stars, astronomers have built up a picture of the distribution and chemical properties of interstellar matter in the solar neighborhood.

Gas and dust are found everywhere in interstellar space. No part of our Galaxy is truly devoid of matter, although the density of the interstellar medium is extremely low. Overall, the gas averages roughly 10^6 atoms per cubic meter—just one atom per cubic centimeter. (For comparison, the best vacuum presently attainable in laboratories on Earth contains about 10^9 atoms per cubic meter.) Interstellar dust is even rarer—about one dust particle for every trillion or so atoms. The space between the stars is populated with matter so thin that harvesting all the gas and dust in an interstellar region the size of Earth would yield barely enough matter to make a pair of dice.

Despite such low densities, over sufficiently large distances interstellar matter accumulates slowly but surely, to the point where it can block visible light and other short-wavelength radiation from distant sources. All told, space in the vicinity of the Sun contains about as much mass in the form of interstellar gas and dust as exists there in the form of stars.

Interstellar matter is distributed very unevenly. In some directions it is largely absent, allowing astronomers to study objects literally billions of parsecs from the Sun. In other directions there are small amounts of interstellar matter, so the obscuration is moderate, preventing us from seeing objects more than a few thousand parsecs away, but allowing us to study nearby stars. Still other regions are so heavily obscured that starlight from even relatively nearby stars is completely absorbed before reaching Earth.

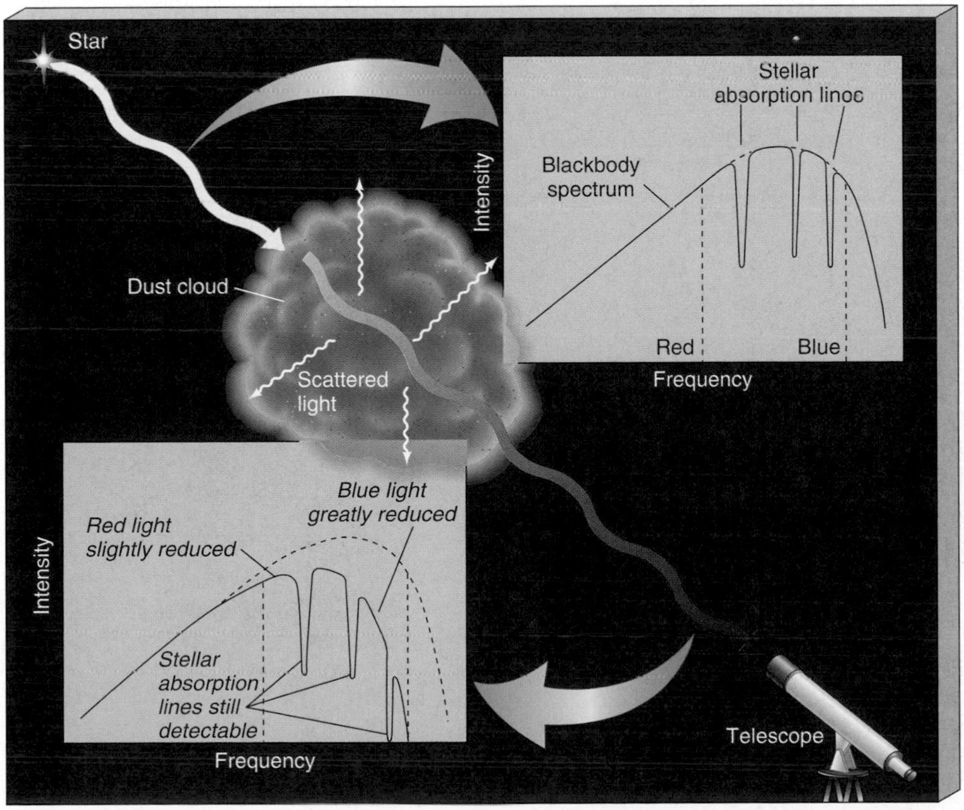

Figure 11.2 Light Reddening Starlight passing through a dusty region of space is both dimmed and reddened, but spectral lines are still recognizable in the light that reaches Earth.

The composition of interstellar gas is reasonably well known from spectroscopic studies of interstellar absorption lines. (Sec. 2.5) Generally speaking, it mirrors the composition of other astronomical objects, such as the Sun, the stars, and the jovian planets: Most of the gas—about 90 percent—is atomic or molecular hydrogen, nine percent is helium, and the remaining one percent consists of heavier elements. The gas is deficient in some heavy elements, such as carbon, oxygen, silicon, magnesium, and iron, most likely because these elements have gone to form the interstellar dust.

In contrast, the composition of the dust is not well known, although there is some infrared evidence for silicates, carbon, and iron, supporting the theory that interstellar dust forms out of interstellar gas. The dust probably also contains some "dirty ice," a frozen mixture of water ice contaminated with trace amounts of ammonia, methane, and other compounds, much like cometary nuclei in our own solar system. (Sec. 4.2)

Concept Check

- If space is a near-perfect vacuum, how can there be enough dust in it to block starlight?

11.2 Interstellar Clouds

Figure 11.3 is a mosaic of photographs showing a region of the sky even larger than that shown in Figure 11.1. The patchy obscuration of background stars by foreground dust is plainly evident. From our vantage point on Earth, this assemblage of stellar and interstellar matter stretches all the way across the sky. On a clear night, it is visible to the naked eye as the Milky Way. In Chapter 14 we will come to recognize this band as the flattened disk, or *plane*, of our own Galaxy.

Star-Forming Regions

2 Figure 11.4 shows a 12°-wide swath of the Milky Way in the general direction of the constellation Sagittarius. The field of view is mottled with stars and interstellar matter. In addition, several large fuzzy patches of light are clearly visible. These fuzzy objects, labeled M8, M16, M17, and M20, correspond to the 8th, 16th, 17th, and 20th objects in a catalog compiled by Charles Messier, an eighteenth-century French astronomer. Today they are known as **emission nebulae**[1]—glowing clouds of hot interstellar matter.

We can gain a better appreciation of these nebulae by examining progressively smaller fields of view. Figure 11.5 is an enlargement of the region near the bottom of Figure 11.4, showing M20 at the top and M8 at the bottom. Figure 11.6 is an enlargement of the top of Figure 11.5, presenting a close-up of M20 and its immediate environment. The total area displayed measures some 10 pc across. Emission nebulae are among the most spectacular objects in the universe, yet they appear only as small, undistinguished patches of light when viewed

[*] Emission nebulae are also often referred to as *HII regions*, where HII refers to the ionized state of hydrogen. Regions of space containing primarily neutral (atomic) hydrogen are known as *HI regions*.

Figure 11.3 **Milky Way Mosaic** The Milky Way photographed almost from horizon to horizon, thus extending over nearly 180°. This band contains high concentrations of stars as well as interstellar gas and dust. The field of view is several times wider than that of Figure 11.1, whose outline is superimposed on this image. *(Palomar Observatory/California Institute of Technology)*

in the larger context of the Milky Way. Perspective is crucial in astronomy.

These nebulae are regions of glowing, ionized gas. At or near the center of each is at least one newly formed hot O- or B-type star producing copious amounts of ultraviolet light. As ultraviolet photons travel outward from the star, they ionize the surrounding gas. As electrons recombine with nuclei, they emit visible radiation, causing the gas to glow. Figure 11.7 shows enlargements of two of the other nebulae visible in Figure 11.4. Notice the hot bright stars embedded within the glowing

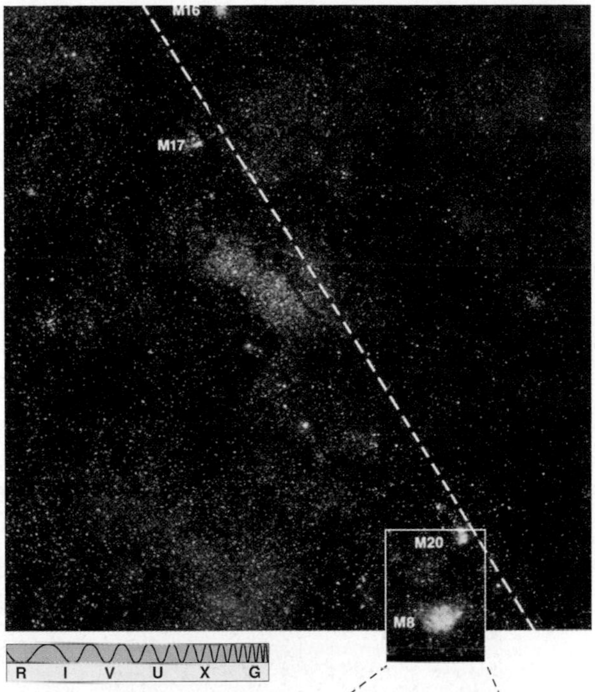

Figure 11.4 Galactic Plane Photo A photograph of a small portion (about 12° across) of the region of the sky shown in Figure 11.1, displaying higher-resolution evidence for stars, gas, and dust as well as several distinct fuzzy patches of light, known as emission nebulae. The center line of the Milky Way is marked with a dashed white line. *(Harvard College Observatory)*

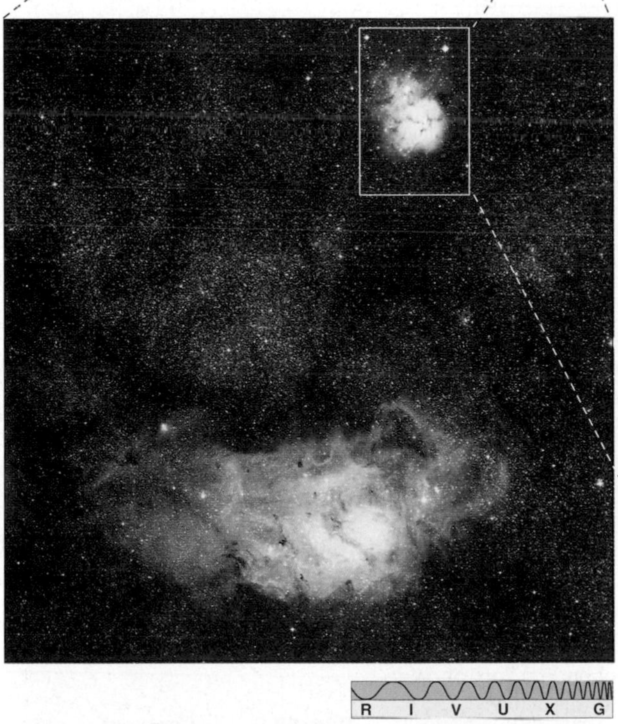

Figure 11.5 M20–M8 Region An enlargement of the bottom of Figure 11.4, showing M20 (top) *(Royal Observatory, Edinburgh/Photo Reachers, Inc.)* and M8 (bottom) more clearly. *(Royal Observatory Edinburgh/Science Photo Library/Photo Reachers, Inc.)*

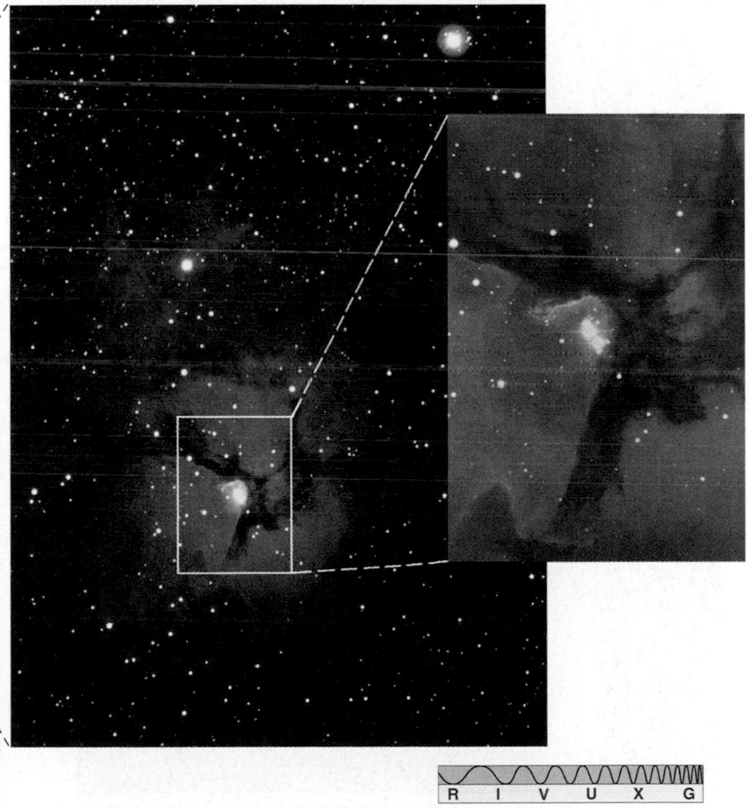

Figure 11.6 Trifid Nebula Further enlargements of the top of Figure 11.5, showing only M20 and its interstellar environment. The nebula (red) is about 4 pc in diameter. It is often called the Trifid Nebula because of the three dust lanes that trisect its midsection (insert). The bluish region immediately above M20 is unrelated to the red emission nebula and is caused by starlight reflected from intervening dust particles. It is called a *reflection nebula. (NOAO; Anglo–Australian Observatory/Royal Observatory Edinburgh. Photograph made from VK Schmidt Plates by David Malin)*

nebular gas, and the predominant red coloration of the emitted radiation—the result of hydrogen atoms emitting light in the red part of the visible spectrum. (Sec. 2.6) The interaction between stars and gas is particularly striking in Figure 11.7(b). The three dark "pillars" visible in this spectacular image are part of the interstellar cloud from which the stars formed. The rest of the cloud in the vicinity of the new stars has already been dispersed by their radiation. The fuzz around the edges of the pillars, especially at top right and center, is the result of this process.

Woven through the glowing nebular gas, and plainly visible in Figures 11.5–11.7, are dark **dust lanes** obscuring the nebular light. These dust lanes are part of the nebulae, and are not just unrelated dust clouds that happen to lie along our line of sight. This relationship is again most evident in Figures 11.7(b) and (d), where the regions of gas and dust are simultaneously silhouetted against background nebular emission and illuminated by foreground nebular stars.

Table 11.1 lists some vital statistics for the nebulae shown in Figure 11.4. Unlike stars, nebulae are large

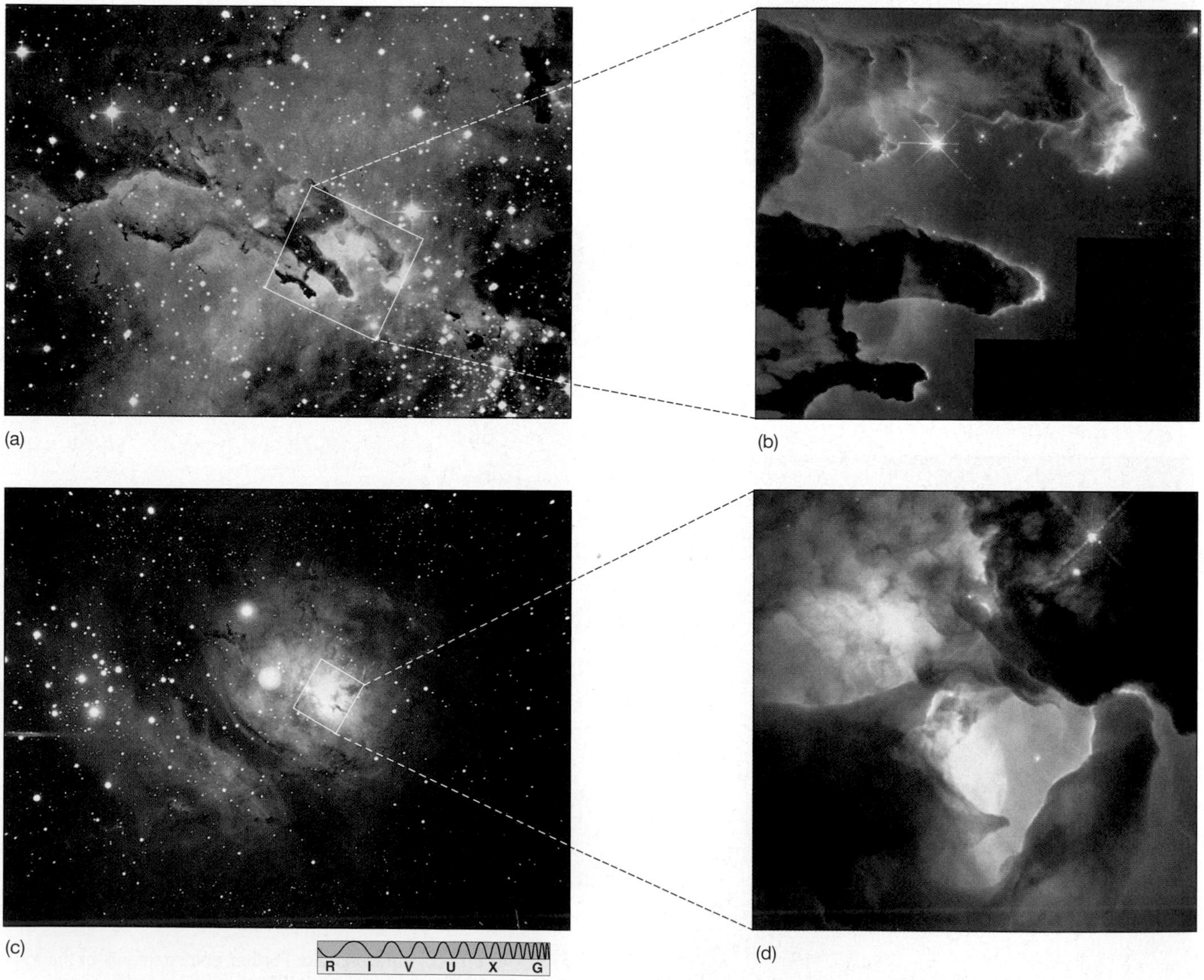

Figure 11.7 Emission Nebulae Enlargements of selected portions of Figure 11.4. (a) M16, the Eagle Nebula. (b) A recent *Hubble Space Telescope* image of huge pillars of cold gas and dust inside M16; delicate sculptures created by the action of stellar ultraviolet radiation on the original cloud. (c) M8, the Lagoon Nebula. (d) A high-resolution view of the core of M8, a region known as the Hourglass. Notice the irregular shape of the emitting regions, the characteristic red color of the light, the bright stars within the gas, and the patches of obscuring dust. *(AURA; NASA; AURA; NASA)*

Animation Video

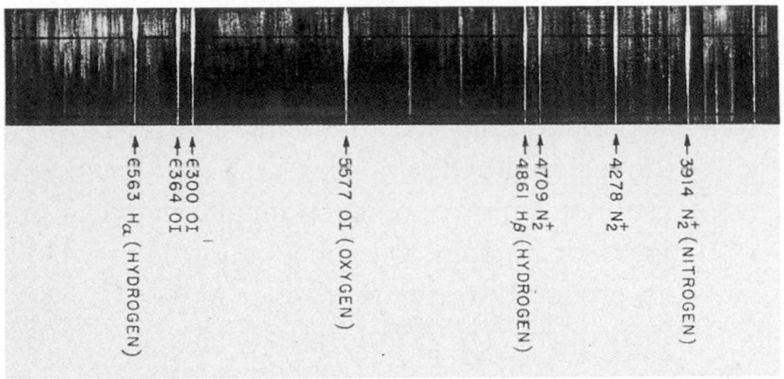

Figure 11.8 Emission Nebula Spectrum The emission spectrum of a typical emission nebula, showing light intensity as a function of frequency over the visible portion of the electromagnetic spectrum. *(Association of Universities for Research in Astronomy)*

enough for their sizes to be measurable by simple geometry. Coupling this size information with estimates of the amount of matter along our line of sight (as revealed by a nebula's total light emission), we can find the nebula's density. Generally, emission nebulae have densities of a few hundred particles, mostly protons and electrons, per cubic centimeter (10^8 per cubic meter).

Figure 11.8 is a typical nebular spectrum spanning part of the visible and near-ultraviolet wavelength interval. ∞ (Sec. 2.5) Numerous emission lines can be seen, and information on the nebula can be extracted from all of them. Analyses of such spectra reveal compositions similar to those found in the Sun and other stars and elsewhere in the interstellar medium. Spectral-line widths imply that the gas atoms and ions have temperatures around 8000 K.

Dark Dust Clouds

3 Emission nebulae are only one small component of interstellar space. Most of space—in fact, more than 99 percent of it—is devoid of such regions and contains no stars. It is simply dark. The average temperature of a typical dark region of interstellar space is about 100 K.

Within these dark voids among the nebulae and the stars lurks another distinct type of astronomical object, the **dark dust cloud**. These clouds are cooler than their surroundings (with temperatures as low as a few tens of kelvins), and thousands or even millions of times denser. In some regions, densities exceeding 10^9 atoms/m^3 (1000 atoms/cm^3) are found. Most of these clouds are bigger than our solar system, and some are many parsecs across (yet, even so, they make up no more than a few percent of the entire volume of interstellar space). Despite their name, these clouds are made up primarily of gas, just like the rest of the interstellar medium. However, their absorption of starlight is due almost entirely to the dust they contain.

Figure 11.9(a) is an optical photograph of a typical dark dust cloud. Pockets of intense blackness mark

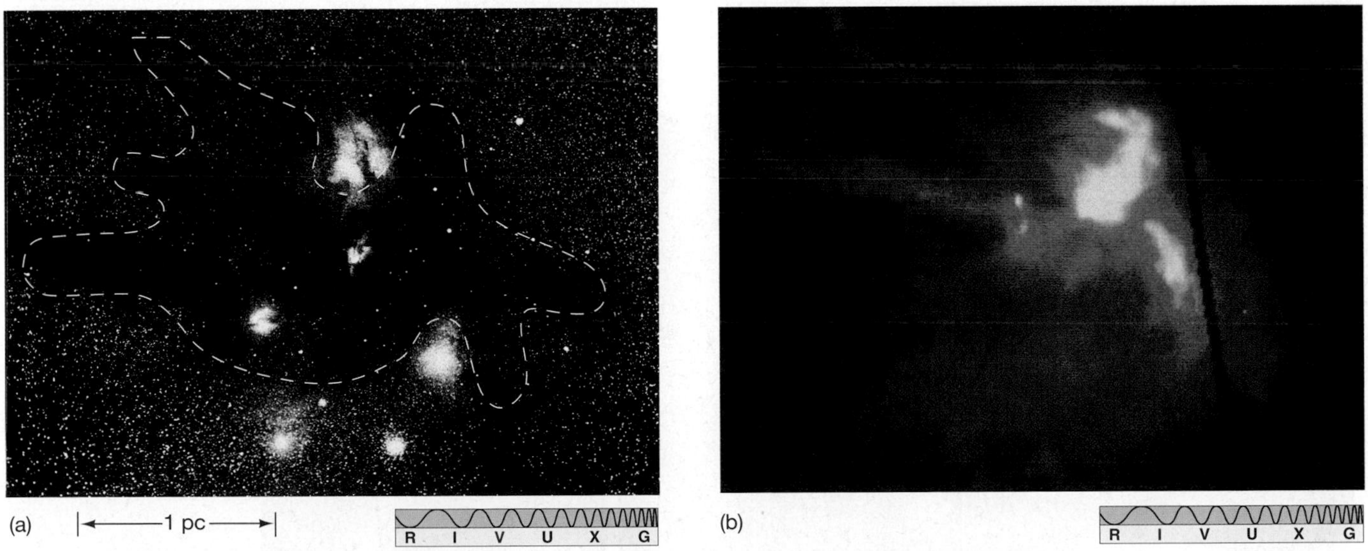

Figure 11.9 Dark Dust Cloud (a) A typical dark dust cloud, known as Rho Ophiuchi, is "visible" only because it blocks the light coming from stars lying behind it. The approximate outline of the cloud is indicated by the dashed line. (b) An infrared map of the same region at roughly the same scale. The infrared emission, and therefore the dust that produces it, display a structure similar to the cloud's visual image. The very bright source of infrared radiation near the top of the cloud comes from a hot emission nebula, which can also be seen in the optical image. (The black streak at right is an instrumental effect.) *(Harvard-Smithsonian Center for Astrophysics; NASA)*

TABLE 11.1 Nebular Properties

OBJECT	APPROX. DISTANCE (pc)	AVERAGE DIAMETER (pc)	DENSITY (10^6 particles/m^3)	MASS (solar masses)	TEMPERATURE (K)
M8	1200	14	80	2600	7500
M16	1800	8	90	600	8000
M17	1500	7	120	500	8700
M20	900	4	100	150	8200

regions where the dust and gas are especially concentrated and the light from background stars is completely obscured. Note the long "streamers" of dust and gas. Most dark dust clouds are very irregularly shaped. The bright patches within the dark region are emission nebulae in the foreground. Some of them are part of the cloud, where newly formed stars near the surface have created a "hot spot" in the cold, dark gas. Others have no connection to the cloud and just happen to lie along our line of sight. Like all dark dust clouds, this cloud is too cold to emit any visible light. However, it does radiate strongly at longer wavelengths. Figure 11.9(b) shows an infrared image of the same region.

Figure 11.10 shows another striking example of a dark dust cloud—the Horsehead Nebula in Orion. This curiously shaped finger of gas and dust projects out from the much larger dark cloud in the bottom half of the image and stands out clearly against the red glow of a background emission nebula.

21-Centimeter Radiation

4 In many cases, dark dust clouds absorb so much light from background stars that they cannot easily be studied by the techniques described in Section 11.1. Fortunately, astronomers have an important alternative means of probing their structure that relies on low-energy radio emissions produced by the interstellar gas itself.

Much of the gas in interstellar space consists of atomic hydrogen. Recall that a hydrogen atom consists of one electron orbiting a single-proton nucleus. ∞ (Sec. 2.6) Besides its orbital motion around the central proton, the electron also has some rotational motion—that is, *spin*—about its own axis. The proton also spins. This model parallels a planetary system, in which, in addition to the orbital motion of a planet about a central star, both the planet (electron) and the star (proton) rotate about their axes.

The laws of physics dictate that there are just two possible spin configurations for a hydrogen atom in its ground state. As illustrated in Figure 11.11, the electron and proton can either rotate in the *same* direction, with their spin axes parallel, or they can rotate in *opposite* directions, with their axes parallel, but oppositely oriented. The former configuration has slightly higher energy than the latter. When a slightly excited hydrogen atom having the electron and proton spinning in the same direction drops down to the less energetic, opposite-spin state, the transition releases a photon with energy equal to the energy difference between the two states.

Figure 11.10 Horsehead Nebula The Horsehead Nebula in Orion is a striking example of a dark dust cloud, silhouetted against the bright background of an emission nebula. The "neck" of the horse is about 0.25 pc across. The dark region lies roughly 1500 pc from Earth. *(Royal Observatory of Belgium; D. Malin/Anglo-Australian Observatory)*

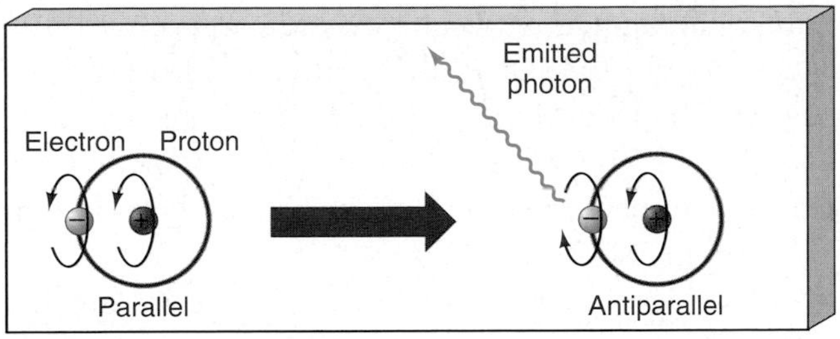

Figure 11.11 **Hydrogen 21cm Emission** A ground-level hydrogen atom changing from a higher-energy state (electron and proton spinning in the same direction) to a lower-energy state (spinning in opposite directions). The emitted photon carries away an energy equal to the energy difference between the two spin states.

Because the energy difference between the two states is very small, the energy of the emitted photon is very low. Consequently, the wavelength of the radiation is long—about 21 cm. That wavelength lies in the radio portion of the electromagnetic spectrum. Researchers refer to the spectral line that results from this hydrogen-spin-flip process as the **21-centimeter line**. It provides a vital probe into any region of the universe containing atomic hydrogen gas. Needing no visible starlight to help calibrate their signals, radio astronomers can observe any interstellar region that contains enough hydrogen to produce a detectable signal. Even the low-density regions between dark dust clouds can be studied.

Of great importance is the fact that the wavelength of 21-cm radiation is much larger than the typical size of interstellar dust particles. Accordingly, this radio radiation reaches Earth completely unaffected by interstellar debris. The opportunity to observe interstellar space well beyond a few thousand parsecs, and in directions lacking detectable background stars, makes 21-cm observations among the most important and useful in all of astronomy.

Molecular Gas

3 4 In certain interstellar regions of cold (typically 20 K) neutral gas, densities can reach as high as 10^{12} particles/m^3. The gas particles in these regions are not atoms, but *molecules*. Because of the predominance of molecules in these dense interstellar regions, they are known as **molecular clouds**.

Only long-wavelength radio radiation can escape from these dense, dusty parts of interstellar space. Molecular hydrogen (H_2) is by far the most common constituent of molecular clouds, but unfortunately this molecule does not emit or absorb radio radiation, so it cannot easily be used as a probe of cloud structure. Nor are 21-cm observations helpful—they are sensitive only to *atomic* hydrogen, not to the *molecular* form of the gas. Instead, astronomers use radio observations of other "tracer" molecules, such as carbon monoxide (CO), hydrogen cyanide (HCN), ammonia (NH_3), water (H_2O), and formaldehyde (H_2CO), to study the dark interiors of these dusty regions. These molecules are produced by chemical reactions within molecular clouds, and have emission lines in the radio part of the spectrum. They are found only in very small quantities—perhaps one part per billion—but when we observe them we know that the regions under study must also contain high densities of molecular hydrogen, dust, and other important constituents.

Figure 11.12 shows a contour map of the distribution of formaldehyde molecules in the immediate vicinity of the M20 nebula. Notice that the amount of formaldehyde (and, we assume, the amount of hydrogen) peaks well away from the visible nebula. Radio maps like this of interstellar gas and infrared maps of interstellar dust reveal that molecular clouds do not exist as distinct and separate objects in space. Rather, they

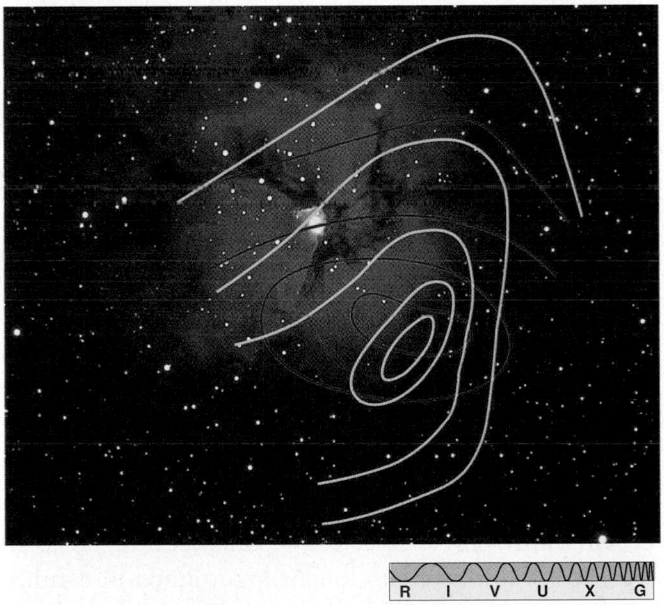

Figure 11.12 **Molecules Near M20** Contour map of the amount of formaldehyde near the M20 nebula, demonstrating how formaldehyde is especially abundant in the darkest interstellar regions. Other kinds of molecules have been found to be similarly distributed. The contour values increase from the outside to the inside, so the maximum density of formaldehyde lies just to the bottom right of the visible nebula. The green and red contours outline the intensity of formaldehyde lines at different frequencies. *(AURA; contours by the authors)*

make up huge **molecular cloud complexes**, some spanning as much as 50 pc across and containing enough gas to make millions of stars like our Sun. About 1000 such complexes are known in our Galaxy.

Why are molecules found only in the densest and darkest of interstellar clouds? One possible reason is that the dust protects the fragile molecules from the harsh interstellar environment—the same absorption that prevents high-frequency radiation from getting out to our detectors also prevents it from getting in to destroy the molecules. Another possibility is that the dust acts as a catalyst that helps form the molecules. The dust grains provide both a place where atoms can stick and react and a means of dissipating any heat associated with the reaction, which might otherwise destroy the newly formed molecules. Probably the dust plays both roles, although the details are still debated.

Concept Check

- If emission nebulae are powered by ultraviolet light from hot stars, why do they appear red?

11.3 The Formation of Stars Like the Sun

5 6 Let us now turn our attention to the connection between the interstellar medium and the stars in our Galaxy. How do stars form? What factors determine their masses, luminosities, and spatial distribution? In short, what basic processes are responsible for the appearance of our night sky?

Gravity and Heat

Simply stated, star formation begins when part of the interstellar medium—one of the cold, dark clouds discussed in the previous section—collapses under its own weight. An interstellar cloud is maintained in equilibrium by a balance between two basic opposing influences: gravity (which is always directed inward) and heat (in the form of outwardly directed pressure). ∞ (*More Precisely 5-2*) However, if gravity somehow begins to dominate over heat, then the cloud can lose its equilibrium and start to contract.

Consider a small portion of a large interstellar cloud. Concentrate first on just a few atoms, as shown in Figure 11.13. Even though the cloud's temperature is very low, each atom has some random motion. ∞ (*More Precisely 2-1*) Each atom is also influenced by the gravitational attraction of all its neighbors. The gravitational force is not large, however, because the mass of each atom is so small. When a few atoms accidentally cluster for an instant, as sketched in Figure 11.13(b), their combined gravity is insufficient to bind them into a lasting, distinct clump of matter. This accidental cluster disperses as quickly as it formed. The effect of heat—the random motion of the atoms—is much stronger than the effect of gravity.

As the number of atoms increases, their gravitational attraction increases too, and eventually the collective gravity of the clump is strong enough to prevent it from dispersing back into interstellar space How many atoms are required for this to be the case? The answer, for a typical cool (100 K) cloud, is about 10^{57}—in other words, the clump must have mass comparable to that of the Sun. If our hypothetical clump is more massive than that, then it will not disperse. Instead, its gravity will cause it to contract, ultimately to form a star.

Of course, 10^{57} atoms don't just clump together by random chance. Rather, star formation is *triggered* when a sufficiently massive pocket of gas is squeezed by some external event. Perhaps it is compressed by the pressure wave produced when a nearby O- or B-type star forms and heats its surroundings, or possibly part of a cloud simply cools below the temperature at which its internal pressure is no longer sufficient to support it against its own gravity. Whatever the cause, theory suggests that once the collapse begins, star formation inevitably follows.

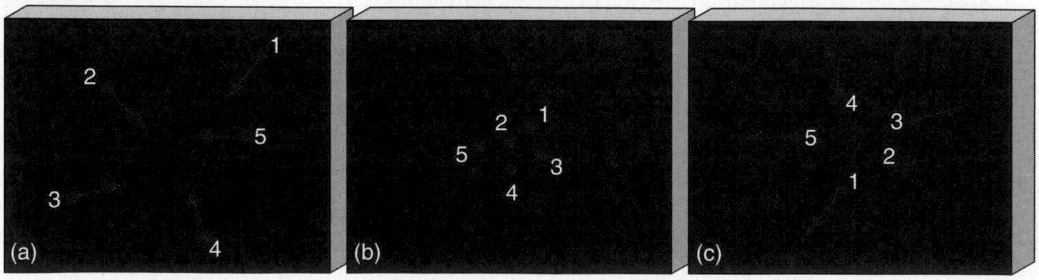

Figure 11.13 Atomic Motions The motions of a few atoms within an interstellar cloud are influenced by gravity so slightly that the atoms' paths are hardly changed (a) before, (b) during, and (c) after an accidental, random encounter.

TABLE 11.2 Prestellar Evolution of a Sunlike Star

STAGE	APPROXIMATE TIME TO NEXT STAGE (yr)	CENTRAL TEMPERATURE (K)	SURFACE TEMPERATURE (K)	CENTRAL DENSITY (particles/m^3)	DIAMETER[1] (km)	OBJECT
1	2×10^6	10	10	10^9	10^{14}	Interstellar cloud
2	3×10^4	100	10	10^{12}	10^{12}	Cloud fragment
3	10^5	10,000	100	10^{18}	10^{10}	Cloud fragment/protostar
4	10^6	1,000,000	3000	10^{24}	10^8	Protostar
5	10^7	5,000,000	4000	10^{28}	10^7	Protostar
6	3×10^7	10,000,000	4500	10^{31}	2×10^6	Star
7	10^{10}	15,000,000	6000	10^{32}	1.5×10^6	Main-sequence star

[1]For comparison, recall that the diameter of the Sun is 1.4×10^6 km and that of the solar system roughly 1.5×10^{10} km.

Table 11.2 lists seven evolutionary stages that an interstellar cloud goes through before becoming a main-sequence star like our Sun. These stages are characterized by different central temperatures, surface temperatures, central densities, and radii of the prestellar object. They trace its progress from a quiescent interstellar cloud to a genuine star. The numbers given in Table 11.2 and the following discussion are valid *only* for stars of approximately the same mass as the Sun. In the next section we will relax this restriction and consider the formation of stars of other masses.

Stage 1—An Interstellar Cloud

The first stage in the star-formation process is a dense interstellar cloud—the core of a dark dust cloud or perhaps a molecular cloud. These clouds are truly vast, sometimes spanning tens of parsecs (10^{14} to 10^{15} km) across. Typical temperatures are about 10 K throughout, with a density of perhaps 10^9 particles/m^3. Stage 1 clouds contain thousands of times the mass of the Sun, mainly in the form of cold atomic and molecular gas. (The dust they contain is important for cooling the cloud as it contracts and also plays a crucial role in planet formation, but it constitutes a negligible fraction of the total mass.) (Sec. 4.3)

The dark region outlined by the red and green radio contours in Figure 11.12 (*not* the emission nebula itself, where stars have already formed) probably represents a stage-1 cloud just starting to collapse. Doppler shifts of the observed formaldehyde lines indicate that it is contracting. Less than a light year across, this region has a total mass over 1000 times the mass of the Sun—considerably greater than the mass of M20 itself.

Once the collapse begins, theory indicates that fragmentation into smaller and smaller clumps of matter naturally follows. As illustrated in Figure 11.14, a typical cloud can break up into tens, hundreds, even thousands, of fragments, each imitating the shrinking behavior of the parent cloud and contracting ever faster. The whole process, from a single quiescent cloud to many collapsing fragments, takes a few million years. Depending on the precise conditions under

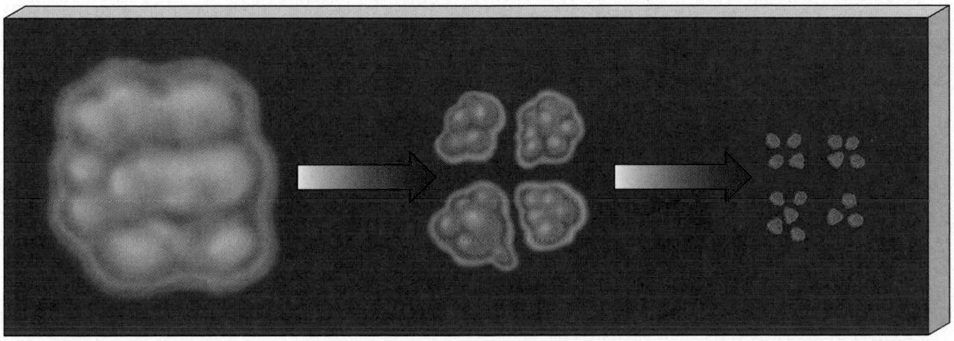

Figure 11.14 Cloud Fragmentation As an interstellar cloud contracts, gravitational instabilities cause it to fragment into smaller pieces. The pieces also contract and fragment, eventually forming many tens or hundreds of individual stars.

which fragmentation takes place, an interstellar cloud can produce either a few dozen stars, each much larger than our Sun, or a collection of hundreds of stars comparable to or smaller than our Sun. There is little evidence for stars born in isolation, one star from one cloud. Most stars—perhaps all stars—appear to originate as members of multiple systems.

The process of continuing fragmentation is eventually stopped by the increasing density within the shrinking cloud. As fragments continue to contract, they eventually become so dense that radiation cannot get out easily. The trapped radiation causes the temperature to rise, the pressure to increase, and the fragmentation to stop. However, the contraction continues.

Stages 2 and 3—A Contracting Cloud Fragment

As it enters stage 2, a fragment destined to form a star like the Sun—the end product of the process sketched in Figure 11.14—contains between one and two solar masses of material. Estimated to span a few hundredths of a parsec across, this fuzzy, gaseous blob is still about 100 times the size of our solar system. Its central density is about 10^{12} particles/m^3.

Even though the fragment has shrunk substantially, its average temperature is not much different from that of its parent cloud. The reason is that the gas constantly radiates large amounts of energy into space. The material of the fragment is so thin that photons produced anywhere within it easily escape without being reabsorbed, so virtually all the energy released in the contraction is radiated away and does not cause any significant increase in temperature. Only at the center, where the radiation must traverse the greatest amount of material in order to escape, is there any appreciable temperature increase. The gas there might be as warm as 100 K by this stage. For the most part, however, the fragment stays cold as it shrinks.

Several tens of thousands of years after it first began contracting, the start of stage 3 occurs when a stage 2 fragment has shrunk to a gaseous sphere with a diameter roughly the size of our solar system (still 10,000 times the size of our Sun). The inner regions have become opaque to their own radiation and have started to heat up considerably, as noted in Table 11.2. The central temperature has reached about 10,000 K—hotter than the hottest steel furnace on Earth. However, the gas near the edge is still able to radiate its energy into space and so remains cool. The central density by this time is approximately 10^{18} particles/m^3 (still only 10^{-9} kg/m^3 or so).

For the first time, our fragment is beginning to resemble a star. The dense, opaque region at the center is called a **protostar**—an embryonic object perched at the dawn of star birth. Its mass increases as more and more material rains down on it from outside, although its radius continues to shrink because its pressure is still unable to overcome the relentless pull of gravity. By the end of stage 3, we can distinguish a "surface" on the protostar—its *photosphere*. Inside the photosphere, the protostellar material is opaque to the radiation it emits. (Note that this is the same operational definition of *surface* we used for the Sun.) ∞ (Sec. 9.1) From here on, the surface temperatures listed in Table 11.2 refer to the photosphere and not to the edge of the collapsing fragment, where the temperature remains low.

Figure 11.15 shows a star-forming region in Orion. Lit from within by several O-type stars, the bright Orion Nebula is partly surrounded by a vast molecular cloud that extends well beyond the roughly 5 × 10 parsec region bounded by the photograph in Figure 11.15(b). The Orion complex harbors several smaller sites of intense radio emission from molecules deep within its core. Shown in Figure 11.15(c), they measure about 10^{10} km, about the diameter of our solar system. Their density is about 10^{15} particles/m^3, much higher than the density of the surrounding cloud. Although the temperatures of these regions cannot be estimated reliably, many researchers regard the regions as objects between stages 2 and 3, on the threshold of becoming protostars. Figures 11.15(d) and (e) are visible-light images of part of the nebula showing other evidence for protostars.

Stages 4 and 5—Protostellar Evolution

As the protostar evolves, it shrinks, its density increases, and its temperature rises, both in the core and at the photosphere. Some 100,000 years after the fragment formed, it reaches stage 4, where its center seethes at about 1,000,000 K. The electrons and protons ripped from atoms whiz around at hundreds of kilometers per second, but the temperature is still short of the 10^7 K needed to ignite the proton–proton nuclear reactions that fuse hydrogen into helium. Still much larger than the Sun, our gassy heap is now about the size of Mercury's orbit. Its surface temperature has risen to a few thousand kelvins.

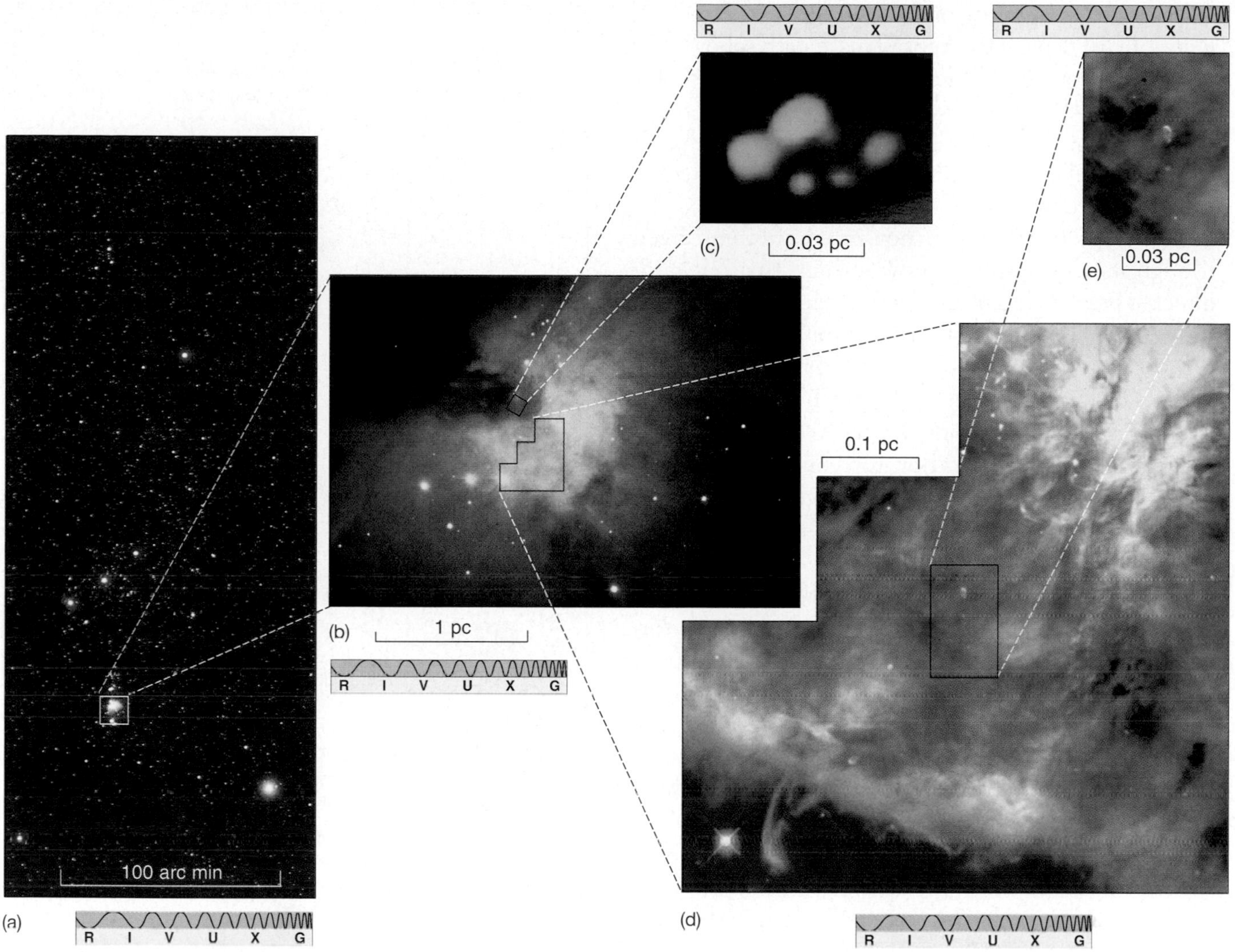

Figure 11.15 Orion Nebula, Up Close (a) The constellation Orion, with the region around its famous emission nebula marked by a rectangle. The Orion Nebula is the middle "star" of Orion's sword. (b) Enlargement of the framed region of part (a), suggesting how the nebula is partly surrounded by a vast molecular cloud. Various parts of this cloud are probably fragmenting and contracting, with even smaller sites forming protostars. (c) and (d) Some of the evidence for those protostars: (c) false-color radio image of some intensely emitting molecular sites, (d) real-color visible image of embedded nebular "knots" thought to harbor protostars, and (e) high-resolution image of several young stars surrounded by disks of gas and dust where planets might ultimately form. *(J. Sanford/Astrostock-Sanford; AURA; James M. Moran; NASA; NASA)*

Video

Knowing the protostar's radius and surface temperature, we can calculate its luminosity using the radius–luminosity–temperature relationship. ∞ (Sec. 10.4) Remarkably, it turns out to be around 1000 times the luminosity of the Sun. Because nuclear reactions have not yet begun in the protostar's core, this luminosity is due entirely to the release of gravitational energy as the protostar continues to shrink and material from the surrounding fragment (which we called the solar nebula back in Chapter 4) rains down on its surface. ∞ (Sec. 4.3)

By the time stage 4 is reached, our protostar's physical properties can be plotted on a Hertzsprung–Russell (H–R) diagram. ∞ (Sec. 10.5) At each phase of a star's evolution, its surface temperature and luminosity can be represented by a single point on the diagram. The motion of that point around the diagram as the star evolves is known as the star's **evolutionary track**. It is a graphical representation of a star's life. The red track on Figure 11.16 depicts the approximate path followed by our interstellar cloud fragment since it became a protostar at the end of stage 3 (which itself lies off the right-hand edge of the figure). Figure 11.17 is an artist's sketch of an interstellar gas cloud proceeding along the evolutionary path outlined so far.

Our protostar is still not in equilibrium. Even though its temperature is now so high that outward-directed pressure has become a powerful countervailing influence against gravity's inward pull, the balance is not yet perfect. The protostar's internal heat gradually diffuses out from the hot center to the cooler surface, where it is radiated away into space. As a result, the contraction slows, but it does not stop completely.

After stage 4, the protostar moves downward on the H–R diagram (toward lower luminosity) and slightly to the left (toward higher temperature), as shown in Figure 11.18. By stage 5, the protostar has shrunk to about 10 times the size of the Sun, its surface temperature is about 4000 K, and its luminosity has fallen to about 10 times the solar value. The central temperature has risen to about 5,000,000 K. The gas is completely ionized by now, but the protons still do not have enough thermal energy for nuclear fusion to begin.

Protostars often exhibit violent surface activity during this phase of their evolution, resulting in ex-

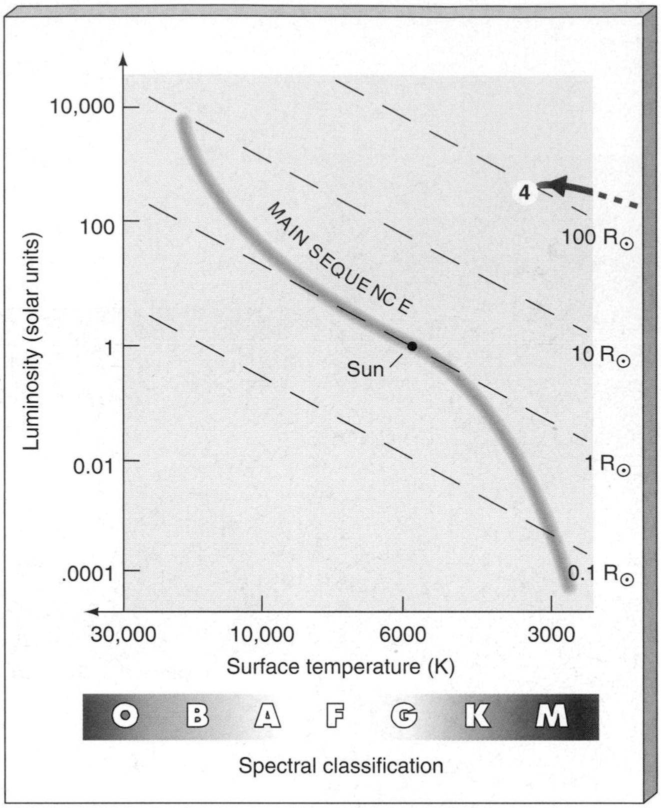

Figure 11.16 Protostar on the H–R Diagram The red arrow indicates the approximate evolutionary track followed by an interstellar cloud fragment before becoming a stage-4 protostar. (The boldface numbers on this and subsequent H–R plots refer to the prestellar evolutionary stages listed in Table 11.2 and described in the text.) Recall from Chapter 10 that "$R_\odot$" means the radius of the Sun.

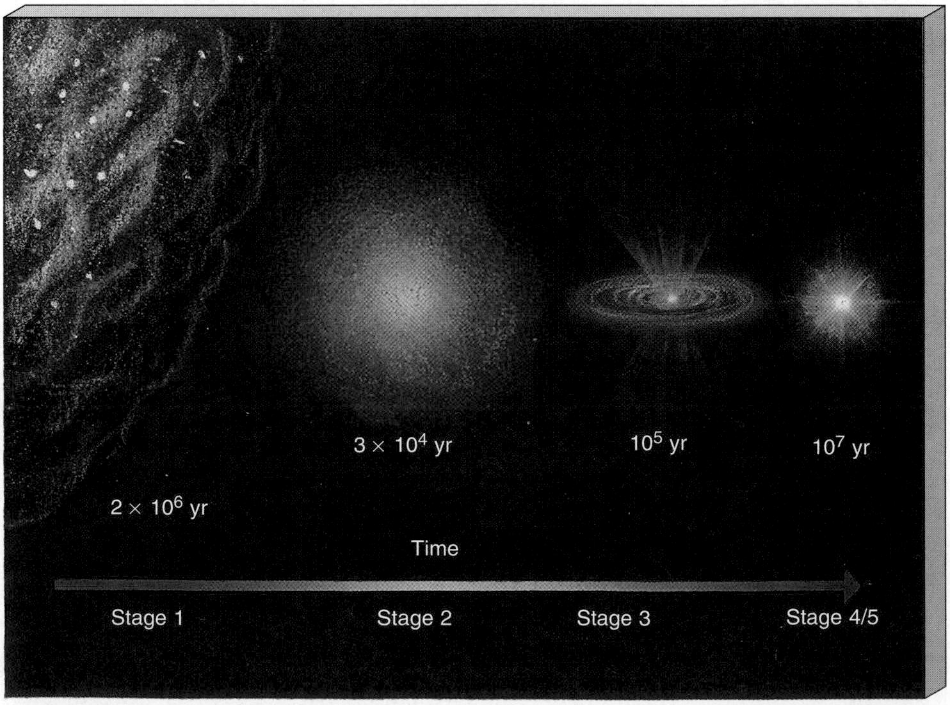

Figure 11.17 Interstellar Cloud Evolution Artist's conception of the changes in an interstellar cloud during the early evolutionary stages outlined in Table 11.2. (Not drawn to scale.) The duration of each stage, in years, is indicated.

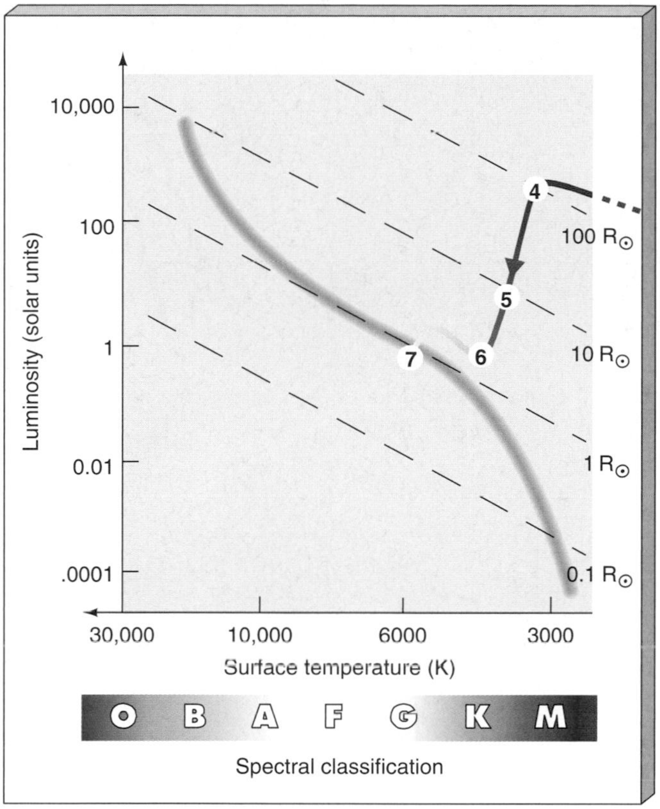

Figure 11.18 Newborn Star on the H–R Diagram The changes in a protostar's observed properties are shown by the path of decreasing luminosity, from stage 4 to stage 6. At stage 7, the newborn star has arrived on the main sequence.

tremely strong protostellar winds, much denser than that of our Sun. This portion of the evolutionary track is often called the **T Tauri phase**, after T Tauri, the first "star" (actually protostar) to be observed in this stage of prestellar development.

Events proceed more slowly as the protostar approaches the main sequence. The initial contraction and fragmentation of the interstellar cloud occurred quite rapidly, but by stage 5, as the protostar nears the status of a full-fledged star, its evolution slows. Its contraction is governed largely by the rate at which it can radiate its internal energy into space. As the luminosity decreases, so too does the contraction rate.

Until the *Infrared Astronomy Satellite* was launched in the early 1980s, astronomers were aware only of very massive stars forming in clouds far away. *IRAS* showed that stars are forming much closer to home, and some of these protostars have masses comparable to that of our Sun. Figure 11.19 shows a premier example of a solar-mass protostar—Barnard 5. Its infrared heat signature is that expected of a stage 5 object.

Stages 6 and 7—A Newborn Star

Some 10 million years after its first appearance, the protostar finally becomes a true star. By stage 6, when our roughly one-solar-mass object has shrunk to a radius of about 1,000,000 km, the contraction has raised the central temperature to 10,000,000 K—enough to ignite nuclear burning. Protons begin fusing into helium nuclei in the core, and a star is born. As shown in Figure 11.18, the star's surface temperature at this point is about 4500 K, still a little cooler than the Sun. Even though the radius of the newly formed star is slightly larger than that of the Sun, its lower temperature means that its luminosity is slightly less than (actually, about two-thirds of) the present solar value.

Indirect evidence for stage-6 stars comes from infrared observations of objects that seem to be luminous hot stars hidden from optical view by surrounding dark clouds. Their radiation is mostly absorbed by a "cocoon" of dust, then reemitted by the dust as infrared radiation. Two considerations support the idea that the hot stars responsible for heating the clouds have only recently

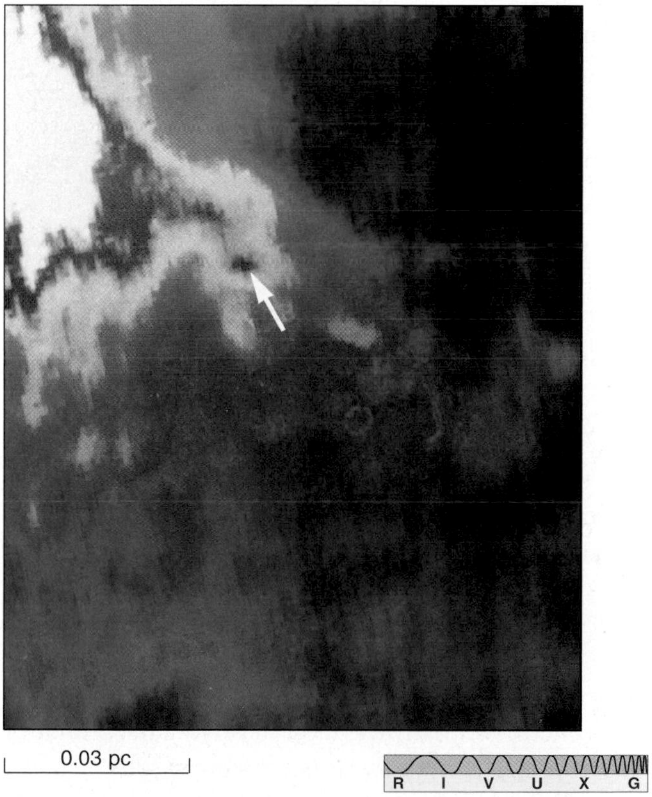

Figure 11.19 Protostar, Imaged An infrared image of the nearby region containing the source Barnard 5 (indicated by the arrow). On the basis of its temperature and luminosity, Barnard 5 appears to be a protostar around stage 5 on the H–R diagram. *(NASA)*

ignited: 1) dust cocoons are predicted to disperse quite rapidly once their central stars form, and 2) these objects are invariably found in the dense cores of molecular clouds, consistent with the star-formation sequence just outlined.

Over the next 30 million years or so, the stage-6 star contracts a little more. Its central density rises to about 10^{32} particles/m^3 (more conveniently expressed as 10^5 kg/m^3), the central temperature increases to 15,000,000 K, and the surface temperature reaches 6000 K. By stage 7, the star finally reaches the main sequence just about where our Sun now resides. Pressure and gravity are finally balanced, and the rate at which nuclear energy is generated in the core exactly matches the rate at which energy is radiated from the surface.

The evolutionary events just described occur over the course of 40 to 50 million years. Although this is a long time by human standards, it is still less than one percent of the Sun's lifetime on the main sequence. Once an object begins fusing hydrogen and establishes a "gravity-in/pressure-out" equilibrium, it burns steadily for a very long time. The star's location on the H–R diagram will remain almost unchanged for the next 10 billion years.

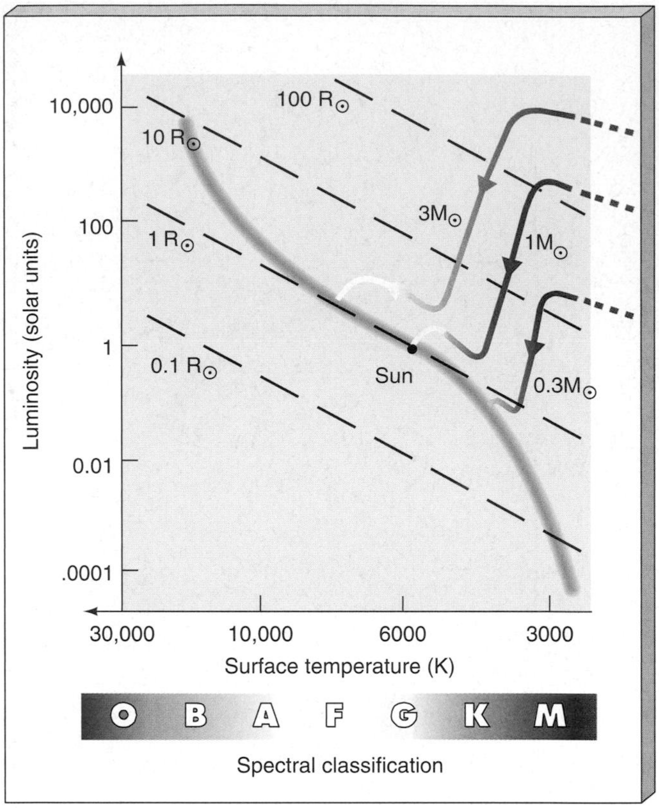

Figure 11.20 Prestellar Evolutionary Tracks Prestellar evolutionary paths for stars more massive and less massive than our Sun.

✓ Concept Check

- What distinguishes a collapsing cloud from a protostar and a protostar from a star?

11.4 Stars of Other Masses

The Zero-Age Main Sequence

7 The numerical values and evolutionary track just described are valid only for one-solar-mass stars. The temperatures, densities, and radii of prestellar objects of other masses exhibit similar trends, but the details differ, in some cases quite considerably. Perhaps not surprisingly, the most massive fragments formed within interstellar clouds tend to produce the most massive protostars and eventually the most massive stars. Similarly, low-mass fragments give rise to low-mass stars.

Figure 11.20 compares the theoretical pre-main-sequence evolutionary track taken by our Sun with the corresponding tracks of a 0.3-solar-mass star and a 3.0-solar-mass star. All three tracks traverse the H–R diagram in the same general manner, but cloud fragments that eventually form stars more massive than the Sun approach the main sequence along a higher track on the diagram, while those destined to form less massive stars take a lower track. The *time* required for an interstellar cloud to become a main-sequence star also depends strongly on its mass. ∞ (Sec. 10.6) The most massive fragments contract into O-type stars in a mere million years, roughly 1/50 the time taken by the Sun. The opposite is the case for prestellar objects having masses much less than our Sun. A typical M-type star, for example, requires nearly a billion years to form.

Whatever the mass, the endpoint of the prestellar evolutionary track is the main sequence. A star is considered to have reached the main sequence when hydrogen burning begins in its core and the star's properties settle down to stable values. The main-sequence band predicted by theory is usually called the **zero-age main sequence** (ZAMS). It agrees quite well with the main sequences observed for stars in the vicinity of the Sun and those observed in more distant stellar systems.

It is important to realize that *the main sequence is not an evolutionary track—stars do not evolve along it.* Rather, it is a "waystation" on the H–R diagram where stars stop and spend most of their lives—low-mass stars at the bottom, high-mass stars at the top. Once on the main sequence, a star stays in roughly the same location in the H–R diagram during its whole time as a stage 7 object. In other words, a star that arrives on the main sequence as, say, a G-type star can never "work its way up" to become a B- or an O-type main-sequence star, nor move down to become an M-type red dwarf. As we will see in Chapter 12, the next stage of stellar evolution occurs when a star leaves the main sequence. When this occurs, the star will have pretty much the same surface temperature and luminosity it had when it arrived on the main sequence millions (or billions) of years earlier.

"Failed" Stars

Some cloud fragments are too small ever to become stars. Pressure and gravity come into equilibrium before the central temperature becomes hot enough to fuse hydrogen, so they never evolve beyond the protostar stage. Rather than turning into stars, such low-mass fragments continue to cool, eventually becoming compact, dark "clinkers"—cold fragments of unburned matter—in interstellar space. Small, faint, and cool (and growing ever colder), these objects are known collectively as **brown dwarfs**. On the basis of theoretical modeling, astronomers believe that the minimum mass of gas needed to generate core temperatures high enough to begin nuclear fusion is about 0.08 solar masses—80 times the mass of Jupiter.

Vast numbers of brown dwarfs may well be scattered throughout the universe—fragments frozen in time somewhere in the cloud-contraction phase. With our current technology we have great difficulty in detecting them, be they planets associated with stars or interstellar cloud fragments alone in space. Conceivably, they might even account for more mass than we observe in the form of stars and interstellar gas combined.

Recent advances in observational hardware and image-processing techniques have identified several likely brown-dwarf candidates, often using techniques similar to those employed in the search for extrasolar planets. ∞ *(Interlude 4-1)* Figure 11.21(a) shows Gliese 623, a binary system containing a brown dwarf candidate originally identified by radial velocity measurements. Figures 11.21(b) and (c) show the binary star system Gliese 229, first identified as a possible brown dwarf by ground-based infrared observations, and more recently imaged from space.

✔ Concept Check

■ Do stars evolve along the main sequence?

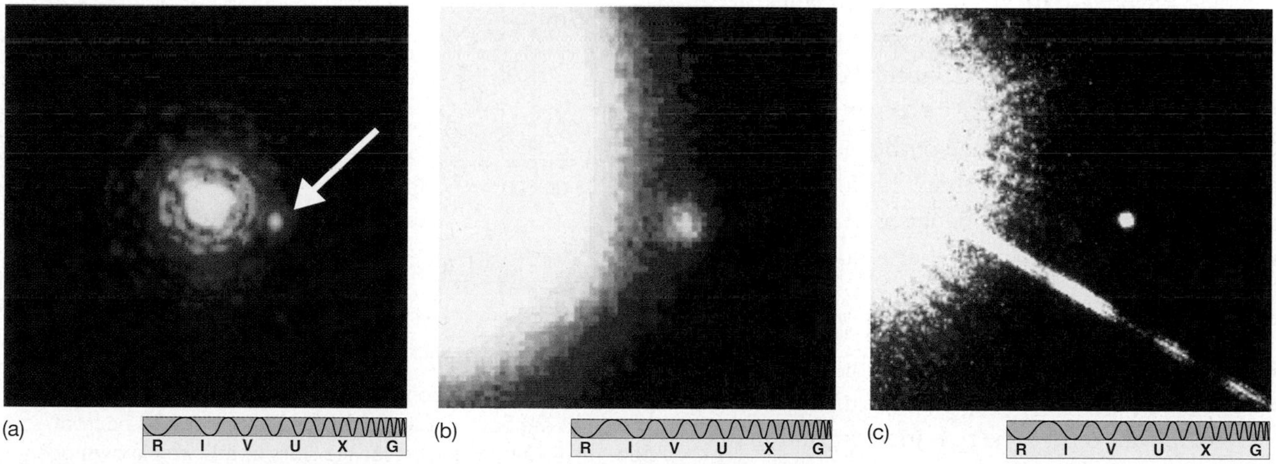

Figure 11.21 Brown Dwarfs, Imaged (a) This *Hubble Space Telescope* image shows Gliese 623, a binary system that may contain a brown dwarf (marked by an arrow). Astronomers hope that continued observations of this system will allow the companion's mass to be measured with sufficient accuracy to determine whether or not it really is a brown dwarf. The "rings" in the image are instrumental artifacts. (b) A ground-based image of the binary star system Gliese 229. The two objects are only 7″ apart; the fainter "star" has a luminosity only a few millionths that of the Sun and an estimated mass about 50 times that of Jupiter. (c) A *Hubble Space Telescope* image of the same system. (The bright diagonal streak in the latter image is caused by a hardware problem in the CCD chip used to record it.) *(NASA)*

11.5 Star Clusters

6 The end result of cloud collapse is a group of stars, all formed from the same parent cloud and lying in the same region of space. Such a collection of stars is called a **star cluster**. Because all the stars formed at the same time out of the same cloud of interstellar gas, and under the same environmental conditions, clusters are near-ideal "laboratories" for stellar studies—not in the sense that astronomers can perform experiments on them, but because the properties of the stars are very tightly constrained. The only factor distinguishing one star from another in the same cluster is mass, so theoretical models of star formation and evolution can be compared with reality without the complications introduced by broad spreads in age, chemical composition, and place of origin.

Clusters and Associations

Figure 11.22(a) shows a small star cluster called the Pleiades, or Seven Sisters, a well-known naked-eye object in the constellation Taurus, lying about 120 pc from Earth. This type of loose, irregular cluster, found mainly in the plane of the Milky Way, is called an **open cluster**. Open clusters typically contain from a few hundred to a few tens of thousands of stars and are a few parsecs across. Figure 11.22(b) shows the H–R diagram for stars in the Pleiades. The cluster contains stars in all parts of the main sequence. The blue stars must be relatively young, for, as we saw in Chapter 10, they burn their fuel very rapidly. ∞ (Sec. 10.6) Thus, even though we have no direct evidence of the cluster's birth, we can estimate its age as less than 20 million years, the lifetime of an O-type star. The wisps of leftover gas evident in the photograph are further evidence of the cluster's youth.

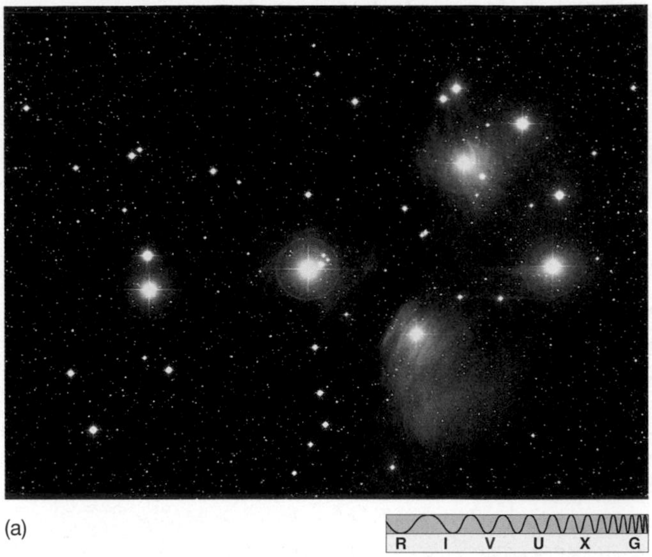

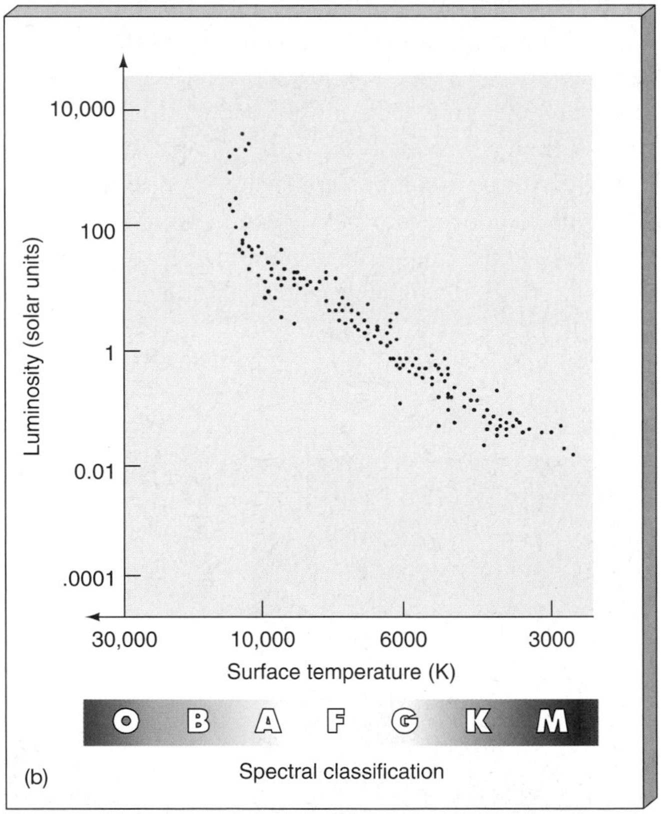

Figure 11.22 Open Cluster (a) The Pleiades cluster (also known as the Seven Sisters or M45) lies about 120 pc from Sun. (b) An H–R diagram for the stars of this well-known open cluster. *(NOAO)*

Less massive, but more extended, clusters are known as **associations**. These typically contain no more than a few hundred stars, but may span many tens of parsecs. Associations tend to be rich in very young stars. They are very loosely bound, and may survive for only a few tens of millions of years before dissolving in the Galactic tidal field. Open clusters are also eventually destroyed, but generally live much longer—hundreds of millions or billions of years. It is quite likely that the main difference between associations and open clusters is simply the efficiency with which stars formed—that is, how large a fraction of the original cloud ended up in the form of stars.

Figure 11.23(a) shows a very different type of star cluster, called a **globular cluster**. All globular clusters are roughly spherical (which accounts for their name), are generally found away from the Milky Way plane, and contain hundreds of thousands, and sometimes millions, of stars spread out over about 50 pc. The H–R diagram

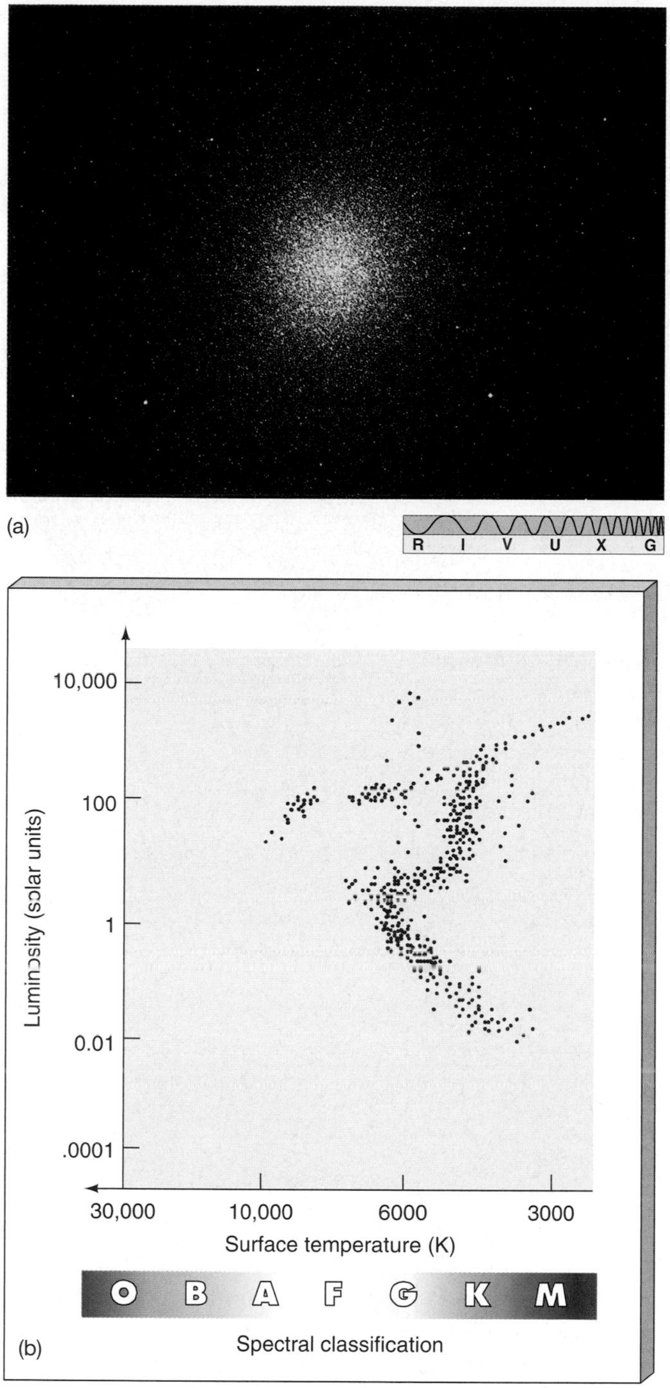

Figure 11.23 **Globular Cluster** (a) The globular cluster Omega Centauri is approximately 5000 pc from Earth and some 40 pc in diameter. (b) An H–R diagram for some of its stars. The distance to this cluster has been determined by a variation on the method of spectroscopic parallax, applied to the entire cluster rather than to individual stars. (Sec. 10.7) *(NOAO)*

for this cluster (called Omega Centauri) is shown in Figure 11.23(b).

The most outstanding feature of globular clusters is their lack of upper main-sequence stars. In fact, globular clusters contain no main-sequence stars with masses greater than about 0.8 times the mass of the Sun. (The A-type stars in this plot are stars at a much later evolutionary stage that happen to be passing through the location of the upper main sequence.) Their more massive O- through F-type stars have long since exhausted their nuclear fuel and disappeared from the main sequence. On the basis of these and other observations, astronomers estimate that all globular clusters are at least 10 billion years old. They contain the oldest known stars in our Galaxy. Astronomers speculate that the 150 or so globular clusters observed today are just the survivors of a much larger population of clusters that formed long ago.

Clusters and Nebulae

Until fairly recently, the existence of star clusters within emission nebulae was largely conjecture—the stars cannot be seen optically because they are obscured by dust. However, infrared observations have clearly demonstrated that stars really are found within star-forming regions. Figure 11.24 compares optical and infrared views of the central regions of the Orion Nebula. The optical image in Figure 11.24(a) shows the Trapezium, the group of four bright stars responsible for ionizing the nebula. However, the false-color infrared image in Figure 11.24(c) reveals an extensive cluster of stars within and behind the visible nebula. The central yellow spot within this region (known as the Becklin–Neugebauer object) is thought to be a dust-shrouded B-type star just beginning to form its own emission nebula. These remarkable images show many stages of star formation. The stars will eventually become part of the general Galactic population when the cluster dissolves.

It seems that interactions among newborn and still forming stars within the cluster play a crucial role in the star-formation process. Careful studies of star-forming regions indicate that the first massive stars to form tend to prevent the growth of additional high-mass stars by disrupting the environment in which other stars are developing. This is one reason why low-mass stars are so much more common than high-mass stars. (Sec. 10.5) It also helps explain the existence of brown dwarfs, by providing a natural way in which star formation can stop before nuclear fusion begins in a growing stellar core.

Concept Check

- If stars in a cluster all start to form at the same time, how can some influence the formation of others?

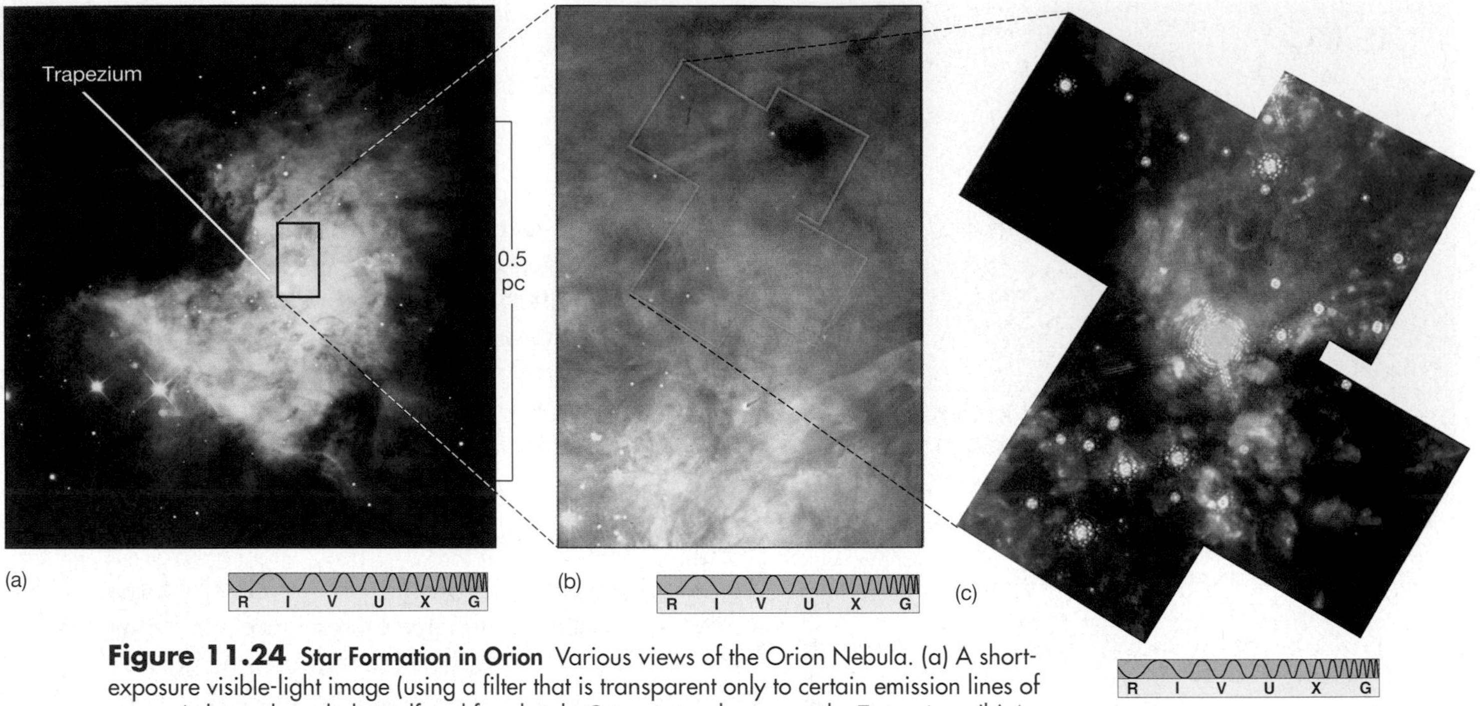

Figure 11.24 Star Formation in Orion Various views of the Orion Nebula. (a) A short-exposure visible-light image (using a filter that is transparent only to certain emission lines of oxygen) shows the nebula itself and four bright O-type stars known as the Trapezium. (b) A magnified view of a smaller part of the nebula shows large quantities of irregular gas and dust, but few obvious stars, which are hidden in the dust. (c) This short-exposure infrared image, acquired by the *Hubble Space Telescope* in 1997, shows several faint red stars emerging from the nebular gas. *(UC/Lick Observatory; NASA; NASA)*

Chapter Review

Summary

The **interstellar medium** (p. 278) occupies the space between stars. It is made up of cold (less than 100 K) gas, mostly atomic or molecular hydrogen and helium, and **dust grains** (p. 278). Interstellar dust is very effective at blocking our view of distant stars, even though the density of the interstellar medium is very low. The spatial distribution of interstellar matter is very patchy. The dust preferentially absorbs short-wavelength radiation, leading to **reddening** (p. 279) of light passing through interstellar clouds.

Emission nebulae (p. 280) are extended clouds of hot, glowing interstellar matter. Associated with star formation, they are caused by hot O- and B-type stars heating and ionizing their surroundings. They are often crossed by dark **dust lanes** (p. 282)—part of the larger molecular cloud from which they formed.

Dark dust clouds (p. 283) are cold, irregularly shaped regions in the interstellar medium that diminish or completely obscure the light from background stars. Astronomers study these clouds by observing their effect on the light from more distant stars. Another way to observe these regions of interstellar space is through spectral analysis of the **21-centimeter line** (p. 285) produced whenever the electron in a hydrogen atom reverses its spin, changing its energy very slightly in the process.

The interstellar medium also contains many cold, dark **molecular clouds** (p. 285). Dust within these clouds probably both protects the molecules and acts as a catalyst to help them form. Astronomers usually study these clouds through observations of other "tracer" molecules that are less common than hydrogen, but much easier to detect. Molecular clouds are likely sites of future star formation. Often, several molecular clouds are found close to one another, forming a **molecular cloud complex** (p. 286) millions of times more massive than the Sun.

Stars form when an interstellar cloud collapses under its own gravity and breaks up into smaller pieces. The evolution of the contracting cloud—the changes in its temperature and luminosity—can be conveniently represented as an **evolutionary track** (p. 289) on the Hertzsprung–Russell diagram. A cold interstellar cloud containing a few thousand solar masses of gas can fragment into tens or hundreds of smaller clumps of matter, from which stars eventually form.

As a collapsing prestellar fragment heats up and becomes denser, it eventually becomes a **protostar** (p. 288)—a warm, very luminous object that emits radiation mainly in the infrared portion of the electromagnetic spectrum. At this stage of its evolution, the protostar is in the **T Tauri phase** (p. 291), named after the first object of this type discovered. Eventually, a protostar's central temperature becomes high enough for hydrogen fusion to begin, and the protostar becomes a star. For a star like the Sun, the whole formation process takes about 50 million years. The **zero-age main sequence** (p. 292) is the main-sequence band predicted by stellar evolutionary theory. It agrees quite well with observed main sequences.

The most massive stars have the shortest formation times and the shortest main-sequence lifetimes. At the other extreme, some low-mass fragments never reach the point of nuclear ignition. Objects not massive enough to fuse hydrogen to helium are called **brown dwarfs** (p. 293). They may be very common in the universe.

A single collapsing and fragmenting cloud can give rise to hundreds or thousands of stars—a **star cluster** (p. 294). Many hundreds of star clusters are known. **Open clusters** (p. 294), which are loose, irregular clusters that typically contain from a few tens to a few thousands of stars, are found mostly in the Milky Way plane. They typically contain many bright blue stars, indicating that they formed relatively recently. **Globular clusters** (p. 294) are roughly spherical, and may contain millions of stars. They include no main-sequence stars more massive than the Sun, indicating that they formed long ago. Globular clusters are believed to date back to the formation of our Galaxy.

Infrared observations have revealed young star clusters in several emission nebulae. Loosely bound groups containing smaller numbers of newborn stars are called stellar **associations** (p. 294). Eventually, star clusters and associations break up into individual stars, although the process may take billions of years to complete.

Review and Discussion

1. What is the composition of interstellar gas? Of interstellar dust?
2. How is interstellar matter distributed through space?
3. What are some methods that astronomers use to study interstellar dust?
4. What is an emission nebula?
5. Why can't 21-centimeter radiation be used to probe the interiors of molecular clouds?
6. If our Sun were surrounded by a cloud of gas, would this cloud be an emission nebula? Why or why not?
7. Briefly describe the basic chain of events leading to the formation of a star like the Sun.
8. What is an evolutionary track?
9. Why do stars tend to form in groups?
10. What critical event must occur in order for a protostar to become a star?
11. What are brown dwarfs?
12. Because stars live much longer than we do, how do astronomers test the accuracy of theories of star formation?
13. At what evolutionary stages must astronomers use radio and infrared radiation to study prestellar objects? Why can't they use visible light?
14. Explain the usefulness of the Hertzsprung–Russell diagram in studying the evolution of stars. Why can't evolutionary stages 1 to 3 be plotted on the diagram?
15. Compare and contrast the properties of open and globular star clusters.

True or False?

____ **1.** Interstellar matter is evenly distributed throughout the entire Milky Way Galaxy.

____ **2.** In the vicinity of the Sun, there is about as much mass in the form of interstellar matter as in the form of stars.

____ **3.** Interstellar gas is deficient in heavy elements because those elements go into making interstellar dust.

____ **4.** Because of the obscuration of visible light by interstellar dust, we can observe stars only within a few thousand parsecs of Earth.

____ **5.** A typical dark dust cloud is many hundreds of parsecs across.

____ **6.** Because of their low temperatures, dark dust clouds radiate mainly in the ultraviolet part of the electromagnetic spectrum.

____ **7.** 21-centimeter radiation provides astronomers with information on interstellar molecular hydrogen gas.

____ **8.** 21-centimeter radiation can pass unimpeded through the entire Milky Way Galaxy.

____ **9.** Emission nebulae display spectra almost identical to those of the stars embedded in them.

____ **10.** Given the typical temperatures found in interstellar space, a cloud containing as few as 1000 atoms has sufficient gravity for it to begin to collapse.

____ **11.** Most stars form as members of clusters.

____ **12.** A stage-5 T Tauri (proto)star has a luminosity about 10 times that of the Sun.

____ **13.** The rate of evolution speeds up dramatically as a protostar approaches the main sequence.

____ **14.** The earliest stages of star formation can be observed using optical telescopes.

____ **15.** Globular clusters generally contain between a few hundred and a few thousand stars.

Fill in the Blank

1. The interstellar medium is made up of matter in the form of ______ and ______.

2. To scatter a beam of radiation most effectively, a particle must be ______ in size to the wavelength of the radiation.

3. 21-centimeter radiation results from a change in the ______ of the electron in a ______ atom.

4. Molecular clouds typically have temperatures of about ______ K.

5. Emissions from molecular clouds are in the ______ part of the electromagnetic spectrum.

6. The most common constituent of molecular clouds is molecular ______.

7. A molecular cloud complex may contain as much as ______ solar masses of gas.

8. The temperature of a typical emission nebula is about ______ K.

9. Astronomers look for emissions at ______ wavelengths to identify interstellar clouds in the early stages of collapse.

10. An ______ plots a star's or protostar's changing location on the H–R diagram as the object evolves.

11. As the fragments of an interstellar cloud contract, their central densities and temperatures ______.

12. Protostars emit a great deal of radiation in the ______ part of the electromagnetic spectrum.

13. When hydrogen is burning stably in the core, the star has reached the ______.

14. It takes a star like the Sun a total of about ______ million years to form.

15. More massive stars evolve more ______.

Problems

1. The average density of interstellar gas within about 100 pc of the Sun is much lower than the value mentioned in the text—in fact, it is just 10^3 hydrogen atoms/m^3. Given that the mass of a hydrogen atom is 1.7×10^{-27} kg, calculate the total mass of interstellar matter contained within a volume equal to that of Earth.

2. Calculate the frequency of 21-cm radiation. If interstellar clouds along our line of sight have radial velocities in the range 75 km/s (receding) to 50 km/s (approaching), calculate the range of frequencies and wavelengths over which the 21-cm line will be observed. ∞ *(More Precisely 2-3)*

3. Calculate the radius of a spherical molecular cloud whose total mass equals the mass of the Sun. Assume a cloud density of 10^{12} hydrogen molecules per cubic meter. Give answer in A.U.

4. A beam of light shining through a dense molecular cloud is diminished in intensity by a factor of 2.5 for every 3 pc it travels. By what total factor is it reduced if the total thickness of the cloud is 30 pc? What is the change, in magnitudes?

5. Estimate the escape speeds near the edges of the four emission nebulae listed in Table 11.1, and compare them with the average speeds of hydrogen nuclei in those nebulae. ∞ *(More Precisely 5-2)* Do you think it is likely that the nebulae are held together by their own gravity?

6. In order for an interstellar gas cloud to contract, the average speed of its constituent atoms must be less than half the cloud's escape speed. Will a molecular hydrogen cloud of mass 1000 solar masses, radius 10 pc, and temperature of 10 K begin to collapse, and why? ∞ *(More Precisely 5-2)*

7. A protostar evolves from a temperature $T = 3500$ K and a luminosity $L = 5000$ times that of the Sun to $T = 5000$ K and $L = 3$ solar units. What is its radius (a) at the start, and (b) at the end of the evolution? ∞ (Sec. 10.4)

8. Through approximately how many magnitudes does a three-solar-mass star decrease in brightness as it evolves from stage 4 to stage 6? (See Figure 11.20.)

9. Use the radius–luminosity–temperature relation to calculate the luminosity, in solar units, of a brown dwarf whose radius is 0.1 solar radii and whose surface temperature is 600 K (0.1 times that of the Sun). ∞ (Sec. 10.4)

10. Approximating the gravitational field of our Galaxy as a mass of 10^{11} solar masses at a distance of 8000 pc (see Chapter 14), estimate the "tidal radius" of a 2000-solar-mass star cluster—that is, the distance from the cluster center outside of which the tidal force due to the Galaxy overwhelms the cluster's gravity. ∞ (*More Precisely 5-1*)

Projects

1. The constellation Orion the Hunter is prominent in the evening sky of winter. Its most noticeable feature is a short, diagonal row of three medium-bright stars—the famous Belt of Orion (Figure 11.15). The line of stars beginning at the middle star of the Belt and extending toward the south represents Orion's sword. Toward the bottom of the sword is the sky's most famous emission nebula, M42, the Orion Nebula. Observe it with your eye, with binoculars, and with a telescope. What is its color? How can you account for this color? With the telescope, try to find the Trapezium, the grouping of four stars in the center of M42. These are hot, young stars; their energy causes the Orion Nebula to glow.

2. The Trifid Nebula, also known as M20, is a place where new stars are forming. It has been called a "dark-night revelation, even in modest apertures." An 8- to 10-inch telescope is needed to see the triple-lobed structure of the nebula. Ordinary binoculars reveal the Trifid as a hazy patch located in the constellation Sagittarius. This nebula is set against the richest part of the Milky Way, the edgewise projection of our own Galaxy around the sky. It is one of many wonders in this region of the heavens. What are the dark lanes in M20? Why are other parts of the nebula bright? There have been reports of large-scale changes occurring in this nebula in the last century and a half. The reports are based on old drawings that show M20 looking slightly different from how it appears today. Do you think it possible that a cloud in space might undergo a change in appearance on a time scale of years, decades, or centuries?

3. Summer is a good time to search with binoculars for open star clusters. Open clusters are generally found in the hazy band of the Milky Way arcing across your night sky. If you are far from city lights and looking at an appropriate time of night and year—you can simply sweep with your binoculars along the Milky Way. Numerous clumps of stars will pop into view. Many will turn out to be open clusters.

4. Globular clusters are harder to find. They are intrinsically larger, but they are also much farther away and therefore appear smaller in the sky. The most famous globular cluster visible from the Northern Hemisphere is M13 in the constellation Hercules, visible on spring and summer evenings. This cluster contains half a million or so of the Galaxy's most ancient stars. Through binoculars, it may be glimpsed as a little ball of light, located in the constellation Hercules, about one-third of the way from the star Eta Herculis to the star Zeta Herculis. Telescopes reveal this cluster as a magnificent, symmetrical grouping of stars.

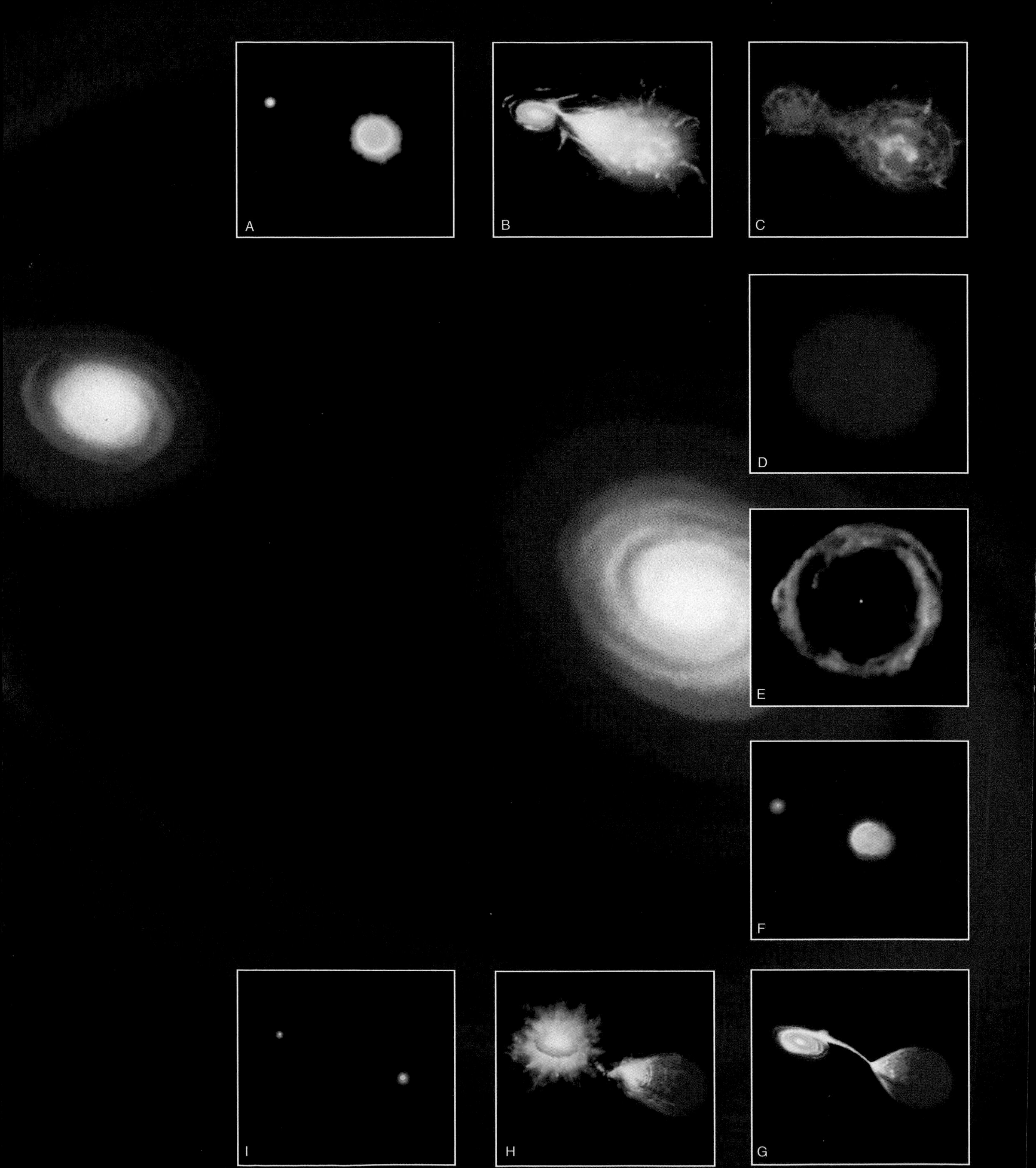
A
B
C
D
E
F
G
H
I

12 STELLAR EVOLUTION

The Lives and Deaths of Stars

LEARNING GOALS

Studying this chapter will enable you to:

1. Explain why stars evolve off the main sequence.
2. Summarize the evolutionary stages followed by a Sun-like star once it leaves the main sequence, and describe the resulting remnant.
3. Explain how white dwarfs in binary systems can become explosively active.
4. Contrast the evolutionary histories of high-mass and low-mass stars.
5. Describe the two types of supernova, and explain how each is produced.
6. Explain the origin of elements heavier than helium, and discuss the significance of these elements for the study of stellar evolution.
7. Discuss the observations that help verify the theory of stellar evolution.

(Opposite page, background) These frames are the conceptions of noted space artist Dana Berry. The ten frames depict a sequence of the birth, evolution, and death of a binary-star system. The sequence starts with the large rendering of a 1-solar-mass star and a 4-solar-mass star in the process of formation and then proceeds clockwise from top left. Some highlights:

(Inset B) Nearing the end of its life, the 4-solar mass star swells, spilling gas onto its companion and forming an accretion disk around it.

(Inset E) The red giant has reached the point where the gravity of the original two stars cannot contain the gas. The result is a gentle expulsion—forming a planetary nebula rich in oxygen gas (hence the green color).

(Inset G) Much later, an accretion bridge again joins the two stars as the lower-mass star becomes a red giant.

(Inset I) The end point of the system is two white dwarfs of roughly equal mass circling each other forevermore. *(D. Berry)*

After reaching the main sequence, a newborn star changes little in outward appearance for more than 90 percent of its lifetime. At the end of this period, as the star begins to run out of fuel and die, its properties once again change greatly as its evolutionary track takes it far from the main sequence. Most stars are destined to end their lives quiescently, their outer layers eventually escaping into interstellar space. However, a few will die violently, in spectacular explosions of almost unimaginable fury, enriching the Galaxy with newly created heavy elements. By continually comparing theoretical calculations with detailed observations of stars of all types, astronomers have refined the theory of stellar evolution into a precise and powerful tool for understanding the universe.

12.1 Leaving the Main Sequence

Most stars spend most of their lives on the main sequence. A star like the Sun, for example, after spending a few tens of millions of years in formation (stages 1–6 in Chapter 11), will reside on or near the main sequence (stage 7) for roughly 10 billion years before evolving into something else. ∞ (Sec. 11.3) That "something else" is the topic of this chapter.

On the main sequence, a star slowly fuses hydrogen into helium in its core. This process is called **core-hydrogen burning**. A star's equilibrium during this phase is the result of a balance between gravity and pressure, in which pressure's outward push exactly counteracts gravity's inward pull (Figure 12.1). Eventually, however, as the hydrogen in the core is consumed, the balance between these opposing forces starts to shift. Both the star's internal structure and its outward appearance begin to change—the star leaves the main sequence. You should keep Figure 12.1 in mind as you study the various stages of stellar evolution described below. Much of a star's complex behavior can be understood in these simple terms.

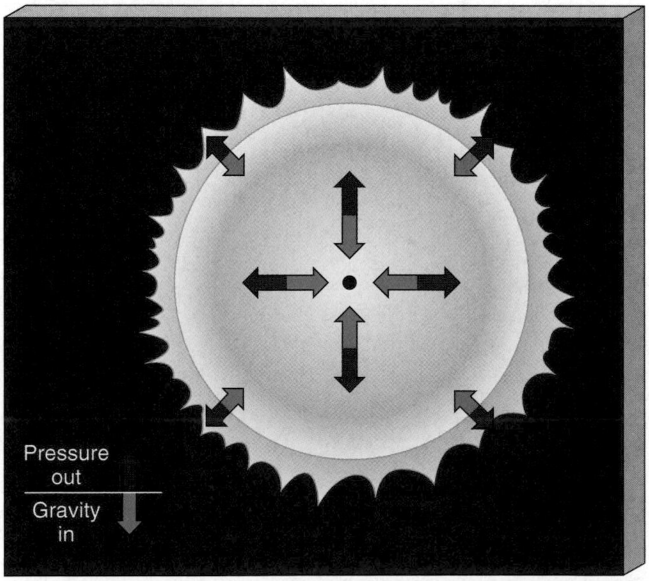

Figure 12.1 Hydrogen-Fusing Star In a steadily burning star on the main sequence, the outward pressure exerted by hot gas balances the inward pull of gravity. This is true at every point within the star, guaranteeing its stability.

The post-main-sequence stages of stellar evolution—the end of a star's life—depend critically on the star's mass. As a rule of thumb, we can say that low-mass stars die gently, while high-mass stars die catastrophically. The dividing line between these two very different outcomes lies around eight times the mass of the Sun, and in this chapter we will refer to stars of more than eight solar masses as "high-mass" stars. Our discussion will concentrate on two fairly representative evolutionary sequences—one specific to a solar-mass star, the other for a "generic" high-mass star much more massive than the Sun.

12.2 Evolution of a Sun-like Star

Figure 12.2 illustrates how the composition of a main-sequence star's interior changes as the star ages. The star's helium content increases fastest at the center, where temperatures are highest and the burning (creating helium from hydrogen) is fastest. ∞ (Sec. 9.5) The helium content also increases near the edge of the core, but more slowly because the burning rate is less rapid there. The inner, helium-rich region becomes larger and more hydrogen deficient as the star continues to shine.

The Red Giant Branch

1 2 Recall from Chapter 9 that a temperature of 10^7 K is needed to fuse hydrogen into helium. Only above that temperature do colliding hydrogen nuclei (that is, protons) have enough speed to overwhelm the repulsive elec-

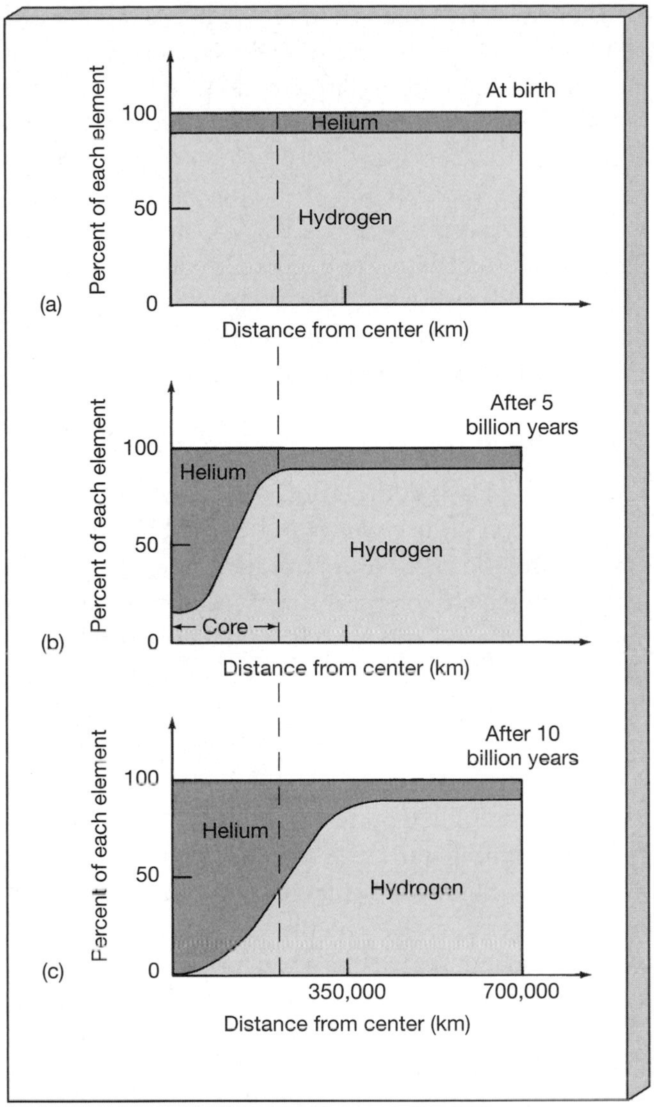

Figure 12.2 Solar Composition Change Theoretical estimates of the changes in a Sun-like star's composition. Hydrogen and helium abundances are shown (a) at birth, just as the star arrives on the main sequence; (b) after five billion years; and (c) after 10 billion years. At stage (b) only about five percent of the star's total mass has been converted from hydrogen to helium. The change speeds up as the nuclear burning rate increases with time.

tromagnetic force between them. (Sec. 9.5) Because helium nuclei (with two protons each, compared to one for hydrogen) carry a greater positive charge, their electromagnetic repulsion is larger, and so higher temperatures are needed to cause them to fuse. The core temperature at this stage is too low for helium fusion to begin. Eventually, hydrogen becomes completely depleted at the center, the nuclear fires there cease, and the location of principal burning moves to higher layers in the core. An inner core of nonburning pure helium starts to grow.

Without nuclear burning to maintain it, the outward-pushing gas pressure weakens in the helium inner core; however, the inward pull of gravity does not. Once the pressure is relaxed—even a little—structural changes in the star become inevitable. As soon as hydrogen becomes substantially depleted, about 10 billion years after the star arrived on the main sequence, the helium core begins to contract.

The shrinkage of the helium core releases gravitational energy, driving up the central temperature and heating the overlying layers, causing the hydrogen there to fuse even more rapidly than before. Figure 12.3 depicts this **hydrogen-shell-burning** stage, in which hydrogen is burning at a furious rate in a relatively thin layer surrounding the nonburning inner core of helium "ash." The hydrogen shell generates energy faster than did the original main-sequence star's hydrogen-burning core, and the shell's energy production continues to increase as the helium core contracts. Strange as it may seem, the star's response to the disappearance of the nuclear fire at its center is to get brighter!

The pressure exerted by this enhanced hydrogen burning causes the star's nonburning outer layers to increase in radius. Not even gravity can stop them. Even while the core is shrinking and heating up, the overlying layers are expanding and cooling. The star, aged and unbalanced, is on its way to becoming a red giant. The

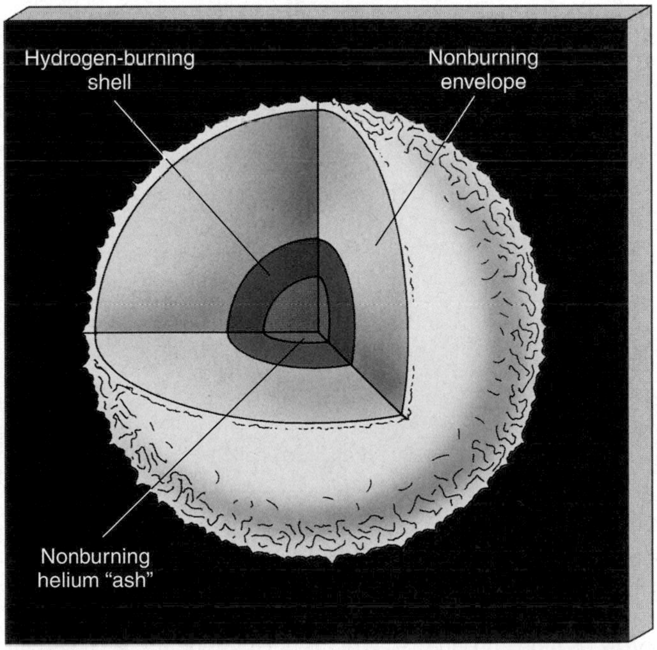

Figure 12.3 Helium-Core Star As a star's core loses more and more of its hydrogen, the hydrogen in the shell surrounding the nonburning helium "ash" burns ever more violently.

change from normal main-sequence star to elderly red giant takes about 100 million years.

We can trace these large-scale changes on an H–R diagram. Figure 12.4 shows the star's path away from the main sequence, labeled as stage 7. ∞ (Sec. 11.3) The star first evolves to the right on the diagram, its surface temperature dropping while its luminosity increases only slightly. The star's roughly horizontal track from its main-sequence location (stage 7) to stage 8 on the figure is called the **subgiant branch**. By stage 8, the star's radius has increased to about three times the radius of the Sun. The nearly vertical (constant temperature) path followed by the star between stages 8 and 9 is known as the **red giant branch** of the H–R diagram.

Figure 12.5 compares the relative sizes of a G-type star like our Sun and a stage 9 red giant, and also indicates the stages through which the star will evolve. The red giant is huge. By stage 9, its luminosity is many hundreds of times the solar value. Its radius is around 100 solar radii—about the size of Mercury's orbit. In contrast, its helium core is surprisingly small, only about 1/1000 the size of the entire star, or just a few times larger than Earth. Continued shrinkage of the red giant's core has compacted its helium gas to approximately 10^8 kg/m^3. Contrast this with the 10^{-3} kg/m^3 in the giant's outermost layers, with the 5500 kg/m^3 average density of Earth, and with the 150,000 kg/m^3 in the present core of the Sun. About 25 percent of the mass of the entire star is packed into its planet-sized core.

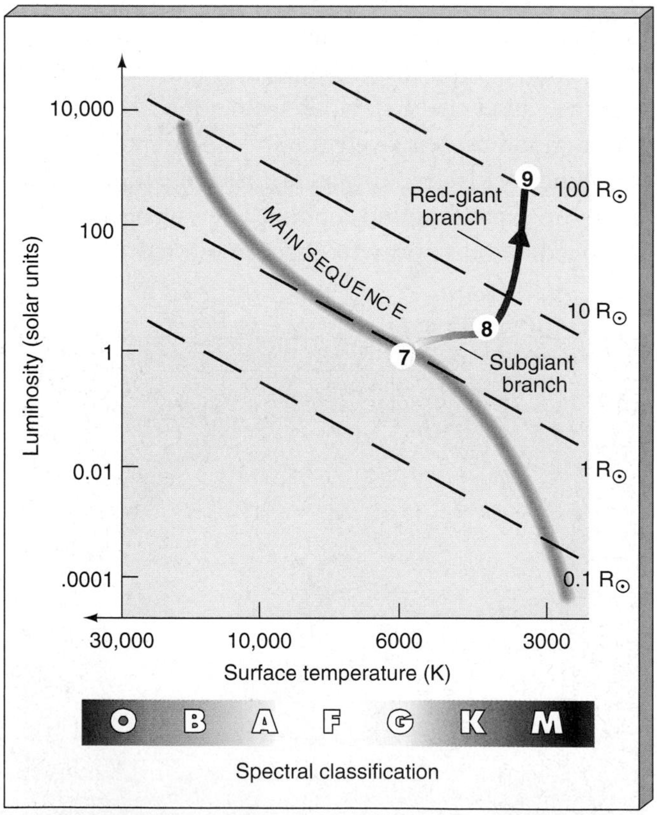

Figure 12.4 Red Giant on the H–R Diagram As its helium core shrinks and its outer envelope expands, the star leaves the main sequence (stage 7). At stage 8, the star is well on its way to becoming a red giant. The star continues to brighten and grow as it ascends the red giant branch to stage 9. As noted in Chapter 10, the dashed diagonal lines are lines of constant radius, allowing us to gauge the changes in the size of our star.

Helium Fusion

This simultaneous contraction of the red giant's core and expansion of its envelope—the nonburning layers surrounding the core—does not continue indefinitely. A few hundred million years after a solar-mass star leaves the main sequence, the central temperature reaches the 10^8 K needed for helium to fuse into carbon, and the nuclear fires reignite.

For stars comparable in mass to the Sun, the high densities and pressures found in the stage-9 core mean that the onset of helium fusion is a very violent event. Once the burning starts, the core cannot respond quickly enough to the rapidly changing conditions within it, and its temperature rises sharply in a runaway explosion called the **helium flash**. For a few hours, the helium burns ferociously, like an uncontrolled bomb. Eventually, the star's structure "catches up" with the flood of energy dumped into it by helium burning. The core expands, its density drops, and equilibrium is restored as the inward pull of gravity and the outward push of gas pressure come back into balance. The core, now stable, begins to burn helium into carbon at temperatures well above 10^8 K.

The helium flash terminates the star's ascent of the red giant branch of the H–R diagram at stage 9 in Figure 12.4. Yet despite the explosive detonation of helium in the core, the flash does *not* increase the star's luminosity. On the contrary, the helium flash produces a rearrangement of the core that ultimately results in a *reduction* in the energy output as the star jumps from stage 9 to stage 10. As indicated in Figure 12.6, the surface temperature is now higher than it was on the red giant branch, although the luminosity is considerably less than at the helium flash. This adjustment in the star's properties occurs quite quickly—in about 100,000 years.

At stage 10 our star is now stably burning helium in its inner core and fusing hydrogen in a shell sur-

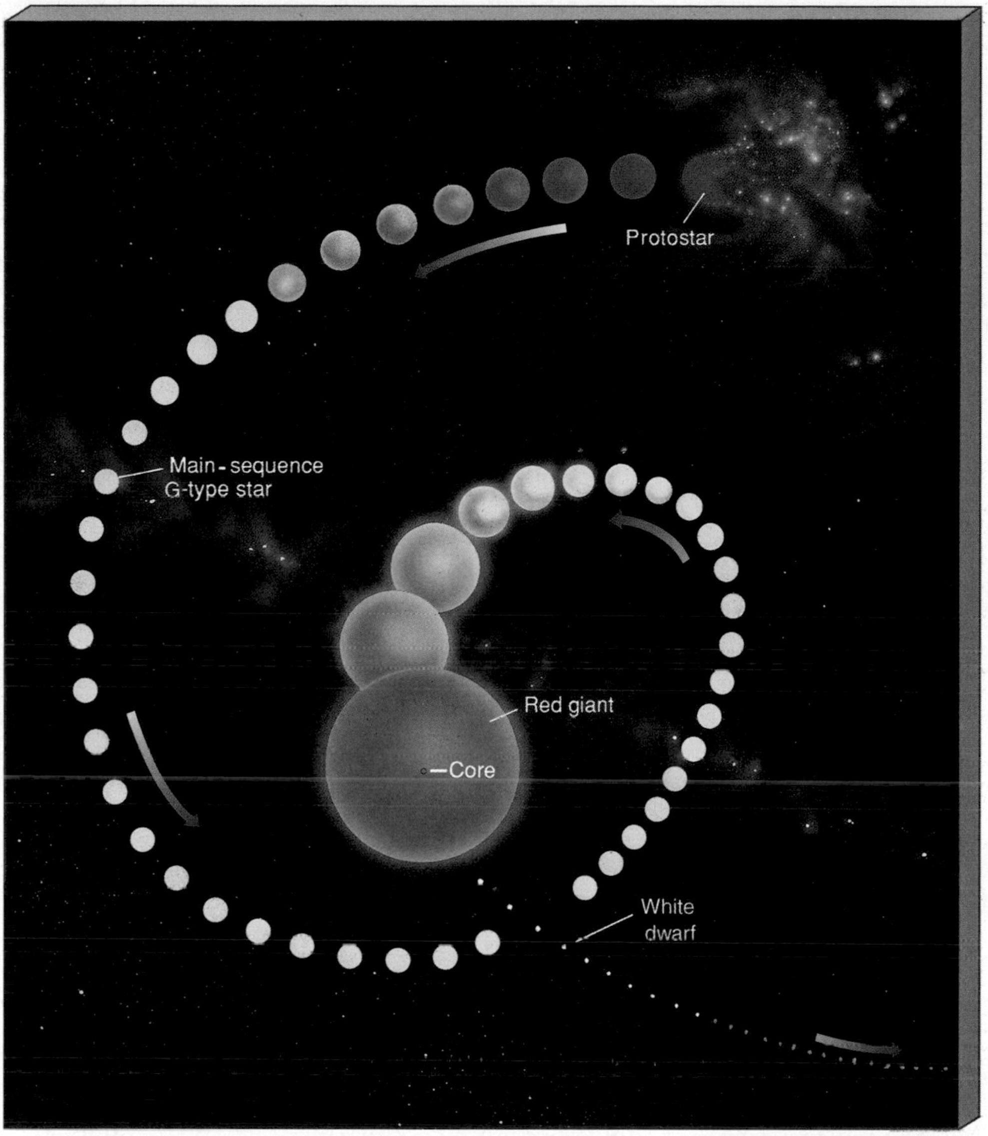

Figure 12.5 G-type Star Evolution Relative sizes and colors of a normal G-type star (such as our Sun) during its formative stages, on the main sequence, and while passing through the red giant and white dwarf stages. At maximum swelling, the red giant is approximately 70 times the size of its main-sequence parent; the core of the giant is about 1/15 the main-sequence size and would be barely discernible if this figure were drawn to scale. The length of time spent in the various stages—protostar, main-sequence star, red giant, and white dwarf—is roughly proportional to the lengths shown in this imaginary trek through space. The star's brief stay on the horizontal branch is not shown.

rounding that core. It resides in a well-defined region of the H–R diagram known as the **horizontal branch**, where core-helium-burning stars remain for a time before resuming their journey around the H–R diagram.

The Carbon Core

Nuclear reactions in stars proceed at rates that increase very rapidly with temperature. At the extremely high temperatures found in the core of a horizontal-branch star, the helium fuel doesn't last long—no more than a few tens of million years after the initial flash.

As helium fuses to carbon, a new inner core of carbon forms and phenomena similar to the earlier buildup of helium begin to occur. Now helium becomes depleted at the center, and eventually fusion ceases there. In response, the nonburning carbon core shrinks and heats up as gravity pulls it inward, causing the hydrogen- and helium-burning rates in the overlying layers to increase. The star now contains a shrinking carbon-ash inner core surrounded by a helium-burning shell, which is in turn surrounded by a hydrogen-burning shell (Figure 12.7). The outer envelope of the star expands, much as it did earlier in the first red giant stage. By the time it reaches stage 11 in Figure 12.8, the star has become a swollen red giant for a second time.

The burning rates at the star's center are much fiercer during its second trip into the red giant region,

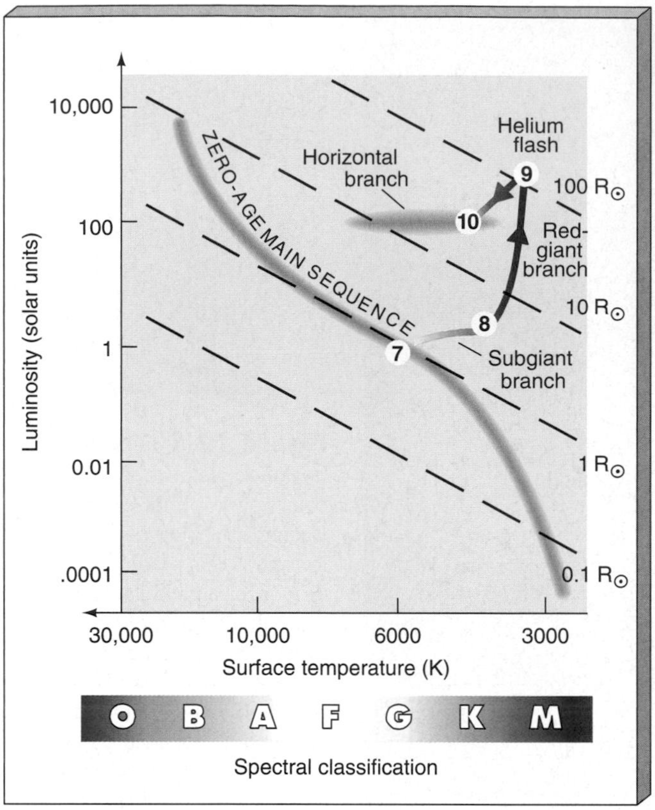

Figure 12.6 **Horizontal Branch on the H–R Diagram** After its large increase in luminosity while ascending the red giant branch is terminated by the helium flash, our star settles down into another equilibrium state at stage 10, on the horizontal branch.

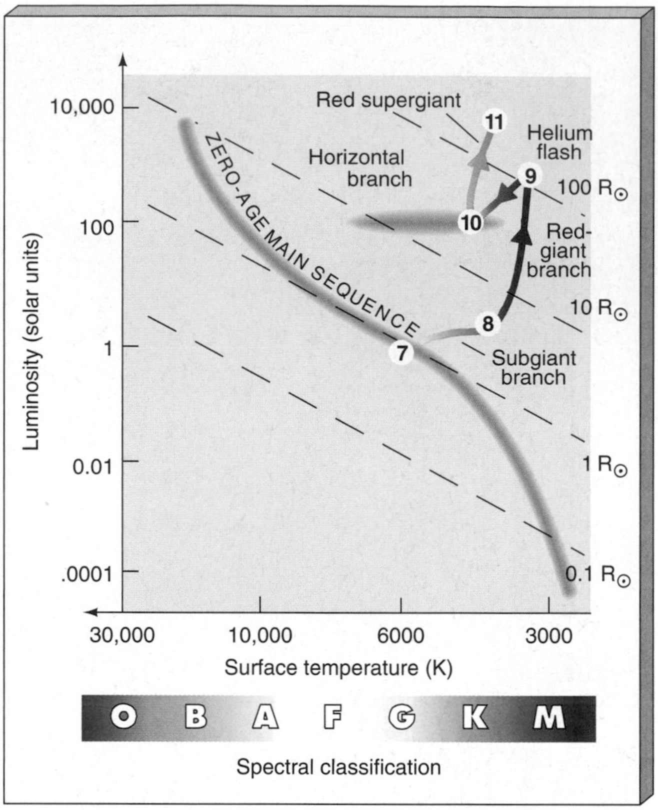

Figure 12.8 **Supergiant on the H–R Diagram** A carbon-core star reenters the giant region of the H–R diagram (stage 11) for the same reason it evolved there the first time around: Lack of nuclear burning in the inner core causes contraction of the core and expansion of the overlying layers.

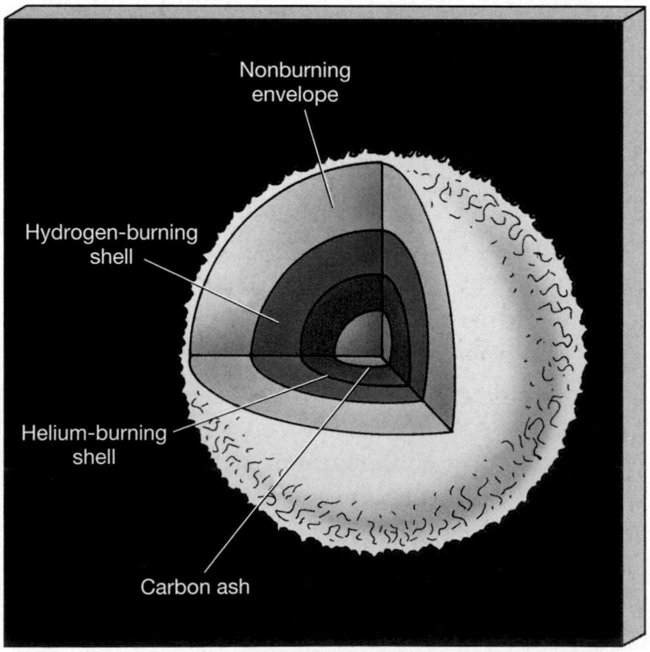

Figure 12.7 **Helium-Fusing Star** Within a few million years after the onset of helium burning (stage 9), carbon accumulates in the star's inner core. Above this core, hydrogen and helium are still burning in concentric shells.

and the radius and luminosity increase to values even greater than those reached during the first visit (stage 9). Our star is now a **red supergiant**. Its carbon core continues to shrink, driving the hydrogen- and helium-burning shells to higher and higher temperatures and luminosities.

Table 12.1 summarizes the key stages through which our solar-mass star evolves. It is a continuation of Table 11.2, except that the density units are now the more convenient kilograms per cubic meter, and we now express radii in units of the solar radius. As in Chapter 11, the numbers in the "Stage" column refer to the evolutionary stages noted in the figures and discussed in the text.

✓ Concept Check

- Why does a star get brighter as it runs out of fuel in its core?

TABLE 12.1 Evolution of a Sun-like Star

STAGE	APPROX. TIME TO NEXT STAGE (yr)	CENTRAL TEMPERATURE (K)	SURFACE TEMPERATURE (K)	CENTRAL DENSITY (kg/m^3)	RADIUS (km)	RADIUS (solar radii)	OBJECT
7	10^{10}	1.5×10^7	6000	10^5	7×10^5	1	Main-sequence star
8	10^8	5×10^7	4000	10^7	2×10^6	3	Subgiant
9	10^5	10^8	4000	10^8	7×10^7	100	Red Giant/ Helium flash
10	5×10^7	2×10^8	5000	10^7	7×10^6	10	Horizontal branch
11	10^4	2.5×10^8	4000	10^8	4×10^8	500	Red supergiant
12	{ 10^5 —	3×10^8 —	100,000 3000	10^{10} 10^{-17}	10^4 7×10^8	0.01 1,000	Carbon core Planetary nebula[1]
13	—	10^8	50,000	10^{10}	10^4	0.01	White dwarf
14	—	Close to 0	Close to 0	10^{10}	10^4	0.01	Black dwarf

[11]Values in columns 2 through 7 refer to the envelope.

12.3 The Death of a Low-Mass Star

2 As our red supergiant moves from stage 10 to stage 11, its envelope swells while its core, too cool for further nuclear burning, continues to contract. If the central temperature could become high enough for carbon fusion to occur, still heavier elements could be synthesized, and the newly generated energy might again support the star, restoring for a time the equilibrium between gravity and pressure. For solar-mass stars, however, this does not occur. The temperature never reaches the 600 million K needed for new nuclear reactions to occur. The red supergiant is now very close to the end of its nuclear-burning lifetime.

Dense Matter

Before the shrinking carbon core can become hot enough for carbon to fuse, the rising central pressure stops the contraction and stabilizes the temperature. However, this pressure is not the "normal" thermal pressure of high-speed particles in a very hot gas. ∞ *(More Precisely 5-2)* Instead, the core enters a state in which the free electrons in the core—a vast sea of charged particles stripped from their parent nuclei by the ferocious heat in the stellar interior—play a critical role in determining the star's future.

The core density at this stage (stage 12 in Table 12.1) is enormous—about 10^{10} kg/m^3. This density is much higher than anything we have seen so far in our study of the cosmos. A single cubic centimeter of core matter would weigh 1000 kg on Earth—a ton of matter compressed into a volume about the size of a grape! Under these extreme conditions, a law of quantum physics known as the *Pauli Exclusion Principle* comes into play, effectively preventing the electrons in the core from being crushed any closer together. (In fact, the core was in a similar high-density state at the moment of helium ignition, causing the helium flash.)

In essence, the electrons behave like tiny rigid spheres that can be squeezed relatively easily up to the point of contact, but become virtually incompressible thereafter. By stage 12, the pressure in the carbon core resisting the force of gravity is supplied almost entirely by tightly packed electrons, and this pressure brings the core back into equilibrium at a central temperature of "only" about 300 million K—too cool to fuse carbon into any of the heavier elements. This stage represents the maximum compression that the star can achieve—there is simply not enough matter in the overlying layers to bear down any harder. Supported by the resistance of its electrons to further compression, the core contraction stops.

Planetary Nebulae

Our aged star is now in quite a predicament. Its inner carbon core is, for all practical purposes, dead. The outer shells continue to burn hydrogen and helium, and as more and more of the inner core reaches its final, high-density state, the burning increases in intensity. Meanwhile, the envelope continues to expand and cool. Eventually, driven by increasing radiation from within and accelerated by the energy released as electrons recombine with nuclei to form atoms, the envelope becomes unstable and is ejected into space at a speed of a few tens of kilometers per second, forming an expanding, cooling shell of matter called a **planetary nebula** (Figure 12.9).

The term *planetary* here is very misleading, for these objects have no association with planets. The name originated in the eighteenth century when, viewed at poor resolution through small telescopes, these shells of gas looked to some astronomers like the circular disks of the planets in our solar system. The term *nebula* is also confusing, as it suggests kinship with the emission nebulae studied in Chapter 11. ∞ (Sec. 11.2) In fact, not only are planetary nebulae much smaller than emission nebulae, they are also associated with much older stars. Emission nebulae are the signposts of recent stellar birth. Planetary nebulae indicate impending stellar death.

The "ring" of a planetary nebula is actually a three-dimensional shell of warm, glowing gas completely surrounding the core. However, as illustrated in Figure 12.9(b), the shell is virtually invisible in the direction of the core, leading to a halo-shaped appearance. In reality, few planetary nebulae are quite as regular as this simple picture suggests (see Figure 12.9c). The star's environment seems to play an important role in determining the nebula's shape and appearance.

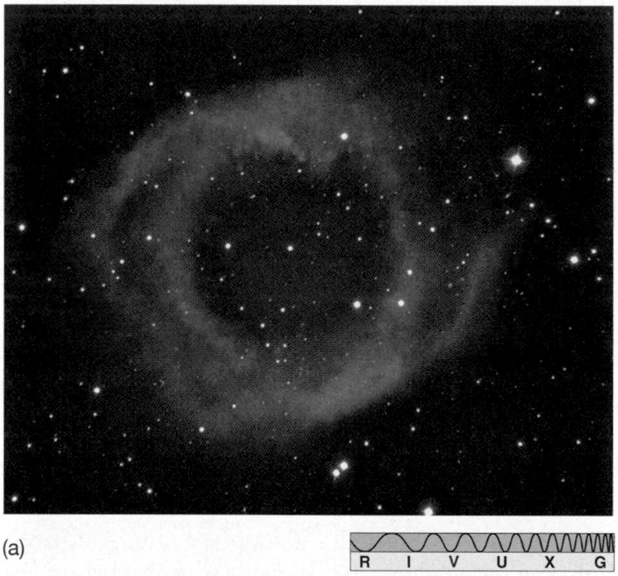

(a)

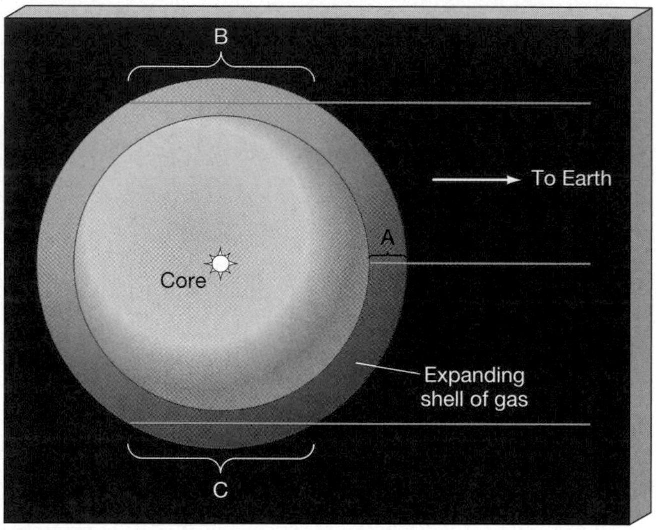

(b)

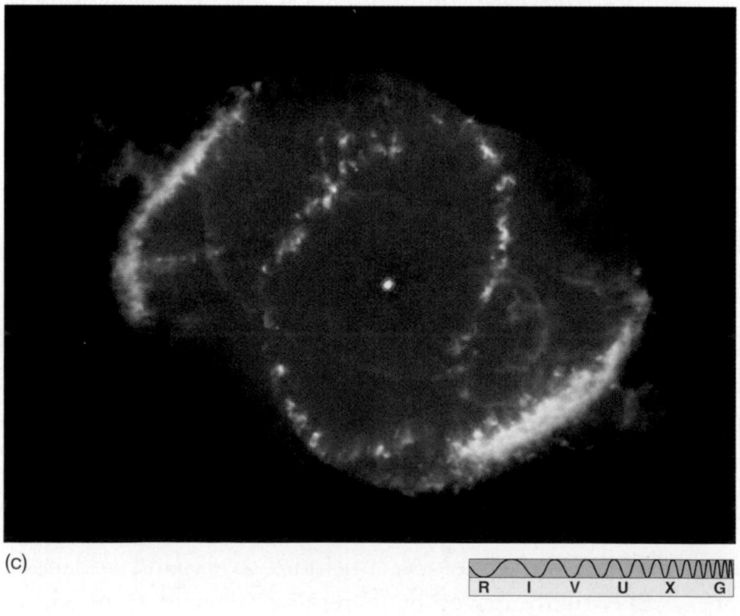

(c)

Animation Video

Figure 12.9 Planetary Nebulae A planetary nebula is an object with a small, dense core (central blue-white star) surrounded by an extended shell (or shells) of glowing matter. (a) The Helix Nebula appears to the eye as a small star with a halo around it. About 140 pc from Earth and 0.6 pc across, its apparent size in the sky is roughly half that of the full Moon. (b) The appearance of a planetary nebula can be explained once we realize that the shell of glowing gas around the central core is quite thin. There is very little gas along the line of sight between the observer and the central star (path A), so that part of the shell is invisible. Near the edge of the shell, however, there is more gas along the line of sight (paths B and C), so the observer sees a glowing ring. (c) The Cat's Eye Nebula is an example of a much more complex planetary nebula. It lies about 1000 pc away. It may have been produced by a pair of binary stars (unresolved at the center) that have both shed planetary nebulae. *(Anglo-Australian Observatory/Royal Observatory Edinburgh. Photograph made from VK Schmidt Plates by D. Malin; J. Harrington & K. Bobrowski/ J. Harrington & K. Bobrowski (University of Maryland/ NASA)*

The planetary nebula continues to spread out with time, becoming more diffuse and cooler, gradually dispersing into interstellar space. In doing so, it enriches the interstellar medium with atoms of helium and carbon dredged up from the depths of the core into the envelope by convection during the star's final years.

White Dwarfs

The carbon core, the stellar remnant at the center of the planetary nebula, continues to evolve. Formerly concealed by the atmosphere of the red giant star, the core becomes visible as a *white dwarf* as the envelope recedes. (Sec. 10.4) The core is very small, about the size of Earth, with a mass about half that of the Sun. Shining only by stored heat, not by nuclear reactions, this small star has a white-hot surface when it first becomes visible, although it appears dim because of its small size. This is stage 13 of Table 12.1. The approximate path followed by the star on the H–R diagram as it evolves from a stage 11 red supergiant to a stage 13 white dwarf is shown in Figure 12.10.

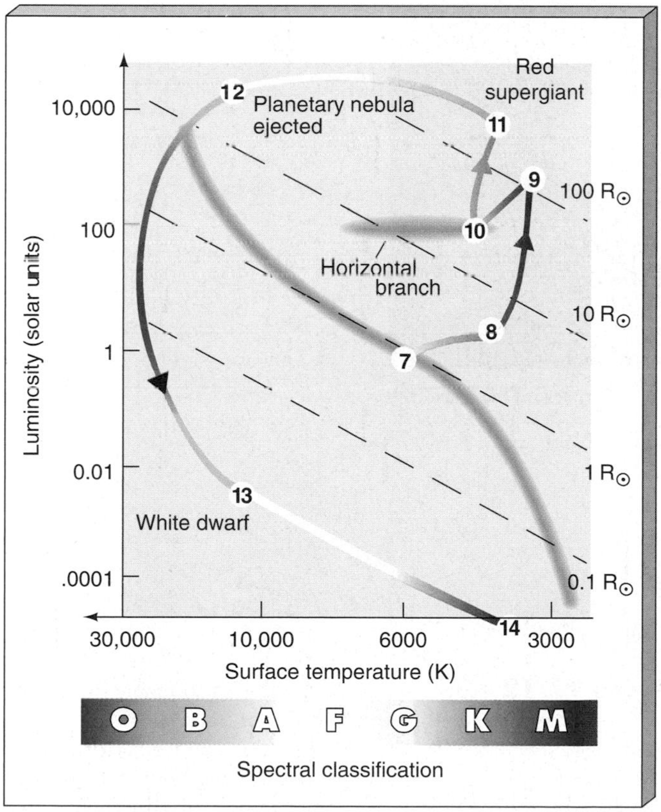

Figure 12.10 White Dwarf on H–R Diagram A star's passage from the horizontal branch (stage 10) to the white dwarf stage (stage 13) creates an evolutionary path that cuts across the H–R diagram.

Animation

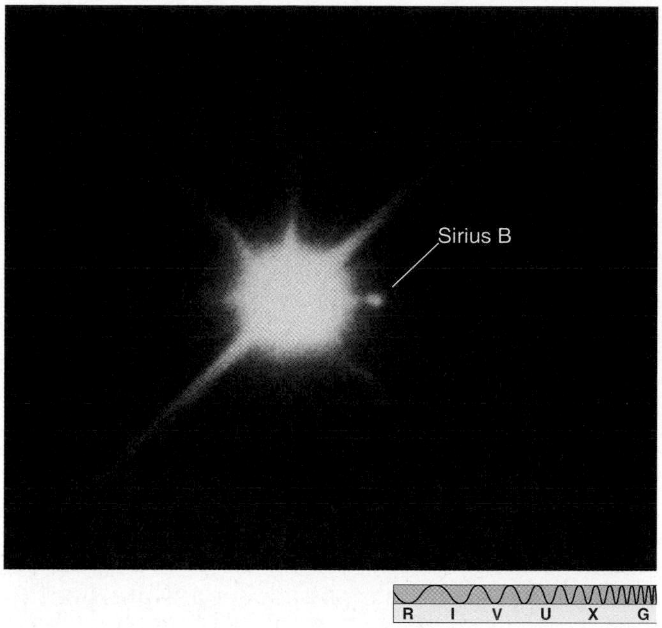

Figure 12.11 Sirius Star System Sirius B (the speck of light to the right of the much larger and brighter Sirius A) is a white dwarf star, a companion to Sirius A. (The "spikes" on the image of Sirius A are not real; they are caused by the support struts of the telescope.) *(Palomar Observatory)*

Not all white dwarfs are seen as the cores of planetary nebulae. Several hundred have been discovered "naked," their envelopes expelled to invisibility long ago. Figure 12.11 shows an example of a white dwarf, Sirius B, that happens to lie particularly close to Earth; it is the faint binary companion of the much brighter Sirius A. (Sec. 10.6) Some of its properties are listed in Table 12.2. With more than the mass of the Sun packed into a volume smaller than Earth, Sirius B's density is about a million times greater than anything familiar to us in the solar system. Sirius B has an unusually high mass for a white dwarf. It is believed to be the evolutionary product of a star roughly four times the mass of the Sun.

TABLE 12.2 Sirius B—A Nearby White Dwarf

Mass	1.1 solar masses
Radius	0.008 solar radii (5500 km)
Luminosity (total)	0.04 solar luminosities
Surface temperature	24,000 K
Average density	3×10^9 kg/m^3

The white dwarf continues to cool and dim with time, following the white–yellow–red line near the bottom of Figure 12.10, eventually becoming a **black dwarf**—a cold, dense, burned-out ember in space. This is stage 14 of Table 12.1, the graveyard of stars. The cooling dwarf does not shrink much as it fades away, however. Even though its heat is leaking away into space, gravity does not compress it further. Its tightly packed electrons will support the star even as its temperature drops (after trillions of years) almost to absolute zero. As the dwarf cools, it remains about the size of Earth.

Novae

3 In some cases, the white dwarf stage does not represent the end of the road for a Sun-like star. Given the right circumstances, it is possible for a white dwarf to become explosively active, in the form of a highly luminous **nova** (plural: novae). The word *nova* means "new" in Latin, and to early observers, these stars did indeed seem new as they suddenly appeared in the night sky. Astronomers now recognize that a nova is what we see when a white dwarf undergoes a violent explosion on its surface, resulting in a rapid, temporary increase in luminosity. Figures 12.12(a) and (b) illustrate the brightening of a typical nova over a period of three days. Figure 12.12(c) shows the nova's light curve, demonstrating how the luminosity rises dramatically in a matter of days, then fades slowly back to normal over the course of several months. On average, two or three novae are observed each year.

What could cause such an explosion on a faint, dead star? The energy involved is far too great to be explained by flares or other surface activity, and as we have just seen, there is no nuclear activity in the dwarf's interior. The answer to this question lies in the white dwarf's surroundings. If the white dwarf is isolated, then it will indeed cool and ultimately become a black dwarf, as just

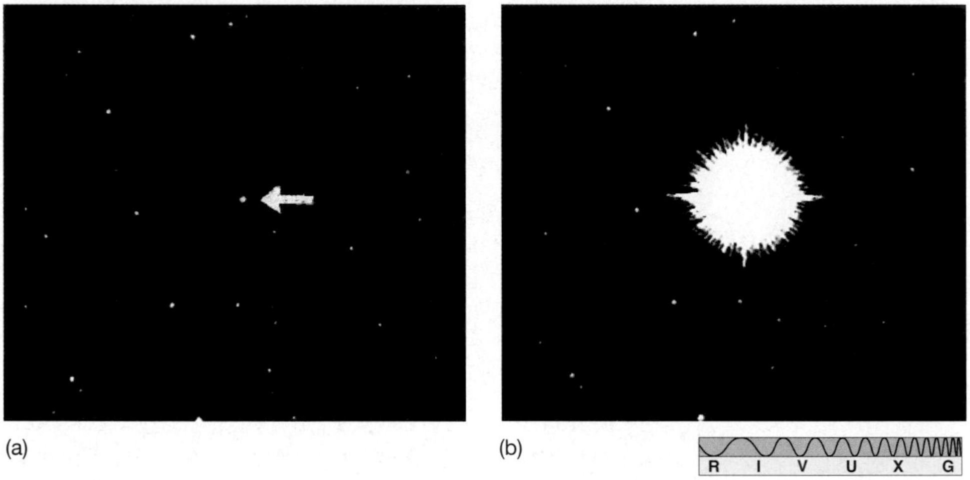

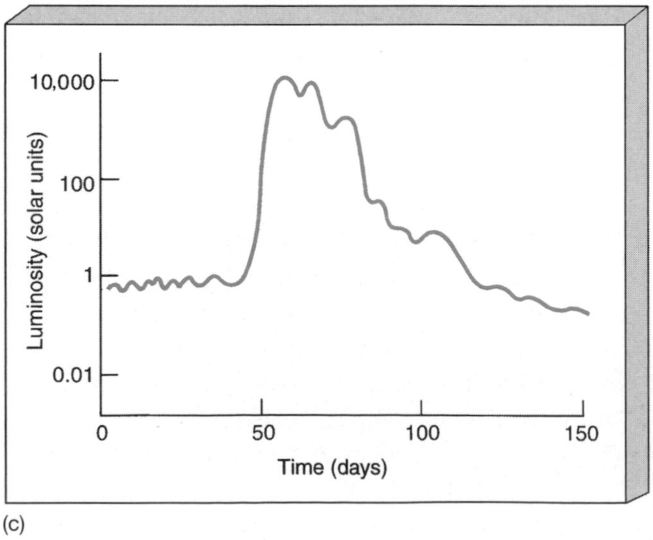

Animation

Figure 12.12 Nova A nova is a star that suddenly increases enormously in brightness, then slowly fades back to its original luminosity. Novae are the result of explosions on the surfaces of faint white dwarf stars, caused by matter falling onto their surfaces from the atmosphere of larger binary companions. Shown is Nova Herculis 1934 in (a) March 1935 and (b) May 1935, after brightening by a factor of 60,000. (c) The light curve of a typical nova. The rapid rise and slow decline in the light received from the star, as well as the maximum brightness attained, are in good agreement with the explanation of the nova as a nuclear flash on a white dwarf's surface. *(UC/Lick Observatory)*

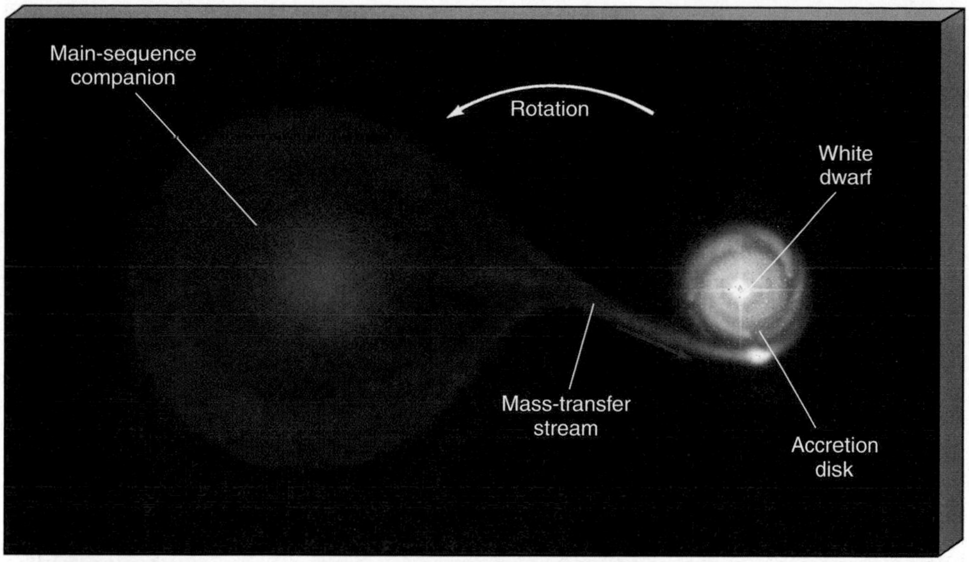

Figure 12.13 Close Binary System If a white dwarf in a binary system is close enough to its companion, its gravitational field can tear matter from the companion's surface. Notice that the matter does not fall directly onto the white dwarf's surface. Instead, it forms an accretion disk of gas spiraling down onto the dwarf.

described. However, should the white dwarf be part of a binary system in which the other star is either a main-sequence star or a giant, an important new possibility exists.

If the distance between the dwarf and the other star is small enough, the dwarf's gravitational field can pull matter—primarily hydrogen and helium—away from the surface of the companion, as illustrated in Figure 12.13. Because of the binary's rotation, material leaving the companion does not fall directly onto the white dwarf. Instead, it "misses" the dwarf, loops around behind it, and goes into orbit around it, forming a swirling, flattened disk of matter called an **accretion disk**, as shown in the figure. As it builds up on the white dwarf's surface, the "stolen" gas becomes hotter and denser. Eventually its temperature exceeds 10^7 K, and the hydrogen ignites, fusing into helium at a furious rate. This surface-burning stage is as brief as it is violent. The star suddenly flares up in luminosity, then fades away as some of the fuel is exhausted and the rest is blown off into space. If the event happens to be visible from Earth, we see a nova.

Once the nova explosion is over and the binary has returned to normal, the mass-transfer process can begin again. Astronomers know of many *recurrent novae*—stars that have been observed to "go nova" several times over the course of a few decades. Such systems can, in principle, repeat their violent outbursts many dozens, if not hundreds, of times.

✓ Concept Check

- Will the Sun ever become a nova?

12.4 Evolution of High-Mass Stars

4 All stars leaving the main sequence on the journey toward the red giant region of the H–R diagram have similar internal structure. Thereafter, however, their evolutionary tracks diverge.

Heavy Element Fusion

Figure 12.14 compares the post-main-sequence evolution of three stars having masses 1, 4, and 15 times the mass of the Sun. Note that, while the solar-mass star ascends the red giant branch almost vertically, the higher-mass stars tend to move nearly horizontally across the H–R diagram after leaving the upper main sequence.

The intermediate 4-solar-mass star loops back and forth across the diagram as it evolves. It does not experience a helium flash when helium fusion begins, and while it does reach temperatures high enough to fuse carbon, nuclear burning subsequently ceases, eventually leading to a white dwarf much as described previously. The 15-solar-mass star, however, can fuse not just hydrogen, helium and carbon, but also oxygen, neon, magnesium, and even heavier elements as its inner core continues to contract and its central temperature continues to rise.

Evolution proceeds so rapidly in the 15-solar-mass star that it doesn't even reach the red giant region before helium fusion begins. The star achieves a central temperature of 10^8 K while still quite close to the main sequence, and its evolutionary track continues smoothly across the H–R diagram, seemingly unaffected by each new phase of burning. The star's luminosity stays roughly constant as its radius increases and its surface temperature drops.

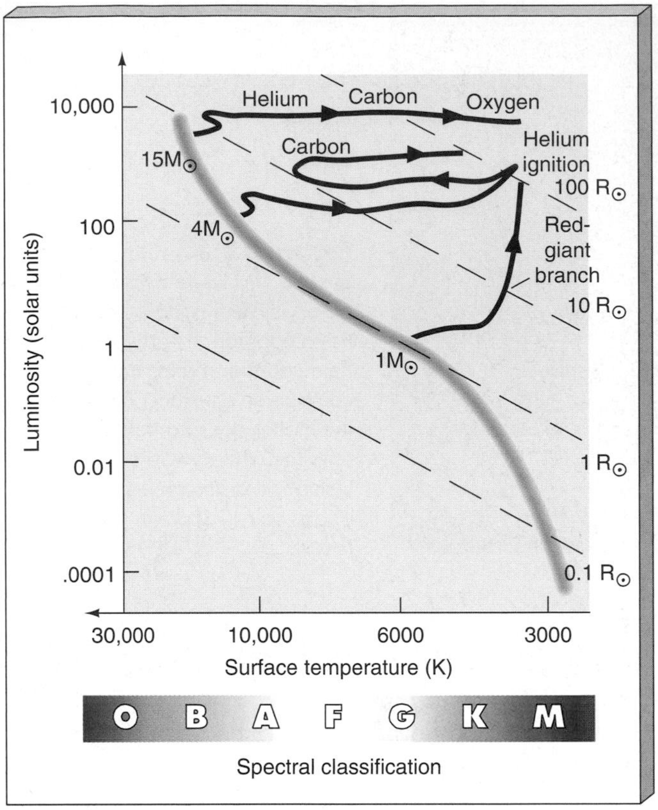

Figure 12.14 High-Mass Evolutionary Tracks Evolutionary tracks for stars of 1, 4, and 15 solar masses (shown only up to helium ignition in the 1-solar-mass case). Stars with masses comparable to the mass of the Sun ascend the giant branch almost vertically, whereas higher-mass stars move roughly horizontally across the H–R diagram from the main sequence into the red giant region. The most massive stars experience smooth transitions into each new burning stage. Some points are labeled with the element that has just started to fuse in the inner core.

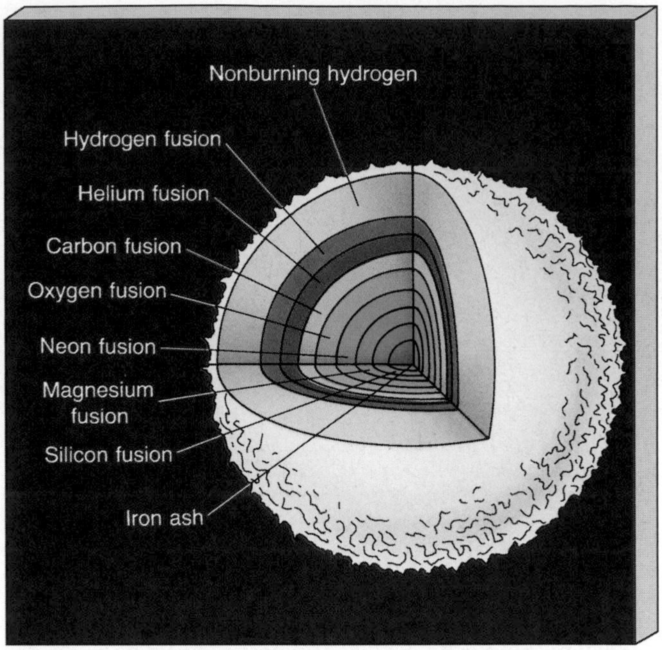

Figure 12.15 Heavy-Element-Fusing Star Cutaway diagram of the interior of a highly evolved high-mass star. The interior resembles the layers of an onion, with shells of progressively heavier elements burning at smaller and smaller radii and at higher and higher temperatures.

Figure 12.15 is a cutaway diagram of an evolved high-mass star. Note the numerous layers where various nuclei burn. As the temperature increases with depth, the product of each burning stage becomes the fuel for the next. At the relatively cool periphery of the core, hydrogen fuses into helium. In the intermediate layers, shells of helium, carbon, and oxygen burn to form heavier nuclei. Deeper down reside neon, magnesium, silicon, and other heavy nuclei, all produced by nuclear fusion in the layers overlying the nonburning inner core. The inner core itself is composed of iron.

As each element is burned to depletion at the center, the core contracts, heats up, and fusion starts again. A new inner core forms, contracts again, heats again, and so on. Through each period of stability and instability, the star's central temperature increases, the nuclear reactions speed up, and the newly released energy supports the star for ever-shorter periods of time. For example, in round numbers, a star 20 times more massive than the Sun burns hydrogen for 10 million years, helium for one million years, carbon for 1000 years, oxygen for one year, and silicon for a week. Its iron core grows less than a day.

The Death of a High-Mass Star

Once the inner core begins to change to iron, our high-mass star is in trouble. Nuclear fusion involving iron does not produce energy, because iron nuclei are so compact that energy cannot be extracted by combining them into heavier elements. In effect, iron plays the role of a fire extinguisher, damping the inferno in the stellar core. With the appearance of substantial quantities of iron, the central fires cease for the last time, and the star's internal support begins to dwindle. The star's foundation is destroyed, and its equilibrium is gone forever. Even though the temperature in the iron core has reached several billion kelvins by this stage, the enormous inward gravitational pull of matter ensures

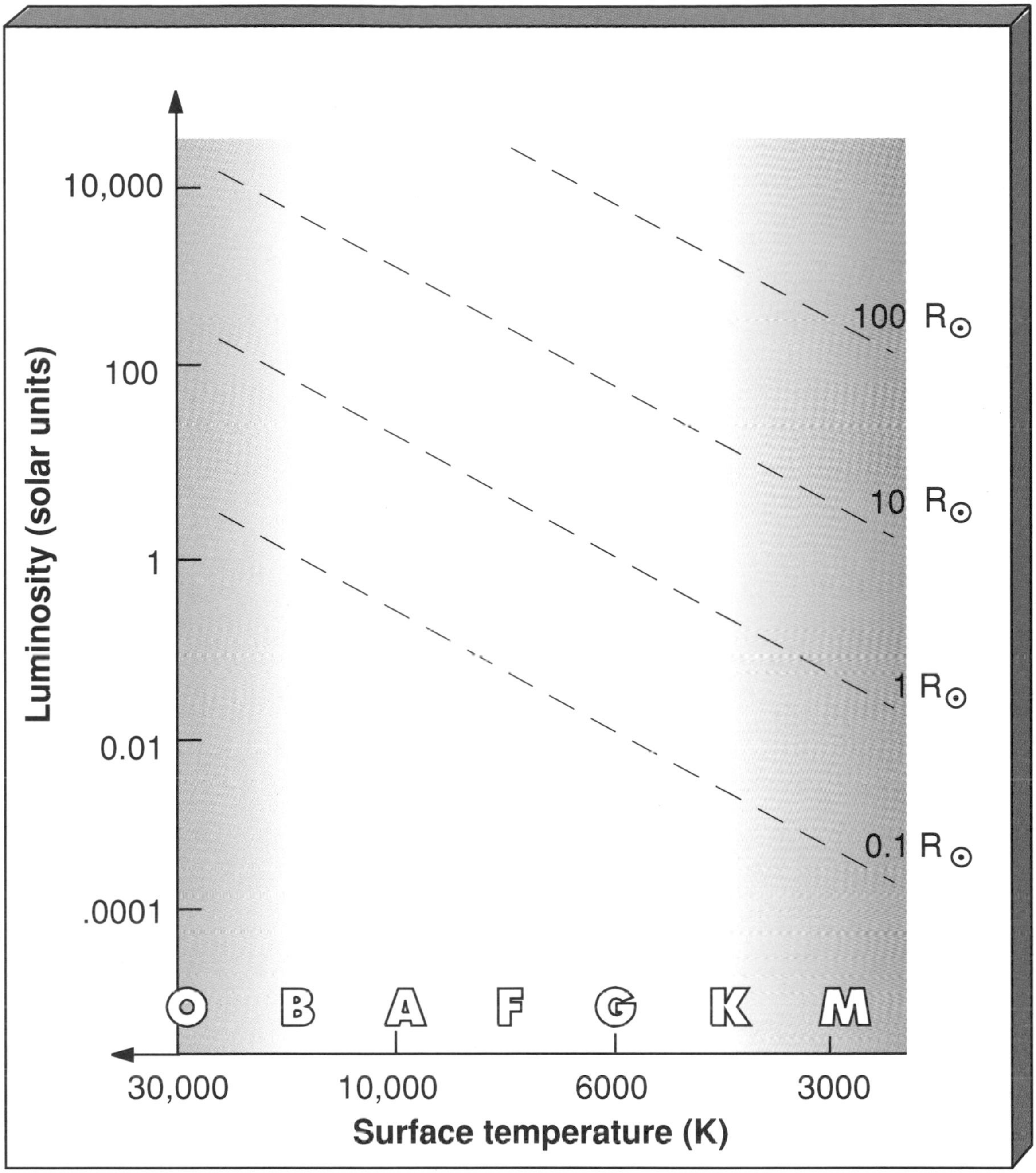

The H R diagram plots stars by luminosity (vertical axis) and temperature, or spectral class (horizontal axis). The dashed diagonal lines are lines of constant radius.

catastrophe in the very near future. Gravity overwhelms the pressure of the hot gas, and the star implodes, falling in on itself.

The core temperature rises to nearly 10 billion K. At these temperatures, individual photons are energetic enough to split iron into lighter nuclei and then break those lighter nuclei apart until only protons and neutrons remain. This process is known as *photodisintegration*. In less than a second, the collapsing core undoes all the effects of nuclear fusion that occurred during the previous 10 million years. But to split iron and lighter nuclei into smaller pieces requires a lot of energy. After all, this splitting is the opposite of the fusion reactions that generated the star's energy during earlier times. The process thus *absorbs* some of the core's heat energy, reducing the pressure and accelerating the collapse.

Now the core consists entirely of electrons, protons, neutrons, and photons at enormously high densities and it is still shrinking. As the density continues to rise, the protons and electrons are crushed together, combining to form more neutrons and releasing neutrinos. Even though the central density by this time may have reached 10^{12} kg/m^3 or more, most of the neutrinos pass through the core as if it weren't there. ∞ (Sec. 9.5) They escape into space, carrying away still more energy as they go.

There is nothing to prevent the collapse from continuing all the way to the point at which the neutrons themselves come into contact with each other, at the incredible density of about 10^{15} kg/m^3. At this density, the neutrons in the shrinking core play a role similar in many ways to that of the electrons in a white dwarf. When far apart, they offer little resistance to compression, but when brought into contact, they produce enormous pressures that strongly oppose further contraction. The collapse finally begins to slow. By the time it is actually halted, however, the core has overshot its point of equilibrium and may reach a density as high as 10^{18} kg/m^3 before beginning to re-expand. Like a fast-moving ball hitting a brick wall, the core becomes compressed, stops, then rebounds—with a vengeance!

The events just described do not take long. Only about a second elapses from the start of the collapse to the "bounce" at nuclear densities. Driven by the rebounding core, an enormously energetic shock wave then sweeps outward through the star at high speed, blasting all the overlying layers—including the heavy elements outside the iron inner core—into space. Although the details of how the shock reaches the surface and destroys the star are still uncertain, the end result is not: The star explodes in one of the most energetic events known in the universe (Figure 12.16). This spectacular death rattle of a high-mass star is known as a **core-collapse supernova**.

✓ Concept Check

- Why does the iron core of a high-mass star collapse?

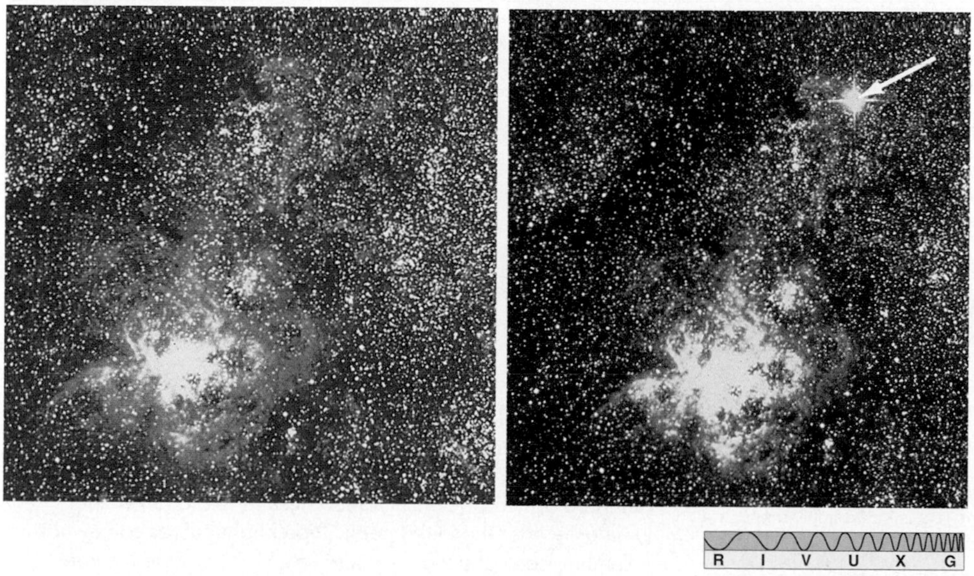

Figure 12.16 Supernova 1987A A supernova called SN1987A (arrow) was exploding near this nebula (30 Doradus) at the moment the photograph on the right was taken. The photograph on the left is the normal appearance of the star field. (See *Interlude 12-1*.) *(Space Telescope Science Institute)*

12.5 Supernova Explosions

Novae and Supernovae

5 Observationally, a *supernova*, like a nova, is a "star" that suddenly increases dramatically in brightness, then slowly dims again, eventually fading from view. However, despite some similarities in their light curves, novae and supernovae are very different phenomena. As we saw in Section 12.2, a nova is a violent explosion on the surface of a white dwarf in a binary system.[1] Supernovae are much more energetic—about a million times brighter than novae—and are driven by very different underlying physical processes. A supernova produces a burst of light billions of times brighter than the Sun, reaching that brightness within just a few hours of the start of the outburst. The total amount of electromagnetic energy radiated by a supernova during the few months it takes to brighten and fade away is roughly the same as the Sun will radiate during its *entire* 10^{10}-year lifetime!

Astronomers divide supernovae into two classes. **Type I supernovae** contain very little hydrogen, according to their spectra, and have light curves (see Figure 12.17) somewhat similar in shape to those of typical novae—a sharp rise in intensity followed by steady, gradual decline. **Type II supernovae** are hydrogen-rich, and usually have a characteristic "plateau" in the light curve a few months after the maximum. Observed supernovae are divided roughly evenly between these two categories.

Type I and Type II Supernovae Explained

How do we explain the two types of supernova just described? In fact, we already have half of the answer—the characteristics of Type II supernovae are exactly consistent with the core collapse supernovae discussed in the previous section. The Type II light curve is in good agreement with computer simulations of a stellar envelope expanding and cooling as it is blown into space by a shock wave sweeping up from below. In addition, since the expanding material consists mainly of unburned hydrogen and helium, it is not surprising that those elements are strongly represented in the supernova's spectrum.

What of Type I supernovae? Is there more than one way for a supernova explosion to occur? The answer is yes. To understand the alternative supernova mechanism, we must reconsider the long-term consequences of the accretion–explosion cycle that causes a nova. A nova explosion ejects matter from a white dwarf's surface, but it does not necessarily expel or burn all the material that has accumulated since the last outburst. In other words, there is a tendency for the dwarf's mass to increase slowly with each new nova cycle. As its mass grows and the internal pressure required to support its weight rises, the white dwarf can enter into a new period of instability—with disastrous consequences.

Recall that a white dwarf is held up by the pressure of electrons that have been squeezed so close together that they have effectively come into contact with one another. However, there is a limit to the mass of a white dwarf, beyond which the electrons cannot support the star against its own gravity. Detailed calculations show that the maximum mass of a white dwarf is about 1.4 solar masses. If an accreting white dwarf exceeds this maximum mass, it immediately starts to collapse. When that occurs, its internal temperature rapidly rises to the point at which carbon can (at last) fuse into heavier elements. Carbon fusion begins everywhere throughout the white dwarf almost simultaneously, and the entire star explodes in a **carbon-detonation supernova**—an event comparable in violence to the core-collapse supernova associated with the death of a high-mass star. In an alternative and (many astronomers think) possi-

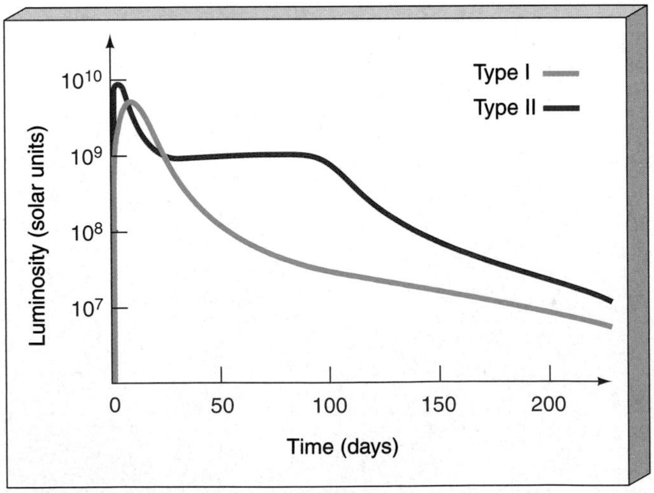

Figure 12.17 Supernova Light Curves The light curves of typical Type I and Type II supernovae. In both cases, the maximum luminosity can sometimes reach that of a billion suns, but there are characteristic differences in the falloff of the luminosity after the initial peak. Type I light curves somewhat resemble those of novae. (Figure 12.12c) Type II curves have a characteristic bump in the declining phase.

[1]Note that, when discussing novae and supernovae, astronomers tend to blur the distinction between the observed event (the sudden appearance and brightening of an object in the sky) and the process responsible for it (a violent explosion in or on a star). The two terms can have either meaning, depending on context.

bly more common scenario, two white dwarfs in a binary system may collide and merge to form a massive, unstable star. The end result is the same—a carbon-detonation supernova.

The detonation of a carbon white dwarf, the descendent of a *low-mass* star, is a Type I supernova. Because the explosion occurs in a system containing virtually no hydrogen, we can readily see why the supernova spectrum shows little evidence of that element. The appearance of the light curve results almost entirely from the radioactive decay of unstable heavy elements produced during the explosion.

Figure 12.18 summarizes the processes responsible for the two types of supernovae. We emphasize that, despite the similarity in the total amounts of energy involved, Type I and Type II supernovae are actually unrelated to one another. They occur in stars of very different types, under very different circumstances. All high-mass stars become Type II (core-collapse) supernovae, but only a tiny fraction of low-mass stars evolve into white dwarfs that ultimately explode as Type I (carbon-detonation) supernovae. However, there are far more low-mass stars than high-mass stars, resulting in the remarkable coincidence that the two types of supernova occur at roughly the same rate.

Supernova Remnants

We have plenty of evidence that supernovae have occurred in our Galaxy. Occasionally, the supernova explosions are visible from Earth. In other cases, we can detect their glowing remains, or **supernova remnants**. One of the best-studied supernova remnants is the Crab Nebula (Figure 12.19a). Its brightness has greatly dimmed now, but the original explosion in the year A.D. 1054 was so brilliant that it is prominently recorded in the manuscripts of Chinese and Middle Eastern astronomers. For nearly a month, this exploded star reportedly could be seen in broad daylight. Even today, the knots and filaments give a strong indication of past violence. The nebula—the envelope of the high-mass star that exploded to create this Type II supernova—is still expanding into space at several thousand kilometers per second.

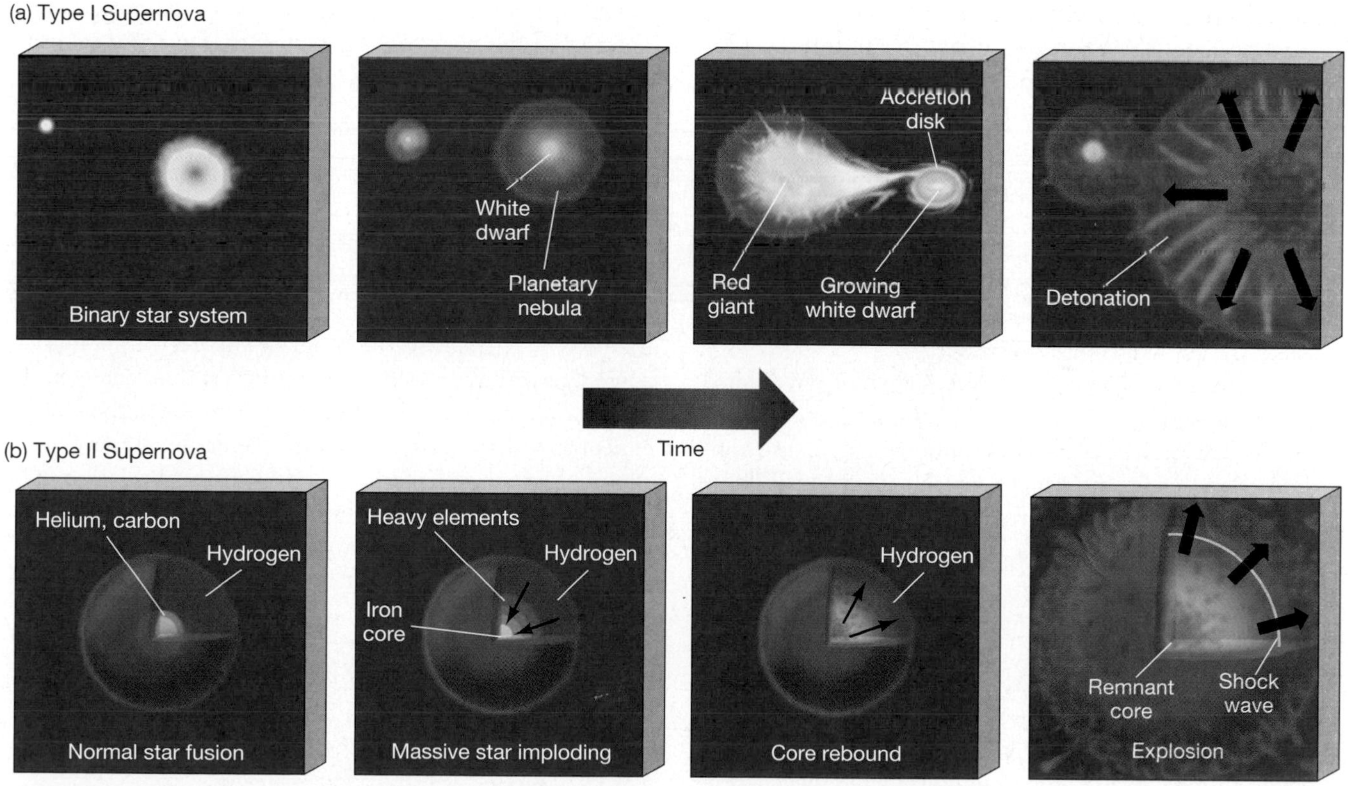

Figure 12.18 Two Types of Supernovae Type I and Type II supernovae have different causes. These sequences depict the evolutionary history of each type. (a) A Type I supernova usually results when a carbon-rich white dwarf pulls matter onto itself from a nearby red giant or main-sequence companion. (b) A Type II supernova occurs when the core of a high-mass star collapses and then rebounds in a catastrophic explosion.

Figure 12.19(b) shows another example (also of Type II). The expansion velocity of the Vela supernova remnant implies that its central star exploded around 9000 B.C. This remnant lies only 500 pc from Earth. Given its proximity, it may have been as bright as the Moon for several months.

Although hundreds of supernovae have been observed in other galaxies during the twentieth century, no one using modern equipment has ever observed one in our own Galaxy (see *Interlude 12-1* for a discussion of a recent supernova in a galaxy very close to our own). A viewable Milky Way star has not exploded since Galileo first turned his telescope to the heavens almost four centuries ago. Based on stellar evolutionary theory, astronomers estimate that an observable supernova ought to occur in our Galaxy every 100 years or so. Our local neighborhood seems long overdue for one. Unless stars explode much less frequently than predicted by theory, we should be treated to a (relatively) nearby version of nature's most spectacular cosmic event any day now.

Supernovae and the Formation of the Heavy Elements

6 From the point of view of life on Earth, probably the most important aspect of supernovae is their role in creating, and then dispersing, the heavy elements out of

INTERLUDE 12-1

Supernova 1987A

Although there have been no observable supernovae in our own Galaxy since the invention of the telescope, astronomers were treated to the next best thing in 1987 when a 15-solar-mass B-type supergiant exploded in the Large Magellanic Cloud (LMC), a small satellite galaxy orbiting our own. (Sec. 15.1) For a few weeks the Type II supernova, designated SN1987A and shown in Figure 12.16, outshone all the other stars in the LMC combined. Because the LMC is relatively close to Earth and the explosion was detected soon after it occurred, SN1987A has provided astronomers with a wealth of detailed information on supernovae, allowing them to make key comparisons between theoretical models and observational reality. By and large, the theory of stellar evolution has held up very well. Still, SN1987A did hold a few surprises.

The light curve of SN1987A, shown below, differed somewhat from the "standard" Type II shape (Figure 12.17). The peak brightness was only about 1/10 the standard value, and occurred much later than expected. These differences are mainly the result of the (relatively) small size of SN1987A's parent star, which was a blue supergiant at the time of the explosion, and not a red supergiant, as indicated by the end of the topmost evolutionary track in Figure 12.14. The reason for this is that the parent star's envelope was deficient in heavy elements, significantly altering the star's evolutionary track. Having looped back toward the main sequence after helium ignition, the star had just turned back to the right on the H–R diagram following the ignition of carbon, with a surface temperature of around 20,000 K, when the rapid chain of events leading to the supernova occurred.

Because the parent star was small and quite tightly bound by gravity, a lot of the energy produced in the form of electromagnetic radiation was used in expanding SN1987A's stellar envelope, so far less was left over to be radiated into space. Thus, SN1987A's luminosity during the first few months was therefore lower than expected, and the early peak evident in Figure 12.17 did not occur. The peak in the SN1987A light curve at about 80 days actually corresponds to the "plateau" in the Type II light curve in Figure 12.17.

About 20 hours before the supernova was detected optically, underground detectors in Japan and the United States recorded a brief (13-second) burst of neutrinos. The neutrinos preceded the light because they escaped during the collapse, whereas the first light of the explosion was emitted only after the supernova shock had plowed through the body of the star to the surface. Theoretical models, consistent with these observations, suggest that many tens of thousands of times more energy was emitted in the form of neutrinos than in any other form. Detection of this neutrino pulse is considered to be a brilliant confirmation of theory and may well herald a new age of astronomy. For the first time, astronomers have received information from beyond the solar system by radiation outside the electromagnetic spectrum.

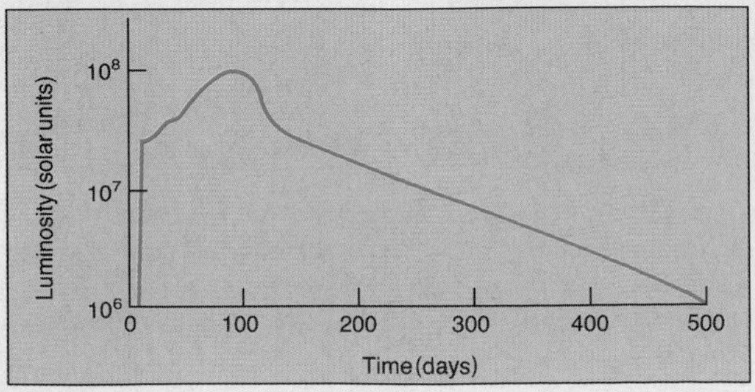

which both our planet and our bodies are made. Since the 1950s, astronomers have come to realize that all of the hydrogen and most of the helium in the universe are *primordial*—that is, they date back to the very earliest times, long before the first stars formed. (Sec. 17.6) All other elements (and, in particular, virtually everything we see around us on Earth) formed later, through stellar evolution.

We have already seen something of how heavy elements are created from light ones by nuclear fusion. This is the basic process that powers all stars. Hydrogen fuses to helium, then helium to carbon. Subsequently, in high-mass stars, carbon fuses to form still heavier elements. Oxygen, neon, magnesium, sulfur, silicon—in fact, all the known elements up to and including iron—were created in turn by fusion reactions in the cores of the most massive stars.

However, all this stops at iron. The fact that iron will not fuse to create more massive nuclei is the basic underlying cause of Type II supernovae. How then were even heavier elements, such as copper, lead, gold, and uranium, formed? The answer is that they were created during supernova explosions (both Type I and Type II), as neutrons and protons, produced when some nuclei were ripped apart by the almost unimaginable violence of the blast, were crammed into other nuclei, creating heavy elements that could not have formed by any other means. The heaviest elements, then, were formed *after*

Theory predicts that the expanding remnant of SN1987A will be large enough to be resolvable by optical telescopes within the next few years. The accompanying photographs show the barely unresolved remnant (at center) surrounded by a much larger shell of glowing gas (in yellow). Scientists reason that the progenitor star expelled this shell during its red giant phase, some 40,000 years before the explosion. The image we see results from the initial flash of ultraviolet light from the supernova hitting the ring, causing it to glow brightly. When the debris from the explosion itself strikes the ring, it will become a temporary but intense source of X rays. As shown in the inset at bottom right, the fastest-moving ejecta have already made it, forming the small (1000 A.U. diameter), glowing region marked by the arrow.

These images also show core debris moving outward toward the ring. The four insets at bottom left (also the sickle-shaped region in the bottom right inset) show material expanding at nearly 3000 km/s. The main image also revealed, to everyone's surprise, two additional faint rings that might be caused by radiation sweeping across an hourglass-shaped bubble of gas. Why the gas should exhibit this odd structure remains unclear.

Buoyed by the success of stellar-evolution theory and armed with firm theoretical predictions of what should happen next, astronomers eagerly await future developments in the story of this remarkable object.

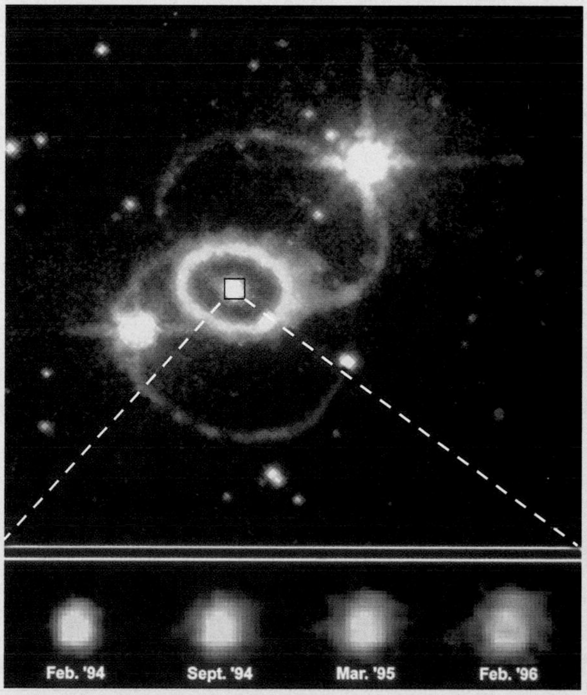

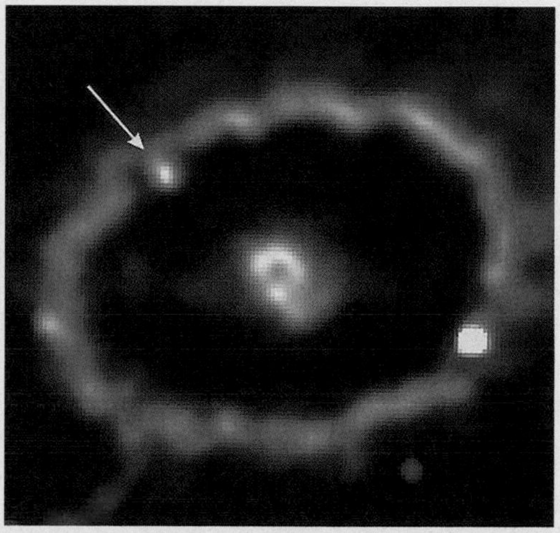

Video

Video

(P. Garnavich (CFA)/NASA) R I V U X G

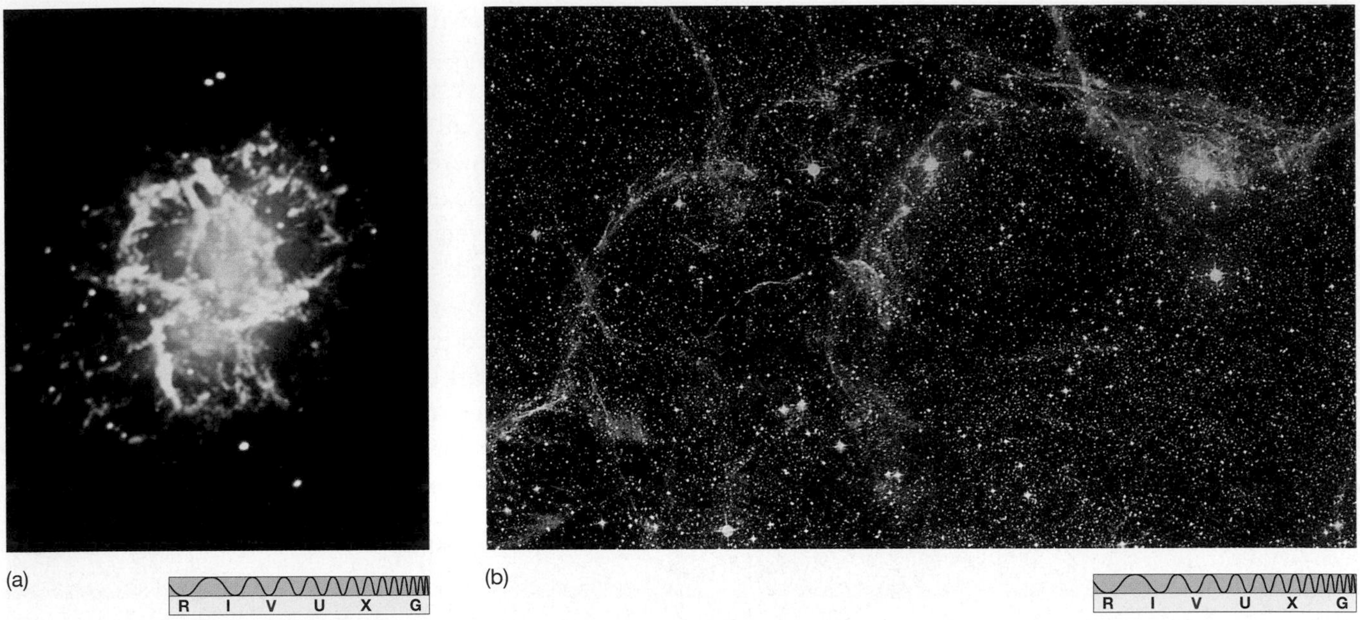

Figure 12.19 **Supernova Remnants** (a) This remnant of an ancient Type II supernova is called the Crab Nebula (M1 in the Messier catalog). It resides about 1800 pc from Earth and has an angular diameter about one-fifth that of the full Moon. Because its debris is scattered over a region of "only" 2 pc, the Crab is considered to be a young supernova remnant. In A.D. 1054 Chinese astronomers observed the supernova explosion. (b) The glowing gases of the Vela supernova remnant are spread across a 6° of the sky. *(AURA; D. Malin/Anglo-Australian Observatory)*

their parent stars had already died, even as the debris from the explosion signaling the stars' deaths was hurled into interstellar space. *More Precisely 12-1* summarizes the process by which stellar evolution steadily enriches the composition of each new generation of stars.

Although no one has ever observed directly the formation of heavy nuclei in stars, astronomers are confident that the chain of events just described really occurs in stars. When nuclear reaction rates (determined from laboratory experiments) are incorporated into detailed computer models of stars and supernovae, the results agree very well, point by point, with the composition of the universe inferred from spectroscopic studies of planets, stars, and the interstellar medium. The reasoning is indirect, but the agreement between theory and observation is so striking that most astronomers regard it as strong evidence supporting the entire theory of stellar evolution.

Concept Check

- Why are supernovae important to life on Earth?

12.6 Observing Stellar Evolution in Star Clusters

7 Star clusters provide excellent test sites for the theory of stellar evolution. Every star in a given cluster formed at the same time, from the same interstellar cloud, with virtually the same composition. (Sec. 11.5) Only the mass varies from one star to another. This uniformity allows us to check the accuracy of our theoretical models in a very straightforward way. Having studied in some detail the evolutionary tracks of individual stars, let us now consider how their collective appearance changes in time.

We begin our study shortly after the cluster's formation, with the high-mass stars already fully formed and burning steadily on the upper main sequence and lower-mass stars just beginning to arrive on the main sequence (Figure 12.20a). The appearance of the cluster at this early stage is dominated by its most massive stars—the bright blue supergiants.

Figure 12.20(b) shows our cluster's H–R diagram after 10 million years. The most massive O-type stars have evolved off the main sequence. Most have already exploded and vanished, but one or two may still be visible as supergiants traversing the top of the diagram. The remaining cluster stars are largely unchanged in appearance. The cluster's H–R diagram has a slightly truncated main sequence. Figure 12.21 shows the twin open clusters h and χ (the Greek letter chi) Persei, along with their observed H–R diagram. Comparing Figure 12.21(b) with Figure 12.20(b), astronomers estimate the age of this double cluster to be about 10 million years.

After 100 million years (Figure 12.20c), stars brighter than type B5 or so (about 4 to 5 solar masses) have left the main sequence, and a few more supergiants are visible. By this time, most of the cluster's low-mass stars have finally arrived on the main sequence, although the dimmest M-type stars may still be in their contraction phase. The appearance of the cluster is now dominated by bright B-type stars and brighter supergiants.

At any time during the evolution, the cluster's original main sequence is intact up to some well-defined stellar mass, corresponding to the stars that are just leaving the main sequence at that instant. We can imagine the main sequence being "peeled away" from the top down, with fainter and fainter stars turning off and heading for the giant branch as time goes on. Astronomers refer to the high-luminosity end of the observed main sequence as the **main-sequence turnoff**. The mass of the star that is just evolving off the main sequence at any moment is known as the *turnoff mass*.

At 1 billion years (Figure 12.20d), the main-sequence turnoff mass is around two solar masses, corresponding roughly to spectral type A2. The subgiant and giant branches associated with the evolution of low-mass stars are just becoming apparent, and the formation of the lower main sequence is now complete.

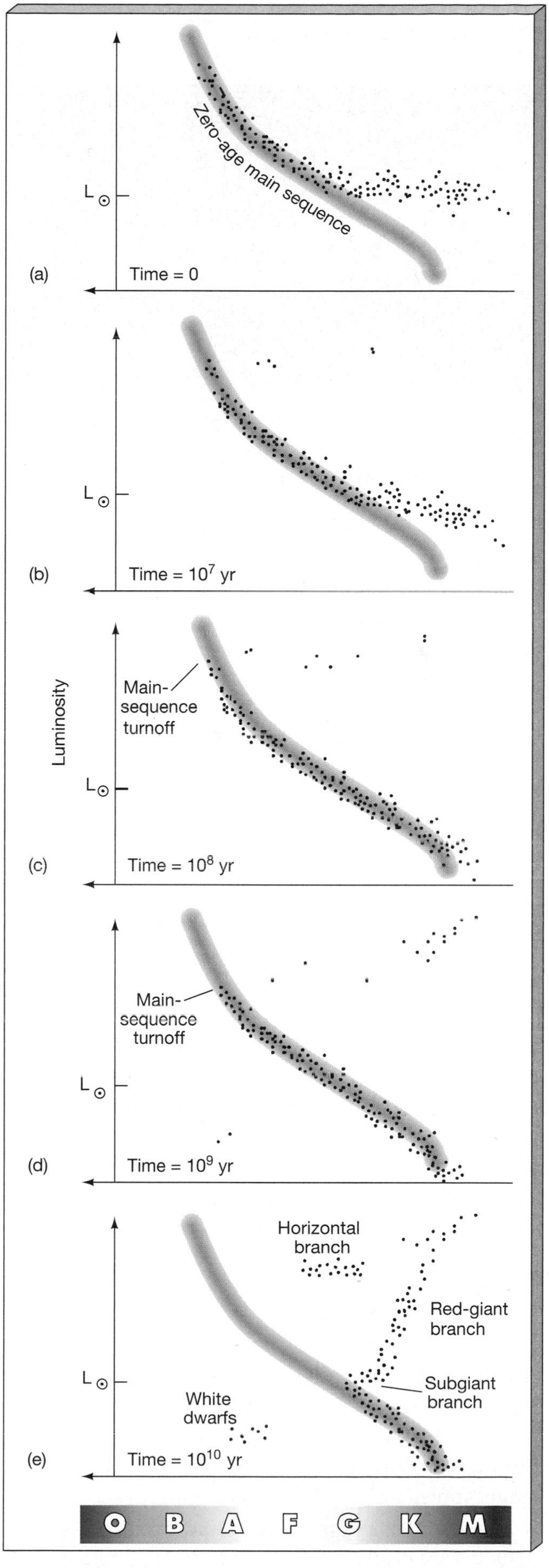

Figure 12.20 Cluster Evolution on the H–R Diagram The changing H–R diagram of a hypothetical star cluster. (a) Initially, stars on the upper main sequence are already burning steadily, while the lower main sequence is still forming. (b) At 10^7 years, O-type stars have already left the main sequence, and a few post-main-sequence supergiants are visible. (c) By 10^8 years, stars of spectral type B have evolved off the main sequence. More supergiants are visible, and the lower main sequence is almost fully formed. (d) At 10^9 years, the main sequence is cut off at about spectral type A. The subgiant and red giant branches are just becoming evident, and the formation of the lower main sequence is complete. A few white dwarfs may be present. (e) At 10^{10} years, only stars less massive than the Sun still remain on the main sequence. The cluster's subgiant, red giant, and horizontal branches are all discernible, and many white dwarfs have now formed.

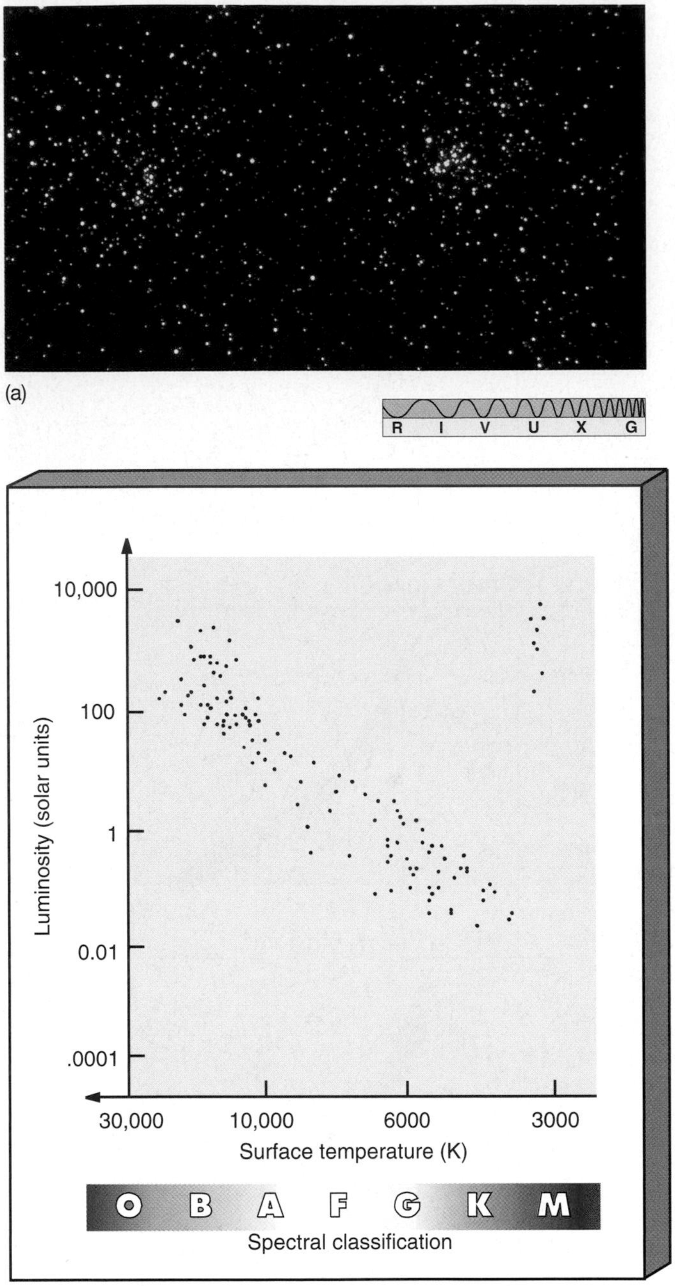

Figure 12.21 Newborn Cluster H–R Diagram (a) The "double cluster" h and χ Persei. (b) The H–R diagram of the pair indicates that the stars are very young—probably only about 10 million years old. *(NOAO)*

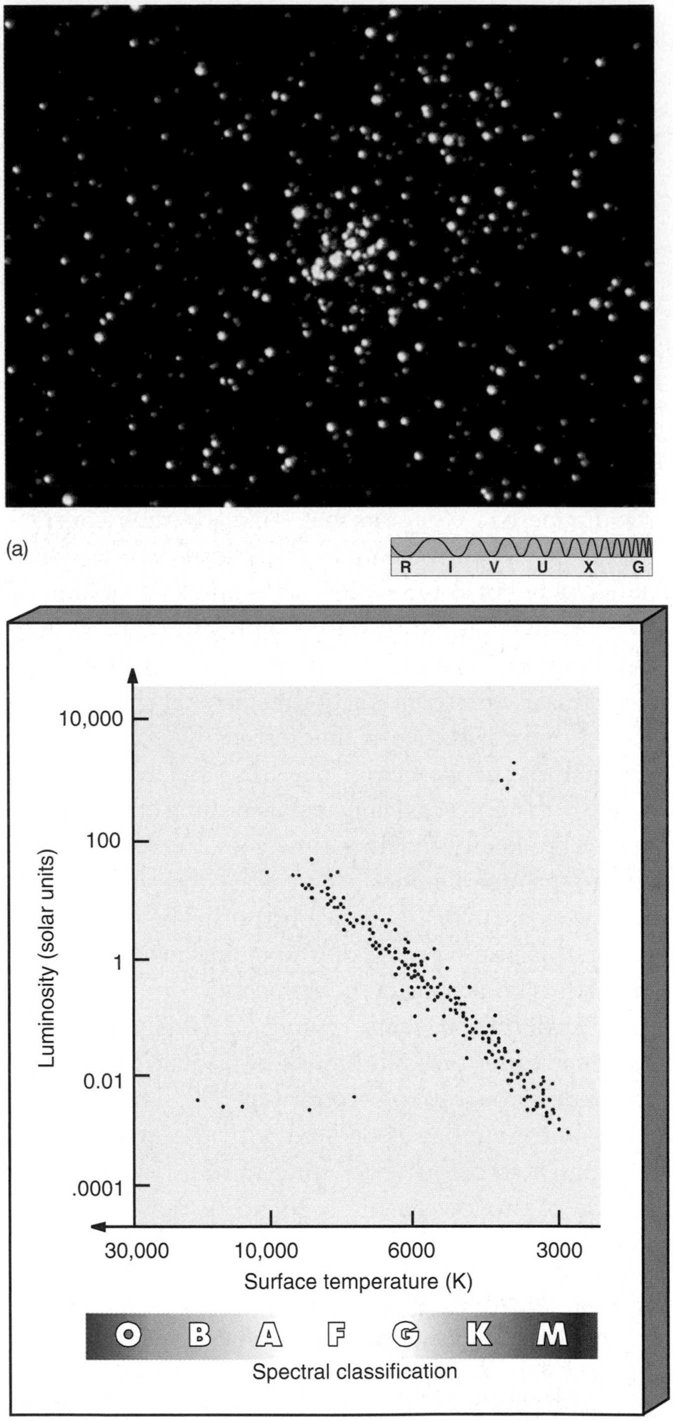

Figure 12.22 Young Cluster H–R Diagram (a) The Hyades cluster, a relatively young group of stars visible to the naked eye. (b) The H–R diagram for this cluster is cut off at about spectral type A, implying an age of about 600 million years. *(NOAO)*

In addition, the first white dwarfs have just appeared, although they are often too faint to be observed at the distances of most clusters. Figure 12.22 shows the Hyades open cluster and its H–R diagram. The H–R diagram appears to lie between Figures 12.20(c) and 12.20(d). More careful measurements yield a cluster age of about 600 million years.

At 10 billion years, the turnoff point has reached solar-mass stars, of spectral type G2. The subgiant and giant branches are now clearly discernible in the H–R diagram (Figure 12.20e), and the horizontal branch appears as a distinct region. Many white dwarfs are also present in the cluster. Figure 12.23 shows a composite H–R diagram for several globular clusters in our Galaxy. The various evolutionary stages predicted by theory are all clearly visible in this figure, although the individual points are shifted somewhat to the left relative to our previous H–R diagrams because of differences in composition between stars like the Sun and stars in globular clusters. (Globular cluster stars tend to be hotter than solar-type stars of the same mass.) The clusters represented in Figure 12.23(b) are quite deficient in heavy elements compared to the Sun. Figure 12.23(a) shows a recent (Hubble) image of the globular cluster M80, which has a similarly low concentration of heavy elements—only about 2 percent the solar value—and whose H–R diagram is expected to look qualitatively very similar to Figure 12.23(b).

By carefully adjusting their theoretical models until the main sequence, subgiant, red giant, and horizontal branches are all well matched, astronomers have determined that this diagram corresponds to a cluster age of roughly 12 billion years, a little older than the hypothetical cluster in Figure 12.20(e). The deficiency of heavy elements is consistent with these clusters being among the first objects to form in our Galaxy (see *More Precisely 12-1*). In fact, globular cluster ages determined this way show a remarkably small spread. Most of the globular clusters in our Galaxy appear to have formed between about 10 and 12 billion years ago. (The "blue stragglers" in the figure are main sequence stars that lie above the turnoff, in apparent contradiction to the theory just described. They are thought to be the results of collisions between lower-mass main sequence stars, most likely in binary systems.)

Stellar evolution is one of the great success stories of astrophysics. Like all good scientific theories, it makes definite, testable predictions about the universe, at the same time remaining flexible enough to incorporate new discoveries as they occur. Theory and observation have advanced hand in hand. At the start of the twentieth century, many scientists despaired of ever knowing even the compositions of the stars, let alone why they shine and how they change. Today, the theory of stellar evolution is a cornerstone of modern astronomy.

✓ Concept Check

- Why are observations of star clusters so important to the theory of stellar evolution?

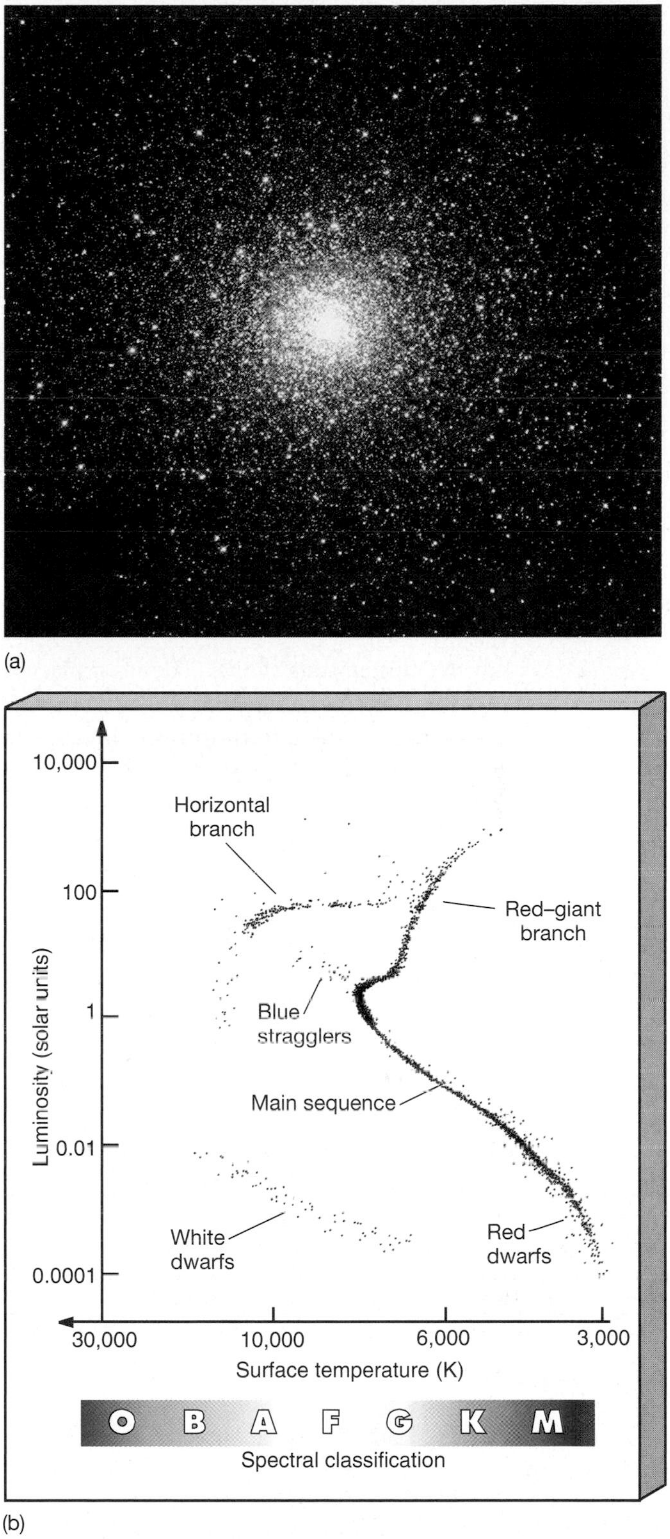

Figure 12.23 **Old Cluster H–R Diagram** (a) The globular cluster M80, which lies some 8 kpc from Earth. (b) Combined H–R diagram, based on ground- and space-based observations, for several globular clusters similar in overall composition to M80. Fitting the main-sequence turnoff and the giant and horizontal branches to theoretical models implies an age of about 12 billion years, making these clusters among the oldest known objects in the Milky Way Galaxy. This is consistent with the low concentrations of heavy elements observed in their spectra. *(NASA; data courtesy of William E. Harris)*

MORE PRECISELY 12-1

The Cycle of Stellar Evolution

We have now seen all the ingredients that make up the complete cycle of star formation and evolution. Let us now briefly summarize that process, which is illustrated in the figure below.

1. Stars form when part of an interstellar cloud is compressed beyond the point at which it can support itself against its own gravity. The cloud collapses and fragments, forming a cluster of stars. The hottest stars heat and ionize the surrounding gas, sending shock waves through the surrounding cloud, possibly triggering new rounds of star formation. ∞ (Sec. 11.3)
2. Within the cluster, stars evolve. The most massive stars evolve fastest, creating heavy elements in their cores and spewing them forth into the interstellar medium in supernova explosions. Low-mass stars take longer to evolve, but they too can create heavy elements and may contribute to the "seeding" of interstellar space when they shed their envelopes as planetary nebulae.
3. The creation and explosive dispersal of new heavy elements are accompanied by further shock waves. The passage of these shock waves through the interstellar medium simultaneously enriches the medium and compresses it into further star formation. Each generation of stars increases the concentration of heavy elements in the interstellar clouds from which the next generation forms. As a result, recently formed stars contain a much greater abundance of heavy elements than stars that formed long ago.

In this way, although some material is used up in each cycle—turned into energy or locked up in low-mass stars—the Galaxy continuously recycles its matter. Each new round of formation creates stars containing more heavy elements than the preceding generation had. From the old globular clusters, which are observed to be deficient in heavy elements relative to the Sun, to the young open clusters, which contain much larger amounts of these elements, we observe this enrichment process in action. Our Sun is the product of many such cycles. We ourselves are another. Without the heavy elements synthesized in the hearts of stars, life on Earth would not exist.

Animation

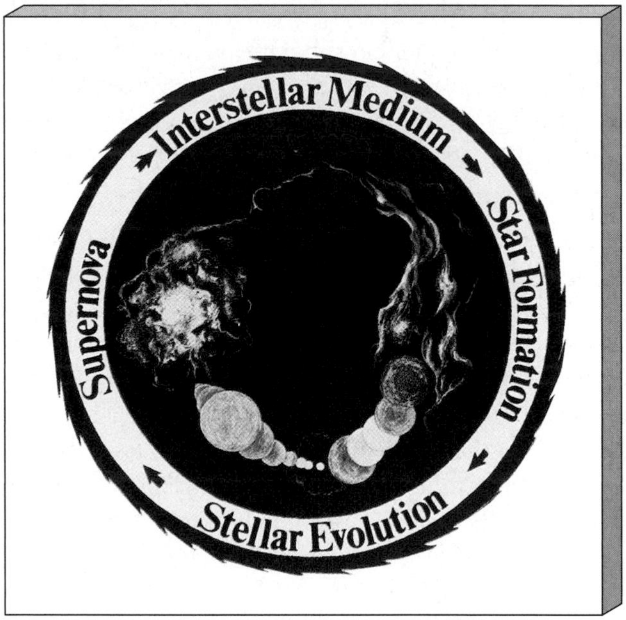

(L. Chaisson)

Chapter Review

Summary

Stars spend most of their lives on the main sequence, in the **core-hydrogen burning** (p. 302) phase of stellar evolution. Stars leave the main sequence when the hydrogen in their cores is exhausted. With no internal energy source, the star's helium core is unable to support itself against its own gravity and begins to shrink. The star at this stage is in the **hydrogen-shell-burning** (p. 303) phase, with nonburning helium at the center surrounded by a layer of burning hydrogen. The energy released by the contracting helium core heats the hydrogen-burning shell, greatly increasing the nuclear reaction rates there. As a result, the star becomes much brighter, while the envelope expands and cools. A solar-mass star moves off the main sequence on the H–R diagram first along the **subgiant branch** (p. 304), then almost vertically up the **red giant branch** (p. 304).

As the helium core contracts, it heats up. Eventually, it reaches the point at which helium begins to fuse into carbon. In a star like the Sun, helium burning begins explosively, in the **helium flash** (p. 304). The flash expands the core and reduces the star's luminosity, sending it onto the **horizontal branch** (p. 305) of the H–R diagram. The star now has a core of burning helium surrounded by a shell of burning hydrogen. An inner core of nonburning carbon forms, shrinks and heats the overlying burning layers, and the star once again becomes a red giant. It reenters the red giant region of the H–R diagram, becoming an extremely luminous **red supergiant** (p. 306) star.

The core of a solar-mass star never becomes hot enough to fuse carbon. Such a star continues to brighten and expand until its envelope is ejected into space as a **planetary nebula** (p. 308). At that point the core becomes visible as a hot, faint, and extremely dense white dwarf. The white dwarf cools and fades, eventually becoming a cold **black dwarf** (p. 310).

A **nova** (p. 310) is a star that suddenly increases greatly in brightness, then slowly fades back to its normal appearance over a period of months. It is the result of a white dwarf in a binary system drawing hydrogen-rich material from its companion. The gas spirals inward in an **accretion disk** (p. 311) and builds up on the white dwarf's surface, eventually becoming hot and dense enough for the hydrogen to burn explosively, temporarily causing a large increase in the dwarf's luminosity.

Stars more massive than about eight solar masses form heavier and heavier elements in their cores, at a more and more rapid pace. As they evolve, their cores form a layered structure consisting of burning shells of successively heavier elements. The process stops at iron, whose nuclei can neither be fused together nor split to produce energy. As a star's iron core grows in mass, it eventually becomes unable to support itself against gravity and begins to collapse. At the high temperatures produced during the collapse, iron nuclei are broken down into protons and neutrons. The protons combine with electrons to form more neutrons. Eventually, when the core becomes so dense that the neutrons are effectively brought into physical contact with one another, their resistance to further squeezing stops the collapse and the core rebounds, sending a violent shock wave out through the rest of the star. The star explodes in a **core-collapse supernova** (p. 313).

Type I supernovae (p. 314) are hydrogen poor and have a light curve similar in shape to that of a nova. **Type II supernovae** (p. 314) are hydrogen rich and have a characteristic bump in the light curve a few months after maximum. A Type II supernova is a core-collapse supernova. A Type I supernova is a **carbon-detonation supernova** (p. 314), which occurs when a white dwarf in a binary system collapses and then explodes as its carbon ignites. We can see evidence for a past supernova in the form of a **supernova remnant** (p. 315), a shell of exploded debris surrounding the site of the explosion and expanding into space at a speed of thousands of kilometers per second.

All elements heavier than helium formed in stars or in supernova explosions. Comparisons between theoretical predictions of element production and observations of element abundances in stars and supernovae provide strong support for the theory of stellar evolution.

The theory of stellar evolution can be tested by observing star clusters. At any instant, stars with masses above the cluster's **main-sequence turnoff** (p. 319) have evolved off the main sequence. By comparing a cluster's main-sequence turnoff mass with theoretical predictions, astronomers can determine the cluster's age.

Review and Discussion

1. For how long can a star like the Sun keep burning hydrogen in its core?
2. Why is the depletion of hydrogen in the core of a star such an important event?
3. What makes an ordinary star become a red giant?
4. How big (in A.U.) will the Sun become when it enters the red giant phase?
5. How long does it take for a star like the Sun to evolve from the main sequence to the top of the red giant branch?
6. What is the helium flash?
7. How do stars of low mass die? How do stars of high mass die?
8. What is a planetary nebula? Why do many planetary nebulae appear as rings?
9. What are white dwarfs? What is their ultimate fate?
10. Under what circumstances will a binary star produce a nova?
11. What occurs in a massive star to cause it to explode?
12. What are the observational differences between Type I and Type II supernovae?
13. How do the mechanisms that cause Type I and Type II supernovae explain their observed differences?
14. What evidence do we have that many supernovae have occurred in our Galaxy?
15. Why do the cores of massive stars evolve into iron and not heavier elements?

True or False?

____ **1.** Low-mass stars are conventionally taken to have masses of less than about eight solar masses.

____ **2.** All the red dwarf stars that ever formed are still on the main sequence today.

____ **3.** Once a star is on the main sequence, gravity is no longer important in determining the star's internal structure.

____ **4.** The Sun will get brighter as it begins to run out of fuel in its core.

____ **5.** As it evolves away from the main sequence, a star gets smaller.

____ **6.** When the Sun becomes a red giant, its core will be smaller than when the Sun was on the main sequence.

____ **7.** When helium starts to fuse inside a solar-mass red giant, it does so very slowly at first; the rate of fusion increases gradually over many years.

____ **8.** A planetary nebula is the disk of matter around a star that will eventually form a planetary system.

____ **9.** A nova is a sudden outburst of light coming from an old main-sequence star.

____ **9.** High-mass stars can fuse carbon in their cores.

____ **11.** It takes less and less time to fuse heavier and heavier elements inside a high-mass star.

____ **12.** In a core-collapse supernova, the outer part of the core rebounds from the high-density inner part of the core, destroying the entire outer part of the star.

____ **13.** The spectrum of a Type I supernova shows the presence of lots of hydrogen.

____ **14.** Stellar evolution can account for the existence of all elements except hydrogen and helium.

____ **15.** A star cluster 100 million years old still contains many O-type stars.

Fill in the Blank

1. A main-sequence star doesn't collapse because of the outward _______ produced by hot gases in the stellar interior.
2. The Sun will leave the main sequence in about _______ years from now.
3. While a star is on the main sequence, _______ is slowly consumed in the core and _______ builds up.
4. A temperature of at least _______ is needed to fuse helium.
5. At the end of its main-sequence lifetime, a star's core starts to _______.
6. The various stages of stellar evolution predicted by theory can be tested using observations of stars in _______.
7. By the time the envelope of a red supergiant is ejected, the core has shrunk down to a diameter of about _______.
8. A typical white dwarf has the following properties: about half a solar mass, fairly _______ surface temperature, small size, and _______ luminosity.
9. As time goes by, the temperature and the luminosity of a white dwarf both _______.
10. In a binary consisting of a white dwarf and a main-sequence or giant companion, matter leaving the companion forms an _______ disk around the dwarf.
11. A nova explosion is due to _______ fusion on the _______ of a white dwarf.
12. When a proton and an electron are forced together, they combine to form a _______ and a _______.

13. A ______ supernova occurs when a white dwarf exceeds 1.4 solar masses.

14. The two types of supernova can be distinguished observationally by their spectra and by their ______.

15. As a star cluster ages, the luminosity of the main-sequence turnoff ______.

Problems

1. Calculate the average density of a red giant core of mass 0.25 solar mass and radius 15,000 km. Compare this with the average density of the giant's envelope, if the mass of the envelope is 0.5 solar mass and its radius is 0.5 A.U. Compare each with the average density of the Sun.

2. A main sequence star at a distance of 20 pc is barely visible through a certain telescope. The star subsequently ascends the giant branch, during which time its temperature drops by a factor of three and its radius increases 100-fold. What is the new maximum distance at which the star would still be visible using the same telescope?

3. If the Sun will reside on the main sequence for 10^{10} years and the luminosity of a main-sequence star is proportional to the cube of the star's mass, what mass star is just now leaving the main sequence in a cluster that formed 400 million years ago?

4. How long will it take the Sun's planetary nebula, expanding at a speed of 50 km/s, to reach the orbit of Neptune? How long to reach the nearest star?

5. What are the escape speed and surface gravity of Sirius B (Table 12.2)? ∞ (*More Precisely* 5-2)

6. A certain telescope could just detect the Sun at a distance of 10,000 pc. What is the maximum distance at which it could detect a nova having a peak luminosity of 10^5 solar luminosities? Repeat the calculation for a supernova having a peak luminosity 10^{10} times that of the Sun.

7. At what distance would the supernova in the previous question look as bright as the Sun? Would you expect a supernova to occur that close to us?

8. A (hypothetical) supernova at a distance of 500 pc has an absolute magnitude of -20. Compare its apparent magnitude with that of (a) the full Moon; (b) Venus at its brightest. ∞ (See Figure 10.6.)

9. The Crab Nebula is now about 1 pc in radius. If it was observed to explode in A.D. 1054, roughly how fast is it expanding? (Assume constant expansion velocity. Is that a reasonable assumption?)

10. A supernova's energy is often compared to the total energy output of the Sun over its lifetime. Using the Sun's current luminosity, calculate the total solar energy output, assuming a 10^{10} year main-sequence lifetime. Using Einstein's formula $E = mc^2$, calculate the equivalent amount of mass, in Earth masses. ∞ (Sec. 9.5)

Projects

1. You can tour the Galaxy without ever leaving Earth, just by looking up. In the winter sky, you'll find the red supergiant Betelgeuse in the constellation Orion. It's easy to see because it's one of the brightest stars visible in our night sky. Betelgeuse is a variable star with a period of about 6.5 years. Its brightness changes as it expands and contracts. At maximum size, Betelgeuse fills a volume of space that would extend from the Sun to beyond the orbit of Jupiter. Betelgeuse is thought to be about 10 to 15 times more massive than our Sun. It is probably between four and 10 million years old—and in the final stages of its evolution. A similar star can be found shining prominently in midsummer. This is the red supergiant Antares in the constellation Scorpius. Can you find these stars? Why are they red? What will happen to them during their next evolutionary stages?

2. In 1758, the French comet-hunter Charles Messier discovered the sky's most legendary supernova remnant, now called M1, or the Crab Nebula. An eight-inch telescope reveals the Crab's oval shape, but it will appear faint. It is located northwest of Zeta Tauri, the star that marks the southern tip of the horns of Taurus the Bull. A 10-inch or larger telescope reveals some of the Nebula's famous filamentary structure.

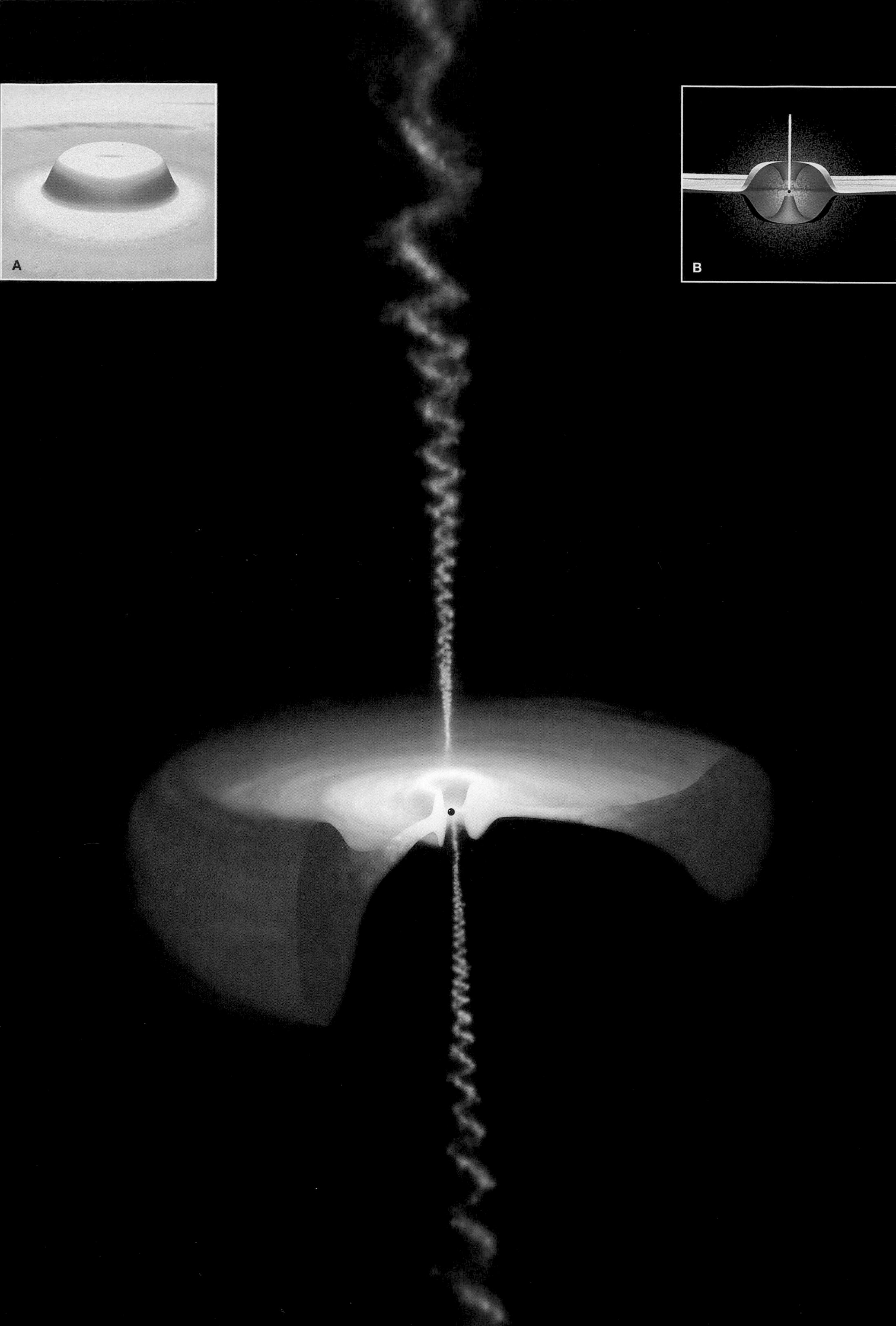
A
B

13 NEUTRON STARS AND BLACK HOLES

Strange States of Matter

LEARNING GOALS

Studying this chapter will enable you to:

1. Describe the properties of neutron stars and explain how these strange objects are formed.
2. Explain the nature and origin of pulsars, and account for their characteristic radiation.
3. List and explain some of the observable properties of neutron-star binary systems.
4. Describe how black holes are formed, and discuss their effects on matter and radiation in their vicinity.
5. Relate the phenomena that occur near black holes to the warping of space around them.
6. Discuss some of the ways in which the presence of a black hole might be detected.

(Opposite page, background) Black holes can be neither seen nor drawn easily. In this artist's conception of the vicinity of a black hole, the disk-shaped region is a rapidly whirling collection of hot gas and dust about to fall into the black hole. The hole itself is represented merely by a black dot at the center of the accretion disk.

(Inset A) A close-up view of the core of the accretion disk, showing a truncated cusp of very hot gas about to make its way down into the central black hole.

(Inset B) The anatomy of a hypothetical black hole would show its accretion disk to the left and right, with the hole itself still only a black point at the very center. The white vertical spike is a geyser (or jet) of matter shot away from near the hole, perpendicular to the innermost part of the disk. *(D. Berry)*

Our study of stellar evolution has led us to some unusual and unexpected objects. Red giants, white dwarfs, and supernovae surely represent extreme states of matter completely unfamiliar to us here on Earth, yet stellar evolution can have even more bizarre consequences. The strangest states of all result from the catastrophic implosion-explosion of stars much more massive than our Sun. The incredible violence of a supernova explosion may bring into being objects so extreme in their behavior that they require us to reconsider some of our most hallowed laws of physics. They open up a science-fiction writer's dream of fantastic phenomena. They may even one day force scientists to construct a whole new theory of the universe.

13.1 Neutron Stars

1 What remains after a supernova explosion? Is the original star blown to bits and dispersed into interstellar space, or does some portion of it survive? For a Type I (carbon-detonation) supernova, most astronomers regard it as unlikely that anything is left behind after the explosion. (Sec. 12.5) The entire star is shattered by the blast. However, for a Type II (core-collapse) supernova, theoretical calculations indicate that part of the star might survive the explosion.

Recall that, in a Type II supernova, the iron core of a massive star collapses until its neutrons effectively come into contact with one another. At that point, the central portion of the core rebounds, creating a powerful shock wave that races outward through the star, violently expelling matter into space. (Sec. 12.4) The key point here is that the shock wave does not start at the very center of the collapsing core. The innermost part of the core—the region that rebounds—remains intact as the shock wave it produces destroys the rest of the star. After the violence of the supernova has subsided, this ultracompressed ball of neutrons is all that is left. Researchers colloquially call this core remnant[1] a **neutron star**.

Neutron stars are extremely small and very massive. Composed purely of neutrons packed together in a tight ball about 20 km across, a typical neutron star is not much bigger than a small asteroid or a terrestrial city (Figure 13.1), yet its mass is greater than that of the Sun. With so much mass squeezed into such a small volume, neutron stars are incredibly dense. Their average density can reach 10^{17} or even 10^{18} kg/m^3, nearly a billion times denser than a white dwarf. (Sec. 12.3) A single thimbleful of neutron-star material would weigh 100 million tons—about as much as a good-sized terrestrial mountain.

Neutron stars are solid objects. Provided that a sufficiently cool one could be found, you might even imagine standing on it. However, this would not be easy, as its gravity is extremely powerful. A 70-kg human would weigh the Earth equivalent of about one billion kg (one million tons). The severe pull of a neutron star's gravity would flatten you much thinner than this piece of paper.

In addition to large mass and small size, newly formed neutron stars have two other very important

Figure 13.1 Neutron Star Neutron stars are not much larger than many of Earth's major cities. In this fanciful comparison, a typical neutron star sits alongside Manhattan Island. *(NASA)*

[1]Astronomers commonly use the term *remnant* to mean whatever is left of a star's inner core after evolution has ended. Such objects are small and compact—no larger than Earth. They should not be confused with *supernova remnants*, which are the aftermath of supernova explosions: glowing clouds of debris scattered across many parsecs of interstellar space. (Sec. 12.4)

properties. First, they *rotate* extremely rapidly, with periods measured in fractions of a second. This is a direct result of the law of conservation of angular momentum (Chapter 4), which tells us that any rotating body must spin faster as it shrinks, and the core of the parent star almost certainly had some rotation before it began to collapse. ∞ (*More Precisely 4-1*) Second, they have very strong *magnetic fields*. The original field of the parent star is amplified as the collapsing core squeezes the magnetic field lines closer together, creating a magnetic field trillions of times stronger than Earth's field.

In time, theory indicates, a neutron star will spin more and more slowly as it radiates its energy into space, and its magnetic field will diminish. However, for a few million years after its birth, these two properties combine to provide the primary means by which this strange object can be detected and studied.

✓ Concept Check

- Are all supernovae expected to lead to neutron stars?

13.2 Pulsars

2 The first observation of a neutron star occurred in 1967, when Jocelyn Bell, a graduate student at Cambridge University, observed an astronomical object emitting radio radiation in the form of rapid *pulses* (Figure 13.2). Each pulse consisted of a roughly 0.01-s burst of radiation, separated by precisely 1.34 s from the next.

Many hundreds of these pulsating objects are now known. They are called **pulsars**. Each has its own characteristic pulse shape and period. Observed pulsar periods are generally quite short, ranging from about a few milliseconds to about a second, corresponding to flashing rates of between one and several hundred times per second. In some cases, the periods are extremely stable—in fact, constant to within a few seconds in a million years—making pulsars the most accurate natural clocks known in the universe, more accurate even than the best atomic clocks on Earth.

When Bell made her discovery in 1967, she did not know what she was looking at. Indeed, no one at the time knew what a pulsar was. The explanation won Bell's thesis advisor, Anthony Hewish, the 1974 Nobel Prize in physics. Hewish reasoned that the only physical mechanism consistent with such precisely timed pulses is a small, rotating source of radiation. Only rotation can cause the high degree of regularity of the observed pulses, and only a small object can account for the sharpness of each pulse (see Section 13.3). The best current model describes a pulsar as a compact, spinning neutron star that periodically flashes radiation toward Earth.

Figure 13.3 outlines the main features of this pulsar model. Two "hot spots," either on the surface of a neutron star or in the magnetosphere just above it, continuously emit radiation in a narrow beam. These spots are most likely localized regions near the neutron star's magnetic poles, where charged particles, accelerated to extremely high energies by the star's rotating magnetic field, emit radiation along the star's magnetic axis. The hot spots radiate more or less steadily, and the resulting beams sweep through space like a revolving lighthouse beacon as the neutron star rotates. Indeed, this pulsar model is often known as the **lighthouse model**. If the neutron star happens to be oriented such that one of the rotating beams sweeps across Earth, then we see the pulses. The period of the pulses is the star's rotation period.

Most pulsars emit pulses in the form of radio radiation. However, some also pulse in the visible, X-ray, and gamma-ray parts of the spectrum. Whatever types of radiation are produced, however, all the electromagnetic flashes from a given pulsar are synchronized, occurring at the same regular intervals, because they all arise from the same object.

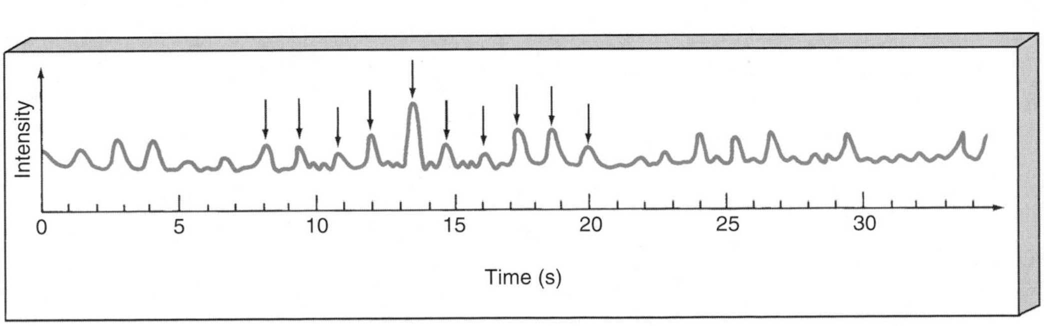

Figure 13.2 Pulsar Radiation Pulsars emit periodic bursts of radiation. This recording shows the regular change in the intensity of the radio radiation emitted by the first such object known. It was discovered in 1967. Some of the pulses are marked by arrows.

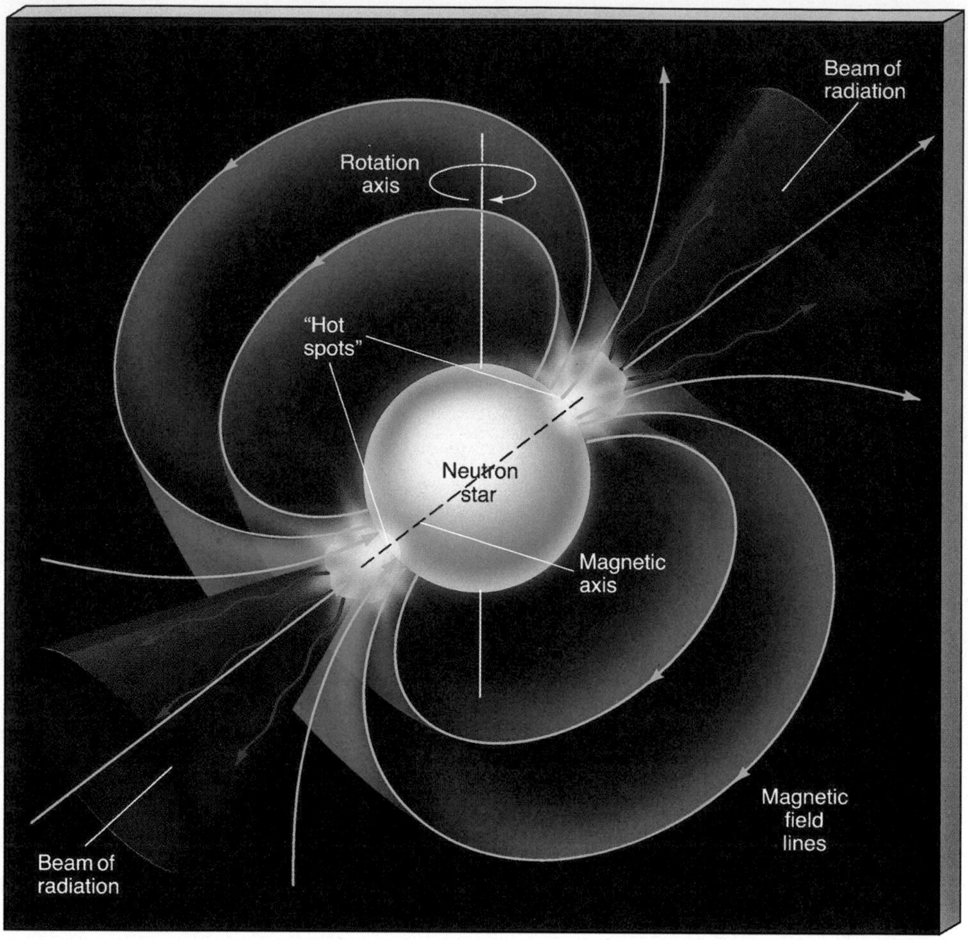

Figure 13.3 Pulsar Model This diagram of the lighthouse model of neutron-star emission accounts for many of the observed properties of pulsars. Charged particles, accelerated by the magnetism of the neutron star, flow along the magnetic field lines, producing radio radiation that beams outward. The beam sweeps across the sky as the neutron star rotates. If it happens to sweep across Earth, we see a pulsar.

All pulsars are neutron stars, but not all neutron stars are pulsars, for two reasons. First, the two ingredients that make the neutron star pulse—rapid rotation and strong magnetic field—both diminish with time, so the pulses gradually weaken and become less frequent. Theory indicates that, within a few tens of millions of years, the pulses have all but stopped. Second, even a young, bright pulsar is not necessarily visible from Earth. The pulsar beam depicted in Figure 13.3 is relatively narrow—perhaps as little as a few degrees across. Only if the neutron star happens to be oriented in just the right way do we see pulses. When we see the pulses from Earth, we call the object a pulsar.

Even though most pulsars are probably not visible from our vantage point, current pulsar observations are consistent with the idea that *every* high-mass star dies in a supernova explosion, leaving a neutron star behind, and that *all* young neutron stars emit beams of radiation, like the pulsars we actually see. A few pulsars are definitely associated with supernova remnants, clearly establishing those pulsars' explosive origin. Figure 13.4 shows optical and X-ray images of the Crab pulsar, at the center of the Crab supernova remnant. ∞ (Sec. 12.5) This is evidently all that remains of the once-massive star whose supernova was observed in 1054.

✓ Concept Check

- Why don't we see pulsars at the centers of all supernova remnants?

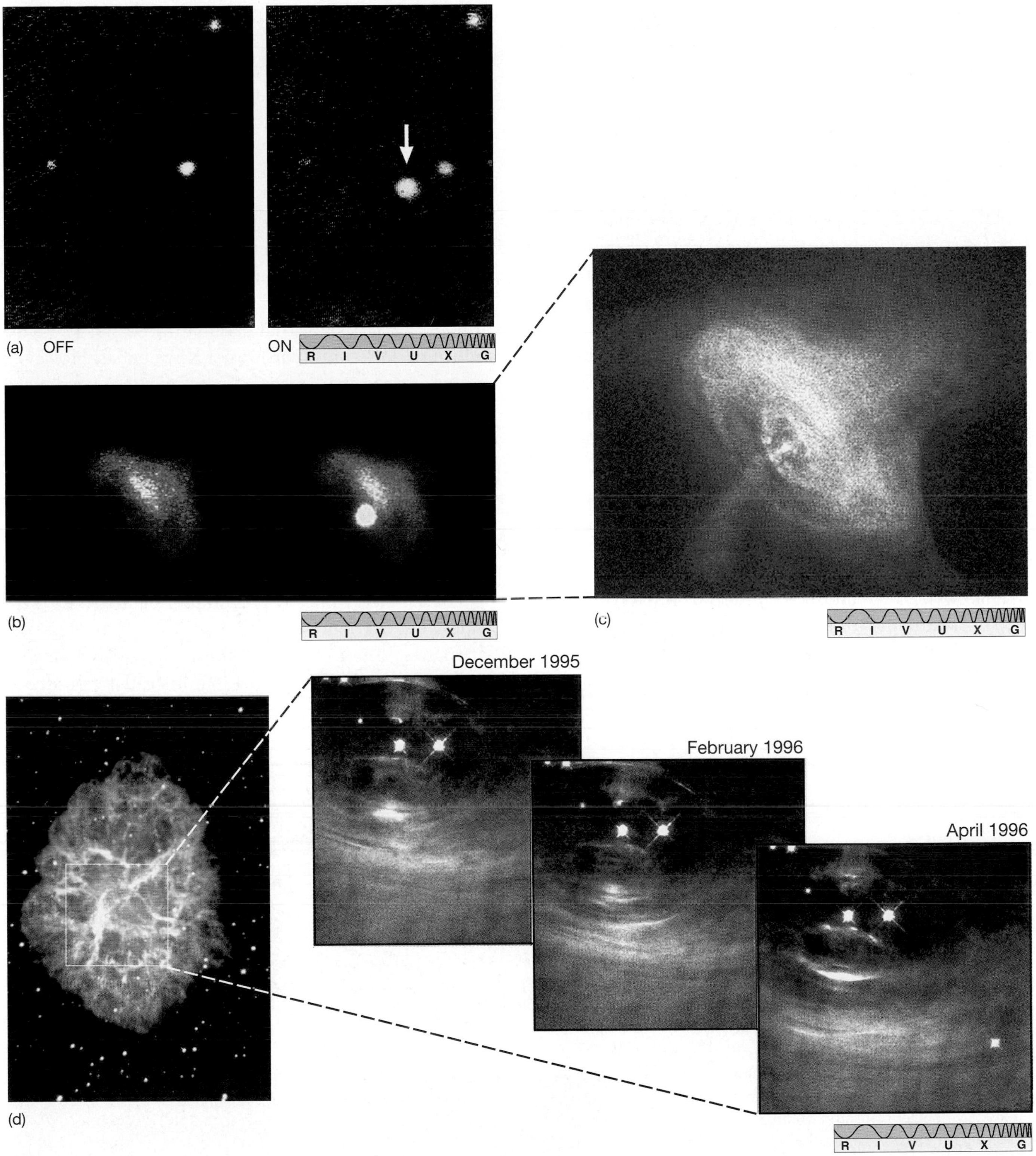

Figure 13.4 Crab Pulsar The pulsar in the core of the Crab Nebula blinks on and off about 30 times each second. (a) In this pair of optical images, the pulsing can be clearly seen. In the left frame, the pulsar is off; in the right frame, it is on. (b) The same phenomenon is also detected in X rays. (c) A recent X-ray image of the Crab, taken by the *Chandra* orbiting telescope, shows more clearly the central pulsar and its accretion disk. (d) These frames show light-year-long "ripples" in the inner portion of the Crab Nebula—shock waves driven outward by the pulsar's intense radiation at nearly half the speed of light. *(UC/Lick Observatory; Max-Planck Institut fur Extraterrestrische Physik; NASA; Palomar Observatory; NASA)*

13.3 Neutron-Star Binaries

X-ray Sources

3 With the launch of the first orbiting X-ray telescopes in the 1970s, numerous X-ray sources were found near the central regions of our Galaxy and also near the centers of a few globular star clusters. ∞ (Sec. 11.5) Some of these sources, known as **X-ray bursters**, emit much of their energy in violent eruptions, each thousands of times more luminous than our Sun but lasting only seconds. A typical burst is shown in Figure 13.5.

The X-ray emission is thought to arise on or near neutron stars that are members of binary systems. Matter torn from the surface of the (main-sequence or giant) companion by the neutron star's strong gravitational pull accumulates on the neutron star's surface. (Figure 12.13 shows the white dwarf equivalent.) ∞ (Sec. 12.2) The gas forms an accretion disk around the neutron star, then slowly spirals inward. The inner portions of the disk become extremely hot, releasing a steady stream of X rays.

As gas builds up on the neutron star's surface, its temperature rises due to the pressure of overlying material. Eventually, it becomes hot enough to fuse hydrogen. The result is a sudden period of rapid nuclear burning that releases a huge amount of energy in a brief but intense X-ray burst. After several hours of renewed accumulation, a fresh layer of matter produces the next burst. Thus, the mechanism is very similar to that responsible for a nova explosion on a white dwarf, except that it occurs on a far more violent scale because of the neutron star's much stronger surface gravity.

Gamma-ray Bursts

At around the same time as the first X-ray bursts were seen, military surveillance satellites discovered **gamma-ray bursts**—bright, irregular flashes of gamma rays typically lasting just a few seconds (Figure 13.6a). Until the 1990s, it was thought that gamma-ray bursts were basically just "scaled-up" versions of X-ray bursts. However, it now appears that this is not the case.

Figure 13.6(b) shows an all-sky plot of the positions of about a thousand gamma-ray bursts detected by the *Compton Gamma-Ray Observatory* (*GRO*) during its first three years of operation. ∞ (Sec. 3.5) On average, *GRO* has detected about one burst per day. Note that the bursts are distributed roughly uniformly over the entire sky, rather than being confined to the relatively narrow band of the Milky Way (compare Figure 3.29). This widespread distribution convinces most astronomers that the bursts do not originate within our own Galaxy, as had previously been assumed, but instead are produced at much greater distances.

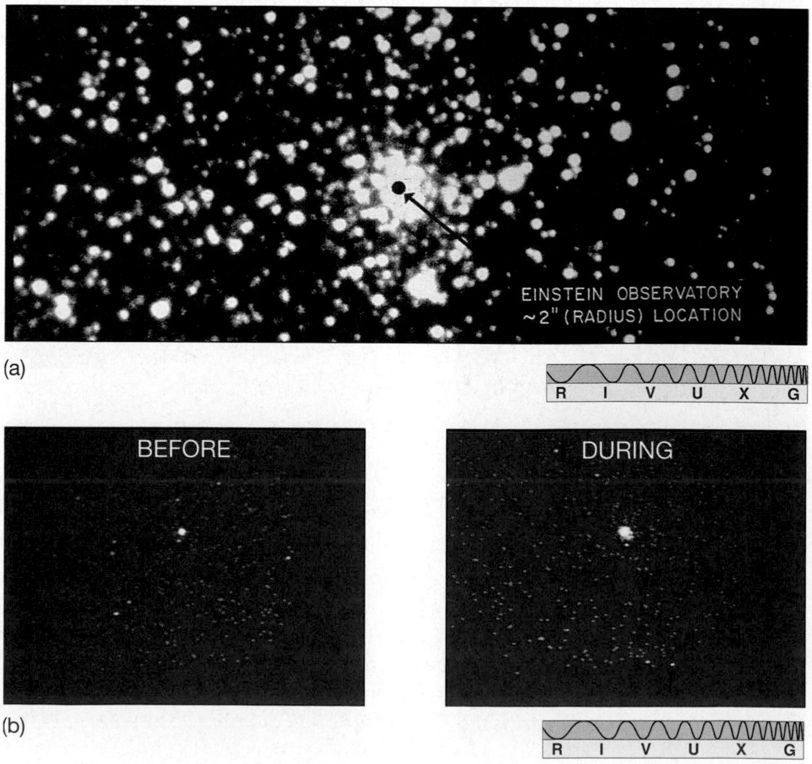

Figure 13.5 X-ray Burster An X-ray burster produces a sudden, intense flash of X rays followed by a period of relative inactivity lasting as long as several hours. Then another burst occurs. The bursts are thought to be caused by explosive nuclear burning on the surface of an accreting neutron star, similar to the explosions on a white dwarf that give rise to novae. (a) An optical photograph of the star cluster Terzan 2, showing a 2″ dot at the center where the X-ray bursts originate. (b) X-ray images taken before and during the outburst. The most intense X rays correspond to the position of the dot shown in (a). *(Harvard-Smithsonian Center for Astrophysics; HEAO-2/NASA)*

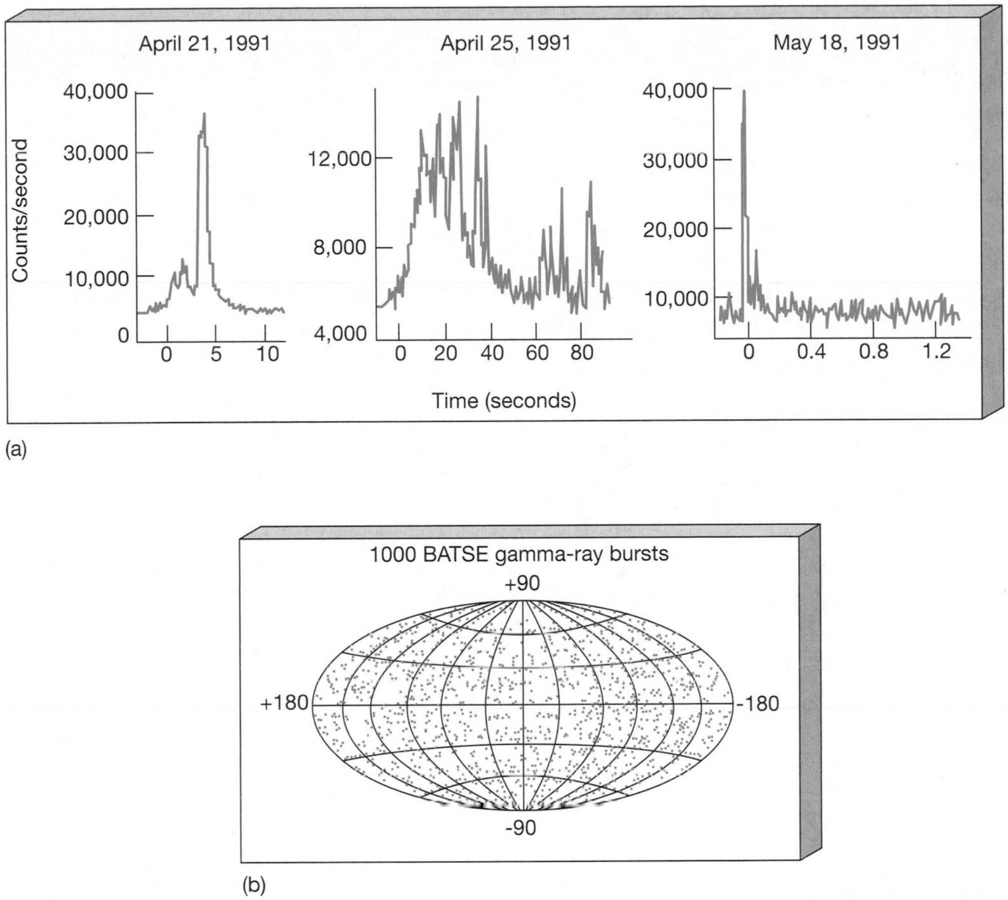

Figure 13.6 Gamma-ray Bursts (a) Plots of intensity versus time (in seconds) for some gamma-ray bursts. Note the substantial differences between them. Some bursts are irregular and spiky, whereas others are much more smoothly varying. Whether this wide variation in burst appearance means that more than one physical process is at work is presently unknown. (b) Positions on the sky of the first 1000 gamma-ray bursts detected by *GRO*; BATSE stands for Burst and Transient Source Experiment. The bursts appear to be distributed uniformly across the entire sky. *(NASA)*

The first direct measurement of the distance to a gamma-ray burst was made in 1997 when astronomers detected the burst's visible "afterglow," and found (using distance-measurement techniques to be discussed in more detail in Chapter 15) that the event occurred more than two *billion* parsecs from Earth. ∞ (Sec. 15.5) Figure 13.7 is an optical photograph of an even more distant burst, almost five billion parsecs away. Some 40 optical or X-ray afterglows of gamma-ray bursts have now been detected, and about a dozen distances are known. All are very large, implying that the bursts must be extremely energetic, since otherwise they wouldn't be detectable by our equipment. ∞ (Sec. 10.2) Each burst apparently generates more energy—and in some cases hundreds of times more energy—than a typical supernova explosion, all in a matter of seconds!

Not only are gamma-ray bursts extremely energetic, they are also very *small*. The millisecond flickering in the bursting gamma rays detected by *GRO* implies that, whatever their origin, all of their energy must come from a volume no larger than a few hundred kilometers across. The reasoning is as follows: If the

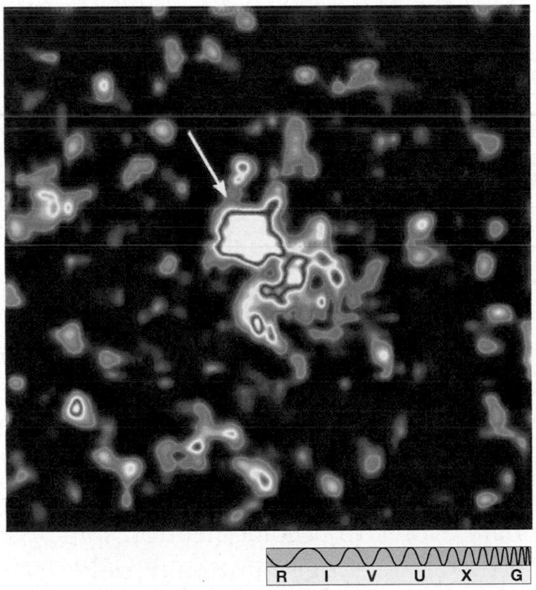

Figure 13.7 Photo of Gamma-ray Burst Optical photograph, in false color, of gamma-ray burst GRB970228. The optical counterpart of the gamma-ray burst (marked by the arrow) is a white blob to the upper left of image center. To the lower right is an extended object thought to be the galaxy in which the burst occurred. The distance to the optical counterpart is almost five billion parsecs. *(NASA)*

INTERLUDE 13-1

Gravity Waves

Electromagnetic waves are common everyday phenomena. They involve periodic changes in the strengths of electric and magnetic fields. (Fig. 2.6) They move through space and transport energy. Any accelerating charged particle, such as an electron in a broadcasting antenna or on the surface of a star, generates electromagnetic waves.

The modern theory of gravity—Einstein's theory of relativity—also predicts waves that move through space. A *gravity wave* is the gravitational counterpart of an electromagnetic wave. *Gravitational radiation* results from changes in the strength of a gravitational field. In theory, any time an object having any mass accelerates, a gravity wave is emitted at the speed of light. The passage of a gravity wave should produce small distortions in the space through which it passes. Gravity is an exceedingly weak force compared with electromagnetism, so these distortions are expected to be very small—in fact, much smaller than the diameter of an atomic nucleus for waves that might be produced by any known astrophysical source. Yet many researchers believe that these tiny distortions should be measurable. No one has yet succeeded in detecting gravity waves, but their detection would provide such strong support for the theory of relativity that scientists are eager to search for them.

Theorists are still debating which kinds of astronomical objects should produce gravity waves detectable on Earth. Leading candidates include (1) the merger of a binary-star system, (2) the collapse of a star into a black hole, and (3) the collision of two black holes or of two neutron stars. Each of these possibilities involves the acceleration of huge masses, resulting in rapidly changing gravitational fields. Of the three, the first probably presents the best chance to detect gravity waves, at least for the present. Binary-star systems should emit gravitational radiation as the component stars orbit one another. As energy escapes in the form of gravity waves, the two stars slowly spiral toward one another, orbiting more rapidly and emitting even more gravitational radiation. This runaway situation can lead to the decay and eventual merger of close binary systems in a relatively short time (which, in this case, means tens or hundreds of millions of years, although most of the radiation is emitted during the last few seconds).

Such a slow but steady decay in the orbit of a binary system has in fact been detected. In 1974, radio astronomer Joseph Taylor and graduate student Russell Hulse at the University of Massachusetts discovered a very unusual binary system. Both components are neutron stars, and one is observable from Earth as a pulsar. This system

emitting region were, say, 300,000 km—one light-second—across, even an instantaneous change in intensity at the source would be smeared out over a time interval of 1 s as seen from Earth, because light from the far side of the object would take 1 s longer to reach us than light from the near side. For the gamma-ray variation not to be blurred by the light-travel time, the source cannot be more than one light-millisecond, or 300 km, in diameter.

The leading explanation of the mechanism for gamma-ray bursts is that they signal the true endpoint of a binary-star system. Suppose that both members of the binary evolve to become neutron stars. As they orbit one another, gravitational radiation (see *Interlude 13-1*) is released and the two ultradense stars spiral together. Once they approach within a few tens of kilometers of one another, coalescence is inevitable. Such a merger will likely produce a violent explosion, comparable in energy to a supernova, and this could conceivably explain the flashes of gamma rays we observe. However, many uncertainties remain—the final word on these enigmatic objects has yet to be written.

Millisecond Pulsars

In the mid-1980s an important new category of pulsars was found—a class of very rapidly rotating objects called **millisecond pulsars**. These objects spin hundreds of times per second (that is, their pulse period is a few milliseconds). This speed is about as fast as a typical neutron star can spin without flying apart. In some cases, the star's equator is moving at more than 20 percent of the speed of light. This suggests a phenomenon bordering on the incredible—a cosmic object of kilometer dimensions, more massive than our Sun, spinning almost at breakup speed, making nearly 1000 complete revolutions *every second*. Yet the observations and their interpretation leave little room for doubt.

The story of these remarkable objects is further complicated by the fact that many of them are found in globular clusters. This is odd because globular clusters are has become known as the *binary pulsar*. Measurements of the periodic Doppler shift of the pulsar's radiation prove that its orbit is slowly shrinking. Furthermore, the rate at which the orbit is shrinking is exactly what would be predicted by relativity theory if the energy were being carried off by gravity waves. Even though the gravity waves themselves have not been detected, the binary pulsar is regarded by most astronomers as very strong evidence supporting general relativity. Taylor and Hulse received the 1993 Nobel Prize in physics for their discovery.

Gravity waves should contain a great deal of information about the physical events in some of the most exotic regions of space. In 1992, funding was approved for an ambitious gravity-wave observatory called *LIGO*—short for *Laser Interferometric Gravity-wave Observatory*. Twin detectors, one in Hanford, Washington (shown in the accompanying figure) and, the other in Livingston, Louisiana, will use laser beams to measure the extremely small distortions of space produced by gravitational radiation. The laser beams will detect the tiny changes (less than one-thousandth the diameter of an atomic nucleus) that will be produced in the lengths of the 4-km-long arms should a gravity wave pass by. The instrument should be capable of detecting gravity waves from many Galactic and extragalactic sources. The detectors were completed in 1999 and will begin taking astronomical data in January 2002. Even more sensitive instruments are planned to come online later in the decade.

If successful, the discovery of gravity waves could herald a new age in astronomy, in much the same way that invisible electromagnetic waves, unknown a century ago, revolutionized classical astronomy and led to the field of modern astrophysics.

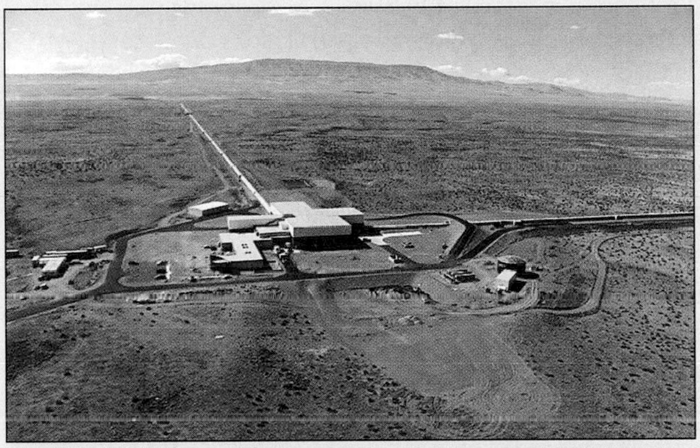

(National Science Foundation/LIGO Project)

known to be at least 10 billion years old, yet Type II supernovae (the kind that create neutron stars) are associated with massive stars that explode within a few tens of *millions* of years after their formation, and no stars have formed in any globular cluster since the cluster came into being. ∞ (Sec. 11.5, 12.4) Thus, no new neutron star has been produced in a globular cluster in a very long time. Furthermore, as mentioned earlier, the pulsar produced in a supernova explosion is expected to slow down and fade away in only a few tens of millions of years—after 10 billion years, its rotation rate should be very slow. The rapid rotation of the pulsars found in globular clusters therefore cannot be a relic of their birth. These pulsars must have been "spun up"—that is, had their rotation rates increased—by some other, much more recent, mechanism.

The most likely explanation for the high rotation rate of millisecond pulsars is that they have been spun up by drawing in matter from a companion star. As matter spirals down onto the neutron star's surface in an accretion disk, it provides the "push" needed to make the neutron star spin faster (Figure 13.8). Theoretical calculations indicate that this process can spin the star up to breakup speed in about a hundred million years. Subsequently, an encounter with another star may eject the pulsar from the binary, or the pulsar's intense radiation may destroy its companion, and an isolated millisecond pulsar results. This general picture is supported by the finding that, of the 40 or so millisecond pulsars seen in globular clusters, roughly half are known currently to be members of binary systems.

Notice that the scenario of accretion onto a neutron star from a binary companion is the same scenario that we just used to explain the existence of X-ray binaries. In fact, the two phenomena are very closely linked. Many X-ray binaries may be on their way to becoming millisecond pulsars.

Pulsar Planets

Radio astronomers can capitalize on the precision with which pulsar signals repeat themselves to make extremely accurate measurements of pulsar motion. In

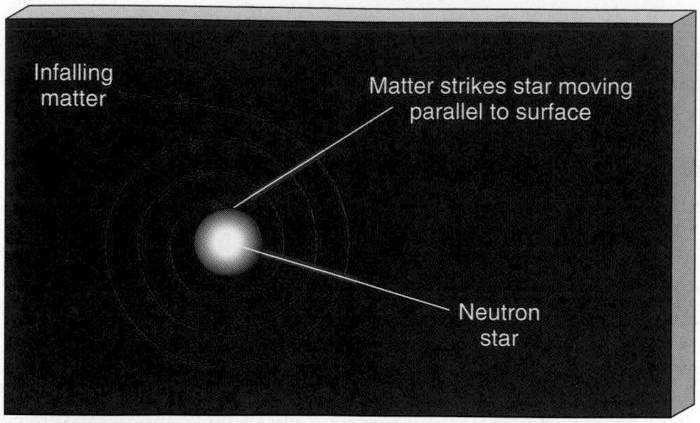

Figure 13.8 Millisecond Pulsar Gas from a companion star spirals down onto the surface of a neutron star. As it strikes the star, the infalling matter moves almost parallel to the surface, so it tends to make the star spin faster. Eventually, this process can result in a millisecond pulsar—a neutron star spinning at the incredible rate of hundreds of revolutions per second.

1992, radio astronomers at the Arecibo Observatory found that the pulse period of a millisecond pulsar lying some 500 pc from Earth exhibits tiny but regular fluctuations (at the level of about one part in 10^7) on two distinct time scales—67 and 98 days.

The most likely explanation for these fluctuations is that they are caused by the Doppler effect as the pulsar wobbles back and forth in space under the combined gravitational pulls of two small bodies, each about three times the mass of Earth. One orbits the pulsar at a distance of 0.4 A.U., the other at a distance of 0.5 A.U., with orbital periods of 67 and 98 days, respectively. Further observations not only confirmed these findings, but also revealed the presence of a third body, with a mass comparable to that of Earth's Moon, orbiting just 0.2 A.U. from the pulsar.

These remarkable results constitute the first definite evidence of planet-sized bodies outside our solar system. A few other millisecond pulsars have since been found with similar behavior. However, it is unlikely that any of these planets formed in the same way as our own. Any planetary system orbiting the pulsar's parent star was almost certainly destroyed in the supernova explosion that created the pulsar. As a result, scientists are still uncertain about how these planets came into being. One possibility involves the binary companion that provided the matter necessary to spin the pulsar up to millisecond speeds. Possibly the pulsar's intense radiation and strong gravity destroyed the companion, then spread its matter out into a disk (a little like the solar nebula) in whose cool outer regions the planets might have condensed.

Astronomers have been searching for decades for planets orbiting main-sequence stars like our Sun, on the assumption that planets are a natural by-product of star formation. ∞ (Sec. 4.3) Only very recently have these searches begun to yield positive results and nothing comparable in size to Earth has yet been detected. ∞ (*Interlude 4-1*) It is ironic that the first and only Earth-sized planets to be found outside the solar system orbit a dead star, and have little or nothing in common with our own world.

✓ Concept Check

- What is the connection between some X-ray sources and millisecond pulsars?

13.4 Black Holes

The Final Stage of Stellar Evolution

4 Neutron stars are supported by the resistance of tightly packed neutrons to further compression. Squeezed together, these elementary particles form a hard ball of ultradense matter that not even gravity can compress further. Or do they? Is it possible that, given enough matter packed into a small enough volume, the collective pull of gravity can eventually crush even a neutron star? Can gravity continue to compress a massive star into an object the size of a planet, a city, a pinhead—even smaller? The answer, apparently, is yes.

Although the precise figure is uncertain, mainly because the behavior of matter at very high densities is not well understood, most researchers concur that the mass of a neutron star cannot exceed about three times the mass of the Sun. This is the neutron-star equivalent of the white dwarf mass limit discussed in the previous chapter. ∞ (Sec. 12.5) Above this mass, not even tightly packed neutrons can withstand the star's gravitational pull. In fact, we know of *no* force that can counteract gravity beyond this point. Thus, if enough material is left behind after a supernova explosion that the central core exceeds this limit, gravity wins the battle with pressure once and for all, and the central core collapses forever. Stellar evolution theory indicates that this is the fate of any star whose main-sequence mass exceeds about 25 times the mass of the Sun.

As the core shrinks, the gravitational pull in its vicinity eventually becomes so great that nothing—not even light—can escape. The resultant object therefore emits no light, no other form of radiation, no information whatsoever. Astronomers call this bizarre endpoint of stellar evolution, in which the core of a very-high-mass star collapses in on itself and vanishes forever, a **black hole**.

Escape Speed

Newtonian mechanics—up to now our reliable and indispensable tool in understanding the universe—cannot adequately describe conditions in or near black holes. ∞ (Sec. 1.4) To comprehend these collapsed objects, we must turn instead to the modern theory of gravity, Einstein's **general theory of relativity** (see *More Precisely 13-1*). Einstein's description of the universe is equivalent to that of Newton for situations encountered in everyday life, but the two theories diverge radically in circumstances where speeds approach the speed of light and in regions of intense gravitational fields.

Despite the fact that general relativity is necessary for a proper description of black-hole properties, we can still usefully discuss some aspects of these strange bodies in more or less Newtonian terms. Let's reconsider the familiar Newtonian concept of escape speed—the speed needed for one object to escape from the gravitational pull of another—supplemented by two key facts from relativity: 1) nothing can travel faster than the speed of light, and 2) all things, *including light*, are attracted by gravity.

Escape speed is proportional to the square root of a body's mass divided by the square root of its radius. ∞ (*More Precisely 5-2*) Earth's radius is 6400 km, and the escape speed from Earth's surface is just over 11 km/s. Now consider a hypothetical experiment in which Earth is squeezed on all sides by a gigantic vise. As our planet shrinks under the pressure, the escape speed increases because the radius is decreasing. Suppose that Earth were compressed to one-fourth its present size. The escape speed would double (because $1/\sqrt{\frac{1}{4}} = 2$). Any object escaping from the surface of this hypothetically compressed Earth would need a speed of at least 22 km/s.

Imagine compressing Earth by an additional factor of 1000, making its radius hardly more than a kilometer. Now a speed of about 700 km/s—more than the escape speed of the Sun—would be needed to escape. Compress Earth still further, and the escape speed continues to rise. If our hypothetical vise were to squeeze Earth hard enough to crush its radius to about a centimeter, the speed needed to escape its surface would reach 300,000 km/s. But this is no ordinary speed. It is the speed of light, the fastest speed allowed by the laws of physics as we currently know them.

Thus, if by some fantastic means the entire planet Earth could be compressed to less than the size of a grape, the escape speed would exceed the speed of light. And because nothing can exceed that speed, the compelling conclusion is that nothing—absolutely nothing—could escape from the surface of such a compressed body. Even radiation—radio waves, visible light, X rays, photons of all wavelengths—would be unable to escape its intense gravity. Our planet would be invisible and uncommunicative—no signal of any sort could be sent to the universe outside. The origin of the term *black hole* becomes clear. For all practical purposes, Earth could be said to have disappeared from the universe. Only its gravitational field would remain behind, betraying the presence of its mass, now shrunk to a point.[2]

The Event Horizon

Astronomers have a special name for the critical radius at which the escape speed from an object would equal the speed of light and within which the object could no longer be seen. It is the **Schwarzschild radius**, after Karl Schwarzschild, the German scientist who first studied its properties. The Schwarzschild radius of any object is proportional to its mass. For Earth, it is 1 cm. For Jupiter, at about 300 Earth masses, it is about 3 m. For the Sun, at 300,000 Earth masses, it is 3 km. For a three-solar-mass stellar core, the Schwarzschild radius is about 9 km. As a convenient rule of thumb, the Schwarzschild radius of an object is simply 3 km multiplied by the object's mass measured in solar masses.

The surface of an imaginary sphere having a radius equal to the Schwarzschild radius and centered on a collapsing star is called the **event horizon**. No event occurring within this region can ever be seen, heard, or known by anyone outside. Even though there is no matter of any sort associated with it, we can think of the event horizon as the "surface" of a black hole.

A 1.4-solar-mass neutron star has a radius of about 10 km and a Schwarzschild radius of 4.2 km. If we were to keep increasing the neutron star's mass, its Schwarzschild radius would increase, although its physical radius would not—the physical radius of a neutron star in

[2]In fact, regardless of the composition or condition of the object that formed the black hole, only three physical properties can be measured from the outside—the hole's mass, charge, and angular momentum. All other information is lost once matter falls into the black hole. Only these three numbers are required to describe completely a black hole's outward appearance. In this chapter, we consider only holes that formed from nonrotating, electrically neutral matter. Such objects are completely specified once their masses are known.

fact *decreases* slightly with increasing mass. By the time our neutron star's mass exceeded about three solar masses, the star would lie within its own event horizon and would collapse of its own accord; it would shrink right past the Schwarzschild radius and disappear forever.

Concept Check

- What happens if you compress an object within its own Schwarzschild radius?

13.5 Black Holes and Curved Space

5 A central concept of general relativity (see *More Precisely 13-1*) is this: Matter—all matter—tends to "warp," or curve, space in its vicinity. At the same time, all objects, such as planets and stars, react to this warping by changing their paths. In the Newtonian view of gravity, particles move on curved trajectories because they feel a gravitational force. In Einsteinian relativity, those same particles move on curved trajectories because they are following the curvature of space produced by some nearby massive object. The greater the mass, the greater the warping. Close to a black hole, the gravitational field becomes overwhelming and the curvature of space extreme. At the event horizon itself, the curvature is so great that space "folds over" on itself, causing objects within to become trapped and disappear.

Some props may help you visualize the curvature of space near a black hole. Bear in mind, however, that these props are in no sense "real"—they are only tools to help you grasp some exceedingly strange concepts.

First, imagine a pool table with the tabletop made of a thin rubber sheet rather than the usual hard felt. As Figure 13.9 suggests, such a rubber sheet becomes distorted when a heavy weight, such as a rock, is placed on it. The otherwise flat sheet sags, especially near the rock. The heavier the rock, the larger the distortion. Trying to play pool, you would quickly find that balls passing near the rock were deflected by the curvature of the tabletop. In much the same way, both matter *and radiation* are deflected by the curvature of space near a star. For example, Earth's orbital path is governed by the relatively gentle curvature of space created by our Sun. The more massive the object, the more the space surrounding it is curved.

Let's consider another analogy. Imagine a large extended family of people living at various locations on a huge rubber sheet—a sort of gigantic trampoline. Deciding to hold a reunion, they converge on a given place at a given time. As shown in Figure 13.10, one person remains behind, not wishing to attend. He keeps in touch with his relatives by means of "message balls" rolled out to him (and back from him) along the surface of the sheet. These message balls are the analog of radiation carrying information through space.

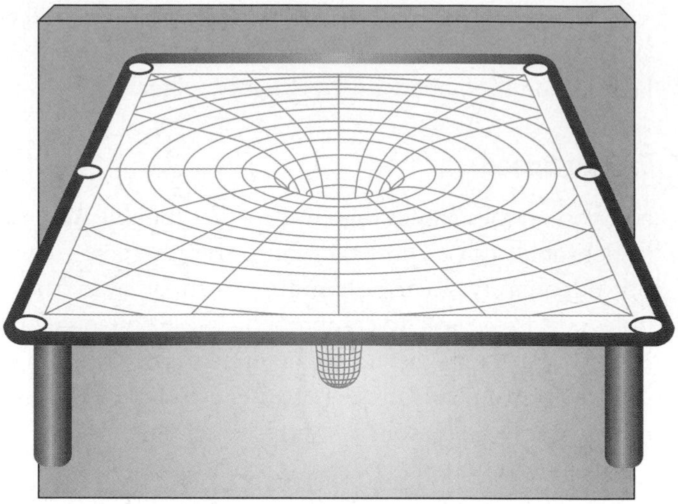

Figure 13.9 **Curved Space** A pool table made of a thin rubber sheet sags when a weight is placed on it. Likewise, space is bent, or warped, in the vicinity of any massive object.

As the people converge, the rubber sheet sags more and more. Their accumulating mass creates an increasing amount of space curvature. The message balls can still reach the lone person far away in nearly flat space, but they arrive less frequently as the sheet becomes more and more warped and stretched and the balls have to climb out of a deeper and deeper well (Figures 13.9b and c). Finally, when enough people have arrived at the appointed spot, the mass becomes too great for the rubber to support. As illustrated in Figure 13.10(d), the sheet pinches off into a "bubble," compressing the people into oblivion and severing their communications with the lone survivor outside. This final stage represents the formation of an event horizon around the party.

Right up to the end—the pinching off of the bubble—two-way communication is possible. Message balls can reach the outside from within (but at a lower and lower rate as the rubber stretches), and messages from outside can get in without difficulty. Once the event horizon (the bubble) forms, balls from the outside can still fall in, but they can no longer be sent back out to the person left behind, no matter how fast they are rolled. They cannot make it past the lip of the bubble in Figure 13.10(d). This anal-

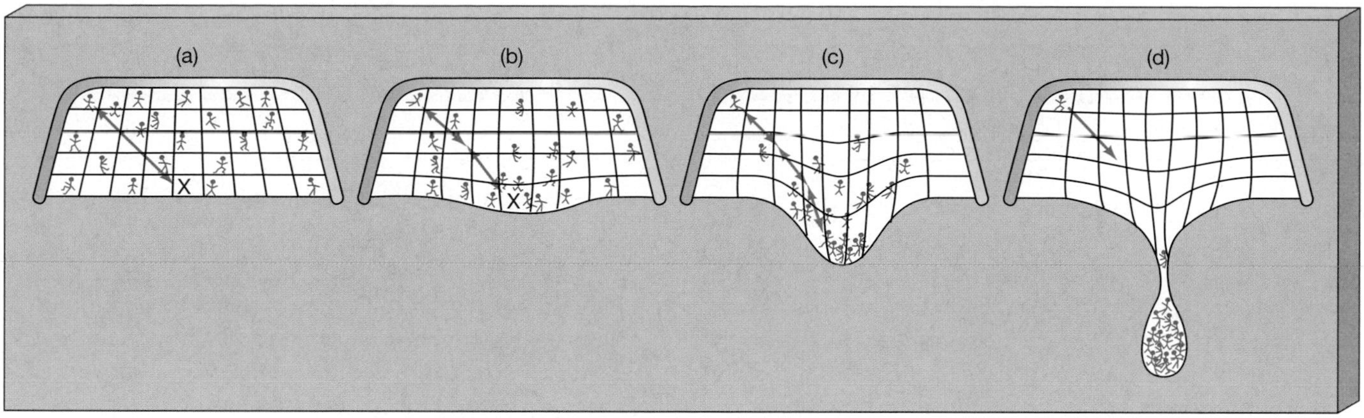

Figure 13.10 **Space Warping** Any mass causes a rubber sheet (space) to be curved. (a–c) As people assemble at the appointed spot on the sheet, the curvature grows progressively larger. The blue arrows represent some directions in which information can be transmitted from place to place. (d) The people are finally sealed inside the bubble, forever trapped and cut off from the outside world.

ogy (very) roughly depicts how a black hole warps space completely around on itself, isolating its interior from the rest of the universe. The essential ideas—the slowing down and eventual cessation of outward-going signals and the one-way nature of the event horizon once it forms—all have parallels in the case of stellar black holes.

✓ Concept Check

- How do Newton's and Einstein's theories differ in their descriptions of gravity?

13.6 Space Travel Near Black Holes

Orbits and Tidal Forces

The orbit of an object near a black hole is essentially the same as its orbit near a "normal" star of the same mass as the black hole. Only if the object happened to pass within a few Schwarzschild radii (perhaps 50 or 100 km for a typical 5–10 solar mass black hole formed in a supernova explosion) of the event horizon would there any significant difference between its actual orbit and the one predicted by Newtonian gravity and described by Kepler's laws. Of course, if some matter did fall into the black hole—that is, if its orbit happened to take it too close to the event horizon—it would be unable to get out. Black holes are like turnstiles, permitting matter to flow in only one direction—inward.

Matter flowing into a black hole is subject to great tidal stress. An unfortunate person falling feet first into a solar-mass black hole would find herself stretched enormously in height and squeezed unmercifully laterally. ∞ (*More Precisely 5-1*) She would be torn apart long before she reached the event horizon, for the pull of gravity would be much stronger at her feet (which are closer to the hole) than at her head. The net result of all this stretching and squeezing is numerous and violent collisions among the resulting debris, causing a great deal of frictional heating of any infalling matter. As illustrated in Figure 13.11, material is simultaneously torn

Figure 13.11 **Black Hole Heating** Any matter falling into the clutches of a black hole will become severely distorted and heated. This sketch shows an imaginary planet being pulled apart by a black hole's gravitational tides.

MORE PRECISELY 13-1

Einstein's Theories of Relativity

Albert Einstein is probably best known for his two *theories of relativity*, the successors to Newtonian mechanics that form the foundation of twentieth-century physics. The *special theory of relativity* (or just *special relativity*), proposed by Einstein in 1905, deals with the preferred status of the speed of light. We have noted that the speed of light, c, is the maximum speed attainable in the universe, but there is more to it than that. In 1887, a fundamental experiment carried out by two American physicists, A. A. Michelson and E. W. Morley, demonstrated a further important and unique aspect of light—that the measured speed of a beam of light is *independent* of the motion of the observer or the source. No matter what our velocity may be relative to the source of the light, we always measure precisely the same value for c—299,792.458 km/s.

A moment's thought leads to the realization that this is a decidedly nonintuitive statement. If we were traveling in a car moving at 100 km/h and we fired a bullet forward with a velocity of 1000 km/h relative to the car, an observer standing at the side of the road would see the bullet pass by at 100 + 1000 = 1100 km/h, as illustrated in the first figure. However, if we were traveling in a rocket ship at 1/10 the speed of light, 0.1c, and we shone a searchlight beam ahead of us, the Michelson–Morley experiment tells us that an outside observer would measure the speed of the beam not as 1.1c, as the preceding example would suggest, but as c. The rules that apply to particles moving at or near the speed of light are different from those we are used to in everyday life.

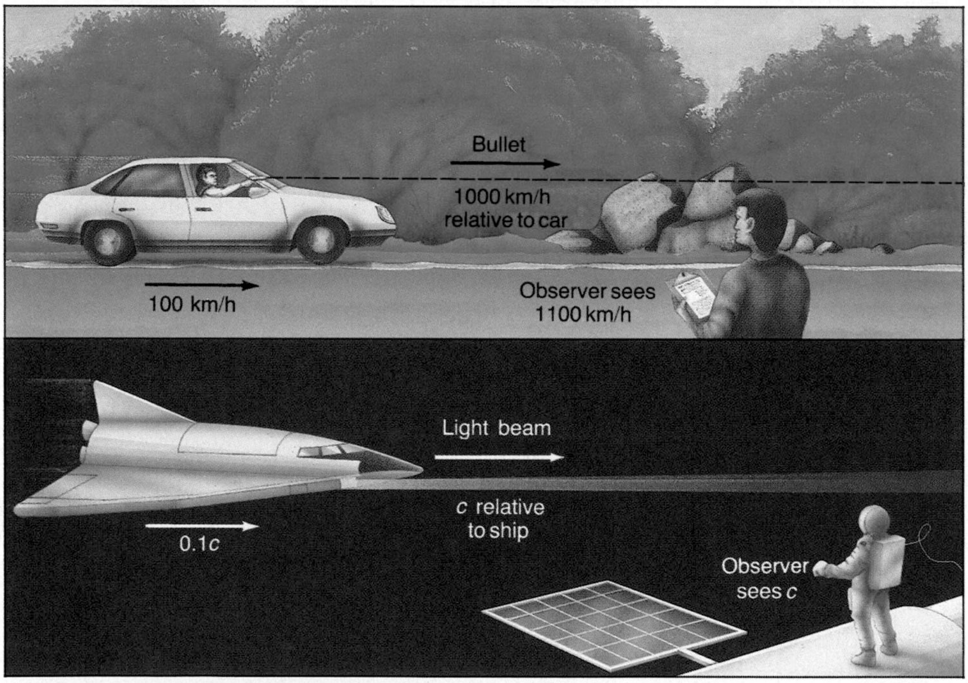

Special relativity is the mathematical framework that allows us to extend the familiar laws of physics from low speeds (that is, speeds much less than c, which are often referred to as *nonrelativistic*) to very high (*relativistic*) speeds comparable to c. Relativity is equivalent to Newtonian mechanics when objects move much

apart and heated to high temperatures as it plunges into the hole.

So efficient is this heating that, prior to reaching the hole's event horizon, matter falling into the hole emits radiation of its own accord. For a black hole of roughly solar mass, the energy is expected to be emitted in the form of X-rays. Thus, contrary to what we might expect from an object whose defining property is that nothing can escape from it, the region surrounding a black hole is expected to be a *source* of energy. Of course, once the hot matter falls below the event horizon, its radiation is no longer detectable—it can never leave the hole.

Approaching the Event Horizon

One safe way to study a black hole would be to go into orbit around it, safely beyond the disruptive influence of the hole's strong tidal forces. After all, Earth and the more slowly than light, but it differs greatly in its predictions at relativistic speeds. For example, special relativity predicts that a rapidly moving spacecraft will appear to contract in the direction of its motion, its clocks will appear to run slow, and its mass will appear to increase. All the theory's predictions have been repeatedly verified to very high accuracy. Today special relativity lies at the heart of all physical science. No scientist seriously doubts its validity.

General relativity is what results when gravity is incorporated into the framework of special relativity. In 1915, Einstein made the connection between special relativity and gravity with the following famous "thought experiment." Imagine that you are enclosed in an elevator with no windows, so that you cannot directly observe the outside world, and that the elevator is floating in space. You are weightless. Now suppose that you begin to feel the floor press up against your feet. Weight has apparently returned. There are two possible explanations for this, shown in the second diagram: A large mass could have come nearby, and you are feeling its downward gravitational attraction, *or* the elevator has begun to accelerate upward and the force you feel is that exerted by the elevator as it accelerates you, too. The crux of Einstein's argument is this: there is *no* experiment that you can perform within the elevator, without looking outside, that will let you distinguish between these two possibilities.

Thus, Einstein reasoned, there is no way to tell the difference between a gravitational field and an accelerated frame of reference (such as the rising elevator in the thought experiment). Gravity can therefore be incorporated into special relativity as a general acceleration of all particles. However, a major modification must be made to special relativity. Central to relativity is the notion that space and time are not separate quantities, but instead must be treated as a single entity—*spacetime*. To incorporate the effects of gravity, the mathematics forces us to the conclusion that spacetime has to be *curved*.

In general relativity, then, gravity is a manifestation of curved spacetime. There is no such thing as a "gravitational field," in the Newtonian sense. Instead, objects move as they do because they follow the curvature of spacetime, and this curvature of spacetime is determined by the amount of matter present. We will explore some of the consequences of this view of gravity in more detail in the text. *More Precisely 13-2* describes some observational tests of the theory.

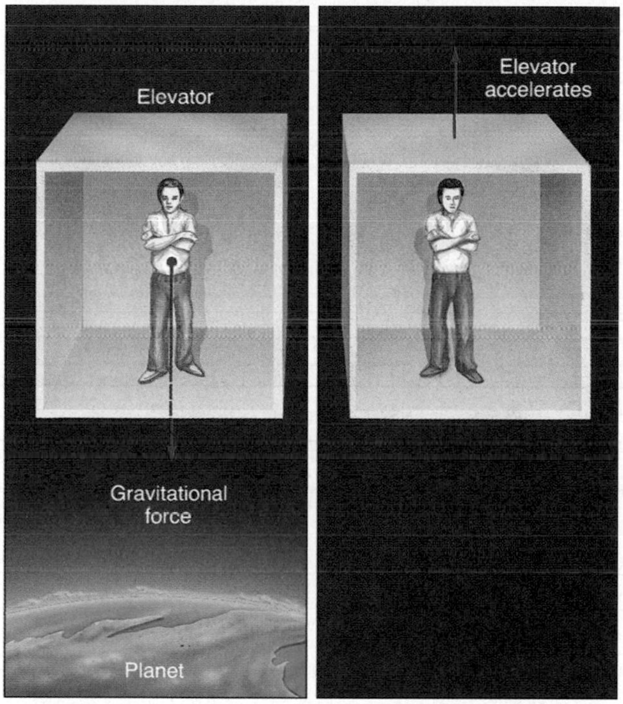

other planets of our solar system all orbit the Sun without falling into it and without being torn apart. The gravity field around a black hole is basically no different. Let's send an imaginary indestructible astronaut—a mechanical robot, say—in a probe toward the center of the hole, as illustrated in Figure 13.12. Watching from a safe distance in our orbiting spacecraft, we can then examine the nature of space and time, at least down to the event horizon. After that boundary is crossed, there is no way for the robot to return any information about its findings.

Suppose our robot has an accurate clock and a light source of known frequency mounted on it. From our safe vantage point far outside the event horizon, we could use telescopes to read the clock and measure the frequency of the light we receive. What might we discover? We would find that the light from the robot would become more and more redshifted—shifted toward

Figure 13.12 Robot–Astronaut A hypothetical robot–astronaut can travel toward a black hole while performing experiments that humans, farther away, can monitor to learn something about space near the event horizon.

longer wavelengths—as the robot neared the event horizon (Figure 13.13). Even if the robot used rocket engines to remain motionless, the redshift would still be detected. The redshift is not caused by motion of the light source—it is not the result of the Doppler effect. ∞ (*More Precisely 2-3*) Rather, it is a result of the black hole's gravitational field, clearly predicted by Einstein's general theory of relativity and known as **gravitational redshift**.

We can explain the gravitational redshift as follows. Photons are attracted by gravity. As a result, in order to escape from a source of gravity, photons must expend some energy—they have to work to get out of the gravitational field. They don't slow down (photons always move at the speed of light); they just lose energy. Because a photon's energy is proportional to the frequency of its radiation, light that loses energy must have its frequency reduced (or, conversely, its wavelength increased).

Thus, as photons traveled from the robot's light source to our orbiting spacecraft, they would become gravitationally redshifted. From our standpoint, a yellow light would appear orange, then red as the robot astronaut approached the black hole. As the robot neared the event horizon, the light would become undetectable with an optical telescope. First infrared and then radio telescopes would be needed to detect it. Theoretically, light emitted *from the event horizon itself* would be gravitationally redshifted to infinitely long wavelengths.

What about the robot's clock? What time does it tell? Is there any observable change in the rate at which the clock ticks while moving deeper into the hole's gravitational field? We would find, from our safely orbiting spacecraft, that any clock close to the hole would appear to tick more *slowly* than an equivalent clock on board our ship. The closer the clock came to the hole, the slower it would appear to run. Upon reaching the event horizon, the clock would seem to stop altogether. All action would become frozen in time. Consequently, we would never actually witness the robot cross the event horizon. The process would appear to take forever.

This apparent slowing down of the robot's clock is known as **time dilation**, and is closely related to gravitational redshift. To see why this is so, think of the

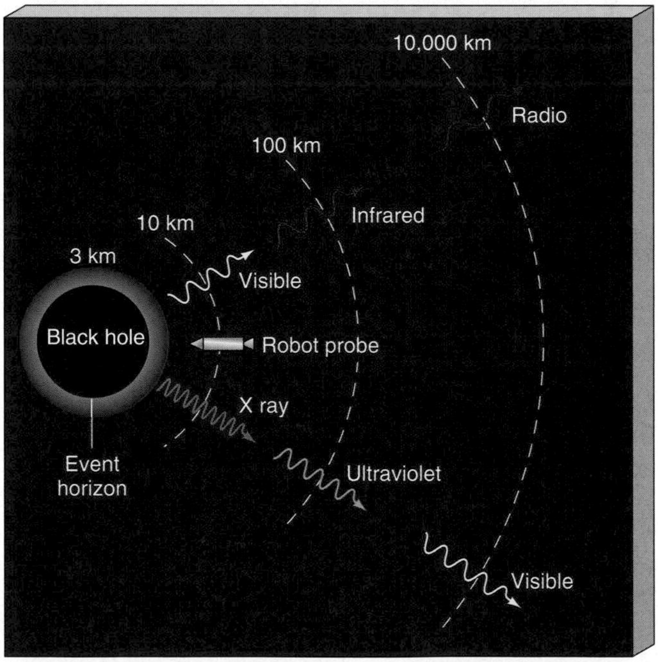

Figure 13.13 Gravitational Redshift As it escapes from the strong gravitational field close to a black hole, a photon must expend energy to overcome the hole's gravity. This energy does not come from a change in the speed at which the photon travels (that speed is always 300,000 km/s, even under these extreme conditions). Rather, the photon "gives up" energy by decreasing its frequency. Thus, the photon's color changes. This figure shows the effect on two beams of radiation, one of yellow light and one of X rays, emitted from a space probe as it nears a one-solar-mass black hole. The beams are shifted to longer and longer wavelengths as they move farther from the event horizon.

MORE PRECISELY 13-2

Tests of General Relativity

Special relativity is the most thoroughly tested and most accurately verified theory in the history of science. General relativity, however, is on somewhat less firm experimental ground. The problem with verifying general relativity is that its effects on Earth and in the solar system—the places where we can most easily perform tests—are very small. Just as special relativity produces major departures from Newtonian mechanics only when speeds approach the speed of light, general relativity predicts large departures from Newtonian gravity only when extremely strong gravitational fields are involved.

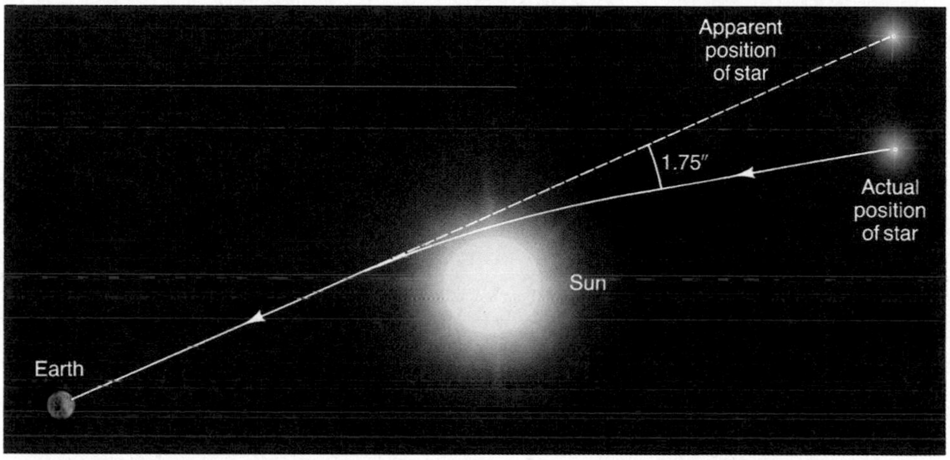

Here we consider just two "classical" tests of the theory—solar-system experiments that helped ensure acceptance of Einstein's theory. Bear in mind, however, that there are no known tests of general relativity in the "strong-field" regime—that part of the theory that predicts black holes, for example—so the full theory has never been experimentally tested.

At the heart of general relativity is the premise that everything, including light, is affected by gravity because of the curvature of spacetime. Shortly after he published his theory in 1915, Einstein noted that light from a star should be deflected by a measurable amount as it passes the Sun. The closer to the Sun the light comes, the more it is deflected. Thus, the maximum deflection should occur for a ray that just grazes the solar surface. Einstein calculated that the deflection angle should be 1.75″—a small but detectable amount.

Of course, it is normally impossible to see stars so close to the Sun. During a solar eclipse, however, when the Moon blocks the Sun's light, the observation does become possible, as illustrated in the accompanying figure. In 1919, a team of observers led by the British astronomer Sir Arthur Eddington succeeded in measuring the deflection of starlight during an eclipse. The results were in excellent agreement with the prediction of general relativity.

Another prediction of general relativity is that planetary orbits should deviate slightly from the perfect ellipses of Kepler's laws. Again, the effect is greatest where gravity is strongest—that is, closest to the Sun. Thus, the largest relativistic effects are found in the orbit of Mercury. Relativity predicts that Mercury's orbit is not exactly an ellipse. Instead, its orbit should rotate slowly, as shown in the (highly exaggerated) diagram below. The amount of rotation is very small—only 43″ per century—but Mercury's orbit is so well charted that even this tiny effect is measurable.

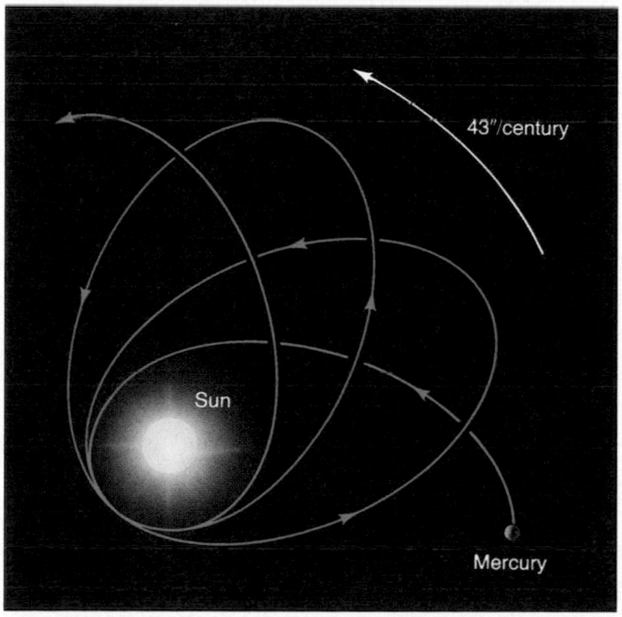

robot's light source as a clock, with the passage of a wave crest (say) constituting a "tick," so the clock ticks at the frequency of the radiation. As the wave is redshifted, the frequency drops, and fewer wave crests pass us each second—the clock appears to slow down. This thought experiment demonstrates that the redshift of the radiation and the slowing of the clock are one and the same thing.

From the point of view of the indestructible robot, relativity theory predicts no strange effects at all. The light source hasn't reddened and the clock keeps perfect time. In the robot's frame of reference, everything is normal. Nothing prohibits it from approaching within the Schwarzschild radius of the hole—no law of physics constrains an object from passing through an event horizon. There is no barrier at the event horizon and no sudden lurch as it is crossed. It is just an imaginary boundary in space. Travelers passing through the event horizon of a sufficiently massive hole (such as might lurk in the heart of our own Galaxy, as we will see) might not even know it—at least until they tried to get out!

Deep Down Inside

No doubt, you are wondering what lies within the event horizon of a black hole. The answer is simple: no one knows. However, the question raises some very fundamental issues that lie at the forefront of modern physics.

General relativity predicts that, without some agent to compete with gravity, the core remnant of a high-mass star will collapse all the way to a point at which both its density and its gravitational field become infinite—a so-called **singularity**. However, we should not take this prediction of infinite density too literally. Singularities always signal the breakdown of the theory producing them. The present laws of physics are simply inadequate to describe the final moments of a star's collapse.

As it stands today, the theory of gravity is incomplete because it does not incorporate a proper description of matter on very small scales. As our collapsing stellar core shrinks to smaller and smaller radii, we eventually lose our ability even to describe, let alone predict, its behavior. Perhaps matter trapped in a black hole never actually reaches a singularity. Perhaps it just approaches this bizarre state, in a manner that we will someday understand as the subject of *quantum gravity*—the merger of general relativity with quantum mechanics—develops.

Having said that, we can at least estimate how small the core can get before current theory fails. It turns out that by the time that stage is reached, the core is already much smaller than any elementary particle. Thus, although a complete description of the endpoint of stellar collapse may well require a major overhaul of the laws of physics, for all practical purposes the prediction of collapse to a point is valid. Even if a new theory somehow succeeds in doing away with the central singularity, it is very unlikely that the external appearance of the hole, or the existence of its event horizon, will change. Any modifications to general relativity are expected to occur only on submicroscopic scales, not on the macroscopic (kilometer-sized) scale of the Schwarzschild radius.

Singularities are places where the rules break down, and some very strange things may occur near them. Many possibilities have been envisaged—gateways to other universes, time travel, or the creation of new states of matter, for instance—but none has been proved, and certainly none has been observed. Because these regions are places where science fails, their presence causes serious problems for many of our cherished laws of physics, including causality (the idea that cause should precede effect, which runs into immediate problems if time travel is possible) and energy conservation (which is violated if material can hop from one universe to another through a black hole). It is unclear whether the removal of the singularity by some future, all-encompassing theory would necessarily also eliminate all of these problematic side effects.

Concept Check

- Why would you never actually witness an infalling object crossing the event horizon of a black hole?

13.7 Observational Evidence for Black Holes

6 Theoretical ideas aside, is there any observational evidence for black holes? What proof do we have that these strange invisible objects really do exist?

Black Holes in Binary Systems

Perhaps the most promising way to find black holes is to look for their effects on other astronomical objects. Our Galaxy harbors many binary-star systems in which only one object can be seen, but we need observe the motion of only one star to infer the existence of an unseen companion and measure some of its properties. ∞ (Sec.

10.6) In the majority of cases, the invisible companion is simply small and dim, nothing more than an M-type star hidden in the glare of an O- or B-type partner, or perhaps shrouded by dust or other debris, making it invisible to even the best available equipment. In either case, the invisible object is not a black hole.

A few close binary systems, however, have peculiarities suggesting that one of their members may be a black hole. Figure 13.14(a) shows the area of the sky in the constellation Cygnus, where the evidence is particularly strong. The rectangle outlines the celestial system of interest, some 2000 pc from Earth. The black-hole candidate is an X-ray source called Cygnus X-1, discovered by the *Uhuru* satellite in the early 1970s. Its visible companion is a blue B-type supergiant. Assuming that it lies on the main sequence, its mass must be around 25 times the mass of the Sun. Spectroscopic observations indicate that the binary system has an orbital period of 5.6 days. Combining this information with spectroscopic measurements of the visible component's orbital speed, astronomers estimate the total mass of the binary system to be around 35 solar masses, implying that Cygnus X-1 has a mass of about 10 times that of the Sun.

X-ray radiation emitted from the immediate neighborhood of Cygnus X-1 indicates the presence of high-temperature gas, perhaps as hot as several million kelvins. Furthermore, the X-ray emission has been observed to vary in intensity on time scales as short as a millisecond. As discussed earlier in the context of gamma-ray bursts, for this variation not to be blurred by the travel time of light across the source, Cygnus X-1 cannot be more than one light-millisecond, or 300 km, in diameter.

These properties suggest that Cygnus X-1 could be a black hole. Figure 13.15 is an artist's conception of this intriguing object. The X-ray-emitting region is likely an accretion disk formed as matter drawn from the visible star spirals down onto the unseen component. The rapid variability of the X-ray emission indicates that the unseen component must be very compact—a neutron star or a black hole. Its large mass, well above the neutron star mass limit discussed earlier, implies the latter.

Have Black Holes Been Detected?

A few other black-hole candidates are known. For example, LMC X-3—the third X-ray source ever discovered in the Large Magellanic Cloud, a small galaxy orbiting our own—is an invisible object that, like Cygnus X-1, orbits a bright companion star. LMC X-3's visible companion seems to be distorted into the shape of an egg by the unseen object's intense gravitational pull. Reasoning similar to that applied to Cygnus X-1 leads to the conclusion that LMC X-3 has a mass nearly 10 times that of the Sun, making it too massive to be anything but a black hole. The Galactic X-ray binary system A0620-00 has been found to contain an invisible compact object of mass 3.8 times the mass of the Sun. In total, there are perhaps half a dozen known objects that may turn out to

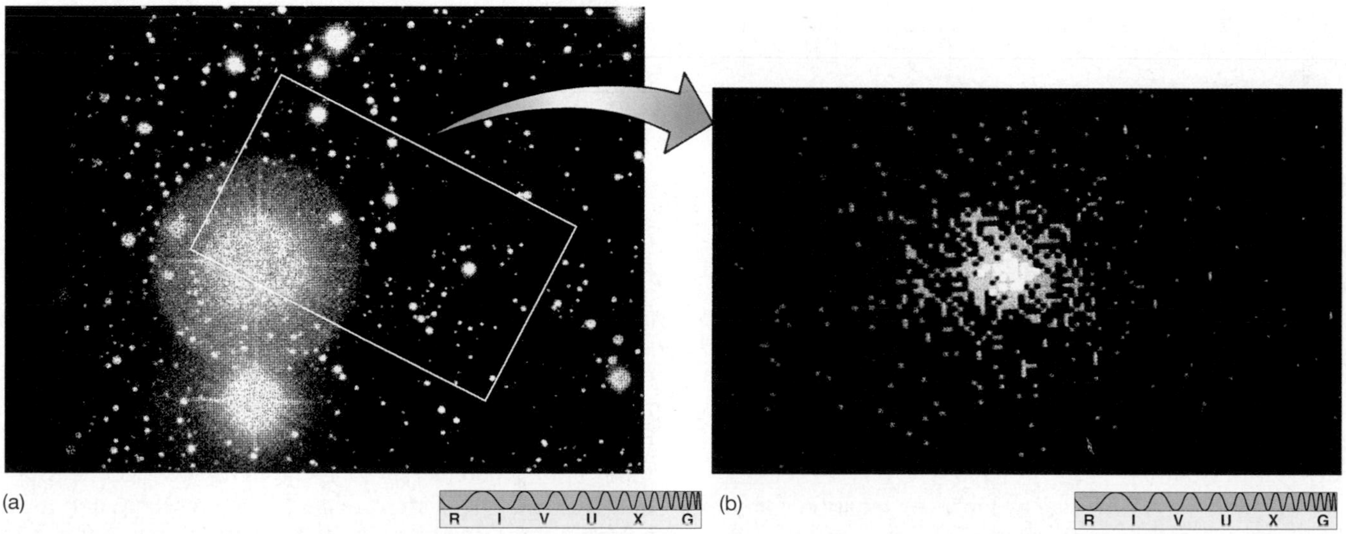

Figure 13.14 **Cygnus X-1** (a) The brightest star in this photograph is a member of a binary system whose unseen companion, called Cygnus X-1, is a leading black-hole candidate. (b) An X-ray image of the field of view outlined by the rectangle in part (a). *(Harvard-Smithsonian Center for Astrophysics; NASA)*

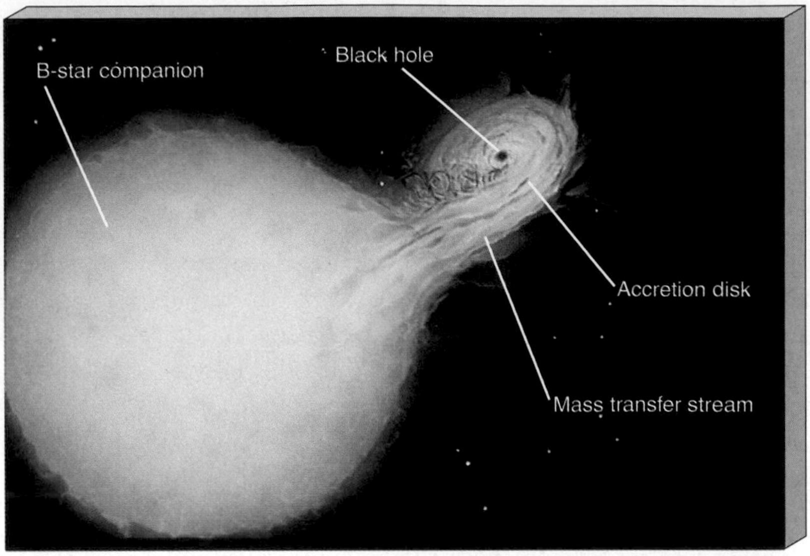

Animation

Figure 13.15 **Black Hole** Artist's conception of a binary system containing a large, bright, visible star and an invisible, X-ray-emitting black hole. This painting is based on data obtained from detailed observations of Cygnus X-1. *(L. Chaisson)*

be black holes, although Cygnus X-1, LMC X-3, and A0620-00 have the strongest claims.

So have stellar black holes really been discovered? The answer is probably yes. Skepticism is healthy in science, but only the most stubborn astronomers (and some do exist!) would take strong issue with the reasoning that supports the case for black holes. Can we guarantee that future modifications to the theory of compact objects will not invalidate our arguments? No, but similar statements could be made in many other areas of astronomy—indeed, about any theory in any area of science. We conclude that, strange as they are, black holes have been detected in our Galaxy. Perhaps some day future generations of space travelers will visit Cygnus X-1 or LMC X-3 and (carefully) test these conclusions firsthand.

✓ Concept Check

- How do astronomers "see" black holes?

Chapter Review www

Summary

A core-collapse supernova may leave behind an ultracompressed ball of material called a **neutron star** (p. 328). According to the **lighthouse model** (p. 329), neutron stars, because they are magnetized and rotating, send regular bursts of electromagnetic energy into space. When we can see the beams from Earth, we call the source neutron star a **pulsar** (p. 329). The pulse period is the rotation period of the neutron star.

A neutron star that is a member of a binary system can draw matter from its companion, forming an accretion disk, which is usually a strong source of X rays. As gas builds up on the star's surface, it eventually becomes hot enough to fuse hydrogen. When hydrogen burning starts on a neutron star, it does so explosively, and an **X-ray burster** (p. 332) results. Even more energetic are **gamma-ray bursts** (p. 332), which may result from the violent merger of neutron stars in very distant binary systems. The rapid rotation of the inner part of the accretion disk causes the neutron star to spin faster as new gas arrives on its surface. The eventual result is a very rapidly rotating neutron star—a **millisecond pulsar** (p. 334). Many millisecond pulsars are found in the hearts of old globular clusters. They cannot have formed recently, and must have been spun up by interactions with other stars. Careful analysis of the radiation received has shown that some millisecond pulsars are orbited by planet-sized objects. The origin of these "pulsar planets" is still uncertain.

The upper limit on the mass of a neutron star is about three solar masses. Beyond that mass, the star can no longer support itself against its own gravity, and it collapses to form a **black hole** (p. 337), a region of space from which nothing can escape. Astronomers believe that the most massive stars, after exploding in a supernova, form black holes rather than neutron stars. Conditions in and near black holes cannot be described by Newtonian mechanics. A proper treatment involves Einstein's **general theory of relativity** (p. 337), which describes gravity in terms of the warping, or bending, of space by the presence of mass. The more mass, the greater the warping. All particles—including photons—respond to that warping by moving along curved paths.

The radius at which the escape speed from a collapsing star equals the speed of light is called the **Schwarzschild radius** (p. 337). The surface of an imaginary sphere centered on the collapsing star and having a radius equal to the star's Schwarzschild radius is called the **event horizon** (p. 337).

To a distant observer, light leaving a spaceship that is falling into a black hole would be subject to **gravitational redshift** (p. 342) as the light climbed out of the hole's intense gravitational field. At the same time, a clock on the spaceship would show **time dilation** (p. 342)—the clock would appear to slow down as the ship approached the event horizon. The observer would never see the ship reach the surface of the hole. Once within the event horizon, no known force can prevent a collapsing star from contracting all the way to a pointlike **singularity** (p. 344), at which point both the density and the gravitational field of the star become infinite. This prediction of relativity theory has yet to be proved. Singularities are places where the known laws of physics break down.

Once matter falls into a black hole, it can no longer communicate with the outside. However, on its way in, it can form an accretion disk and emit X-rays. The best place to look for a black hole is in a binary system in which one component is a compact X-ray source. Cygnus X-1, a well-studied X-ray source in the constellation Cygnus, is a long-standing black-hole candidate. Studies of orbital motions imply that some binaries contain compact objects too massive to be neutron stars, leaving black holes as the only alternative.

Review and Discussion

1. How does the way in which a neutron star forms determine some of its most basic properties?
2. What would happen to a person standing on the surface of a neutron star?
3. Why aren't all neutron stars seen as pulsars?
4. What are X-ray bursters?
5. Why do astronomers think that gamma-ray bursts are very energetic?
6. What is the favored explanation for the rapid spin rates of millisecond pulsars?
7. Why do you think astronomers were surprised to find a pulsar with a planetary system?
8. What does it mean to say that the measured speed of a light beam is independent of the motion of the observer?
9. Use your knowledge of escape speed to explain why black holes are said to be "black."
10. What is an event horizon?
11. Why is it so difficult to test the predictions of the theory of general relativity? Describe two tests of the theory.
12. What would happen to someone falling into a black hole?
13. What makes Cygnus X-1 a good black hole candidate?
14. Imagine that you had the ability to travel at will through the Milky Way Galaxy. Explain why you would discover many more neutron stars than those known to observers on Earth. Where would you be most likely to find these objects?
15. Do you think that planet-sized objects discovered in orbit around a pulsar should be called planets? Why or why not?

True or False?

____ 1. The density of a typical neutron star is about 100,000 times denser than a white dwarf.

____ 2. As a result of their high mass and small size, neutron stars have only a weak gravitational pull at their surface.

____ 3. Newly formed neutron stars have extremely strong magnetic fields.

____ 4. A millisecond pulsar is actually a very old neutron star that has been recently sped up by interaction with a neighbor.

____ 5. Millisecond pulsars are found only in globular clusters.

____ 6. Planet-sized bodies will never be found around a pulsar because the supernova that formed the pulsar would have destroyed any planets in the system.

____ 7. Nothing can travel faster than the speed of light.

____ 8. All things, except light, are attracted by gravity.

____ 9. Loosely speaking, a black hole is an object whose escape speed equals or exceeds the speed of light.

____ **10.** Although visible light cannot escape from a black hole, high-energy radiation, such as gamma rays, can escape.

____ **11.** If you could touch it, the surface of a black hole, the event horizon, would be very hard.

____ **12.** The laws of physics break down near the center of a black hole.

____ **13.** If the Sun were suddenly to turn into a black hole, Earth's orbit would immediately change.

____ **14.** Thousands of black holes have now been identified.

____ **15.** Neutrinos are emitted by matter accreting onto a stellar-mass black hole.

Fill in the Blank

1. No neutron star remains after the explosion of a _______ supernova.

2. A typical neutron star is about _______ km in diameter.

3. Neutron stars may be characterized as having _______ rotation rates and _______ magnetic fields.

4. Pulsars were discovered through observations in the _______ part of the electromagnetic spectrum.

5. Observed pulsar periods range from _______ to _______. (Give approximate numbers and units.)

6. The pulse period of pulsar radiation tells us the _______ of the neutron star emitting the radiation.

7. All millisecond pulsars either are now, or once were, members of _______ star systems.

8. X-ray bursters result when material from a binary companion accretes onto a _______ star.

9. Colliding neutron stars would be expected to release a lot of energy in the form of _______.

10. According to the general theory of relativity, space is warped, or curved, by _______.

11. Photons _______ energy as they escape from a gravitational field.

12. The region of extremely high density at the center of a black hole is called a _______.

13. Matter accreting onto a stellar-mass black hole emits radiation in the form of _______.

14. Black holes are believed to have been discovered in several _______ systems.

15. Scientists infer that the energy-emitting region in Cygnus X-1 must be very small because of the rapid _______ of the radiation received.

Problems

1. The angular momentum of a spherical body is proportional to the body's angular speed times the square of its radius. Using the law of conservation of angular momentum, estimate how fast a collapsed stellar core would spin if its initial spin rate was one revolution per day and its radius decreased from 10,000 km to 10 km. ∞ *(More Precisely 4-1)*

2. What would your mass be if you were composed entirely of neutron-star material, of density 3×10^{17} kg/m^3? (Assume that your average density is 1000 kg/m^3.) Compare this with the mass of (a) the Moon; (b) a typical 1-km diameter asteroid.

3. Calculate the surface gravity (relative to Earth's gravity of 9.8 m/s^2) and the escape speed of 1.4–solar mass neutron star with a radius of 10 km. What would be the escape speed of a 1-solar-mass object with a radius of 3 km? ∞ *(More Precisely 5-2)*

4. Use the radius–luminosity–temperature relation to calculate the luminosity of a 10-km-radius neutron star for temperatures of 10^5 K, 10^7 K, and 10^9 K. What do you conclude about the visibility of neutron stars? Could the coolest of them be plotted on our H–R diagram?

5. A gamma-ray detector of area 0.5 m^2 observing a gamma-ray burst records photons having total energy 10^{-8} joules. If the burst occurred 1000 Mpc away, calculate the total amount of energy it released (assuming that the energy was emitted equally in all directions). How would this figure change if the burst occurred 10,000 pc away instead, in the halo of our Galaxy? What if it occurred within the Oort cloud of our own solar system, at a distance of 50,000 A.U.?

6. A 10-km-radius neutron star is spinning 1000 times per second. Calculate the speed of a point on its equator, and compare your answer with the speed of light. (Consider the equator as the circumference of a circle, and recall that circumference = $2\pi \times$ radius.) Also calculate the orbital speed of a particle in a circular orbit just above the neutron star's surface. (*More Precisely 1-2*)

7. Supermassive black holes are believed to exist in the centers of some galaxies. What would be the Schwarzschild radii of black holes of one million and one billion solar masses? How does the first black hole compare in size with the Sun? How does the second compare in size with the solar system?

8. Calculate the tidal acceleration on a 2-m-tall human falling feet first into a one-solar-mass black hole—that is, compute the difference in the accelerations (forces per unit mass) on his head and his feet just before his feet cross the event horizon. (*More Precisely 5-1*) Repeat the calculation for a one million-solar-mass black hole and for a one billion-solar-mass black hole (see the previous question). Compare these accelerations with the acceleration due to gravity on Earth (9.8 m/s^2).

9. Endurance tests suggest that the human body cannot withstand stress greater than about 10 times the acceleration due to gravity on Earth's surface. At what distance from a one-solar-mass black hole would the human in the previous question be torn apart?

10. Using the data given in the text (assume the upper limit on the stated range for the black-hole mass), calculate the orbital separation of Cygnus X-1 and its B-type stellar companion. (*More Precisely 1-2*)

Projects

1. Many amateur astronomers enjoy turning their telescopes on the ninth-magnitude companion to Cygnus X-1, the sky's most famous black-hole candidate. Because none of us can see X-rays, no sign of anything unusual can be observed. Still, it's fun to gaze toward this region of the heavens and contemplate Cygnus X-1's powerful energy emission and strange properties. Even without a telescope, it is easy to locate the region of the heavens where Cygnus X-1 resides. The constellation Cygnus contains a recognizable star pattern, or asterism, in the shape of a large cross. This asterism is called the Northern Cross. The star in the center of the crossbar is called Sadr, and the star at the bottom of the cross is called Albireo. Approximately midway along an imaginary line between Sadr and Albireo lies the star Eta Cygni. Cygnus X-1 is located slightly less than 0.5 degree from this star. With or without a telescope, sketch what you see.

2. Set up a demonstration of the densities of various astronomical objects—an interstellar cloud, a star, a terrestrial planet, a white dwarf, and a neutron star. Select a common object that is easily held in your hand, something familiar to anyone—an apple, for example. For the lowest densities, calculate how large a volume would contain the object's equivalent mass. For high densities, calculate how many of the objects would have to be fit into a standard volume, such as 1 cm^3. Present your demonstration to your class or to some other group of students. Tell them about each astronomical object and how it comes by its density.

A
B
C
D

14 THE MILKY WAY GALAXY

A Grand Design

LEARNING GOALS

Studying this chapter will enable you to:

1. Describe the overall structure of the Milky Way Galaxy, and specify how the various regions differ from one another.
2. Explain the importance of variable stars in determining the size and shape of our Galaxy.
3. Describe the orbital paths of stars in different regions of the Galaxy, and explain how these motions are accounted for by our understanding of how the Galaxy formed.
4. Discuss some possible explanations for the existence of the spiral arms observed in our own and many other galaxies.
5. Explain what studies of galactic rotation reveal about the size and mass of our Galaxy, and discuss the possible nature of dark matter.
6. Describe some of the phenomena observed at the center of our Galaxy.

(Opposite page, background) The varying interrelationships among the many components of matter in our Milky Way Galaxy comprise a sort of "galactic ecosystem." Its evolutionary balance may be as complex as that of life in a tidal pool or a tropical rainforest. Here, stars abound throughout the Lagoon Nebula, a rich stellar nursery about 1200 pc from Earth. *(AURA)*

(Inset A) An emission nebula, the North American Nebula, glowing amidst a field of stars, its red color produced by the emission of light from vast clouds of hydrogen atoms. *(J. Sanford/Astrostock-Sanford)*

(Inset B) An open cluster, the Jewel Box, containing many young, blue stars. *(AURA)*

(Inset C) A typical globular cluster, 47 Tucanae. This true-color image reveals its dominant member stars to be elderly red giants. *(Space Telescope Science Institute)*

(Inset D) A true-color *Hubble Space Telescope* image of part of the Bubble Nebula (upper left), a 3-pc-long shell ionized by a star (at left center) having about 10–20 times the mass of our Sun. *(NASA)*

Looking up on a dark, clear night, we are struck by two aspects of the night sky. The first is a fuzzy band of light—the Milky Way—that stretches across the heavens. From the Northern Hemisphere, this band is most easily visible in the summertime, arcing high above the horizon. Its full extent forms a great circle that encompasses the entire celestial sphere. Away from that glowing band, however, our second impression is that the nighttime sky seems more or less the same in all directions. Bunches of stars cluster here and there, but overall, apart from the band of the Milky Way, the evening sky looks pretty uniform. Yet this is only a local impression. Ours is a rather provincial view. When we consider much larger volumes of space, on scales far, far greater than the distances between neighboring stars, a new level of organization becomes apparent as the large-scale structure of the Milky Way Galaxy is revealed.

14.1 Our Parent Galaxy

1 A **galaxy** is a gargantuan collection of stellar and interstellar matter—stars, gas, dust, neutron stars, black holes—isolated in space and held together by its own gravity. Astronomers are aware of literally millions of galaxies beyond our own. The particular galaxy we happen to inhabit is known as the **Milky Way Galaxy**, or just "the Galaxy," with a capital "G."

Our Sun lies in a part of the Galaxy known as the **Galactic disk**, an immense, circular, flattened region containing most of our Galaxy's luminous stars and interstellar matter (and virtually everything we have studied so far in this book). Figure 14.1 illustrates how, viewed from within, the Galactic disk appears as a band of light—the *Milky Way*—stretching across our nighttime sky. As indicated in the figure, if we look in a direction away from the Galactic disk (red arrows), relatively few stars lie in our field of view. However, if our line of sight happens to lie within the disk (white and blue arrows), we see so many stars that their light merges into a continuous blur.

Paradoxically, although we can study individual stars and interstellar clouds near the Sun in great detail, our location within the Galactic disk makes deciphering our Galaxy's large-scale structure from Earth a very difficult task. In some directions, the interpretation of what we see is ambiguous and inconclusive. In others, foreground objects completely obscure our view of what lies beyond. As a result, astronomers who study the Milky Way Galaxy are often guided in their efforts by comparisons with more distant, but much more easily observable, systems.

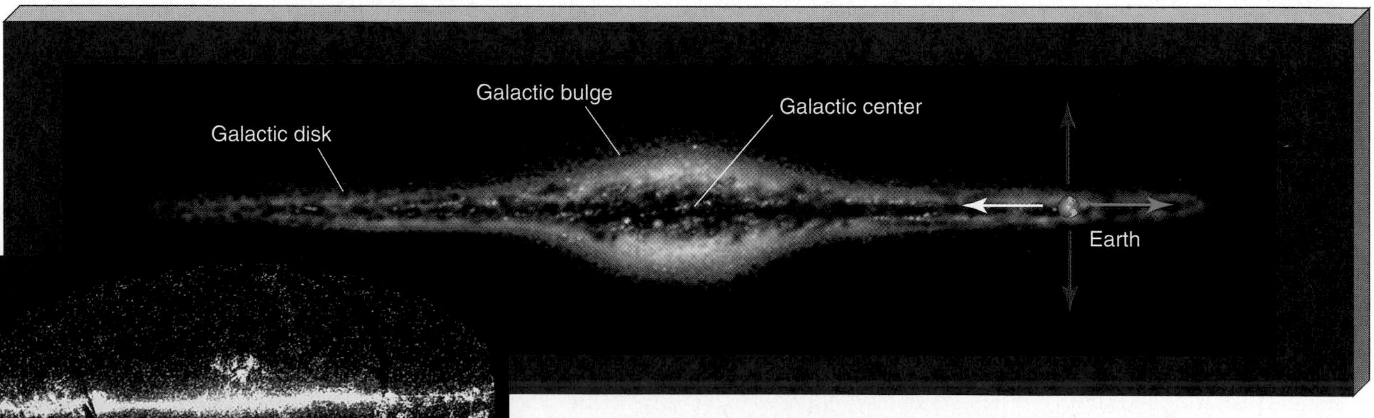

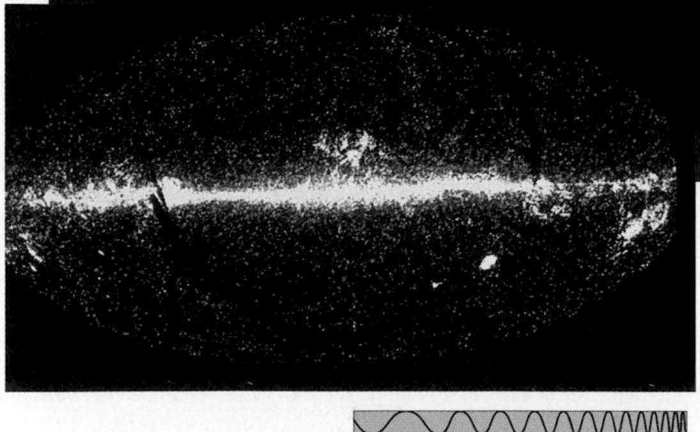

Figure 14.1 Galactic Plane Gazing from Earth toward the Galactic center (white arrow), we see myriad stars stacked up within the thin band of light known as the Milky Way. Looking in the opposite direction (blue arrow), we still see the Milky Way band, but now it is fainter because our position is far from the Galactic center, with the result that we see more stars when looking toward the center than when looking in the opposite direction. Looking perpendicular to the disk (red arrows), we see far fewer stars. The inset is an enhanced satellite view of the sky all around us. The white band is the disk of our Milky Way Galaxy, which can be seen with the naked eye from very dark locations on Earth. *(NASA)*

Figure 14.2 Spiral Galaxies (a) The Andromeda Galaxy probably resembles fairly closely the overall layout of our own Milky Way Galaxy. The disk and bulge are clearly visible in this image, which is about 30,000 pc across. The faint halo stars cannot be seen. The white stars sprinkled all across this image are not part of Andromeda's halo. They are foreground stars in our own Galaxy, lying in the same region of the sky as Andromeda but about a thousand times closer. (b) This galaxy, catalogued as NGC 6744 and seen nearly face-on, is somewhat similar in overall structure to our own Milky Way Galaxy and Andromeda. (c) The galaxy NGC 891 happens to be oriented in such a way that we see it edge-on, allowing us to see clearly its disk and central bulge. *(NASA; NOAO/AURA; Palomar Observatory/California Institute of Technology)*

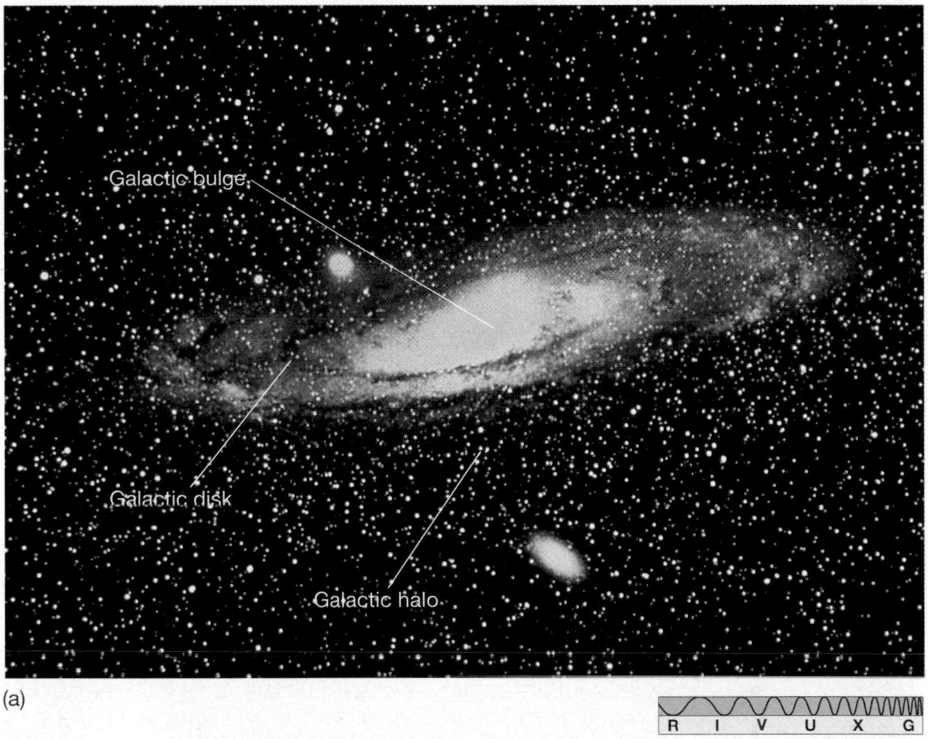

(a)

(b)

(c)

Figure 14.2 shows three **spiral galaxies** thought to resemble our own in overall structure. Figure 14.2(a) is the Andromeda Galaxy, the nearest major galaxy to the Milky Way Galaxy, lying about 900 kpc (nearly three million light-years) away. Andromeda's apparent elongated shape is a consequence of the angle at which we happen to view it. In fact, this galaxy, like our own, consists of a circular **galactic disk** of matter that fattens to a **galactic bulge** at the center. The disk and bulge are embedded in a roughly spherical ball of faint old stars known as the **galactic halo**. These three basic galactic regions are indicated on the figure (the halo stars are so faint that they cannot be discerned here; see Figure 14.9). Figures 14.2(b) and (c) show views of two other galaxies—one seen face-on, the other edge-on—that illustrate these points more clearly.

Concept Check

- Why do we see the Milky Way as a band of light across the sky?

14.2 Measuring the Milky Way

Before the twentieth century, astronomers' conception of the cosmos differed markedly from the modern view. The growth in our knowledge of our Galaxy, as well as the realization that there are many other distant galaxies similar to our own, has gone hand in hand with the development of the cosmic distance scale.

Star Counts

In the late eighteenth century, long before the distances to any stars were known, the English astronomer William Herschel tried to estimate the size and shape of our Galaxy simply by counting how many stars he could see in different directions in the sky. Assuming that all stars were of about equal brightness, he concluded that the Galaxy was a somewhat flattened, roughly disk-shaped collection of stars lying in the plane of the Milky Way, with the Sun at its *center* (Figure 14.3). Subsequent refinements to this approach led to essentially the same picture. Early in the twentieth century, some workers went so far as to estimate the dimensions of this "Galaxy" as about 10 kpc in diameter by 2 kpc thick.

In fact, the Milky Way Galaxy is several tens of kiloparsecs across, and the Sun lies far from the center. The flaw in the above reasoning is that the observations were made in the visible part of the electromagnetic spectrum, and astronomers failed to take into account the (then unknown) absorption of visible light by interstellar gas and dust. ∞ (Sec. 11.1) Only in the 1930s did astronomers begin to realize the true extent and importance of the interstellar medium.

The apparent falloff in the density of stars with distance in the plane of the Milky Way is not a real thinning of their numbers in space, but simply a consequence of the murky environment in the Galactic disk. Objects in the disk lying more than a few kiloparsecs away are hidden from our view by the effects of interstellar absorption. The long "fingers" in Herschel's map are directions where the obscuration happens to be a little less severe than in others. However, perpendicular to the disk, where there is less obscuring gas and dust along the line of sight, the falloff is real.

Observations of Variable Stars

2 While astronomers' attempts to probe the Galactic disk by optical means are frustrated by the effects of the interstellar medium, stars are visible to much greater distances in directions out of the Milky Way plane. An important by-product of the laborious effort to catalog stars around the turn of the twentieth century was the systematic study of **variable stars**. These are stars whose luminosity changes with time—some quite erratically, others more regularly. Only a small fraction of stars fall into this category, but those that do are of great astronomical significance.

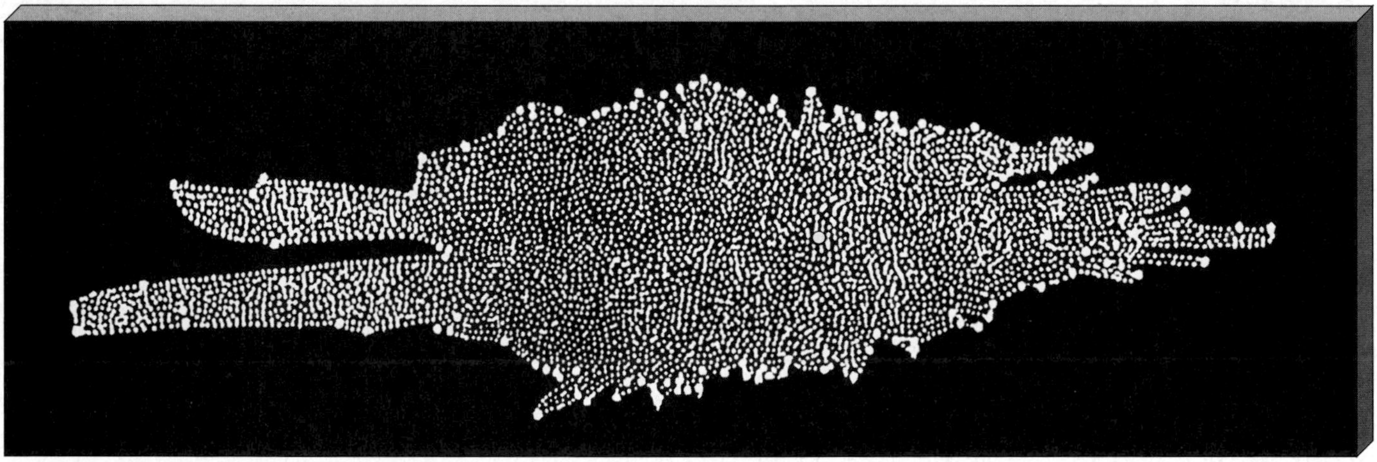

Figure 14.3 Early Galaxy Model Eighteenth-century English astronomer William Herschel constructed this "map" of the Galaxy by counting the numbers of stars he saw in different directions in the sky. He assumed that all stars were of roughly equal luminosity and, within the confines of the Galaxy, were uniformly distributed in space. Our Sun (marked by the large yellow dot) appears to lie near the center of the distribution, and the long axis of the diagram lies in the plane of the Galactic disk.

We have encountered several examples of variable stars in earlier chapters. Often, the variability is the result of membership in a binary system. Eclipsing binaries and novae are cases in point. However, sometimes the variability is an intrinsic property of the star itself, and is not dependent on its being a part of a binary. Particularly important to Galactic astronomy are the **pulsating variable stars,**[1] which vary cyclically in luminosity in very characteristic ways. Two types of pulsating variable stars that have played central roles in revealing both the true extent of our Galaxy and the distances to our galactic neighbors are the **RR Lyrae** and **Cepheid** variables. (Following long-standing astronomical practice, the names come from the first star of each class to be discovered—in this case the variable star labeled RR in the constellation Lyra and the variable star Delta Cephei, the fourth brightest star in the constellation Cepheus.)

RR Lyrae and Cepheid variable stars are recognizable by the characteristic shapes of their light curves. RR Lyrae stars all pulsate in essentially similar ways (Figure 14.4a), with only small differences in period between one RR Lyrae variable and another. Observed periods range from about 0.5 to 1 day. Cepheid variables also pulsate in distinctive ways (the regular "sawtooth" pattern in Figure 14.4b), but different Cepheids can have very different pulsation periods, ranging from about 1 to 100 days. In either case, the stars can be recognized and identified *just by observing the variations in the light they emit.*

Pulsating variable stars are normal stars experiencing a brief period of instability as a natural part of stellar evolution. The conditions necessary to cause pulsations are not found in main-sequence stars. They occur in post-main-sequence stars as they evolve through a region of the Hertzsprung–Russell diagram known as the *instability strip* (Figure 14.5). When a star's temperature and luminosity place it in this strip, the star becomes internally unstable, and both its temperature and its radius vary in a regular way, causing the pulsations we observe. As the star brightens, its surface becomes hotter and its radius shrinks; as its luminosity decreases, the star expands and cools. Cepheid variables are high-mass stars evolving across the upper part of the H–R diagram. (Sec. 12.4) The less luminous RR Lyrae variables are lower-mass horizontal-branch stars that happen to lie within the lower portion of the instability strip. (Sec. 12.3)

[1]Note that these stars have *nothing* whatsoever to do with pulsars.

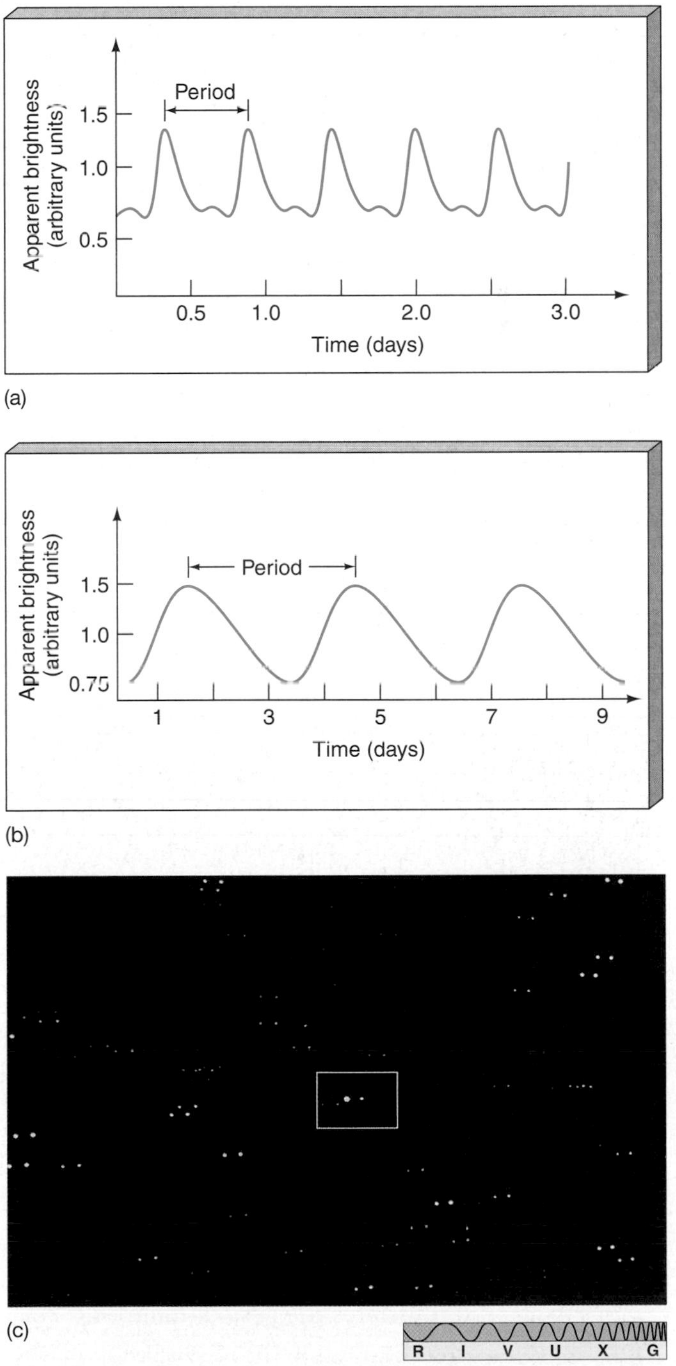

Figure 14.4 **Variable Stars** (a) Light curve of the pulsating variable star RR Lyrae. All RR Lyrae-type variables have essentially similar light curves, with periods of less than a day. (b) The light curve of a Cepheid variable star called WW Cygni. (c) This Cepheid is shown here (boxed) on successive nights, near its maximum and minimum brightness; two photos, one from each night, were superposed and then slightly displaced. *(California Institute of Technology/Palomar/Hale Observatory)*

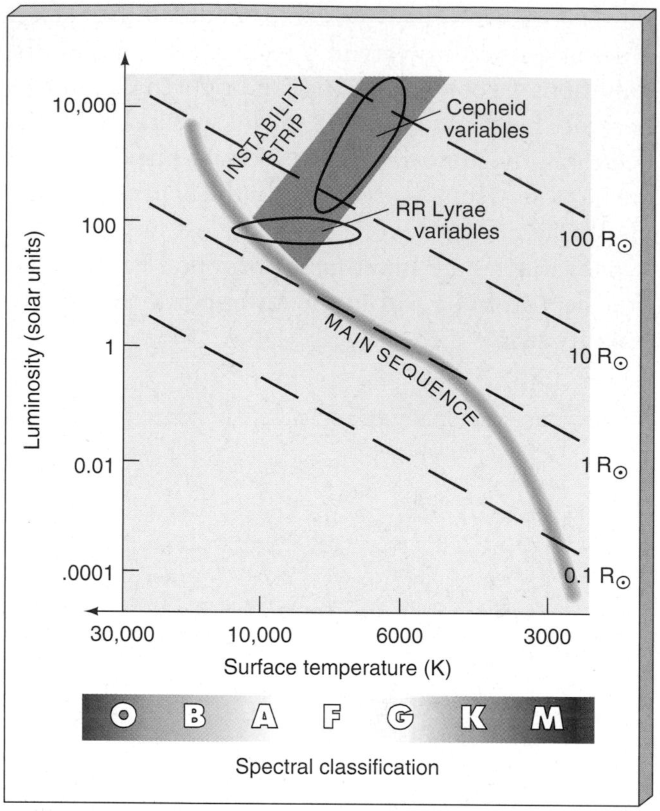

Figure 14.5 Variable Stars on H–R Diagram Pulsating variable stars are found in the instability strip of the H–R diagram. As a high-mass star evolves through the strip, it becomes a Cepheid variable. Low-mass horizontal-branch stars in the instability strip are RR Lyrae variables.

A New Yardstick

Astronomers can use pulsating variables as a means of determining distances, both within our own Galaxy and far beyond. Once we recognize a star as being of the RR Lyrae or Cepheid type, we can infer its luminosity. Then, by comparing the star's (known) luminosity with its (observed) apparent brightness, we can estimate its distance using the inverse-square law. ∞ (Sec. 10.2)

How do we infer a variable star's luminosity? For RR Lyrae variables, this is simple. All such stars have basically the same luminosity (averaged over a complete pulsation cycle)—about 100 times that of the Sun. For Cepheids, we make use of a close correlation between average luminosity and pulsation period, discovered in 1908 by Henrietta Leavitt of Harvard University and known simply as the **period–luminosity relationship**. Cepheids that vary slowly—that is, have long periods—have high luminosities, while short-period Cepheids have low luminosities.

Figure 14.6 illustrates the period–luminosity relationship for Cepheids found within a thousand parsecs or so of Earth. Astronomers can plot such a diagram for relatively nearby stars because they can measure their distances, and hence their luminosities, using stellar or spectroscopic parallax. Thus, a simple measurement of a Cepheid variable's pulsation period immediately tells us its luminosity—we just read it off the graph in Figure 14.6. (The roughly constant luminosities of the RR Lyrae variables are also indicated in the figure.)

This distance-measurement technique works well provided the variable star can be clearly identified and its pulsation period measured. With Cepheids, this method allows astronomers to estimate distances out to about 15 million parsecs, more than enough to take us to the nearest galaxies. Indeed, the existence of galaxies beyond our own was first established in the late 1920s, when American astronomer Edwin Hubble observed Cepheids in the Andromeda Galaxy and thereby succeeded in measuring its distance. The less luminous RR Lyrae stars are not as easily detectable as Cepheids, so their usable range is not as great. However, they are much more common, so, within this limited range, they are actually more useful than Cepheids.

Figure 14.7 extends our cosmic distance ladder, begun in Chapter 1 with radar ranging in the solar system and expanded in Chapter 10 to include stellar and spectroscopic parallax, by adding variable stars as a fourth method of determining distance.

The Size and Shape of Our Galaxy

Many RR Lyrae variables are found in globular clusters, those tightly bound swarms of old, reddish stars we first met in Chapter 11. ∞ (Sec. 11.5) Early in the twenti-

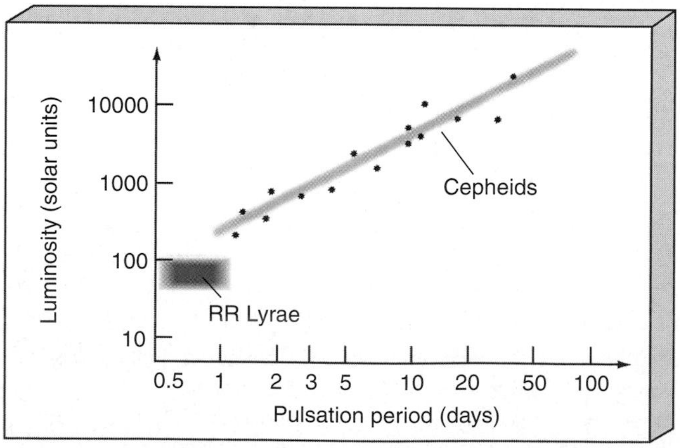

Figure 14.6 Period–Luminosity Plot A plot of pulsation period versus average absolute brightness (that is, luminosity) for a group of Cepheid variable stars. The two properties are quite tightly correlated. The pulsation periods of some RR Lyrae variables are also shown.

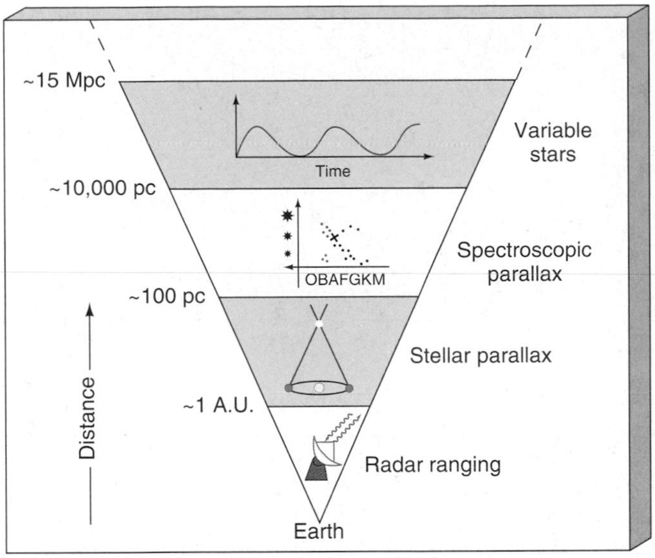

Figure 14.7 Variable Stars on Distance Ladder Application of the period–luminosity relationship for Cepheid variable stars allows us to determine distances out to about 15 Mpc with reasonable accuracy.

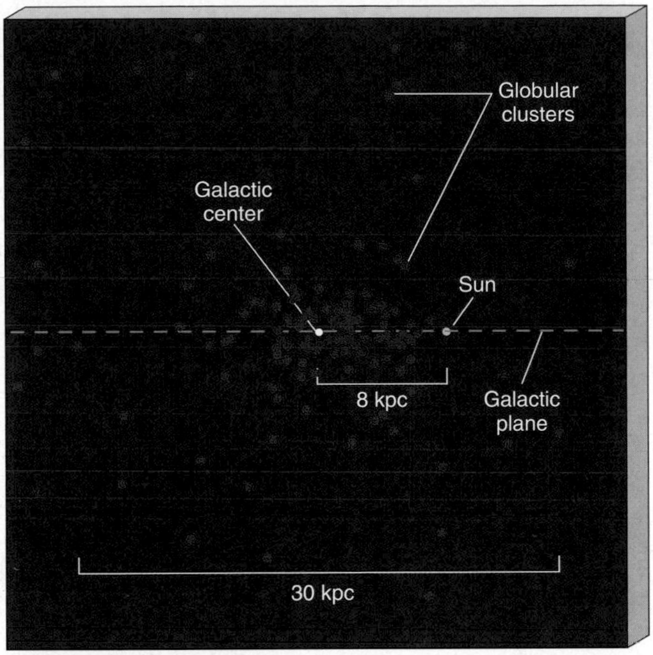

Figure 14.8 Globular Cluster Map Our Sun does not coincide with the center of the very large collection of globular clusters. Instead, more globular clusters are found in one direction than in any other. The Sun resides closer to the edge of the collection, which measures roughly 30 kpc across. We now know that the globular clusters outline the true distribution of stars in the Galactic halo.

eth century, the American astronomer Harlow Shapley used observations of RR Lyrae stars to make two very important discoveries about the Galactic globular cluster system. First, he showed that most globular clusters reside at great distances—many thousands of parsecs—from the Sun. Second, by measuring the direction and distance of each cluster, he was able to determine their three-dimensional distribution in space (Figure 14.8). In this way, Shapley demonstrated that the globular clusters map out a truly gigantic, and roughly *spherical*, volume of space, about 30 kpc across.[2] However, the center of the distribution lies nowhere near our Sun. It is located nearly 8 kpc away from us, in the direction of the constellation Sagittarius.

In a brilliant intellectual leap, Shapley realized that the distribution of globular clusters maps out the true extent of stars in the Milky Way Galaxy—the region that we now call the Galactic halo. The hub of this vast collection of matter, 8 kpc from the Sun, is the **Galactic center**. As illustrated in Figure 14.9, we live in the suburbs of this huge ensemble, in the Galactic disk—the thin sheet of young stars, gas, and dust that cuts through the center of the halo. Since Shapley's time, astronomers have identified many individual stars—that is, stars not belonging to any globular cluster—within the Galactic halo.

Shapley's bold interpretation of the globular clusters as defining the overall distribution of stars in our Galaxy was an enormous step forward in human understanding of our place in the universe. Five hundred years ago, Earth was considered the center of all things. Copernicus argued otherwise, demoting our planet to an undistinguished place removed from the center of the solar system. In Shapley's time, the prevailing view was that our Sun was the center not only of the Galaxy but also of the universe. Shapley showed otherwise. With his observations of globular clusters, he simultaneously increased the size of our Galaxy by almost a factor of 10 over earlier estimates and banished our parent Sun to its periphery, virtually overnight!

[2]The Galactic globular cluster system and the Galactic halo of which it is a part are somewhat flattened in the direction perpendicular to the disk, but the degree of flattening is uncertain. The halo is certainly much less flattened than the disk, however.

✓ Concept Check

- Can variable stars be used to map out the structure of the Galactic disk?

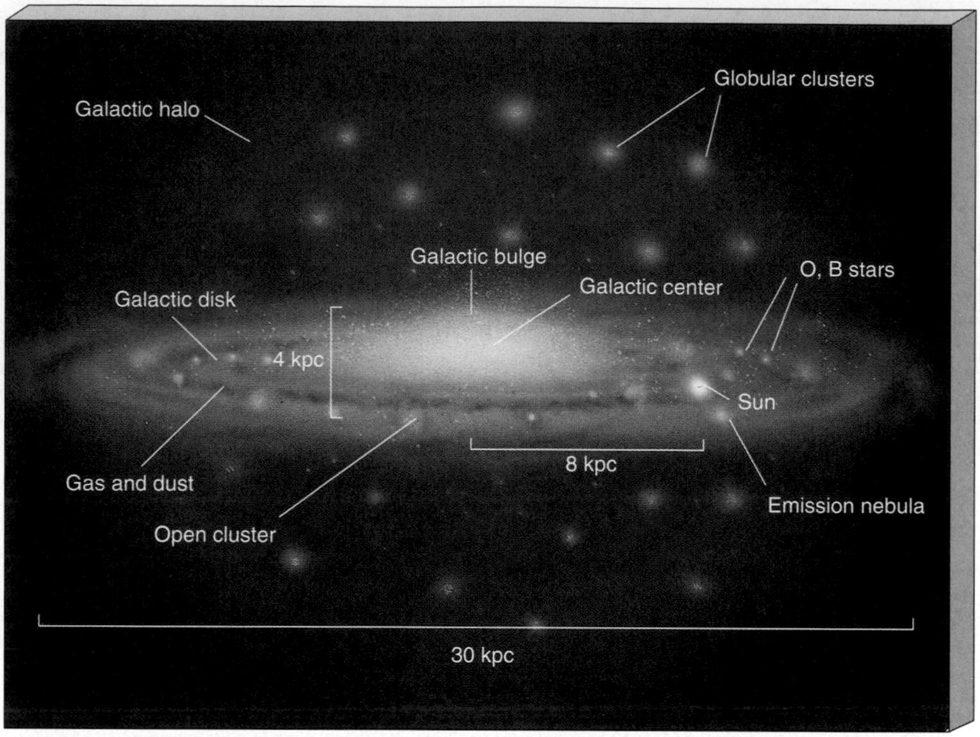

Figure 14.9 Stellar Populations in Our Galaxy Artist's conception of a (nearly) edge-on view of the Milky Way Galaxy, showing the distributions of young blue stars, open clusters, old red stars, and globular clusters.

14.3 Large-scale Structure

Mapping Our Galaxy

1 Based on optical, infrared, and radio studies of our galaxy, the very different spatial distributions of the disk, bulge, and halo components of the Milky Way Galaxy are shown in Figure 14.9. The extent of the halo is based largely on optical observations of globular clusters and other halo stars. However, as we have seen, optical techniques can cover only a small portion of the dusty Galactic disk. Much of our knowledge of the structure of the disk on larger scales is based on radio observations, particularly of the 21-cm radio emission line produced by atomic hydrogen. ∞ (Sec. 11.2) Because long-wavelength radio waves are largely unaffected by interstellar dust, and hydrogen is by far the most abundant element in interstellar space, the 21-cm signals are strong enough that virtually the entire disk can be observed in this way.

According to radio studies, the center of the gas distribution coincides roughly with the center of the globular cluster system, lying about 8 kpc from the Sun. In fact, this figure is derived most accurately from radio observations of Galactic gas. The densities of both stars and gas in the disk decline quite rapidly beyond about 15 kpc from the Galactic center (although some radio-emitting gas has been observed out to at least 50 kpc). Perpendicular to the Galactic plane, the disk in the vicinity of the Sun is "only" about 300 pc thick, or about 1/100 of the 30 kpc Galactic diameter. Don't be fooled, though. Even if you could travel at the speed of light, it would take you 1000 years to traverse the thickness of the Galactic disk. The disk may be thin compared with the Galactic diameter, but it is huge by human standards.

Also shown in Figure 14.9 is our Galaxy's central bulge, measuring roughly 6 kpc across in the plane of the Galactic disk by 4 kpc perpendicular to that plane. Obscuration by interstellar dust makes it difficult to study the bulge at optical wavelengths. However, at longer wavelengths, which are less affected by interstellar matter, a much clearer picture emerges (Figure 14.10). Detailed measurements of the motion of gas and stars in and near the bulge imply that it is actually football shaped, about half as wide as it is long, with the long axis of the football lying in the Galactic plane.

Stellar Populations

Aside from their shapes, the three components of the Galaxy—disk, bulge, and halo—have other properties that distinguish them from one another. First, the halo contains essentially *no* gas or dust—just the opposite of the disk and bulge, in which interstellar matter is common. Second, there are clear differences in *color* between disk, bulge, and halo stars. Stars in the Galactic bulge and halo are distinctly redder than those found in the disk. Observations of other spiral galaxies also show this trend—the blue-white tint of the disk and the yellowish coloration of the bulge are evident in Figures 14.2(a) and (b).

All the bright, blue stars visible in our night sky are part of the Galactic disk, as are the young open star clusters and star-forming regions. In contrast, the cooler, redder stars—including those found in the old globular clusters—are more uniformly distributed throughout the disk, bulge, and halo. Galactic disks appear bluish because main-sequence O- and B-type blue supergiants are very much brighter than G-, K-, and M-type dwarfs, even though the dwarfs are present in far greater numbers.

The explanation for the marked difference in stellar content between disk and halo is that, whereas the gas-rich Galactic disk is the site of ongoing star formation and so contains stars of all ages, all the stars in the Galactic halo are *old*. The absence of dust and gas in the halo means that no new stars are forming there, and star formation apparently ceased long ago—at least 10 billion years in the past, judging from the ages of the globular clusters. ∞ (Sec. 12.6) The gas density is very high in the inner part of the Galactic bulge, making this region the site of vigorous ongoing star formation, and both very old and very young stars mingle there. The bulge's gas-poor outer regions have properties more similar to those of the halo.

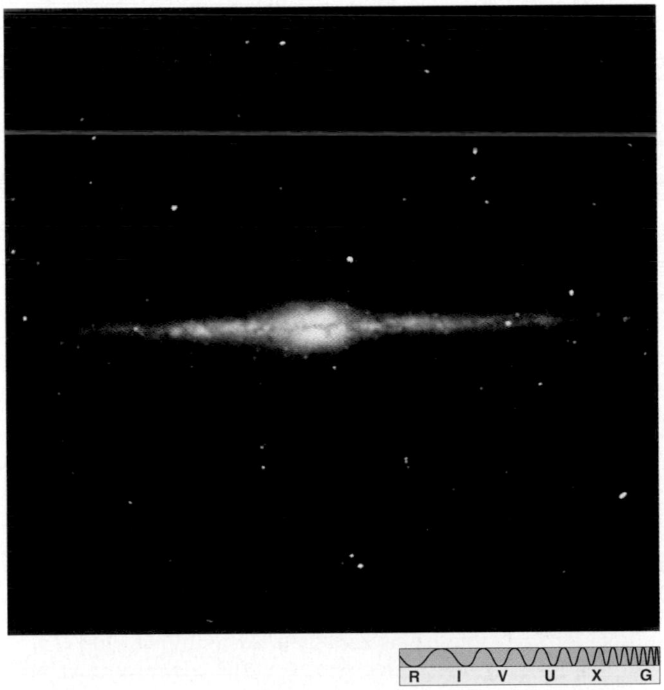

Figure 14.10 Infrared View of Galaxy A wide-angle infrared image of the disk and bulge of the Milky Way Galaxy, as observed by the *Cosmic Background Explorer* (*COBE*) satellite. Compare Figure 14.2(c). *(National Optical Astronomy Observatories)*

Spectroscopic studies indicate that halo stars are far less abundant in heavy elements (that is, elements heavier than helium) than are nearby stars in the disk. Each successive cycle of star formation and evolution enriches the interstellar medium with the products of stellar evolution, leading to a steady increase in heavy elements with time. ∞ (*More Precisely 12-1*) Thus, the scarcity of these elements in halo stars is consistent with the view that the halo formed long ago.

Astronomers often refer to young disk stars as *Population I* stars and old halo stars as *Population II* stars. The idea of two stellar "populations" dates back to the 1930s, when the differences between disk and halo stars first became clear. It represents something of an oversimplification, as there is actually a continuous variation in stellar ages throughout the Milky Way Galaxy, and not a simple division of stars into two distinct "young" and "old" categories. Nevertheless, the terminology is still widely used.

Orbital Motion

3 Are the internal motions of our Galaxy's members chaotic and random, or are they part of some gigantic "traffic pattern"? The answer depends on our perspective. The motion of stars and clouds we see on small scales (within a few tens of parsecs of the Sun) seems random, but on larger scales (hundreds or thousands of parsecs) the motion is much more orderly.

As we study the Doppler shifts of radiation received from stars and gas at different locations in the Galactic disk, a clear pattern of motion emerges. The entire disk is *rotating* about the Galactic center (Figure 14.11). In the vicinity of the Sun, 8 kpc from the center, the orbital speed is about 220 km/s, so material takes about 225 million years to complete one circuit. At other distances from the center, the rotation period is different—shorter closer to the center, longer at greater distances—that is, the Galactic disk rotates not as a solid object, but *differentially*.

This picture of orderly circular orbital motion about the Galactic center applies only to the Galactic disk. The old globular clusters in the halo and the faint, reddish individual stars in both the halo and the bulge do *not* share the disk's well-defined rotation. Instead, as

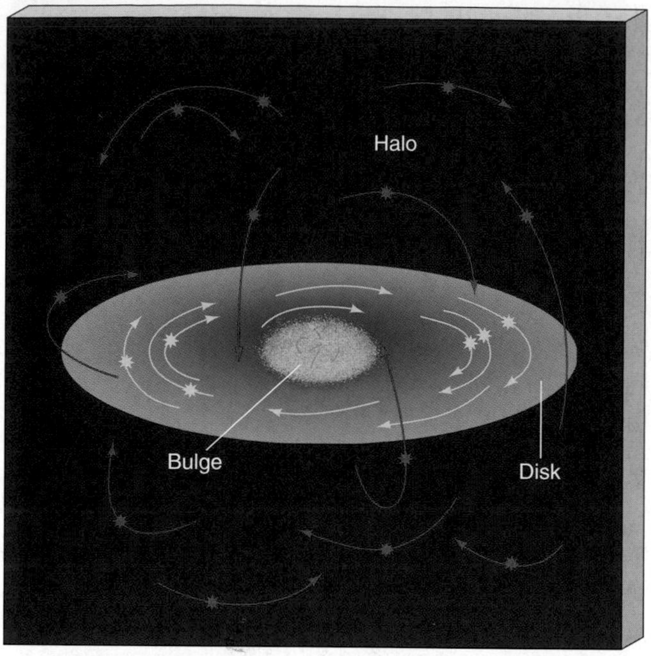

Figure 14.11 Stellar Orbits in Our Galaxy Stars in the Galactic disk move in orderly, circular orbits about the Galactic center. In contrast, halo stars have orbits with largely random orientations and eccentricities. The orbit of a typical halo star takes it high above the Galactic disk, then down through the disk plane, then out the other side and far below the disk. The orbital properties of bulge stars are intermediate between those of disk stars and those of halo stars.

shown in Figure 14.11, their orbits are largely random.[3] Although they do orbit the Galactic center, they move in all directions, their paths filling an entire three-dimensional volume rather than a nearly two-dimensional disk. At any given distance from the Galactic center, bulge or halo stars move at speeds comparable to the disk's rotation speed at that radius but in *all* directions, not just one. Their orbits carry these stars repeatedly through the disk plane and out the other side.

Table 14.1 compares some key properties of the three basic components of the Galaxy.

[3]Halo stars may in fact have some small net rotation about the Galactic center, but the rotational component of their motion is overwhelmed by the much larger random component. The motion of bulge stars also has a rotational component, larger than that of the halo but still smaller than the random component of stellar motion within the bulge.

The Formation of the Milky Way

3 Figure 14.12 illustrates the current view of our Galaxy's evolution. It explains, in broad terms at least, the Galactic structure we see today. Not unlike the star-formation scenario outlined in Chapter 11, it starts from a contracting cloud of pregalactic gas. ∞ (Sec. 11.3)

When the first Galactic stars and globular clusters formed, some 12 billion years ago, the gas in our Galaxy had not yet accumulated into a thin disk. Instead, it was spread out over an irregular, and quite extended, region of space, spanning many tens of kiloparsecs in all directions. When the first stars formed, they were distributed throughout this volume. Their distribution today (the Galactic halo) reflects that fact—it is an imprint of their birth. Many astronomers believe that the very first stars formed even earlier, in smaller systems that later merged to create our Galaxy. The present-day halo would look much the same in either case. ∞ (Sec. 15.4)

TABLE 14.1 Overall Properties of the Galactic Disk, Halo, and Bulge

GALACTIC DISK	GALACTIC HALO	GALACTIC BULGE
highly flattened	roughly spherical—mildly flattened	somewhat flattened—elongated in the plane of the disk ("football shaped")
contains both young and old stars	contains old stars only	contains both young and old stars; more old stars at greater distances from the center
contains gas and dust	contains no gas and dust	contains gas and dust, especially in the inner regions
site of ongoing star formation	no star formation during the last 10 billion years	ongoing star formation in the inner regions
gas and stars move in circular orbits in the Galactic plane	stars have random orbits in three dimensions	stars have largely random orbits, but with some net rotation about the Galactic center
spiral arms (Sec. 14.4)	no obvious substructure	ring of gas and dust near center; central Galactic nucleus (Sec. 14.6)
overall white coloration, with blue spiral arms	reddish in color	yellow-white

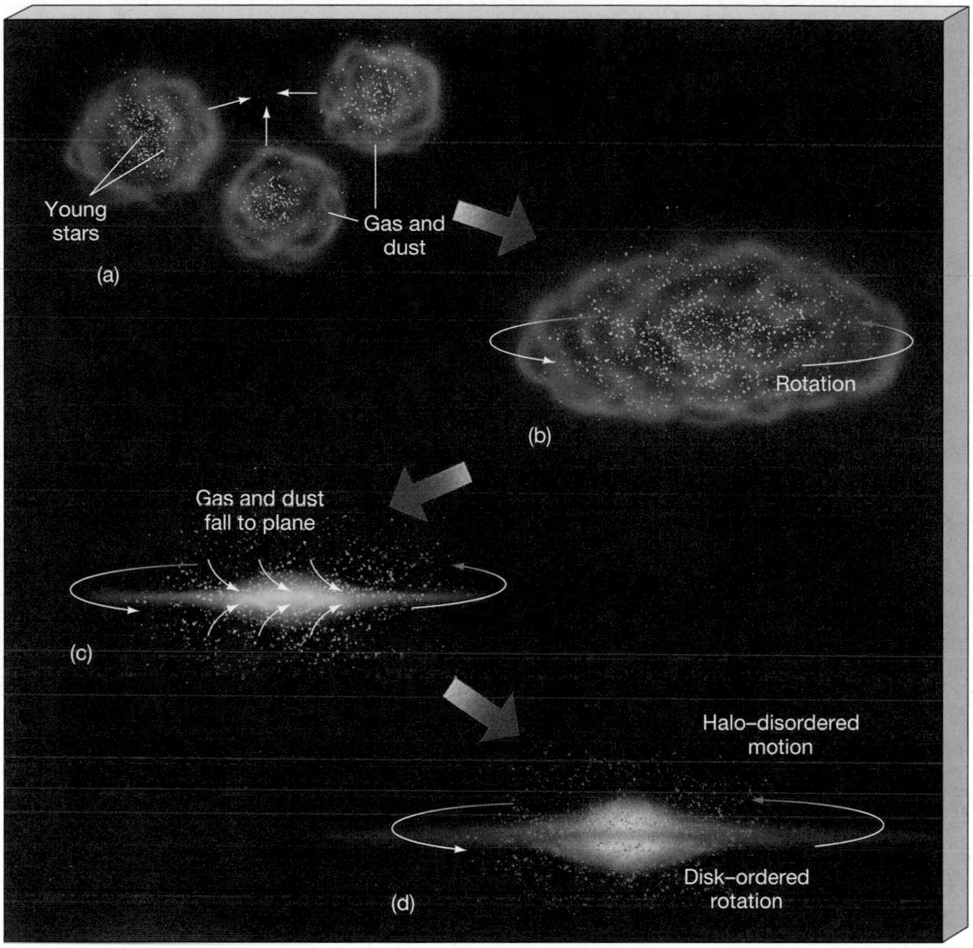

Figure 14.12 Milky Way Formation (a) The Milky Way Galaxy possibly formed through the merger of several smaller systems. (b) Astronomers reason that, early on, our Galaxy was irregularly shaped, with gas distributed throughout its volume. When stars formed during this stage, there was no preferred direction in which they moved and no preferred location in which they were found. (c) In time, rotation caused the gas and dust to fall to the Galactic plane and form a spinning disk. The stars that had already formed were left behind, forming the halo. (d) New stars forming in the disk inherit its overall rotation and so orbit the Galactic center on ordered, circular orbits.

During the past 12 billion years, rotation has flattened the gas in our Galaxy into a relatively thin disk. Physically, this process is similar to the flattening of the solar nebula during the formation of the solar system, except on a vastly larger scale. ∞ (Sec. 4.3) Star formation in the halo ceased billions of years ago when the raw materials fell to the Galactic plane. Ongoing star formation in the disk gives it its bluish tint, but the halo's short-lived blue stars have long since burned out, leaving only the long-lived red stars that give it its characteristic pinkish glow. The Galactic halo is ancient, whereas the disk is full of youthful activity.

The chaotic orbits of the halo stars are also explained by this theory. When the halo developed, the irregularly shaped Galaxy was rotating only very slowly, so there was no strongly preferred direction in which matter tended to move. As a result, halo stars were free to travel along nearly any path once they formed (or when their parent systems merged). As the Galactic disk formed, however, conservation of angular momentum caused it to spin more rapidly. Stars forming from the gas and dust of the disk inherit its rotational motion and so move on well-defined, circular orbits.

✓ Concept Check

- Why are there no young halo stars?

14.4 Galactic Spiral Arms

4 Radio studies provide perhaps the best direct evidence that we live in a spiral galaxy. Figure 14.13 is an artist's conception (based on observational data) of the appearance of our Galaxy as seen from far above the disk, clearly showing our Galaxy's **spiral arms**, pinwheel-like structures originating close to the Galactic bulge and extending outward throughout much of the Galactic disk. One of these arms, as far as we can tell, wraps around a large part of the disk and contains our Sun. Notice, incidentally, the scale markers on Figures 14.8, 14.9, and 14.13: The Galactic globular-cluster distribution (shown

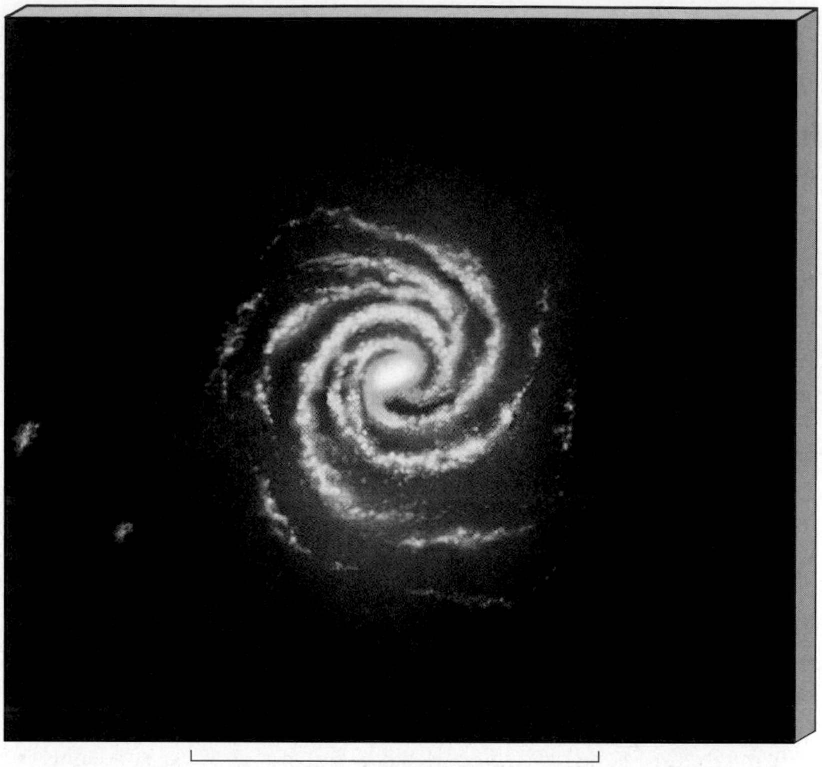

Figure 14.13 Milky Way, Rendering An artist's conception of our Milky Way Galaxy seen face-on. This illustration is based on data accumulated by legions of astronomers during the past few decades, including radio maps of gas density in the Galactic disk. Painted from the perspective of an observer 100 kpc above the Galactic plane, the spiral arms are at their best-determined positions. All the features are drawn to scale (except for the oversized yellow dot near the top, which represents our Sun). The two small blotches to the left are dwarf galaxies, called the Magellanic Clouds. We study them in Chapter 15. *(L. Chaisson)*

in Figure 14.8), the luminous stellar component of the disk (Figure 14.9), and the known spiral structure (Figure 14.13) all have roughly the same diameter—about 30 kpc.

The spiral arms in our Galaxy are made up of much more than just interstellar gas and dust. Studies of the Galactic disk within a kiloparsec or so of the Sun indicate that young stellar and prestellar objects—emission nebulae, O- and B-type stars, and recently formed open clusters—are also distributed in a spiral pattern that closely follows the distribution of interstellar clouds. The obvious conclusion is that the spiral arms are the part of the Galactic disk where star formation takes place. The brightness of the young stellar objects just listed is the main reason that the spiral arms of other galaxies are easily seen from afar (Figure 14.2b).

A central problem facing astronomers trying to understand spiral structure is how that structure persists over long periods of time. The basic issue is simple: Differential rotation makes it impossible for any large-scale structure "tied" to the disk material to survive. Figure 14.14 shows how a spiral pattern consisting always of the same group of stars and gas clouds would necessarily "wind up" and disappear within a few hundred million years. How then do the Galaxy's spiral arms retain their structure over long periods of time in spite of differential rotation? A leading explanation for the existence of spiral arms holds that they are **spiral density waves**—coiled waves of gas compression that move through the Galactic disk, squeezing clouds of interstellar gas and triggering the process of star formation as they go. (Sec. 11.3) The spiral arms we observe are defined by the denser than normal clouds of gas the density waves create and by the new stars formed as a result of the spiral waves' passage.

This explanation of spiral structure avoids the problem of differential rotation because the wave pattern is not tied to any particular piece of the Galactic disk. The spirals we see are merely patterns moving through the disk, not great masses of matter being transported from place to place. The density wave moves through the collection of stars and gas making up the disk just as a sound wave moves through air or an ocean wave passes through water, compressing different parts of the disk at different times. Even though the rotation rate of the disk material varies with distance from the Galactic center, the wave itself remains intact, defining the Galaxy's spiral arms.

Over much of the visible portion of the Galactic disk (within about 15 kpc of the center), the spiral wave pattern is predicted to rotate *more slowly* than the stars and gas. Thus, as shown in Figure 14.15, Galactic material catches up with the wave, is temporarily slowed

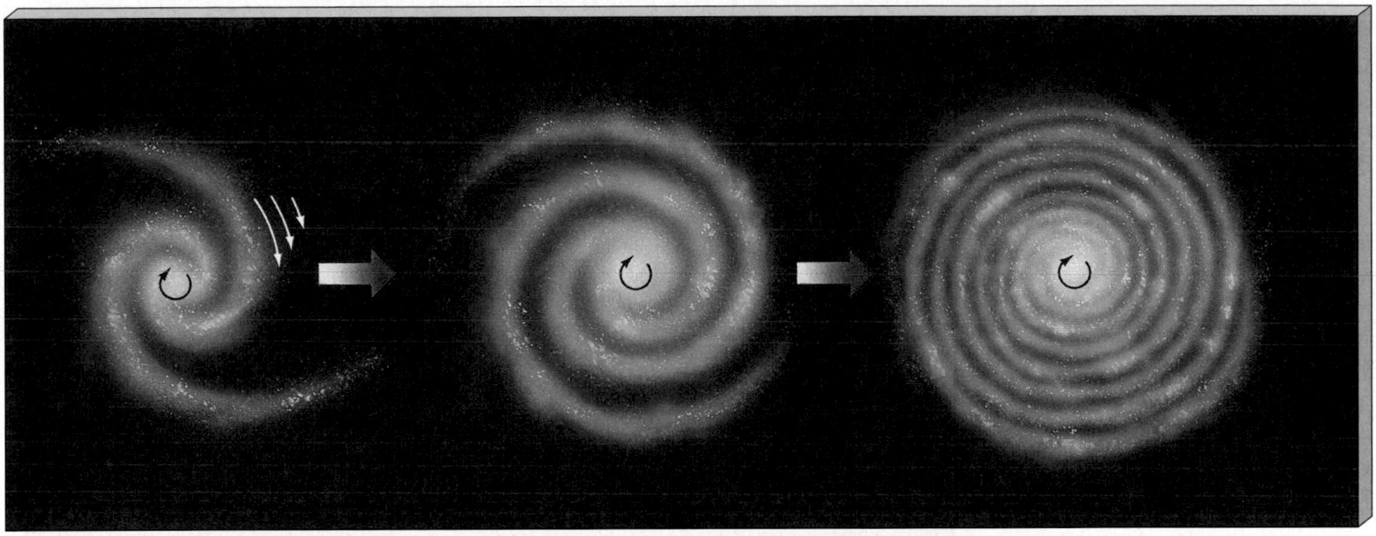

Figure 14.14 Differential Galactic Rotation The disk of our Galaxy rotates differentially—stars close to the center take less time to orbit the Galactic center than those at greater distances. If spiral arms were somehow tied to the material of the Galactic disk, this differential rotation would cause the spiral pattern to wind up and disappear in a few hundred million years. Spiral arms would be too short-lived to be consistent with the numbers of spiral galaxies we observe today.

down and compressed as it passes through, then continues on its way. (For a more down-to-earth example of an analogous process, see *Interlude 14-1*.)

As gas enters the arm from behind, the gas is compressed and forms stars. Dust lanes mark the regions of highest-density gas. The most prominent stars—the bright O- and B-type blue giants—live for only a short time, so emission nebulae and young star clusters are found only within the arms, near their birthsites, just ahead of the dust lanes. Their brightness emphasizes the spiral structure. Further downstream, ahead of the spiral arms, we see mostly older stars and star clusters, which

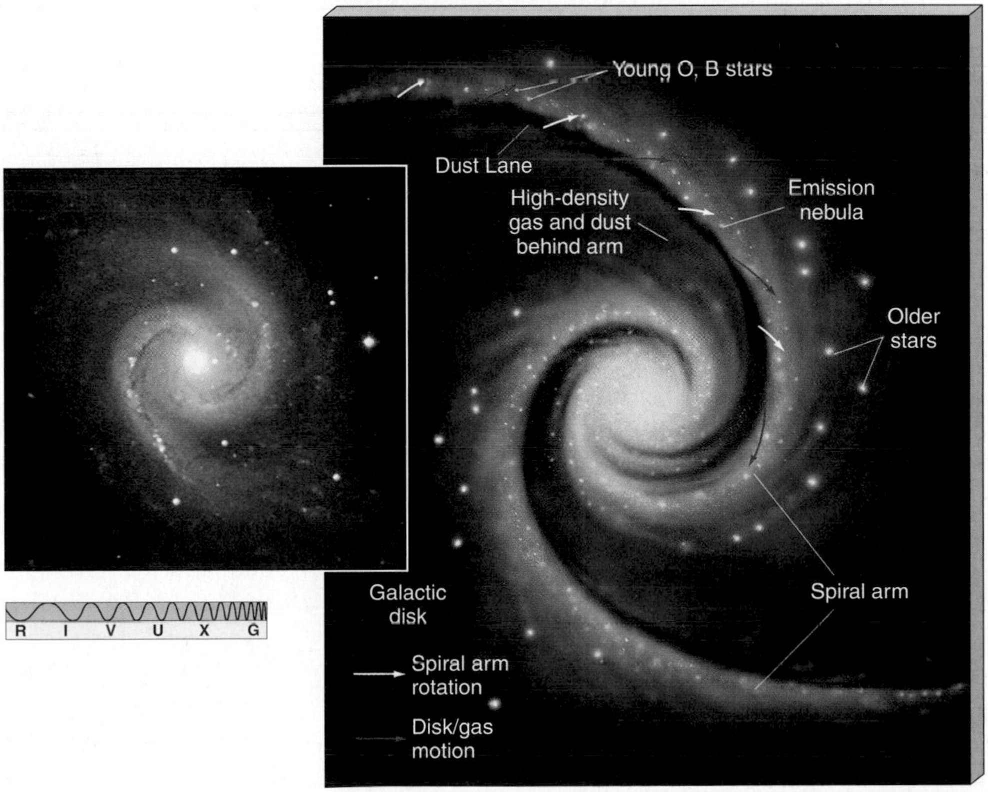

Figure 14.15 Density-wave Theory This theory holds that the spiral arms seen in our own and many other galaxies are waves of gas compression and star formation moving through the material of the Galactic disk. In the painting at right, gas motion is indicated by red arrows, and arm motion by white arrows. Gas enters an arm from behind, is compressed, and forms stars. The spiral pattern is delineated by dust lanes, regions of high gas density, and newly formed bright stars. The inset shows the spiral galaxy NGC 1566, which displays many of the features just described. *(AURA)*

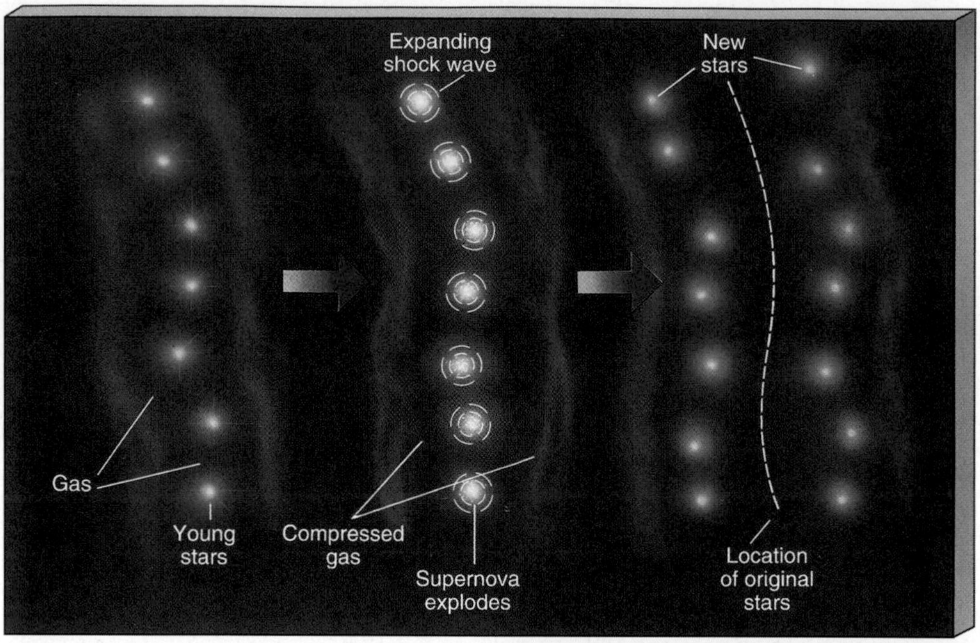

(a)

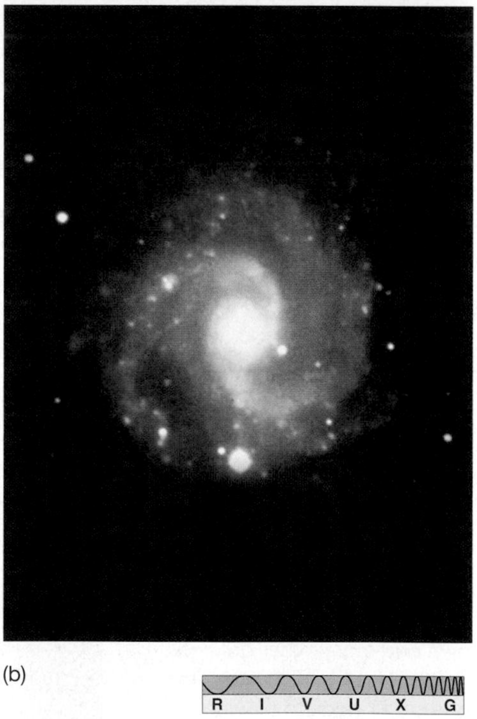

(b)

Figure 14.16 Self-propagating Star Formation (a) In this theory of the formation of spiral arms, the shock waves produced by the formation and later evolution of a group of stars provide the trigger for new rounds of star formation. We have used supernova explosions to illustrate the point here, but the formation of emission nebulae and planetary nebulae are also important. (b) This process may well be responsible for the partial spiral arms seen in some galaxies, such as NGC 3184, shown here in true color. The distinct blue appearance derives from the vast numbers of young stars that pepper its ill-defined spiral arms. *(NASA)*

have had enough time since their formation to outdistance the wave and pull away from it.

An alternative possibility is that the formation of stars drives the waves, instead of the other way around. Imagine a row of newly formed massive stars somewhere in the disk. The emission nebula created when these stars form, and the supernovae when they die, send shock waves through the surrounding gas, possibly triggering new star formation. ∞ (*More Precisely 12-1*) Thus, as illustrated in Figure 14.16(a), the formation of one group of stars provides the mechanism for the creation of more stars. Computer simulations suggest that it is possible for the "wave" of star formation thus created to take on the form of a partial spiral and for this pattern to persist for some time. However, this process, known as **self-propagating star formation**, can produce only pieces of spirals, as are seen in some galaxies (Figure 14.16b). It apparently cannot produce the galaxy-wide spiral arms seen in other galaxies and present in our own. It may well be that there is more than one process at work in the spectacular spirals we see.

An important question, but one that unfortunately is not answered by either of the two theories just described, is: Where do these spirals come from? What was responsible for generating the density wave in the first place or for creating the line of newborn stars whose evolution drives the advancing spiral arm? Scientists speculate that 1) instabilities in the gas near the Galactic bulge, 2) the gravitational effects of nearby galaxies, or 3) the elongated shape of the bulge itself may have had a big enough influence on the disk to get the process going. The fact is that we still don't know for sure how galaxies—including our own—acquire such beautiful spiral arms.

✓ Concept Check

- Why can't spiral arms simply be clouds of gas and young stars orbiting the Galactic center?

INTERLUDE 14-1

Density Waves

In the late 1960s, American astrophysicists C. C. Lin and Frank Shu proposed a way in which spiral arms in the Galaxy could persist for many Galactic rotations. They argued that the arms themselves contain no "permanent" matter. They should not be viewed as assemblages of stars, gas, and dust moving intact through the disk because such structures would quickly be destroyed by differential rotation. Instead, as described in the text, a spiral arm should be envisaged as a *density wave*—a wave of alternating compression and expansion sweeping through the Galaxy.

A wave in water builds up material temporarily in some places (crests) and lets it down in others (troughs). Similarly, as Galactic matter encounters a spiral density wave, the matter is compressed to form a region of slightly higher than normal density. The matter enters the wave, is temporarily slowed down and compressed as it passes through, then continues on its way. This compression triggers the formation of new stars and nebulae. In this way, the spiral arms are formed and reformed repeatedly, without wrapping up.

The accompanying figure illustrates the formation of a density wave in a more familiar context—a traffic jam triggered by the presence of a repair crew moving slowly down the road. Cars slow down temporarily as they approach the crew, then speed up again as they pass the worksite and continue on their way. The result observed by a traffic helicopter flying overhead is a region of high traffic density concentrated around the work crew and moving with it. An observer on the side of the road, however, sees that the jam never contains the same cars for very long. Cars constantly catch up to the bottleneck, move slowly through it, then speed up again, only to be replaced by more cars arriving from behind.

The traffic jam is analogous to the region of high stellar density in a Galactic spiral arm. Just as the traffic density wave is not tied to any particular group of cars, the spiral arms are not attached to any particular piece of disk material. Stars and gas enter a spiral arm, slow down for a while, then exit the arm and continue on their way around the Galactic center. The result is a moving region of high stellar and gas density, involving different parts of the disk at different times. Notice also that, just as in our Galaxy, the traffic jam wave moves more slowly than, and independently of, the overall traffic flow.

We can extend our traffic analogy a little further. Most drivers are well aware that the effects of a such a tie-up can persist long after the road crew has stopped work and gone home for the night. Similarly, spiral density waves can continue to move through the disk even after the disturbance that originally produced them has long since subsided. According to spiral density wave theory, this is precisely what has happened in the Milky Way Galaxy. Some disturbance in the past produced a wave that has been moving through the Galactic disk ever since.

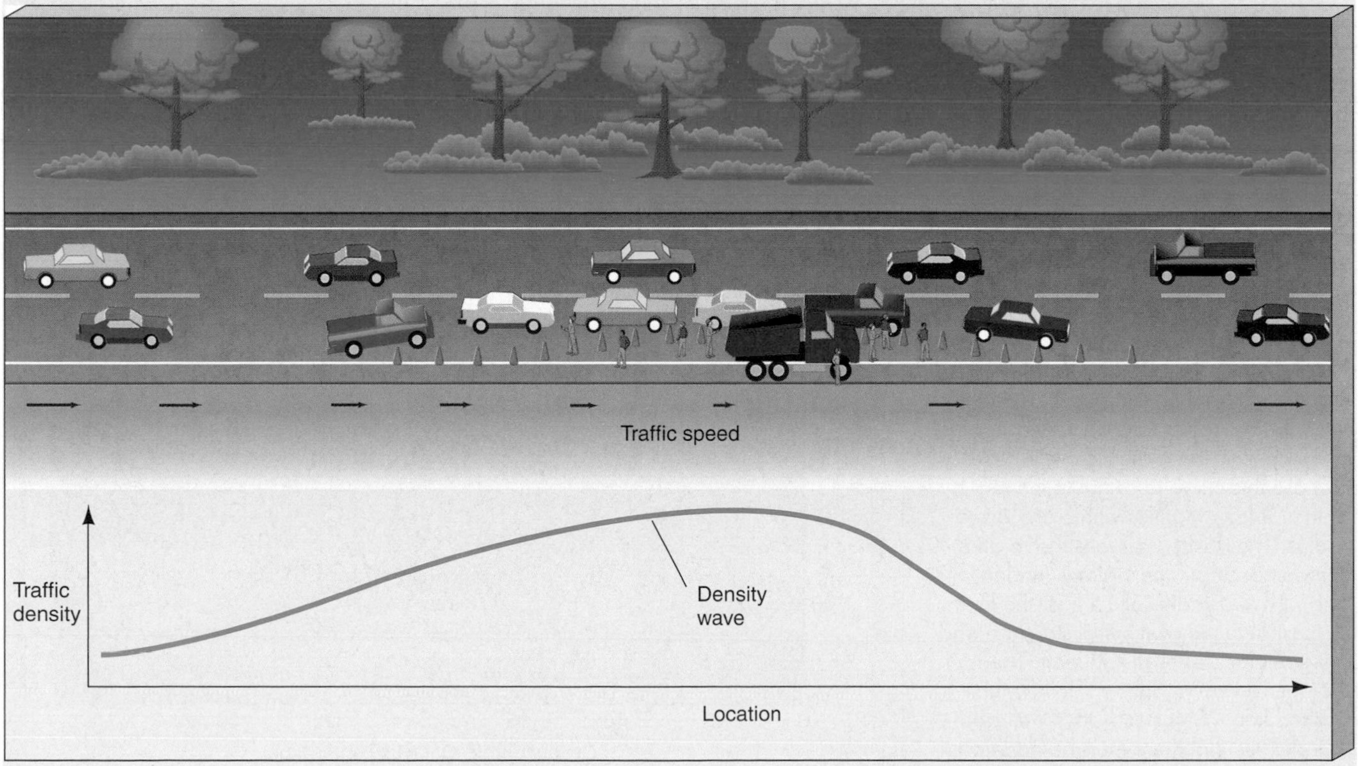

14.5 The Mass of the Milky Way

Galactic Rotation

5 We can measure our Galaxy's mass by studying the motions of gas clouds and stars in the Galactic disk. Recall from Chapter 1 that Kepler's third law (as modified by Newton) connects the orbital period, orbit size, and masses of any two objects in orbit around one another: (Sec. 1.4)

$$\text{total mass (solar masses)} = \frac{\text{orbit size (A.U.)}^3}{\text{orbit period (years)}^2}.$$

The distance from the Sun to the Galactic center is about 8 kpc and the Sun's orbital period is 225 million years. Substituting these numbers into the equation, we obtain a mass of almost 10^{11} solar masses—100 *billion* times the mass of our Sun.

But what mass have we just measured? When we performed the analogous calculation in the case of a planet orbiting the Sun, there was no ambiguity: Neglecting the planet's mass, the result of our calculation was the mass of the Sun. (*More Precisely 1-2*) However, the Galaxy's matter is not concentrated at the Galactic center (as the Sun's mass is concentrated at the center of the solar system). Instead, Galactic matter is distributed over a large volume of space. What portion of the Galaxy's mass controls the Sun's orbit? Isaac Newton answered this question three centuries ago: The Sun's orbital period is determined by the portion of the Galaxy that lies *within the orbit of the Sun*. This is the mass computed from the above equation.

Dark Matter

Based largely on data obtained from radio observations, Figure 14.17 shows the Galactic **rotation curve**, which plots the rotation speed of the Galactic disk against distance from the Galactic center. Using this graph, we can repeat our earlier calculation to compute the total mass that lies within any given distance from the Galactic center. We find, for example, that the mass within about 15 kpc from the center—the volume defined by the globular clusters and the known spiral structure—is roughly 2×10^{11} solar masses, about twice the mass contained within the Sun's orbit. Does most of the matter in the Galaxy "cut off" at this point, where the luminosity drops off sharply? Surprisingly, the answer is no.

If all of the mass of the Galaxy were contained within the edge of the visible structure, Newton's laws of motion predict that the orbital speed of stars and gas beyond 15 kpc would decrease with increasing distance from the Galactic center, just as the orbital speeds of the planets diminish as we move outward from the Sun. (*More Precisely 1-2*) The dashed line in Figure 14.17 indicates how the rotation curve should look in that case. However, the actual rotation curve is quite different. Far from declining at larger distances, it *rises* slightly out to the limits of our measurement capabilities. This implies that the amount of mass contained within successively larger radii continues to grow beyond the orbit of the Sun, apparently out to a distance of at least 50 kpc.

According to the above equation, the amount of mass within 50 kpc is approximately 6×10^{11} solar masses. Since 2×10^{11} solar masses lies within 15 kpc of the

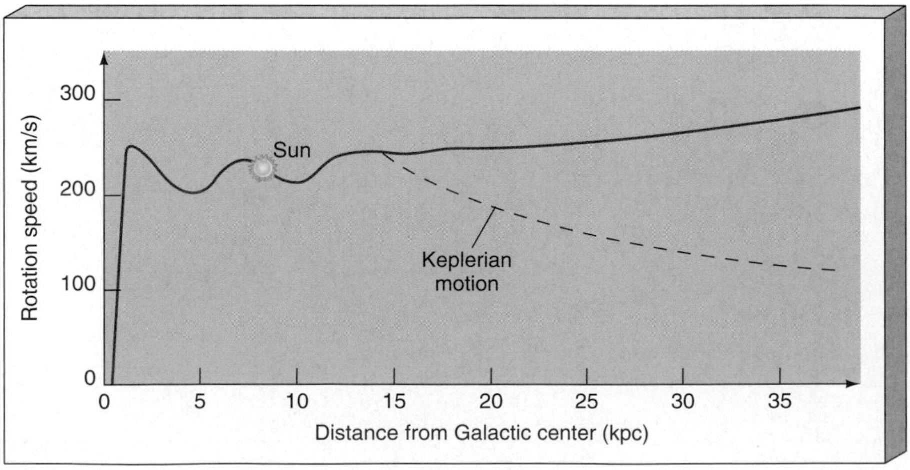

Figure 14.17 Galaxy Rotation Curve The rotation curve for the Milky Way Galaxy plots rotation speed against distance from the Galactic center. We can use this curve to compute the mass of the Galaxy that lies within any given radius. The dashed curve is the rotation curve expected if the Galaxy "ended" abruptly at a radius of 15 kpc, the limit of most of the known spiral structure and the globular cluster distribution. The fact that the red curve does not follow this dashed line, but instead stays well above it, indicates that there must be additional matter beyond that radius.

Galactic center, we have to conclude that at least twice as much mass lies *outside* the luminous part of our galaxy—the part made up of stars, star clusters, and spiral arms—as lies inside!

Based on these observations, astronomers now believe that the luminous portion of the Milky Way Galaxy—the region outlined by the globular clusters and by the spiral arms—is merely the "tip of the Galactic iceberg." Our Galaxy is in reality very much larger. The luminous region is surrounded by an extensive, invisible **dark halo**, which dwarfs the inner halo of stars and globular clusters and extends well beyond the 15-kpc radius once thought to represent the limit of our Galaxy. But what is this dark halo made of? We do not detect enough stars or interstellar matter to account for the mass that our computations tell us must be there. We are inescapably drawn to the conclusion that most of the mass in our Galaxy exists in the form of invisible **dark matter**, which we presently simply do not understand.

The term *dark* here does not refer just to matter undetectable in visible light. The material has (so far) escaped detection at *all* electromagnetic wavelengths, from radio to gamma rays. Only by its gravitational pull do we know of its existence. Dark matter is not hydrogen gas (atomic or molecular), nor is it made up of ordinary stars. Given the amount of matter that must be accounted for, we would have been able to detect it with present-day equipment if it were in either of those forms. Its nature and its consequences for the evolution of galaxies and the universe are among the most important questions in astronomy today.

Many candidates have been suggested for this dark matter, although none is proven. Among the strongest "stellar" contenders are brown dwarfs, white dwarfs, and faint, low-mass red dwarfs. These objects could in principle exist in great numbers throughout the Galaxy, yet would be exceedingly hard to see. A radically different alternative is that the dark matter is made up of exotic *subatomic particles* that pervade the entire universe. Many theoretical astrophysicists believe that these particles could have been produced in abundance during the very earliest moments of our universe. If they survived to the present day, there might be enough of them to account for all the dark matter we believe must be out there. This idea is hard to test, however, because such particles would be very difficult to detect. Several detection experiments have been attempted, so far without success.

The Search for Stellar Dark Matter

Recently, researchers have obtained insight into the distribution of stellar dark matter by using a key element of Albert Einstein's theory of general relativity—the prediction that a beam of light can be deflected by a gravitational field, which has already been verified in the case of starlight passing close to the Sun. ∞ (*More Precisely 13-2*) Although this effect is small, it has the potential for making otherwise invisible stellar objects observable from Earth. Here's how.

Imagine looking at a distant star as a faint foreground object (such as a brown dwarf or a white dwarf) happens to cross your line of sight. As illustrated in Figure 14.18, the intervening object deflects a little more starlight than usual toward you, resulting in a temporary, but quite substantial, *brightening* of the distant star. In some ways, the effect is like the focusing of light by a lens, and the process is known as **gravitational lensing**. The foreground object is referred to as a *gravitational lens*. The amount of brightening and the duration of the effect depend on the mass, distance, and speed of the lensing object. Typically, the apparent brightness of the background star increases by a factor of two to five for a period of several weeks. Thus, even though the foreground object cannot be seen directly, its effect on the light of the background star makes it detectable.

Of course, the probability of one star passing almost directly in front of another, as seen from Earth, is extremely small. However, by observing millions of stars every few days over a period of years (using automated telescopes and high-speed computers to reduce the burden of coping with so much data), astronomers have so far seen 15 such events since these studies began in 1993. The observations are consistent with lensing by low-mass white dwarf stars and suggest that such stars exist in great numbers in the halo, but it now seems unlikely that they can account for all the dark matter inferred from dynamical studies.

The identity of the dark matter is not necessarily an "all-or-nothing" proposition. It is perfectly conceivable that more than one type of dark matter exists. For example, it is quite possible that most of the dark matter in the inner (visible) parts of galaxies could be in the form of low-mass stars, while the dark matter farther out might be primarily in the form of exotic particles. We will return to this perplexing problem in later chapters.

✓ Concept Check

- In what sense is dark matter "dark"?

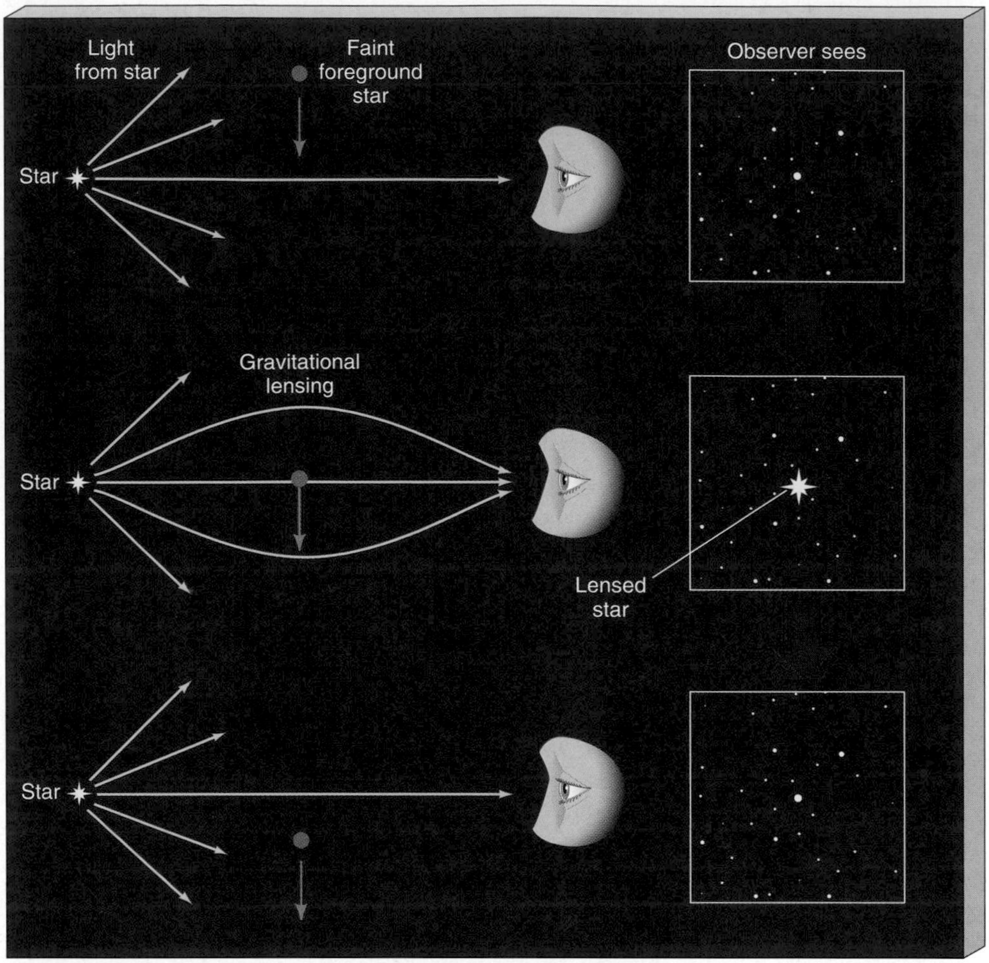

Figure 14.18 **Gravitational Lensing** Gravitational lensing by a faint foreground object (such as a brown dwarf) can temporarily cause a background star to brighten significantly, providing a means of detecting otherwise invisible stellar dark matter.

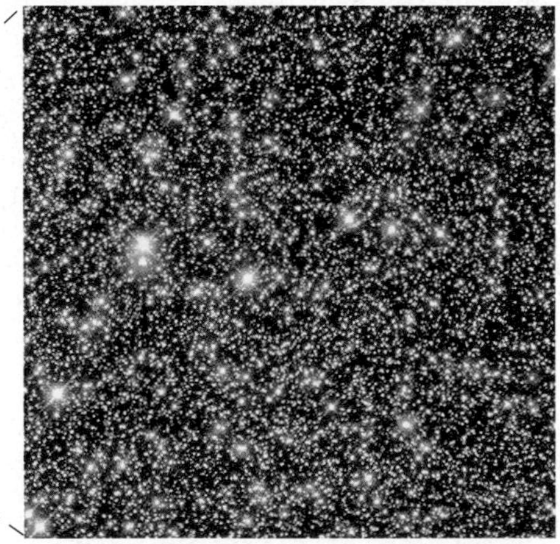

Figure 14.19 **Visible Image of the Galactic Center** A photograph of stellar and interstellar matter in the direction of the Galactic center. Because of heavy obscuration, even the largest optical telescopes can see no farther than one-tenth the distance to the center. The M8 nebula (arrow) can be seen at extreme top center. The field is roughly 20°, top to bottom, and is a continuation of the bottom part of Figure 11.4. The box indicates the location of the center of our Galaxy and corresponds to the same area boxed in Figure 14.20(a). The inset shows the best optical view of the area near the Galactic center, which is still heavily confused by stars along the line of sight. *(Space Telescope Science Institute; NASA)*

14.6 The Galactic Center

6 Theory predicts that the Galactic bulge, and especially the region close to the Galactic center, should be densely populated with billions of stars. However, we are unable to see this region of our Galaxy—the interstellar medium in the Galactic disk shrouds what otherwise would be a stunning view. Figure 14.19 shows the (optical) view we do have of the region of the Milky Way toward the Galactic center, in the general direction of the constellation Sagittarius.

We can peer more deeply into the central regions of our Galaxy using infrared and radio techniques. Infrared observations (Figure 14.20a) indicate that the heart of our Galaxy harbors roughly 50,000 stars per cubic parsec. That's a stellar density about a million times greater than in our solar neighborhood, high enough that stars must experience frequent close encounters and even collisions. Infrared radiation has also been detected from what appear to be huge clouds rich in dust. In addition, radio observations indicate a ring of molecular gas nearly 400 pc across, containing some 30,000 solar masses of material and rotating at about 100 km/s. The origin of this ring is unclear, but researchers suspect that the gravitational influence of our Galaxy's elongated, rotating bulge may well be involved. The ring surrounds a bright radio source that marks the Galactic center.

High-resolution radio observations show more structure on small scales. Figure 14.20(b) shows the bright radio source Sagittarius A, which lies at the center of the boxed regions in Figures 14.19 and Figure 14.20(a), and, we think, at the center of our Galaxy. On

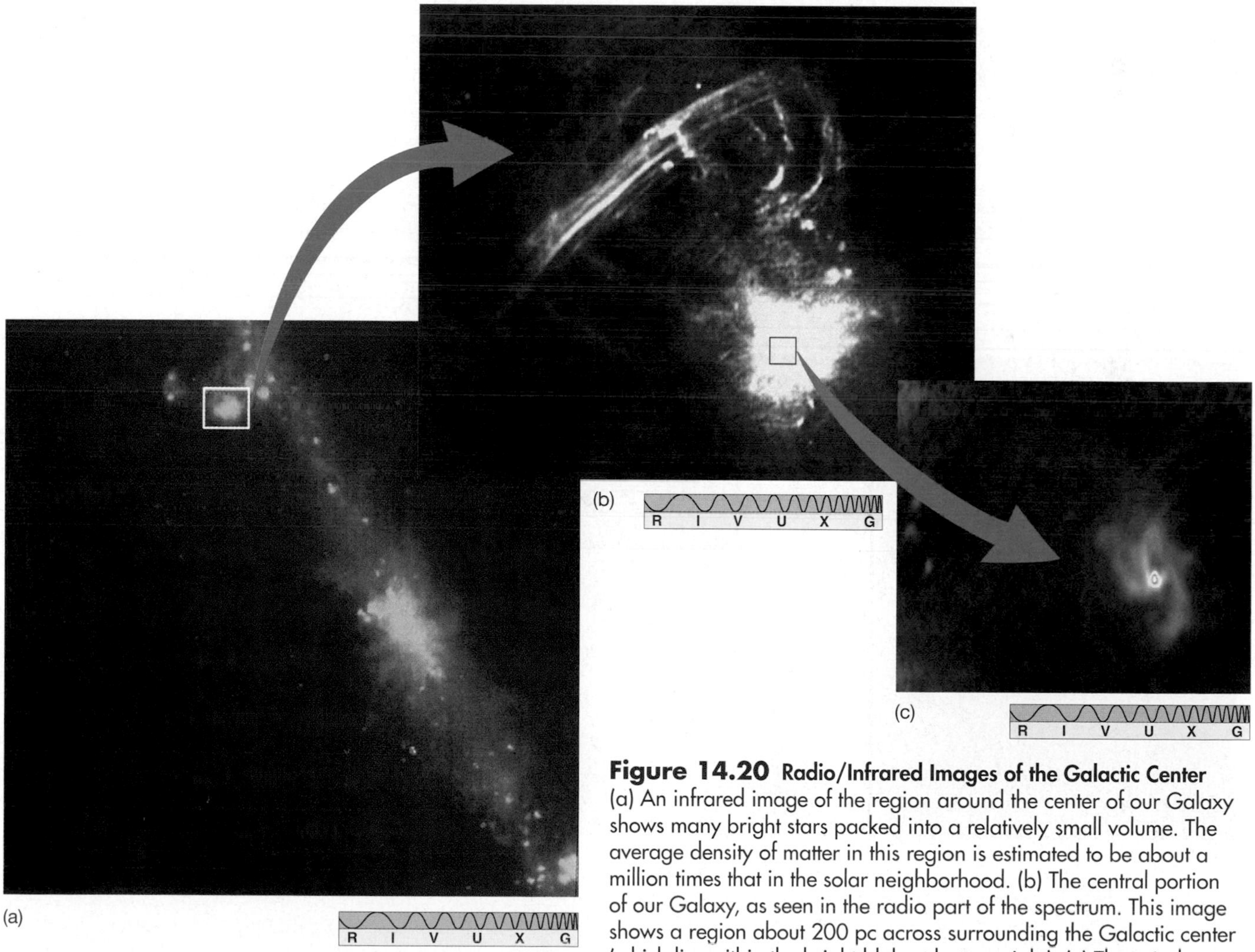

Figure 14.20 **Radio/Infrared Images of the Galactic Center** (a) An infrared image of the region around the center of our Galaxy shows many bright stars packed into a relatively small volume. The average density of matter in this region is estimated to be about a million times that in the solar neighborhood. (b) The central portion of our Galaxy, as seen in the radio part of the spectrum. This image shows a region about 200 pc across surrounding the Galactic center (which lies within the bright blob at bottom right). (c) The spiral pattern of radio emission arising from the very center of the Galaxy. The data suggest a rotating ring of matter only 5 pc across. *(NASA; NRAO)*

a scale of about 100 pc, extended filaments can be seen. Their presence suggests to many astronomers that strong magnetic fields operate in the vicinity of the center, creating structures similar in appearance (but much larger) to those observed on the active Sun. On even smaller scales (Figure 14.20c), the observations indicate a rotating ring or disk of matter only a few parsecs across.

At the very center of our Galaxy is a remarkable object with the odd-sounding name Sgr A* (pronounced "saj-ay-star"). Its total energy output (at all wavelengths) is estimated to be about 10^{33} W, more than a million times the luminosity of the Sun. Most of its energy is emitted in the radio and infrared part of the spectrum, although it is also detectable at X- and gamma-ray wavelengths.

What could cause all this activity? The leading candidate is a massive black hole at the Galactic center. VLBI observations imply that Sgr A* is small—no more than 10 A.U. across—and infrared spectroscopic observations of stars and gas within a few arc seconds of Sgr A* indicate that the orbital speed increases closer to the center, just as one would expect for matter orbiting a very massive object. The accumulated data imply that Sgr A* contains between two and three million solar masses. However, even with this large mass, if Sgr A* is a genuine black hole, the size of its event horizon is still less than 0.05 A.U. Such a small region, 8 kpc away, is currently unresolvable with any telescope now in existence.

The hole itself is not the source of the energy, of course. Instead, the radiation we see comes from a vast accretion disk of matter spiraling into the hole. ∞ (Sec. 13.3, 13.7) An adaptive-optics infrared imaging camera has spotted a bright source very close to Sgr A* that seems to vary with a 10-minute period, and could be a hot spot on the accretion disk. The strong magnetic fields are thought to be generated by the swirling motion within the accretion disk itself. In the late 1990s, the *Compton Gamma Ray Observatory* found indirect evidence for a fountain of antimatter, possibly produced by violent processes close to the event horizon, gushing

Figure 14.21 Galactic Center Zoom Six artist's conceptions, each centered on the Galactic center and each increasing in resolution by a factor of 10. Frame (a) shows the same scene as Figure 14.13. Frame (f) is a rendition of a vast whirlpool within the innermost parsec of our Galaxy. The data imaged in Figure 14.20 do not closely match these artistic renderings because the Figure 14.20 view is parallel to the Galactic disk, whereas these six paintings portray an idealized version perpendicular to that disk. *(L. Chaisson)*

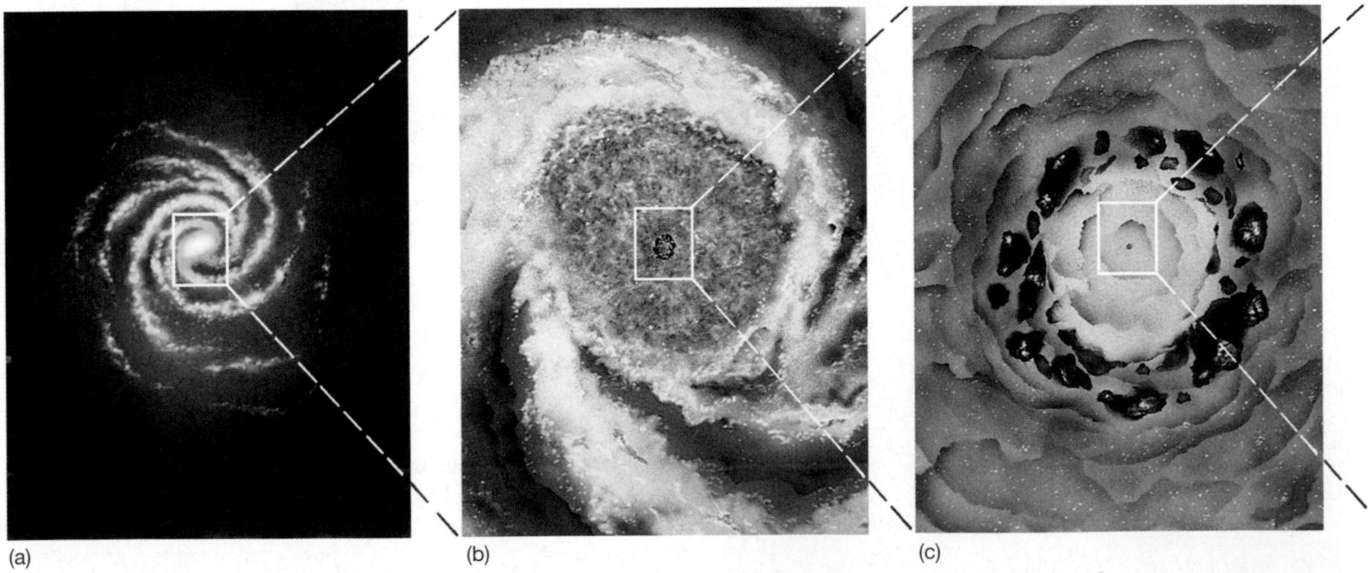

into the halo more than a thousand parsecs beyond the Galactic center.

Figure 14.21 places these findings into a simplified perspective. Each frame is centered on the Galaxy's core, and each increases in resolution by a factor of 10. Frame (a) renders the Galaxy's overall shape, as painted in Figure 14.13. The scale of this frame measures about 100 kpc from top to bottom. Frame (b) spans a distance of 10 kpc from top to bottom and is nearly filled by the great circular sweep of the innermost spiral arm. Moving in to a 1-kpc span, frame (c) depicts the 400-pc ring of matter mentioned earlier. The dark blobs represent giant molecular clouds, the pink patches emission nebulae associated with star formation within those clouds.

In frame (d), at 100 pc, a pinkish region of ionized gas surrounds the reddish heart of the Galaxy. The source of energy producing this vast ionized cloud is assumed to be related to the activity in the Galactic center. Frame (e), spanning 10 pc, depicts the tilted, spinning whirlpool of hot (10^4 K) gas that marks the center of our Galaxy. The innermost part of this gigantic whirlpool is painted in frame (f), in which a swiftly spinning, white-hot disk of gas with temperatures in the millions of kelvins nearly engulfs a massive black hole too small in size to be pictured (even as a minute dot) on this scale.

If our knowledge of the Galaxy's center seems sketchy, that's because it is sketchy. Astronomers are still deciphering the clues hidden within its invisible radiation. We are only beginning to appreciate the full magnitude of this strange realm deep in the heart of the Milky Way.

✓ Concept Check

- What is the most likely explanation of the energetic events observed at the Galactic center?

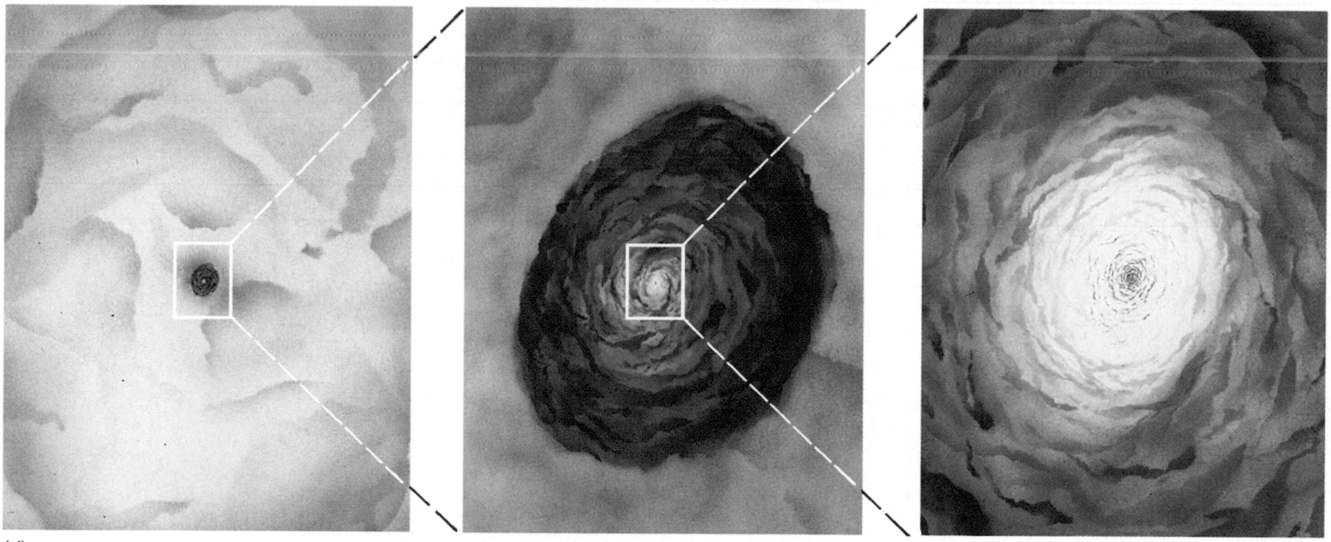

(d) (e) (f)

Chapter Review

Summary

A **galaxy** (p. 352) is a huge collection of stellar and interstellar matter isolated in space and bound together by its own gravity. Because we live within it, the **Galactic disk** (p. 352) of our own **Milky Way Galaxy** (p. 352) appears as a broad band of light across the sky—the Milky Way. Near the center, the Galactic disk thickens into the **Galactic bulge** (p. 353). The disk is surrounded by a roughly spherical **Galactic halo** (p. 353) of old stars and star clusters. Our Galaxy, like many others visible in the sky, is a **spiral galaxy** (p. 353).

The halo can be studied using **variable stars** (p. 354), whose luminosity changes with time. **Pulsating variable stars** (p. 355) vary in brightness in a repetitive and predictable way. Two types of pulsating variable stars of great importance to astronomers are **RR Lyrae variables** (p. 355) and **Cepheid variables** (p. 355). All RR Lyrae stars have roughly the same luminosity. For Cepheids, astronomers can determine the luminosity by measuring the pulsation period and using the **period–luminosity relationship** (p. 356). The brightest Cepheids can be seen at distances of millions of parsecs, extending the cosmic distance ladder well beyond our own Galaxy.

In the early twentieth century, Harlow Shapley used RR Lyrae stars to determine the distances to many of the Galaxy's globular clusters. He found that the clusters have a roughly spherical distribution in space, but the center of the sphere lies far from the Sun. The globular clusters are now known to map out the true extent of the luminous portion of the Milky Way Galaxy. The center of their distribution is close to the **Galactic center** (p. 357), which lies about 8 kpc from the Sun.

Disk and halo stars differ in their spatial distributions, ages, colors, and orbital motion. The luminous portion of our Galaxy has a diameter of about 30 kpc. In the vicinity of the Sun, the Galactic disk is about 300 pc thick. The halo lacks gas and dust, so no stars are forming there. All halo stars are old. The gas-rich disk is the site of current star formation and contains many young stars. Stars and gas within the Galactic disk move on roughly circular orbits around the Galactic center. Stars in the halo and bulge move on largely random three-dimensional orbits that pass repeatedly through the disk plane but have no preferred orientation. Halo stars appeared early on, before the Galactic disk took shape, when there was still no preferred orientation for their orbits. As the gas and dust formed a rotating disk, stars that formed in the disk inherited its overall spin and so moved on circular orbits in the Galactic plane, as they do today.

Radio observations clearly reveal the extent of our Galaxy's **spiral arms** (p. 361), regions of the densest interstellar gas where star formation is taking place. The spirals cannot be "tied" to the disk material, as the disk's differential rotation would have wound them up long ago. Instead, they may be **spiral density waves** (p. 362) that move through the disk, triggering star formation as they pass by. Alternatively, the spirals may arise from **self-propagating star formation** (p. 364), when shock waves produced by the formation and evolution of one generation of stars trigger the formation of the next.

The Galactic **rotation curve** (p. 366) plots the orbital speed of matter in the disk against distance from the Galactic center. By applying Newton's laws of motion, astronomers can determine the mass of the Galaxy. They find that the Galactic mass continues to increase beyond the radius defined by the globular clusters and the spiral structure we observe, indicating that our Galaxy has an invisible **dark halo** (p. 367). The **dark matter** (p. 367) making up this dark halo is of unknown composition. Leading candidates include low-mass stars and exotic subatomic particles. Recent attempts to detect stellar dark matter have used the fact that a faint foreground object can occasionally pass in front of a more distant star, deflecting the star's light and causing its apparent brightness to increase temporarily. This deflection is called **gravitational lensing** (p. 367).

Astronomers working at infrared and radio wavelengths have uncovered evidence for energetic activity within a few parsecs of the Galactic center. The leading explanation is a massive black hole at the heart of our Galaxy.

Review and Discussion

1. What do the globular clusters tell us about our galaxy?
2. How are Cepheid variables used in determining distances?
3. Roughly how far out into space can we use Cepheids to measure distance?
4. What important discoveries were made early in this century using RR Lyrae variables?
5. Why can't we study the central regions of the Galaxy using optical telescopes?
6. Of what use is radio astronomy in the study of Galactic structure?
7. Contrast the motions of disk and halo stars.

8. Explain why galactic spiral arms are believed to be regions of recent and ongoing star formation.

9. Describe the motion of interstellar gas as it passes through a spiral density wave.

10. What is self-propagating star formation?

11. What do the red stars in the Galactic halo tell us about the history of the Milky Way Galaxy?

12. What does the rotation curve of our Galaxy tell us about its total mass?

13. What evidence is there for dark matter in the Galaxy?

14. What are some possible explanations for dark matter?

15. Why do astronomers believe that a supermassive black hole lies at the center of the Milky Way Galaxy?

True or False?

____ **1.** Globular clusters trace out the large-scale structure of the Galactic disk.

____ **2.** Cepheid variables can be used to determine the distances to the nearest galaxies.

____ **3.** RR Lyrae stars are eclipsing binary systems.

____ **4.** The Galactic halo contains about as much gas and dust as the Galactic disk.

____ **5.** The Galactic disk contains only old stars.

____ **6.** Population I objects are found only in the Galactic halo.

____ **7.** Up until the 1930s, the main error made in determining the size of the Galaxy was the neglect of interstellar absorption.

____ **8.** Astronomers use 21-cm radiation to study Galactic molecular clouds.

____ **9.** Radio techniques are capable of mapping the entire Galactic disk.

____ **10.** In the neighborhood of the Sun, the Galaxy's spiral density waves rotate more slowly than the overall Galactic rotation.

____ **11.** The mass of the Galaxy is determined by counting stars.

____ **12.** Most of the mass of our Galaxy exists in the form of dark matter.

____ **13.** Dark matter is now known to be due to large numbers of black holes.

____ **14.** A million-solar-mass black hole could account for the unusual properties of the Galactic center.

____ **15.** The Galactic center has been extensively studied at visible and ultraviolet wavelengths.

Fill in the Blank

1. One difficulty in studying our Galaxy in its entirety is that the Sun lies within the _______.

2. Herschel's attempt to map the Milky Way by counting stars led to an estimate of the Galaxy's size that was too _______.

3. The highly flattened, circular part of the Galaxy is called the Galactic _______.

4. The roughly spherical region of faint old stars in which the rest of the Galaxy is embedded is the Galactic _______.

5. The _______ of Cepheids and RR Lyrae stars are observed to vary in a periodic fashion.

6. Cepheid pulsation periods range from _______ to _______.

7. Cepheids and RR Lyrae variables lie in a region of the H–R diagram called the _______.

8. According to the period–luminosity relationship, the longer the pulsation period of a Cepheid, the _______ its luminosity.

9. Harlow Shapley used _______ to determine distances to the globular clusters.

10. The Sun lies roughly _______ pc from the Galactic center.

11. The orbital speed of the Sun around the Galactic center is _______ km/s.

12. The direction of motion of halo objects at any particular location is _______.

13. The original cloud of gas from which the Galaxy formed probably had a size and shape similar to those of the present Galactic _______.

14. Rotational speeds in the outer part of the Galaxy are _______ than would be expected on the basis of observed stars and gas, indicating the presence of _______.

15. Spectroscopic observations of gas and stars near the Galactic center indicate that material there is orbiting at extremely _______ speeds.

Problems

1. At one time some astronomers claimed that the Andromeda "nebula" (Figure 14.2a) was a star-forming region lying within our own Galaxy. Calculate the angular diameter of a prestellar nebula of radius 100 A.U., lying 100 pc from Earth. ∞ (Sec. P.5) How close would the nebula have to lie to Earth for it to have the same angular diameter (roughly 6°) as Andromeda?
2. What is the greatest distance at which an RR Lyrae star of absolute magnitude 0 could be seen by a telescope capable of detecting objects as faint as 20th magnitude? ∞ (Sec. 10.2)
3. A typical Cepheid variable is 100 times brighter than a typical RR Lyrae star. On average, how much farther away than RR Lyrae stars can Cepheids be used as distance-measuring tools?
4. The *Hubble Space Telescope* can just detect a star like the Sun at a distance of 100,000 pc. The brightest Cepheids have luminosities 30,000 times greater than that of the Sun. How far away can *HST* see these Cepheids?
5. Calculate the proper motion (in arc seconds/year) of a globular cluster with a transverse velocity (relative to the Sun) of 200 km/s and a distance of 3 kpc. Do you think that this motion might be measurable? ∞ (Sec. 10.1)
6. Calculate the total mass of the Galaxy lying within 20 kpc of the Galactic center if the rotation speed at that radius is 240 km/s.
7. Using the data presented in Figure 14.17, calculate how long it takes the Sun to "lap" stars orbiting 15 kpc from the Galactic center. How long does it take matter at 5 kpc to lap us?
8. A density wave made up of two spiral arms is moving through the Galactic disk. At the 8-kpc radius of the Sun's orbit around the Galactic center, the wave's speed is 120 km/s, and the Galactic rotation speed is 220 km/s. Calculate how many times the Sun has passed through a spiral arm since the Sun formed 4.6 billion years ago.
9. Given the data in the previous question and the fact that O-type stars live at most 10 million years before exploding as supernovae, calculate the maximum distance at which an O-type star (orbiting at the Sun's distance from the Galactic center) can be found from the density wave in which it formed.
10. Material at an angular distance of 0.2″ from the Galactic center is observed to have an orbital speed of 1100 km/s. If the Sun's distance to the Galactic center is 8 kpc, and the material's orbit is circular and is seen edge-on, calculate the mass of the object around which the material is orbiting.

Project

1. If you are far from city lights, look for a hazy band of light arching across the sky. This is our edgewise view of the Milky Way Galaxy. The Galactic center is located in the direction of the constellation Sagittarius, highest in the sky during the summer but visible from spring through fall. Look at the band making up the Milky Way and notice dark regions. These are relatively nearby dust clouds. Sketch what you see. Look for faint fuzzy spots in the Milky Way and note their positions in your sketch. Draw in the major constellations for reference. Compare your sketch with a map of the Milky Way in a star atlas. How many dust clouds did you discover? Can you identify the faint fuzzy spots?

15 NORMAL GALAXIES

The Large-Scale Structure of the Universe

LEARNING GOALS

Studying this chapter will enable you to:

1. Describe the basic properties of the main types of normal galaxies.
2. Discuss the distance-measurement techniques that enable astronomers to map the universe beyond our Milky Way.
3. Summarize what is known about the large-scale distribution of galaxies in the universe.
4. Describe some of the methods used to determine the masses of distant galaxies.
5. Explain why astronomers think that most of the matter in the universe is invisible.
6. Discuss some theories of how galaxies form and evolve.
7. State Hubble's law and explain how it is used to derive distances to the most remote objects in the observable universe.

This image is known as the Hubble Deep Field-South, a long (approximately 100-hour) exposure taken in 1998 by the *Hubble Space Telescope* over the course of 10 days of orbits. Like the original Hubble Deep Field-North (reproduced as Figure 15.22), this new image is a very narrow "core sample" of assorted cosmic objects distributed along a thin corridor of space approximately 3000 Mpc in length, or 10 billion light-years. By "deep," we mean dim and distant, for this is the faintest photograph ever made; it reaches 30th magnitude, or about four billion times fainter than can be seen by the human eye. In all, the image—which takes us far out in space and far back in time—implies that about 40 billion galaxies inhabit the observable universe. *(HDF–S Team/NASA)*

We know of literally millions of galaxies beyond our own—many smaller than the Milky Way, some comparable in size, a few much larger. All are vast, gravitationally bound assemblages of stars, gas, dust, dark matter, and radiation separated from us by almost incomprehensibly large distances. Even a modest-sized galaxy harbors more stars than the number of people who have ever lived on Earth. The light we receive tonight from the most distant galaxies was emitted long before Earth existed. By comparing and classifying the properties of galaxies, astronomers have begun to understand their complex dynamics. By mapping out their distribution in space, astronomers trace out the immense realms of the universe. The galaxies remind us that our position in the universe is no more special than that of a boat adrift at sea.

15.1 Hubble's Galaxy Classification

1 Figure 15.1 shows a vast expanse of space lying about 100 million pc from Earth. Almost every patch or point of light in this figure is a separate galaxy—several hundred can be seen in just this one photograph. Some of the blobs of light in Figure 15.1 are spiral galaxies like the Milky Way Galaxy and Andromeda. Others, however, are definitely not spirals—no disks or spiral arms can be seen. Even when we take into account their different orientations in space, galaxies do *not* all look the same.

The American astronomer Edwin Hubble was the first to categorize galaxies in a comprehensive way. Working with the then recently completed 2.5-m optical telescope on Mount Wilson in California in 1924, he classified the galaxies he saw into four basic types—*spirals*, *barred spirals*, *ellipticals*, and *irregulars*—solely on the basis of appearance. Many modifications and refinements have been incorporated over the years, but the basic **Hubble classification scheme** is still widely used today.

(a) (b)

Figure 15.1 **Coma Cluster** (a) A collection of many galaxies, each consisting of hundreds of billions of stars. Called the Coma Cluster, this group of galaxies lies more than 100 million pc from Earth. (The blue spiked object at top right is a nearby star; virtually every other object visible is a galaxy.) (b) A recent *Hubble Space Telescope* image of part of the Coma Cluster. *(AURA; NASA)*

Spirals

We saw several examples of **spiral galaxies** in Chapter 14—for example, our own Milky Way Galaxy and our neighbor Andromeda. All galaxies of this type contain a flattened galactic disk in which spiral arms are found, a central galactic bulge with a dense nucleus, and an extended halo of faint, old stars. ∞ (Sec. 14.3)

In Hubble's scheme, a spiral galaxy is denoted by the letter S and classified as type a, b, or c according to the size of its central bulge—Type Sa galaxies having the largest bulges, Type Sc the smallest (Figure 15.2). The tightness of the spiral pattern is quite well correlated with the size of the bulge (although the correspondence is not perfect). Type Sa spiral galaxies tend to have tightly wrapped, almost circular, spiral arms, Type Sb galaxies typically have more open spiral arms, while Type Sc spirals often have loose, poorly defined spiral structure. The arms also tend to become more "knotty," or clumped, in appearance as the spiral pattern becomes more open.

The bulges and halos of spiral galaxies contain large numbers of reddish old stars and globular clusters, similar to those observed in our own Galaxy and in Andromeda. ∞ (Sec. 14.1) Most of the light from spirals, however, comes from A- through G-type stars in the disk, giving these galaxies an overall whitish glow. Like the disk of the Milky Way, the flat galactic disks of typical spiral galaxies are rich in gas and dust. The 21-cm radio radiation emitted by spirals betrays the presence of interstellar gas, and obscuring dust lanes are clearly visible in many systems (see Figures 15.2b and c). Stars are forming within the spiral arms, which contain numerous emission nebulae and newly formed O- and B-type stars. ∞ (Sec. 14.5) The arms appear bluish because of the presence of bright blue O- and B-type stars there. Type Sc galaxies contain the most interstellar gas and dust, Sa galaxies the least.

Most spirals are not seen face-on as in Figure 15.2. Many are tilted with respect to our line of sight, making their spiral structure hard to discern, However, we do not need to see spiral arms to classify a galaxy as a spiral. The presence of the disk, with its gas, dust, and newborn stars, is sufficient. For example, the galaxy shown in Figure 15.3 is classified as a spiral because of the clear line of obscuring dust seen along its midplane.

A variation of the spiral category in Hubble's classification scheme is the **barred-spiral galaxy**. Barred spirals differ from ordinary spirals mainly by the presence of an elongated "bar" of stellar and interstellar matter passing through the center and extending beyond the bulge, into the disk. The spiral arms project from near the ends of the bar rather than from the bulge (as they do in normal spirals). Barred spirals are designated by the letters SB and are subdivided, like the ordinary spirals, into categories SBa, SBb, and SBc, depending on the size of the bulge. Again like ordinary spirals, the

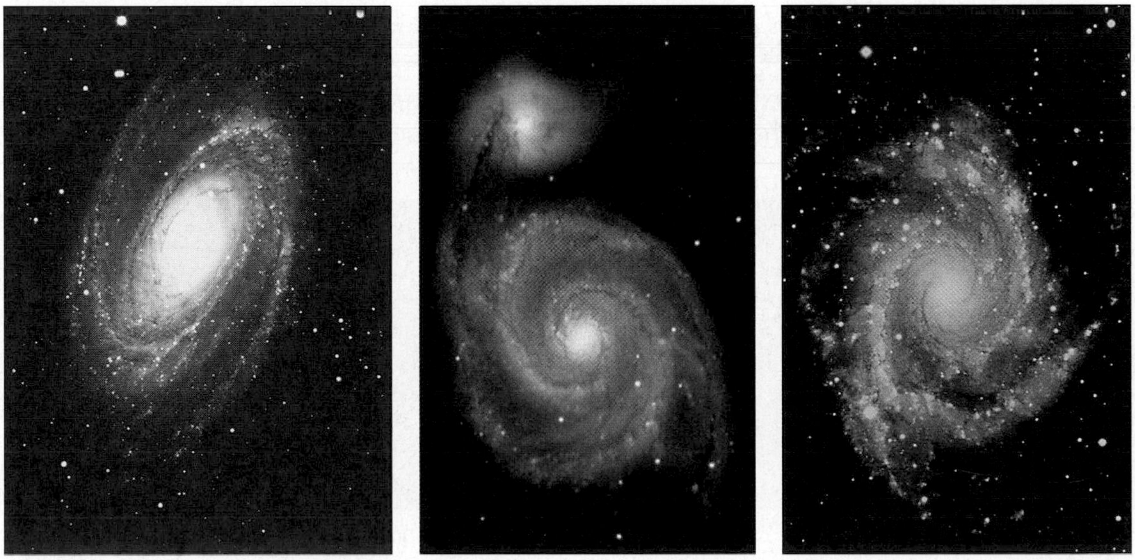

M81 Type Sa M51 Type Sb NGC 2997 Type Sc

R I V U X G

Figure 15.2 Spiral Galaxy Shapes Variation in shape among spiral galaxies. As we progress from type Sa to Sb to Sc, the bulges become smaller while the spiral arms tend to become less tightly wound. *(NOAO; Anglo-Australian Observatory, Photograph by D. Malin)*

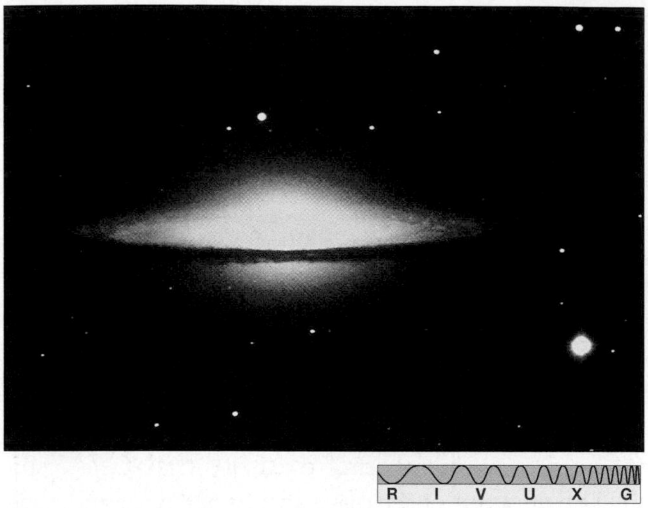

Figure 15.3 Sombrero Galaxy The Sombrero Galaxy, a spiral system seen edge-on. Officially cataloged as M104, this galaxy has a dark band composed of interstellar gas and dust. The large size of this galaxy's central bulge marks it as type Sa, even though its spiral arms cannot be seen. *(AURA)*

tightness of the spiral pattern is correlated with bulge size. Figure 15.4 shows the variation among barred-spiral galaxies. In the case of the SBc category, it is often hard to tell where the bar ends and the spiral arms begin.

The recent findings that the bulge of our own Galaxy is elongated suggest that the Milky Way is a barred spiral, of type SBb or SBc. ∞ (Sec. 14.3)

Ellipticals

Unlike the spirals, **elliptical galaxies** have no spiral arms and, in most cases, no obvious galactic disk—in fact, other than a dense central nucleus, they often exhibit little internal structure of any kind. Denoted by the letter E, these systems are subdivided according to how elliptical they are. The most circular are designated E0, slightly flattened systems are labeled E1, and so on, all the way to the most elongated ellipticals, of type E7 (Figure 15.5).

There is a large range in both the size and the number of stars contained in elliptical galaxies. The largest elliptical galaxies are much larger than our own Milky Way Galaxy. These *giant ellipticals* can range up to a few megaparsecs across and contain trillions of stars. At the other extreme, *dwarf ellipticals* may be as small as 1 kpc in diameter and contain fewer than a million stars. The significant observational differences between giant and dwarf ellipticals have led many astronomers to conclude that these galaxies are members of separate classes, with quite different formation histories and stellar content. The dwarfs are by far the most common type of ellipticals, outnumbering their brighter counterparts by about ten to one. However, most of the *mass* that exists in the form of elliptical galaxies is contained in the larger systems.

Lack of spiral arms is not the only difference between spirals and ellipticals. Most ellipticals also contain little or no gas and dust and display no evidence of young stars or ongoing star formation. Like the halo of our own Galaxy, ellipticals are made up mostly of old, reddish, low-mass stars. Indeed, having no disk, gas, or dust, elliptical galaxies are, in a sense, "all halo." Again like the halo of our Galaxy, the orbits of stars in ellipticals are disordered, exhibiting little or no overall rotation; objects move in all directions, not in regular, circular paths as in our Galaxy's disk. Apparently all, or nearly all, of the interstellar gas within elliptical galaxies was swept up into stars—or out of the galaxy—long ago, before a disk had a chance to

Figure 15.4 Barred-spiral Galaxy Shapes Variation in shape among barred-spiral galaxies. The variation from SBa to SBc is similar to that for the spirals in Figure 15.2, except that now the spiral arms begin at either end of a bar through the galactic center. *(Smithsonian Astrophysical Observatory; University of Hawaii Board of Regents; W. Keel)*

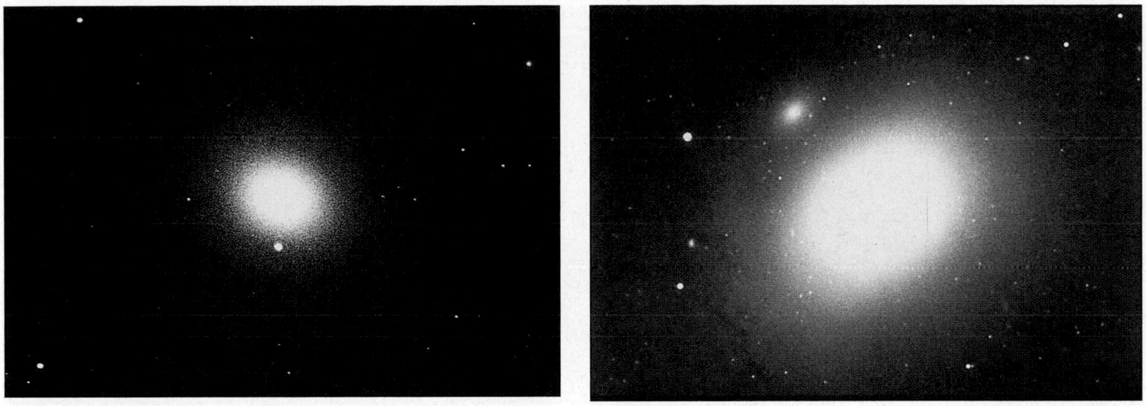

(a) M49 Type E1 (b) M84 Type E3

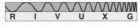

Figure 15.5 Elliptical Galaxy Shapes Variation in shape among elliptical galaxies. (a) The E1 galaxy M49 is nearly circular in appearance. (b) M84 is a slightly more elongated elliptical galaxy. It is classified as E3. Both of these galaxies lack spiral structure, and neither shows evidence of interstellar matter. *(AURA)*

form, leaving stars in random orbits, with no loose gas and dust for the creation of future stellar generations.

Some giant ellipticals are exceptions to many of the foregoing general statements about elliptical galaxies, as they have been found to contain disks of gas and dust in which stars are forming. Some astronomers speculate that these galaxies may really be otherwise "normal" ellipticals that have collided and merged with a companion spiral system.

Intermediate between the E7 ellipticals and the Sa spirals in the Hubble classification is a class of galaxies that show evidence of a thin disk and a flattened bulge but that contain no gas and no spiral arms. Two such objects are shown in Figure 15.6. These galaxies are known as **S0 galaxies** if no bar is evident and **SB0 galaxies** if a bar is present. They look a little like spirals whose dust and gas have been stripped away, leaving behind just a stellar disk. Observations in recent years have shown that many normal elliptical galaxies have faint disks within them, like the S0 galaxies. As with the S0s, the origin of these disks is uncertain, but some researchers suspect that S0s and ellipticals may be closely related.

Irregulars

The final galaxy class identified by Hubble is a catchall category—**irregular galaxies**—so named because their visual appearance does not allow us to place them into any of the other categories just discussed. Irregulars tend to be rich in interstellar matter and young, blue stars, but they lack any regular structure, such as well-defined spiral arms or central bulges. They are divided into two subclasses—Irr I galaxies and Irr II galaxies. The Irr I galaxies often look like misshapen spirals. The much rarer Irr II galaxies (Figure 15.7), in addition to their irregular shape, have other peculiarities, often exhibiting a distinctly explosive or filamentary appearance. Their appearance once led astronomers to suspect that violent

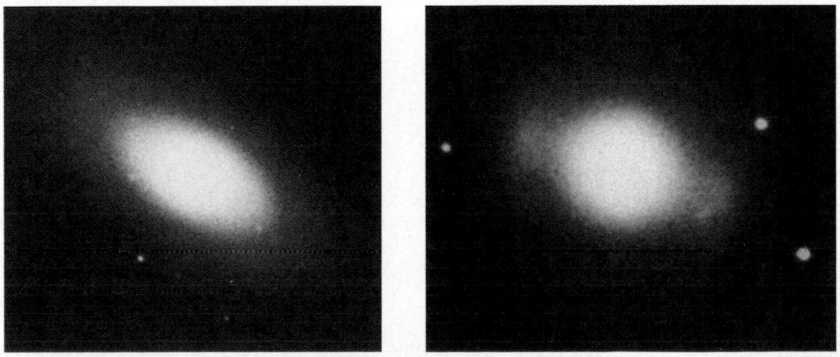

(a) NGC 1201 Type S0 (b) NGC 2859 Type SB0

Figure 15.6 S0 Galaxies (a) S0 galaxies contain a disk and a bulge but no interstellar gas and no spiral arms. They are in many respects intermediate between E7 ellipticals and Sa spirals in their properties. (b) SB0 galaxies are similar to S0 galaxies, except for a bar of stellar material extending beyond the central bulge. *(Palomar Observatory/California Institute of Technology)*

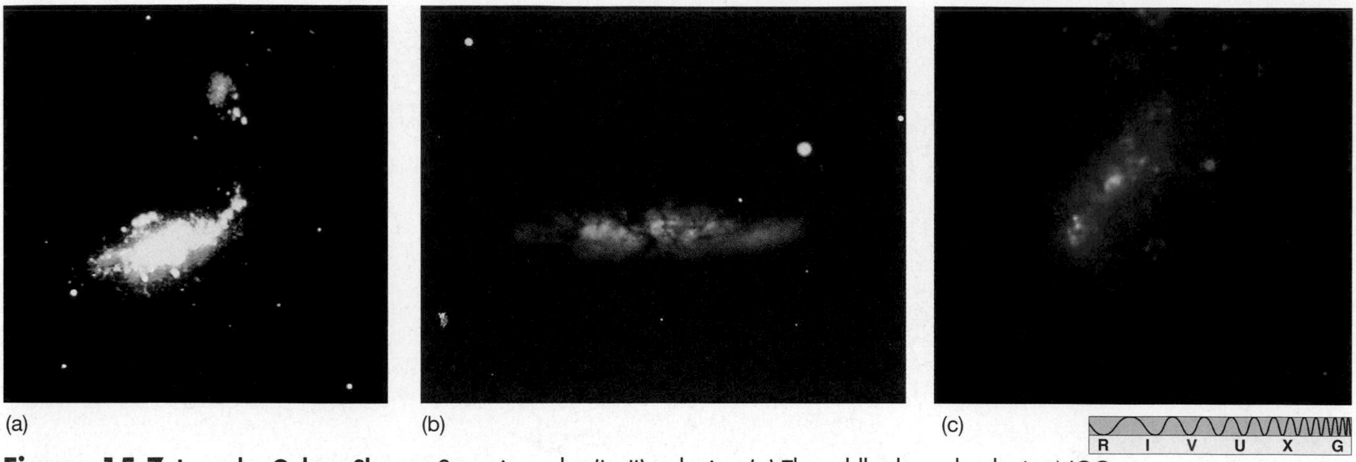

Figure 15.7 **Irregular Galaxy Shapes** Some irregular (Irr II) galaxies. (a) The oddly shaped galaxies NGC 4485 and NGC 4490 may be close to one another and interacting gravitationally. (b) The galaxy M82 seems to show an explosive appearance, although interpretations remain uncertain. (c) Many irregular galaxies are small and dim, but this one, NGC 4449, is comparable in both size and luminosity to the Milky Way Galaxy. *(Association of Universities for Rearch in Astronomy; Palomar Observatory/California Institute of Technology; NASA)*

events had occurred within them. However, it now seems more likely that in some (but probably not all) cases, we are seeing the result of a close encounter or collision between two previously "normal" systems.

Irregular galaxies tend to be smaller than spirals but somewhat larger than dwarf ellipticals. They typically contain between 10^8 and 10^{10} stars. The smallest are called *dwarf irregulars*. As with elliptical galaxies, the dwarf type is the most common irregular. Dwarf ellipticals and dwarf irregulars occur in approximately equal numbers and together make up the vast majority of galaxies in the universe. They are often found close to a larger "parent" galaxy.

Figure 15.8 shows the **Magellanic Clouds**, a famous pair of Irr I galaxies that orbit the Milky Way Galaxy. They are shown to proper scale in Figure 14.13. Studies of

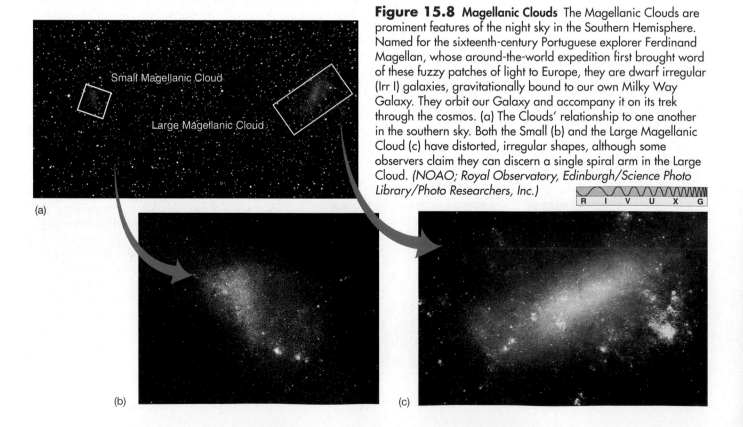

Figure 15.8 **Magellanic Clouds** The Magellanic Clouds are prominent features of the night sky in the Southern Hemisphere. Named for the sixteenth-century Portuguese explorer Ferdinand Magellan, whose around-the-world expedition first brought word of these fuzzy patches of light to Europe, they are dwarf irregular (Irr I) galaxies, gravitationally bound to our own Milky Way Galaxy. They orbit our Galaxy and accompany it on its trek through the cosmos. (a) The Clouds' relationship to one another in the southern sky. Both the Small (b) and the Large Magellanic Cloud (c) have distorted, irregular shapes, although some observers claim they can discern a single spiral arm in the Large Cloud. *(NOAO; Royal Observatory, Edinburgh/Science Photo Library/Photo Researchers, Inc.)*

TABLE 15.1 Basic Galaxy Properties by Type

	SPIRAL/BARRED SPIRAL (S/SB)	ELLIPTICAL (E)	IRREGULAR (Irr)
Shape and structural properties	Highly flattened disk of stars and gas, containing spiral arms and thickening to central bulge. Sa and SBa galaxies have largest bulges, the least obvious spiral structure, and roughly spherical stellar halos. SB galaxies have an elongated central "bar" of stars and gas.	No disk. Stars smoothly distributed through an ellipsoidal volume ranging from nearly spherical (E0) to very flattened (E7) in shape. No obvious substructure other than a dense central nucleus.	No obvious structure. Irr II galaxies often have "explosive" appearance.
Stellar content	Disks contain both young and old stars; halos consist of old stars only.	Contain old stars only.	Contain both young and old stars.
Gas and dust	Disks contain substantial amounts of gas and dust; halos contain little of either.	Contain little or no gas and dust.	Very abundant in gas and dust.
Star formation	Ongoing star formation in spiral arms.	No significant star formation during the last 10 billion years.	Vigorous ongoing star formation.
Stellar motion	Gas and stars in disk move in circular orbits around the galactic center; halo stars have random orbits in three dimensions.	Stars have random orbits in three dimensions.	Stars and gas have very irregular orbits.

Cepheid variables within the Clouds show them to be approximately 50 kpc from the center of our Galaxy. ∞ (Sec. 14.2) The Large Cloud contains about six billion solar masses of material and is a few kiloparsecs across. Both Magellanic Clouds contain lots of gas, dust, and blue stars, indicating current star formation. Both also contain many old stars and several old globular clusters, so we know that star formation has been going on there for a very long time.

An H–R Diagram for Galaxies?

Table 15.1 summarizes the basic characteristics of the various galaxy types. When he first developed his classification scheme, Hubble arranged the galaxy types into the "tuning fork" diagram shown in Figure 15.9. His aim in doing this was simply to indicate similarities in appearance among galaxies, not to suggest that any connection might exist, but some astronomers have since

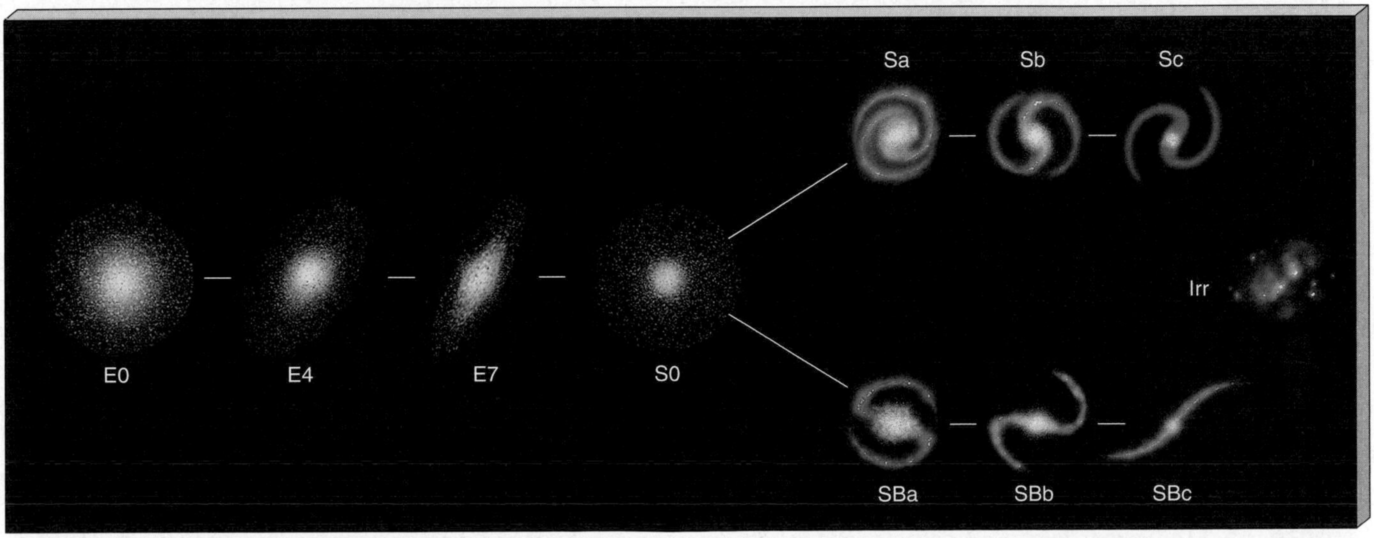

Figure 15.9 Galaxy "Fork" Diagram Hubble's "tuning fork" diagram, showing his basic galaxy classification scheme. The placement of the four basic galaxy types—ellipticals, spirals, barred spirals, and irregulars—in the diagram is suggestive, but no "evolutionary track" along the sequence (in either direction) is proven.

speculated that perhaps the diagram has deeper significance. Does some sort of evolutionary connection exist among galaxies of different types? The answer, as far as we can tell, is no. Isolated normal galaxies do *not* evolve from one type to another. Spirals are not ellipticals with arms, nor are ellipticals spirals that have somehow lost their star-forming disks. In short, astronomers know of no parent-child relationship among normal galaxies.

✓ Concept Check

- In what sense are large spirals like the Milky Way and Andromeda not representative of galaxies as a whole?

15.2 The Distribution of Galaxies in Space

Now that we have seen some of their basic properties, let us ask how galaxies are spread through the expanse of the universe beyond the Milky Way Galaxy. To begin to answer this question, we must first know the *distances* to the galaxies.

Extending the Distance Scale

2 Although some reside close enough (within about 15 Mpc) for the Cepheid variable technique to work, most known galaxies lie much farther away. ∞ (Sec. 14.2) Cepheid variables in very distant galaxies simply cannot be observed well enough, even through the world's most sensitive telescopes, to allow us to measure their luminosity and period. To extend our distance-measurement ladder, therefore, we must find some new object to study.

One way in which researchers have tackled this problem is through observations of **standard candles**—easily recognizable astronomical objects whose luminosities are confidently known. The basic idea is very simple. Once an object is identified as a standard candle—by its appearance or by the shape of its light curve, say—its luminosity can be estimated. Comparison of the luminosity with the apparent brightness then gives the object's distance, and hence the distance to the galaxy in which it resides. (Note that, other than the way in which the luminosity is determined, the Cepheid variable technique relies on identical reasoning.)

To be most useful, a standard candle must (1) have a narrowly defined luminosity, so that the uncertainty in estimating its brightness is small and (2) be bright enough to be seen at great distances. In recent years, Type I supernovae have proved particularly reliable as standard candles. ∞ (Sec. 12.5) They have remarkably consistent peak luminosities and are bright enough to be identified and measured out to distances of many hundreds of megaparsecs.

An important alternative to standard candles is known as the **Tully–Fisher relation**. Observations of spiral galaxies within a few tens of megaparsecs of the Milky Way reveal that their rotation speeds are closely

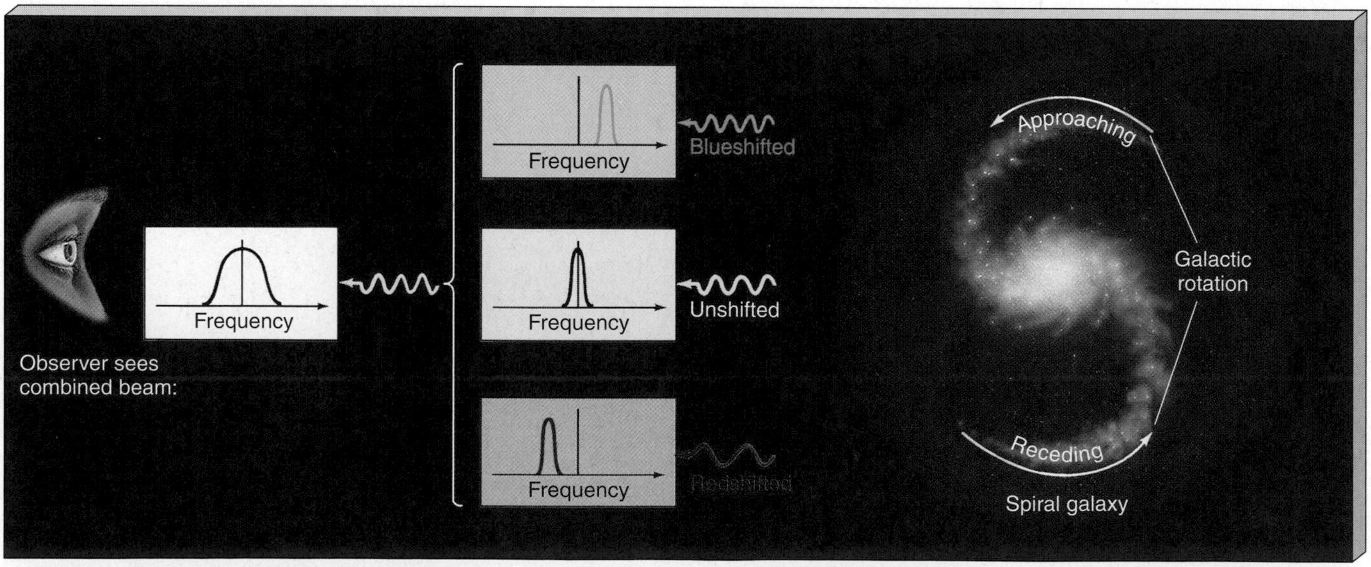

Figure 15.10 Galaxy Rotation A galaxy's rotation causes some of the radiation it emits to be blueshifted and some to be redshifted (relative to what the emission would be from an unmoving source). From a distance, when the radiation from the galaxy is combined into a single beam and analyzed spectroscopically, the redshifted and blueshifted components combine to produce a broadening of the galaxy's spectral lines. The amount of broadening is a direct measure of the rotation speed of the galaxy.

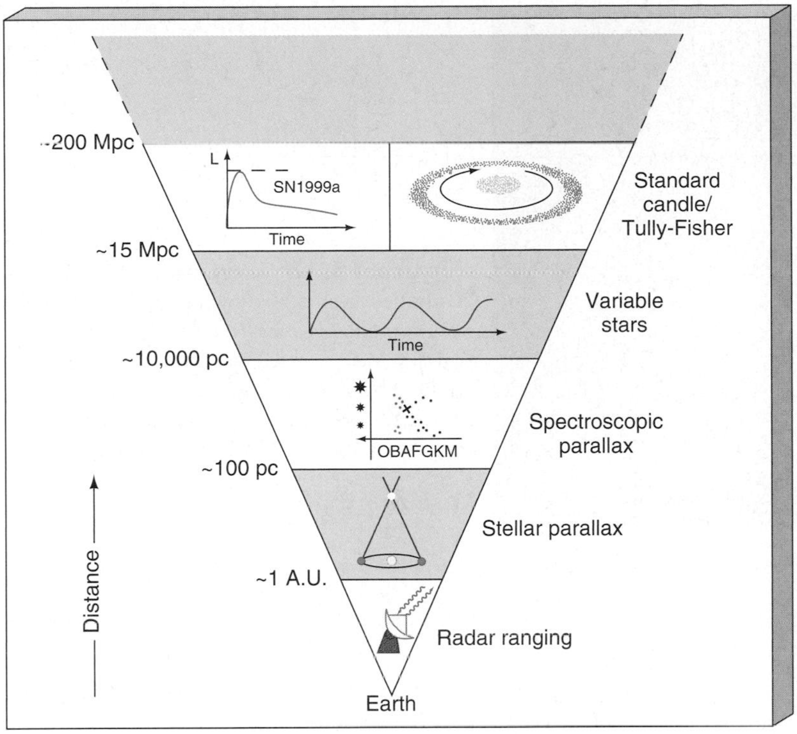

Animation

Figure 15.11 Supernova Distance Ladder An inverted pyramid summarizes the distance techniques used to study different realms of the universe. The techniques shown in the bottom four layers—radar ranging, stellar parallax, spectroscopic parallax, and variable stars—take us as far as the nearest galaxies. To go farther, we must use new techniques, the method of standard candles and the Tully–Fisher relation, each based on distances determined by the four lowest techniques.

correlated with their luminosities. Rotation speed is a measure of a spiral galaxy's total mass, so it is perhaps not surprising that this property is related to luminosity. ∞ (Sec. 14.5) What *is* surprising is just how tight the correlation is. The Tully-Fisher relation allows astronomers to obtain a remarkably accurate estimate of a spiral galaxy's luminosity simply by observing how fast it rotates. As illustrated in Figure 15.10, the rotation speed is usually determined by measuring the "smearing," or broadening, it causes in a galaxy's spectral lines. ∞ (Sec. 2.6) The Tully–Fisher relation then tells us the galaxy's luminosity, and hence (comparing the known luminosity with the measured apparent magnitude) its distance.

The Tully–Fisher relation can be used to measure distances to spiral galaxies out to about 200 Mpc, beyond which the line broadening becomes increasingly difficult to measure accurately. A somewhat similar connection, relating line broadening to a galaxy's *diameter*, exists for elliptical galaxies. Once the galaxy's diameter and angular size are known, its distance can be computed from elementary geometry. ∞ (*More Precisely P-2*) These methods bypass many of the standard candles often used by astronomers and so provide independent means of determining distances to faraway objects.

To emphasize their independence, standard candles and the Tully–Fisher relation share the fifth rung of our cosmic distance ladder (Figure 15.11). Just as with the lower rungs, the properties of these new techniques are calibrated using distances measured by more local means. However, in the process, the errors and uncertainties in each step accumulate, so the distances to the farthest objects are the least well known.

Galaxy Clusters

3 Figure 15.12 sketches the locations of all the known major astronomical objects within about 1 Mpc of the Milky Way Galaxy. Our Galaxy appears with its three

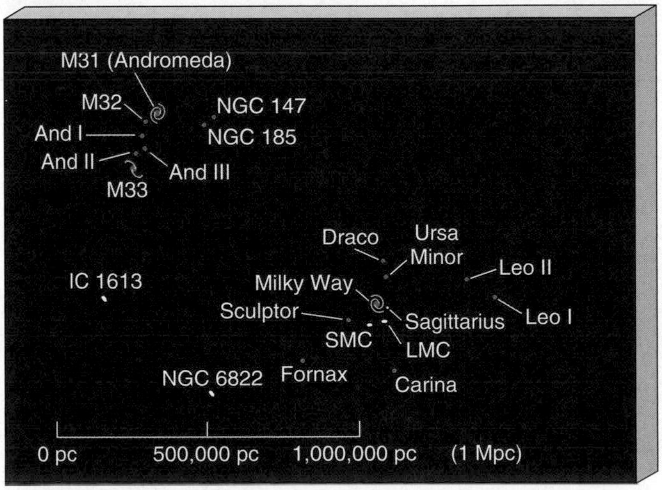

Figure 15.12 Local Group Diagram of the Local Group, made up of about 20 galaxies within approximately 1 Mpc of our Milky Way Galaxy. Only a few are spirals; most of the rest are dwarf elliptical or irregular galaxies. Spirals are shown in blue, ellipticals in pink, and irregulars in white.

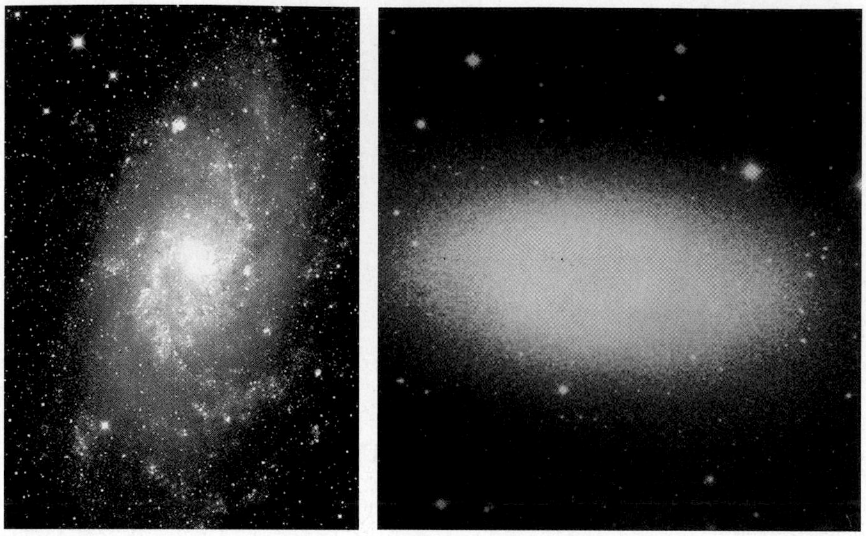

(a) (b)

Figure 15.13 Andromeda's Nearby Galaxies Two well-known neighbors of the Andromeda Galaxy (M31): (a) the spiral galaxy M33 and (b) the dwarf elliptical galaxy M32 (also visible in Figure 14.2a, a larger-scale view of the Andromeda system). *(Palomar Observatory/NASA)*

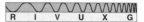

R I V U X G

known satellite galaxies—the two Magellanic Clouds and a more recently discovered companion (labeled "Sagittarius" in the figure) lying almost within our own Galactic plane. The Andromeda Galaxy, lying 900 kpc from us, is also shown. Two galactic neighbors of Andromeda are shown in Figure 15.13. M33 is a spiral, while M32 is a dwarf elliptical, easily seen in Figure 14.2(a) to the bottom right of Andromeda's central bulge.

All told, some 20 galaxies populate our Galaxy's neighborhood. Three of them (the Milky Way, Andromeda, and M33) are spirals; the rest are dwarf irregulars and ellipticals. Together, they form a small **galaxy cluster** called the *Local Group* (Figure 15.12), a new level of structure in the universe above the scale of our Galaxy. Its diameter is roughly 1 Mpc. The Milky Way Galaxy and Andromeda are by far its largest members. The Local Group is held together by the combined gravity of the galaxies it contains, like stars in a star cluster, but on a millionfold larger scale.

Moving beyond the Local Group, the next large concentration of galaxies we come to is the *Virgo Cluster* (Figure 15.14). It lies some 17 Mpc from the Milky Way. The Virgo Cluster does not contain a mere 20 galaxies, however. Rather, it houses approximately 2500 galaxies, bound by gravity into a tightly knit group about 3 Mpc across. Figure 15.15 illustrates the approximate locations of Virgo and several other galaxy clusters in our cosmic neighborhood, within about 30 Mpc of the Milky Way.

Many thousands of galaxy clusters have now been identified and cataloged, and they come in many shapes and sizes. Large, "rich" clusters like Virgo contain thousands of individual galaxies distributed fairly smoothly in space. Small clusters, such as the Local Group, contain only a few galaxies and are quite irregular in shape. A few galaxies are not members of any cluster. They are apparently isolated systems, moving alone through intercluster space.

R I V U X G

Figure 15.14 Virgo Cluster The central region of the Virgo Cluster of galaxies, about 17 Mpc from Earth. Several large spiral and elliptical galaxies can be seen. The galaxy near the center is a giant elliptical known as M86. *(National Optical Astronomy Observatories)*

Clusters of Clusters

3 Most astronomers believe that the galaxy clusters themselves are clustered, forming titanic agglomerations of matter known as **superclusters**.

Figure 15.16 shows the *Local Supercluster*, containing the Local Group, the Virgo Cluster, and most of the other clusters shown in Figure 15.15. (The direc-

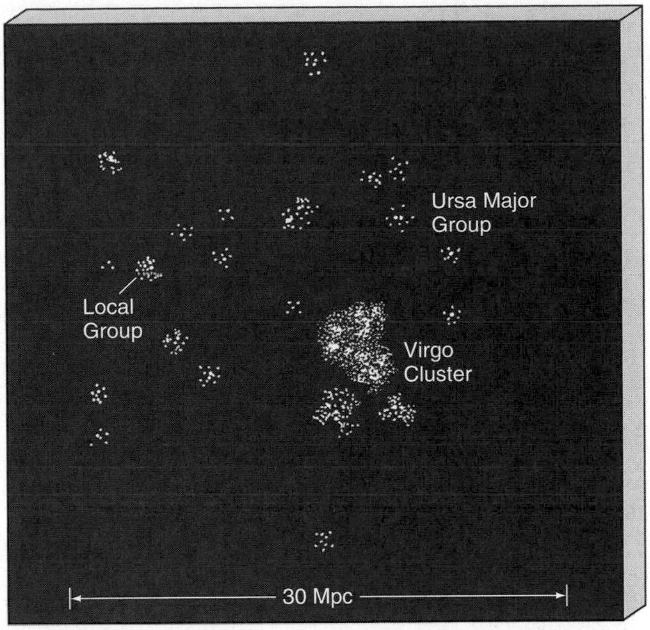

Figure 15.15 Galaxy Supercluster Schematic diagram of the location of several galaxy clusters in our part of the universe. Our Milky Way Galaxy is just one of these dots, and our Local Group is just one of the clusters.

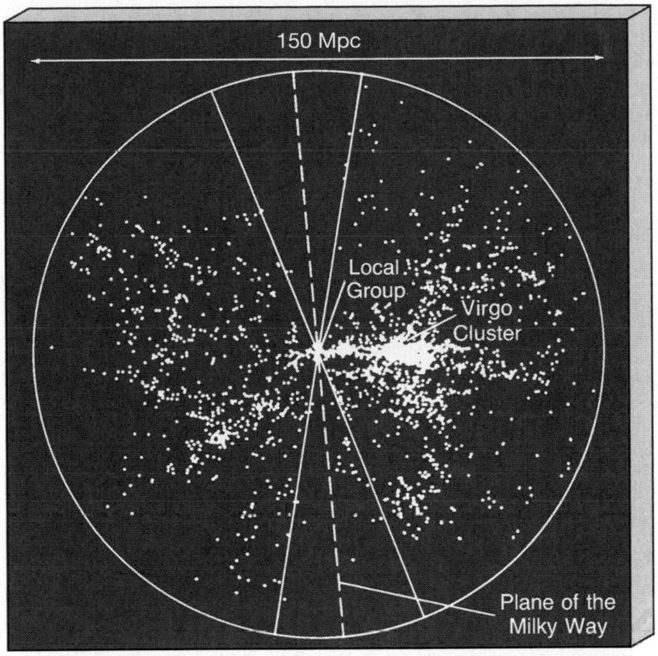

Figure 15.16 Local Supercluster The Local Supercluster. Each of the 2200 points shown represents a galaxy, and the Milky Way Galaxy is at the center of the diagram. The Virgo Cluster and the plane of our own Galaxy are marked. Our Galaxy is seen edge-on. Its dust obscures our views to the top and the bottom, resulting in two empty V-shaped regions on the map. The circle shown here is about 150 Mpc across.

tion of view is roughly perpendicular to that shown in the previous figure, so many of the clusters noted there are difficult to discern here.) Each point represents a galaxy, and the diagram is centered on the Milky Way Galaxy. The perspective is such that the disk of our Galaxy is seen edge-on and runs vertically up the page. The two nearly empty V-shaped regions at the top and bottom of the figure are not devoid of galaxies—they are simply obscured from view by dust in the plane of our Galaxy.

The Local Supercluster is some 30–40 Mpc across and contains about 10^{15} solar masses of material. Significantly flattened and somewhat irregular in shape, its center lies within the Virgo Cluster. By now it should perhaps come as no surprise that the Local Group is not found at its heart—we live far off in the periphery, nearly 20 Mpc from the center.

The Local Supercluster contains tens of thousands of individual galaxies, yet most known galaxy clusters and superclusters lie far beyond its edge. Figure 15.17 is

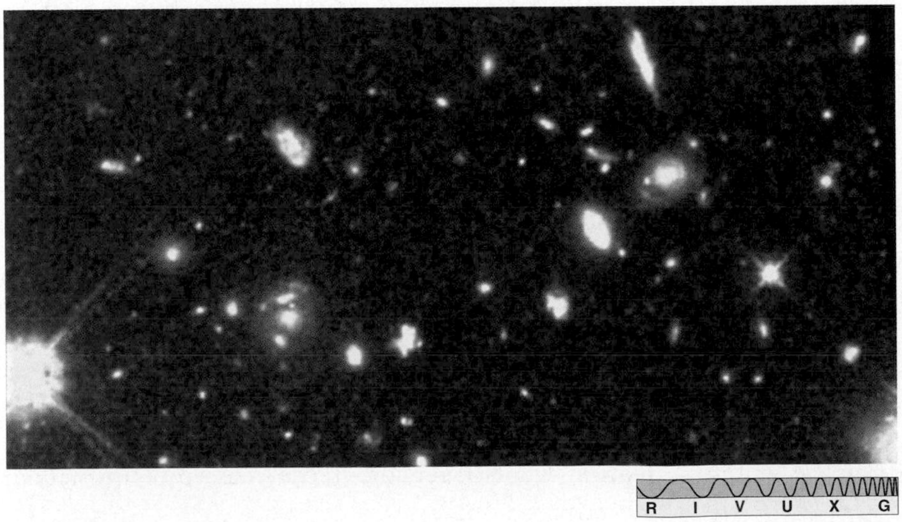

Figure 15.17 Distant Galaxy Cluster The galaxy cluster CL1358162 contains huge numbers of galaxies and resides roughly a billion parsecs from Earth. Every patch of light in this photograph is a separate galaxy. Thanks to the high resolution of the *Hubble Space Telescope*, we can now discern, even at this great distance, spiral structure in some of the galaxies. We also see many galaxies in collision—some tearing matter from one another, others merging into single systems. *(NASA)*

a long-exposure photograph of one such remote cluster. This rich cluster is far outside the limit of the circle shown in Figure 15.16. It is only one of many large and distant groups of galaxies scattered throughout the observable universe. The Coma Cluster, shown in Figure 15.1, is another. On and on, the picture is much the same. The farther we peer into deep space, the more galaxies, clusters of galaxies, and superclusters we see.

Is there structure on even larger scales? As we will see in a moment, the answer is still yes. However, before we make our final leap in distance to the limits of the visible universe, let's pause for a moment to take stock of some important properties of galaxies and galaxy clusters and consider a few current ideas on how galaxies form and evolve.

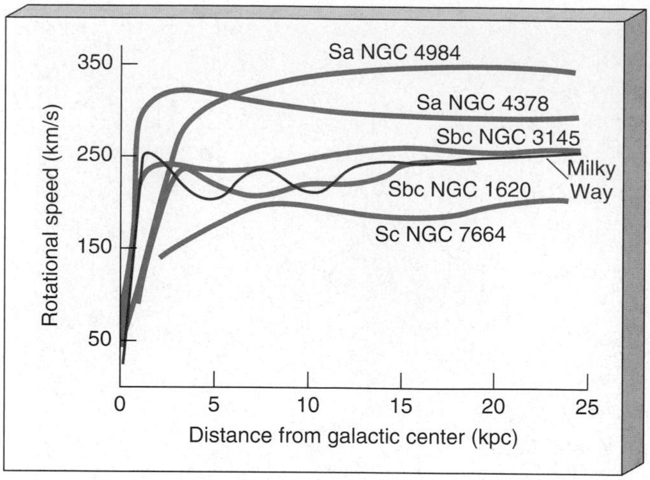

Figure 15.18 Galaxy Rotation Curves Rotation curves for some nearby spiral galaxies indicate masses of a few hundred billion times the mass of the Sun. The corresponding curve for our own Galaxy (from Figure 14.17) is marked in red for comparison.

✓ Concept Check

- Why are distances to faraway galaxies uncertain by about a factor of two?

15.3 Galaxy Masses

4 Despite their enormous sizes, galaxies and galaxy clusters obey the same physical laws as the planets in our own solar system. To determine galaxy masses, we turn as usual to Newton's law of gravity.

Mass Measurement

Astronomers can calculate the masses of some spiral galaxies by determining their *rotation curves*, which plot rotation speed versus distance from the galactic center. The mass within any given radius then follows directly from Newton's laws. ∞ (Sec. 14.5) Rotation curves for a few nearby spirals are shown in Figure 15.18. They imply masses ranging from about 10^{11} to 5×10^{11} solar masses within about 25 kpc of the center—comparable to the results obtained for our own Galaxy using the same technique. For galaxies too distant for detailed rotation curves to be drawn, the rotation speed can still be inferred from the broadening of spectral lines, as described in the previous section.

This approach is useful for measuring the mass lying within about 50 kpc of a galaxy's center (the extent of the electromagnetic emission from stellar and interstellar material). To probe farther from the centers of galaxies, astronomers turn to *binary* galaxy systems (Figure 15.19a), whose components may lie hundreds of kiloparsecs apart. The orbital period of such a system is typically billions of years, far too long for the orbit to be accurately measured. However, by estimating the period and semi-major axis using the information available—the line-of-sight velocities and angular separation of the components—an approximate total mass can be derived. ∞ (*More Precisely 1-2*)

Galaxy masses obtained in this way are fairly uncertain. However, by combining many such measurements, astronomers can obtain quite reliable *statistical* information about galaxy masses. Most normal spirals (the Milky Way Galaxy included) and large ellipticals contain between 10^{11} and 10^{12} solar masses of material. Irregular galaxies often contain less mass, about 10^8 to 10^{10} times that of the Sun. Dwarf ellipticals and dwarf irregulars can contain as little as 10^6 or 10^7 solar masses of material.

We can use another statistical technique to derive the combined mass of all the galaxies within a galaxy cluster. As depicted in Figure 15.19(b), each galaxy within a cluster moves relative to all other cluster members, and we can estimate the cluster's mass simply by asking how massive it must be in order to bind its galaxies gravitationally. Typical cluster masses obtained in this way lie in the range of 10^{13}–10^{14} solar masses. Notice that this calculation gives us no information whatsoever about the masses of individual

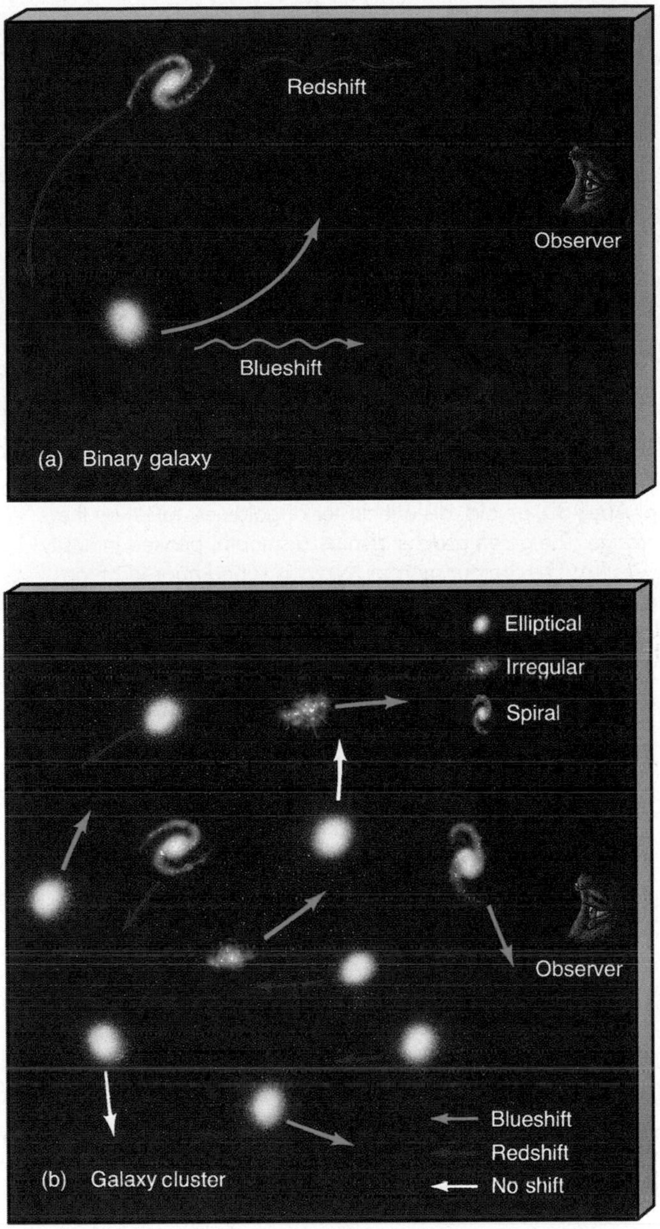

Figure 15.19 Galaxy Masses (a) In a binary galaxy system, galaxy masses may be estimated by observing the orbit of one galaxy about the other. (b) In a galaxy cluster, cluster masses are measured by observing the motion of many galaxies in the cluster and then estimating how much mass is needed to prevent the cluster from flying apart.

galaxies. It tells us only about the *total* mass of the entire cluster.

Dark Matter in the Universe

3 5 The rotation curves of the spiral galaxies shown in Figure 15.18 remain flat (that is, do not decline and may even rise slightly) far beyond the visible images of the galaxies themselves, implying that these galaxies—and perhaps all spiral galaxies—have invisible dark halos similar to that surrounding the Milky Way. ∞ (Sec. 14.5) Spiral galaxies seem to contain from three to 10 times more mass than can be accounted for in the form of visible matter.

Astronomers find an even greater discrepancy when they study galaxy clusters. Calculated cluster masses range from 10 to nearly 100 times the mass suggested by the light emitted by individual cluster galaxies. A lot more mass is needed to bind galaxy clusters than we can see. Thus the problem of dark matter exists not just in our own Galaxy, but also in other galaxies and, to an even greater degree, in galaxy clusters too. In that case, we are compelled to accept the fact that *upward of 90 percent of the universe is composed of dark matter*. As noted in Chapter 14, this matter is not just dark in the visible portion of the spectrum—it is undetected at *any* electromagnetic wavelength. ∞ (Sec. 14.5)

The dark matter in clusters apparently cannot simply be the accumulation of dark matter within individual galaxies. Even including the galaxies' dark halos, we still cannot account for all the dark matter in galaxy clusters. As we look on larger and larger scales, we find that a larger and larger fraction of the matter in the universe is dark.

Could the additional dark matter be diffuse intergalactic matter existing among the galaxies within the clusters—that is, intracluster gas? Satellites orbiting above Earth's atmosphere have detected substantial amounts of X-ray radiation from many galaxy clusters. Figure 15.20 shows false-color X-ray images of two such clusters. The X-ray-emitting region is centered on, and comparable in size to, the visible image of the cluster. These X-ray observations demonstrated for the first time the existence of large amounts of invisible hot (10 million K) gas within galaxy clusters. At least as much matter—and in a few cases substantially more—exists within clusters in the form of hot gas as is visible in the form of stars. This is a lot of material, but it still doesn't solve the dark-matter problem. To account for the total masses of galaxy clusters implied by dynamical studies, we would have to find from 10 to 100 times more mass in gas than in stars.

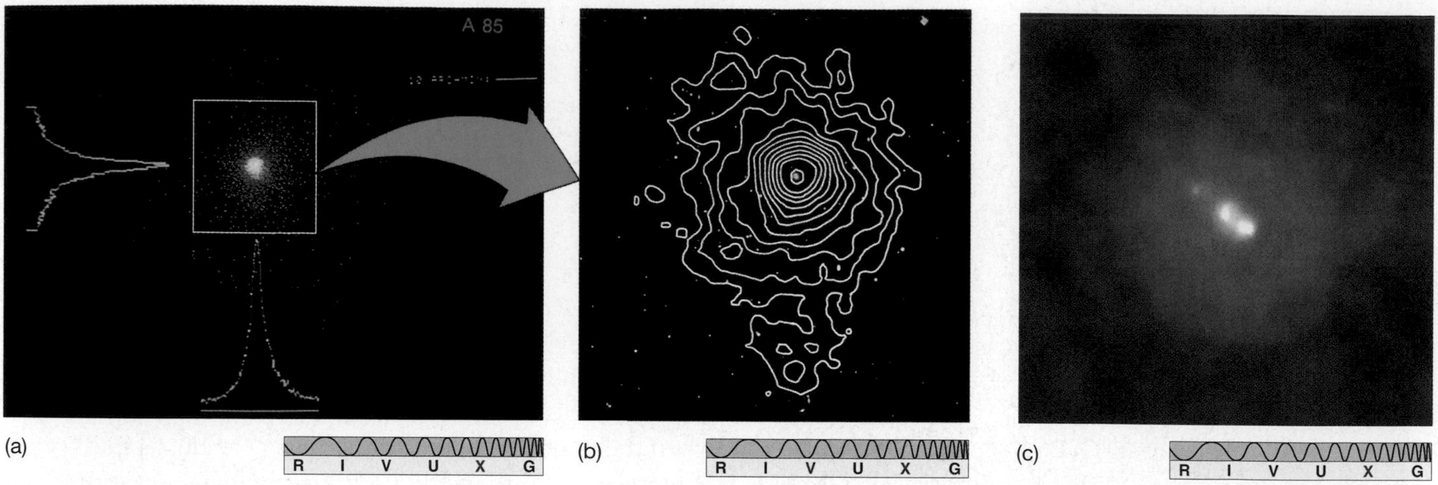

Figure 15.20 Galaxy Cluster X-Ray Emission (a) X-ray image of Abell 85, an old, distant cluster of galaxies, taken by the *Einstein* X-ray Observatory. The cluster's X-ray emission is shown in orange. The green graphs display a smooth, peaked intensity profile centered on the cluster but not associated with individual galaxies. (b) The contour map of X rays is superimposed on an optical photo, showing its X rays peaked on Abell 85's central supergiant galaxy. Images like these demonstrated that the space between the galaxies within galaxy clusters is filled with superheated gas. (c) A *Chandra* X-ray image of hot gas within another cluster of galaxies (called 3C295). The cluster is nearly 2 billion parsecs from Earth and measures several Mpc in diameter. *(NASA; AURA; NASA)*

Concept Check

- Why do astronomers think that most of the matter in the universe is dark?

15.4 Galaxy Formation and Evolution

6 How did the different galaxy types come into being? Astronomers know of no evolutionary connection among the various categories in the Hubble classification scheme. To address this question, we must understand how galaxies formed.

Unfortunately, the theory of galaxy formation is still very much in its infancy. Galaxies are much more complex than stars, they are harder to observe, and the observations are harder to interpret. In addition, we have only a partial understanding of conditions in the universe immediately preceding galaxy formation, quite unlike the corresponding situation for stars. ∞ (Sec. 11.2) Finally, whereas stars almost never collide with one another, with the result that most single stars and binaries evolve in isolation, galaxies may suffer numerous collisions and mergers during their lives, making it much harder to decipher their pasts. Nevertheless, some general ideas have begun to gain widespread acceptance, and we can offer a few insights into the processes responsible for the galaxies we see.

Mergers and Acquisitions

The seeds of galaxy formation were sown in the very early universe, when small density fluctuations in the primordial matter began to grow. ∞ (Sec. 17.2) Our discussion begins with these "pregalactic" blobs of gas already formed. The masses of these fragments were quite small—only a few million solar masses, comparable to the masses of the smallest present-day dwarf galaxies, which may in fact be remnants of this early time. Most astronomers believe that galaxies grew by repeated *merging* of smaller objects, as illustrated in Figure 15.21(a). Contrast this with the process of star formation, in which a large cloud fragments into smaller pieces that eventually become stars. ∞ (Sec. 11.3)

Theoretical evidence for this picture is provided by computer simulations of the early universe, which clearly show merging taking place. Further strong support comes from observations that indicate that galaxies at large distances from us (meaning that the light we see was emitted long ago) appear distinctly smaller and more irregular than those found nearby. Figure 15.21(b) and 15.22 show some of these images. The vague bluish patches are separate small galaxies, each containing only a few percent of the mass of the Milky Way Galaxy. Their irregular shape is thought to be the result of galaxy

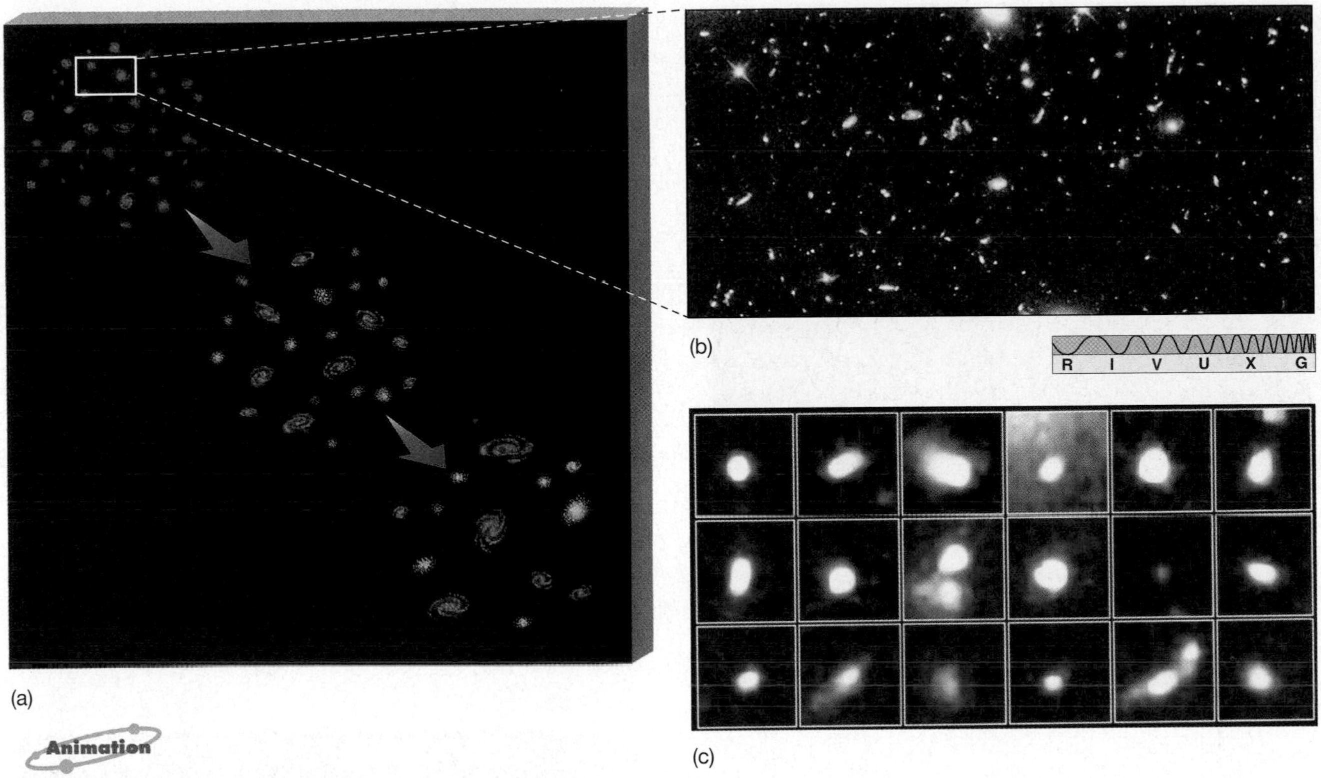

Figure 15.21 **Galaxy Formation** (a) The present view of galaxy formation holds that large systems were built up from smaller ones through collisions and mergers, as shown schematically in this drawing. (b) This photograph, one of the deepest ever taken of the universe, provides "fossil evidence" for hundreds of galaxy shards and fragments, up to 3000 Mpc distant. (c) Enlargements of selected portions of (b) reveal rich (billion-star) "star clusters," all lying within a relatively small volume of space (some 600 kpc across). Their proximity to one another suggests that we may be seeing a group of pregalactic fragments about to merge to form a galaxy. The events pictured took place about 10 billion years ago. *(NASA)*

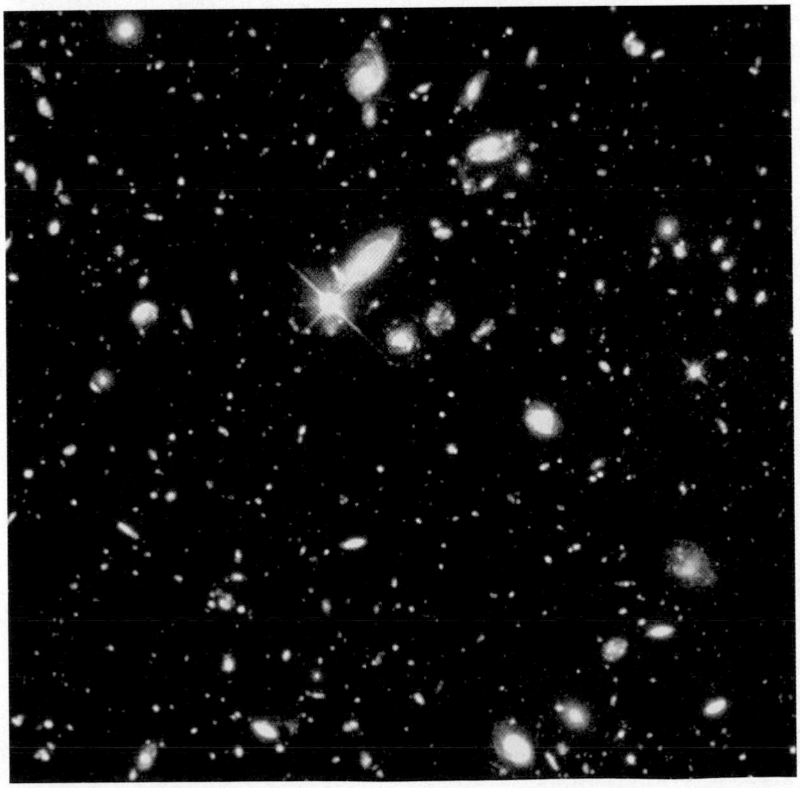

Video

Figure 15.22 **Hubble Deep Field** Numerous small, irregularly shaped young galaxies can be seen in this very deep optical image. Known as the Hubble Deep Field-North, this image, made with an exposure of approximately 100 hours, captured objects as faint as the 30th magnitude. Redshift measurements (see Section 15.5) indicate that some of the galaxies lie well over 1000 Mpc from Earth. Their size, color, and irregular appearance support the theory that galaxies grew by mergers and were smaller and less regular in the past. The field of view is about two arc minutes across. (See also the image at the opening of this chapter.) *(R. Williams; NASA)*

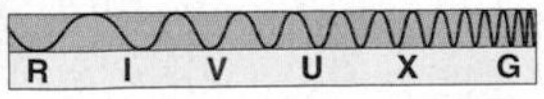

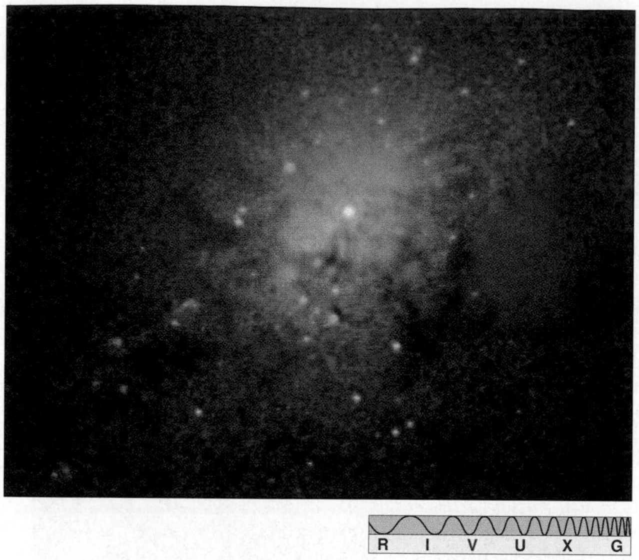

Figure 15.23 Peculiar Galaxy The peculiar (Irr II) galaxy NGC 1275 contains a system of long filaments that seem to be exploding outward into space. Its blue blobs, as revealed by the *Hubble Space Telescope,* are probably young globular clusters formed by the collision of two galaxies. *(NASA)*

mergers; the bluish coloration comes from young stars that formed during the merger process.

Figure 15.21(c) shows details of some of the objects in Figure 15.21(b). Each frame spans about one-tenth the size of the Milky Way Galaxy. Each blob seems to contain several billion stars spread throughout a distorted spheroid about a kiloparsec across—just about the size and scale expected for a pregalactic fragment.

How might we account for the differences between spirals and ellipticals? The answer remains unclear, although there are strong hints that the environment plays an important role. For example, spiral galaxies are relatively rare in the dense central parts of rich galaxy clusters. Is this because they simply tended not to form there, or is it because their disks are easily destroyed by collisions and mergers, which are more common in denser regions?

Computer simulations suggest that collisions between spiral galaxies can indeed destroy the spirals' disks, ejecting much of the gas into intergalactic space (in the process creating the hot intracluster gas noted in Section 15.3) and leaving behind objects that look very much like ellipticals. Observations of interacting galaxies, such as those shown in Figure 15.23, appear to support this scenario. Additional evidence comes from the observed fact that spirals are more common at large distances, which implies that their numbers are decreasing with time, presumably as the result of collisions.

However, nothing in this area of astronomy is clear cut. We know of numerous isolated elliptical galaxies in low-density regions of the universe, which are hard to explain as the result of mergers. Apparently some, but *not* all, ellipticals formed in this way.

Galaxy Interactions

Astronomers have ample evidence that galaxies interact with one another (see *Interlude 15-1*). Many spiral galaxies have huge, invisible dark halos surrounding them, and astronomers strongly suspect that all galaxies have similar halos. These halos make galaxies much "bigger" than the luminous objects we see, greatly increasing the chance of an encounter or collision with a neighbor.

Figure 15.24 Galactic Cannibalism This computer-enhanced, false-color composite optical photograph of the galaxy cluster Abell 2199 is thought to show an example of galactic cannibalism. The large central galaxy of the cluster (itself 120 kpc along its long axis) is displayed with a superimposed "window." (This results from a shorter time exposure, which shows only the brightest objects that fall within the frame.) Within the core of the large galaxy are several smaller galaxies (the three bright yellow images at center) apparently already "eaten" and now being "digested" (that is, being torn apart and becoming part of the larger system). Other small galaxies swarm on the outskirts of the swelling galaxy, almost certainly to be eaten, too. *(Smithsonian Astrophysical Observatory)*

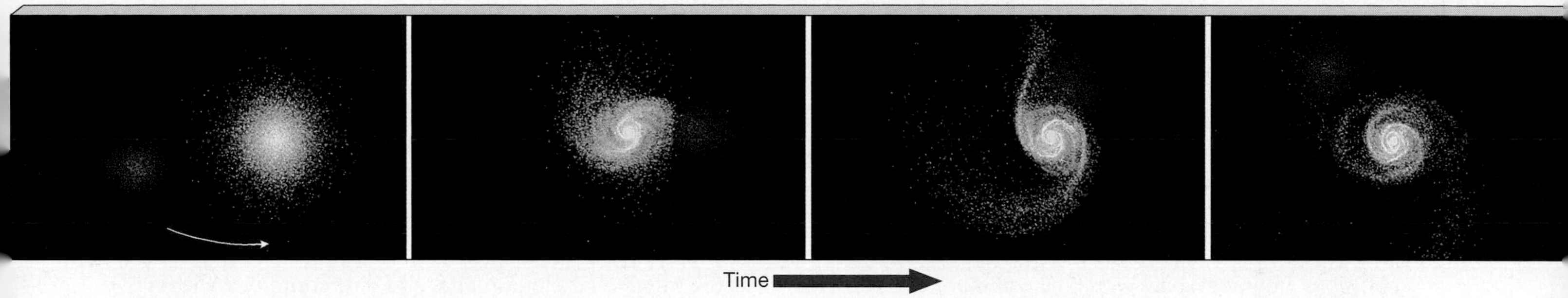

Figure 15.25 **Galactic Environmental Interaction** Galaxies can change their shapes long after their formation. In this computer-generated sequence, two galaxies closely interact over several hundred million years. The smaller galaxy, in red, has gravitationally disrupted the larger galaxy, in blue, changing it into a spiral galaxy. Compare the result of this supercomputer simulation with Figure 15.2b, a photograph of M51 and its small companion. *(J. Barnes & L. Hernquist)*

Consider two galaxies orbiting one another—a binary galaxy system. As they orbit, the galaxies interact with each other's dark halos, one galaxy stripping halo material from the other by tidal forces. The freed matter is either redistributed between the galaxies or is entirely lost from the binary system. In either case, the interaction changes the orbits of the galaxies, causing them to spiral toward one another and eventually to merge.

If one galaxy of the pair happens to have a much lower mass than the other, the process is colloquially termed *galactic cannibalism*. Such cannibalism might explain why supermassive galaxies are often found at the cores of rich galaxy clusters. Having dined on their companions, they now lie at the center of the cluster, waiting for more food to arrive. Figure 15.24 is a remarkable combination of images that has apparently captured this process at work. Closer to home, the Magellanic Clouds (Figure 15.8) will eventually suffer a similar fate at the center of the Milky Way.

Now consider two interacting disk galaxies, one a little smaller than the other but each having a mass comparable to the Milky Way Galaxy. As shown in the computer-generated frames of Figure 15.25, the smaller galaxy can substantially distort the larger one, causing spiral arms to appear where none existed before. The entire event requires several hundred million years—a span of evolution that a supercomputer can model in minutes. The final frame of Figure 15.25 looks remarkably similar to the double galaxy shown in Figure 15.2(b), demonstrating how the two galaxies might have interacted millions of years ago, and how spiral arms might have been created or enhanced as a result. Astronomers have cataloged numerous **starburst galaxies** (Figure 15.26; see also *Interlude 15-1*), where violent events, possibly a near-collision with a neighbor, appear to have rearranged the galaxy's internal structure and triggered a sudden, intense burst of star formation in the recent past.

Close encounters are random events and do not seem to represent any genuine evolutionary sequence linking all spirals to all ellipticals and irregulars. However, it *is* clear that many galaxies and galaxy clusters have evolved greatly since they first formed long ago.

Concept Check

- Other than scale, in what important way does galaxy formation differ from star formation?

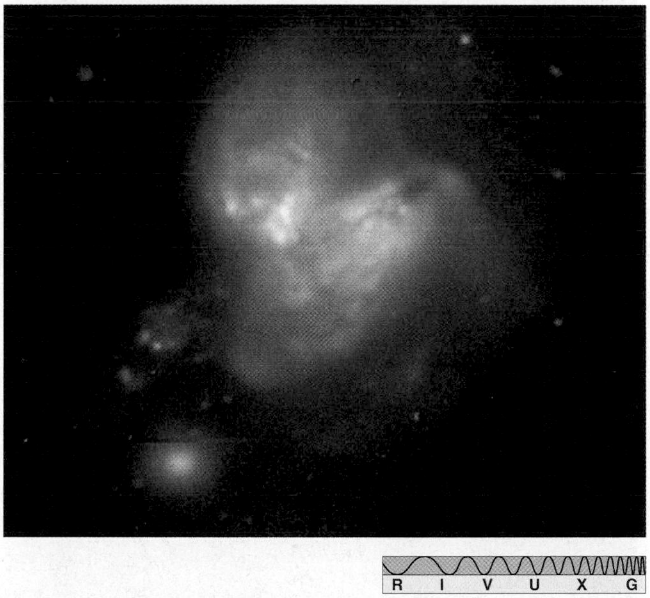

Figure 15.26 **Interacting Galaxies** This interacting galaxy pair (IC 694, at left center, and NGC 3690) shows starbursts now under way in both galaxies—hence the bluish tint. Such intense, short-lived bursts probably last for no more than a few tens of millions of years—a small fraction of a typical galaxy's lifetime. *(NASA)*

INTERLUDE 15-1
Colliding Galaxies

Contemplating the congested confines of a rich galaxy cluster (such as Virgo or Coma, with thousands of member galaxies orbiting within a few megaparsecs), we might expect that collisions among galaxies would be common. Gas particles collide in our atmosphere, and hockey players collide in the rink—do galaxies in clusters collide, too? The answer is yes.

This stunning image captures two spiral galaxies apparently passing each other like majestic ships in the night. It is unclear if they are only experiencing a close-encounter, or are about to have a head-on collision. The larger and more massive galaxy on the left is NGC 2207; the smaller one on the right is IC 2163. The strong tidal forces of the former are severely distorting parts of the latter, causing stars and gas to be ejected along streamers that stretch nearly 100,000 light-years toward the right-hand edge of the image. Analysis of this image suggests that IC 2163 is now swinging past NGC 2207 in a counterclockwise direction, having made a close approach some 40 million years ago. The two galaxies seem destined to undergo further close encounters, as IC 2163 apparently does not have enough energy to escape the gravitational pull of NGC 2207. Each time the two experience a very close brush, bursts of star formation erupt in both galaxies, as the interstellar clouds of gas and dust contained within each are pushed, shoved, and shocked. In roughly a billion years, these two galaxies will probably merge into a single, more massive galaxy, much the way several other grand galaxies, including our Milky Way, is thought to have been assembled by a similar process of coalescence of smaller galaxies.

No human will ever witness an entire galaxy collision, for it would last many millions of years. However, computer simulations display formations remarkably similar to observations. The particular simulation shown in the left-hand box on the facing page began with two spiral galaxies, but the details of the original structure have been largely obliterated by the collision. Notice the similarity to the real image of NGC 4038/4039 (the black and white image at right), the so-called Antennae galaxies, which show extended tails as well as double galactic centers only a few

(NASA)

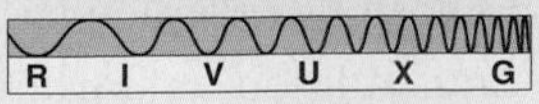

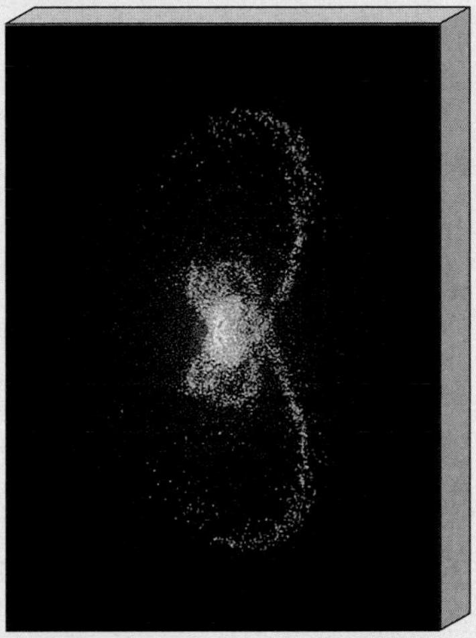

(J. Barnes)

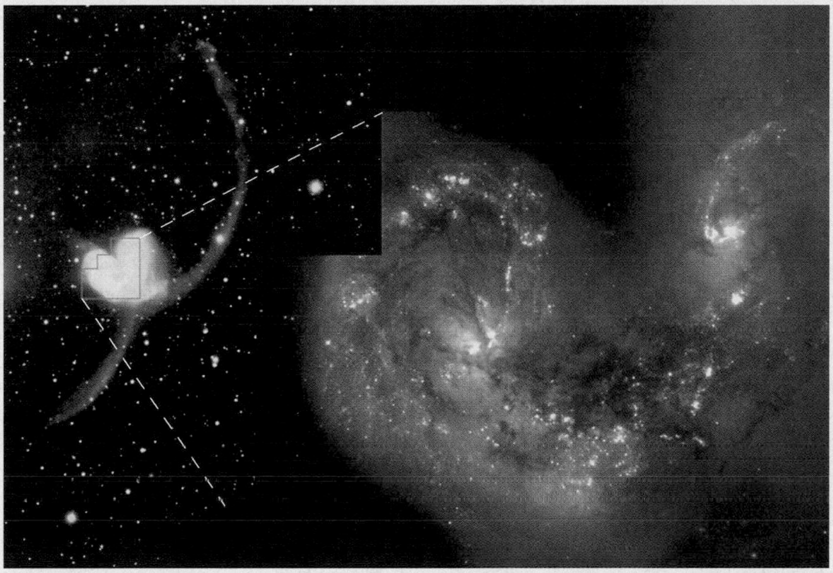

(B. Whitmore (STSC); NASA)

R I V U X G

hundred parsecs across. Star formation induced by the collision is clearly traced by the blue light from thousands of young, hot stars, evident in the high-resolution *Hubble Space Telescope* image at right. (The Hubble field of view is outlined on the black and white image.) The simulations suggest that ultimately the two galaxies will merge into one.

Direct observational evidence now indicates that galaxies in clusters apparently collide quite often. In the smaller groups, the galaxies' speeds are low enough that interacting galaxies tend to "stick together," and mergers, as shown in the computer simulation, are a very common outcome. In larger groups, galaxies move faster and tend to pass through one another without sticking. Since the early 1980s, it has become increasingly clear, on the basis of both observations and numerical simulations, that collisions can have very large effects on the galaxies involved, rearranging the stellar and interstellar contents of each. Some researchers go so far as to suggest that most galaxies have been strongly influenced by collisions, in many cases in the relatively recent past.

Curiously, although a collision may wreak havoc on the large-scale structure of the galaxies involved, it has essentially no effect on the individual stars they contain. The stars within each galaxy just glide past one another. Although we have plenty of photographic evidence for galaxy collisions, no one has ever witnessed or photographed a collision between two stars. Stars do collide in other circumstances—in the dense central cores of galactic nuclei and globular clusters, or as a result of stellar evolution in binary systems—but stellar collisions are a very rare consequence of galaxy interactions.

To understand why individual stars do not collide when galaxies collide, recall that the galaxies within a typical cluster are bunched together fairly tightly. The distance between adjacent galaxies in a given cluster averages a few hundred thousand parsecs, which is only about 10 times greater than the size of a typical galaxy. Galaxies simply do not have much room to roam around without bumping into one another. By contrast, the average distance between stars within a galaxy is a few parsecs, which is millions of times greater than the size of a typical star. When two galaxies collide, the star population merely doubles for a time, and the stars continue to have so much space that they do not run into each other. The stellar and interstellar contents of each galaxy are certainly rearranged, and the resultant burst of star formation may indeed be spectacular from afar, but from the point of view of the stars, it's clear sailing.

15.5 Hubble's Law

2 7 Within a galaxy cluster, individual galaxies move more or less randomly. You might expect that, on even larger scales, the clusters themselves would also have random, disordered motion—some clusters moving this way, some that. In fact, this is not the case. On the largest scales, galaxies and galaxy clusters alike move in a very *ordered* way.

Universal Recession

In 1912, the American astronomer Vesto M. Slipher, working under the direction of Percival Lowell, discovered that virtually every spiral galaxy he observed had a redshifted spectrum—it was *receding* from our Galaxy. (*More Precisely 2-3*) In fact, with the exception of a few nearby systems, *every* known galaxy is part of a general motion away from us in all directions. Individual galaxies that are not part of galaxy clusters are steadily receding. Galaxy clusters too have an overall recessional motion, although their individual member galaxies move randomly with respect to one another. (Consider a jar full of fireflies that has been thrown into the air. The fireflies within the jar, like the galaxies within the cluster, have random motions due to their individual whims, but the jar as a whole, like the galaxy cluster, has some directed motion as well.)

Figure 15.27 shows the optical spectra of several galaxies, arranged in order of increasing distance from the Milky Way Galaxy. These spectra are redshifted, indicating that the galaxies are receding. Furthermore, the extent of the redshift increases progressively from top to bottom in the figure. There is a connection between

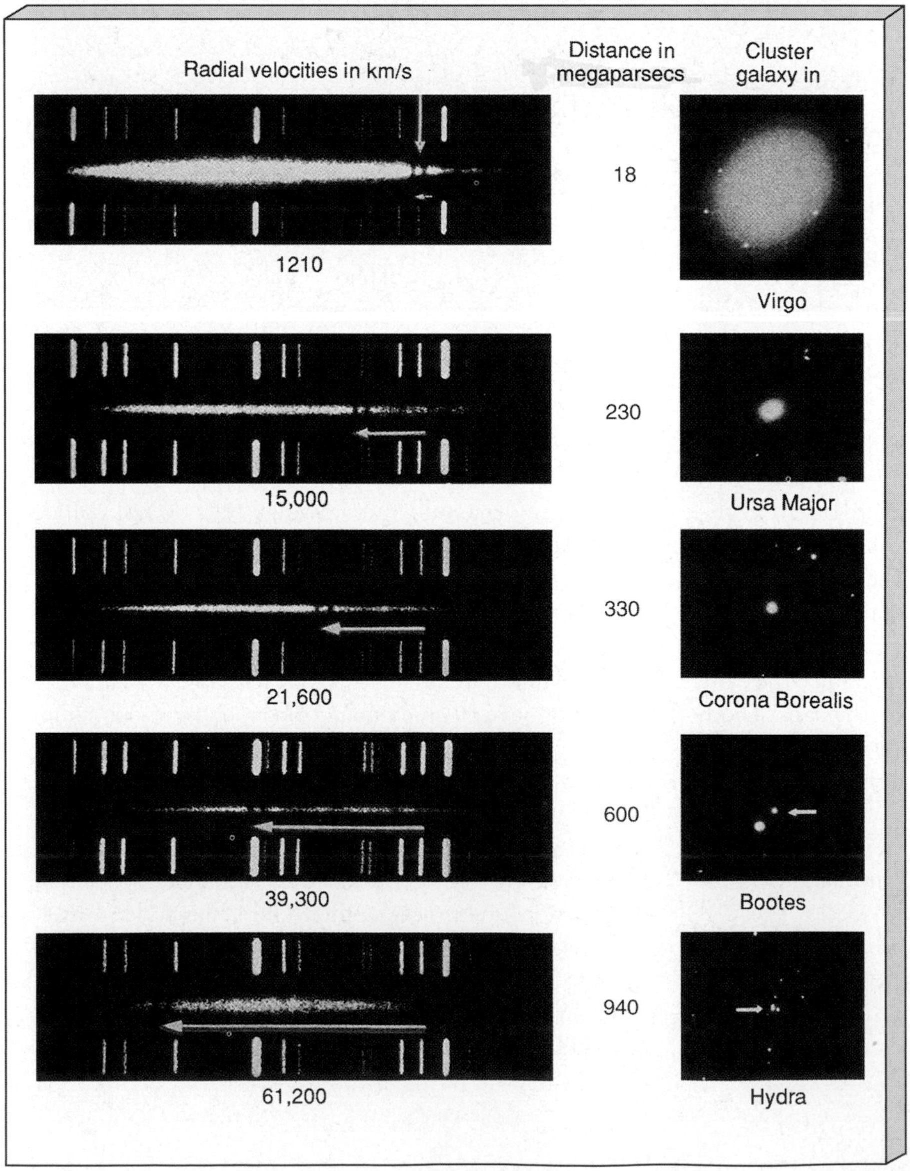

Figure 15.27 Galaxy Spectra Optical spectra of several galaxies. Both the extent of the redshift (denoted by the horizontal yellow arrows) and the distance from the Milky Way Galaxy to each galaxy (numbers in center column) increase from top to bottom. The vertical yellow arrow in the top spectrum highlights a particular spectral feature (a pair of dark absorption lines). The horizontal yellow arrows indicate how this feature shifts to longer wavelengths in spectra of more distant galaxies. The lines at the top and bottom of each spectrum are laboratory references. *(Palomar Observatory/California Institute of Technology)*

R I V U X G

Doppler shift and distance: the greater the distance, the greater the redshift. This trend holds for nearly all galaxies in the universe. (Two galaxies within our Local Group, including Andromeda, and a few galaxies in the Virgo Cluster display blueshifts and so are moving toward us, but this results from their random motions within their parent clusters. Recall the fireflies in the jar.)

Figure 15.28(a) shows recessional velocity plotted against distance for the galaxies of Figure 15.27. Figure 15.28(b) is a similar plot for some more galaxies within about one billion parsecs of Earth. Plots like these were first made by Edwin Hubble in the 1920s and now bear his name—*Hubble diagrams*. The data points generally fall close to a straight line, indicating that the rate at which a galaxy recedes is *directly proportional* to its distance from us. This rule is called **Hubble's law**. We could construct such a diagram for any group of galaxies, provided we could determine their distances and velocities. The universal recession described by the Hubble diagram is sometimes called the *Hubble flow*.

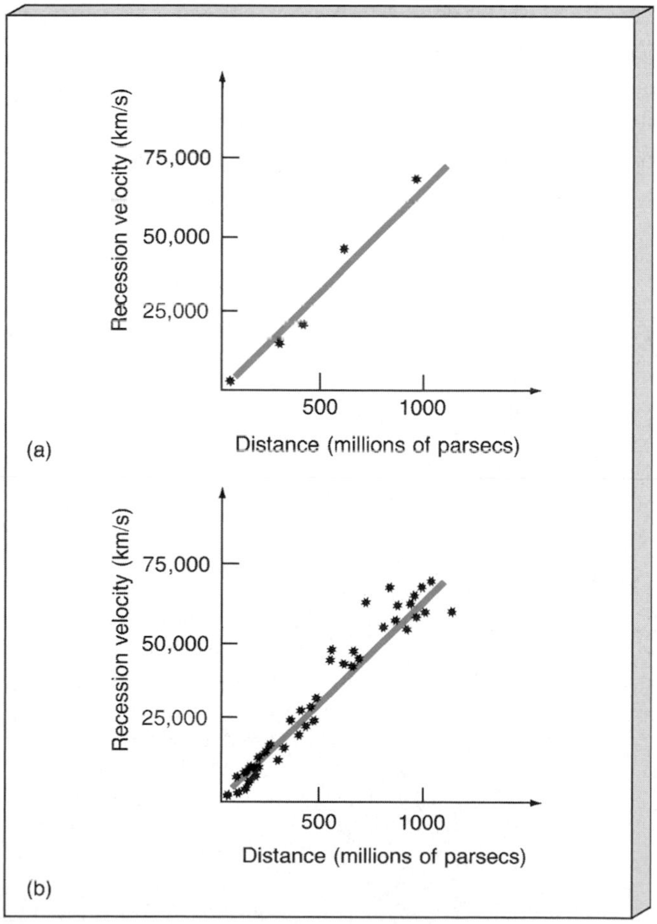

Figure 15.28 Hubble's Law Plots of recessional velocity against distance (a) for the galaxies shown in Figure 15.27 and (b) for numerous other galaxies within about one billion pc of Earth.

The recessional motions of the galaxies prove that the cosmos is not steady and unchanging on the largest scales. The universe is *expanding*. However, let's be clear on just *what* is expanding and what is not. Hubble's law does not mean that humans, Earth, the solar system, or even individual galaxies are physically increasing in size. These groups of atoms, rocks, planets, and stars are held together by their own internal forces and are not themselves getting bigger. Only the largest framework of the universe—the ever-increasing distances separating the galaxies and the galaxy clusters—is expanding.

To distinguish recessional redshift from redshifts caused by motion *within* an object—for example, galaxy orbits within a cluster or explosive events in a galactic nucleus—the redshift resulting from the Hubble flow is called the **cosmological redshift**. Objects that lie so far away that they exhibit a large cosmological redshift are said to be at *cosmological distances*—distances comparable to the scale of the universe itself.

Hubble's law has some fairly dramatic implications. If nearly all galaxies show recessional velocity according to Hubble's law, then doesn't that mean that they all started their journey from a single point? If we could run time backward, wouldn't all the galaxies fly back to this one point, perhaps the site of some explosion in the remote past? In Chapter 17 we explore the ramifications of the Hubble flow for the past and future evolution of our universe. For the remainder of this chapter, however, we set aside its cosmic implications and use Hubble's law simply as a convenient distance-measuring tool.

Hubble's Constant

The constant of proportionality between recessional velocity and distance in Hubble's law is known as **Hubble's constant** and denoted by the symbol H_0. The data shown in Figure 15.28 then obey the equation:

$$\text{recessional velocity} = H_0 \times \text{distance}.$$

The value of Hubble's constant is the slope of the straight line—recessional velocity divided by distance—in Figure 15.28(b). Reading the numbers off the graph, this comes to roughly 65,000 km/s divided by 1000 Mpc, or 65 km/s/Mpc (kilometers per second per megaparsec, the most commonly used unit for H_0). Astronomers continually strive to refine the accuracy of the Hubble diagram and the resulting estimate of H_0 because Hubble's constant is one of the most fundamental quantities of nature—it specifies the rate of expansion of the entire cosmos.

The precise value of Hubble's constant is the subject of considerable debate. In the 1970s, astronomers obtained a value of around 50 km/s/Mpc, using a chain of standard candles to extend their observations to large distances. However, in the early 1980s, when the Tully–Fisher technique had become fairly well established, other researchers used it to obtain a measurement of H_0 that was largely independent of methods relying on standard candles. From observations of galaxies within about 150 Mpc, the latter group deduced a value of $H_0 = 90$ km/s/Mpc, a result inconsistent with the earlier measurements (even allowing for the uncertainties in the methods used). For some reason, the distances obtained using the Tully–Fisher method were only about half those determined using standard candles, and so the measured value of Hubble's constant nearly doubled using this approach.

The most recent measurements of H_0 by a number of researchers, using different galaxies and a variety of distance-measurement techniques, give results mostly within the range 50–80 km/s/Mpc. Most astronomers would be quite surprised if the true value of H_0 turned out to lie outside this range. However, the width of the quoted range is not the result of measurement uncertainties in any one method—there remain real, and as yet unresolved, inconsistencies between the different techniques currently in use. For now, astronomers must live with this uncertainty. We will adopt $H_0 = 65$ km/s/Mpc as the best current estimate of Hubble's constant for the remainder of the text.

The Cosmic Distance Scale

Using Hubble's law, we can derive the distance to a remote object simply by measuring the object's recessional velocity and dividing by Hubble's constant. (Notice, however, that the uncertainty in Hubble's constant translates directly into a similar uncertainty in the distance determined by this method.) Using Hubble's law in this way tops our inverted pyramid of distance-measurement techniques (Figure 15.29). This sixth method simply assumes that Hubble's law holds. If this assumption is correct, Hubble's law enables us to measure great distances in the universe—so long as we can obtain an object's spectrum, we can determine how far away it is.

Many redshifted objects have recessional motions that are a substantial fraction of the speed of light. The most distant object thus far observed in the universe has the catalog name 0140+326RD1. Its extremely high redshift implies a recessional velocity 95 percent that of light. At that speed, ultraviolet spectral lines are Doppler shifted all the way into the far infrared! Hubble's law implies that, solely on the basis of its observed redshift, 0140+326RD1 lies some 5500 Mpc away from us (see *More Precisely 16-1* for a fuller discussion of redshifts and recession velocities comparable to the speed of light). This object resides as close to the limits of the observable universe as astronomers have yet been able to probe.

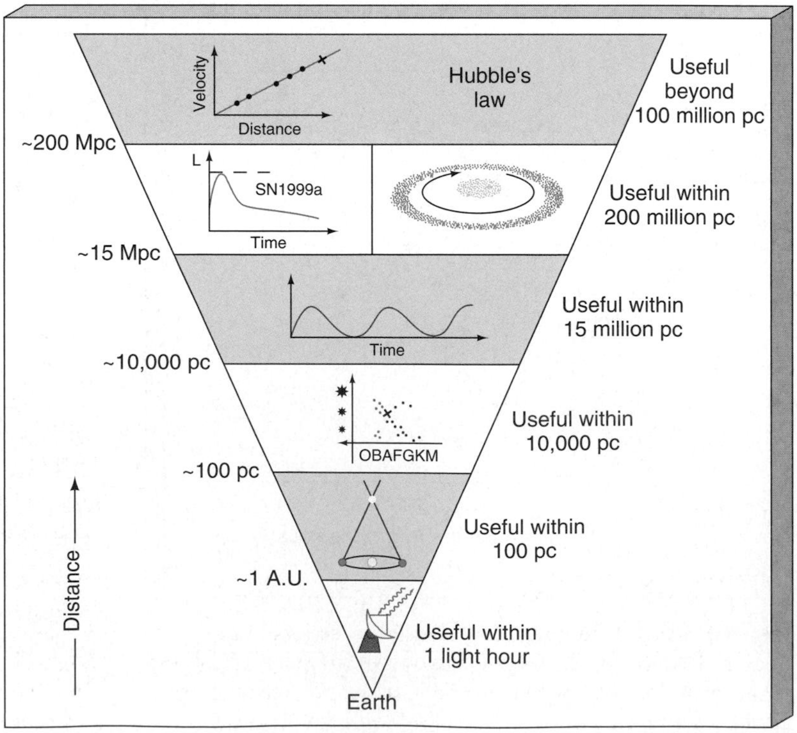

Figure 15.29 Galaxies on Distance Ladder Hubble's law tops the hierarchy of distance-measurement techniques. It is used to find the distances of astronomical objects all the way out to the limits of the observable universe.

Large-Scale Structure in the Universe

3 Figure 15.30 shows part of an extensive survey of the universe performed by astronomers at Harvard University. Using Hubble's law as a distance indicator, the team systematically mapped out the locations of galaxies within about 200 Mpc of the Milky Way in a series of wedge-shaped "slices," each 6° thick, starting in the northern sky. The first slice (Figure 15.30) covered a region of the sky containing the Coma Cluster (see Figure 15.1), which happens to lie in a direction almost perpendicular to our Galaxy's plane.

The most striking feature of maps such as this is that the distribution of galaxies on very large scales is decidedly nonrandom. The galaxies appear to be arranged in a network of strings, or filaments, surrounding large, relatively empty regions of space known as **voids**. The biggest voids measure some 100 Mpc across. The most likely explanation for the voids and filamentary structure in Figure 15.30 is that the galaxies and galaxy clusters are spread across the surfaces of vast "bubbles" in space. The voids are the interiors of these gigantic bubbles. The galaxies seem to be distributed like beads on strings only because of the way our slice of the universe cuts through the bubbles. Like suds on soapy water, these bubbles fill the entire universe. The densest clusters and superclusters lie in regions where several bubbles meet.

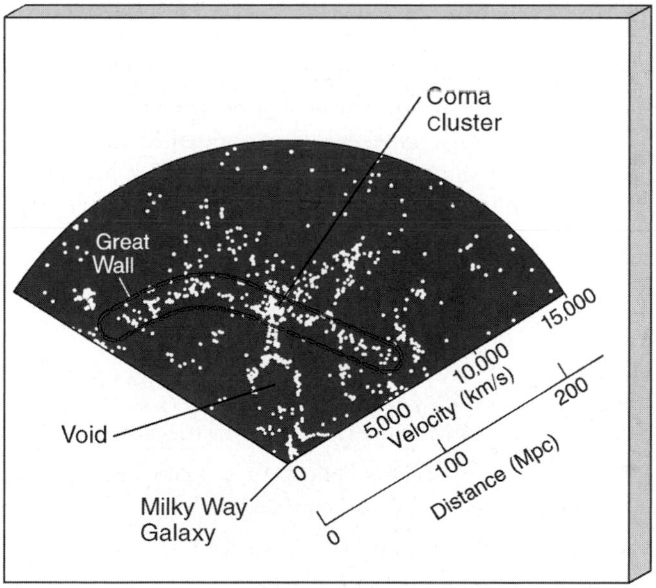

Figure 15.30 Galaxy Survey The first slice of a survey of the universe, covering 1057 galaxies out to an approximate distance of 200 Mpc, clearly shows that galaxies and clusters are not randomly distributed on large scales. Instead, they appear to have a filamentary structure, surrounding vast, nearly empty voids. We assume H_0 = 65 km/s/Mpc. The slice covers 6° of the sky in the direction out of the plane of the paper.

The idea that the filaments are the intersection of the survey slice with much larger structures (the bubble surfaces) was confirmed when the next three slices of the survey, lying above and below the first, were completed. The region of Figure 15.30 indicated by the red outline was found to continue through both the other slices. This extended sheet of galaxies, which has come to be known as the *Great Wall*, measures at least 70 Mpc (out of the plane of the page) by 200 Mpc (across the page). It is one of the largest known structures in the universe. Figure 15.31 combines the survey data with results of other surveys of the southern sky. The Great Wall can be seen arcing around the left side of the figure.

Most theorists believe that this "frothy" distribution of galaxies, and in fact all structure on scales larger than a few megaparsecs, traces its origin directly to the conditions found in the very earliest stages of the universe (Chapter 17). As such, studies of large-scale structure are vital to our efforts to understand the origin and nature of the cosmos itself.

✓ Concept Check

- How does the expansion of the universe provide a means of measuring distances to galaxies?

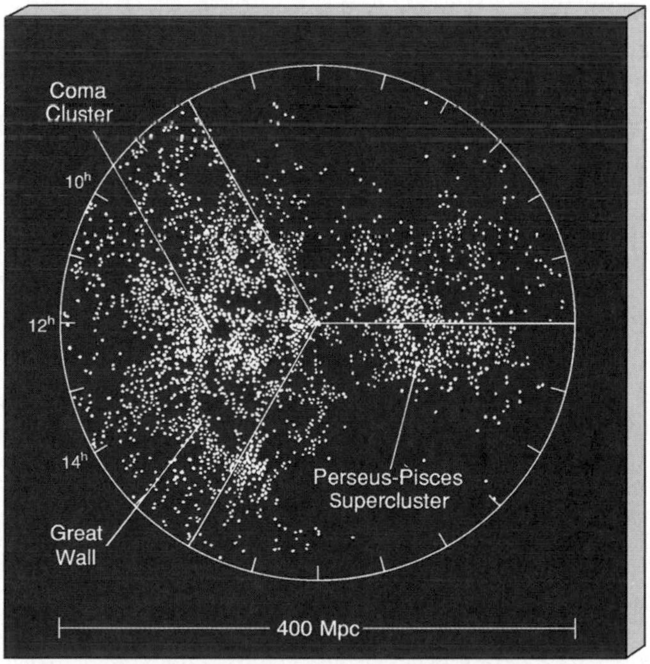

Figure 15.31 Distant Galaxy Survey Combination of data from several redshift surveys of the universe reveal the extent of large-scale structure within 200–300 Mpc of the Sun. The arc on the left is the Great Wall. The empty regions are mostly areas obscured by our Galaxy. Positions for more than 4500 galaxies are plotted here (H_0 = 65 km/s/Mpc).

Chapter Review

Summary

The **Hubble classification scheme** (p. 378) divides galaxies into several classes, depending on their appearance. **Spiral galaxies** (p. 379) have flattened disks, central bulges, and spiral arms. They are further subdivided on the basis of the size of the bulge and the tightness of the spiral structure. The halos of these galaxies consist of old stars, whereas the gas-rich disks are the sites of ongoing star formation. **Barred-spiral galaxies** (p. 379) contain an extended "bar" of material projecting beyond the central bulge. **Elliptical galaxies** (p. 380) have no disk and contain no gas or dust. In most cases, they consist entirely of old stars. They range in size from dwarf ellipticals, which are much less massive than the Milky Way Galaxy, to giant ellipticals, which may contain trillions of stars.

S0 and **SB0 galaxies** (p. 381) are intermediate in their properties between ellipticals and spirals. They have extended halos and stellar disks and bulges (and bars, in the SB0 case) but little or no gas and dust. **Irregular galaxies** (p. 381) are galaxies that do not fit into either of the other categories. Some may be the result of galaxy collisions or close encounters. Many irregulars are rich in gas and dust and are the sites of vigorous star formation. The **Magellanic Clouds** (p. 382), two small systems that orbit the Milky Way Galaxy, are examples of this type of galaxy.

Astronomers often use **standard candles** (p. 384) as distance-measuring tools. These are objects that are easily identifiable and whose luminosities lie in some reasonably well-defined range. Comparing luminosity and apparent brightness, astronomers determine the distance using the inverse-square law. An alternative approach is the **Tully–Fisher relation** (p. 384), an empirical correlation between rotational velocity and luminosity in spiral galaxies.

The Milky Way, Andromeda, and several other smaller galaxies form the Local Group, a small **galaxy cluster** (p. 386). Galaxy clusters consist of a collection of galaxies orbiting one another, bound together by their own gravity. The nearest large galaxy cluster to the Local Group is the Virgo Cluster. Galaxy clusters themselves tend to clump together into **superclusters** (p. 386). The Virgo Cluster, the Local Group, and several other nearby clusters form the Local Supercluster.

The masses of nearby spiral galaxies can be determined by studying their rotation curves. Astronomers also use studies of binary galaxies and galaxy clusters to obtain statistical mass estimates of the galaxies involved. The measurements reveal the presence of large amounts of dark matter. The fraction of dark matter grows as the scale under consideration increases. Large amounts of hot X-ray-emitting gas have been detected among the galaxies in many clusters, but not enough to account for the dark matter inferred from dynamical studies.

Collisions and mergers play important roles in galaxy evolution. Large galaxies probably formed by mergers of smaller ones, and interactions between galaxies are common. A **starburst galaxy** (p. 393) may result when a galaxy experiences a close encounter with a neighbor.

Distant galaxies are receding from the Milky Way at speeds proportional to their distances from us. This relationship is called **Hubble's law** (p. 397). The constant of proportionality in the law is **Hubble's constant** (p. 397). Its value is believed to lie between 50 and 80 km/s/Mpc. Astronomers use Hubble's law to determine distances to the most remote objects in the universe. The redshift associated with the Hubble expansion is called the **cosmological redshift** (p. 397).

On very large scales, galaxies and galaxy clusters are arranged on the surfaces of enormous "bubbles" of matter surrounding vast low-density regions called **voids** (p. 399). The origin of this structure is thought to be related to conditions in the very earliest epochs of the universe.

Review and Discussion

1. In what sense are elliptical galaxies "all halo"?
2. Describe the four rungs in the distance-measurement ladder used to determine the distance to a galaxy lying 5 Mpc away.
3. Describe the contents of the Local Group. How much space does it occupy compared with the volume of the Milky Way Galaxy?
4. How does the Tully–Fisher relation allow astronomers to measure the distances to galaxies?
5. What is the Virgo Cluster?
6. Describe two techniques for measuring the mass of a galaxy.
7. Why do astronomers believe that galaxy clusters contain more mass than we can see?
8. Why do clusters of galaxies emit X-rays?
9. What evidence do we have that galaxies collide with one another?
10. Describe the role of collisions in the formation and evolution of galaxies.
11. Do you think that collisions between galaxies constitute "evolution" in the same sense as the evolution of stars?

12. What is Hubble's law?

13. How is Hubble's law used by astronomers to measure distances to galaxies?

14. What is the most likely range of values for Hubble's constant? Why is the exact value uncertain?

15. What are voids? What is the distribution of galactic matter on very large (more than 100 Mpc) scales?

True or False?

_____ **1.** Barred-spiral galaxies have similar properties to normal spirals, except for the "bar" feature.

_____ **2.** Elliptical galaxies do not contain a flattened disk.

_____ **3.** Elliptical galaxies contain substantial amounts of interstellar gas but no interstellar dust.

_____ **4.** Most ellipticals contain only old stars.

_____ **5.** Irregular galaxies, although small, have lots of star formation taking place in them.

_____ **6.** Most galaxies are spirals.

_____ **7.** Isolated spiral galaxies evolve into ellipticals.

_____ **8.** Type I supernovae can be used to determine distances to galaxies.

_____ **9.** Every galaxy is a member of some galaxy cluster.

_____ **10.** Galaxy collisions can occur, but are extremely rare.

_____ **11.** Galaxy collisions destroy most of the stars in the galaxies involved.

_____ **12.** Distant galaxies appear to be much larger than those nearby.

_____ **13.** A typical galaxy cluster has a mass of about 10^6 solar masses.

_____ **14.** Most galaxies appear to be receding from the Milky Way Galaxy.

_____ **15.** Hubble's law can be used to determine distances to the farthest objects in the universe.

Fill in the Blank

1. Galaxies are classified by their type in the _______ classification scheme.

2. Spiral galaxies that have tightly wrapped spiral arms tend to have _______ central bulges.

3. Galaxies of type _______ have the least amount of gas; type _______ have the most.

4. An E5 galaxy is more _______ than an E1 galaxy.

5. The Milky Way Galaxy, the Andromeda Galaxy, and about 20 others form a small cluster known as the _______.

6. In the Tully–Fisher relation, a galaxy's luminosity is found to be related to the _______ of its spectral lines.

7. Galaxy clusters are themselves clustered, forming _______.

8. When galaxies collide, the star formation rate often _______.

9. Galaxy mass determinations from rotation curves, line broadening, and binary galaxies all make use of the law of _______.

10. Intergalactic gas in galaxy clusters emits large amounts of energy in the form of _______.

11. More than 90 percent of the mass in galaxy clusters exists in the form of _______.

12. Galaxies form by the process of _______ (small objects accumulating into large); they do not form by _______ (large objects breaking up into small).

13. Hubble's law is a correlation between the redshifts and the _______ of galaxies.

14. Hubble's constant is believed to lie in the range _______ to _______ km/s/Mpc.

15. The largest known structures in the universe, such as voids and the Great Wall, have sizes on the order of _______.

Problems

1. A supernova of absolute magnitude −18 is used as a standard candle to measure the distance to a faraway galaxy. From Earth, the supernova appears as bright as the Sun would appear from a distance of 10 kpc. What is the distance to the galaxy?

2. A Cepheid variable star in the Virgo cluster has an apparent magnitude of 26 and is observed to have an pulsation period of 20 days. Use these figures and Figure 14.6 to estimate the distance to the Virgo cluster.

3. The Andromeda Galaxy is approaching our Galaxy with a radial velocity of 266 km/s. Given the galaxies' present separation of 930 kpc, and neglecting both the transverse component of the velocity and the effect of gravity in accelerating the motion, estimate when the two galaxies will collide.

4. Calculate the average speed of hydrogen nuclei (protons) in a gas of temperature 10 million K. ∞ (*More Precisely 5-2*) Compare this with the typical orbital speed of 300–500 km/s of galaxies in a large galaxy cluster.

5. Two galaxies are orbiting one another at a separation of 500 kpc. Their orbital period is estimated to be 30 billion years. Use Kepler's third law to find the total mass of the pair. ∞ (Sec. 14.5)

6. Use Kepler's third law to estimate the mass required to keep a galaxy moving at 400 km/s in a circular orbit of radius 1 Mpc around the center of a galaxy cluster. ∞ (*More Precisely 1-2*)

7. In a galaxy collision, two similar-sized galaxies pass through each other with a combined relative velocity of 1500 km/s. If each galaxy is 100 kpc across, how long does the event last?

8. According to Hubble's law, with $H_0 = 65$ km/s/Mpc, what is the recessional speed of a galaxy at a distance of 200 Mpc? How far away is a galaxy whose recessional speed is 4000 km/s? How would these answers change if (a) $H_0 = 50$ km/s/Mpc, (b) $H_0 = 80$ km/s/Mpc?

9. If $H_0 = 65$ km/s/Mpc, how long will it take for the distance from the Milky Way Galaxy to the Virgo Cluster to double?

10. A Type-I supernova has absolute magnitude -18 and a recession velocity of 13,000 km/s. If $H_0 = 65$ km/s/Mpc, calculate the supernova's apparent magnitude.

Projects

1. Look for a copy of the *Atlas of Peculiar Galaxies* by Halton Arp. It is available in book form or on laser disk. Search for examples of interacting galaxies of various types: 1) tidal interactions, 2) starburst galaxies, 3) collisions between two spirals, and 4) collisions between a spiral and an elliptical. For (1), look for galactic material pulled away from a galaxy by a neighboring galaxy. Is the latter galaxy also tidally distorted? In (2), the surest sign of starburst activity are bright knots of star formation. In what type(s) of galaxies do you find starburst activity? For (3) and (4), how do collisions differ depending on the types of galaxies involved?

2. Look for the Virgo Cluster. An 8-inch telescope is a perfect size for this project, although a smaller telescope will also work. The constellation Virgo is visible from the United States during much of fall, winter, and spring. To locate the center of the cluster, first find the constellation Leo. The eastern part of Leo is composed of a distinct triangle of stars—Denebola (β), Chort (θ), and Zosma (δ). Move your eye eastward from Chort to Denebola and then continue along that same line until you have moved a distance equal to the distance between Chort and Denebola. You are now looking at the approximate middle of the Virgo Cluster. Look for the following Messier objects that make up some of the brightest galaxies in the cluster. M49, 58, 59, 60, 84, 86, 87 (which is a giant elliptical thought to have a massive black hole at its center), 89, and 90. Examine each galaxy for unusual features; some have very bright nuclei.

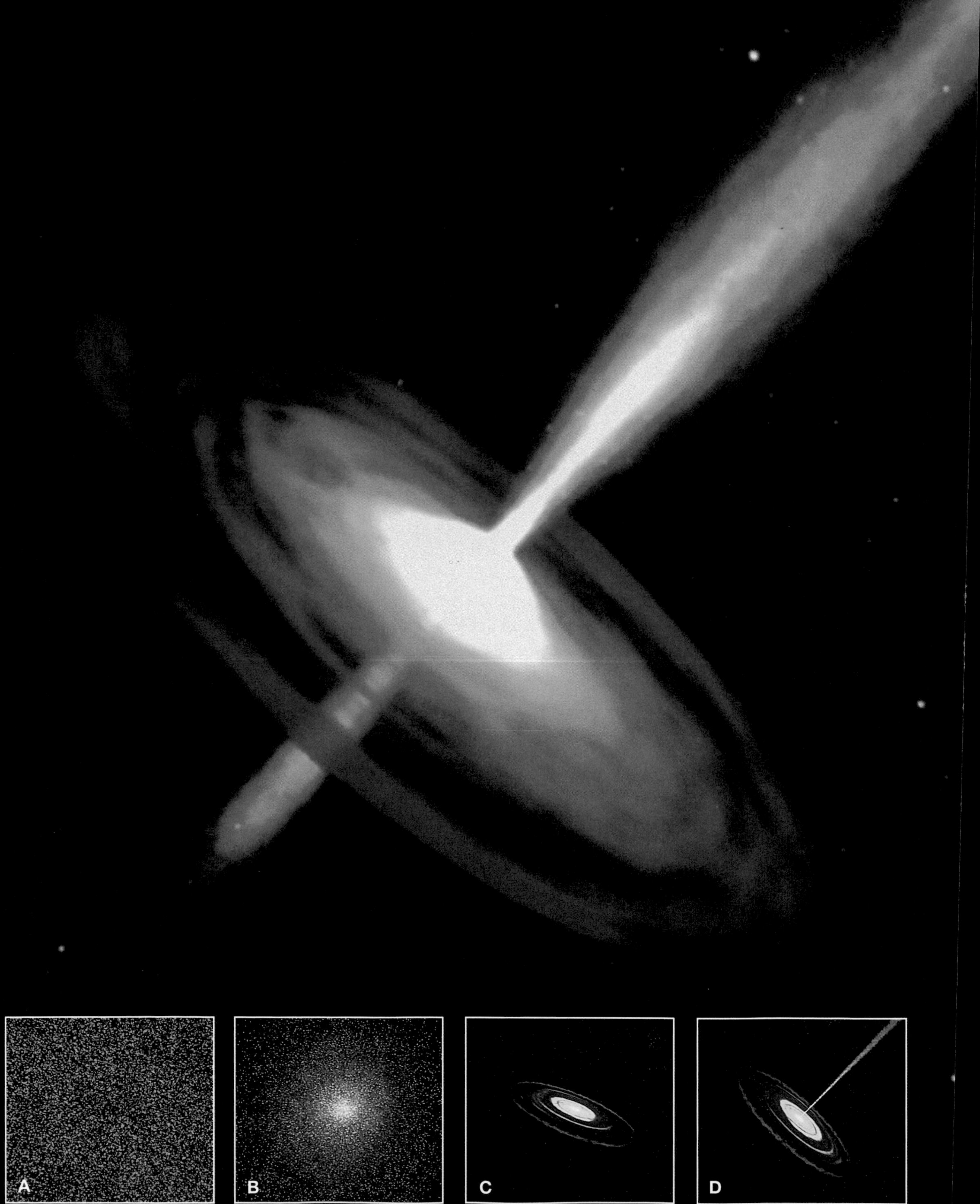
A
B
C
D

16 ACTIVE GALAXIES AND QUASARS

Limits of the Observable Universe

WWW

LEARNING GOALS

Studying this chapter will enable you to:

1 Specify the basic differences between active and normal galaxies.

2 Describe the important features of Seyfert and radio galaxies.

3 Explain what drives the central engine thought to power all active galaxies.

4 Describe the observed properties of quasars and discuss the special properties of the radiation they emit.

5 Discuss the place of active galaxies in current theories of galactic evolution.

(Opposite page, background) This artist's conception depicts one possible scenario for the "central engine" of an active galaxy. Two jets of matter are shown moving outward perpendicular to a flattened accretion disk surrounding a supermassive black hole.

(Inset A) Numerical simulations performed on supercomputers can help us visualize key events in the universe. Here a uniform distribution of about a thousand stars is spread over a dimension of several hundred parsecs.

(Inset B) As the simulation proceeds, the hypothetical stars congregate preferentially toward the core of the star cluster, forming a "luminosity cusp" of light.

(Inset C) If we could see inside the cusp, an accretion disk would probably be evident, swirling around a giant black hole.

(Inset D) As the accretion disk rotates, tilts, and acquires more matter, jets of matter emerge to cool the environment surrounding the black hole. *(D. Berry)*

Our journey from the Milky Way to the Great Wall in the past two chapters has widened our cosmic field of view by a factor of 10,000, yet the galaxies that make up the structures we see show remarkable consistency in their properties. Most fit neatly into the Hubble classification scheme, showing few, if any, unusual characteristics. However, sprinkled throughout the mix of normal galaxies, even relatively close to the Milky Way Galaxy, are some that are decidedly abnormal in their properties. Although their optical appearance is often quite ordinary, these abnormal galaxies emit huge amounts of energy—far more than normal galaxies—mostly in the nonvisible part of the electromagnetic spectrum. Observing such objects at great distances, we may be seeing some of the formative stages of our own Galactic home.

16.1 Beyond the Local Realm

1 Astronomers estimate that some 100 billion galaxies exist in the observable universe. Most lie thousands of megaparsecs from Earth—too far for their Hubble types to be reliably determined with current telescopes—yet, to the extent that their properties can be measured, even very distant galaxies seem basically "normal." Their luminosities and spectra are generally consistent with the standard categories in the Hubble classification scheme. ∞ (Sec. 15.1) However, scattered throughout the universe, a few galaxies differ considerably from the norm. They are far more luminous than even the brightest spiral or elliptical galaxies discussed in the previous chapter. Having luminosities sometimes thousands of times greater than that of the Milky Way, they are known collectively as **active galaxies**.

In addition to their greater overall luminosities, active galaxies differ fundamentally from normal galaxies in the *character* of the radiation they emit. Most of a normal galaxy's energy is emitted in or near the visible portion of the electromagnetic spectrum, much like the radiation from stars. Indeed, to a large extent, the light we see from a normal galaxy *is* just the accumulated light of its many component stars. By contrast, as illustrated schematically in Figure 16.1, the radiation from active galaxies does *not* peak in the visible. Most active galaxies do emit substantial amounts of visible radiation, but far more energy is emitted at longer wavelengths. Put another way, the radiation from active galaxies is *inconsistent* with what we would expect if it were the combined radiation of myriad stars. Their radiation is said to be *nonstellar*.

Two important categories of active galaxies are *Seyfert galaxies* and *radio galaxies*. More extreme in their properties are the even more luminous *quasars*. Astronomers conventionally distinguish between active galaxies and quasars based on their appearance, spectra, and distance from us. At visible wavelengths, active galaxies typically *look* like normal galaxies. Indeed, we can think of active galaxies as being otherwise "normal" systems that happen also to be extremely intense sources of radio and/or infrared radiation. Quasars, on the other hand, are mostly so far away that little internal structure can be discerned. However, the distinction between quasars and active galaxies may not be so clear-cut. Most astronomers now believe that quasars are simply an early stage of galaxy formation, and that the same basic processes power all of these active objects.

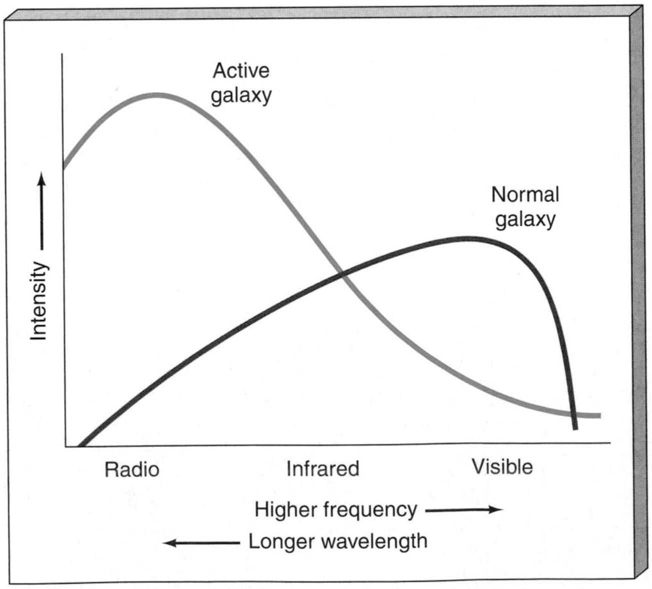

Figure 16.1 Galaxy Energy Spectrum The energy emitted by a normal galaxy differs from that emitted by an active galaxy. This plot illustrates the general run of intensity for all galaxies of a particular type and does not represent any one individual galaxy.

Some active galaxies are found locally, scattered among the normal galaxies that make up most of our cosmic neighborhood, but as a general rule, active galaxies are more common at greater distances, and the most active objects—the quasars—lie farthest from Earth. Since light travels at a finite speed, looking out to great distances in space is equivalent to looking back in time. The tremendous power and nonstellar radiation of these faraway objects suggest to many astronomers that the universe was once a much more violent place than it is today.

✓ Concept Check

- What distinguishes active galaxies from normal galaxies?

16.2 Seyfert Galaxies

2 In 1943 Carl Seyfert, an American optical astronomer studying spiral galaxies from Mount Wilson Observatory, discovered the type of active galaxy that now bears his name. **Seyfert galaxies** are a class of astronomical objects whose properties lie between those of normal galaxies and those of the most violent active galaxies known. Their spectral lines are usually substantially redshifted, telling us that most Seyferts reside at large distances (hundreds of megaparsecs) from us. (Sec. 15.5) However, a few lie relatively nearby—just 20 or 30 Mpc away.

Superficially, Seyferts resemble normal spiral galaxies (Figure 16.2a). Indeed, the stars in a Seyfert's galactic disk and spiral arms produce about the same amount of visible radiation as the stars in a normal spiral galaxy. However, closer study reveals some peculiarities not found in normal spirals. Nearly all of a Seyfert's energy is emitted in the form of radio and infrared radiation from a small central region known as the **galactic nucleus**—the center of the overexposed white patch in Figure 16.2(a), shown in more detail in Figure 16.2(b). A Seyfert nucleus may be quite similar to the center of a normal galaxy (the Milky Way, for example), but with one very important difference: The nucleus of a Seyfert is some 10,000 times brighter than the center of our Galaxy. In fact, the brightest Seyfert nuclei are 10 times more energetic than the *entire* Milky Way Galaxy.

Seyfert spectral lines have many similarities to those observed toward the center of our own Galaxy. (Sec. 14.6) The lines are very broad, most likely indicating rapid (1000 km/s) motion within the nuclei. In addition, the energy emission often varies in time

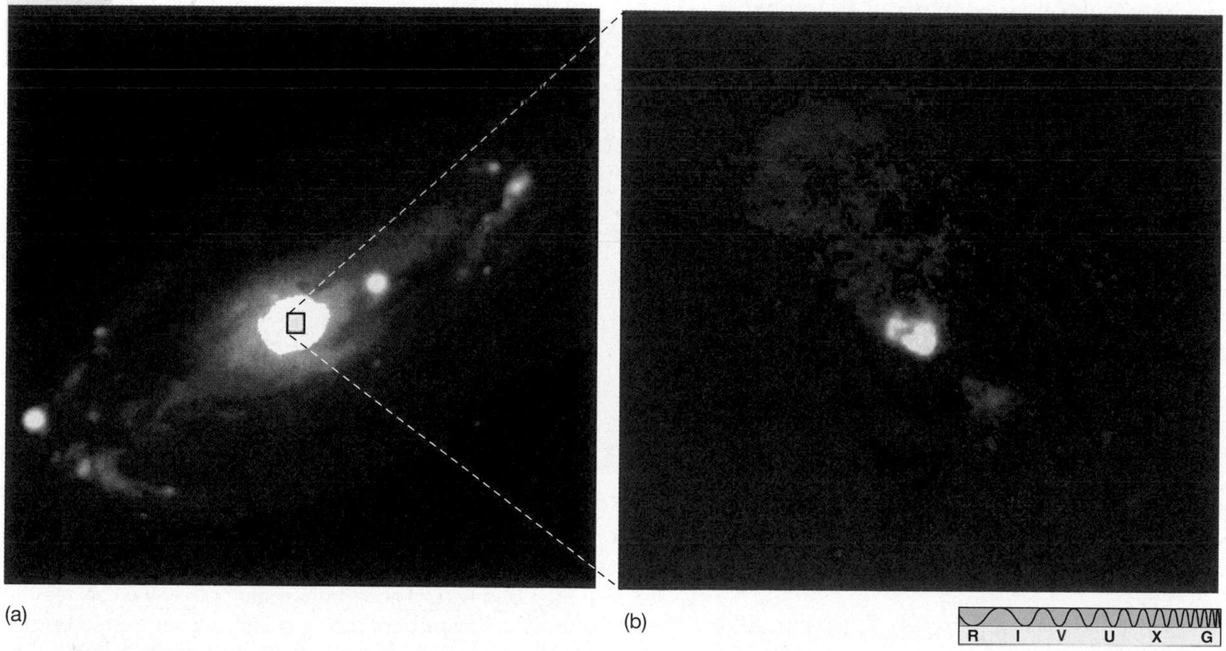

Figure 16.2 Seyfert Galaxy Photographs of a Seyfert galaxy (NGC 5728) (a) from the ground and (b) from Earth orbit. The enlarged view in (b) of two cone-shaped beams of light shows a group of glowing "blobs" near the galaxy's core, perhaps illuminated by radiation arising from the accretion disk of a black hole. This object is about 40 Mpc away. *(AURA; NASA)*

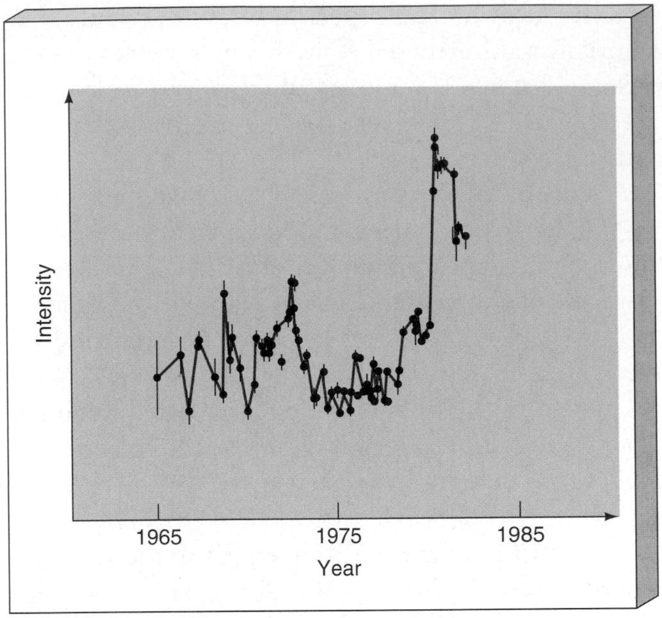

Figure 16.3 Seyfert Time Variability This graph illustrates the irregular variations of a particular Seyfert galaxy's luminosity over two decades. Because this Seyfert, called 3C 84, emits most strongly in the radio part of the electromagnetic spectrum, these observations were made with large radio telescopes. The optical and X-ray luminosities vary as well.

(Figure 16.3). A Seyfert's luminosity can double or halve within a fraction of a year. These rapid fluctuations in luminosity lead us to conclude that the source of energy emissions must be quite compact—simply put, an object cannot "flicker" in less time than radiation takes to cross it. ∞ (Sec. 13.4) The emitting region must therefore be less than one light-year across—an extraordinarily small region, considering the amount of energy emanating from it. High-resolution interferometric radio maps of Seyfert nuclei generally confirm this reasoning. ∞ (Sec. 3.4)

Concept Check

- How does a Seyfert nucleus compare with the center of the Milky Way Galaxy?

16.3 Radio Galaxies

2 As the name suggests, **radio galaxies** are active galaxies that emit most of their energy in the radio portion of the electromagnetic spectrum. They differ from Seyferts not only in the wavelengths at which they radiate most strongly, but also in both the appearance and the extent of their emitting regions. Seyferts radiate at shorter radio wavelengths and in the infrared, with virtually all of the nonstellar emission coming from a small central nucleus. Radio galaxies radiate at longer radio wavelengths, and their emitting regions can be very much larger—hundreds of kiloparsecs across in some cases.

Core–Halo Radio Galaxies

One common type of radio galaxy is often called a *core–halo* radio galaxy. As illustrated in Figure 16.4, the radio energy from such an object comes mostly from a small central nucleus (which radio astronomers refer to as the *core*) less than 1 pc across, with weaker radio emission coming from an extended *halo* surrounding the nucleus. The halo typically measures about 50 kpc across, comparable in size to the visible galaxy,[1] which is usually elliptical and often quite faint. The radio luminosity from the nucleus can be as great as 10^{37} W—about the same as the total emission from a Seyfert nucleus, and comparable to the output from the Milky Way Galaxy at all wavelengths.

[1]The term "visible galaxy" is commonly used to refer to the components of an active galaxy that emit visible "stellar" radiation, as opposed to the nonstellar "active" component of the galaxy's emission.

Figure 16.4 Core–Halo Radio Galaxy On this radio contour map of a typical core–halo radio galaxy we can see that the radio emission from such a galaxy comes from a bright central nucleus, which is surrounded by an extended, less intense radio halo. The radio map is superimposed on an optical image of the galaxy and some of its neighbors, shown previously in Figure 15.14. *(Harvard-Smithsonian Center for Astrophysics)*

Figure 16.5 presents several images of a nearby core–halo radio galaxy, the giant elliptical galaxy M87, a prominent member of the Virgo Cluster and one of the closest active galaxies. ∞ (Sec. 15.2) A long time exposure (Figure 16.5a) shows a large, fuzzy ball of light—a fairly normal-looking E1 galaxy about 100 kpc across. A shorter-time exposure of M87 (Figure 16.5b), capturing only the galaxy's bright inner regions, reveals a long thin *jet* of matter ejected from M87's center. The jet is about 2 kpc long and is traveling outward at very high speed, possibly as much as half the speed of light. Computer enhancement shows that it is made up of a series of distinct "blobs" more or less evenly spaced along its length, suggesting that the material was ejected during bursts of activity. The jet has also been imaged in the radio and infrared regions of the spectrum (Figures 16.5c and d). Jets such as this are a very common feature of active galaxies.

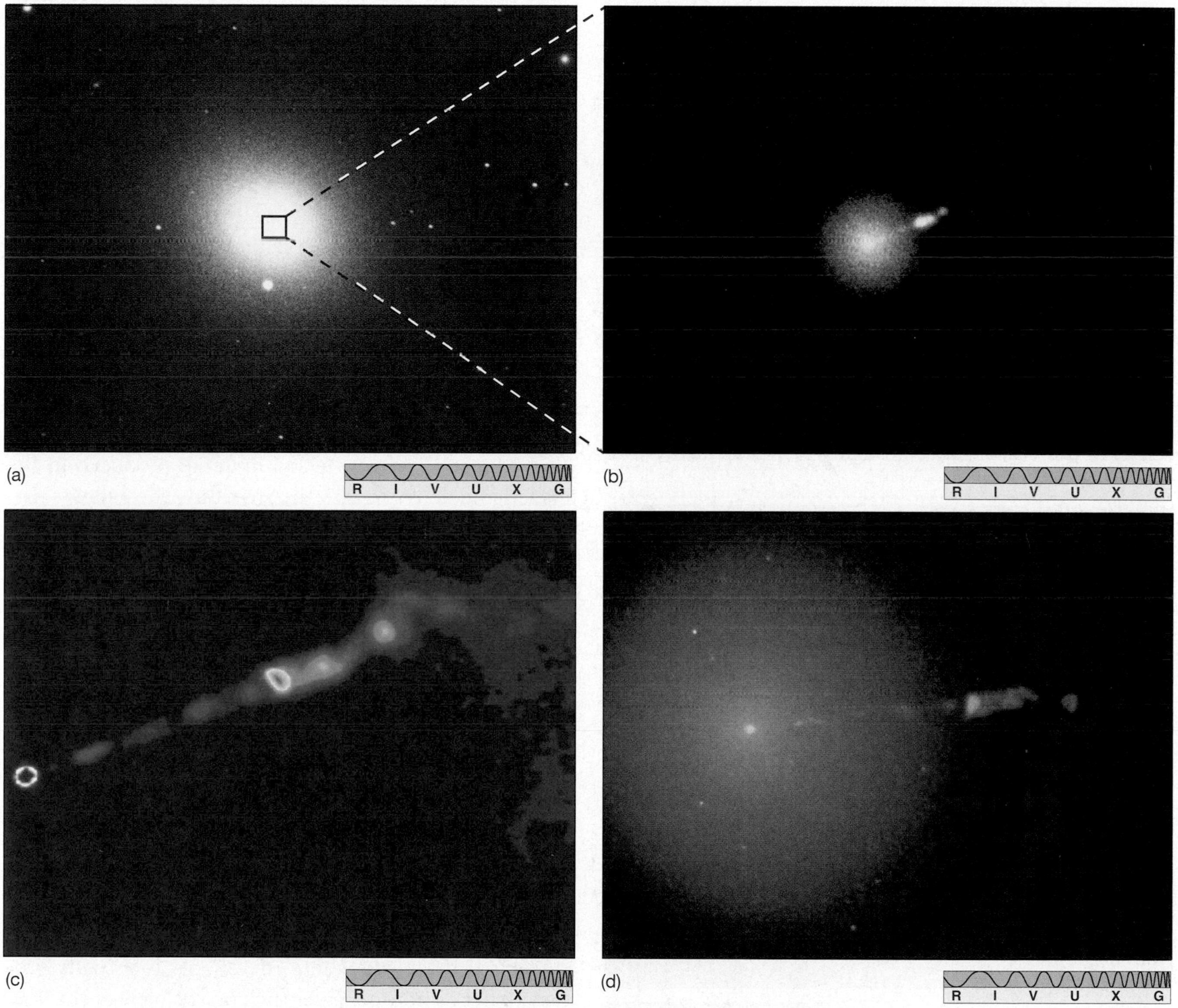

Figure 16.5 **M87 Jet** A core–halo radio galaxy—the giant elliptical galaxy M87 (also called Virgo A)—is displayed here at several wavelengths. (a) A long optical exposure, showing mainly the galaxy's outer halo. (b) A shorter optical exposure of the galaxy's nucleus and an intriguing jet of matter, at a smaller scale than (a). (c) A radio image of the jet, at a scale larger than that in (b). The red dot at left marks the bright nucleus of the galaxy; the red and yellow blob near the center of the image corresponds to the bright area visible in the jet in (b). (d) A near-infrared image of the jet, at roughly the same scale as (c). *(NOAO; AURA; NRAO; NASA)*

Lobe Radio Galaxies

Many radio galaxies are not of the core–halo type. Very little of their radio emission arises from the compact central nucleus, but instead comes from huge extended regions called **radio lobes**—roundish clouds of gas up to 1 Mpc across—lying well beyond the visible part of the galaxy. For this reason, these objects are known as *lobe radio galaxies*.

The radio lobes of lobe radio galaxies are truly enormous. From end to end, an entire lobe radio galaxy typically is more than 10 times the size of the Milky Way Galaxy, comparable in size to the entire Local Group. The radio luminosity of the lobes can range from 10^{36} to 10^{38} W—between 1/10 and 10 times the total energy emitted by our Galaxy. One such system, known as Centaurus A, is shown in Figure 16.6. It lies only 4 Mpc from Earth.

Figure 16.7 shows the relationship between the galaxy's visible and radio emission. In visible light, Centaurus A is apparently a large E2 galaxy some 500 kpc in diameter, bisected by an irregular band of dust. Numerical simulations suggest that this system is probably the result of a merger between an elliptical galaxy and a smaller spiral galaxy about 500 million years ago. The radio lobes are roughly symmetrically placed, jutting out from the center of the visible galaxy, roughly perpendicular to the dust lane, suggesting that they consist of material ejected in opposite directions from the galactic nucleus. This conclusion is strengthened by the presence of a pair of smaller secondary lobes (marked in Figure 16.7) closer to the visible galaxy, and by the presence of a roughly 1-kpc-long jet in the galactic center, all aligned with the main lobes.

If the material was ejected from the nucleus at close to the speed of light, then Centaurus A's outer lobes were created a few hundred million years ago, quite possibly around the time of the elliptical/spiral merger thought to be responsible for the galaxy's odd optical appearance. The secondary lobes were expelled more recently. Apparently some violent process at the center of Centaurus A—most probably triggered by the merger—started up around that time and has been intermittently firing jets of matter out into intergalactic space ever since.

Additional evidence supporting this view comes from another radio galaxy, called Cygnus A (Figure 16.8). The high-resolution radio map in Figure 16.8(b) clearly shows the jets joining Cygnus A's radio lobes to the center of the visible galaxy (the dot at the center of the radio image). Notice also that, as with Centaurus A, the optical image of the galaxy appears to show a galaxy collision in progress. These observations strongly suggest that radio lobes are formed when two oppositely directed, narrow jets of material produced in the galactic nucleus run into the gas filling the galaxy's parent cluster. (Sec. 15.3) In some systems, the motion of the galaxy through the intracluster gas causes the lobes to be swept back behind the main part of the galaxy, forming a *head–tail* radio galaxy such as that shown in Figure 16.9.

It may well be that all radio galaxies have jets and lobes, and that the differences we have just described between core–halo and lobe systems are largely a matter of perspective. As illustrated in Figure 16.10, if we view the jets and lobes from the side, we see a lobe radio galaxy. However, if we view the jet almost head-on—in other words, looking *through* the lobe—we see a core–halo system. If so, then in *all* cases studied so far—both Seyferts and radio galaxies—the central compact nucleus is the place where the energy is actually produced. Let us now consider the current view of the "engine" that powers all this activity.

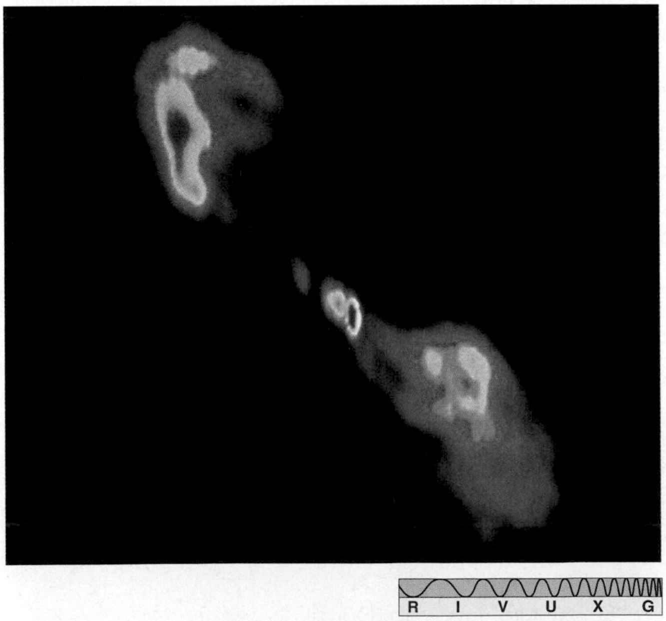

Figure 16.6 Centaurus A Radio Lobes Lobe radio galaxies, such as Centaurus A shown here, have huge radio-emitting regions extending a million parsecs or more beyond the center of the galaxy. The lobes emit no visible light and so are observable only with radio telescopes. The lobes are shown here in false color, with decreasing intensity from red to yellow to green to blue. *(NRAO)*

Concept Check

- What evidence do we have that the energy emitted by lobe radio galaxies actually originates in the galactic nucleus?

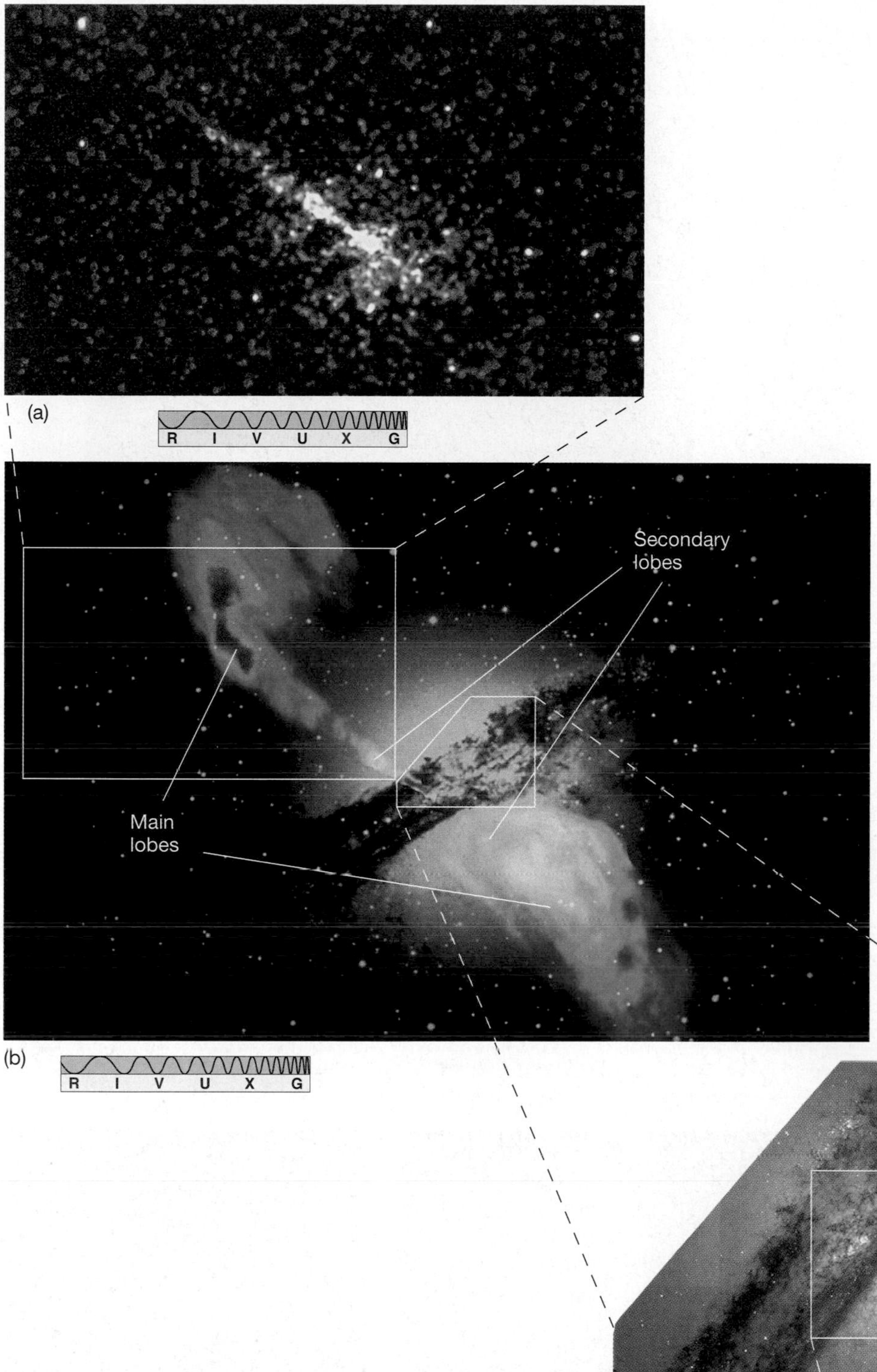

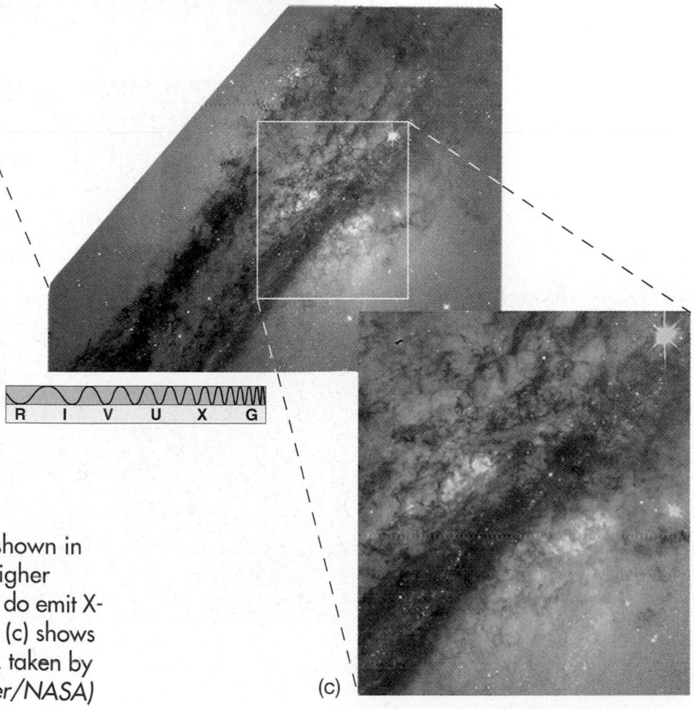

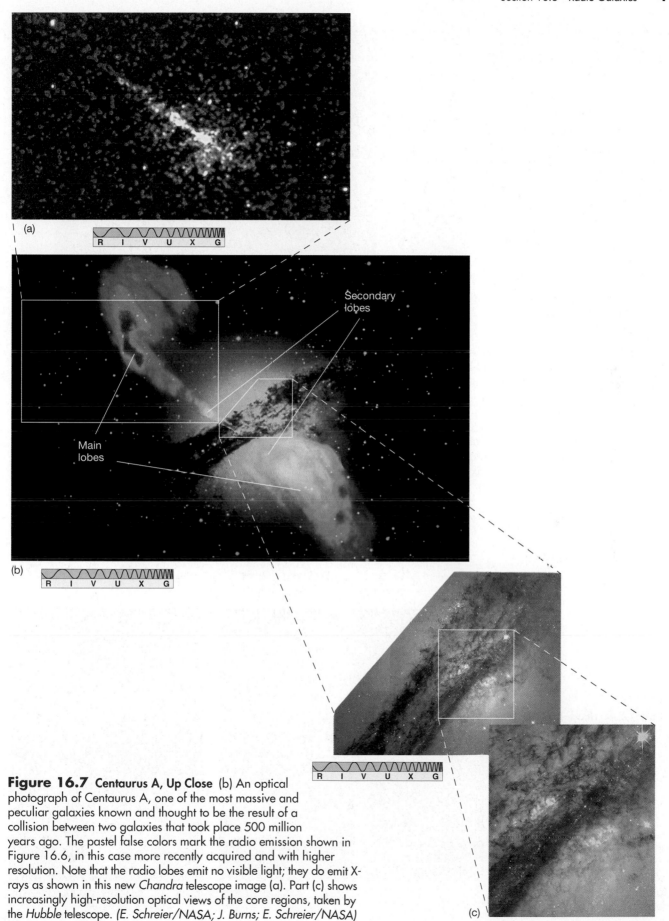

Figure 16.7 Centaurus A, Up Close (b) An optical photograph of Centaurus A, one of the most massive and peculiar galaxies known and thought to be the result of a collision between two galaxies that took place 500 million years ago. The pastel false colors mark the radio emission shown in Figure 16.6, in this case more recently acquired and with higher resolution. Note that the radio lobes emit no visible light; they do emit X-rays as shown in this new *Chandra* telescope image (a). Part (c) shows increasingly high-resolution optical views of the core regions, taken by the *Hubble* telescope. *(E. Schreier/NASA; J. Burns; E. Schreier/NASA)*

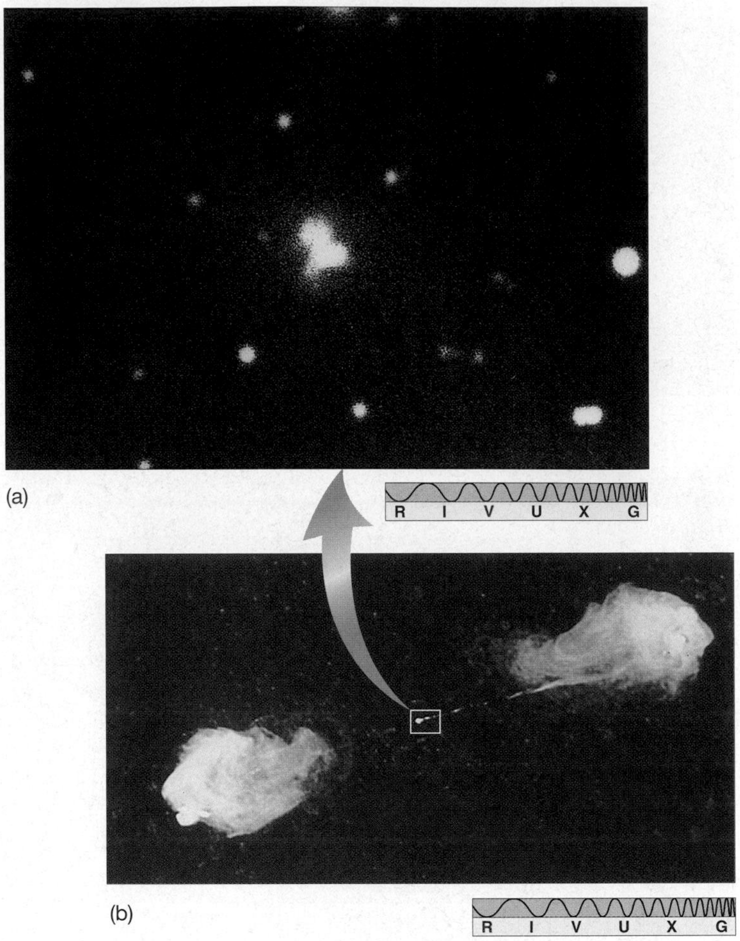

Figure 16.8 Cygnus A Radio Galaxy (a) Cygnus A also appears to be two galaxies in collision, although it is not completely clear what is really happening. (b) A radio image of Cygnus A shows the radio-emitting lobes on either side of the visible center. To put these images into proper perspective, the optical galaxy in (a) is about the size of the small dot at the center of the radio image in (b). Notice in particular the thin line of radio-emitting material joining the right lobe to the galaxy center. *(NOAO; NRAO)*

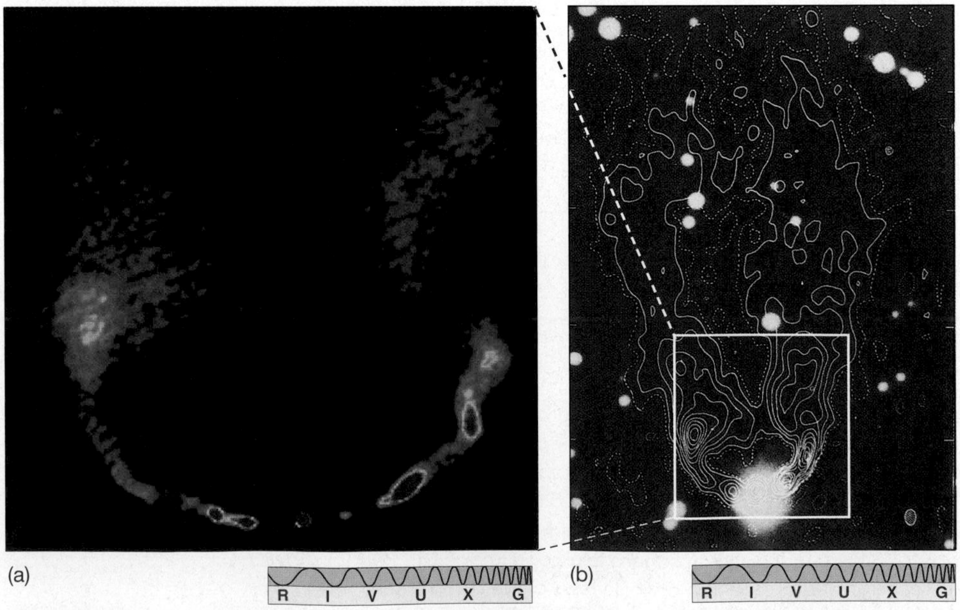

Figure 16.9 Head–Tail Radio Galaxy (a) Radio image, in false color, of the head–tail radio galaxy NGC 1265. (b) The same radio data, on a somewhat larger scale and in contour form, superposed on the optical image of the galaxy and its surroundings. Astronomers reason that this object is moving rapidly through space, the motion forcing the lobes to trail behind the central part of the galaxy, forming the tail. *(NRAO; Palomar Observatory/Caltech)*

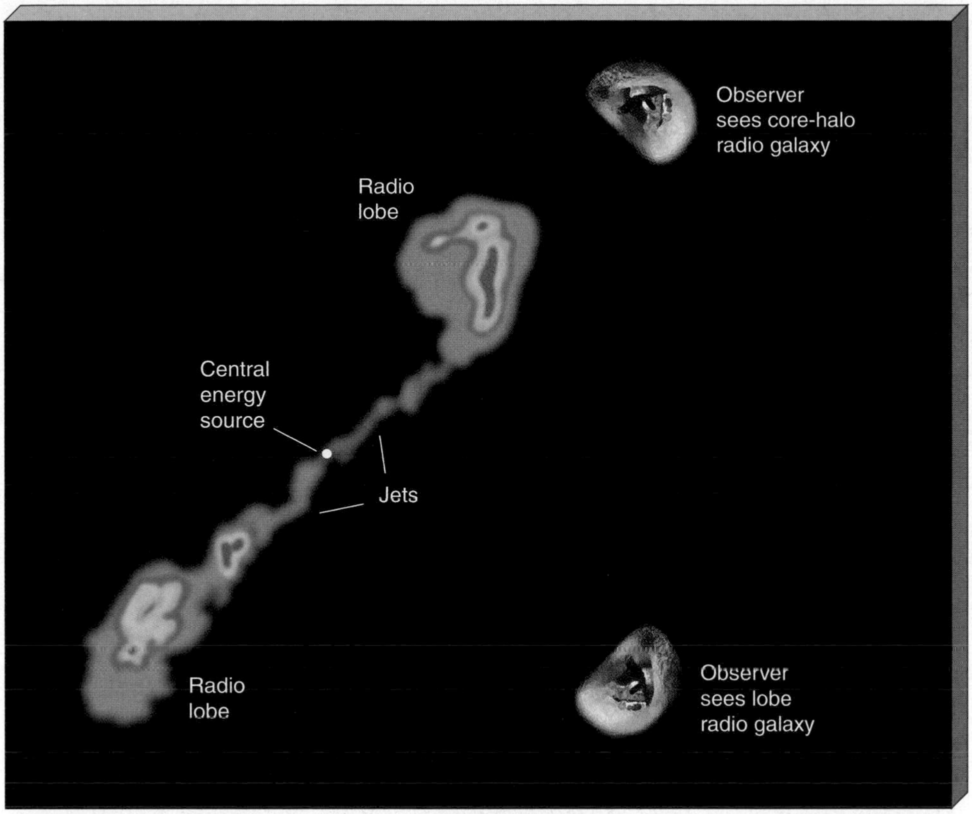

Figure 16.10 Radio Galaxy Appearance A central energy source produces high-speed jets of matter that interact with intergalactic gas to form radio lobes. The system may appear to us as either a lobe or a core–halo radio galaxy, depending on our location with respect to the jets and lobes.

16.4 The Central Engine of an Active Galaxy

3 The present consensus among astronomers is that, despite great differences in appearance, Seyferts and radio galaxies may share a common energy-generation mechanism. As a class, active galaxies (and quasars too, as we will see) have some or all of the following properties:

1. They have *high luminosities*, generally greater than the 10^{37} W, characteristic of a fairly bright normal galaxy.
2. Their energy emission is mostly *nonstellar*—it cannot be explained as the combined radiation of even trillions of stars.
3. Their energy output can be highly *variable*, implying that it is emitted from a small central nucleus much less than a parsec across.
4. They often exhibit *jets* and other signs of explosive activity.
5. Their optical spectra may show broad emission lines, indicating *rapid internal motion* within the energy-producing region.

The principal questions then are: How can such vast quantities of energy arise from these relatively small regions of space? Why is so much of the energy radiated at long wavelengths, in the radio and infrared? And what is the origin of the extended radio-emitting lobes and jets? We first consider how the energy is *produced*, then turn to the question of how it is actually *emitted* into intergalactic space.

Energy Production

To develop a feeling for the enormous emissions of active galaxies, consider for a moment an object having a luminosity of 10^{38} W. In and of itself, this energy output is not inconceivably large. The brightest giant ellipticals are comparably powerful. Thus, some 10^{12} stars—a few normal galaxies' worth of material—could *equivalently* power a typical active galaxy. However, in an active galaxy this energy production is packed into a region much less than a parsec in diameter.

The twin requirements of large energy generation and small physical size bring to mind our discussion of X-ray sources in Chapter 13. ∞ (Sec. 13.3, 13.7) The presence of the jet in M87 and the radio lobes in Centaurus A and Cygnus A strengthen the connection, as similar phenomena have also been observed in some stellar X-ray-emitting systems. Recall that the best current explanation for those "small-scale" phenomena involves the accretion of material onto a compact object—a neutron star or a black hole. Large amounts of energy are produced as matter spirals down onto the central object. In Chapter 14 we suggested that a similar mechanism, involving a *supermassive black hole*—having a mass of around a million suns—may also be responsible for the energetic radio and infrared emission observed at the center of our own Galaxy. ∞ (Sec. 14.6)

As illustrated in Figure 16.11, the leading model for the central engine of active galaxies is essentially a scaled-up version of the same accretion process, only now the black holes involved are millions or even *billions* of times more massive than the Sun. As with this model's smaller-scale counterparts, infalling gas forms an accretion disk and spirals down toward the black hole. It is heated to high temperatures by friction within the disk and emits large amounts of radiation as a result. In this case, however, the origin of the accreted gas is not a binary companion, as in stellar X-ray sources, but entire stars and clouds of interstellar gas that come too close to the hole and are torn apart by its strong gravity.

Accretion is extremely efficient at converting infalling mass (in the form of gas) into energy (in the form of electromagnetic radiation). Detailed calculations indicate that as much as 10 or 20 percent of the total mass–energy of the infalling matter can be radiated away before it crosses the hole's event horizon and is lost forever. ∞ (Sec. 13.4) Since the total mass–energy of a star like the Sun—the mass times the speed of light

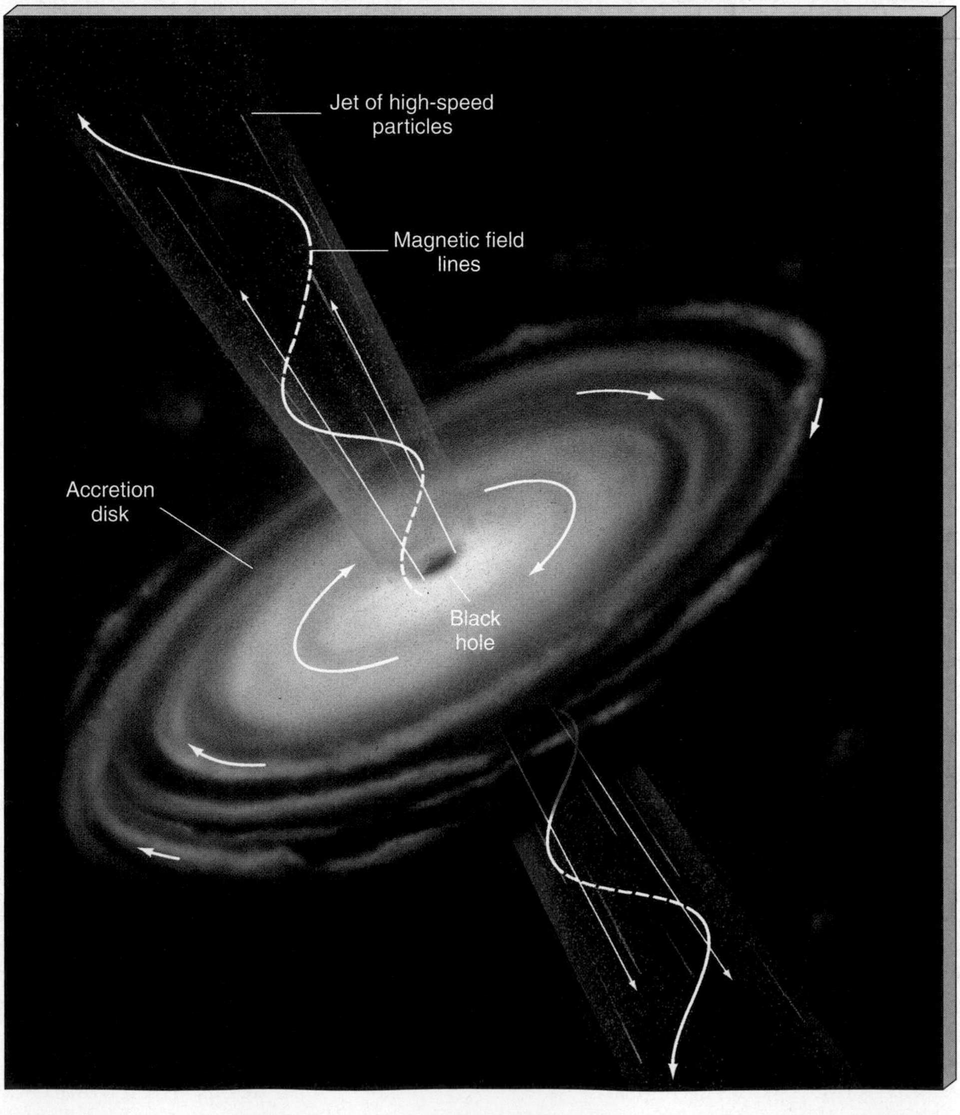

Animation

Figure 16.11 Active Galactic Nucleus The leading theory for the energy source in active galactic nuclei holds that these objects are powered by material accreting onto a supermassive black hole. As matter spirals toward the black hole, the matter heats up, producing large amounts of energy. At the same time, high-speed jets of gas may be ejected perpendicular to the accretion disk, giving rise to the jets and lobes seen in many active objects. The jets carry magnetic fields generated in the disk out to the radio lobes, where they play a crucial role in producing the radiation we observe.

squared—is about 2×10^{47} J, it follows that the 10^{38} W luminosity of a bright active galaxy can be accounted for by the consumption of only one solar mass of gas per decade by a billion-solar-mass black hole. Less luminous active galaxies would require correspondingly less fuel—for example, the central black hole of a 10^{36}-W Seyfert galaxy would devour only one Sun's worth of material every thousand years.

The small size of the emitting region is a direct consequence of the compact nature of the central black hole. Even a billion-solar-mass black hole has a radius of only 3×10^{9} km, or 10^{-4} pc—about 20 A.U.—and theory suggests that the part of the accretion disk responsible for most of the emission would be much less than 1 pc across. Instabilities in the accretion disk can cause fluctuations in the energy released, leading to the variability observed in many objects. The broadening of the spectral lines observed in the nuclei of many active galaxies results from the rapid orbital motion of the gas in the black hole's intense gravity.

The jets consist of material (mainly protons and electrons) blasted out into space—and completely out of the visible portion of the galaxy—from the inner regions of the disk. The details of how jets form remain uncertain, but there is a growing consensus among theorists that jets are a common feature of accretion flows, large and small. They are most likely formed by strong magnetic fields produced within the accretion disk itself, which accelerate charged particles to nearly the speed of light and eject them parallel to the disk's rotation axis. Figure 16.12 shows a *Hubble Space Telescope* image of a disk of gas and dust at the core of the radio galaxy NGC 4261 in the Virgo Cluster. Consistent with the theory just described, the disk is perpendicular to the huge jets emanating from the galaxy's center.

Figure 16.13 shows imaging and spectroscopic data from the center of M87 (Figure 16.5), suggesting a rapidly rotating disk of matter orbiting the galaxy's center, again perpendicular to the jet. Measurements of the gas velocity on opposite sides of the disk indicate that the mass within a few parsecs of the center is approximately 3×10^{9} solar masses—we assume that this is the mass of the central black hole. At M87's distance, *HST*'s resolution of 0.05 arc second corresponds to a scale of about 5 pc, so we are still far from seeing the (solar-system-sized) central black hole itself, but the improved "circumstantial" evidence has convinced many astronomers of the basic correctness of the theory.

Perhaps the most compelling evidence for supermassive black holes at the centers of galaxies comes from radio studies. Using the Very Long Baseline Array, a continent-wide network of 10 radio telescopes, a U.S.-Japanese team has achieved angular resolution hundreds of times better than that attainable with *HST*. Observations of NGC 4258, a spiral galaxy about 6 Mpc away, have uncovered a group of molecular clouds swirling in an organized fashion about the galaxy's center. As depicted in Figure 16.14, Doppler measurements reveal a slightly warped, spinning disk centered precisely on the galaxy's heart. The rotation speeds imply the presence of more than 40 million solar masses packed into a region less than 0.2 pc across.

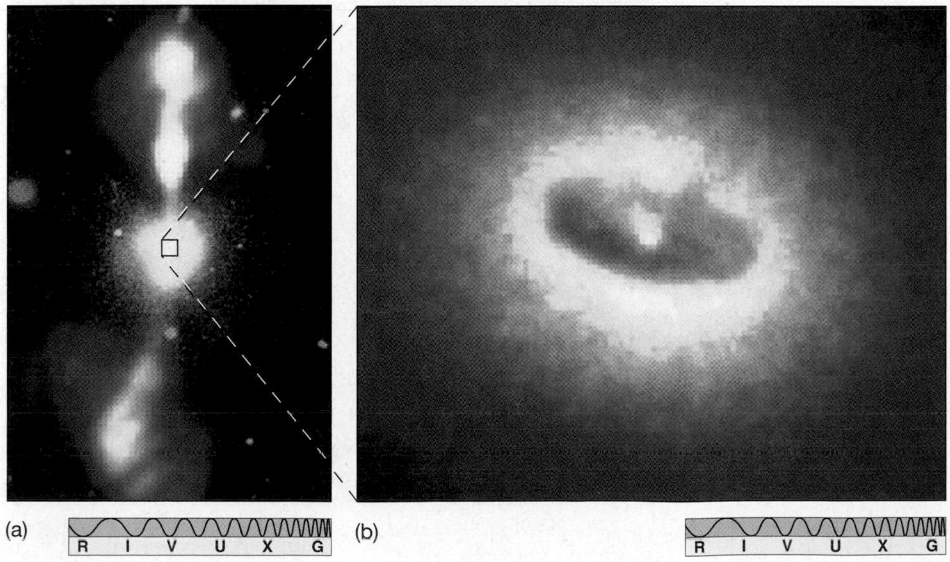

Animation

Figure 16.12 Giant Elliptical Galaxy (a) A combined optical/radio image of the giant elliptical galaxy NGC 4261, in the Virgo Cluster, shows a white visible galaxy at center, from which red-orange (false color) radio lobes extend for about 60 kpc. (b) A close-up photograph of the galaxy's nucleus reveals a 100-pc-diameter disk surrounding a bright hub thought to harbor a black hole. *(NASA)*

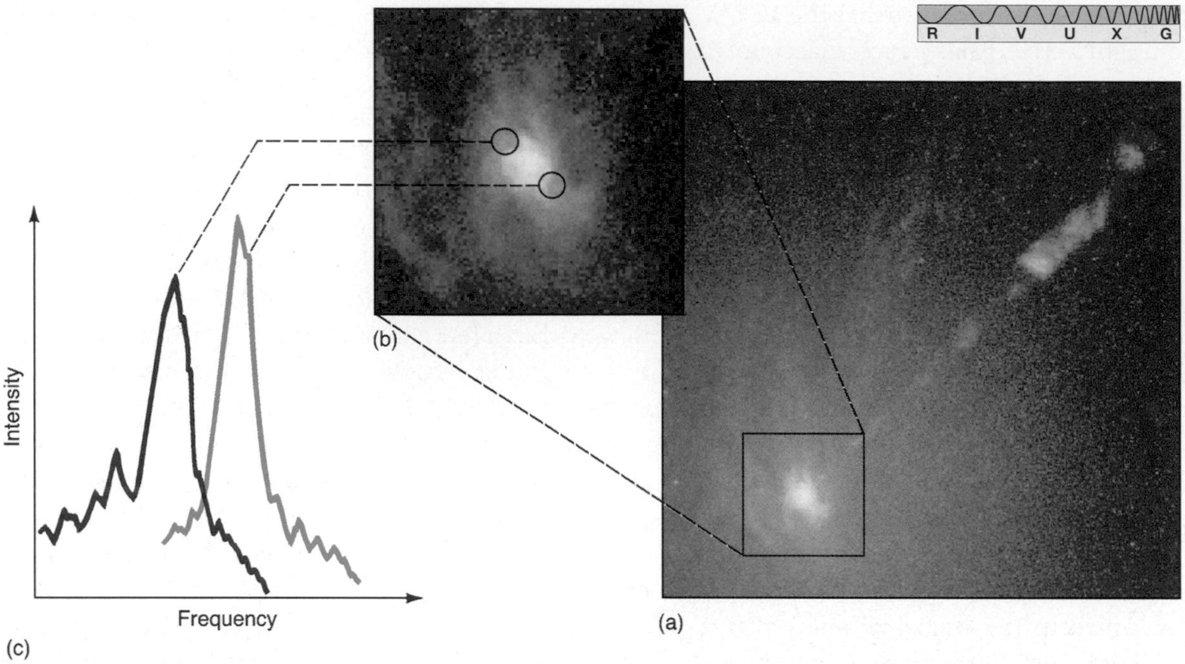

Figure 16.13 **M87 Disk** Recent imaging and spectroscopic observations of M87 support the idea of a rapidly whirling accretion disk at its heart. (a) An image of the central region of M87, similar to that shown in Figure 16.5(d), shows its bright nucleus and jet. The scale is comparable to the scale of Figure 16.5(c). (b) A magnified view of the nucleus suggests a spiral swarm of stars, gas, and dust. (c) Spectral-line features observed on opposite sides of the nucleus show opposite Doppler shifts, implying that material on one side of the nucleus is coming toward us and material on the other side is moving away from us. The strong implications are that an accretion disk spins perpendicular to the jet and that at its center is a black hole having three billion times the mass of the Sun. *(NASA)*

Animation

Energy Emission

In general, the theoretically predicted radiation spectra for matter accreting onto a black hole do not match well with the spectra observed in most active galaxies. In Seyfert galaxies, it appears that the energy emitted from the accretion disk itself is "reprocessed"—that is, absorbed and reemitted at infrared and longer wavelengths—by gas and dust surrounding the nucleus before eventually reaching our detectors. A different reprocessing mechanism operates in radio galaxies. It involves the *magnetic fields* produced within the accretion disk and transported by the jets into intergalactic space (Figure 16.11).

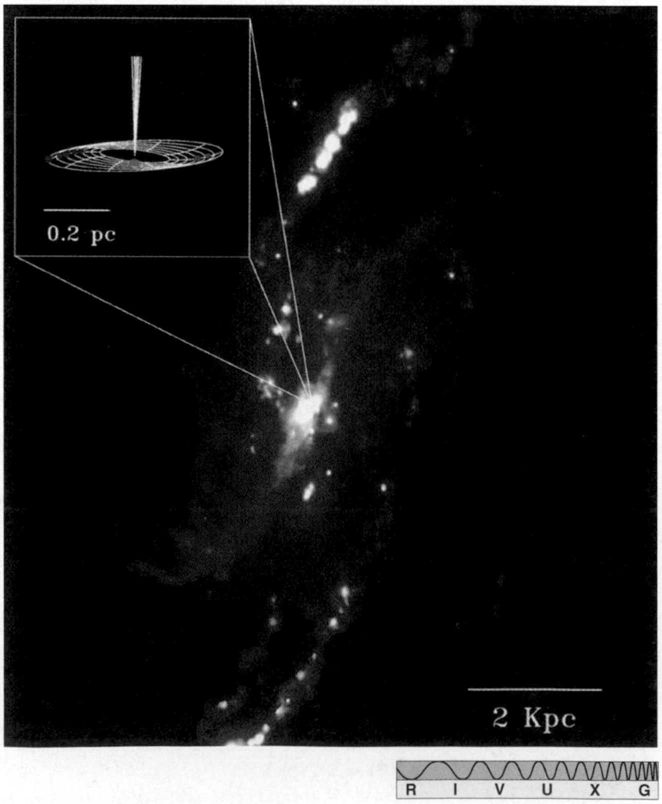

Figure 16.14 **Galaxy Black Hole** A network of radio telescopes has probed the nucleus of the spiral galaxy NGC 4258, shown here in the light of mostly hydrogen emission. Within the innermost 0.2 pc (inset showing an artist's conception), observations of Doppler-shifted molecular clouds (designated by red, green, and blue dots) show that they obey Kepler's third law perfectly and reveal a slightly warped disk of rotating gas. At the center of the disk presumably lurks a huge black hole. *(James Moran)*

As sketched in Figure 16.15(a), whenever a charged particle (here an electron) encounters a magnetic field, the particle tends to spiral around the magnetic field lines. We have already encountered this idea in the discussions of Earth's magnetosphere and solar activity. ∞ (Sec. 5.7, 9.4) As the particles whirl around, they emit electromagnetic radiation. ∞ (Sec. 2.2) The faster the particles move, or the stronger the magnetic field, the greater the amount of energy radiated. In most cases, the electrons are the fastest-moving particles, and they are responsible for virtually all of the radiation we observe.

The electromagnetic radiation produced in this way—called **synchrotron radiation**, after the type of particle accelerator in which it was first observed—is *nonthermal* in nature: There is no link between the emission and the temperature of the radiating object, so the radiation is not described by a blackbody curve. Instead, its intensity decreases with increasing frequency, as shown in Figure 16.15(b). This is just what is needed to explain the overall spectrum of radiation recorded from active galaxies (compare Figure 16.15b with Figure 16.1). Observations of the radiation received from the jets and radio lobes of active galaxies are completely consistent with this process.

Eventually, the jet is slowed and stopped by the intergalactic medium, the flow becomes turbulent, and the magnetic field grows tangled. The result is a gigantic radio lobe emitting virtually all of its energy in the form of synchrotron radiation. Thus, even though the radio *emission* comes from an enormously extended volume of space that dwarfs the visible galaxy, the *source* of the energy is still the accretion disk—a billion billion times smaller in volume than the radio lobe—lying at the galactic center. The jets serve merely as a conduit to transport energy from the nucleus, where it is generated, into the lobes, where it is finally radiated into space.

As we have seen, jet formation seems to be an intermittent process, and many nearby active galaxies (Centaurus A and Cygnus A, for example) appear to have been "caught in the act" of interacting with another galaxy, suggesting that in many, if not all, cases, the fuel supply can be turned on by a companion. Just as tidal forces can trigger star formation in starburst galaxies, they may also divert gas and stars into the galactic nucleus, triggering an outburst that may last for millions or even billions of years. ∞ (Sec. 15.4)

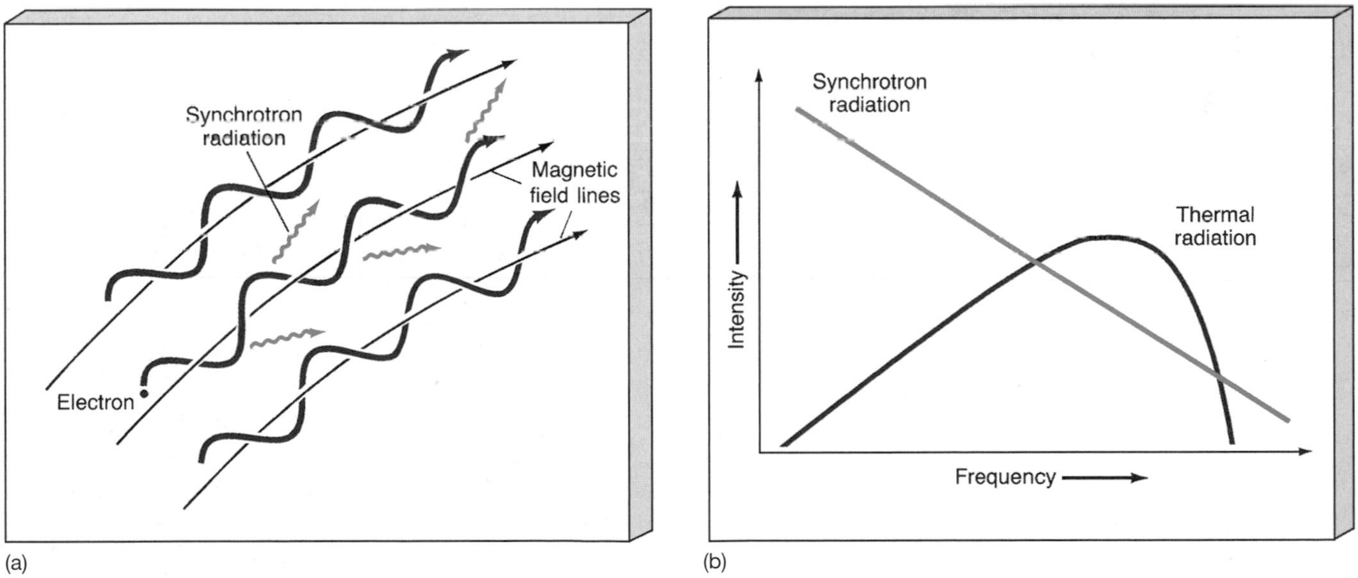

(a) (b)

Figure 16.15 Nonthermal Radiation (a) Charged particles, especially fast-moving electrons, emit synchrotron radiation while spiraling in a magnetic field. This process is not confined to active galaxies. It occurs, on smaller scales, when charged particles interact with magnetism in Earth's Van Allen belts ∞ (Sec. 5.6), when charged matter arches above sunspots on the Sun ∞ (Sec. 9.4), in the vicinity of neutron stars ∞ (Sec. 13.2), and at the center of our own Galaxy ∞ (Sec. 14.7). (b) How the intensity of thermal and synchrotron (nonthermal) radiation varies with frequency. Thermal radiation, described by a blackbody curve, peaks at some frequency that depends on the temperature of the source. Nonthermal synchrotron radiation, by contrast, is most intense at low frequencies and is independent of the temperature of the emitting object. Compare this figure with Figure 16.1.

Concept Check

- How does accretion onto a supermassive black hole power the energy emission from a lobe radio galaxy?

16.5 Quasi-Stellar Objects

The Discovery of Quasars

4 In the early days of radio astronomy, many radio sources were detected for which no corresponding visible object was known. By 1960, several hundred such sources were listed in the *Third Cambridge Catalog* and astronomers were scanning the skies in search of visible counterparts to these radio sources. Their job was made difficult both by the low resolution of the radio observations (which meant that the observers did not know exactly where to look) and by the faintness of these objects at visible wavelengths.

In 1960, astronomers detected what appeared to be a faint blue star at the location of the radio source 3C 48 (the 48th object on the third *Cambridge* list) and obtained its spectrum. Containing many unknown broad emission lines, the unusual spectrum defied interpretation. 3C 48 remained a unique curiosity until 1962, when another similar-looking, and similarly mysterious, faint blue object with "odd" spectral lines was discovered and identi-

MORE PRECISELY 16-1

Relativistic Redshifts and Look-Back Time

When discussing very distant objects, astronomers usually talk about their redshifts rather than their distances. Indeed, it is very common for researchers to speak of an event occurring "at" a certain redshift—meaning that the light received today from that event is redshifted by the specified amount. Of course, because of Hubble's law, redshift and distance are equivalent to one another. However, redshift is the preferred quantity because it is a directly observable property of an object, whereas distance is derived from redshift using Hubble's constant, whose value is not accurately known. (In the next chapter, we will see another reason why astronomers favor the use of redshift in studies of the cosmos.)

The redshift of a beam of light is, by definition, the *fractional* increase in its wavelength resulting from the recessional motion of the source. (*More Precisely 2-3*) Thus, a redshift of 1 corresponds to a *doubling* of the wavelength. Using the formula for the Doppler shift given previously, the redshift of radiation received from a source moving away from us with speed v is given by

$$\text{redshift} = \frac{\text{observed wavelength} - \text{true wavelength}}{\text{true wavelength}}$$

$$= \frac{\text{recessional velocity, } v}{\text{speed of light, } c}.$$

Let's illustrate this with two examples, rounding the speed of light, c, to 300,000 km/s. A galaxy at a distance of 100 Mpc has a recessional speed (by Hubble's law) of 65 km/s/Mpc × 100 Mpc = 6,500 km/s. Its redshift therefore is 6,500 km/s ÷ 300,000 km/s = 0.022. Conversely, an object that has a redshift of 0.05 has a recessional velocity of 0.05 × 300,000 km/s = 15,000 km/s and hence a distance of 15,000 km/s ÷ 65 km/s/Mpc = 230 Mpc.

Unfortunately, while the foregoing equation is correct for low speeds, it does not take into account the effects of relativity. As we saw in Chapter 13, the rules of everyday physics have to be modified when speeds begin to approach the speed of light, and the formula for the Doppler shift is no exception. (*More Precisely 13-1*). In particular, while our formula is valid for speeds much less than the speed of light, when $v = c$ the redshift is not 1, as the equation suggests, but is in fact *infinite*. Radiation received from an object moving away from us at nearly the speed of light is redshifted to almost infinite wavelength.

Thus do not be alarmed to find that many quasars have redshifts greater than 1. This does not mean that they are receding faster than light! It simply means that their recessional speeds are relativistic—comparable to the speed of light—and the preceding simple formula is not applicable. Table 16.1 presents a conversion chart relating redshift, recession speed, and present distance. The column headed "v/c" gives recession velocities based on the Doppler effect, taking relativity properly into account (but see Section 17.2 for a more correct interpretation of the redshift). All values are based on reasonable assumptions and are usable even for $v \approx c$. We take Hubble's constant to be 65 km/s/Mpc and assume a critical-density, flat universe (see Section 17.3). The conversions in the table are used consistently throughout this text.

fied with the radio source 3C 273. Several more of these peculiar objects are shown in Figure 16.16.

The following year saw a breakthrough when astronomers realized that the strongest unknown lines in 3C 273's spectrum were simply familiar spectral lines of hydrogen redshifted by a very unfamiliar amount—about 16 percent, corresponding to a recession velocity of 48,000 km/s. Figure 16.17 shows the spectrum of 3C 273. Some prominent emission lines and the extent of their redshift are marked on the diagram. Once the nature of the strange spectral lines was known, astronomers quickly found a similar explanation for the spectrum of 3C 48. Its 37 percent redshift implied that it is receding from Earth at almost one-third the speed of light.

These huge speeds mean that neither of these two objects can possibly be members of our Galaxy. Applying Hubble's law with our adopted value of H_0 = 65 km/s/Mpc, we obtain distances of 660 Mpc for 3C 273 and 1340 Mpc for 3C 48. *More Precisely 16-1* discusses in more detail how these distances are determined and what they mean. Clearly not stars (because of such enormous redshifts), these objects became known as *quasi-stellar radio sources* (quasi-stellar means "starlike"), or **quasars**. Because we now know that not all such highly redshifted, starlike objects are strong radio sources, the term **quasi-stellar object** (or QSO) is more common today. However, the name quasar persists, and we will continue to use it here.

Because the universe is expanding, the "distance" to a galaxy is not very well defined—do we mean the distance when the galaxy emitted the light we see today, or the present distance (as presented in the table, even though we do not see the galaxy as it is today), or is some other measure more appropriate? Largely because of this ambiguity, astronomers prefer to work in terms of a quantity known as the *look-back time* (shown in the last column of Table 16.1), which is simply how long ago an object emitted the radiation we see today. While astronomers talk frequently about redshifts and sometimes about look-back times, they hardly ever talk of distances to high-redshift objects.

For nearby sources, the look-back time is numerically equal to the distance in light-years—the light we receive tonight from a galaxy at a distance of 100 million light-years was emitted 100 million years ago. However, for more distant objects, the look-back time and the present distance in light-years differ because of the expansion of the universe, and the divergence increases dramatically with increasing redshift (see Section 17.2). For example, a galaxy now located 15 billion light-years from Earth was much closer to us when it emitted the light we now see. Consequently, its light has taken considerably less than 15 billion years—in fact, only about 8.8 billion years—to reach us.

TABLE 16.1 Redshift, Distance, and Look-Back Time

REDSHIFT	v/c	PRESENT DISTANCE (Mpc)	PRESENT DISTANCE (10^6 light-years)	LOOK-BACK TIME (millions of years)
0.000	0.000	0	0	0
0.010	0.010	46	149	149
0.025	0.025	113	369	365
0.050	0.049	222	725	708
0.100	0.095	429	1401	1336
0.200	0.180	804	2622	2400
0.250	0.220	974	3177	2853
0.500	0.385	1693	5523	4570
0.750	0.508	2251	7346	5697
1.0	0.600	2702	8815	6483
1.5	0.724	3390	11,062	7492
2.0	0.800	3899	12,720	8099
3.0	0.882	4612	15,048	8775
4.0	0.923	5099	16,637	9132
5.0	0.946	5459	17,809	9347
10.0	0.984	6443	21,022	9754
50.0	0.999	7933	25,882	10,001
100.0	0.9998	8307	27,101	10,019
∞	1.000	9224	30,096	10,029

Animation

Figure 16.16 Distant Quasar (a) 3C 275 was one of the first quasars discovered. Its starlike appearance shows no obvious structure and gives little outward indication of this object's enormous luminosity. However, 3C 275 has a much larger redshift than any of the other stars or galaxies in this image; it is about two billion parsecs away. (b) A field of quasars (marked), including QSO 1229+204, one of the most powerful quasars known, shown enlarged in (c). Like 3C 275, its distance from Earth is about 2000 Mpc. *(NOAO; Palomar Observatory/California Institute of Technology; NASA)*

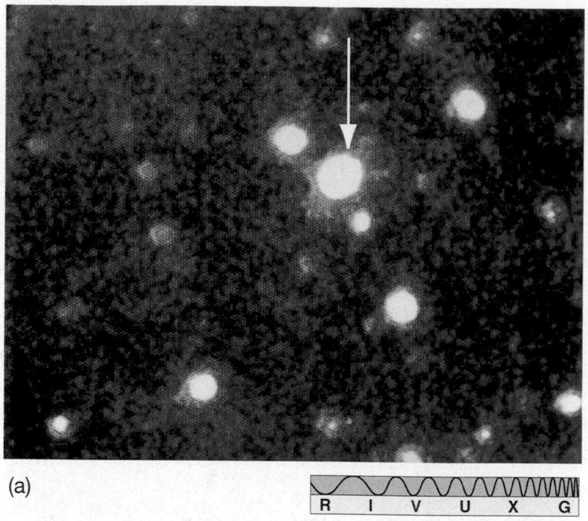

(a)

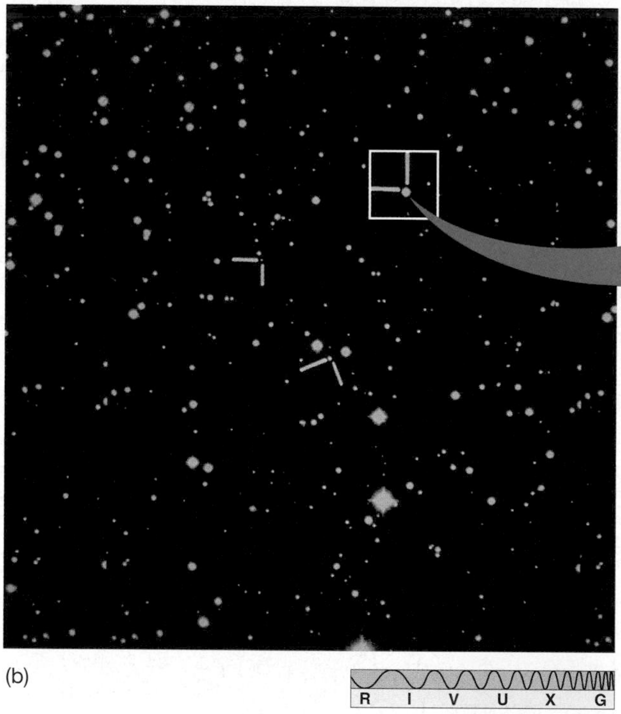

(b)

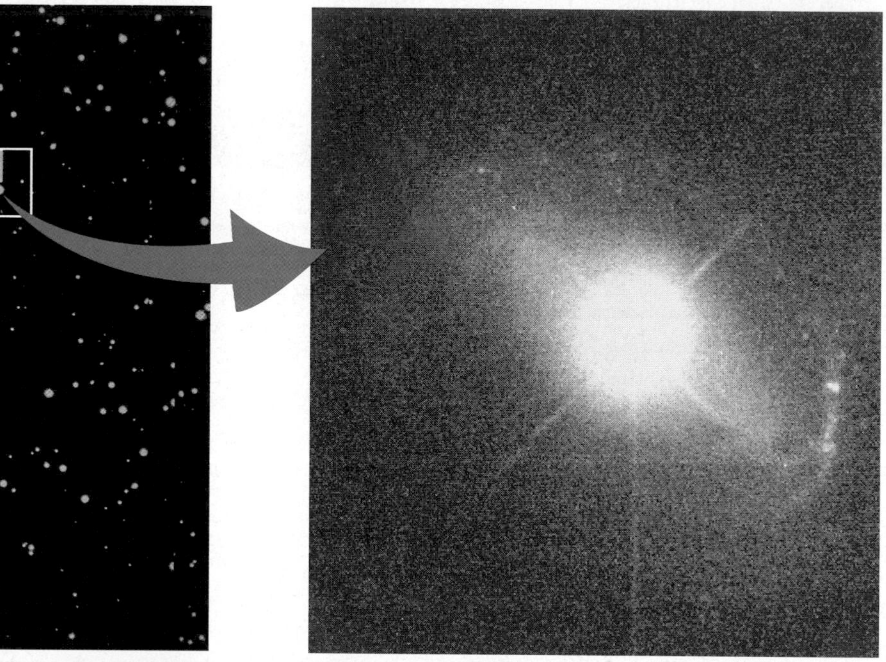

(c)

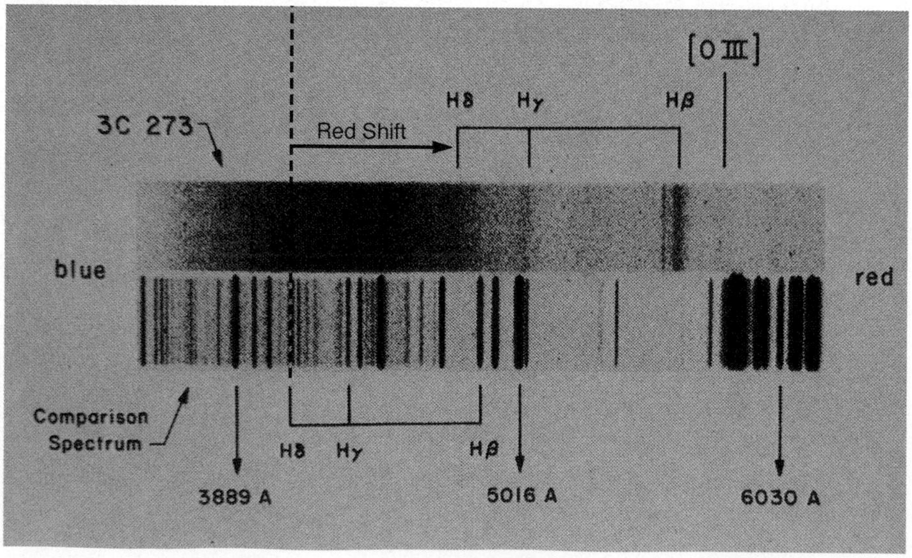

Figure 16.17 Quasar Spectrum In this optical spectrum of the distant quasar 3C 273, notice both the redshift and the widths of the three hydrogen spectral lines marked as Hβ, Hγ, and Hδ. The redshift indicates the quasar's enormous distance. The width of the lines implies rapid internal motion within the quasar. (Note that, in this figure, red is to the right and blue is to the left.) *(Palomar Observatory/California Institute of Technology)*

Observed Properties of Quasars

The most striking characteristic of the thousands of quasars now known is that their spectra all show large redshifts, ranging from 0.06 (that is, a 6 percent increase in wavelength) up to the current maximum of just under 6. (See again *More Precisely 16-1* for an explanation of redshifts greater than 1.) Thus, *all* quasars lie at large distances from us—the closest is 240 Mpc away, the farthest nearly 5500 Mpc (Table 16.1). Most quasars lie much more than 1000 Mpc from Earth—they represent the universe as it was in the distant past.

Figure 16.18, compares a quasar and a spiral galaxy that happen to lie close to one another on the sky. Despite their unimpressive optical appearance, the large distances implied by quasar redshifts mean that these faint "stars" are in fact the brightest known objects in the universe! 3C 273, for example, has a luminosity of about 10^{40} W. More generally, quasars range in luminosity from around 10^{38} W—about the same as the brightest radio galaxies—up to nearly 10^{42} W. A value of 10^{40} W, comparable to 20 trillion Suns or 1000 Milky Way Galaxies, is fairly typical.

Quasars have many of the same general properties as active galaxies. Their radiation is nonthermal, may vary irregularly in brightness over periods of months, weeks, days, or (in some cases) even hours, and some show evidence of jets and extended emission features. Figure 16.19 is an optical photograph of 3C 273. Notice the jet of luminous matter, reminiscent of the jet in M87, extending nearly 3 kpc from the center of the quasar. Often, as shown in Figure 16.20, quasar radio radiation arises from regions lying beyond the bright nucleus, much like the emission from core–halo and lobe radio galaxies. In other cases, the radio emission is confined to the central part of the visible image. Quasars have been observed in all parts of the electromagnetic spectrum, although most emit most of their energy in the infrared.

In addition to their own strongly redshifted spectra, many quasars also show additional absorption features that are redshifted by substantially *less* than the lines from the quasar itself. For example, the quasar PHL 938 has an emission-line redshift of 1.955, placing it at a distance of some 3400 Mpc, but it also shows absorption lines having redshifts of just 0.613. These are interpreted as arising from intervening gas that is much closer to us (only about 1700 Mpc away) than the quasar itself. Most probably this gas is part of an otherwise invisible galaxy lying along the line of sight, affording astronomers an important means of probing previously undetected parts of the universe.

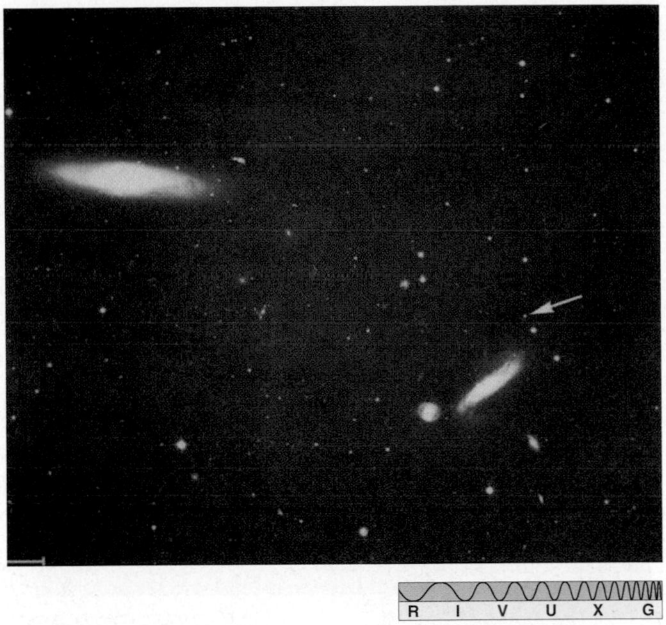

Figure 16.18 Typical Quasar Although quasars are the most luminous objects in the universe, they are often unimpressive in appearance. In this optical image, a distant quasar (marked by an arrow) is seen close (on the sky) to a nearby spiral galaxy. The quasar's much greater distance makes it appear much fainter than the galaxy. *(AURA)*

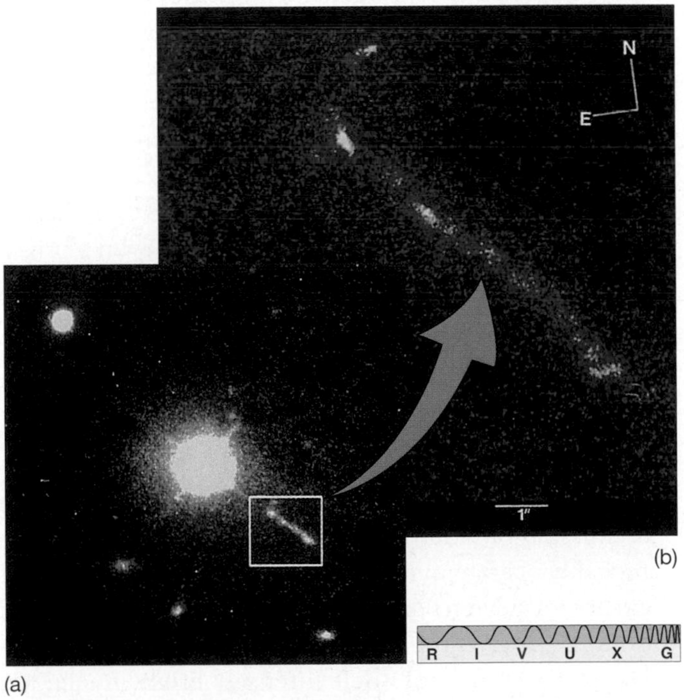

Figure 16.19 Quasar 3C 273 (a) The bright quasar 3C 273 displays a luminous jet of matter, but the main body of the quasar is starlike in appearance. (b) The jet extends for about 30 kpc and can be seen better in this high-resolution image. *(AURA)*

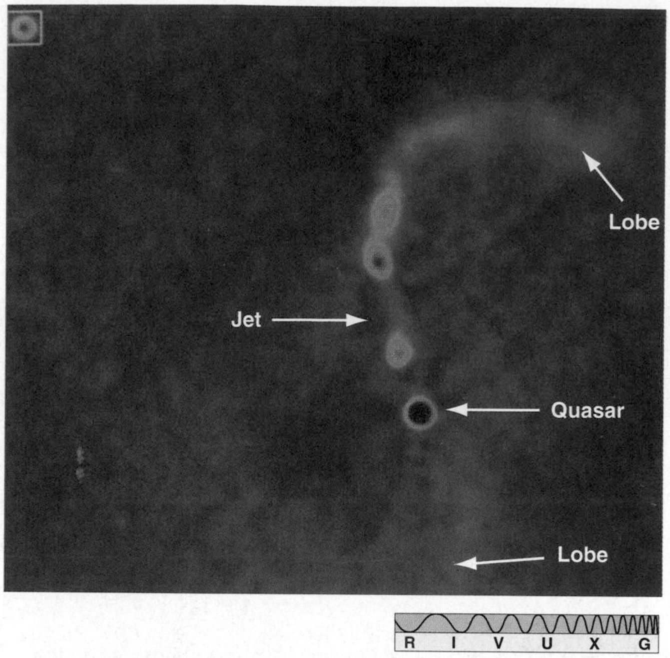

Figure 16.20 Quasar Jets Radio image of the quasar 2300-189 showing radio jets feeding faint radio lobes. The bright (red) central object is the quasar, some 400 Mpc away from Earth. *(NRAO)*

Quasar Energy Generation and Lifetimes

5 Quasars have all the general properties described earlier for active galaxies, so it should come as no surprise that the current best explanation of the quasar engine is basically a scaled-up version of the mechanism powering lower-luminosity active galaxies—accretion onto a central, supermassive black hole.

A black hole having a mass of 10^8 or 10^9 solar masses can emit enough energy to power even a bright (10^{38} W) radio galaxy by swallowing stars and gas at the relatively modest rate of one star every 10 years. To power a 10^{40}-W quasar, which is 100 times brighter, the black hole simply consumes 100 times more fuel—10 stars per year. The reprocessing mechanisms that convert the quasar's energy into the radiation we detect—namely, the ejection of matter in jets and lobes and the reemission of radiation by surrounding gas and dust—probably operate in much the same manner as described earlier for Seyferts and radio galaxies. At the distances of most quasars, the galaxies themselves cannot easily be seen. Often, only their intensely bright nuclei are visible from Earth.

By this account, the brightest known quasars devour about 1000 solar masses of material every year. A simple estimate of the total amount of fuel required to sustain this rate of consumption for 10 billion years (the approximate age of the universe, based on globular cluster ages), together with the prevalence of quasars at high redshifts, suggests to most astronomers that a typical quasar spends only a fairly short period of time in this highly luminous phase—perhaps a few tens of millions of years—before running out of fuel. ∞ (Sec. 12.6)

The most likely explanation for the high luminosities of young quasars is simply that there was more fuel available at very early times, perhaps left over from the formation of the galaxies in which the quasars reside, or (as with nearby active galaxies) the result of galaxy interactions in the early universe.

Quasar "Mirages"

In 1979, astronomers were surprised to discover what appeared to be a binary quasar—two quasars with exactly the same redshift and very similar spectra, separated by only a few arc seconds on the sky. Remarkable as the discovery of such a binary would have been, the truth about this pair of quasars turned out to be even more amazing. Closer study of the quasars' radio emission revealed that they were *not* two distinct objects. Instead, they were two separate images of the *same* quasar! Optical views of such a *twin quasar* are shown in Figure 16.21.

What could produce such a "doubling" of a quasar image? The answer is gravitational lensing—the deflection and focusing of light from a background object by the gravity of some foreground body (Figure 16.22). In Chapter 14 we saw how lensing by compact objects in the halo of the Milky Way Galaxy may amplify the light from a distant star, allowing astronomers to detect otherwise invisible stellar dark matter. ∞ (Sec. 14.5) In the case of quasars, the idea is the same, except that the foreground lensing object is an entire galaxy or galaxy cluster, and the deflection of the light is so great (a few arc seconds) that several separate images of the quasar may be formed, as shown in Figure 16.23. About two dozen such gravitational lenses are known.

The existence of these multiple images provides astronomers with a number of useful observational tools. First, lensing by a foreground galaxy tends to amplify the light of the quasar, as just mentioned, making it easier to observe. At the same time, *microlensing* by individual stars within the galaxy may cause large fluctuations in the quasar's brightness, allowing astronomers to study the galaxy's stellar content.

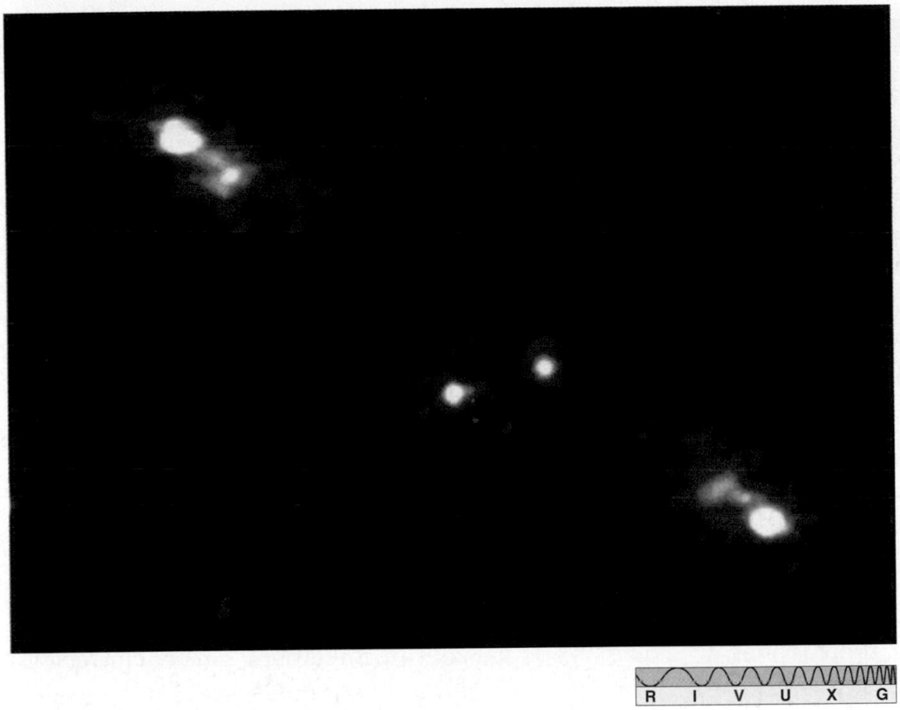

Figure 16.21 **Twin Quasar** This twin quasar (designated AC114 and located about two billion parsecs away) is not two separate objects at all. Instead, the two large "blobs" (at upper left and lower right) are images of the same object, created by a gravitational lens. The lensing galaxy itself is probably not visible in this image—the two objects near the center of the frame are thought to be unrelated galaxies in a foreground cluster. *(NASA)*

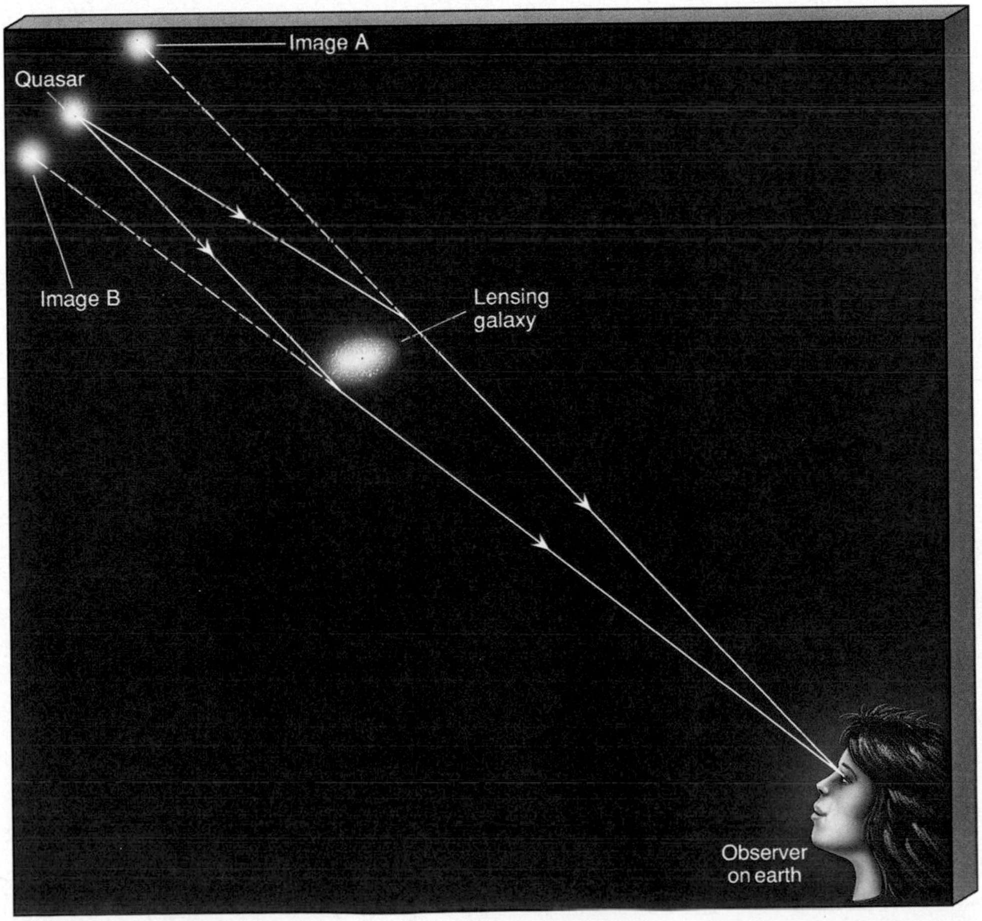

Figure 16.22 **Gravitational Lens** When light from a distant object passes close to a galaxy or cluster of galaxies along the line of sight, the image of the background object (here, the quasar) can sometimes be split into two or more separate images (A and B). The foreground object is a gravitational lens.

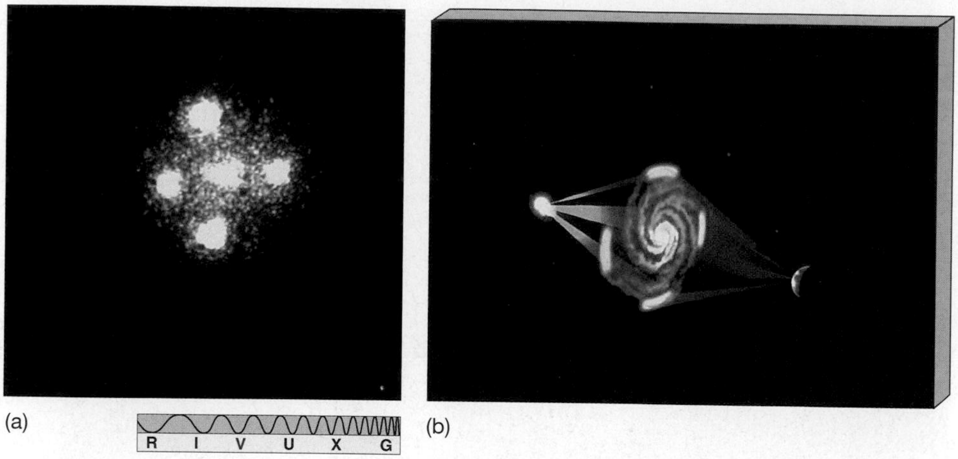

Figure 16.23 **Einstein's Cross** (a) The Einstein Cross, a multiply imaged quasar. In this *Hubble Space Telescope* view, spanning only a couple of arc seconds, four images of the same quasar have been produced by the galaxy at the center. (b) An artist's conception of what might be occurring here, with Earth at right and the distant quasar at left. *(NASA; D. Berry)*

Second, because the light rays forming the images usually follow paths of different lengths, there is often a time delay, ranging from days to years, between them. This delay provides advance notice of interesting events, such as sudden changes in the quasar's brightness—if one image flares up, in time the other(s) will too, giving astronomers a second chance to study the event. The time delay between these events also allows astronomers to determine the distance to the lensing galaxy. If enough lenses can be found, this method may provide an alternative means of measuring Hubble's constant that is independent of any of the techniques discussed in the previous chapter. (Sec. 15.5)

Finally, by studying the lensing of background quasars and galaxies by foreground galaxy clusters, astronomers can obtain a better understanding of the distribution of dark matter in those clusters. (Sec. 15.3) Figure 16.24 shows how the images of faint, background galaxies are bent into arcs by the gravity of a nearby galaxy cluster. The degree of bending allows the total mass of the cluster (*including* the dark matter) to be measured.

Concept Check

- Why did astronomers initially have difficulty recognizing quasars as very luminous, very distant objects?

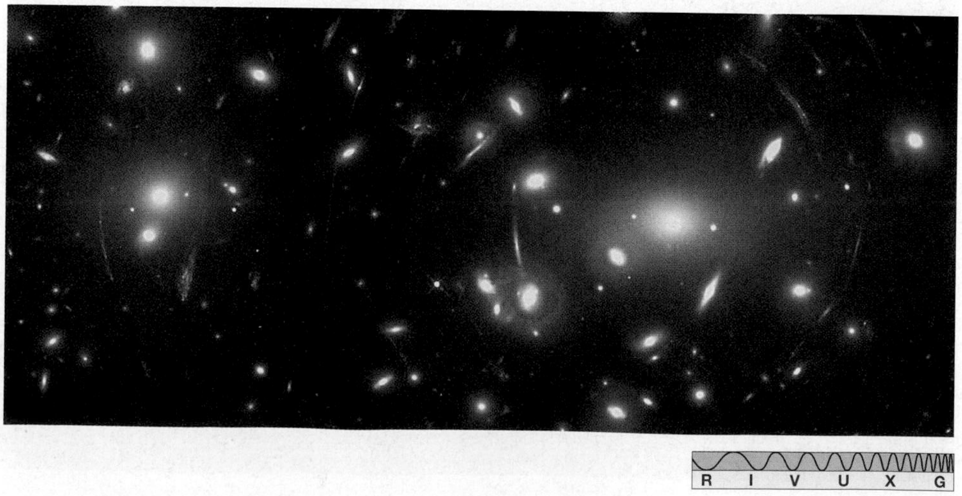

Figure 16.24 **Galaxy Cluster Lens** A spectacular example of gravitational lensing by a galaxy cluster. This wispy "spider's web" of more than a hundred faint arcs around a foreground galaxy cluster (A2218; itself one billion parsecs away from us) is an illusion caused by the cluster's gravitational field, which deflects and distorts the light from very distant background galaxies. By measuring the extent of this distortion, astronomers can estimate the mass of the cluster. *(NASA)*

16.6 Active Galaxy Evolution

5 In previous chapters we addressed the issue of evolutionary change in normal galaxies. (Sec. 14.3, 15.4) Let us now briefly consider the possibility of evolutionary links among active galaxies and between normal and active galaxies.

Most quasars are very distant, indicating that they were more common in the past than they are today. At the same time, normal galaxies seem to have been less common in the distant past. These two pieces of evidence suggest to many astronomers that, when galaxies first formed, they probably were quasars. This view is strengthened by the fact that the same black hole energy-generation mechanism can account for the luminosity of quasars, active galaxies, and the central regions of normal galaxies. Large black holes do not simply vanish. Thus, the presence of supermassive black holes in the centers of many normal galaxies is consistent with the idea that they started off as quasars, then "wound down" to become the relatively quiescent objects we see today. Some possible evolutionary connections between active and normal galaxies are illustrated in Figure 16.25.

We can thus construct the following (speculative) scenario for the evolution of galaxies in the universe. Galaxies began to form some 9–10 billion years ago (corresponding to a redshift of 5, the upper limit on the measured redshifts of known quasars—see Table 16.1). The first massive stars to form in a galaxy may have given rise to many stellar-mass black holes which sank to the galactic center and merged into a supermassive black hole there. Alternatively, the supermassive hole may have formed directly by gravitational collapse of the galaxy's dense central regions. Whatever their precise origin, large black holes appeared at the centers of many galaxies at a time when there was still plenty of fuel available to power them, resulting in many highly luminous quasars. The brightest quasars—the ones we now see from Earth—were those with the greatest fuel supply.

Figure 16.25 Galaxy Evolution A possible evolutionary sequence for galaxies, beginning with the highly luminous quasars, decreasing in violence through the radio and Seyfert galaxies, and ending with normal spirals and ellipticals. The central black holes that powered the early activity are still there at later times but they run out of fuel as time goes on.

Young galaxies at this early stage were much fainter than their bright quasar cores. As a result, until quite recently, astronomers were hard-pressed to discern any galactic structure in quasar images. Since 1996, several groups of astronomers have used the *Hubble Space Telescope* to search for quasar "host" galaxies. After removing the bright quasar core from the *HST* images and carefully analyzing the remnant light, the researchers have reported that, in every case studied—several dozen quasars so far—a normal host galaxy can be seen enveloping the quasar. Figure 16.26 shows some of the longest quasar exposures ever taken. Even without sophisticated computer processing, the hosts are clearly visible.

As a young galaxy developed and its central black hole used up its fuel, the luminosity of the nucleus diminished. While still active, this galaxy no longer completely overwhelmed the emission from the surrounding stars. The result was an active galaxy—either radio or Seyfert—still emitting a lot of energy, but now with a definite "stellar" component in its spectrum. The central activity continued to decline. Eventually, only the surrounding galaxy could be seen—a normal galaxy, like the majority of those we now see around us. Today, the black holes lie dormant in galactic nuclei, producing only a relative trickle of radiation. Occasionally, two normal galaxies may interact with one another, causing a flood of new fuel to be directed toward the central black hole of one or both. The engine starts up again for a while, giving rise to the nearby active galaxies we observe.

Should this picture be correct—and the confirmation of host galaxies surrounding many quasars lends strong support to this view—then many normal galaxies, including perhaps our own Milky Way Galaxy, were once brilliant quasars. Perhaps some alien astronomer, thousands of megaparsecs away, is at this very moment observing our Galaxy—seeing it as it was billions of years ago—and is commenting on its enormous luminosity, nonstellar spectrum, and high-speed jets, and wondering what exotic physical process could possibly account for its violent activity!

When they were first discovered, active galaxies and quasars seemed to present astronomers with insurmountable problems. For a time, their dual properties of enormous energy output and small size appeared incompatible with the known laws of physics and threatened to overturn our modern view of the universe. Yet the problems were eventually solved and the laws of physics remain intact. Far from jeopardizing our knowledge of the cosmos, these violent phenomena have become part of the thread of understanding that binds our own Galaxy to the earliest epochs of the universe we live in.

✓ Concept Check

- Do we live in a burned-out quasar?

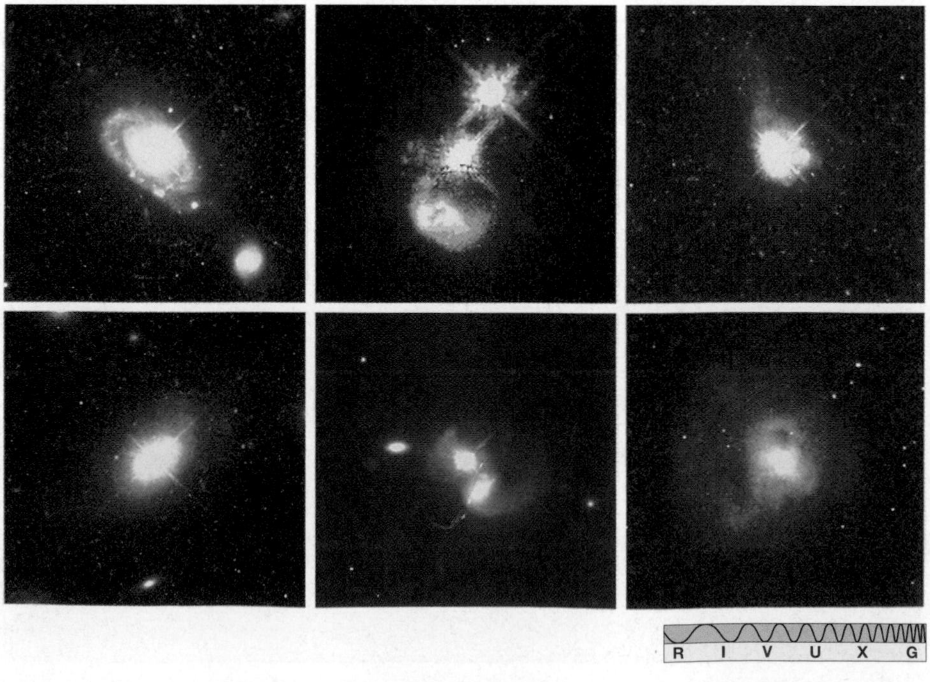

Figure 16.26 Quasar Host Galaxies These long-exposure *Hubble Space Telescope* images of distant quasars clearly show the young host galaxies in which the quasars reside, lending strong support to the view that quasars represent an early, very luminous phase of galactic evolution. The quasar at top left is the best example, having the catalog name PG0052+251 and residing roughly 430 Mpc from Earth. *(NASA)*

Chapter Review

Summary

Active galaxies (p. 406) are much more luminous than normal galaxies and have nonstellar spectra, emitting most of their energy in the radio and infrared parts of the electromagnetic spectrum.

A **Seyfert galaxy** (p. 407) looks like a normal spiral except that the Seyfert has an extremely bright central **galactic nucleus** (p. 407). Spectral lines from Seyfert nuclei are very broad, indicating rapid internal motion, and rapid variability implies that the source of the radiation is much less than one light-year across. **Radio galaxies** (p. 408) emit most of their energy in the radio part of the spectrum. They are generally comparable to Seyferts in total energy output. The corresponding visible galaxy is usually elliptical. In a core–halo radio galaxy, most of the energy is emitted from a small central nucleus. In a lobe radio galaxy, the energy comes from enormous **radio lobes** (p. 410) that lie far beyond the visible portion of the galaxy.

Many active galaxies have high-speed, narrow jets of matter shooting out from their central nuclei. In lobe radio galaxies, astronomers believe that the jets transport energy from the nucleus, where it is generated, to the lobes, where it is radiated into space. The jets often appear to be made up of distinct "blobs" of gas, suggesting that the process generating the energy is intermittent.

The generally accepted explanation for the observed properties of active galaxies is that the energy is generated by accretion of galactic gas onto a supermassive (billion-solar-mass) black hole lying at the center of the nucleus. The small size of the accretion disk explains the compact extent of the emitting region, and the high-speed orbit of gas in the black hole's intense gravity accounts for the rapid motion observed. Typical active galaxy luminosities require the consumption of about one solar mass of material every few years. Some of the infalling matter is blasted out into space, producing magnetized jets that create and feed the radio lobes. Charged particles spiraling around the magnetic field lines produce **synchrotron radiation** (p. 417) whose spectrum is consistent with the nonstellar radiation observed in radio galaxies and Seyferts.

Quasars (p. 419), or **quasi-stellar objects** (p. 419), are starlike radio sources with spectra redshifted to wavelengths much longer than normal. All quasars have large redshifts, indicating that we see them as they were in the distant past. Even the closest quasars lie at great distances from us. Quasars are the most luminous objects known. They exhibit the same basic features as active galaxies, and astronomers believe that their power source is also basically the same—a supermassive black hole consuming many stars per year. The brightest quasars consume so much fuel that their energy-emitting lifetimes must be quite short.

Some quasars have been observed to have double or multiple images due to gravitational lensing, where the gravitational field of a foreground galaxy or galaxy cluster bends and focuses the light from the more distant quasar. Analysis of this bending provides a means of determining the masses of galaxy clusters—including the dark matter—far beyond the region defined by the optical images of the galaxies.

Quasars, active galaxies, and normal galaxies may represent an evolutionary sequence. Quasars probably represent a brief early phase of violent galactic activity. As the fuel supply diminishes, the quasar dims, and the host galaxy becomes visible as an active galaxy. When the fuel supply declines to the point where the nucleus becomes virtually inactive, a normal galaxy is all that remains.

Review and Discussion

1. Name two basic differences between normal galaxies and active galaxies.
2. Describe some of the basic properties of Seyfert galaxies.
3. What distinguishes a core–halo radio galaxy from a lobe radio galaxy?
4. What is the evidence that the radio lobes of some active galaxies consist of material ejected from the galaxy's nucleus?
5. What conditions result in a head–tail radio galaxy?
6. Briefly describe the leading model for the central engine of an active galaxy.
7. How do astronomers know that the energy-producing region of an active galaxy must be very small?
8. What is synchrotron radiation, and what does it tell us about energy emission from active galaxies?
9. What evidence do we have that the energy production in active galaxies is intermittent?
10. What was it about the spectra of quasars that was so unexpected and surprising?

11. Why do astronomers prefer to speak in terms of redshifts rather than distances to faraway objects?

12. How do we know that quasars are extremely luminous?

13. How are the spectra of distant quasars used to probe the space between us and them?

14. What evidence do we have that quasars represent an early stage of galactic evolution?

15. If many galaxies were once quasars, what has happened to the energy sources at their centers?

True or False?

____ **1.** Active galaxies can emit thousands of times more energy than our own Galaxy.

____ **2.** The spectrum of a typical active galaxy is well described by a blackbody curve.

____ **3.** The "extra" radiation emitted by active galaxies is due to the tremendous number of stars they contain.

____ **4.** Active galaxies emit most radiation at visible wavelengths.

____ **5.** Most radio galaxies are ellipticals.

____ **6.** The diameter of a billion-solar-mass black hole is about 20 A.U.

____ **7.** Nearby active galaxies are most likely the result of galaxy interactions.

____ **8.** A redshift greater than 1 means a recessional velocity greater than the speed of light.

____ **9.** All active galaxies are far away.

____ **10.** All quasars are far away.

____ **11.** Other than a small amount of visible light, quasars emit all of their radiation at radio wavelengths.

____ **12.** Many nearby normal galaxies may once have been quasars.

____ **13.** Quasars emit about as much energy as normal galaxies.

____ **14.** Astronomers have no direct evidence that quasars are found in the hearts of young galaxies.

____ **15.** The quasar stage of a galaxy ends because the central black hole uses up all the matter around itself.

Fill in the Blank

1. Active galaxies generally emit most of their radiation at _______ wavelengths.

2. A Seyfert galaxy looks like a normal spiral, but with a very bright galactic _______.

3. In a core–halo radio galaxy, most of the radio radiation is emitted from the _______.

4. Lobe radio galaxies emit radio radiation from regions that are typically much _______ in size than the visible galaxy.

5. Radio lobes are always found aligned with the _______ of the visible galaxy.

6. For all types of active galaxies, the original source of the tremendous energy emitted is the galactic _______.

7. The energy source of an active galaxy is unusual in that a large amount of energy is emitted from a region much less than _______ in diameter. (Give size and unit.)

8. The mass of the black hole responsible for energy production in the active galaxy M87 is thought to be approximately equal to _______ solar masses.

9. The amount of mass that must be consumed by a supermassive black hole to provide the energy for an active galaxy is about _______ per _______.

10. Quasars are also known as quasi-stellar objects because of their _______ appearance at visible wavelengths.

11. Quasar spectra were understood when it was discovered that their radiation is _______ by an unexpectedly large amount.

12. The distance to a quasar in light-years is not simply equal to the time in years since the quasar emitted the light we see because of the _______ of the universe.

13. The fact that a typical quasar would consume an entire galaxy's worth of mass in 10 billion years suggests that quasar lifetimes are relatively _______.

14. The image of a distant quasar can be split into several images by gravitational lensing, produced by a foreground _______ along the line of sight.

15. Quasar host galaxies are hard to see because they are so much _______ than the quasar itself.

Problems

1. A Seyfert galaxy is observed to have broadened emission lines indicating a speed of 1000 km/s at a distance of 1 pc from its center. Assuming circular orbits, use Kepler's laws (Section 14.6) to estimate the mass within this 1-pc radius. ∞ (Sec. 14.5)

2. Centaurus A—from one radio lobe to the other—spans about 1 Mpc. It lies at a distance of 4 Mpc from Earth. What is the angular size of Centaurus A? Compare this value with the angular diameter of the Moon.

3. Assuming a jet speed of 0.75 *c*, calculate the time taken for material in Cygnus A's jet to cover the 500 kpc between the galaxy's nucleus and its radio-emitting lobes.

4. Assuming the same efficiency as indicated in the text, calculate the amount of energy an active galaxy would generate if it consumed one Earth mass of material every day. Compare this value with the luminosity of the Sun.

5. Based on the data presented in the text, calculate the orbital speed of material orbiting at a distance of 0.5 pc from the center of M87.

6. A certain quasar has a redshift of 0.25 and an apparent magnitude of 13. Using the data from Table 16.1, calculate the quasar's absolute magnitude, and hence its luminosity. ∞ (*More Precisely 10-1*) Compare the apparent brightness of the quasar, viewed from a distance of 10 pc, with that of the Sun as seen from Earth.

7. Assuming an energy-generation efficiency (that is, the ratio of energy released to total mass—energy available) of 10 percent, calculate how long a 10^{41}-W quasar can shine if a total of 10^{10} solar masses of fuel is available.

8. The spectrum of a quasar with a redshift of 0.20 contains two sets of absorption lines, redshifted by 0.15 and 0.155, respectively. If $H_0 = 65$ km/s/Mpc, estimate the distance between the intervening galaxies responsible for the two sets of lines.

9. Light from a distant quasar passes 200 kpc from the center of an intervening galaxy cluster before being deflected to a detector on Earth. If Earth, the cluster, and the quasar are all aligned, the quasar is 2500 Mpc away, and the cluster lies midway between Earth and the quasar, calculate the angle through which the galaxy cluster bends the quasar's light.

10. Light from a distant star is deflected by 1.75″ as it grazes the Sun. ∞ (*More Precisely 13-2*) Given that the deflection angle is proportional to the mass of the gravitating body and is inversely proportional to the minimum distance between the light ray and the body, calculate the mass of the galaxy cluster in the previous question.

Projects

Here are three observing projects that are increasingly challenging. If you can find the three objects listed here, you have started to become an accomplished observer!

1. In Project 2 of Chapter 15, you were given directions for finding the Virgo Cluster of galaxies. ∞ (Chapter 15, Project 2) M87, in the central part of this cluster, is the core–halo radio galaxy nearest Earth and has coordinates RA = 12^h 30.8^m, declination = +12° 249. At magnitude 8.6, it should not be difficult to find in an 8-inch telescope. Its distance is roughly 20 Mpc. Describe its nucleus; compare what you see with the appearance of other nearby ellipticals in the Virgo Cluster.

2. NGC 4151 is the brightest Seyfert galaxy. Its coordinates are RA = 12^h 10.5^m, dec = +39° 24′, and it can be found below the Big Dipper in Canes Venatici. At magnitude 10–12 (it is variable), it should be visible in an 8-inch telescope but will be challenging to find. Its distance from Earth is 13.5 Mpc. Describe its nucleus and compare its appearance with what you have seen for other galaxies.

3. 3C 273 is the nearest and brightest quasar. However, that does not mean it will be easy to find and see. Its coordinates are RA = 12^h 29.2^m, dec = +12° 03′. It is located in the southern part of the Virgo Cluster but is not associated with that cluster. At magnitude 12–13, it may require a 10- or 12-inch telescope to see, but try first with an 8-inch. 3C 273 should appear as a very faint star. The significance of seeing this object is that it is 640 Mpc distant. The light you are seeing left this object over two billion years ago. 3C 273 is the most distant object observable with a small telescope.

B1
A1
B2
A2
B3
A3

17 COSMOLOGY

The Big Bang and the Fate of the Universe

WWW

LEARNING GOALS

Studying this chapter will enable you to:

1. State the cosmological principle and explain its significance.
2. Explain how the age of the universe is determined and discuss the uncertainties involved.
3. Summarize the leading evolutionary models of the universe and discuss the factors that determine whether or not the universe will expand forever.
4. Describe the cosmic microwave background radiation and explain its importance to our understanding of cosmology.
5. Explain how nuclei and atoms emerged from the primeval fireball.
6. Summarize the horizon and flatness problems, and discuss the theory of cosmic inflation as a possible solution to these problems.
7. Explain the formation of large-scale structure in the cosmos and discuss the observational evidence for the leading theory of structure formation.

(Opposite page, background) By virtually all accounts, the universe began in a fiery explosion some 10 to 20 billion years ago. Out of this maelstrom emerged all the energy that would later form galaxies, stars, and planets (depicted here in an artist's rendering). The story of the origin and fate of all these systems—and especially of the entire universe itself—comprises the subject of cosmology. *(D. Berry)*

(Inset A) These three images of spiral galaxies lying at different distances from Earth capture representative views of what the universe was like at much earlier times. At top, about 10 billion years ago, the galaxy's spiral shape is barely recognizable. At middle, about 8 billion years ago, the spiral is still vague, though starburst activity is evident in the outer regions. At bottom, some 5 billion years ago, the galaxy's spiral features are more prominent, with much star formation in its arms. *(NASA)*

(Inset B) Elliptical galaxies also suggest evolution with time—indeed, evolution more rapid than that of spiral galaxies. These images of different galaxies (from top to bottom, 10, 8, and 5 billion years ago) suggest that many ellipticals had already taken on their eventual shapes early in the universe. In contrast to spirals, ellipticals seem to have made all their stars long, long ago. *(NASA)*

Our field of view now extends for billions of parsecs into space and billions of years back in time. We have asked and answered many questions about the structure and evolution of planets, stars, and galaxies. At last we are in a position to address the central issues of the biggest puzzle of all. How big is the universe? How long has it been around, and how long will it last? What was its origin, and what will be its fate? Many cultures have asked these questions, in one form or another, and have developed their own cosmologies—theories about the nature, origin, and destiny of the universe—to answer them. In this chapter, we see how modern scientific cosmology addresses these important issues, and what it has to tell us about the universe we inhabit. After more than 10,000 years of civilization, science may be ready to provide some insight into the ultimate origin of all things.

17.1 The Universe on the Largest Scales

1 We saw in Chapter 15 how galaxy surveys have revealed the existence of structures in the universe up to 200 Mpc across. (Sec. 15.5) Are there even larger structures? Does the hierarchy of clustering of stars into galaxies and galaxies into galaxy clusters, superclusters, and filaments continue forever, on larger and larger scales? Perhaps surprisingly, given the trend we have just described, most astronomers think the answer is *no*.

Figure 17.1 shows a survey of galaxies similar to that presented in Figure 15.30, except that it includes nearly 24,000 galaxies within almost 1000 Mpc of the Milky Way, and so covers a much greater volume of space than the earlier figure. Numerous voids and "Great Wall-like" filaments can be seen but, apart from the general falloff in numbers of galaxies at large distances—basically because more distant galaxies are harder to see—there is no obvious evidence for any structures on scales larger than about 200 Mpc. Careful statistical analysis confirms this impression.

The results of this and other large-scale surveys strongly suggest that the universe is **homogeneous** (the same everywhere) on scales greater than a few hundred megaparsecs. In other words, if we took a huge cube—300 Mpc on a side, say—and placed it anywhere in the universe, its overall contents would look much the same no matter where it was centered. The universe also appears to be **isotropic** (the same in all directions) on these scales. Excluding directions that are obscured by our Galaxy, we count roughly the same number of galaxies per square degree in any patch of the sky we choose to observe, provided we look deep (far) enough that local inhomogeneities don't distort our sample.

In the science of **cosmology**—the study of the structure and evolution of the entire universe—researchers generally assume that the universe is homogeneous and isotropic on sufficiently large scales. No one knows if these assumptions are precisely correct, but we can at least say that they are consistent with current observations. In this chapter we simply assume that they hold. The twin assumptions of homogeneity and isotropy are known as the **cosmological principle**.

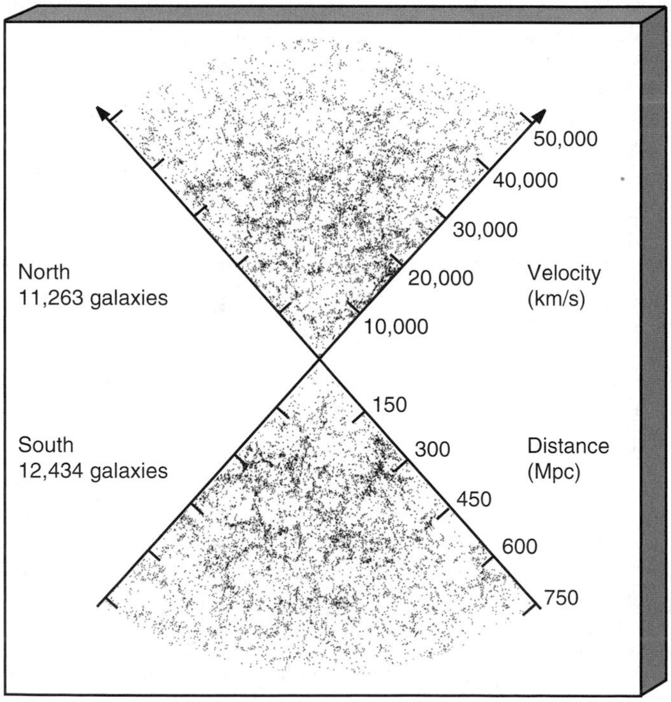

Figure 17.1 Galaxy Survey This large-scale galaxy survey, carried out at the Las Campanas Observatory in Chile, consists of 23,697 galaxies within about 1000 Mpc, in two 80° × 4.5° wedges of the sky. Many voids and "walls" on scales of up to 100–200 Mpc can be seen, but no larger structures are evident, suggesting that the universe is roughly homogeneous on sufficiently large scales.

The cosmological principle has some very far-reaching implications. For example, it implies that there can be no *edge* to the universe because that would violate the assumption of homogeneity. And there can be no *center* because then the universe would not look the same in all directions from any noncentral point, violating the assumption of isotropy. This is the familiar Copernican principle expanded to truly cosmic proportions—not only are we not central to the universe, but *no one* can be central, because *the universe has no center*!

Concept Check

- In what sense is the universe homogeneous and isotropic?

17.2 Cosmic Expansion

Olbers's Paradox

Every time you go outside at night and notice that the sky is dark, you are making a profound cosmological observation. Here's why.

Let's assume that, in addition to being homogeneous and isotropic, the universe is also infinite in spatial extent and unchanging in time. In that case, as illustrated in Figure 17.2, when you look up at the night sky, your line of sight must *eventually* encounter a star. The star may lie at an enormous distance, in some remote galaxy, but the laws of probability dictate that, sooner or later, any line drawn outward from Earth will run into a bright stellar surface. This fact has a dramatic implication: No matter where you look, the sky should be as bright as the surface of a star—the entire night sky should be as brilliant as the surface of the Sun! The obvious difference between this prediction and the actual appearance of the night sky is known as **Olbers's paradox**, after the nineteenth-century German astronomer Heinrich Olbers, who popularized the idea.

Why is the sky dark at night? Given that the universe is homogeneous and isotropic, then one (or both) of the other two assumptions must be false. Either the universe is finite in extent, or it evolves in time. In fact, the answer involves a little of each, and is intimately tied to the behavior of the universe on the largest scales.

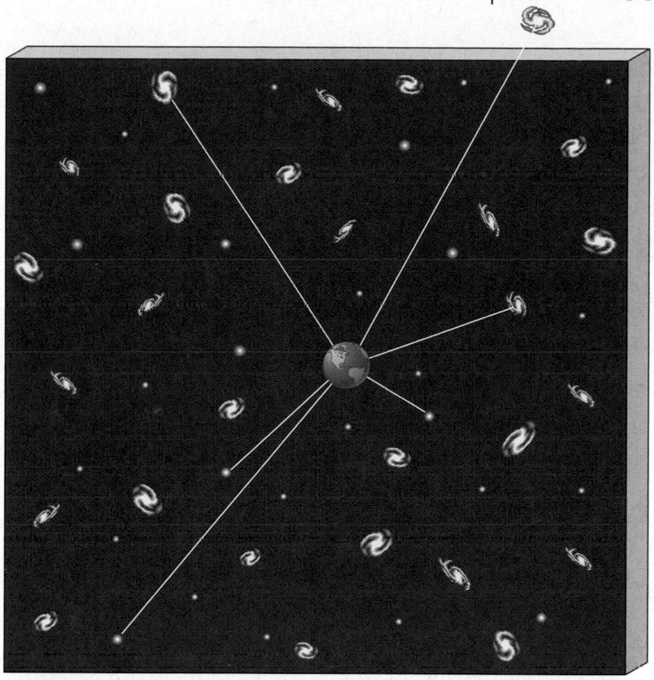

Figure 17.2 **Olber's Paradox** If the universe were homogeneous, isotropic, infinite in extent, and unchanging, then any line of sight from Earth should eventually run into a star and the entire night sky should be bright. This obvious contradiction of the facts is known as Olbers's paradox.

The Birth of the Universe

2 We have seen that all the galaxies in the universe are rushing away from us as described by Hubble's law:

$$\text{recession velocity} = H_0 \times \text{distance},$$

where we take Hubble's constant H_0 to be 65 km/s/Mpc. (Sec. 15.5) We have used this relationship as a convenient means of determining the distances to galaxies and quasars, but it is much more than that.

Assuming that all velocities have remained constant in time, we can ask how long it has taken for any given galaxy to reach its present distance from us. The answer follows from Hubble's law. The time taken is simply the distance traveled divided by the velocity:

$$\begin{aligned}\text{time} &= \frac{\text{distance}}{\text{velocity}}\\ &= \frac{\text{distance}}{H_0 \times \text{distance}} \quad \text{(using Hubble's law for the velocity)}\\ &= \frac{1}{H_0}.\end{aligned}$$

For H_0 = 65 km/s/Mpc, this time is about 15 billion years. Notice that it is *independent* of the distance—galaxies twice as far away are moving twice as fast, so the time they took to cross the intervening distance is the same.

Hubble's law therefore implies that, at some time in the past—15 billion years ago, according to the foregoing simple calculation—*all* the galaxies in the universe lay right on top of one another. In fact, astronomers believe that *everything* in the universe—matter and radiation alike—was confined to a single point at that instant. Then the point exploded, flying apart at high speeds. The present locations and velocities of the galaxies are a direct consequence of that primordial blast. This gargantuan explosion, involving everything in the cosmos, is known as the **Big Bang**. It marked the beginning of the universe.

The Big Bang provides the resolution of Olbers's paradox. Whether the universe is actually finite or infinite in extent is irrelevant, at least as far as the appearance of the night sky is concerned. The key point is that we see only a *finite* part of it—the region lying within roughly 15 billion light-years of us. What lies beyond is unknown—its light has not had time to reach us.

Where Was the Big Bang?

Now we know *when* the Big Bang occurred. Is there any way of telling *where*? The observed recession of the galaxies described by Hubble's law implies that all the galaxies exploded from a point some time in the past. Wasn't that point, then, different from the rest of the universe, violating the assumption of homogeneity expressed in the cosmological principle? The answer is a definite *no*!

To understand why there is no center to the expansion, we must make a great leap in our perception of the universe. If we were to imagine the Big Bang as simply an enormous explosion that spewed matter out into space, ultimately to form the galaxies we see, then the foregoing reasoning would be quite correct. The universe would have a center and an edge, and the cosmological principle would not apply. But the Big Bang was *not* an explosion in an otherwise featureless, empty universe. The only way that we can have Hubble's law *and* retain the cosmological principle is to realize that the Big Bang involved the entire universe—not just the matter and radiation within it, but the universe *itself*. In other words, the galaxies are not flying apart into the rest of the universe. The universe itself is expanding. Like raisins in a loaf of raisin bread that move apart as the bread expands in an oven, the galaxies are just along for the ride.

Let's reconsider some of our earlier statements in this new light. We now recognize that Hubble's law describes the expansion of the universe itself. Apart from small scale individual random motions, galaxies are not moving with respect to the fabric of space. The component of the galaxies' motion that makes up the Hubble flow is really an expansion of space itself. The expanding universe remains homogeneous at all times. There is no "empty space" beyond the galaxies into which they rush. At the time of the Big Bang, the galaxies did not reside at a point located at some well-defined place within the universe. The *entire universe* was a point. The Big Bang happened *everywhere* at once.

To illustrate these ideas, imagine an ordinary balloon with coins taped to its surface, as shown in Figure 17.3. (Better yet, do the experiment yourself.) The coins represent galaxies and the two-dimensional surface of the balloon represents the "fabric" of our three-

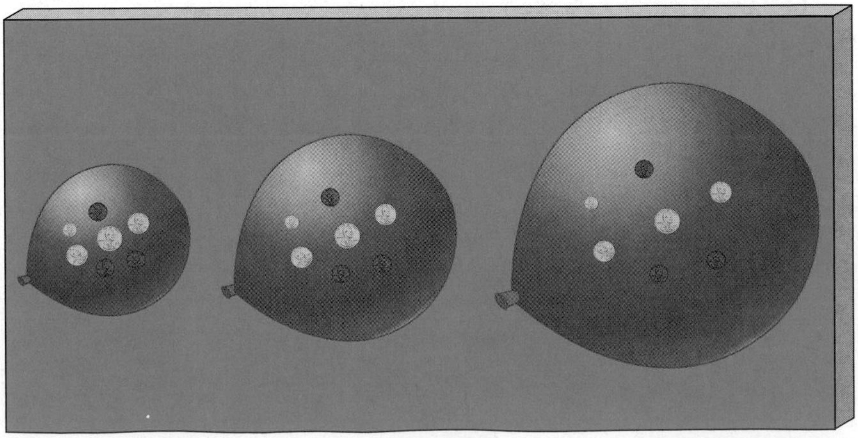

Figure 17.3 **Receding Galaxies** Coins taped to the surface of a spherical balloon recede from one another as the balloon inflates (left to right). Similarly, galaxies recede from one another as the universe expands. As the coins recede, the distance between any two of them increases, and the rate of increase of this distance is proportional to the distance between them. Thus, the balloon expands according to Hubble's law.

dimensional universe. The cosmological principle applies here because every point on the balloon looks pretty much the same as every other. Imagine yourself as a resident of one of the coin "galaxies" on the leftmost balloon, and note your position relative to your neighbors. As the balloon inflates (that is, as the universe expands), the other galaxies recede from you and more distant galaxies recede more rapidly.

Regardless of which galaxy you chose to consider, you would see all the other galaxies receding from you, but nothing is special or peculiar about this fact. There is no center to the expansion and no position that can be identified as the location from which the universal expansion began. Everyone sees an overall expansion described by Hubble's law, with the same value of Hubble's constant in all cases. Now imagine letting the balloon deflate, running the "universe" backward from the present time to the moment of the Big Bang. *All* the galaxies (coins) would arrive at the same place at the same time at the instant the balloon reached zero size, but no single point on the balloon could be said to be *the* place where that occurred.

The Cosmological Redshift

Up to now, we have explained the cosmological redshift of galaxies as a Doppler shift, a consequence of their motion relative to us. ∞ (Sec. 15.5) However, we have just argued that the galaxies are not moving with respect to the universe, in which case the Doppler interpretation is incorrect. The true explanation is that, as a photon moves through space, its wavelength is influenced by the expansion of the universe. Think of the photon as being attached to the expanding fabric of space, so its wavelength expands along with the universe, as illustrated in Figure 17.4.

It is standard practice in astronomy to refer to the cosmological redshift in terms of recessional velocity, but bear in mind that, strictly speaking, this is not the right thing to do. The cosmological redshift is a consequence of the changing size of the universe, and is *not* related to velocity at all.

The redshift of a photon measures the amount by which the universe has expanded since that photon was emitted. For example, when we measure the light from a quasar to have been redshifted by a factor of five, that means the light was emitted at a time when the universe was just one-fifth its present size (and we are observing the quasar as it was at that time). In general, the larger a photon's redshift, the smaller the universe was at the time the photon was emitted, and so the longer ago that emission occurred. ∞ (*More Precisely 16-1*)

✓ Concept Check

- Why does Hubble's Law imply a Big Bang?

17.3 The Fate of the Universe

3 Will the universe expand forever? This fundamental question about the fate of the universe has been at the heart of cosmology since Hubble's law was first discovered. Until the late 1990s, the prevailing view among cosmologists was that the answer would most likely be found by determining the extent to which gravity would slow, and perhaps ultimately reverse, the current expansion. However, it now appears that the answer is rather more subtle—and perhaps a lot more profound in its implications—than was hitherto believed.

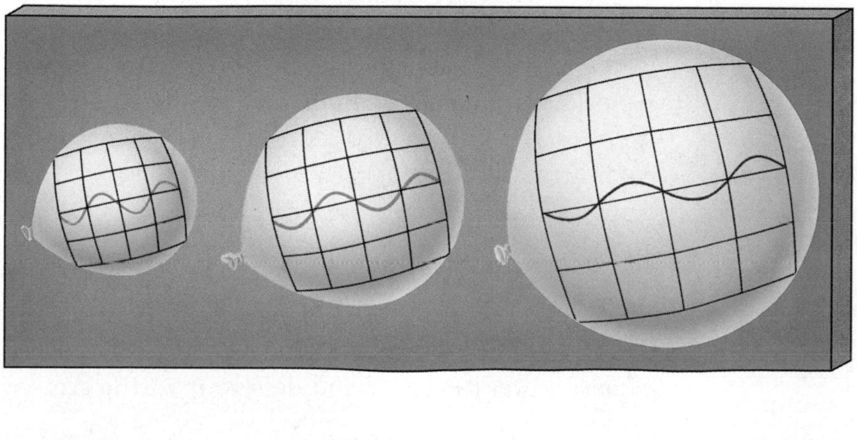

Figure 17.4 Cosmological Redshift As the universe expands, photons of radiation are stretched in wavelength, giving rise to the cosmological redshift.

Critical Density

Let's begin with the conventional view, assuming that gravity is the only force that affects the cosmic expansion, and consider as an analogy a rocket ship launched from the surface of a planet. What is the likely outcome of that motion? There are basically just two possibilities, depending on the launch speed of the ship. If the speed is high enough, it will exceed the planet's escape speed and the ship will never return to the surface. ∞ (*More Precisely 5-2*) The spacecraft leaves the planet on an *unbound* trajectory, as illustrated in Figure 17.5(a). Alternatively, if the launch speed is lower than the escape speed, the ship will reach a maximum distance from the planet, then fall back to the surface. Its *bound* trajectory is shown in Figure 17.5(b).

Similar reasoning applies to the expansion of the universe. Imagine two galaxies at some known distance from one another, their present relative velocity given by Hubble's law. The same possibilities exist for these galaxies as for our rocket ship—the distance between them can increase forever, or it can increase for a while and then start to decrease. What's more, the cosmological principle says that, whatever the outcome, it must be the same for *any* two galaxies—in other words, the same statement applies to the universe *as a whole*. Thus, as illustrated in Figure 17.6, the universe can continue to expand forever—an *unbound* universe—or the present expansion will someday stop and turn around into a contraction—a *bound* universe.

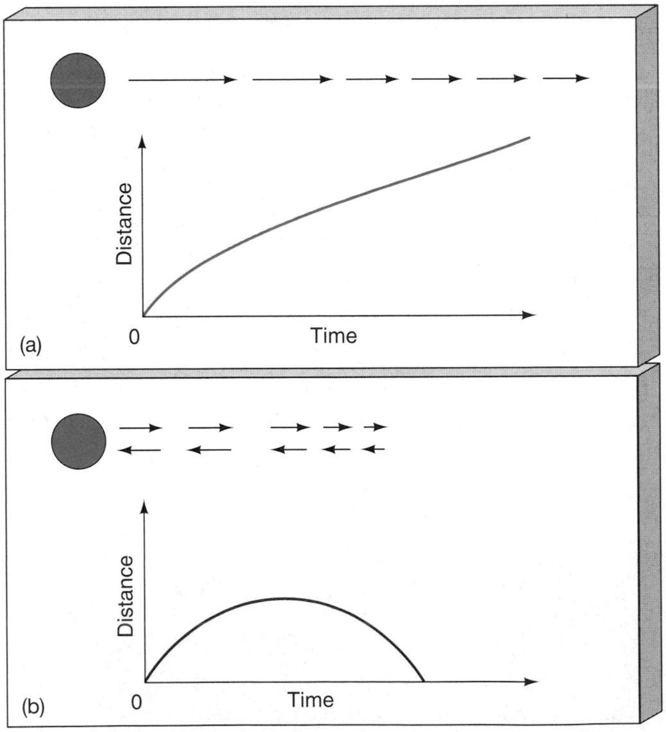

Figure 17.5 Escape Velocity (a) A rocket ship (arrow) leaving a planet (blue ball) with a speed greater than the escape speed follows an unbound trajectory. The graph shows the distance between the ship and the planet as a function of time. (b) If the launch speed is less than the escape speed, the ship eventually drops back to the planet. Its distance from the planet first rises, then falls.

The middle curve on Figure 17.6 marks the dividing line between these two possibilities. It shows a *marginally bound* universe that expands forever, but at a decreasing rate, analogous to our rocket ship's leaving the planet with precisely the escape speed. The three curves are drawn so that they all pass through the same point at the present time. All are possible descriptions of the universe given its present size and expansion rate.

What determines which of these possibilities will actually occur? In the case of a rocket ship of fixed launch speed, the *mass* of the planet determines whether or not escape will occur—a more massive planet has a higher escape speed, making it less likely that the rocket can escape. For the universe, the corresponding factor is the *density* of the cosmos. A high-density universe contains enough mass to stop the expansion and eventually cause a recollapse. Such a bound universe is destined ultimately to shrink toward a superdense, superhot singularity much like the one from which it originated. Some astronomers call this the "Big Crunch." A low-density universe, conversely, is unbound and will expand forever. In that case, in time, even with the most powerful telescopes, we will see no galaxies in the sky beyond the Local Group (which is not itself expanding). The rest of the observable universe will appear dark, the distant galaxies too faint to be seen.

The density of a marginally bound universe, in which gravity is just sufficient to halt the present expansion, is called the **critical density**. For H_0 = 65 km/s/Mpc, the critical density is about 8×10^{-27} kg/m^3. That's an extraordinarily low density—just five hydrogen atoms per cubic meter, a volume the size of a typical household closet. In more "cosmological" terms, it corresponds to about 0.1 Milky Way Galaxies (including the dark matter) per cubic megaparsec.

The Density of the Universe

Cosmologists conventionally call the ratio of the actual density of the universe to the critical value the *cosmic density parameter* and denote it by the symbol

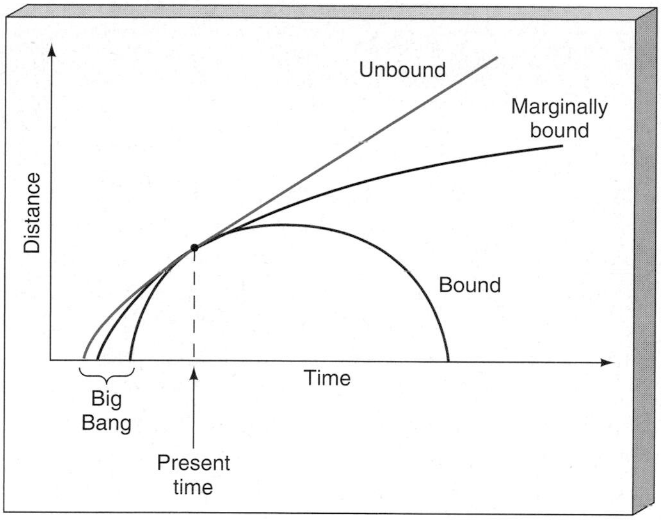

Figure 17.6 Model Universes Distance between two galaxies as a function of time in each of the possible universes discussed in the text, illustrating the competition between gravity and the cosmic expansion. The point where the three curves touch represents the present time.

Ω_0 ("omega nought"). In terms of this quantity, then, a critical universe has $\Omega_0 = 1.0$. A universe with Ω_0 less than 1.0 has insufficient gravity to halt and reverse the present cosmic expansion.

How can we determine the average density of the universe? On the face of it, it would seem simple—just measure the average mass of the galaxies residing within a large parcel of space, calculate the volume of that space, and compute the total mass density. When astronomers do this, they usually find a little less than 10^{-28} kg/m^3 in the form of luminous matter. Largely independent of whether the chosen region contains only a few galaxies or a rich galaxy cluster, the resulting density is about the same, within a factor of two or three. Comparing this density with the critical value, we find that galaxy counts yield a value of Ω_0 of about 0.01.

But there is an important additional consideration. Most of the matter in the universe is *dark*—it exists in the form of invisible material that has been detected only through its gravitational effect in galaxies and galaxy clusters. ∞ (Sec. 14.5, 15.3) We currently do not know *what* the dark matter is, but we *do* know that it is there. Galaxies may contain as much as 10 times more dark matter than luminous material, and the figure for galaxy clusters is even higher—perhaps as much as 95 percent of the total mass in clusters is invisible. Even though we cannot see it, dark matter contributes to the average density of the universe and plays its part in opposing the cosmic expansion. Including all the dark matter that is known to exist in galaxies and galaxy clusters increases the value of Ω_0 to about 0.2 or 0.3.

Unfortunately, the distribution of dark matter on larger scales is not very well known. We can infer its presence in galaxies and galaxy clusters, but we are largely ignorant of its extent in superclusters, voids, or other larger structures. One of the few mass estimates for an object much larger than a supercluster comes from optical and infrared observations of the overall motion of galaxies within the Local Supercluster. The measured velocities suggest the presence of a nearby huge accumulation of mass known as the *Great Attractor*, with a total mass of about 10^{17} solar masses and a size of 100–150 Mpc. Its average density may be quite close to the critical value.

However, it seems that, when averaged over the entire universe, even objects like the Great Attractor don't raise the overall cosmic density by much. In short, there doesn't seem to be much additional dark matter "tucked away" on very large scales. Most cosmologists agree that Ω_0 is less than 1.0, and probably not much greater than 0.3. Thus, as best we can tell given the current data, the universe is destined to expand forever.

An Accelerating Universe?

Determining the density of the universe is an example of a *local* measurement that provides an estimate of Ω_0. But, as we have just seen, there are many uncertainties in the result, especially on large scales. In an attempt to get around this problem, astronomers have devised alternative methods that rely instead on *global* measurements, covering much larger regions of the observable universe. In principle, such global tests should indicate the universe's overall density, not just its value in our cosmic neighborhood.

One such global method is based on observations of Type I (carbon detonation) supernovae. ∞ (Sec. 12.5) Recall that these objects are both very bright and have a remarkably narrow spread in luminosities, making them particularly useful as standard candles. ∞ (Sec. 15.2) They can be used as probes of the universe because, by measuring their distances (*without* using Hubble's law) and their redshifts, we can determine the rate of cosmic expansion in the distant past. Here's how the method works.

Suppose the universe is decelerating, as we would expect if gravity were slowing its expansion. Then, because the expansion rate is decreasing, objects at great distances—that is, objects that emitted their radiation long ago—should appear to be receding *faster* than

INTERLUDE 17-1

The Cosmological Constant

Even the greatest minds are fallible. The first scientist to apply general relativity to the universe was (not surprisingly), the theory's inventor, Albert Einstein. When he derived and solved the equations describing the behavior of the universe, Einstein discovered that they predicted a universe that evolved in time. But in 1917 neither he nor anyone else knew about the expansion of the universe, as described by Hubble's law, which would not be discovered for another 10 years. (Sec. 15.5) At the time Einstein, like most scientists, believed that the universe was static—that is, unchanging and everlasting. The discovery that there was no static solution to his equations seemed to Einstein a near-fatal flaw in his new theory.

To bring his theory into line with his beliefs, Einstein tinkered with his equations, introducing a "fudge factor" now known as the cosmological constant. The accompanying figure shows the effect of introducing a cosmological constant into the equations describing the expansion of a critical-density universe. One new possible solution is a "coasting" universe, whose radius does indeed remain constant for an indefinite period of time. Einstein took this to be the static universe he expected. Later, when the expansion of the universe was discovered and Einstein's equations—without the cosmological constant—were found to describe it perfectly, he declared that the cosmological constant was the biggest mistake of his scientific career.

For many researchers—Einstein included—the main problem with the cosmological constant was (and still is) the fact that it had no clear physical interpretation. Einstein introduced it to fix what he thought was a problem with his equations, but he discarded it immediately once he realized that no problem actually existed. Scientists are very reluctant to introduce unknown quantities into their equations purely to make the results "come out right." As a result, the cosmological constant fell out of favor among astronomers for many years. However, as described in the text, it may well be making a comeback. Recent observations suggesting that the expansion of the universe is accelerating are readily explained by models including a cosmological constant. (See, for example, the light-green curve in the figure.)

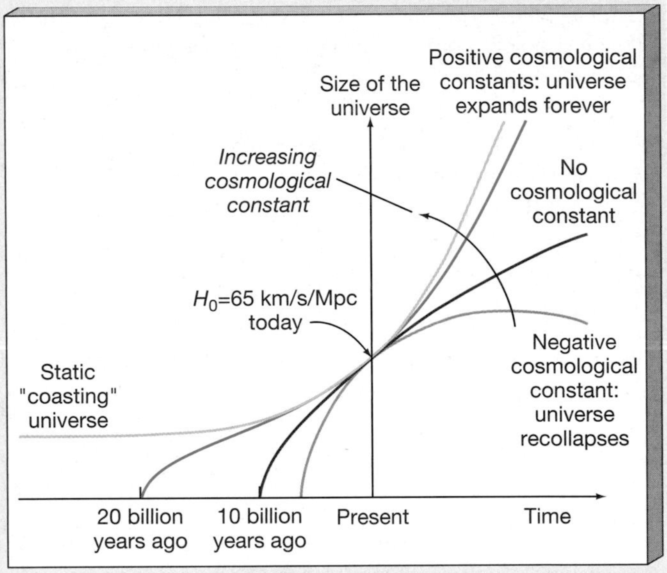

Hubble's law predicts. Figure 17.7 illustrates this concept. If the universal expansion were constant in time, recessional velocity and distance would be related by the black line in the figure. (The line is not quite straight because it takes the expansion of the universe properly into account in computing the distance.) (*More Precisely 16-1*) In a decelerating universe, the velocities of distant objects should lie *above* the black curve, and the deviation from that curve is greater for a denser universe, in which gravity has been more effective at slowing the expansion.

How does theory compare with reality? In the late 1990s, two groups of astronomers announced the results of independent, systematic surveys of distant supernovae. Some of their data are shown on Figure 17.7. Far from clarifying the picture of cosmic deceleration, however, these findings seem to indicate that the expansion of the universe is not slowing, but actually accelerating! According to the supernova data, galaxies at large distances are receding *less* rapidly than Hubble's law would predict. The deviations from the decelerating curves appear small in the figure, but they are statistically very significant, and both groups report similar findings. These observations are inconsistent with the standard Big Bang model just described, and may necessitate a major revision of our view of the cosmos.

What could cause an overall acceleration of the universe? Cosmologists do not know, although several possibilities have been suggested. One leading candi-

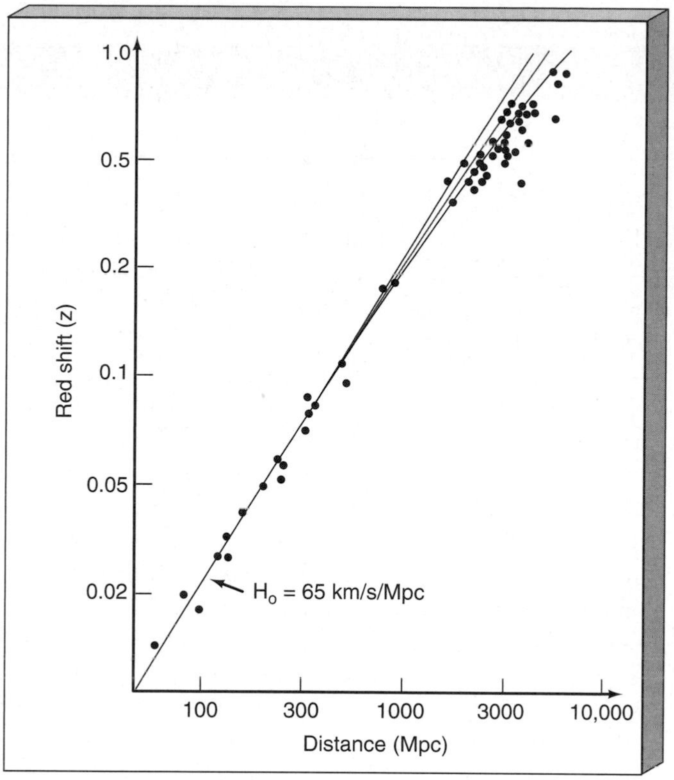

Figure 17.7 Accelerating Universe? Observations of distant supernovae allow astronomers to measure changes in the expansion rate of the universe. In a decelerating universe (purple and red curves), redshifts of distant objects are greater than would be predicted from Hubble's Law (black curve). The reverse is true for an accelerating universe. The points showing observations of some 50 supernovae strongly suggest that the cosmic expansion is accelerating. The vertical scale shows redshift; for small velocities, redshift is just velocity divided by the speed of light. ∞ (More Precisely 16-1)

date is an additional "vacuum pressure" force associated with empty space and effective only on very large scales. It is known simply as the **cosmological constant** (*Interlude 17-1*). Models including this force can fit the observational data but, at present, astronomers have no clear physical interpretation of what it actually means—it is neither required nor explained by any known law of physics.

Whatever it is, the mysterious cosmic field causing the universe to accelerate is neither matter nor radiation. It is sometimes referred to as *dark energy*. If confirmed, the magnitude of the cosmic acceleration implies that the amount of dark energy in the universe may exceed the total mass-energy of matter (luminous and dark) by a wide margin. By opposing the attractive force of gravity, the repulsive effect of the dark energy strengthens our earlier conclusion that the universe will expand forever.

The Age of the Universe

In Section 17.2, when we estimated the age of the universe from the accepted value of Hubble's constant, we made the assumption that the expansion speed of the cosmos was constant in the past. However, as we have just seen, this is probably not the case. The effects of gravity have tended to slow the universe's expansion, particularly at early times, while the cosmic expansion rate today may actually be increasing. In the absence of a cosmological constant, the universe expanded *faster* in the past than it does today, so the assumption of constant expansion rate leads to an overestimate of the universe's age—such a universe is younger than the 15 billion years we calculated earlier. How much younger depends on how much deceleration has occurred. Conversely, a repulsive cosmological constant tends to increase the age of the cosmos.

Figure 17.8 illustrates these points. It is similar to Figure 17.6, except that here we have added two lines, one corresponding to constant expansion rate at the present value, the other to the accelerating universe that best fits the supernova data. The age of a critical-density universe with no cosmological constant is about 10 billion years. The age corresponding to the accelerating universe is about 14 billion years, coincidentally quite close to the value for constant expansion.

✓ Concept Check

- Why do astronomers think the universe will expand forever?

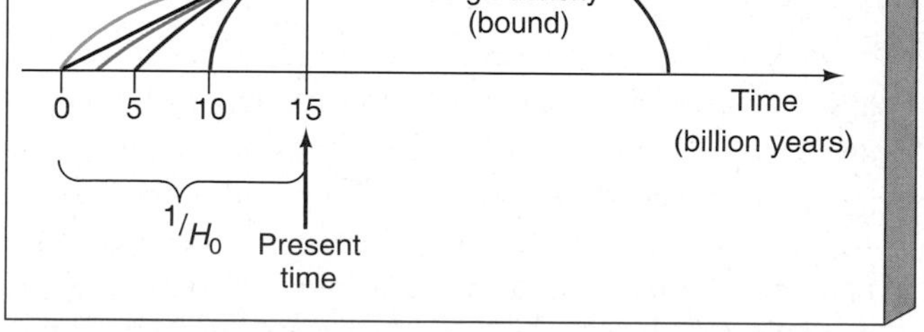

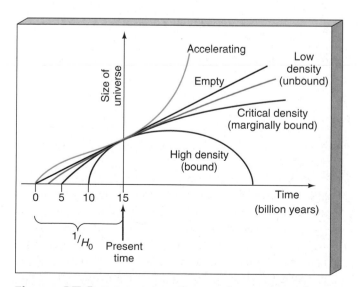

Figure 17.8 Universe Age The age of a universe without a cosmological constant is always less than $1/H_0$, and decreases for larger values of the present-day density. The existence of a repulsive cosmological constant increases the age of the cosmos.

MORE PRECISELY 17-1

Curved Space

Euclidean geometry is the geometry of flat space—the geometry taught in high schools everywhere. Set forth by Euclid, one of the most famous of the ancient Greek mathematicians, it is the geometry of everyday experience. Houses are built with flat floors. Writing tablets and blackboards are also flat. We work easily with flat, straight objects because a straight line is the shortest distance between any two points.

In constructing houses or any other straight-walled buildings on the surface of Earth, some other basic axioms of Euclid's geometry also apply: parallel lines never meet even when extended to infinity; the angles of any triangle always sum to 180°; the circumference of a circle equals π times its diameter. If these rules were not obeyed, walls and roof would never meet to form a house!

In reality, though, the geometry of Earth's surface is not flat. It is curved. We live on a sphere, and on its surface Euclidean geometry breaks down. Instead, the rules are those of *Riemannian geometry*, named after the nineteenth-century German mathematician Georg Friedrich Riemann. There are no parallel lines (or curves) on a sphere's surface—any "straight" lines drawn on the surface and around the full circumference will eventually intersect. The sum of a triangle's angles always *exceeds* 180°. (For example, in the 90°-90°-90° triangle shown in the figure, the sum is 270°.) And the circumference of a circle is *less* than π times its diameter.

Riemannian geometry on the curved surface of a sphere differs greatly from the flat-space geometry of Euclid. The two are approximately the same only if we confine ourselves to a small patch on the surface. If the patch is much smaller than the sphere's radius, the surface surrounding any point looks flat, and Euclidean geometry is approximately valid. This is why we can draw a usable map of our home, our city, even our state, on a flat sheet of paper, but an accurate map of the entire Earth must be drawn on a globe.

When we work with larger portions of Earth, we must abandon Euclidean geometry. World navigators are well aware of this. Aircraft do not fly along what you might regard as a straight-line path from one point to another. Instead they follow a *great circle*, which, on the curved surface of a sphere, is the shortest distance between two points. For example, a flight from Los Angeles to London does not proceed directly across the United States and the Atlantic Ocean as you might expect from looking at a flat map. Instead, it goes far to the north, over Canada and Greenland, above the Arctic Circle, finally coming in over Scotland for a landing at London. This is the shortest path between the two cities. Look at a globe of Earth (or see the accompanying figure) to verify that this is indeed the case.

The *positively curved* space of Riemann is not the only possible departure from flat space. Another is the *negatively curved* space first studied by Nikolai Ivanovich Lobachevsky, a nineteenth-century Russian mathematician. In this geometry, there are an *infinite* number of lines through any given point parallel to another line, the sum of a triangle's angles is *less* than 180° (see the figure), and the circumference of a circle is *greater* than π times its diameter. Instead of the flat surface of a plane or the curved surface of a sphere, this type of space is described by the surface of a curved saddle. It is a difficult geometry to visualize!

Even though the universe on the largest scales may be curved as we have just described, most of the local realm of the *three*-dimensional universe (including the solar system, the neighboring stars, and even our Milky Way Galaxy) is correctly described by Euclidean geometry.

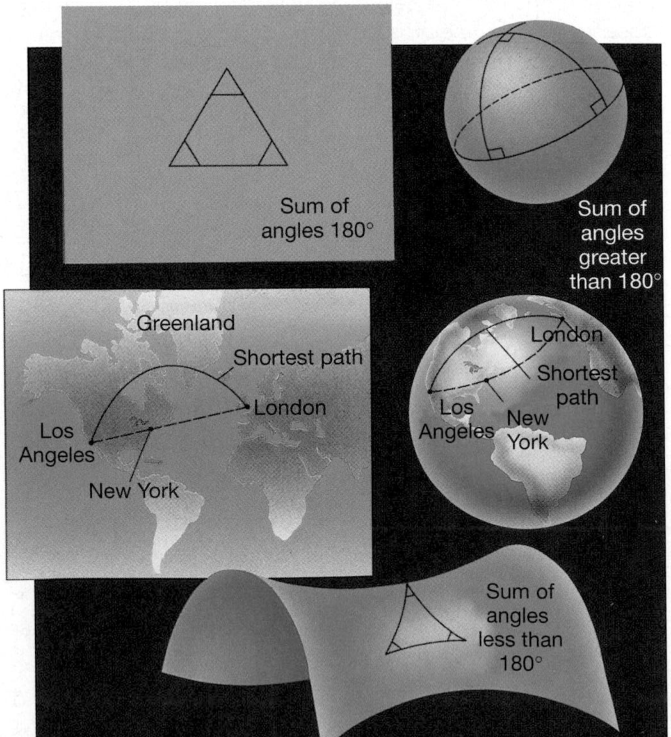

17.4 The Geometry of Space

4 The idea of the entire universe expanding from a point—with *nothing*, not even space and time, outside—takes a lot of getting used to. Nevertheless, it lies at the heart of modern cosmology, and few modern astronomers seriously doubt it. But this description of the universe itself (not just its contents) as a dynamic, evolving object is far beyond the capabilities of Newtonian mechanics, which we have used almost everywhere in this book. (Sec. 1.4) Instead, the more powerful techniques of Einstein's theory of general relativity, with its built-in notions of warped space and dynamical spacetime, are needed. (Sec. 13.5)

The theory of general relativity states that mass (or, equivalently, energy) curves, or "warps" space in its vicinity. The greater the *total* density of the cosmos—including not just matter, but also radiation and dark energy (if any), suitably converted to mass units using the relation $E = mc^2$—the greater the curvature. (Sec. 9.5) In the universe, the curvature must be the same everywhere (assuming homogeneity), so there are really only three possibilities for the large-scale geometry of space. For more information on the different types of geometry involved, see *More Precisely 17-1*.

If the density is above the critical value, space is curved so much that it bends back on itself and closes off, making the universe *finite* in size. Such a universe is known as a **closed universe**. It is difficult to visualize a three-dimensional volume uniformly arching back on itself in this way, but the two-dimensional version is well known: It is the surface of a sphere, like the balloon we discussed earlier. Figure 17.3, then, is the two-dimensional likeness of a three-dimensional closed universe. Like the surface of a sphere, a closed universe has no boundary, yet it is finite in extent.[1] One remarkable property of a closed universe is illustrated in Figure 17.9: A flashlight beam shone in some direction in space might eventually traverse the entire universe and return from the opposite direction.

The surface of a sphere curves, loosely speaking, "in the same direction" no matter which way we move from a given point. A sphere is said to have *positive curvature*. However, if the density of the universe is below the critical value, then the curvature of space is qualitatively quite different from that of a sphere. The two-dimensional surface corresponding to such a space curves like a saddle and is said to have *negative curvature*. Most people have seen a saddle it curves "up" in one direction and "down" in another, but no one has ever seen a uniformly negatively curved surface, for the good reason that it cannot be constructed in three-dimensional space! It is just "too big" to fit. A low-density saddle-curved universe is infinite in extent and is usually called an **open universe**.

Figure 17.9 Einstein's Curve Ball In a closed universe, a beam of light launched in one direction might return someday from the opposite direction after circling the universe, just as motion in a "straight line" upon Earth's surface will eventually encircle the globe.

The intermediate case, when the density is precisely equal to the critical density, is the easiest to visualize. This **critical universe** has no curvature. It is said to be "flat," and it is infinite in extent. In this case, and only in this case, the geometry of space on large scales is precisely the familiar Euclidean geometry taught in high school. Apart from its overall expansion, this is basically the universe that Newton knew.

Euclidean geometry—the geometry of flat space—is familiar to most of us because it is a good description of space in the vicinity of Earth. It is the geometry of everyday experience. Does this mean that the universe is flat, which would in turn mean that it has exactly the critical density? The answer is no. Just as a flat street

[1]Notice that, for the sphere analogy to work, we must imagine ourselves as two-dimensional "flatlanders" who cannot visualize or experience in any way the third dimension perpendicular to the sphere's surface. Flatlanders are confined to the sphere's surface, just as we are confined to the three-dimensional volume of our universe.

map is a good representation of a city, even though we know Earth is really a sphere, Euclidean geometry is a good description of space within the solar system, or even the Galaxy, because the curvature of the universe is negligible on scales smaller than about 1000 Mpc. Only on the very largest scales would the geometrical effects we have just described become evident.

Concept Check

- How is the curvature of space related to the density of the universe?

17.5 Back to the Big Bang

We now turn from studies of the far future of our universe to the quest to understand its distant past. Just how far back in time can we probe? Is there any way to study the universe beyond the most remote quasar? How close can we come to perceiving directly the edge of time, the very origin of the universe?

The Cosmic Microwave Background

4 A partial answer to these questions was discovered by accident in 1964, during an experiment designed to improve America's telephone system. As part of a project to identify and eliminate unwanted interference in satellite communications, Arno Penzias and Robert Wilson, scientists at Bell Telephone Laboratories in New Jersey, were studying the Milky Way's emission at microwave wavelengths. In their data, they noticed a bothersome background "hiss"—a little like the background static on an AM radio station. Regardless of where and when they pointed their antenna, the hiss persisted. Never diminishing or intensifying, the weak signal was detectable at any time of the day and on any day of the year, apparently filling all of space.

Eventually, after discussions with colleagues at Bell Labs and theorists at nearby Princeton University, the two experimentalists realized that the origin of the mysterious static was nothing less than the fiery creation of the universe itself. The radio hiss that Penzias and Wilson detected is now known as the **cosmic microwave background**. Their discovery won them the 1978 Nobel Prize in physics.

In fact, theorists had predicted the existence and general properties of the microwave background well before its discovery. In addition to being extremely dense, they reasoned, the early universe must also have been very *hot*, and so, shortly after the Big Bang, the universe must have been filled with extremely high-energy thermal radiation—gamma rays of very short wavelength. As illustrated in Figure 17.10, this intensely hot primordial radiation has been redshifted by the expansion of the universe from gamma ray, through optical and infrared wavelengths, all the way into the radio (microwave) range of the electromagnetic spectrum. (Sec. 2.4) The radiation observed today is the "fossil remnant" of the primeval fireball that existed at those very early times.

Researchers at Princeton confirmed the existence of the microwave background and estimated its temperature at about 3 K. However, this part of the electromagnetic spectrum happens to be very difficult to observe from the ground, and it was 25 years before astronomers could demonstrate conclusively that the radiation was indeed described by a blackbody curve. (Sec. 2.3) In 1989, the *Cosmic Background Explorer* satellite—*COBE* for short—measured the intensity of the microwave background at wavelengths straddling the peak, from half a millimeter to about 10 cm. The results are shown in Figure 17.11. The solid line is the blackbody curve that best fits the *COBE* data. The near-perfect fit corresponds to a universal temperature of just over 2.7 K.

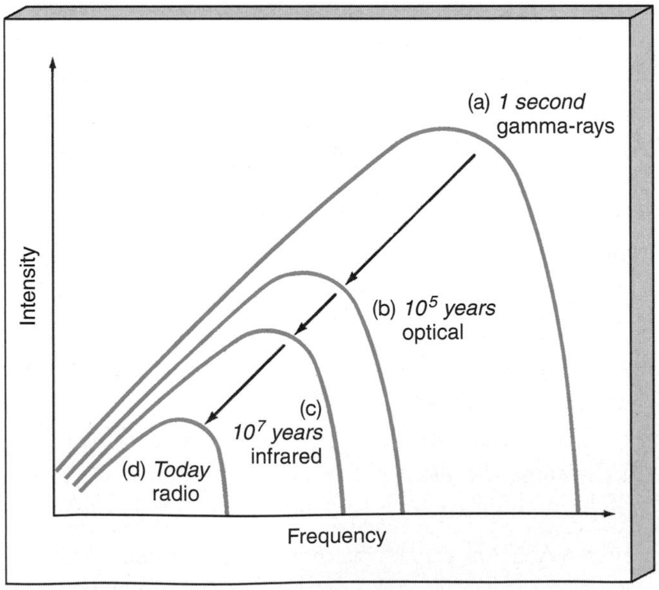

Figure 17.10 Universe Planck Curves Theoretically derived blackbody curves for the entire universe (a) one second after the Big Bang, (b) 100,000 years after the Big Bang, (c) 10 million years after the Big Bang, and (d) at present, approximately 15 billion years after the Big Bang.

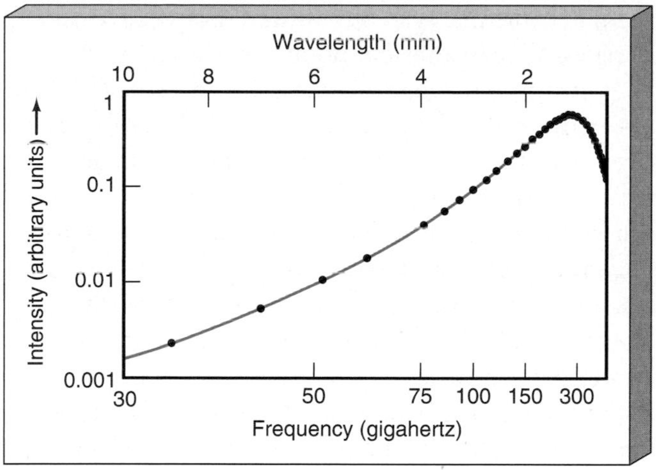

Figure 17.11 Cosmic Microwave Background The intensity of the cosmic background radiation, as measured by the *COBE* satellite, agrees very well with theory. The curve is the best fit to the data, corresponding to a temperature of 2.735 K. The experimental errors in this remarkably accurate observation are smaller than the dots representing the data points.

A striking aspect of the cosmic microwave background is its high degree of *isotropy*. When we correct for Earth's motion through space (which causes the microwave background to appear a little hotter than average in front of us and slightly cooler behind, by the Doppler effect), the intensity of the radiation is virtually constant (to about one part in 10^5) from one direction in the sky to another.

Remarkably, the cosmic microwave background contains more energy than has been emitted by all the stars and galaxies that have ever existed in the history of the universe. Stars and galaxies, though very intense sources of radiation, occupy only a small fraction of space. When their energy is averaged out over the volume of the entire cosmos, it falls short of the energy in the microwave background by at least a factor of 10. For our present purposes, then, the cosmic microwave background is the only significant form of radiation in the universe.

Matter and Radiation

Is matter the dominant constituent of the present universe, or does radiation also play an important role on the largest scales? We can answer this question by comparing the energy contained in the microwave background with the total energy in the form of matter. To make the comparison we must first convert the energy in the microwave background into an equivalent mass density by calculating the number of microwave background photons in any cubic meter of space, then converting the total energy of these photons to a mass using $E = mc^2$. When we do this, we find that the density of the microwave background is about 5×10^{-30} kg/m^3. The overall density of matter in the universe is not known with certainty, but it is thought to be fairly close to the critical density of 8×10^{-27} kg/m^3. Thus, *at the present moment* the density of matter in the universe greatly exceeds the density of radiation. In cosmological terminology (which predates the possibility of dark energy in the cosmos), we say that we live in a **matter-dominated** universe.

Was the universe always matter-dominated? To address this question, we must ask how the densities of both matter and radiation change as the universe expands. Both decrease, as the expansion dilutes the numbers of atoms and photons alike, but the radiation is also diminished in energy by the cosmological redshift, so its equivalent density falls faster than the density of matter as the universe grows (Figure 17.12). Thus, even though

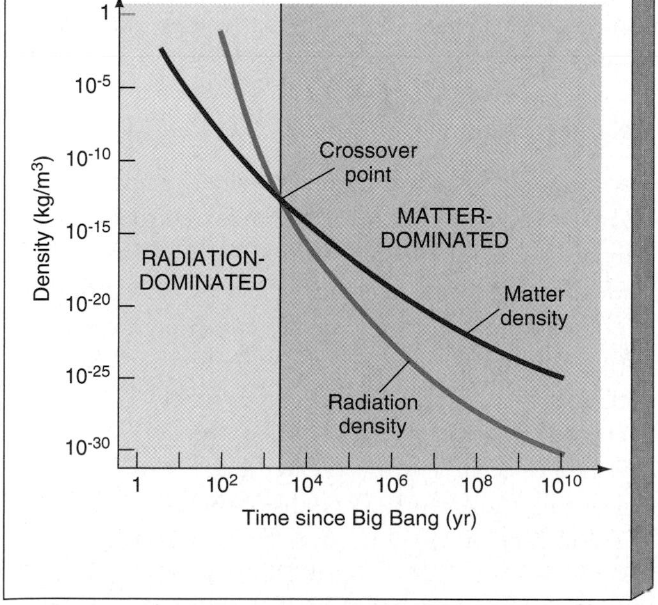

Figure 17.12 Radiation–Matter Dominance As the universe expanded, the number of both matter particles and photons per unit volume decreased. However, the photons were also reduced in energy by the cosmological redshift, reducing their equivalent mass, and hence their density, still further. As a result, the density of radiation fell faster than the density of matter as the universe grew. Tracing the curves back from the densities we observe today, we see that radiation must have dominated matter at early times—that is, at times to the left of the crossover point.

today the radiation density is much less than the density of matter, there must have been a time in the past when they were equal. Before that time, radiation was the main constituent of the cosmos. The universe is said to have been **radiation-dominated** then. (The density associated with the dark energy, if it exists, remains constant as the universe expands, and was therefore negligible at early times.)

✓ Concept Check

- When was the cosmic microwave background formed?

17.6 The Formation of Nuclei and Atoms

5 At the very earliest times, theorists believe, the cosmos consisted entirely of radiation. During the first minute or so of the universe's existence, temperatures were high enough that individual photons of radiation had sufficient energy to transform themselves into matter, in the form of elementary particles. This period saw the creation of all the basic building blocks of matter we know today—both "normal" matter, such as protons and neutrons, and electrons, and the "exotic" particles that may comprise the dark matter in galaxies and galaxy clusters. ∞ (Sec. 14.5) Since then, matter has evolved, clumping together into more and more complex structures forming the nuclei, atoms, planets, stars, galaxies and large-scale structure we know today, but no *new* matter has been created. *Everything we see around us was created out of radiation as the early universe expanded and cooled.*

Helium Formation

Astronomers find that, no matter where they look, and no matter how low a star's abundance of heavy elements may be, there is a minimum amount of helium—a little under 25 percent by mass—in all stars. This is the case even in stars containing very few heavy elements, indicating that the helium already existed when those ancient stars formed. ∞ (*More Precisely 12-1*) The accepted explanation is that this base level of helium is *primordial*—that is, it was created not by fusion in stars but during the early, hot epochs of the universe, before any stars had formed. The production of elements heavier than hydrogen by nuclear fusion shortly after the Big Bang is called **primordial nucleosynthesis**.

As the temperature of the universe fell below about 900 million K, roughly 2 minutes after the Big Bang, conditions became favorable for protons and neutrons to fuse to form deuterium.[2] Prior to that time, deuterium nuclei were broken apart by high-energy gamma rays as quickly as they were created. Once the deuterium was at last able to survive the radiation background, other fusion reactions quickly converted it into heavier elements, especially helium-4. In just a few minutes, most of the available neutrons were consumed, leaving a universe whose "normal" matter content was primarily hydrogen and helium.

This early burst of fusion did not last long. Conditions in the cooling, thinning universe became less and less conducive to further fusion reactions as time went by. For all practical purposes, primordial nucleosynthesis stopped at helium-4. By about 15 minutes after the Big Bang, the cosmic elemental abundance was set, and helium accounted for approximately one-quarter of the total mass of matter in the universe. The remaining 75 percent of the matter in the universe was hydrogen. It would be almost a billion years before nuclear reactions in stars would change these figures.

Nucleosynthesis and the Density of the Universe

Although most deuterium was quickly burned into helium as soon as it formed, a small amount was left when the primordial nuclear reactions ceased. Detailed calculations indicate that the amount of deuterium remaining is a very sensitive indicator of the present-day density of matter in the universe: The *denser* the universe is today, the more matter (protons and neutrons) there was at those early times to react with the deuterium, and the *less* deuterium was left when nucleosynthesis ended. Comparison between theoretical calculations of the production of deuterium in the early universe and the observed abundance of deuterium in stars and the interstellar medium implies a present-day cosmic density of *at most* three or four percent of the critical value.

Primordial nucleosynthesis depends only on the presence of *protons* and *neutrons* in the early universe.

[2]Recall from Chapter 9 that deuterium is simply a heavy form of hydrogen. Its nucleus contains one proton and one neutron. ∞ (Sec. 9.5)

Thus, measurements of the abundance of helium and deuterium tell us only about the density of "normal" matter—matter made up of protons and neutrons—in the cosmos. This has an important implication for the overall composition of the universe. Since Ω_0, the ratio of the total cosmic mass density to the critical density, seems to be around 0.3, we are forced to conclude that not only is most of the matter in the universe dark, but most of the dark matter is *not* composed of protons and neutrons. The bulk of the matter in the universe apparently exists in the form of elusive subatomic particles whose nature we do not fully understand and whose very existence has yet to be conclusively demonstrated in laboratory experiments. (Sec. 14.5)

For the sake of brevity, we will adopt from here on the convention that the term "dark matter" refers only to these unknown particles, and not to "stellar" dark matter, such as brown or white dwarfs, which are made of normal matter.

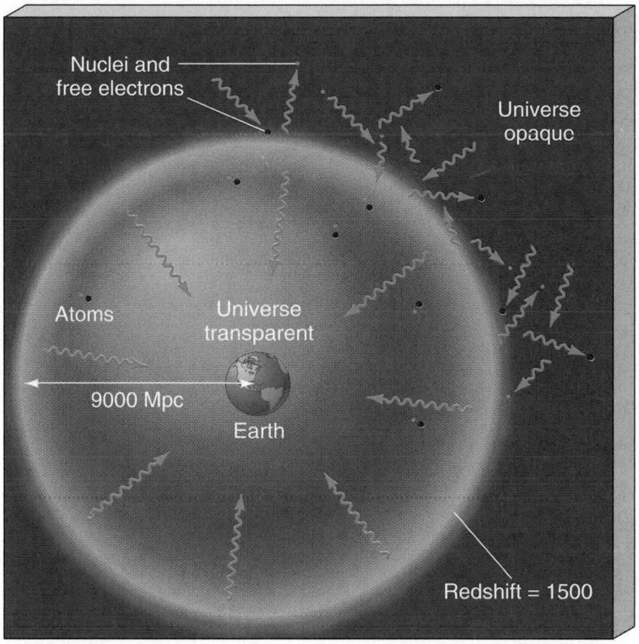

Figure 17.13 Radiation–Matter Decoupling When atoms formed, the universe became virtually transparent to radiation. Thus, observations of the cosmic background radiation allow us to study conditions in the universe around a time at a redshift of 1500, when the temperature dropped below about 4500 K.

The Formation of Atoms

When the universe was a few thousand years old, matter (consisting of electrons, protons, helium nuclei, and dark matter) began to dominate over radiation. The temperature at that time was several tens of thousands of kelvins—too hot for atoms of hydrogen to exist, although some helium ions may already have formed. During the next few hundred thousand years, the universe expanded by another factor of 10, the temperature dropped to a few thousand kelvins, and electrons and nuclei combined to form neutral atoms. By the time the temperature had fallen to 4500 K, the universe consisted of atoms, photons, and dark matter.

The period during which nuclei and electrons combined to form atoms is often called the epoch of **decoupling**, for it was during this period that the radiation background parted company with normal matter. At early times, when matter was ionized, the universe was filled with large numbers of free electrons, which interacted frequently with electromagnetic radiation of all wavelengths. A photon could not travel far before encountering an electron and scattering off it. The universe was *opaque* to radiation. However, when the electrons combined with nuclei to form atoms of hydrogen and helium, only certain wavelengths of radiation—those corresponding to the spectral lines of those atoms—could interact with matter. (Sec. 2.6) Light of all other wavelengths could travel virtually forever without being absorbed. The universe became nearly *transparent*. After that time, as the universe expanded, the radiation simply cooled, eventually to become the microwave background we observe today.

The microwave photons now detected on Earth have been traveling through the universe ever since they decoupled. As illustrated in Figure 17.13, their last interaction with matter (at the epoch of decoupling) occurred when the universe was a few hundred thousand years old and roughly 1500 times smaller (and hotter) than it is today—that is, at a redshift of 1500. Thus, by observing the microwave background, we are probing conditions in the universe almost all the way back in time to the Big Bang.

✓ Concept Check

- How do we know that most of the dark matter in the universe is not composed of "normal" matter?

17.7 Cosmic Inflation

The Horizon and Flatness Problems

6 In the late 1970s, cosmologists were confronted with two nagging problems that had no easy explanation within the standard Big Bang model. The first is known as the **horizon problem**. Imagine observing the microwave background in two opposite directions on the sky, as illustrated in Figure 17.14. As we have just seen, in doing so we are actually observing two distant regions of the universe, marked A and B on the figure, where the radiation background last interacted with matter. The fact that the background is known to be *isotropic* to high accuracy means that regions A and B must have had very similar densities and temperatures at the time the radiation we see left them.

The problem is, within the Big Bang theory as just described, there is no particular reason *why* these regions should have been so similar. At the time in question, a few hundred thousand years after the Big Bang, regions A and B were separated by many megaparsecs, and no signal—sound waves, light rays, or anything else—had had time to travel from one to the other. In cosmological terminology, each region lay outside the other's *horizon*. But in that case, with no possibility of information or energy exchange between these regions, there was no way for them to "know" that they were supposed to look the same. The only alternative is that they simply started off looking alike—an assumption that cosmologists are very unwilling to make.

The second problem with the standard Big Bang model is called the **flatness problem**. Whatever the exact value of the cosmic density, it appears to be fairly close to the critical value. In terms of space-time curvature, the universe is remarkably close to being *flat*. But again, there is no good reason *why* the universe should have formed with a density very close to critical. Why not a millionth or a million times that value? Furthermore, as can be seen in Figure 17.15, a universe that starts off close to, but not exactly on, the critical curve soon deviates greatly from it, so if the universe is close to critical now, it must have been *extremely* close to critical in the past. For example, if $\Omega_0 = 0.1$ today, the departure from the critical density at the time of nucleosynthesis would have been only one part in 10^{15} (a thousand trillion).

These observations constitute "problems" because cosmologists want to be able to *explain* the present condition of the universe, not just accept it "as is." They

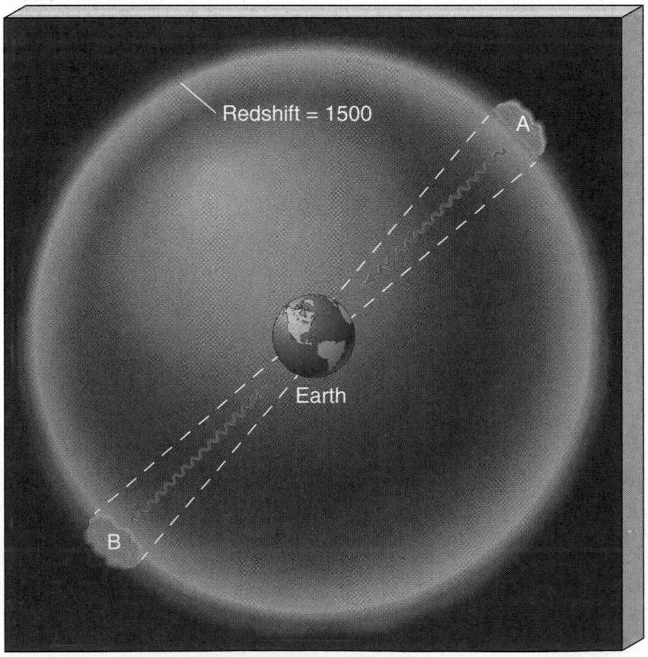

Figure 17.14 Horizon Problem The isotropy of the microwave background indicates that regions A and B in the universe were very similar to one another when the radiation we observe left them, but there has not been enough time since the Big Bang for them ever to have interacted with one another. Why, then, should they look the same?

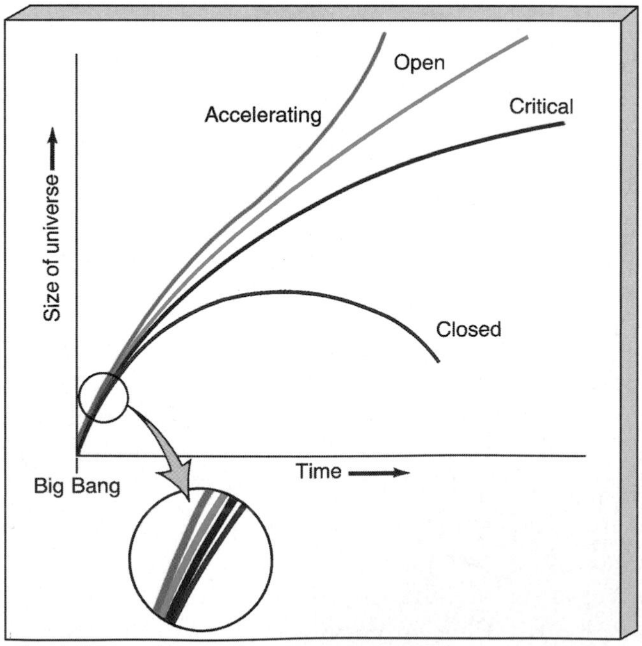

Figure 17.15 Flatness Problem If the universe deviates even slightly from critical density, that deviation grows rapidly in time. For the universe to be as close to critical as it is today, it must have differed from the critical density in the past by only a tiny amount.

would prefer to resolve the horizon and flatness problems in terms of physical processes that could have taken a universe with no special properties and caused it to evolve into the cosmos we now see. The resolution of both problems takes us back in time even earlier than nucleosynthesis or the formation of any of the elementary particles we know today—back, in fact, almost to the instant of the Big Bang itself.

The Epoch of Inflation

In the 1970s and 1980s, theoretical physicists succeeded in combining, or *unifying*, the three nongravitational forces in the universe—electromagnetism and the strong and weak forces—into a single all-encompassing "superforce." ∞ (*More Precisely 9-1*) A general prediction of the **Grand Unified Theories** (GUTs) that describe this superforce is that the three forces are unified and indistinguishable from one another *only* at enormously high energies—corresponding to temperatures in excess of 10^{28} K. At lower temperatures, the superforce splits into three, revealing its separate electromagnetic, strong, and weak characters.

In the early 1980s, cosmologists discovered that Grand Unified Theories had some remarkable implications for the very early universe. About 10^{-34} s after the Big Bang, as temperatures fell below 10^{28} K and the basic forces of nature became reorganized—a little like a gas liquefying or water freezing as the temperature drops—the universe briefly entered a very odd, and unstable, high-energy state that physicists call the "false vacuum." In essence, the universe remained in the "unified" condition a little too long, like water that has been cooled below freezing but has not yet turned to ice. This had dramatic consequences. For a short while, empty space acquired an enormous *pressure* (similar in effect but much greater in magnitude than the pressure associated with the cosmological constant), which temporarily overcame the pull of gravity and accelerated the expansion of the universe at an enormous rate. The pressure remained constant as the cosmos expanded, and the acceleration grew more and more rapid with time—in fact, the size of the universe *doubled* every 10^{-34} s or so. This period of unchecked cosmic expansion, illustrated in Figure 17.16, is known as the epoch of **inflation**.

Eventually, the universe returned to the lower-energy "true vacuum" state. Regions of normal space began to appear within the false vacuum and rapidly spread to include the entire cosmos. With the return of the true vacuum, inflation stopped, and the universe resumed its normal, relatively leisurely expansion. The whole episode lasted a mere 10^{-32} s, but during that time the universe swelled in size by the incredible factor of about 10^{50}

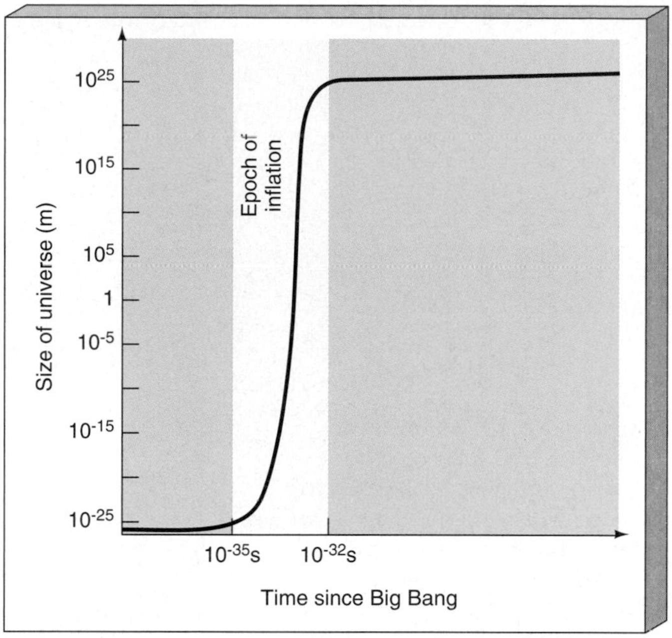

Figure 17.16 Cosmic Inflation During the period of inflation the universe expanded enormously in a very short time. Afterward, it resumed its earlier "normal" expansion rate, except that now the size of the cosmos was about 10^{50} times bigger than it was before inflation.

Implications for the Universe

Inflation provides a natural solution for the horizon and flatness problems. The horizon problem is solved because inflation took regions of the universe that had already had time to communicate with one another—and so had established similar physical properties—and then dragged them far apart, well out of communication range of one another. In effect, the universe expanded much faster than the speed of light during the inflationary epoch, so what was once well within the horizon now lies far beyond it. (Relativity restricts matter and energy to speeds less than the speed of light, but imposes no such limit on the universe as a whole.) Regions A and B in Figure 17.15 have been out of contact with one another since 10^{-32} s after the Big Bang, but they *were* in contact before then. Their properties are the same today because they were the same long ago, before inflation separated them.

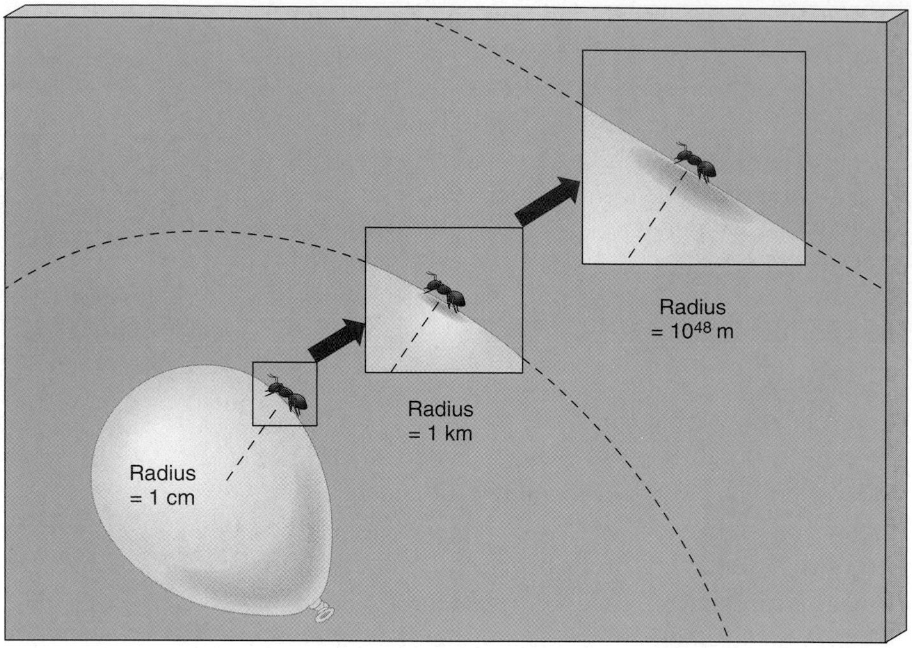

Figure 17.17 Inflation Conceptualized Inflation solves the flatness problem by taking a curved surface, here represented by the surface of the expanding balloon, and expanding it enormously. To an ant on the surface, the balloon looks perfectly flat when the expansion is complete.

Figure 17.17 illustrates how inflation solves the flatness problem. Imagine that you are a 1-mm-long ant sitting on the surface of an expanding balloon. When the balloon is just a few centimeters across, you can easily see that its surface is curved because its circumference is comparable to your own size. However, as the balloon expands, the curvature becomes less pronounced. When the balloon is a few kilometers across, your "ant-sized" patch of the surface looks quite flat, just as Earth's surface looks flat to us. Now imagine that the balloon expands 100 trillion trillion trillion trillion times. Your local patch of the surface would be completely indistinguishable from a perfectly flat plane. Thus, any curvature the universe may have had before inflation has been expanded so much that space is now perfectly flat on all scales we can ever hope to observe.

This resolution to the flatness problem—observations indicate that the universe is close to being flat because the universe *is* flat, to very high accuracy—has a very important consequence. Because the universe is flat, the *total* density (including matter, radiation, and dark energy) must be exactly critical. As we have seem. observations of galaxies and clusters imply that $\Omega_0 =$ 0.3. Thus, not only is most matter dark (Section 17.6), but most of the cosmic density isn't made up of matter. The best match to the supernova data presented in Section 17.3 is consistent with a critical-density universe comprising 30 percent (mostly dark) matter and 70 percent dark energy. In a sense, this is the ultimate statement of the Copernican principle. Not only is Earth in no way central to the cosmos, but the stuff of which we are made is the minority constituent of the universe!

Concept Check

- Why does the theory of inflation imply that much of the energy density of the universe may be neither matter nor radiation?

17.8 The Formation of Large-Scale Structure in the Universe

7 As far as we can tell, all of the present large-scale structure in the universe grew from small *inhomogeneities*—slight deviations from perfectly uniform density—that existed in the early universe. Denser than average clumps of matter contracted under the influence of gravity and merged with other clumps, eventually reaching the point where stars began to form and luminous galaxies appeared. (Sec. 15.4)

Most cosmologists believe that the intense background radiation in the early universe effectively prevented clumps of normal matter from forming or growing. Just as in a star, the pressure of the radiation stabilized the clumps against the inward pull of their own gravity. ∞ (Sec. 12.1) Instead, the first structures to appear did so in the *dark* matter that comprises the bulk of matter in the universe. While the true nature of dark matter remains uncertain, its defining property is the fact that it interacts only very weakly with both normal matter and radiation. Consequently, clumps of dark matter were unaffected by the radiation background and began to grow as soon as matter first began to dominate the cosmic density.

Thus, as illustrated in Figure 17.18, dark matter determined the overall distribution of mass in the universe and clumped to form the observed large-scale structure. Then, at later times, normal matter was drawn by gravity into the regions of highest density, eventually forming galaxies and galaxy clusters. This picture explains why so much dark matter is found outside the visible galaxies. The luminous material is strongly concentrated near the density peaks and dominates the dark matter there, but the rest of the universe is essentially devoid of normal matter. Like foam on the crest of an ocean wave, the universe we see is only a tiny fraction of the total.

Figure 17.19 shows the results of a supercomputer simulation of a (mainly) dark-matter universe. Compare these images with the real observations of nearby structure shown in Figures 15.30 and 15.31. Although calculations like this do not prove that dark-matter models are the correct description of the universe, the similarities between the models and reality are certainly very striking.

Although dark matter does not interact directly with photons, the background radiation is influenced slightly by the *gravity* of the growing dark clumps. This causes a slight gravitational redshift in the background radiation that varies from place to place depending on the dark-matter density. As a result, dark-matter models predict that there should be tiny "ripples" in the microwave background—temperature variations of only a few parts per million from place to place on the sky. In 1992, after almost two years of careful observation, the *COBE* team announced that the expected ripples had been detected. The temperature variations are tiny—only 30–40 millionths of a kelvin from place to place in the sky—but

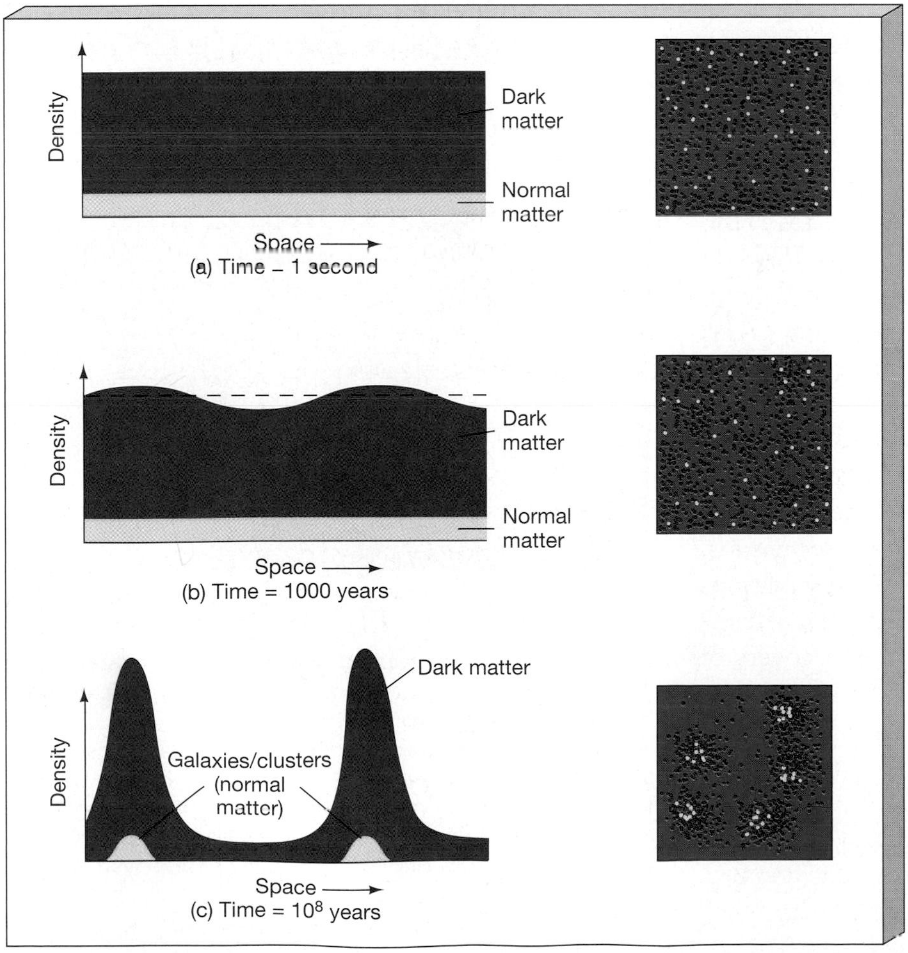

Figure 17.18 Structure Formation (a) The universe started out as a mixture of (mostly) dark and normal matter. (b) A few thousand years after the Big Bang, the dark matter began to clump. (c) Eventually the dark matter formed large structures (represented here by the two high-density peaks) into which normal matter flowed, ultimately to form the galaxies we see today. The three frames at right represent the densities of dark matter (red) and normal matter (yellow) graphed at left.

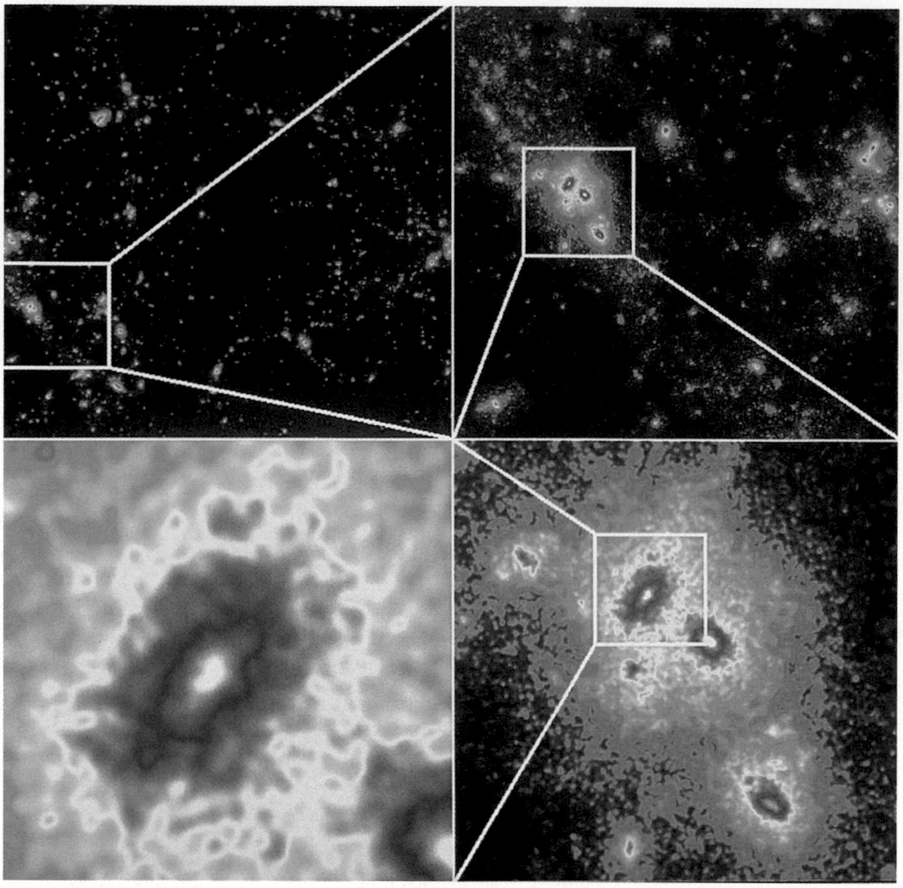

Figure 17.19 Structure Simulated Successively magnified views of a 100 × 100 × 100 Mpc cube in a simulated dark-matter universe in which $\Omega_0 = 1$, showing the present-day structure that results from the growth of small density inhomogeneities in the very early universe. Colors represent mass density, ranging from the cosmic mean (dark blue), through green, yellow, and red, to 100 times the mean (white). The enlargements zoom in on one particular galaxy in one particular small group of galaxies. The last frame (at bottom left) is roughly 1.5 Mpc across. Notice both the large-scale filamentary structure evident in the top two frames and the extensive dark-matter halos surrounding individual galaxies (which are roughly the white regions in the bottom two frames). *(Edmund Bertschinger, MIT)*

they are there. The *COBE* results are displayed as a temperature map of the microwave sky in Figure 17.20.

The ripples seen by *COBE*, combined with computer simulations such as that shown in Figure 17.19, predict present-day structure that is quite consistent with the superclusters, voids, filaments and Great Walls we actually see around us. Detailed analysis of the ripples also supports the key prediction of inflation theory—that the universe is of exactly critical density and hence spatially flat. For these reasons, most cosmologists now regard the *COBE* observations as striking confirmation of a central prediction of dark-matter theory. They rank alongside the discovery of the microwave background itself in terms of their importance to the field of cosmology.

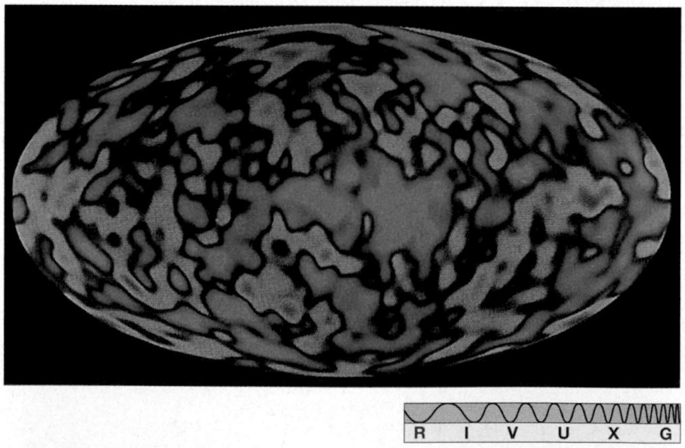

Figure 17.20 Cosmic Microwave Background Map *COBE* map of temperature inhomogeneities in the cosmic microwave background over the entire sky. Hotter than average regions are shown in red, cooler than average regions in blue. The total temperature range shown is ±200 millionths of a kelvin. The temperature variation due to Earth's motion has been subtracted out, as has the radio emission from the Milky Way Galaxy, and temperature deviations from the average are displayed. *(NASA)*

✓ Concept Check

- Why is dark matter important to galaxy formation?

Chapter Review

Summary

On scales larger than a few hundred megaparsecs, the universe appears roughly **homogeneous** (p. 432) (the same everywhere) and **isotropic** (p. 432) (the same in all directions). In **cosmology** (p. 432)—the study of the universe as a whole—researchers usually assume that the universe is homogeneous and isotropic on large scales. This assumption is known as the **cosmological principle** (p. 432).

If the universe were homogeneous, isotropic, infinite, and unchanging, then the night sky would be bright because any line of sight would eventually intercept a star. The fact that the night sky is instead dark is called **Olbers's paradox.** (p. 433) Its resolution is that we see only a finite part of the universe—the region within about 15 billion light-years, beyond which light has not yet had time to reach us.

Tracing the observed motions of galaxies back in time implies that, about 15 billion years ago, the universe consisted of a single point that expanded rapidly in the **Big Bang** (p. 434). Space itself was compressed to a point at that instant—the Big Bang happened everywhere at once. The cosmological redshift occurs as a photon's wavelength is "stretched" by cosmic expansion. The extent of the observed redshift is a direct measure of the expansion of the universe since the photon was emitted.

If the density of the universe is less than the **critical density** (p. 436), then there is insufficient cosmic matter to stop the expansion and cause the universe to recollapse. Luminous matter by itself contributes only about one percent of the critical density. When dark matter in galaxies and clusters is taken into account, the figure rises to 20 or 30 percent. The fraction of dark matter on larger scales is uncertain, but it may be even greater than is found in clusters. Most astronomers believe that the present mass density of the universe lies somewhere around 30 or 40 percent of the critical value. Recent observations of distant supernovae suggest that the expansion of the universe may be accelerating, possibly driven by the effects of the **cosmological constant** (p. 439), a repulsive force that exists throughout all space. Its physical nature is unknown.

General relativity provides a description of the geometry of the universe on the largest scales. The curvature of spacetime in a high-density **closed universe** (p. 441) is sufficiently large that the universe "bends back" on itself and is finite in extent, somewhat like the surface of a sphere. A low-density **open universe** (p. 441) has curvature a little like that of a two-dimensional saddle surface and is infinite in extent. The intermediate-density **critical universe** (p. 441) is spatially flat.

The **cosmic microwave background** (p. 442) is isotropic blackbody radiation that fills the entire universe. Its present temperature is about 3 K. Its existence is evidence that the universe expanded from a hot, dense state. As the universe has expanded, the initially high-energy radiation has been redshifted to lower and lower temperatures. At the present time, the density of matter in the universe greatly exceeds the equivalent mass density of radiation. We live in a **matter-dominated** (p. 443) universe. The matter density was much greater in the past, when the universe was smaller. However, because radiation is redshifted as the universe expands, the density of radiation was greater still. The early universe was **radiation dominated** (p. 444).

Most of the helium observed in the universe today is primordial, created by fusion between protons and neutrons in the early universe; this is known as **primordial nucleosynthesis** (p. 444). Detailed studies of this process indicate that "normal" matter can account for at most three percent of the critical density. By the time the universe was about 1500 times smaller than it is today, the temperature had become low enough for atoms to form. At that time, the (then visible) radiation background **decoupled** (p. 445) from the matter. The photons that now make up the microwave background have been traveling freely through space ever since.

According to modern **Grand Unified Theories** (p. 447), the three nongravitational forces of nature first began to display their separate characters about 10^{-34} s after the Big Bang. A brief period of rapid cosmic expansion called the epoch of **inflation** (p. 447) ensued, during which the size of the universe increased by a factor of about 10^{50}. The **horizon problem** (p. 446) is the fact that, according to the standard (that is, noninflationary) Big Bang model, there is no good reason for widely separated parts of the universe to be as similar as they are. Inflation solves the horizon problem by taking a small homogeneous patch of the early universe and expanding it enormously. Inflation also solves the **flatness problem** (p. 446), which is the fact that there is no obvious reason why the density of the universe is so close to critical. Inflation implies that the cosmic density is almost exactly critical.

The large-scale structure observed in the universe today formed when density inhomogeneities in the dark matter clumped and grew to create the "skeleton" of the structure now observed. Normal matter then flowed into the densest regions of space, eventually forming the galaxies we now see. "Ripples" in the microwave background are the imprint of these early inhomogeneities on the radiation field.

Review and Discussion

1. What evidence do we have that there is no structure in the universe on very large scales? How large is "very large"?
2. What is the cosmological principle?
3. What is Olbers's paradox? How is it resolved?
4. Explain how an accurate measure of Hubble's constant can lead to an estimate of the age of the universe.
5. Why isn't it correct to say that the expansion of the universe involves galaxies flying outward into empty space?
6. Why are recent observations of distant supernovae so important to cosmology?
7. Where and when did the Big Bang occur?
8. How does the cosmological redshift relate to the expansion of the universe?
9. What is the cosmic microwave background, and why is it so significant?
10. Why do all stars, regardless of their abundance of heavy elements, contain at least 25 percent helium by mass?
11. What is Ω_0? How do measurements of the cosmic deuterium abundance provide a lower limit on Ω_0?
12. When did the universe become transparent to radiation?
13. What is cosmic inflation? How does inflation solve the horizon problem? The flatness problem?
14. What is the connection between dark matter and the formation of large-scale structure in the universe?
15. What did dark-matter models predict that was later found by the *COBE* satellite?

True or False?

____ 1. Cosmic homogeneity can be tested observationally, but cosmic isotropy has no observational test.

____ 2. Olbers's paradox asks, "Why is the night sky dark?"

____ 3. The Big Bang is an expansion only of matter, not of space.

____ 4. All points in space appear to be at the center of the expanding universe.

____ 5. The cosmological redshift is a direct measure of the expansion of the universe.

____ 6. As it travels through space, a photon's wavelength expands at the same rate as the universe expands.

____ 7. Observations of supernovae imply that the expansion of universe is slowing down faster than can be accounted for by gravity alone.

____ 8. The cosmic microwave background is the highly redshifted radiation of the early Big Bang.

____ 9. The light emitted from all the stars in the universe now far outshines the cosmic microwave background.

____ 10. Most of the helium in the universe is primordial.

____ 11. The present-day microwave background radiation was created at the time of decoupling.

____ 12. After decoupling, neutral atoms could no longer exist.

____ 13. The horizon problem has to do with the isotropy found in the microwave background radiation.

____ 14. The flatness problem is the fact that the observed density of matter is unexpectedly different from the critical density.

____ 15. If the inflation model is correct, then the density of the universe is exactly equal to the critical density.

Fill in the Blank

1. Deep surveys of the universe suggest that the largest structures in space are no larger than about _______ Mpc across.
2. Homogeneous means "the same _______."
3. Isotropic means "the same in all _______."
4. Together, the assumptions of cosmic homogeneity and isotropy are known as the _______.
5. If the universe had an edge, this would violate the assumption of _______.
6. If the universe had a center, this would violate the assumption of _______.
7. A value of 65 km/s/Mpc for Hubble's constant implies an age for the universe of roughly _______ billion years.
8. Luminous matter makes up at most _______ percent of the critical density.
9. The surface of a sphere is a two-dimensional example of a _______ universe.
10. The present average temperature of the universe, as measured by the cosmic microwave background, is _______ K.
11. Comparing the mass density of radiation and matter, we find that, at the present time, _______ dominates.
12. When the universe was a few minutes old, nuclear fusion produced _______ and _______.

13. Elements heavier than these were not formed primordially because the density and temperature of the universe became too ________after helium formed.

14. If the inflation model is correct, then the density of the universe is ________ (greater than, equal to, less than) the critical density.

15. Theory predicted tiny inhomogeneities in the ________ of the microwave background; the ________ satellite found them.

Problems

1. What is the greatest distance at which a galaxy survey sensitive to objects as faint as 20th magnitude could detect a galaxy as bright as the Milky Way (absolute magnitude −20)? ∞ *(More Precisely 10-1)*

2. If the entire universe were filled with Milky Way-like galaxies, with an average density of 0.1 per cubic megaparsec, calculate the total number of galaxies observable by the survey in the previous question if it covered the entire sky.

3. Eight galaxies are located at the corners of a cube. The present distance from each galaxy to its nearest neighbor is 10 Mpc, and the entire cube is expanding according to Hubble's law, with $H_0 = 65$ km/s/Mpc. Calculate the recession velocity (magnitude *and* direction) of one corner of the cube relative to the opposite corner.

4. According to the standard Big Bang theory (neglecting any effects of cosmic acceleration), what is the maximum possible age of the universe if $H_0 = 50$ km/s/Mpc? 65 km/s/Mpc? 80 km/s/Mpc?

5. For $H_0 = 65$ km/s/Mpc, the critical density is 8×10^{-27} kg/m^3. (a) How much mass does that correspond to in a volume of 1 A.U.^{3}? (b) How large a cube would be required to enclose one Earth mass of material?

6. The Virgo Cluster is observed to have a recessional velocity of 1200 km/s. Assuming $H_0 = 65$ km/s/Mpc and a critical-density universe, calculate the total mass contained within a sphere centered on Virgo and just enclosing the Milky Way Galaxy. What is the escape speed from the surface of this sphere? ∞ *(More Precisely 5-2)*

7. Assuming critical density and using the distances presented in Table 16.1, estimate the total mass of the observable universe. Express your answer (a) in kilograms, (b) in solar masses, and (c) in "galaxies," where 1 galaxy $= 10^{12}$ solar masses.

8. What were the matter density and the equivalent mass density of the cosmic radiation field when the universe was one-thousandth its present size? Assume critical density, and don't forget the cosmological redshift.

9. How many times did the universe double in size during the inflationary period if its final size was 10^{50} larger than when it started?

10. The "blobs" evident in Figure 17.20 are about 10° across. If those blobs represent clumps of matter around the time of decoupling (redshift = 1500), and assuming Euclidean geometry, estimate the size of the clumps (at the time of decoupling). ∞ *(More Precisely 16-1)*

Projects

1. Make a model of a two-dimensional universe and examine Hubble's law on it. Find a balloon that will blow up into a nice large sphere. Blow it up about halfway and mark dots all over its surface; the dots represent galaxies. Arbitrarily choose one dot as your home galaxy. Using a cloth measuring tape, measure the distances from your home galaxy to various other galaxies, numbering the dots so you do not confuse them later. Now blow the balloon up to full size and measure the distances again. Calculate the change in the distances for each galaxy; this is a measure of their speed (= change in position/change in time; the time is the same for each and is arbitrary). Plot their speeds against their new distances as in Figure 15.28. Do you get a straight-line correlation (that is, a plot that obeys Hubble's law)? Try this again using a different dot for your home galaxy. Do you still get a plot that obeys Hubble's law? Does it matter which dot you choose as home?

2. Write a paper on the philosophical differences between living in an open, closed, or flat universe. It is quite possible that astronomers may finally determine which (if any) of these is correct within your lifetime. Does it really matter? Are there aspects of any of these three possibilities that are uncomfortable to accept? Do you have a preference?

3. Go to your library and read about the model called the *steady-state universe*, which enjoyed some measure of popularity in the 1950s and 1960s. How does it differ from the standard Big Bang model? Why do you think the steady-state model is not accepted by most cosmologists today?

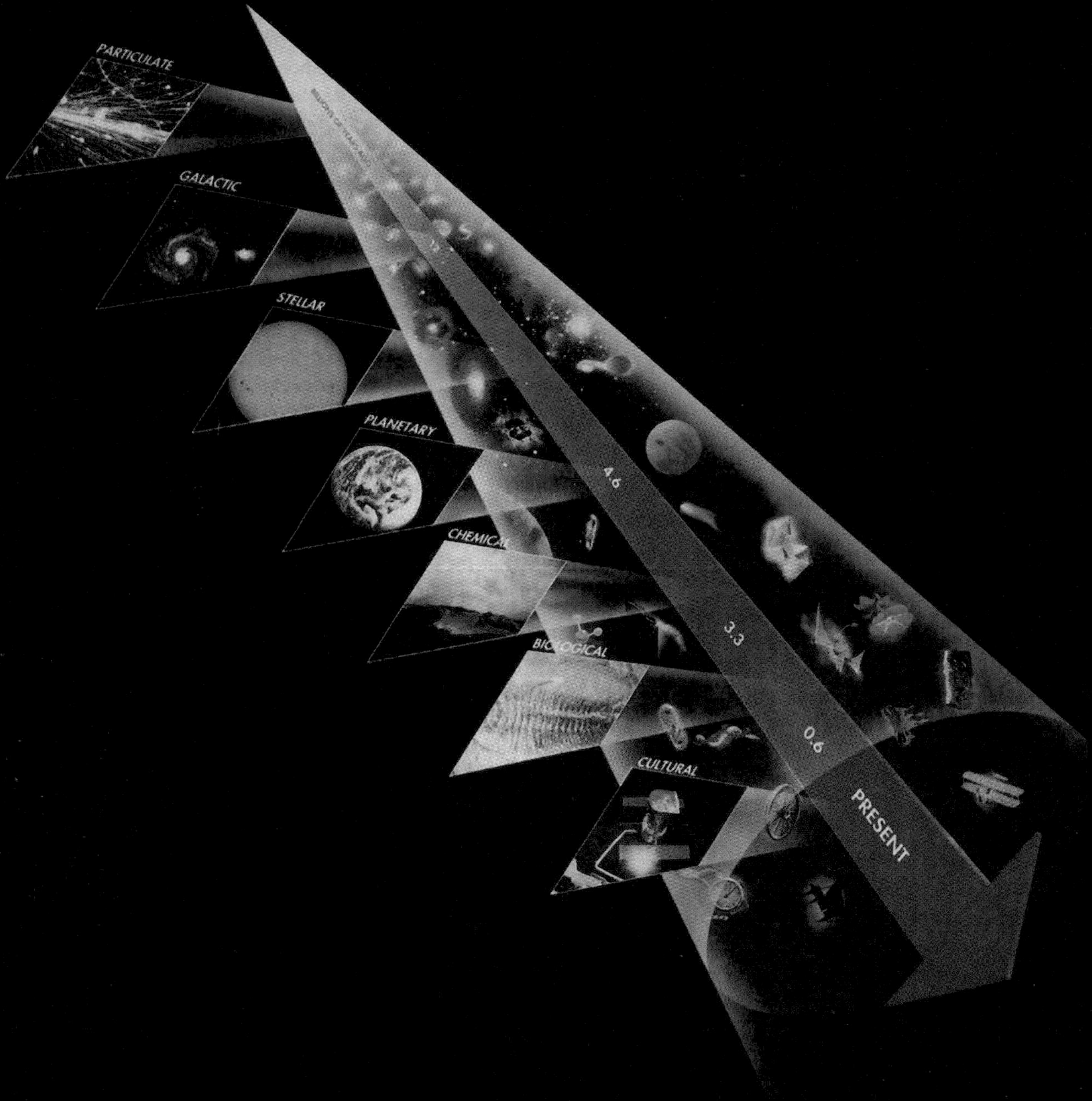
PARTICULATE
GALACTIC
STELLAR
PLANETARY
CHEMICAL
BIOLOGICAL
CULTURAL
BILLIONS OF YEARS AGO
4.6
3.3
0.6
PRESENT

18 LIFE IN THE UNIVERSE

Are We Alone? WWW

LEARNING GOALS

Studying this chapter will enable you to:

1 Summarize the process of cosmic evolution as it is currently understood.

2 Evaluate the chances of finding life in the solar system.

3 Summarize the various probabilities used to estimate the number of advanced civilizations that might exist in our Galaxy.

4 Discuss some of the techniques we might use to search for extraterrestrials and to communicate with them.

(Opposite page) The arrow of time, from the origin of the universe to the present and beyond, spans several major epochs of history. Cosmic evolution is the study of the many varied changes in the assembly and composition of energy, matter, and life in the thinning and cooling universe. Inserted along the arrow of time are seven "windows" outlining some of the key events in the history of the universe. *(Art by D. Berry)*

- evolution of primal energy into elementary particles
- evolution of atoms into galaxies and stars
- evolution of stars into heavy elements
- evolution of elements into solid, rocky planets
- evolution of those elements into the molecular building blocks of life
- evolution of those molecules into life itself
- evolution of advanced life forms into intelligence, culture, and technological civilization

Are we unique? Is life on our planet the only example of life in the universe? These are difficult questions, for the subject of extraterrestrial life is one on which we have no data, but they are important questions, with profound implications for the human species. Earth is the only place in the universe where we know for certain that life exists. In this chapter we take a look at how humans evolved on Earth and then consider whether those evolutionary steps might have happened elsewhere. Having done that, we assess the likelihood of our having Galactic neighbors and consider how we might learn about them if they exist.

18.1 Cosmic Evolution

1 Figure 18.1 (see also the chapter opening art) identifies seven major phases in the history of the universe: *particulate*, *galactic*, *stellar*, *planetary*, *chemical*, *biological*, and *cultural* evolution. Together, these evolutionary stages make up the grand sweep of **cosmic evolution**—the continuous transformation of matter and energy that has led to the appearance of life and civilization on Earth. The first four phases represent, in reverse order, the contents of this book. We now expand our field of view beyond astronomy to include the other three.

From the Big Bang to the evolution of intelligence and culture, the universe has evolved from simplicity to complexity. We are the result of an incredibly complex chain of events that spanned billions of years. Were those events random, making us unique, or are they in some sense *natural* and technological civilization inevitable? Put another way, are we alone in the universe, or are we just one among countless other intelligent life forms?

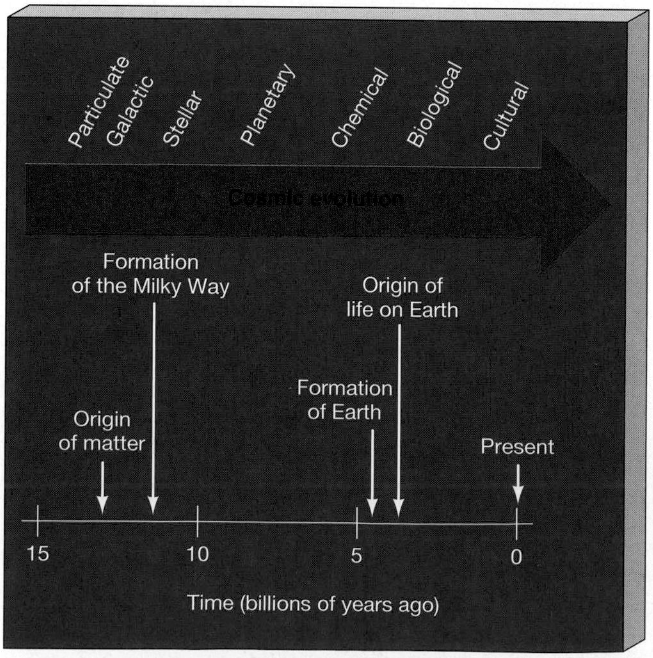

Figure 18.1 Arrow of Time Some highlights of cosmic history are indicated along this arrow of time, reaching from the beginning of the universe to the present. Noted along the top of the arrow are the major phases of cosmic evolution.

Life in the Universe

The distinction between the living and the nonliving is perhaps not as obvious as we might at first think. Generally speaking, scientists regard the following as characteristics of living organisms: 1) they can *react* to their environment and can often heal themselves when damaged; 2) they can *grow* by taking in nourishment from their surroundings; 3) they can *reproduce*, passing along some of their own characteristics to their offspring; and 4) they have the capacity for genetic change and can therefore *evolve* from generation to generation to adapt to a changing environment.

These rules are not hard and fast, and a good working definition of life is elusive. Stars, for example, react to the gravity of their neighbors, grow by accretion, generate energy, and "reproduce" by triggering the formation of new stars, but no one would suggest that they are alive. A virus is crystalline and inert when isolated from living organisms, but, once inside a living system, it exhibits all the properties of life, seizing control of the cell and using the cell's own genetic machinery to grow and reproduce. Most researchers believe that the distinction between living and nonliving is more one of structure and complexity than a simple checklist of rules.

The general case in favor of extraterrestrial life is summed up in what are sometimes called the *assumptions of mediocrity*: 1) because life on Earth depends on just a few basic molecules, 2) because the elements that make up these molecules are (to a greater or lesser extent) common to all stars, and 3) if the laws of science we know apply to the entire universe (as we have supposed throughout this book), then—given sufficient time—life must have originated elsewhere in the cosmos. The opposing view maintains that intelligent life on Earth is the product of a series of extremely fortunate accidents—

astronomical, geological, chemical, and biological events unlikely to have occurred anywhere else in the universe. The purpose of this chapter is to examine some of the arguments for and against these viewpoints.

Life on Earth

What information do we have about the earliest stages of planet Earth? Unfortunately, not very much. Geological hints about the first billion years or so were largely erased by violent surface activity as volcanoes erupted and meteorites bombarded our planet; subsequent erosion by wind and water has seen to it that little evidence has survived to the present day. Scientists believe, however, that the early Earth was barren, with shallow lifeless seas washing upon grassless treeless continents. Outgassing from our planet's interior through volcanoes, fissures, and geysers produced an atmosphere rich in hydrogen, nitrogen, and carbon compounds and poor in free oxygen. As Earth cooled, ammonia, methane, carbon dioxide, and water formed. The stage was set for the appearance of life.

The surface of the young Earth was a very violent place. Natural radioactivity, lightning, volcanism, solar ultraviolet radiation and meteoritic impacts all provided large amounts of energy that eventually shaped the ammonia, methane, carbon dioxide and water into more complex molecules known as **amino acids** and **nucleotide bases**—organic (carbon-based) molecules that are the building blocks of life as we know it. Amino acids build *proteins*, and proteins control *metabolism*, the daily utilization of food and energy by means of which organisms stay alive and carry out their vital activities. Sequences of nucleotide bases form *genes*—parts of the DNA molecule—which direct the synthesis of proteins and thus determine the characteristics of organisms. These same genes also carry an organism's hereditary characteristics from one generation to the next. In all living creatures on Earth—from bacteria to amoebas to humans—genes mastermind life, while proteins maintain it.

The first experimental demonstration that complex molecules could have evolved naturally from simpler ingredients found on the primitive Earth came in 1953. Scientists Harold Urey and Stanley Miller, using laboratory equipment somewhat similar to that shown in Figure 18.2, took a mixture of the materials thought to be present on Earth long ago—a "primordial soup" of water, methane, carbon dioxide, and ammonia—and energized it by passing an electrical discharge ("lightning") through it. After a few days, they analyzed their mixture and found that it contained many of the same amino acids found today in all living things on Earth. About a decade later, scientists succeeded in constructing nucleotide bases in a similar manner. These experiments have been repeated in many different forms, with more realistic mixtures of gases and a variety of energy sources, but always with the same basic outcomes.

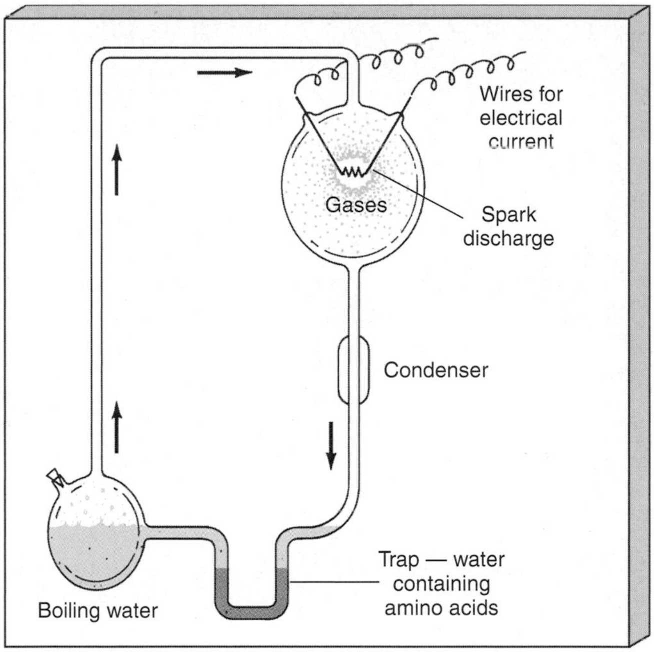

Figure 18.2 Miller-Urey Experiment This chemical apparatus is designed to synthesize complex biochemical molecules by energizing a mixture of simple chemicals. A mixture of gases (ammonia, methane, carbon dioxide, water vapor) is placed in the upper bulb to simulate the primordial Earth atmosphere and then energized by spark-discharge electrodes. After about a week, amino acids and other complex molecules are found in the trap at the bottom, which simulates the primordial oceans into which heavy molecules produced in the overlying atmosphere would have fallen.

Although none of these experiments has ever produced a living organism, or even a single strand of DNA, they do demonstrate conclusively that "biological" molecules can be synthesized by strictly nonbiological means, using raw materials available on the early Earth. More advanced experiments have fashioned proteinlike blobs (Figure 18.3a) that behave to some extent like true biological cells. They resist dissolution in water and tend to cluster into small droplets called *microspheres*—a little like oil globules floating on the surface of water. The walls of these laboratory-made droplets permit the inward passage of small molecules, which then combine within the droplet to produce more complex molecules too large to pass back out through the walls. As the droplets "grow," they tend to "reproduce," forming smaller droplets.

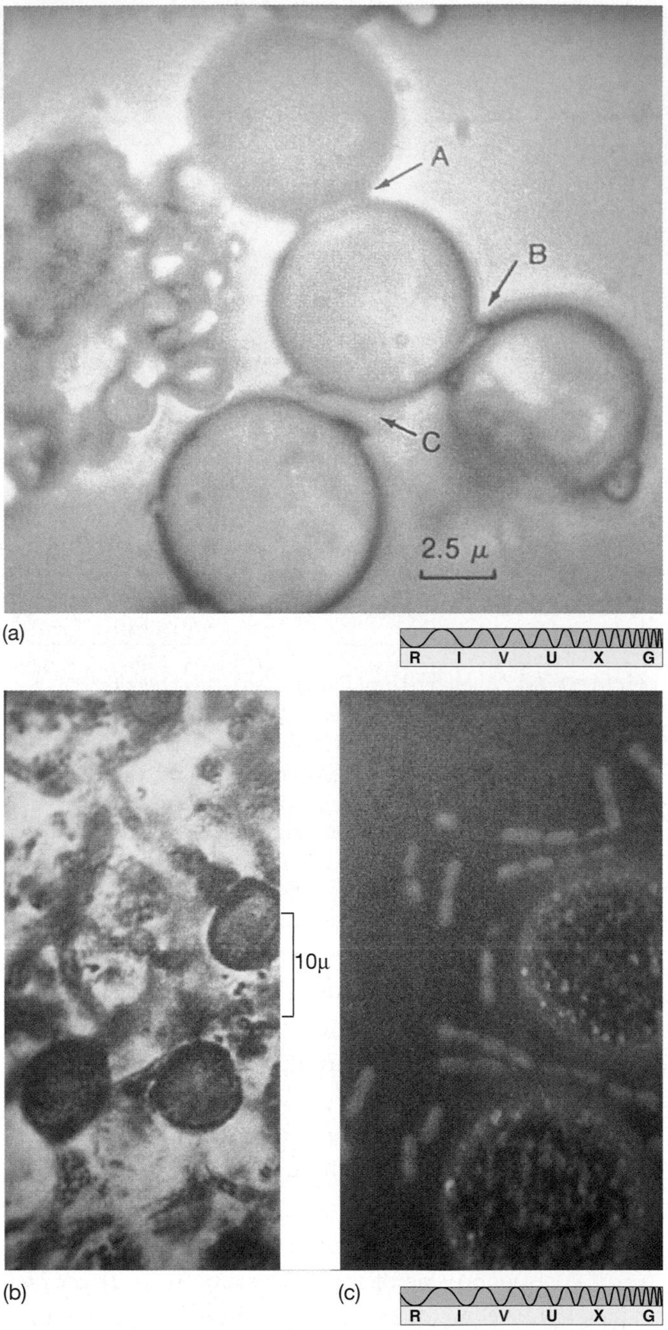

Figure 18.3 Chemical Evolution (a) These carbon-rich, proteinlike droplets display the clustering of as many as a billion amino acid molecules in a liquid. Droplets can "grow," and parts of droplets can separate from the "parent" to become new individual droplets (such as at A, B, C). The scale of 2.5 microns noted here is 1/4000 of a centimeter. (b) This photograph, taken through a microscope, shows a fossilized organism found in sediments radioactively dated as being two billion years old. This primitive system possesses concentric spheres or walls connected by smaller spheroids. The roundish fossils here measure about a thousandth of a centimeter in diameter. (c) For comparison, modern blue-green algae. *(Sidney Fox; Elso Barghoorn)*

Can we consider these proteinlike microspheres to be alive? Almost certainly not. The microspheres contain many of the basic ingredients needed to form life but are not life itself. They lack the hereditary molecule DNA. However, they do have similarities to ancient cells found in the fossil record (Figure 18.3b). Thus, while no actual living cells have yet been created "from scratch," many biochemists feel that *chemical evolution*—the chain of events leading from simple non-biological molecules almost to the point of life itself—has been amply demonstrated.

Some scientists do not agree that Earth's primitive atmosphere was a suitable environment for the production of complex molecules. They argue that there may not have been sufficient energy available to power the chemical reactions, and the early atmosphere may not have contained enough raw material for the reactions to occur at a significant rate in any case. Instead, they suggest, most of the organic material that combined to form the first living cells was produced in *interstellar space*, and subsequently arrived on Earth in the form of interplanetary dust and meteors that did not burn up during their descent through the atmosphere.

Interstellar molecular clouds are known to contain many complex molecules, and large amounts of organic material were detected on comets Halley and Hale-Bopp when they last visited the inner solar system, so the idea that organic matter is constantly raining down on Earth from space in the form of interplanetary debris is quite plausible. ∞ (Sec. 4.2) However, whether this was the *primary* means by which complex molecules appeared on Earth remains unclear.

Diversity and Culture

However the basic materials appeared on Earth, we know that life *did* appear. The fossil record chronicles how life on Earth became widespread and diversified over time. The study of fossil remains shows the initial appearance about 3.5 billion years ago of simple one-celled organisms, such as blue-green algae. These were followed about two billion years ago by more complex one-celled creatures like the amoeba. Multicellular organisms did not appear until about one billion years ago, after which a wide variety of increasingly complex organisms flourished—among them insects, reptiles, mammals, and humans.

The fossil record leaves no doubt that biological organisms have changed over time—all scientists accept the reality of *biological evolution*. As conditions on Earth have shifted and Earth's surface has evolved, those organisms that could best take advantage of their new surroundings succeeded and thrived—often at the expense of those organisms that could not make the necessary adjustments and consequently became extinct.

Many anthropologists believe that, like any other highly advantageous trait, intelligence is strongly favored by natural selection. The social cooperation that went with coordinated hunting efforts was an important competitive advantage that developed as brain size increased. Perhaps most important of all was the development of language. Experience and ideas, stored in the brain as memories, could be passed down from one generation to the next. A new kind of evolution had begun, namely, *cultural evolution*—the changes in the ideas and behavior of society. Our more recent ancestors created, within about 10,000 years, the entirety of human civilization.

✓ Concept Check

- Has chemical evolution been verified in the laboratory?

18.2 Life in the Solar System

2 Simple one-celled life forms reigned supreme on Earth for most of our planet's history. It took time—a great deal of time—for life to emerge from the oceans, to evolve into simple plants, to continue to evolve into complex animals, and to develop intelligence, culture, and technology. Have those (or similar) events occurred elsewhere in the solar system? Let's try to assess what little evidence we have on the subject.

Life as We Know It

"Life as we know it" is generally taken to mean carbon-based life that originated in a liquid-water environment—in other words, life on Earth. Might such life exist on some other body in our planetary system?

The Moon and Mercury lack liquid water, protective atmospheres, and magnetic fields, and so are subjected to fierce bombardment by solar ultraviolet radiation, the solar wind, meteoroids, and cosmic rays. Simple molecules could not survive in such hostile environments. On the other hand, Venus has far too much protective atmosphere. Its dense, dry, scorchingly hot atmospheric blanket effectively rules it out as a possible abode for life.

The jovian planets have no solid surfaces, and Pluto and most of the moons of the outer planets are too cold. However, the possibility of liquid water below Europa's icy surface has refueled speculation about the development of life there, making this moon of Jupiter a prime candidate for future exploration. ∞ (Sec. 8.1) Saturn's moon Titan, with its atmosphere of methane, ammonia, and nitrogen gases, and possibly with some liquid on its surface, is conceivably another site where life might have arisen, although the results of the 1980 *Voyager 1* flyby suggest that Titan's frigid surface conditions are inhospitable for anything familiar to us. ∞ (Sec. 8.2)

What about the cometary and meteoritic debris that orbits within our solar system? Comets contain many of the basic ingredients for life, but they spend almost all of their time frozen, far from the Sun. A small fraction of the meteorites that survive the plunge to Earth's surface also contain organic compounds. A meteorite fell near Murchison, Australia, in 1969 that was located soon after crashing to the ground. This well-studied meteorite contains many of the amino acids normally found in living cells (Figure 18.4). The moderately large molecules found in meteorites and in interstellar clouds are our only evidence that chemical evolution has occurred elsewhere in the universe. However, most researchers regard this organic matter as prebiotic—matter that could eventually lead to life, but that has not yet done so.

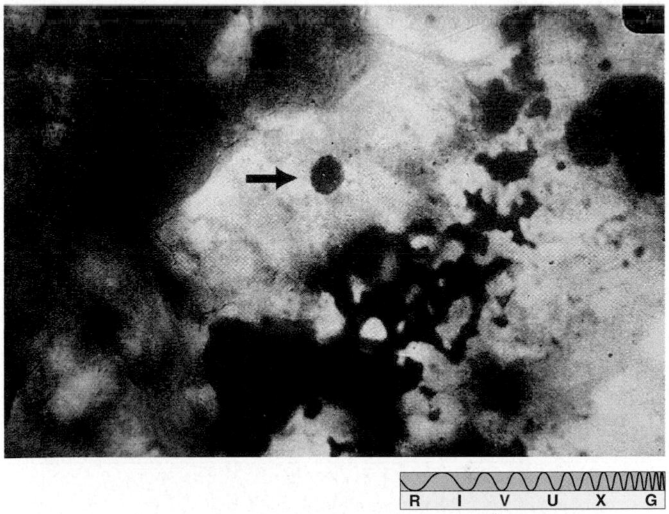

Figure 18.4 Murchison Meteorite The Murchison meteorite contains relatively large amounts of amino acids and other organic material, indicating that chemical evolution of some sort has occurred beyond our own planet. In this magnified view of a fragment from the meteorite, the arrow points to a microscopic sphere of organic matter. *(Harvard-Smithsonian Center for Astrophysics)*

The planet most likely to harbor life (or to have harbored it in the past) seems to be Mars. This planet is harsh by Earth standards—liquid water is scarce, the atmosphere is thin, and the lack of magnetism and of an ozone layer allows solar high-energy particles and ultraviolet radiation to reach the surface unabated. But the Martian atmosphere was thicker, and the surface warmer and much wetter, in the past. ∞ (Sec. 6.8) The *Viking* landers found no evidence of life, but they landed on the safest Martian terrain, not in the most interesting regions, such as near the moist polar caps.

Some scientists think that a different type of biology might be operating, or might have operated, on the Martian surface. They suggest that Martian microbes capable of digesting oxygen-rich compounds in the Martian soil could explain the *Viking* results. This speculation will be greatly strengthened if recent announcements of fossilized bacteria in meteorites originating on Mars are confirmed. ∞ (*Interlude 6-1*) The consensus among biologists and chemists today is that Mars does not house any life similar to that on Earth, but a solid verdict regarding present or past life on Mars will likely not be reached until we have thoroughly explored our intriguing neighbor.

Alternative Biochemistries

Conceivably, some types of biology may be so different from life on Earth that we do not know how to test for them. Some researchers have pointed out that the abundant element silicon has chemical properties somewhat similar to those of carbon and have suggested silicon as a possible alternative to carbon as the basis for living organisms. Ammonia (made of the common elements hydrogen and nitrogen) is sometimes put forward as a possible liquid medium in which life might develop, at least on a planet cold enough for ammonia to exist in the liquid state. Together or separately, these alternatives would surely give rise to organisms having biochemistries radically different from those we know on Earth. Conceivably, we might have difficulty even recognizing these organisms as being alive.

While the possibility of such alien life forms is a fascinating scientific problem, most biologists would argue that chemistry based on carbon and water is the one most likely to give rise to life. Silicon's chemical bonds are weaker than those of carbon and may not be able to form complex molecules—an apparently essential aspect of carbon-based life. Also, the low temperatures necessary for ammonia to be liquid might inhibit or even prevent completely the chemical reactions leading to the equivalent of amino acids and nucleotide bases. Still, we must admit that we know next to nothing about non-carbon, non-water biochemistries, for the good reason that there are no examples of them to study experimentally. We can speculate about alien life forms and try to make general statements about their properties, but we can say little of substance about them.

✓ Concept Check

- Which solar system bodies (other than Earth) are the leading candidates in the search for extraterrestrial life?

18.3 Intelligent Life in the Galaxy

3 With humans apparently the only intelligent life in the solar system, we must broaden our search for extraterrestrial intelligence to other stars, perhaps even other galaxies. At such distances, though, we have little hope of detecting life with current equipment. Instead, we must ask: "How likely is it that life in any form—carbon based, silicon based, water based, ammonia based, or something we cannot even dream of—exists?" Let's look at some numbers to develop statistical estimates of the probability of life elsewhere in the universe.

The Drake Equation

An early approach to this statistical problem is known as the **Drake equation**, after Frank Drake, the U.S. astronomer who pioneered this analysis:

number of technological, intelligent civilizations now present in the Galaxy	=	rate of star formation, averaged over the lifetime of the Galaxy	×	fraction of stars having planetary systems	×	average number of habitable planets within those planetary systems	×	fraction of those habitable planets on which life arises	×	fraction of those life-bearing planets on which intelligence evolves	×	fraction of those intelligent-life planets that develop technological society	×	average lifetime of a technologically competent civilization.

Several of the terms in this equation are largely a matter of opinion. We do not have nearly enough information to determine—even approximately—all of the terms, so the Drake equation cannot give us a hard-and-fast answer. Its real value is that it subdivides a large and very difficult question into smaller pieces that we can attempt to answer separately. Figure 18.5 illustrates how, as our requirements become more and more stringent, only a small fraction of star systems in the Milky Way Galaxy have the advanced qualities specified by the combination of terms on the right-hand side of the equation.

Let's examine the terms in the equation one by one and make some educated guesses about their values. Bear in mind, though, that if you ask two scientists for their best estimates of any given term, you will likely get two very different answers!

Rate of Star Formation

We can estimate the average number of stars forming each year in the Galaxy simply by noting that at least 100 billion stars now shine in it. Dividing this number by the roughly 10-billion-year lifetime of the Galaxy, we obtain a formation rate of 10 stars per year. This may be an overestimate because astronomers believe that fewer stars are forming now than formed during earlier epochs, when more interstellar gas was available. However, we do know that stars are forming today, and our

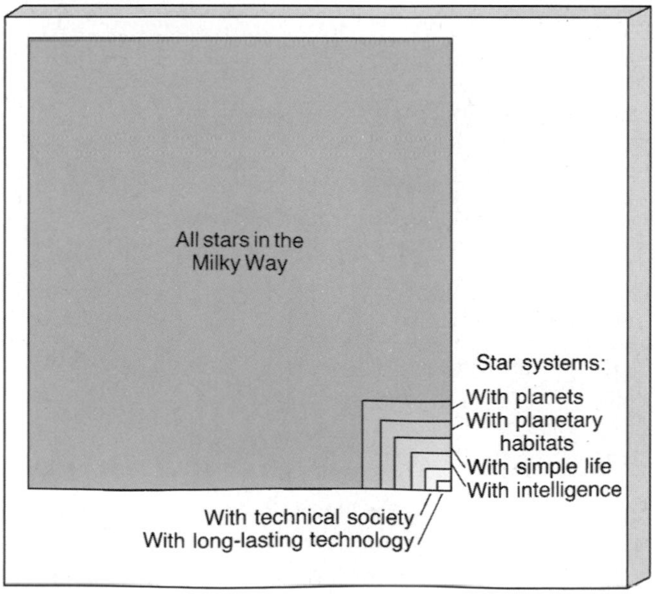

Figure 18.5 Graphical Drake Equation Of all the star systems in our Milky Way Galaxy (represented by the largest box), progressively fewer and fewer have each of the qualities typical of a long-lasting technological society (represented by the smallest box at the lower right corner).

estimate does not include stars that formed in the past and have since died, so our value of 10 stars per year is probably reasonable when averaged over the lifetime of the Milky Way Galaxy.

Fraction of Stars Having Planetary Systems

Many astronomers believe planet formation to be a natural result of the star-formation process. If the condensation theory is correct, and if there is nothing special about our Sun, as we have argued throughout this book, we would expect many stars to have at least one planet. ∞ (Sec. 4.3)

No planets like our own have yet been *seen* orbiting any other star. The light reflected by an Earth-like planet circling even the closest star would be too faint to detect with current equipment, although planned orbiting telescopes may soon be able to detect Jupiter-sized and smaller planets orbiting the nearest stars. However, there is good *indirect* evidence for planets orbiting other stars in the back-and-forth or side-to-side motions of a parent star, which betray an orbiting planet's gravitational pull. ∞ (*Interlude 4-1*) The planets found so far are Jupiter-sized rather than Earth-sized, and most move on orbits quite unlike those found in our solar system, but astronomers are confident that Earth-like planets will one day be detected by one of these means.

Accepting the condensation theory and its consequences, and without being either too conservative or naively optimistic, we assign a value near 1 to this term—that is, we believe that essentially all stars have planetary systems.

Number of Habitable Planets per Planetary System

Temperature, more than any other single quantity, determines the feasibility of life on a given planet. The surface temperature of a planet depends on two things: the planet's distance from its parent star and the thickness of its atmosphere. Planets with a nearby parent star (but not too close) and some atmosphere (though not too thick) should be reasonably warm, like Earth or Mars. Planets far from the star and with no atmosphere, like Pluto, will surely be cold by our standards. And planets too close to the star and with a thick atmosphere, like Venus, will be very hot indeed.

Figure 18.6 illustrates how a three-dimensional zone of "comfortable" temperatures—often called a *habitable zone*—surrounds every star. It represents the range of distances within which a planet of mass and composition similar to those of Earth would have a surface

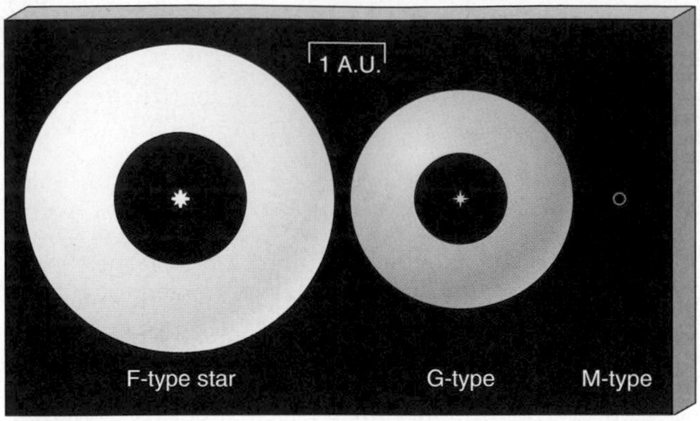

Figure 18.6 Habitable Zones The extent of the habitable zone is much larger around a hot star than around a cool one. For a star like the Sun (a G-type star), the zone extends from about 0.85 A.U. to 2.0 A.U. For an F-type star, the range is 1.2 to 2.8 A.U. For a faint M-type star, only planets orbiting between about 0.02 and 0.06 A.U. would be habitable.

temperature between the freezing and boiling points of water. (Our Earth-based bias is plainly evident here!) The hotter the star, the larger this zone. For example, an A- or an F-type star has a rather large habitable zone. G-, K-, and M-type stars have successively smaller zones. O- and B-type stars are not considered here because they are not expected to last long enough for life to develop, even if they do have planets.

Three planets—Venus, Earth, and Mars—reside within the habitable zone surrounding our Sun. Venus is too hot because of its thick atmosphere and proximity to the Sun. Mars is a little too cold because its atmosphere is too thin and it is too far from the Sun. But if Venus had Mars's thin atmosphere and if Mars had Venus's thick atmosphere, both of these nearby planets might conceivably have surface conditions resembling those on Earth.

To estimate the number of habitable planets per planetary system, we first take inventory of how many stars of each type shine in our Galaxy and calculate the sizes of their habitable zones. Then we eliminate binary-star systems because a planet's orbit within the habitable zone of a binary would be unstable in many cases, as illustrated in Figure 18.7. Given the known properties of binaries in our Galaxy, habitable planetary orbits would probably be unstable in most cases, so there would not be time for life to develop.

Taking all these factors into account, we assign a value of 1/10 to this term in our equation. In other words, we believe that, on average, there is one potentially habitable planet for every 10 planetary systems that might exist in our Galaxy. Single F-, G-, and K-type stars are the best candidates. However, if the current inventory of extrasolar planets turns out to be typical, then planetary systems with roughly circular orbits (required to remain within the habitable zone) may be a distinct *minority*, and our estimate of this term will have to be substantially reduced. ∞ (*Interlude 4-1*)

Fraction of Habitable Planets on Which Life Arises

The number of possible combinations of atoms is incredibly large. If the chemical reactions that led to the complex molecules that make up living organisms occurred completely at random, then it is extremely unlikely that those molecules could have formed at all. In that case, life is extraordinarily rare, this term is close to zero, and we are probably alone in the Galaxy, perhaps even in the entire universe.

However, laboratory experiments (like the Urey-Miller experiment described earlier) seem to suggest that

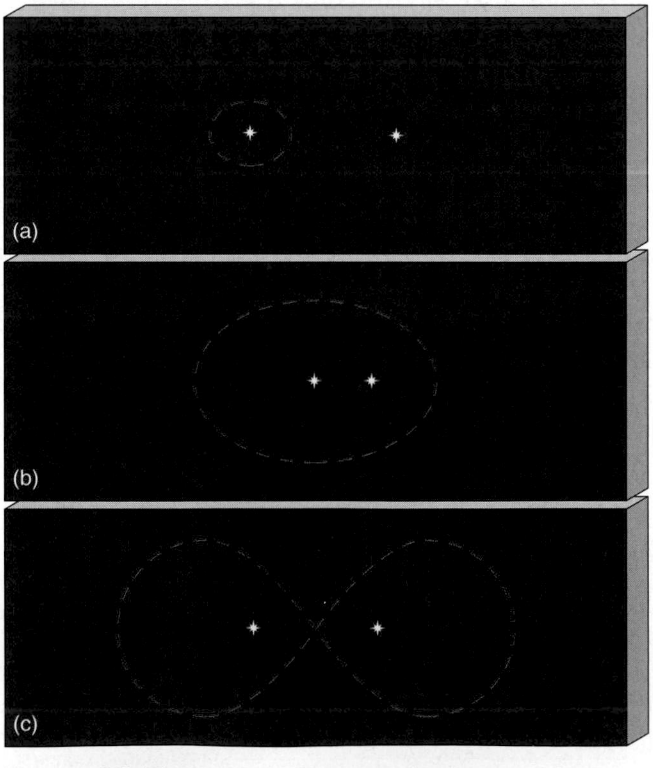

Figure 18.7 Binary-star Planets In binary-star systems, planets are restricted to only a few kinds of orbits that are gravitationally stable. (a) This orbit is stable only if the planet lies very close to its parent star, so the gravity of the other star is negligible. (b) A planet circulating at a great distance about both stars in an elliptical orbit. This orbit is stable only if it lies far from both stars. (c) Another possible but unstable path interweaves between the two stars in a figure-eight pattern.

certain chemical combinations are strongly favored over others—that is, the reactions are not random. Of the billions upon billions of basic organic groupings that could possibly occur on Earth from the random combination of all sorts of simple atoms and molecules, only about 1500 actually do occur. Furthermore, these 1500 organic groups of terrestrial biology are made from only about 50 simple "building blocks" (including the amino acids and nucleotide bases mentioned earlier). This suggests that molecules critical to life may not be assembled by pure chance. Apparently, additional factors are at work at the microscopic level. If a relatively small number of chemical "evolutionary tracks" is likely to exist, then the formation of complex molecules—and hence, we assume, life—becomes much more likely, given sufficient time.

To assign a very low value to this term in the equation is to believe that life arises randomly and rarely. To assign a value close to 1 is to believe that life is inevitable, given the proper ingredients, a suitable environment, and a long enough period of time. No easy experiment can distinguish between these extreme alternatives, and there is little or no middle ground. To many researchers, the discovery of life—past or present—on Mars, Europa, Titan, or any other object in our solar system would convert the appearance of life from an unlikely miracle to a virtual certainty throughout the Galaxy. We will take the optimistic view and adopt a value of 1.

Fraction of Life-Bearing Planets on Which Intelligence Arises

As with the evolution of life, the appearance of a well-developed brain is a very unlikely event if only random chance is involved. However, biological evolution through natural selection is a mechanism that generates apparently highly improbable results by singling out and refining useful characteristics. Organisms that profitably use adaptations can develop more complex behavior, and complex behavior provides organisms with the *variety* of choices needed for more advanced development.

One school of thought maintains that, given enough time, intelligence is inevitable. In this view, assuming that natural selection is a universal phenomenon, at least one organism on a planet will always rise to the level of "intelligent life." If this is correct, then the fifth term in the Drake equation equals or nearly equals 1.

Others argue that there is only one known case of intelligence, and that case is life on Earth. For 2.5 billion years—from the start of life about 3.5 billion years ago to the first appearance of multicellular organisms about one billion years ago—life did not advance beyond the one-celled stage. Life remained simple and dumb, but it survived. If this latter view is correct, then the fifth term in our equation is very small, and we are faced with the depressing prospect that humans may be the smartest form of life anywhere in the Galaxy. As with the previous term, we will be optimistic and adopt a value of 1 here.

Fraction of Planets on Which Intelligent Life Develops and Uses Technology

To evaluate the sixth term of our equation, we need to estimate the probability that intelligent life eventually develops technological competence. Should the rise of technology be inevitable, this term is close to 1, given long enough periods of time. If it is not inevitable—if intelligent life can somehow "avoid" developing technology—then this term could be much less than 1. The latter possibility envisions a universe possibly teeming with civilizations but very few among them ever becoming technologically competent. Perhaps only one managed it—ours.

Again, it is difficult to decide conclusively between these two views. We don't know how many (if any) prehistoric Earth cultures either failed to develop technology or rejected its use. We do know that tool-using societies arose independently at several places on Earth, including Mesopotamia, India, China, Egypt, Mexico, and Peru. Because so many of these ancient "technological" cultures originated *independently* at about the same time, we conclude that the chances are good that some sort of technological society will inevitably develop, given some basic intelligence and enough time. Accordingly, we take this term to be close to 1.

Average Lifetime of a Technological Civilization

The reliability of these estimates declines markedly from the first to the last term of the Drake equation. Our knowledge of astronomy allows us to make a reasonably good stab at the first term, but it is much harder to evaluate some of the later terms, and this term is totally unknown. There is only one example of such a civilization that we are aware of—humans on planet Earth. Our own civilization has presently survived in its technological state for only about 100 years, and how long we will be around before a human-made catastrophe or a planetwide natural disaster (see *Interlude 18-1*) ends it all is impossible to tell.

One thing is certain: if the correct value for *any one term* in the equation is very small, then few technological civilizations now exist in the Galaxy. If the pessimistic view of the development of life or of intelligence is correct, then we are unique, and that is the end of our story. However, if both life and intelligence are inevitable consequences of chemical and biological evolution, as many scientists believe, and if intelligent life always becomes technological, then we can plug the higher, more optimistic values into the Drake equation. In that case, combining our estimates for the other six terms (and noting that $10 \times 1 \times 1/10 \times 1 \times 1 \times 1 = 1$), we can say

number of technological, intelligent civilizations now present in the Milky Way Galaxy	=	average lifetime of a technologically competent civilization, in years.

Thus, if advanced civilizations typically survive for 1000 years, there should be 1000 of them currently in existence scattered throughout the Galaxy. If they survive for a million years, on average, we would expect there to be a million of them in the Galaxy, and so on.

✓ Concept Check

- How does the Drake equation assist astronomers in refining their search for extraterrestrial life?

18.4 The Search for Extraterrestrial Intelligence

4 Let's continue our optimistic assessment of the prospects for life and assume that civilizations enjoy a long stay on their parent planet once their initial technological "teething problems" are past. In that case, they are likely to be plentiful in the Galaxy. How might we become aware of their existence?

Meeting Our Neighbors

For definiteness, let us assume that the average lifetime of a technological civilization is one million years—only one percent of the reign of the dinosaurs, but 100 times longer than human civilization has survived thus far. Given the size and shape of our Galaxy, we can then estimate the average *distance* between these civilizations to be some 30 pc, or about 100 light-years. Thus, any two-way communication with our neighbors—using signals traveling at or below the speed of light—will take at least 200 years (100 years for the message to reach the planet and another 100 years for the reply to travel back to us).

Might we hope to visit our neighbors by developing the capability of traveling far outside our solar system? This may never be a practical possibility. At a speed of 50 km/s, the speed of the current fastest space probes, the round trip to even the nearest Sun-like star, Alpha Centauri, would take about 50,000 years. The journey to the nearest technological neighbor (assuming a distance of 100 pc) and back would take one million years—the entire lifetime of our civilization! Interstellar travel at these speeds is clearly not feasible. Speeding up our ships to near the speed of light would reduce the travel time, but this is far beyond our present technology.

Actually, our civilization has already launched some interstellar probes, although they have no specific stellar destination. Figure 18.8 is a reproduction of a plaque mounted on board the *Pioneer 10* spacecraft launched in the mid-1970s and now well beyond the orbit of Pluto, on its way out of the solar system. Similar information was also included aboard the *Voyager* probes launched in 1978. While these spacecraft would be incapable of reporting back to Earth the news that

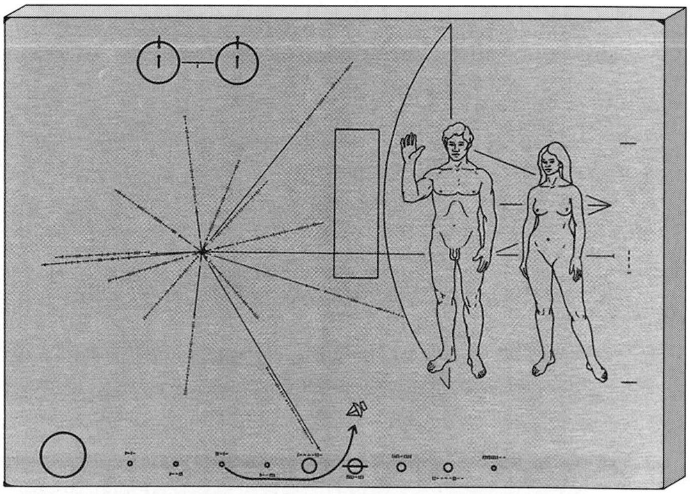

Figure 18.8 ***Pioneer 10* Plaque** A replica of a plaque mounted on board the *Pioneer 10* spacecraft showing the important features of the plaque. Included are a scale drawing of the spacecraft, a man, and a woman; a diagram of the hydrogen atom undergoing a change in energy (top left); a starburst pattern representing various pulsars and the frequencies of their radio waves that can be used to estimate when the craft was launched (middle left); and a depiction of the solar system, showing that the spacecraft departed the third planet from the Sun and passed the fifth planet on its way into outer space (bottom). All the drawings have computer-coded (binary) markings from which actual sizes, distances, and times can be derived. *(NASA)*

they had encountered an alien culture, scientists hope that the civilization on the other end would be able to unravel most of its contents using the universal language of mathematics. The caption to Figure 18.8 notes how the aliens might discover from where and when the *Pioneer* and *Voyager* probes were launched.

Radio Communication

A cheaper, and much more practical, alternative is to try to make contact with extraterrestrials using electromagnetic radiation, the fastest known means of transferring information from one place to another. Because light and other short-wavelength radiation are heavily scattered while moving through dusty interstellar space, long-wavelength radio radiation seems to be the best choice. ∞ (Sec. 2.3) We do not attempt to broadcast to all nearby candidate stars, however—that would be far too expensive and inefficient. Instead, radio telescopes on Earth listen *passively* for radio signals emitted by other civilizations.

In what direction should we aim our radio telescopes? The answer to this question at least is fairly easy. On the basis of our earlier reasoning, we should target all F-, G-, and K-type stars in our vicinity. But are extraterrestrials broadcasting radio signals? If they are not, this search technique will obviously fail. If they are, how do we distinguish their artificially generated radio signals from signals emitted naturally by interstellar gas clouds? This depends on whether the signals are produced deliberately or are simply "waste radiation" escaping from a planet.

Consider first how "waste" radio wavelengths originating on Earth would look to extraterrestrials. Figure 18.9 shows the pattern of radio signals we broadcast into space. From the viewpoint of a distant observer, the

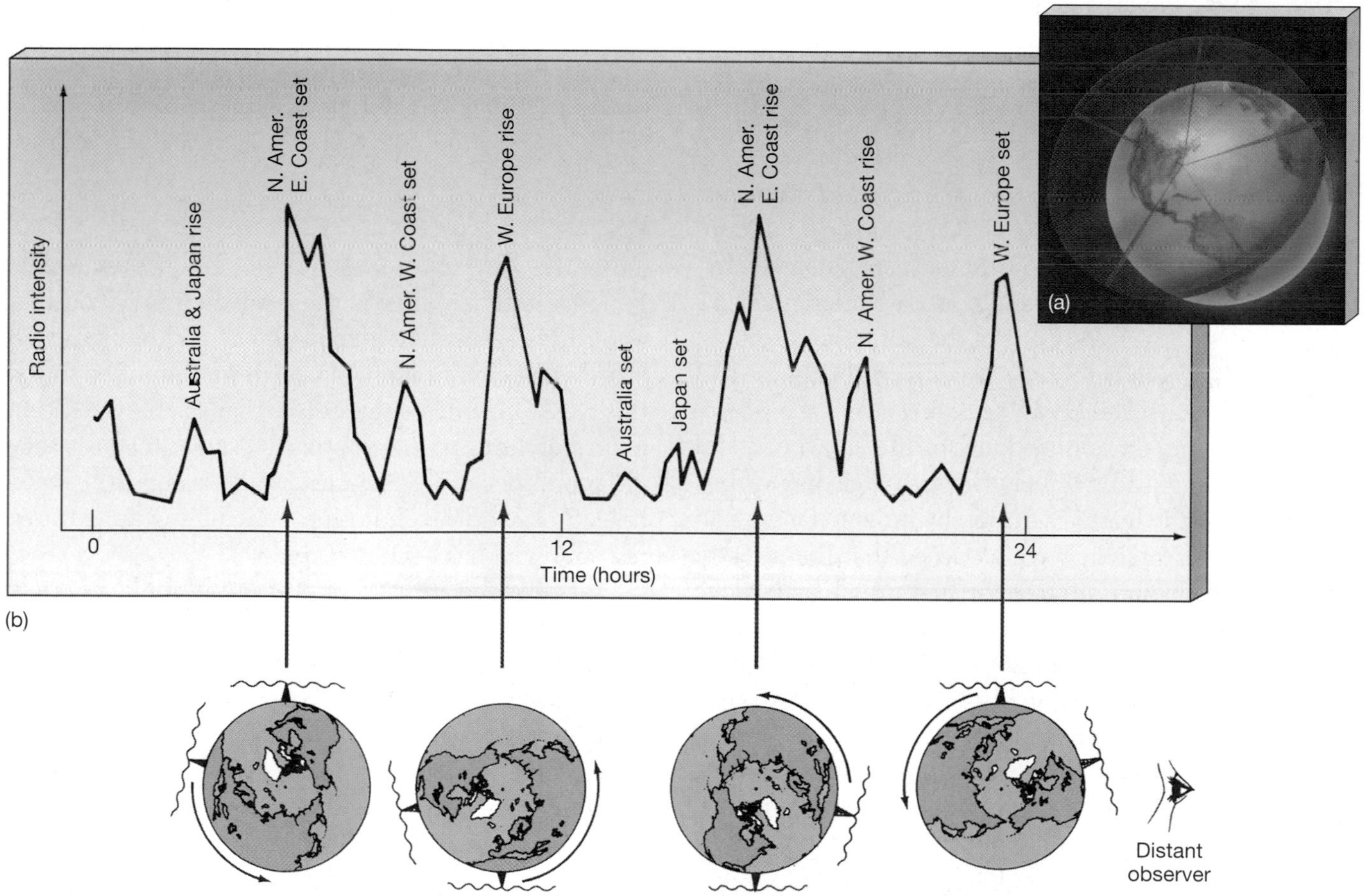

Figure 18.9 Earth's Radio Leakage Radio radiation now leaks from Earth into space because of the daily activities of our technological civilization. (a) FM radio and television transmitters broadcast their energy parallel to Earth's surface, so they send a great "sheet" of electromagnetic radiation into interstellar space, producing the strongest signal in any given direction when they happen to lie on Earth's horizon, as seen from that direction. (The more common AM broadcasts are trapped below our ionosphere, so those signals never leave Earth.) (b) Because the great majority of transmitters are clustered in the eastern United States and western Europe, a distant observer would detect blasts of radiation from Earth as our planet rotates each day.

spinning Earth emits a bright flash of radio radiation every few hours as the most technological regions of our planet rise or set. In fact, Earth is now a more intense radio emitter than the Sun. Our radio and television transmissions race out into space, and have been doing so since the invention of these technologies nearly seven decades ago. Another civilization as advanced as ours might have constructed devices capable of detecting this radiation. If any sufficiently advanced (and sufficiently interested) civilization resides within about 65 light-years (20 pc) of Earth, then we have already announced our presence to them.

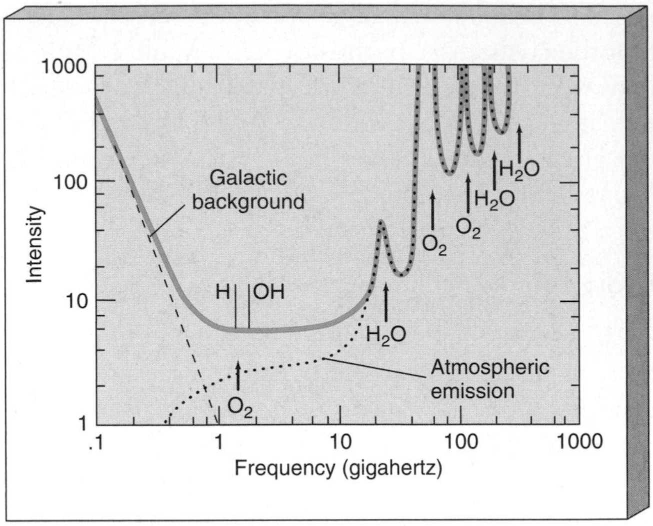

Figure 18.10 Water Hole The "water hole" is bounded by the natural emission frequencies of the hydrogen (H) atom (21-cm wavelength) and the hydroxyl (OH) molecule (18-cm wavelength). The topmost solid (blue) curve sums the natural emissions of our Galaxy (dashed line on left side of diagram labeled "Galactic background") and Earth's atmosphere (dotted line on right side of diagram denoted by various chemical symbols). This sum is minimized near the water hole frequencies. Perhaps all intelligent civilizations conduct their interstellar communications within this quiet "electromagnetic oasis."

The Water Hole

Now let us suppose that a civilization has decided to assist searchers by actively broadcasting its presence to the rest of the Galaxy. At what frequency should we listen for such an extraterrestrial beacon? The electromagnetic spectrum is enormous; the radio domain alone is vast. To hope to detect a signal at some unknown radio frequency is like searching for a needle in a haystack. Are some frequencies more likely than others to carry alien transmissions?

Some basic arguments suggest that civilizations might well communicate at a wavelength near 20 cm. The basic building blocks of the universe, namely, hydrogen atoms, radiate naturally at a wavelength of 21 cm. ∞ (Sec. 11.2) Also, one of the simplest molecules, hydroxyl (OH), radiates near 18 cm. Together, these two substances form water (H_2O). Arguing that water is likely to be the interaction medium for life anywhere, and that radio radiation travels through the disk of our Galaxy with the least absorption by interstellar gas and dust, some researchers have proposed that the interval between 18 cm and 21 cm is the best wavelength range for civilizations to transmit or listen. Called the **water hole**, this radio interval might serve as an "oasis" where all advanced Galactic civilizations would gather to conduct their electromagnetic business.

This water-hole frequency interval is only a guess, of course, but its use is supported by other arguments. Figure 18.10 shows the water hole's location in the electromagnetic spectrum and plots the amount of natural emission from our Galaxy and from Earth's atmosphere. The 18- to 21-cm range lies within the quietest part of the spectrum, where the Galactic "static" from stars and interstellar clouds happens to be minimized. Furthermore, the atmospheres of typical planets are also expected to interfere least at these wavelengths. Thus, the water hole seems like a good choice for the frequency of an interstellar beacon, although we cannot be sure of this reasoning until contact is actually achieved. Some radio searches are now in progress, simultaneously sampling millions of frequencies in and around the water hole. So far, however, nothing resembling an extraterrestrial signal has been detected.

The space surrounding all of us could be, right now, flooded with radio signals from extraterrestrial civilizations. If only we knew the proper direction and frequency, we might be able to make one of the most startling discoveries of all time. The result would provide whole new opportunities to study the cosmic evolution of energy, matter, and life throughout the universe.

✓ Concept Check

- Why do many researchers regard the water hole as a likely place to search for extraterrestrial signals?

INTERLUDE 18-1

What Killed the Dinosaurs?

The name *dinosaur* derives from the Greek words *deinos* (terrible) and *sauros* (lizard). Dinosaurs were no ordinary reptiles. In their prime, roughly 100 million years ago, they were the all-powerful rulers of Earth. Their fossilized remains have been uncovered on all the world's continents. Yet despite their dominance, according to the fossil record, these creatures vanished from Earth quite suddenly about 65 million years ago. What happened to them?

Many explanations have been offered to account for the dinosaurs' sudden and complete disappearance: devastating plagues, magnetic field reversals, increased geological activity, severe climate changes, even supernova explosions. In the 1980s, it was suggested that the cause was a huge extraterrestrial object that collided with Earth 65 million years ago. (An ancient, heavily eroded crater in the Yucatan peninsula in Mexico is often cited as a possible point of impact.) According to this idea, illustrated in the accompanying figure, a 10- to 15-km-wide asteroid or comet struck Earth, releasing as much energy as 10 million or more of the largest hydrogen bombs humans have ever constructed and propelling huge quantities of dust high into the atmosphere. The dust may have shrouded our planet for many years, virtually extinguishing the Sun's rays during this time. On the darkened surface, most plants could not survive. The entire food chain was disrupted, and the dinosaurs, at the top of that chain, eventually died out.

Although we have no direct astronomical evidence to confirm or refute this idea, we can estimate the chances of a large asteroid or comet striking the Earth today, on the basis of observations of the number of objects presently on Earth-crossing orbits. The second figure shows the likelihood of an impact as a function of the energy released by the collision, measured in *megatons* of TNT. (The megaton—4.2×10^{16} joules, the explosive yield of a large nuclear warhead—is the only common terrestrial measure of energy adequate to describe the violence of these occurrences.) Large meteoroids or asteroids are relatively rare, so 100-million-megaton events like the planetwide catastrophe that supposedly wiped out the dinosaurs are very infrequent, occurring only once every 10 million years or so. ∞ (Sec. 4.2) However, smaller impacts, equivalent to "only" a few tens of kilotons of TNT, could happen every few years—we may be long overdue for one. The most recent large impact was in 1908, the Tunguska explosion in Siberia which packed a roughly 1-megaton punch (Figure 4.16).

While this theory is (arguably) the leading explanation for the demise of the dinosaurs, it is by no means universally accepted. Perhaps predictably, geologists tend to be less convinced than astronomers, citing numerous pieces of geological data for which the impact theory offers only partial explanation and presenting alternative explanations for many of the facts used to support the impact theory. Whatever killed the dinosaurs, however, dramatic environmental change of some sort was almost surely responsible. It is important that we continue the search for the cause of their extinction, for there is no telling if and when that sudden change might strike again. As the dominant species on Earth, we are now the ones with the most to lose.

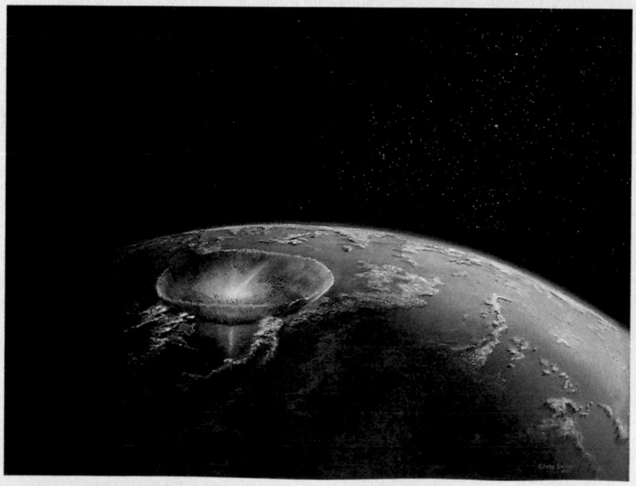

(C. Butler/Astrostock-Sanford)

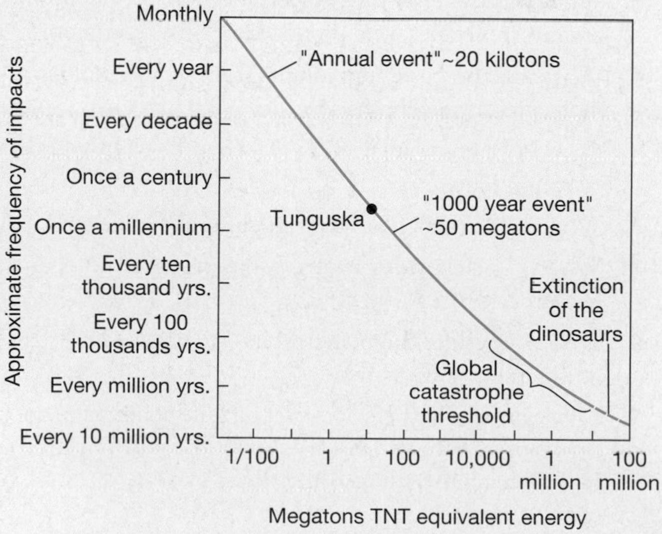

Chapter Review

Summary

Cosmic evolution (p. 456) is the continuous process that has led to the appearance of galaxies, stars, planets, and life on Earth. Living organisms may be characterized by their ability to react to their environment, to grow by taking in nutrition from their surroundings, to reproduce, passing along some of their own characteristics to their offspring, and to evolve in response to a changing environment. Organisms that can best take advantage of their new surroundings succeed at the expense of those organisms that cannot make the necessary adjustments. Intelligence is strongly favored by natural selection.

Powered by natural energy sources, reactions between simple molecules in the oceans of the primitive Earth may have led to the formation of **amino acids** (p. 457) and **nucleotide bases** (p. 457), the basic molecules of life. Alternatively, some complex molecules may have been formed in interstellar space and then delivered to Earth by meteors or comets. The best hope for life beyond Earth in the solar system is the planet Mars, although no evidence for living organisms has been found. Jupiter's Europa and Saturn's Titan may also be possibilities, but conditions on both those bodies are harsh by terrestrial standards.

The **Drake equation** (p. 460) provides a means of estimating the probability of other intelligent life in the Galaxy. The astronomical terms in the equation are the Galactic star-formation rate, the likelihood of planets, and the number of habitable planets. Chemical and biological terms are the probability of life appearing and the probability that it subsequently develops intelligence. Cultural and political terms are the probability that intelligence leads to technology and the lifetime of a technological civilization. Taking an optimistic view of the development of life and intelligence leads to the conclusion that the total number of technologically competent civilizations in the Galaxy is approximately equal to the lifetime of a typical civilization, expressed in years.

A technological civilization would probably "announce" itself to the universe by the radio and television signals it emits into space. Observed from afar, our planet would appear as a radio source with a 24-hour period, as different regions of the planet rise and set. The **water hole** (p. 466) is a region in the radio range of the electromagnetic spectrum, near the 21-cm line of hydrogen and the 18-cm line of hydroxyl, where natural emissions from the Galaxy happen to be minimal. Many researchers regard this as the best part of the spectrum for communications purposes.

Review and Discussion

1. Why is life difficult to define?
2. What is chemical evolution?
3. What is the Urey-Miller experiment? What important organic molecules were produced in this experiment?
4. What are the basic ingredients from which biological molecules formed on Earth?
5. How do we know anything at all about the early episodes of life on Earth?
6. What is the role of language in cultural evolution?
7. Where else, besides Earth, have organic molecules been found?
8. Where—besides the planet Mars—might we find signs of life in our solar system?
9. Do we know whether Mars ever had life at any time during its past? What argues in favor of the position that it may once have harbored life?
10. What is generally meant by "life as we know it"? What other forms of life might be possible?
11. How many of the terms in the Drake equation are known with any degree of certainty? Which factor is least well known?
12. What is the relationship between the average lifetime of Galactic civilizations and the possibility of our someday communicating with them?
13. How would Earth appear, at radio wavelengths, to extraterrestrial astronomers?
14. What are the advantages in using radio waves for communication over interstellar distances?
15. What is the water hole? What advantage does it have over other parts of the radio spectrum?

True or False?

____ **1.** The assumptions of mediocrity favor the existence of extraterrestrial life.

____ **2.** The definition of life requires only that, to be considered "alive," you must be able to reproduce.

____ **3.** The Urey-Miller experiment produced biological molecules from nonbiological molecules.

____ **4.** Organic molecules important for life could have reached Earth's surface via comets.

____ **5.** Organic molecules exist only on Earth.

____ **6.** Laboratory experiments have created living cells from nonbiological molecules.

____ **7.** For most of the history of Earth, life consisted of only single-celled life forms.

____ **8.** The *Viking* landers on Mars discovered microscopic evidence of life but found no large fossil evidence.

____ **9.** The rate of star formation in the Galaxy is reasonably well known.

____ **10.** As yet there is no direct evidence that Earth-like planets orbit other stars.

____ **11.** In estimating whether intelligence arises and develops technology, we have only life on Earth as an example.

____ **12.** Dinosaurs existed on Earth for 1000 times longer than human civilization has existed to date.

____ **13.** Our civilization has launched probes into interstellar space.

____ **14.** One disadvantage of interstellar radio communication is that we can do it only with another civilization that has a technology less advanced than our own.

____ **15.** We have already broadcast our presence to our interstellar neighbors.

Fill in the Blank

1. Amino acids are the building blocks of _______.

2. The naturally occurring molecules present on the young Earth included water, carbon dioxide, _______, and _______.

3. Two sources of energy for chemical reactions on the young Earth were _______ and _______.

4. The fossil record clearly shows evidence of life dating back _______ years.

5. Multicellular organisms did not appear on Earth until about _______ years ago.

6. The Murchison meteorite was discovered to contain relatively large amounts of _______.

7. The Drake equation estimates the number of _______ in the Milky Way Galaxy.

8. Planets in binary-star systems are not considered habitable because the planetary orbits are usually _______.

9. The probabilities of the development of life and intelligence on Earth are extremely _______ if random chance is the only evolutionary factor involved.

10. Direct contact between extraterrestrial lifeforms may be impractical because of the large _______ between civilizations.

11. Radio communication over interstellar distances is practical because the signals travel at the speed of _______.

12. Radio waves can travel throughout the Galaxy because they are not blocked by interstellar _______.

13. Radio waves leaking away from Earth have now traveled a distance of _______ light-years.

14. Radio wavelengths between 18 cm and 21 cm are referred to as the _______.

15. A two-way communication with another civilization at a distance of 100 light-years will require _______ years.

Problems

1. As a way of visualizing the period of time since our planet formed, imagine that Earth's 4.6-billion-year age is compressed to just 46 years. What would be your age, in seconds? How long ago (in seconds) was the end of World War II, the Declaration of Independence, Columbus's discovery of the New World? How long ago did the extinction of the dinosaurs occur (in days)?

2. According to the inverse-square law, a planet receives energy from its parent star at a rate proportional to the star's luminosity and inversely proportional to the square of the planet's distance from the star. ∞ (Sec. 10.2) According to Stefan's law, the rate at which the planet radiates energy into space is proportional to the fourth power of its surface temperature. ∞ (Sec. 2.4) In equilibrium, the two rates are equal. Based on this information and the data presented in Figure 18.6, estimate the extent of the habitable zone surrounding a K-type star of luminosity $\frac{1}{10}$ the luminosity of the Sun.

3. Based on the numbers presented in the text, and assuming an average lifetime of five billion years for suitable stars, estimate the total number of habitable planets in the Galaxy.

4. A planet orbits one component of a binary star system at a distance of 1 A.U. (see Figure 18.7a). If both stars have the same mass, and their orbit is circular, estimate the minimum distance between the stars for the tidal force due to the companion not to exceed a "safe" 0.01 percent of the gravitational force between the planet and its parent star.

5. Suppose that each of the "fraction" terms in the Drake equation turns out to have a value of $\frac{1}{10}$, that stars form at an average rate of 20 per year, and that each star with a planetary system has exactly one habitable planet orbiting it. Estimate the present number of technological civilizations in the Milky Way Galaxy if the average lifetime of a civilization is (a) 100 years (b) 10,000 years (c) one million years.

6. Adopting the estimate from the text that the number of technological civilizations in the Milky Way Galaxy is equal to the average lifetime of a civilization, in years, it follows that the distance to our nearest neighbor decreases as the average lifetime increases. Assuming that civilizations are uniformly spread over a two-dimensional Galactic disk of radius 15 kpc, and all have the same lifetime, calculate the minimum lifetime for which a two-way radio communication with our nearest neighbor would be possible before our civilization ends. Repeat the calculation for a round-trip personal visit, using current-technology spacecraft that travel at 50 km/s.

7. How fast would a spacecraft have to travel in order to complete the trip from Earth to Alpha Centauri (a distance of 1.3 pc) and back in less than an average human lifetime (80 years, say)?

8. Assuming that there are 10,000 FM radio stations on Earth, each transmitting at a power level of 50 kW, calculate the total radio luminosity of Earth in the FM band. Compare this value with the roughly 10^6 W radiated by the Sun in the same frequency range.

9. Convert the water-hole's wavelengths to frequencies. For practical reasons, any search of the water hole must be broken up into channels, much like you find on a television, except these channels are very narrow in radio frequency, about 100 Hz wide. How many channels must astronomers search in the water hole?

10. There are 20,000 stars within 100 light years that are to be searched for radio communications. How long will the search take if one hour is spent looking at each star? What if one day is spent per star?

Projects

1. Some people suggest that if extraterrestrial life is discovered, it will have a profound effect on people. Interview as many people as you can and ask the following two questions: 1) Do you believe that extraterrestrial life exists? 2) Why? From your results, try to decide whether there will be a profound effect on people if extraterrestrial life is discovered.

2. Conduct another poll, or do it at the same time you do the first one. Ask the following question: What one question would you like to ask an extraterrestrial life form in a radio communication? How many responses do you receive that indicate the person is very "Earth centered" in thinking? How many responses suggest a lack of understanding of how alien an extraterrestrial life-form might be?

3. The Drake equation should be able to "predict" at least one civilization in our Galaxy: us. Try changing the values of various factors so that you end up with at least one. What do these various combinations of factors imply about how life arises and develops? Are there some combinations that just don't make any sense?

APPENDIX 1

Scientific Notation

The objects studied by astronomers range in size from the smaller particles to the largest expanse of matter we know—the entire universe. Subatomic particles have sizes of about 0.000000000000001 meter, while galaxies (like that shown in Figure P.3) typically measure some 1,000,000,000,000,000,000,000 meters across. The most distant known objects in the universe lie on the order of 100,000,000,000,000,000,000,000,000 meters from Earth.

Obviously, writing all those zeros is both cumbersome and inconvenient. More important, it is also very easy to make an error—write down one zero too many or too few and your calculations become hopelessly wrong! To avoid this, scientists always write large numbers using a shorthand notation in which the number of zeros following or preceding the decimal point is denoted by a superscript power, or *exponent*, of 10. The exponent is simply the number of places between the first significant (nonzero) digit in the number (reading from left to right) and the decimal point. Thus 1 is 10^0, 10 is 10^1, 100 is 10^2, 1000 is 10^3, and so on. For numbers less than 1, with zeros between the decimal point and the first significant digit, the exponent is negative: 0.1 is 10^{-1}, 0.01 is 10^{-2}, 0.001 is 10^{-3}, and so on. Using this notation we can shorten the number describing subatomic particles to 10^{-15} meter and write the number describing the size of a galaxy as 10^{21} meters.

More complicated numbers are expressed as a combination of a power of 10 and a multiplying factor. This factor is conventionally chosen to be a number between 1 and 10, starting with the first significant digit in the original number. For example, 150,000,000,000 meters (the distance from Earth to the Sun, in round numbers) can be more concisely written at 1.5×10^{11} meters, 0.000000025 meters as 2.5×10^{-8} meter, and so on. The exponent is simply the number of places the decimal point must be moved *to the left* to obtain the multiplying factor.

Some other examples of scientific notation are:

- the approximate distance to the Andromeda Galaxy = 3,000,000 light years = 3×10^6 light-years
- the size of a hydrogen atom = 0.00000000005 meter = 5×10^{-11} meter
- the diameter of the Sun = 1,392,000 kilometers = 1.392×10^6 kilometers
- the U.S. national debt (as of midnight, July 1, 2000) = \$5,800,945,000,000.00 = \$5.800945 trillion = 5.800945×10^{12} dollars.

In addition to providing a simpler way of expressing very large or very small numbers, this notation also makes it easier to do basic arithmetic. The rule for multiplication of numbers expressed in this way is simple: Just multiply the factors and add the exponents. Similarly for division: Divide the factors and subtract the exponents. Thus, 3.5×10^{-2} multiplied by 2.0×10^3 is simply $(3.5 \times 2.0) \times 10^{-2+3} \times = 7.0 \times 10^1$—that is, 70. Again, 5×10^6 divided by 2×10^4 is just $(5/2) \times 10^{6-4}$, or 2.5×10^2 (= 250). Applying these rules to unit conversions, 200,000 nanometers is $200{,}000 \times 10^{2-9}$ meter (since 1 nanometer = 10^{-9} meter; see Appendix 2), or $(2 \times 10^5) \times 10^{-9}$ meter, or $2 \times 10^{5-9} = 2 \times 10^{-4}$ meter = 0.2 mm. Verify these rules yourself with a few examples of your own. The advantages of this notation when considering astronomical objects will soon become obvious.

Scientists often use "rounded-off" versions of numbers, both for simplicity and for ease of calculation. For example, we will usually write the diameter of the Sun as 1.4×10^6 kilometers, instead of the more precise number given earlier. Similarly, the diameter of Earth is 12,756 kilometers, or 1.2756×10^4 kilometers, but for "ballpark" estimates, we really don't need so many digits, and the more approximate number 1.3×10^4 kilometers will suffice. Very often, we perform rough calculations using only the first one or two significant digits in a number, and that may be all that is necessary to make a particular point. For example, to support the statement, "The Sun is much larger than Earth," we need only say that the ratio of the two diameters is roughly 1.4×10^6 divided by 1.3×10^4. Since 1.4/1.3 is close to 1, the ratio is approximately $10^6/10^4 = 10^2$, or 100. The essential fact here is that the ratio is much larger than 1; calculating it to greater accuracy (to get 109.13) would give us no additional *useful* information. This technique of stripping away the arithmetic details to get to the essence of a calculation is very common in astronomy, and we use it frequently throughout this text.

APPENDIX 2

Astronomical Measurement

Astronomers use many different kinds of units in their work, simply because no single system of units will do. Rather than the *Système Internationale* (SI), or meter-kilogram-second (MKS), metric system used in most high school and college science classes, many professional astronomers still prefer the older centimeter-gram-second (CGS) system. However, astronomers also commonly introduce new units when convenient. For example, when discussing stars, the mass and radius of the Sun are often used as reference points. The solar mass, written as $M_\odot$, is equal to 2.0×10^{33} g, or 2.0×10^{30} kg (since 1 kg = 1000 g). The solar radius, $R_\odot$, is equal to 700,000 km, or 7.0×10^8 m (1 km = 1000 m). The subscript $\odot$ always stands for Sun. Similarly, the subscript $\oplus$ always stands for Earth. In this book, we try to use the units that astronomers commonly use in any given context, but we also give the "standard" SI equivalents where appropriate.

Of particular importance are the units of length astronomers use. On small scales, the *angstrom* (1 Å = 10^{-10} m = 10^{-8} cm), the *nanometer* (1 nm = 10^{-9} m = 10^{-7} cm), and the *micron* (1 μm = 10^{-6} m = 10^{-4} cm) are used. Distances within the solar system are usually expressed in terms of the *astronomical unit* (A.U.), the mean distance between Earth and the Sun. One A.U. is approximately equal to 150,000,000 km, or 1.5×10^{11} m. On larger scales, the *light-year* (1 ly = 9.5×10^{15} m = 9.5×10^{12} km) and the *parsec* (1 pc = 3.1×10^{16} m = 3.1×10^{13} km = 3.3 ly) are commonly used. Still larger distances use the regular prefixes of the metric system: *kilo* for one thousand and *mega* for one million. Thus 1 kiloparsec (kpc) = 10^3 pc = 3.1×10^{19} m, 10 megaparsecs (Mpc) = 10^7 pc = 3.1×10^{23} m, and so on.

Astronomers use units that make sense within a context, and as contexts change, so do the units. For example, we might measure densities in grams per cubic centimeter (g/cm^3), in atoms per cubic meter (atoms/m^3), or even in solar masses per cubic megaparsec ($M_\odot$/Mpc^3), depending on the circumstances. The important thing to know is that once you understand the units, you can convert freely from one set to another. For example, the radius of the Sun could equally well be written as $R_\odot = 6.96 \times 10^8$ m, or 6.96×10^{10} cm, or 109 $R_\oplus$, or 4.65×10^{23} A.U., or even 7.36×10^{-8} ly—whichever happens to be most useful. Some of the more common units used in astronomy, and the contexts in which they are most likely to be encountered, are listed below.

Length:

1 angstrom (Å)	= 10^{-18} m	atomic physics,
1 nanometer (nm)	= 10^{-9} m	spectroscopy
1 micron (mm)	= 10^{-6} m	interstellar dust and gas
1 centimeter (cm)	= 0.01 m	in widespread use
1 meter (m)	= 100 cm	throughout all
1 kilometer (km)	= 1000 m = 10^5 cm	astronomy
Earth radius ($R_\oplus$)	= 6378 km	planetary astronomy
Solar radius ($R_\odot$)	= 6.96×10^8 m	solar system, stellar evolution
1 astronomical unit (A.U.)	= 1.496×10^{11} m	
1 light-year (ly)	= 9.46×10^{15} m = 63,200 A.U.	
1 parsec (pc)	= 3.09×10^{16} m = 3.26 ly	galactic astronomy, stars and star clusters
1 kiloparsec (kpc)	= 1000 pc	galaxies, galaxy clusters,
1 megaparsec (Mpc)	= 1000 kpc	cosmology

Mass:

1 gram (g)		in widespread use in
1 kilogram (kg)	= 1000 g	many different areas
Earth mass ($M_\oplus$)	= 5.98×10^{24} kg	planetary astronomy
Solar mass ($M_\odot$)	= 1.99×10^{30} kg	"standard" unit for all mass scales larger than Earth

Time:

1 second (s)		in widespread use throughout astronomy
1 hour (h)	= 3600 s	planetary and stellar
1 day (d)	= 86,400 s	scales
1 year (yr)	= 3.16×10^7 s	virtually all processes occurring on scales larger than a star

APPENDIX 3

Tables

TABLE 1 Some Useful Constants and Physical Measurements[1]

astronomical unit	1 A.U. = 1.496×10^8 km (1.5×10^8 km)
light-year	1 ly = 9.46×10^{12} km (10^{13} km; about 6 trillion miles)
parsec	1 pc = 3.09×10^{13} km = 3.3 ly
speed of light	c = 299,792.458 km/s (3×10^5 km/s)
Stefan–Boltzmann Constant	$\sigma = 5.67 \times 10^{-8}$ W/m^{2}K^4
gravitational constant	G = 6.67×10^{-11} m^3/kg/s^2
mass of Earth	$M_\oplus = 5.97 \times 10^{24}$ kg (6×10^{24} kg; 6000 billion billion tons)
radius of Earth	$R_\oplus$ = 6378 km (6500 km)
mass of the Sun	$M_\odot = 1.99 \times 10^{30}$ kg (2×10^{30} kg)
radius of the Sun	$R_\odot = 6.96 \times 10^5$ km (7×10^5 km)
luminosity of the Sun	$L_\odot = 3.90 \times 10^{26}$ W (4×10^{26} W)
effective temperature of the Sun	$T_\odot$ = 5778 K (5800 K)
Hubble constant	$H_0 \approx$ 65 km/s/Mpc
mass of an electron	$m_e = 9.11 \times 10^{-31}$ kg
mass of a proton	$m_p = 1.67 \times 10^{-27}$ kg

[1]The rounded-off values used in the text are shown above in parentheses.

CONVERSIONS BETWEEN COMMON ENGLISH AND METRIC UNITS

English	Metric
1 inch	= 2.54 centimeters (cm)
1 foot (ft)	= 0.3048 meters (m)
1 mile	= 1.609 kilometers (km)
1 pound (lb)	= 453.6 grams (g) or 0.4536 kilograms (kg) (on Earth)

TABLE 2A Planetary Data: Orbital Properties

Planet	SEMI-MAJOR AXIS (A.U.)	SEMI-MAJOR AXIS (10^6 km)	SIDEREAL PERIOD (tropical years)	MEAN ORBITAL SPEED (km/s)	ORBITAL ECCENTRICITY	INCLINATION TO THE ECLIPTIC (degrees)
Mercury	0.39	57.9	0.24	47.9	0.206	7.00
Venus	0.72	108.2	0.612	35.0	0.007	3.39
Earth	1.00	149.6	1.00	29.8	0.017	0.01
Mars	1.52	227.9	1.88	24.1	0.093	1.85
Jupiter	5.20	778.4	11.86	13.1	0.048	1.31
Saturn	9.54	1427	29.42	9.65	0.054	2.49
Uranus	19.19	2871	83.75	6.80	0.047	0.77
Neptune	30.07	4498	163.7	5.43	0.009	1.77
Pluto	39.48	5906	248.0	4.74	0.249	17.2

TABLE 2B Planetary Data: Physical Properties

PLANET	EQUATORIAL RADIUS (km)	EQUATORIAL RADIUS (Earth = 1)	MASS (kg)	MASS (Earth = 1)	MEAN DENSITY (kg/m^3)	SIDEREAL ROTATION PERIOD (solar days)[1]
Mercury	2440	0.38	3.30×10^{23}	0.055	5430	58.6
Venus	6052	0.95	4.87×10^{24}	0.82	5240	−243.0
Earth	6378	1.00	5.97×10^{24}	1.00	5520	0.9973
Mars	3394	0.53	6.42×10^{23}	0.11	3930	1.026
Jupiter	71,492	11.21	1.90×10^{27}	317.8	1330	0.41
Saturn	60,268	9.45	5.68×10^{26}	95.2	690	0.44
Uranus	25,559	4.01	8.68×10^{25}	14.5	1270	−0.72
Neptune	24,766	3.88	1.02×10^{26}	17.1	1640	0.67
Pluto	1137	0.18	1.27×10^{22}	0.0021	2060	−6.39

PLANET	AXIAL TILT (degrees)	SURFACE GRAVITY (Earth = 1)	ESCAPE SPEED (km/s)	SURFACE TEMPERATURE[2] (K)	NUMBER OF MOONS
Mercury	0.0	0.38	4.2	100 to 700	0
Venus	177.4	0.91	10.4	730	0
Earth	23.5	1.00	11.2	290	1
Mars	24.0	0.38	5.0	180 to 270	2
Jupiter	3.1	2.53	60	124	16
Saturn	26.7	1.07	36	97	18
Uranus	97.9	0.91	21	58	17
Neptune	29.6	1.14	24	59	8
Pluto	118	0.07	1.2	40 to 60	1

[1]A negative sign indicates retrograde rotation.

[2]Temperature is effective temperature for jovian planets.

TABLE 3 The Twenty Brightest Stars

NAME	STAR	SPECTRAL TYPE[1] A	SPECTRAL TYPE[1] B	PARALLAX (arc seconds)	DISTANCE (pc)	APPARENT VISUAL MAGNITUDE[1] A	APPARENT VISUAL MAGNITUDE[1] B
Sirius	α CMa	A1V	wd[2]	0.37	2.7	−1.46	+8.7
Canopus	α Car	F01b-II		0.033	30	−0.72	
Rigel Kentaurus	α Cen	G2V	K0V	0.77	1.3	−0.01	+1.3
Arcturus	α Boo	K2IIIp		0.091	11	−0.06	
Vega	α Lyr	A0V		0.13	8.0	+0.04	
Capella	α Aur	GIII	M1V	0.071	14	+0.05	+10.2
Rigel	β Ori	B8 Ia	B9	—	250	+0.14	+6.6
Procyon	α CMi	F5IV-V	wd	0.29	3.5	+0.37	+10.7
Betelgeuse	α Ori	M2Iab		—	150	+0.41	
Achernar	α Eri	B5V		0.050	20	+0.51	
Hadar	β Cen	B1III	?	0.011	90	+0.63	+4
Altair	α Aql	A7IV-V		0.20	5.1	+0.77	
Acrux	α Cru	B1IV	B3	0.008	120	+1.39	+1.9
Aldebaran	α Tau	K5III	M2V	0.063	16	+0.86	+13
Spica	α Vir	B1V		0.013	80	+0.91	
Antares	α Sco	M1Ib	B4V	0.008	120	+0.92	+5.1
Pollux	β Gem	K0III		0.083	12	+1.16	
Fomalhaut	α PsA	A3V		0.14	7.0	+1.19	+6.5
Deneb	α Cyg	A2Ia		—	430	+1.26	
Mimosa	β Cru	B1IV		—	150	+1.28	

NAME	LUMINOSITY (Sun = 1) A	LUMINOSITY (Sun = 1) B	ABSOLUTE VISUAL MAGNITUDE[1] A	ABSOLUTE VISUAL MAGNITUDE[1] B	PROPER MOTION (arc seconds/yr)	TRANSVERSE VELOCITY (km/s)	RADIAL VELOCITY (km/s)
Sirius	23.5	0.04	+1.4	+11.6	1.33	17.0	−7.6[3]
Canopus	1510		−3.1		0.02	2.8	+20.5
Rigel Kentaurus	1.56	0.46	+4.4	+5.7	3.68	22.7	−24.6
Arcturus	115		−0.3		2.28	119	−5.2
Vega	55.0		+0.5		0.34	12.9	−13.9
Capella	166	0.01	−0.7	+9.5	0.44	29	+30.2[3]
Rigel	4.6×10^4	126	−6.8	−0.4	0.00	1.2	+20.7[3]
Procyon	7.7	0.0006	+2.6	+13.0	1.25	20.7	−3.2[3]
Betelgeuse	1.4×10^4		−5.5		0.03	21	+21.0[3]
Achernar	219		−1.0		0.10	9.5	+19
Hadar	3800	182	−4.1	−0.8	0.04	17	−12[3]
Altair	11.5		+2.2		0.66	16	−26.3
Acrux	3470	2190	−4.0	−3.5	0.04	24	−11.2
Aldebaran	105	0.0014	−0.2	+12	0.20	15	+54.1
Spica	2400		−3.6		0.05	19	+1.0[3]
Antares	5500	115	−4.5	−0.3	0.03	17	−3.2
Pollux	41.7		+0.8		0.62	35	+3.3
Fomalhaut	13.8	0.10	+2.0	+7.3	0.37	12	+6.5
Deneb	5.0×10^4		−6.9		0.003	6	−4.6[3]
Mimosa	6030		−4.6		0.05	36	

[1]A and B columns identify individual components of binary systems.
[2]"wd" stands for "white dwarf."
[3]Average value of variable velocity.

TABLE 4 The Twenty Nearest Stars

NAME	SPECTRAL TYPE[1] A	SPECTRAL TYPE[1] B	PARALLAX (arc seconds)	DISTANCE (pc)	APPARENT VISUAL MAGNITUDE[1] A	APPARENT VISUAL MAGNITUDE[1] B
Sun	G2V				−26.72	
Proxima Cen	M5e		0.772	1.30	+11.05	
Rigel Kentaurus	G2V	K0V	0.750	1.33	−0.01	+1.33
Barnard's Star	M5V		0.545	1.83	+9.54	
Wolf 359	M8V		0.421	2.38	+13.53	
BD 1 36°2147	M2V		0.397	2.52	+7.50	
Luyten 726-8	M5.5V	M5.5V	0.387	2.58	+12.52	+13.02
Sirius	A1V	wd[2]	0.377	2.65	−1.46	+8.3
Ross 154	M4.5V		0.345	2.90	+10.45	
Ross 248	M6V		0.314	3.18	+12.29	
ε Eridani	K2V		0.303	3.30	+3.73	
Ross 128	M5V		0.298	3.36	+11.10	
61 Cygni	K5V	K7V	0.294	3.40	+5.22	+6.03
ε Indi	K5V		0.291	3.44	+4.68	
BD 1 43°44	M1V	M6V	0.290	3.45	+8.08	+11.06
Luyten 789-6	M6V		0.290	3.45	+12.18	
Procyon	F5IV-V	wd	0.285	3.51	+0.37	+10.7
BD 1 59°1915	M4V	M5V	0.285	3.55	+8.90	+9.69
CD 2 36°15693	M2V		0.279	3.58	+7.35	
G51-15	MV		0.278	3.60	+14.81	

NAME	LUMINOSITY (Sun = 1) A	LUMINOSITY (Sun = 1) B	ABSOLUTE VISUAL MAGNITUDE[1] A	ABSOLUTE VISUAL MAGNITUDE[1] B	PROPER MOTION (arc seconds/yr)	TRANSVERSE VELOCITY (km/s)	RADIAL VELOCITY (km/s)
Sun	1.0		+4.85				
Proxima Cen	0.00006		+15.5		3.86	23.8	−16
Rigel Kentaurus	1.6	0.45	+4.4	+5.7	3.68	23.2	−22
Barnard's Star	0.00045		+13.2		10.34	89.7	−108
Wolf 359	0.00002		+16.7		4.70	53.0	+13
BD 1 36°2147	0.0055		+10.5		4.78	57.1	−84
Luyten 726-8	0.00006	0.00004	+15.5	+16.0	3.36	41.1	+30
Sirius	23.5	0.04	+1.4	+11.2	1.33	16.7	−8
Ross 154	0.00048		+13.3		0.72	9.9	−4
Ross 248	0.00011		+14.8		1.58	23.8	−81
ε Eridani	0.30		+6.1		0.98	15.3	+16
Ross 128	0.00036		+13.5		1.37	21.8	−13
61 Cygni	0.082	0.039	+7.6	+8.4	5.22	84.1	−64
ε Indi	0.14		+7.0		4.69	76.5	−40
BD 1 43°44	0.0061	0.00039	+10.4	+13.4	2.89	47.3	+17
Luyten 789-6	0.00014		+14.6		3.26	53.3	−60
Procyon	7.65	0.00055	+2.6	+13.0	1.25	2.8	−3
BD 1 59°1915	0.0030	0.0015	+11.2	+11.9	2.28	38.4	+5
CD 2 36°15693	0.013		+9.6		6.90	117	+10
G51-15	0.00001		+17.0		1.26	21.5	—

[1]A and B columns identify individual components of binary systems.

[2]"wd" stands for "white dwarf."

GLOSSARY

A

A ring One of three Saturnian rings visible from Earth. The A ring is farthest from the planet and is separated from the B ring by the Cassini division. (p. 218)

absolute brightness The apparent brightness a star would have if it were placed at a standard distance of 10 parsecs from Earth. (p. 257)

absolute magnitude The apparent magnitude a star would have if it were placed at a standard distance of 10 parsecs from Earth. (p. 259)

absorption line A dark line in an otherwise continuous bright spectrum, where light has been removed from one or more narrow frequency ranges. (p. 56)

acceleration The rate of change of velocity of a moving object. (p. 36)

accretion Gradual growth of bodies, such as stars or planets, by the accumulation of gas or other smaller bodies. (p. 118)

accretion disk Flat disk of matter spiraling down onto the surface of a star or black hole. Often, the matter originated on a companion star in a binary-star system. (p. 311)

active galaxies The most energetic galaxies, which can emit hundreds or thousands of times more energy per second than does the Milky Way. (p. 406)

active optics Collection of techniques now being used to increase the resolution of ground-based telescopes. Minute modifications are made to the overall configuration of an instrument as its temperature and orientation change to maintain the best possible focus at all times. (p. 82)

active region Region of the photosphere of the Sun surrounding a sunspot group, which can erupt violently and unpredictably. During sunspot maximum the number of active regions is also a maximum. (p. 243)

active Sun The unpredictable aspects of the Sun's behavior, such as sudden explosive outbursts of radiation in the form of prominences and flares. (p. 239)

adaptive optics Technique used to increase the resolution of a telescope by deforming the shape of the mirror's surface under computer control while a measurement is being made, in order to undo the effects of atmospheric turbulence. (p. 82)

amino acids Organic (carbon-based) molecules, forming the basis for building proteins, which direct metabolism in living creatures. (p. 457)

amplitude The maximum deviation of a wave above or below the zero point. (p. 46)

angular momentum A measure of a rotating object's tendency to keep spinning. Mathematically, an object's angular momentum is equal to the product of its mass, angular speed, and the square of its radius. (p. 119)

angular resolution The ability of a telescope to distinguish between adjacent objects in the sky. (p. 77)

annular eclipse Solar eclipse occurring at a time when the Moon is far enough from Earth so that it does not cover the disk of the Sun completely, leaving a ring of sunlight visible around its edge. (p. 14)

aphelion A planet's greatest distance from the Sun. (p. 32)

apparent brightness The brightness that a star appears to have, as measured by an observer on Earth. (p. 257)

apparent magnitude The apparent brightness of a star expressed using the magnitude scale. (p. 258)

association Small grouping of stars (typically 100 or less), spanning up to a few tens of parsecs across, usually rich in very young stars. (p. 294)

asteroid belt Region of the solar system between the orbits of Mars and Jupiter in which most asteroids are found. (pp. 103, 228)

asteroid One of thousands of very small members of the solar system orbiting the Sun between the orbits of Mars and Jupiter. Asteroids are often referred to as "minor planets." (pp. 103, 228)

astronomical unit (A.U.) The average distance of Earth from the Sun. Precise radar measurements yield a value for the A.U. of 149,603,500 km. (p. 33)

astronomy Branch of science dedicated to the study of everything in the universe that lies above Earth's atmosphere. (p. 2)

atmosphere Layer of gas confined close to a planet's surface by the force of gravity. (p. 128)

atom Building block of matter composed of positively charged protons and neutral neutrons in the nucleus surrounded by negatively charged neutrons. (p. 59)

aurora Event which occurs when atmospheric molecules, excited by incoming charged particles from the solar wind, emit energy as they fall back to their ground states. Aurorae generally occur at high latitudes, near the north and south magnetic poles. (p. 148)

autumnal equinox Date on which the Sun crosses the celestial equator moving southward, occurring on or near September 22. (p. 9)

B

B ring One of three Saturnian rings visible from Earth. The B ring is the brightest of the three and lies just within the Cassini division, closer to the planet than the A ring. (p. 218)

barred-spiral galaxy Spiral galaxy containing a central elongated "bar" of stars and gas, with the spiral arms beginning near the ends of the bar. (p. 379)

baseline The distance between two observing locations used for the purpose of triangulation. In general, the larger the baseline, the better the attainable resolution. (p. 17)

belt Dark, low-pressure region in the atmosphere of a jovian planet where gas flows downward (toward the planet's center). (p. 190)

Big Bang Event that cosmologists consider the beginning of the universe in which all space, time, matter, and radiation in the universe came into being. (p. 434)

binary-star system A system consisting of two stars in orbit about their common center of mass and held together by their mutual gravitational attraction. Most stars are found in binary-star systems. (p. 267)

blackbody curve The characteristic distribution of the radiation emitted by a hot object. The frequency at which the emitted radiation peaks is an indicator of the object's temperature. Also referred to as the Planck curve. (p. 52)

black dwarf The ultimate endpoint of the evolution of an isolated, low-mass star. Theoretically, after the white dwarf stage, a star will slowly cool to the point where it eventually becomes a dark "clinker" in interstellar space. (p. 310)

black hole A region of space where the pull of gravity is so great that nothing—not even light—can escape. A possible outcome of the evolution of a very massive star. (p. 337)

blue giant Large, hot, bright star at the upper left end of the main sequence on the Hertzsprung–Russell diagram. Its name comes from its color and size. (p. 266)

blue shift Any Doppler shift of radiation toward shorter wavelengths (higher frequencies); the result of relative motion of the observer toward the source. (p. 64)

blue supergiant The very largest of the large, hot, bright stars at the uppermost left end of the main sequence on the Hertzsprung–Russell diagram. (p. 266)

Bohr model First theory of the hydrogen atom to explain that element's observed spectral lines. This model rests on three ideas: there is a state of lowest energy for the electron, there is a maximum energy beyond which the electron is no longer bound to the nucleus, and between these two energies the electron can exist only in certain well-defined energy levels. (p. 59)

brown dwarf Remnant of a fragment of collapsing gas and dust that did not contain enough mass to begin core nuclear fusion. Such objects are "frozen" somewhere along their pre-main-sequence contraction phases, slowly cooling into compact dark objects. Brown dwarfs are extremely difficult to detect observationally because of their small sizes and low temperatures. (p. 293)

brown oval Feature of Jupiter's atmosphere that appears only at latitudes near 20° N. Such structures are long-lived holes in the clouds that allow us to look down into Jupiter's lower atmosphere. (p. 192)

C

C ring One of three Saturnian rings visible from Earth. The C ring is the faintest of the three and lies between the B ring and the planet's surface. (p. 218)

carbon-detonation supernova See Type I supernova. (p. 314)

Cassegrain telescope A type of reflecting telescope in which incoming light hits the primary mirror and is reflected upward toward the prime focus, where a secondary mirror reflects the light back down through a small hole in the main mirror into a detector or eyepiece. (p. 75)

Cassini Division A relatively empty gap in Saturn's rings, lying between the A and B rings, discovered in 1675 by Giovanni Cassini. It is now known to contain a number of thin ringlets. (p. 218)

celestial coordinates Pair of quantities—right ascension and declination—similar to longitude and latitude on Earth, used to specify the locations of objects on the celestial sphere. (p. 6)

celestial equator The projection of Earth's equator onto the celestial sphere. (p. 6)

celestial sphere Imaginary sphere surrounding Earth to which all objects in the sky were once considered to be attached. (p. 5)

center of mass The "average" position in space of a collection of massive bodies, taking their masses into account. This point moves with constant velocity in an isolated system, according to Newtonian mechanics. (p. 39)

Cepheid variable Star whose luminosity varies in a characteristic way, a rapid rise in brightness is followed by a slower decline. The luminosity of a Cepheid variable star is directly related to its period, so a determination of the period can be used to obtain an estimate of the star's distance. (p. 355)

charge-coupled device (CCD) Electronic device used for data acquisition composed of many tiny pixels, each of which records a buildup of charge proportional to the amount of light striking it. (p. 81)

chromosphere The Sun's lower atmosphere lying just above the visible photosphere. (p. 233)

closed universe Geometry that the universe as a whole would have if the total density exceeds the critical value. A closed universe is finite in extent and has no edge, like the surface of a sphere. In the absence of a cosmological constant, such a universe has enough mass to stop the present expansion and will eventually recollapse. (p. 441)

collecting area The total area of a telescope that is capable of capturing incoming radiation. The larger the telescope, the greater its collecting area and the fainter the objects it can detect. (p. 76)

coma The brightest part of a comet, often referred to as the "head." (p. 107)

comet A small body, composed mainly of ice and dust, in an elliptical orbit about the Sun. As it comes close to the Sun some of its material is vaporized to form a gaseous coma and extended tail. (pp. 103, 232)

condensation nuclei Dust grains in the interstellar medium that act as seeds around which other material can coagulate. The presence of dust was very important in causing matter to clump during the formation of the solar system. (p. 118)

condensation theory Currently favored model of solar system formation that combines features of the old nebular theory with new information about interstellar dust grains, which acted as condensation nuclei. (p. 117)

constellation A grouping of stars in the night sky seen as a recognizable pattern by humans. (p. 4)

continuous spectrum Spectrum in which the radiation is distributed over all frequencies, not just a few specific frequency ranges. A prime example is the blackbody radiation emitted by a hot, dense body. (p. 55)

convection Churning motion resulting from the constant upwelling of warm fluid and the concurrent downward flow of cooler material to take its place. (p. 133)

convection zone Region of the Sun's interior, lying just below the surface, where the material of the Sun is in constant convective motion. This region extends into the solar interior to a depth of about 200,000 km. (p. 233)

Copernican revolution The realization toward the end of the sixteenth century that Earth is not at the center of the universe. (p. 26)

core, Earth The central region of Earth, lying below the mantle. (p. 128)

core, Sun The central region of the Sun, where nuclear fusion occurs. (p. 233)

core-halo galaxy Radio galaxy in which the observed emission comes from a central bright core surrounded by a diffuse halo. (p. 408)

core-hydrogen burning The energy burning stage for main-sequence stars in which helium is produced by hydrogen fusion in a stellar core. A typical star spends up to 90 percent of its lifetime in a state of equilibrium resulting from the balance between gravity and the energy generated by core-hydrogen burning. (p. 302)

core-collapse supernova See Type II supernova. (p. 313)

corona The tenuous outer atmosphere of the Sun which lies just above the chromosphere, and at great distances becomes the solar wind. (p. 233)

coronae Large, roughly circular regions on the surface of Venus thought to have been caused by upwelling mantle material causing the planet's crust to bulge outward. (p. 165)

coronal hole Vast regions of the Sun's atmosphere where the density of matter is about 10 times lower than average. There the gas streams freely into space at high speeds, escaping the Sun completely. (p. 239)

cosmic distance scale Collection of direct and indirect distance-measurement techniques that astronomers use to measure the scale of the universe. (p. 17)

cosmic evolution The seven major phases in the history of the universe: galactic, stellar, planetary, chemical, biological, cultural, and future evolution. (p. 456)

cosmic microwave background The almost perfectly isotropic radio signal that is the electromagnetic remnant of the Big Bang. (p. 442)

cosmological constant Term originally introduced by Einstein into general relativity to make his equations describe a static universe. Now it is one of several candidates for the repulsive force responsible for the observed cosmic acceleration. (p. 439)

cosmological principle Two assumptions that form the foundation of modern cosmology; the universe is homogeneous and isotropic on sufficiently large scales. (p. 432)

cosmological redshift The component of the redshift of an object that is due to the expansion of the universe. (p. 397)

cosmology The study of the structure and evolution of the entire universe. (p. 432)

crater Bowl-shaped depression on the surface of a planet or moon resulting from a collision with interplanetary debris. (p. 143)

critical density The cosmic matter density corresponding to the dividing line between a universe that recollapses and one that expands forever; the total matter density dividing an open universe from a closed one. (p. 436)

critical universe Geometry that the universe would have if the total density is exactly equal to the critical density. Such a universe is infinite in extent and has zero curvature. In the absence of a cosmological constant, the present expansion will continue forever, but will approach an expansion speed of zero. (p. 441)

crust Layer of Earth containing the solid continents and the seafloor. (p. 128) The solid surface of any planet or moon.

D

D ring Collection of very faint, thin rings of Saturn extending down from the inner edge of the C ring almost to the planet's cloud tops. This region contains so few particles that it is completely invisible from Earth. (p. 220)

dark dust cloud A large interstellar cloud, often many parsecs across, containing gas and dust in a ratio of about 10^{12} gas atoms for every dust particle. Typical densities are a few tens or hundreds of millions of particles per cubic meter. Such objects are very effective at blocking visible light and ultraviolet radiation. (p. 283)

dark halo Region of a galaxy beyond the visible halo where dark matter predominates. (p. 367)

dark matter Term used to describe the mass in galaxies and clusters whose existence is inferred from rotation curves and other techniques, but that is not detectable at any electromagnetic wavelength. (p. 367)

declination Celestial coordinate used to measure "latitude" above or below the celestial equator on the celestial sphere. (p. 6)

decoupling Epoch in the early universe when atoms first formed, after which photons could propagate freely through space. (p. 445)

deferent A construct of the geocentric model of the solar system needed to explain observed planetary motions. A deferent is a large circle encircling Earth on which an epicycle moves. (p. 24)

density A measure of the compactness of the matter within an object computed by dividing the object's mass by its volume. Units are kilograms per cubic meter (kg/m^3), or grams per cubic centimeter (g/cm^3). (p. 102)

differential rotation The tendency for a gaseous sphere, such as a jovian planet or the Sun, to rotate at a different rate at the equator than at the poles, or for the rotation rate to vary with depth. For a galaxy or other object, it is a condition where the angular speed varies with location within the object. (p. 188)

differentiation Variation with depth of the density and composition of a body such as Earth, with low-density material on the surface and higher-density material in the core. (p. 138)

diffraction The tendency of waves to bend around corners. The diffraction of light establishes its nature as a wave. (p. 77)

Doppler effect Motion-induced change in a wave's properties. Relative motion of the source away from the observer leads to an increase in the observed wavelength (or decreased frequency); approaching motion leads to a decrease in wavelength (or increased frequency). (p. 64)

Drake equation Expression estimating the probability that intelligence exists elsewhere in the Galaxy based on a number of supposedly necessary conditions for intelligent life to develop. (p. 460)

dust grain An interstellar dust particle, roughly 10^{-8} m in size, comparable to the wavelength of visible light. (p. 278)

dust lane A lane of dark, obscuring interstellar dust in an emission nebula or galaxy. (p. 282)

dust tail The component of a comet's tail that is composed of dust particles. (p. 107)

dwarf Any star with radius comparable to, or smaller than, that of the Sun (including the Sun itself). (p. 263)

E

E ring A faint ring lying well outside the main ring system of Saturn discovered by *Voyager 2* and believed to be associated with volcanism on the moon Enceladus. (p. 220)

Earth-crossing asteroid An asteroid whose orbit crosses that of Earth. Earth-crossing asteroids are also called Apollo asteroids after the first of the type discovered. (p. 103)

earthquake A sudden dislocation of rocky material near Earth's surface. (p. 135)

eccentricity A measure of the flatness of an ellipse that is numerically equal to the distance between the two foci divided by the length of the major axis. (p. 32)

eclipse Event during which one body passes in front of another and blocks the light from the occulted body. (p. 11)

eclipsing binary Rare binary-star system aligned in such a way that from Earth we periodically observe one star pass in front of the other, eclipsing it. (p. 267)

ecliptic The apparent path of the Sun, relative to the stars on the celestial sphere, over the course of a year. (p. 9)

electric field A field extending outward in all directions from a charged particle, such as a proton or an electron. The electric field determines the electric force exerted by the particle on all other charged particles in the universe; the strength of the electric field decreases with increasing distance from the charge according to an inverse-square law. (p. 47)

electromagnetic radiation Another term for light, electromagnetic radiation transfers energy and information from one place to another, even through the vacuum of empty space. (p. 44)

electromagnetic spectrum The complete range of electromagnetic radiation, from radio waves to gamma rays, including the visible spectrum. All types of electromagnetic radiation are basically the same phenomenon, differing only in wavelength or frequency; all move at the speed of light. (p. 49)

electromagnetism The union of electricity and magnetism, which do not exist as independent quantities but are in reality two aspects of a single physical phenomenon. (p. 48)

electron An elementary particle with a negative electric charge; one of the components of the atom. (p. 47)

element Matter made up of one particular atom. The number of protons in the nucleus of an atom determines which element it represents. (p. 62)

ellipse Geometric figure resembling a flattened circle. An ellipse is characterized by its degree of flattening, or eccentricity, and by the length of its long (major) axis. In general, bound orbits of objects moving under gravity are ellipses. (p. 32)

elliptical galaxy Category of galaxy which appears elliptical on the sky, ranging from highly elongated to nearly circular in shape. (p. 380)

emission line Bright line at a specific point in the spectrum of an object corresponding to emission of light at a certain well-defined frequency. A heated gas in a glass container produces an emission-line spectrum. (p. 55)

emission nebula A glowing cloud of hot interstellar gas. The gas glows as a result of one or more nearby young stars ionizing the gas. Since the gas is mostly hydrogen, the reemitted radiation falls predominantly in the red region of the spectrum because of the dominant Hα hydrogen emission line. (p. 280)

emission spectrum The pattern of spectral emission lines produced by an element. Each element has its own unique emission spectrum. (p. 55)

Encke Division A small gap in Saturn's A ring. (p. 218)

epicycle A construct of the geocentric model of the solar system that was necessary to explain observed planetary motions. Each planet was thought to ride on a small epicycle whose center in turn traversed a larger circle (the deferent). (p. 24)

equinox See autumnal equinox and vernal equinox. (p. 9)

escape speed The speed necessary for one object to escape the gravitational pull of another. An object moving away from another with more than the escape speed will never return. (p. 128)

event horizon Imaginary spherical surface, having radius equal to the Schwarzschild radius, surrounding a collapsing star within which no event can be seen, heard, or known about by an outside observer. (p. 337)

evolutionary theory A theory that explains observations as a series of gradual steps, understandable in terms of well-established physical principles. (p. 26)

evolutionary track A graphical representation of a star's life as a path on the Hertzsprung–Russell diagram. (p. 289)

excited state State of an atom when one of its electrons is in a higher energy orbital than the ground state. Atoms can become excited by absorbing a photon of a specific energy or by colliding with a nearby atom. (p. 59)

F

F ring Faint narrow outer ring of Saturn, discovered by *Pioneer 11* in 1979. The F ring lies just inside the Roche limit of Saturn, and was shown by the *Voyager* spacecraft to be made up of several ring strands, apparently braided together. (p. 220)

field line Imaginary line in space that indicates the direction of the local gravitational, electric, or magnetic field. (p. 146)

flare Explosive event occurring in or near an active region on the Sun. (p. 244)

flatness problem One of two conceptual problems with the standard Big Bang model—there is no natural way to explain why the density of the universe is so close to the critical value. (p. 446)

focus One of two special points within an ellipse whose separation determines the eccentricity of the figure. In a bound orbit, objects move in ellipses with the center of mass of the system at one focus. (p. 32)

focus Point where light rays converge after reflection (or refraction) by the mirror (or lens) in a telescope. (p. 72)

force Action on an object that causes its momentum to change. The rate at which the momentum changes is numerically equal to the force. (p. 35)

fragmentation The breaking up of a large object into many smaller pieces (for example, as the result of high-speed collisions between planetesimals and protoplanets in the early solar system). (p. 118)

frequency The number of wave crests passing any given point per unit time. (p. 46)

full Moon Phase of the Moon in which it appears as a complete circular disk in the sky. (p. 17)

fusion Combining two light nuclei to form a heavier nucleus, usually releasing energy in the process. (p. 11)

G

galactic bulge Thick distribution of warm gas and stars around the galactic center. (p. 353)

galactic center The center of the Milky Way or any other galaxy. The point around which the disk of a spiral galaxy rotates. (p. 357)

galactic disk Flattened region of gas and dust that bisects the galactic halo in a spiral galaxy. This is the region of active star formation. (p. 352)

galactic halo Region of a galaxy, extending far above and below the galactic disk, where globular clusters and other old stars reside. (p. 353)

galactic nucleus Small, high-density region near a galactic center. Nearly all of the radiation from an active galaxy is generated within the nucleus. (p. 407)

galaxy Gravitationally bound collection of a large number of stars. The Sun is a star in the Milky Way Galaxy. (p. 352)

galaxy cluster A collection of galaxies held together by their mutual gravitational attraction. (p. 386)

Galilean moons The four large satellites of Jupiter—Io, Europa, Ganymede, and Callisto. (p. 184)

gamma ray Region of the electromagnetic spectrum, toward the short-wavelength side of the X-ray range, corresponding to radiation of very high frequency and very short wavelength. (p. 45)

gamma-ray burst Object that emits a large amount of energy in the form of a brief burst of gamma rays. Gamma-ray burst sources are thought to lie at very large distances from our Galaxy, but the precise nature of these violent events is currently unknown. (p. 332)

general theory of relativity Einstein's theory of gravity that reinterprets the force of gravity as a curvature of space-time in the vicinity of a massive object. (p. 337)

geocentric model A model of the solar system with the Earth at the center of the universe and all other bodies revolving around it. The earliest theories of the solar system were geocentric. (p. 24)

giant Star having a radius between 10 and 100 times that of the Sun. (p. 263)

globular cluster Tightly bound, roughly spherical collection of hundreds of thousands or millions of stars spanning about 50 parsecs. Globular clusters are distributed in the halos of the Milky Way and other galaxies. (p. 294)

Grand Unified Theories Theories describing the behavior of the single force that results from unification of the strong, weak, and electromagnetic forces. (p. 447)

granulation Mottled appearance of the solar surface caused by rising (hot) and falling (cool) material in convective cells just below the photosphere. (p. 236)

gravitational field Field created by any object with mass, extending outward in all directions, that determines the influence of that object on all others. The strength of the gravitational field decreases as the square of the distance. (p. 38)

gravitational force The force exerted by one massive object on another, according to Newton's law of gravity. (p. 36)

gravitational lensing Effect produced in the image of a distant object by a massive foreground object. Light from the distant background object may be bent into two or more separate images. (p. 367)

gravitational redshift A prediction of Einstein's general theory of relativity. Photons lose energy as they escape from the gravitational field of a massive object. Because a photon's energy is proportional to its frequency, a photon that loses energy suffers a decrease in frequency, corresponding to an increase, or redshift, in wavelength. (p. 342)

gravity The attractive effect that any massive object has on all other massive objects. The greater the mass of the object, the stronger its gravitational attraction. (p. 36)

Great Dark Spot Prominent (but now vanished) storm system in the atmosphere of Neptune that was located near the planet's equator and comparable in size to Earth. (p. 197)

Great Red Spot Large, high-pressure, long-lived storm system visible in Jupiter's atmosphere. The Red Spot is roughly twice the size of Earth. (p. 190)

greenhouse effect The partial trapping of solar radiation by a planetary atmosphere; similar to the trapping of heat in a greenhouse. (p. 134)

ground state The lowest energy state that an electron can have within an atom. (p. 59)

H

Head–tail radio galaxy Radio galaxy whose radio lobes have been "swept back" by the galaxy's motion through the intergalactic medium in its parent cluster, causing most of the radio emission to be observed to one side of the visible galaxy. (p. 410)

heliocentric model Model of the solar system centered on the Sun with Earth and the other planets orbiting it. (p. 26)

helioseismology The study of conditions far below the Sun's surface through the analysis of internal "sound" waves that repeatedly cross the solar interior. (p. 234)

helium flash An explosive event in the post-main-sequence evolution of a low-mass star. When helium fusion begins in a dense stellar core the burning is explosive in nature. It continues until the energy released is enough to expand the core, at which point the star regains stable equilibrium. (p. 304)

Hertzsprung–Russell (H–R) diagram A plot of luminosity (or magnitude) against temperature (or spectral class) for a group of stars. (p. 265)

high-energy telescope Telescope designed to detect radiation in the form of X rays or gamma rays. (p. 90)

highlands Relatively light-colored regions on the surface of the Moon that are elevated several kilometers above the maria. Also called terrae. (p. 142)

homogeneous The same everywhere. In a homogeneous universe, the number of galaxies in an imaginary large cube is the same no matter where in the universe the cube is placed. (p. 432)

horizon problem One of two conceptual problems with the standard Big Bang model—some regions of the universe with very similar properties are too far apart to have exchanged information within the age of the universe, so there is no obvious reason why they should look so similar. (p. 446)

horizontal branch Region of the Hertzsprung–Russell diagram where post-main-sequence stars again reach hydrostatic equilibrium. At this point, the star is burning helium in its core and hydrogen in a shell surrounding the core. (p. 305)

Hubble classification scheme Method of classifying galaxies according to their appearance developed by Edwin Hubble. (p. 378)

Hubble's constant The constant of proportionality giving the relation between recessional velocity and distance in Hubble's law. (p. 397)

Hubble's law Law relating a galaxy's observed recession velocity to its distance from us. The recession velocity of a galaxy is directly proportional to its distance. (p. 397)

hydrogen envelope Invisible region engulfing the coma of a comet, usually distorted by the solar wind and extending across millions of kilometers of space. (p. 107)

hydrogen-shell burning Fusion of hydrogen in a shell, driven by contraction and heating of the helium core. Once hydrogen is depleted in the core of a star, hydrogen burning stops and the core contracts due to gravity. This causes the temperature to rise, heating the surrounding layers of hydrogen in the star and increasing the burning rate there. (p. 303)

hydrosphere Layer of Earth containing the liquid oceans and accounting for roughly 70 percent of Earth's total surface area. (p. 128)

I

image Representation of an object produced when light from the object is reflected or refracted by a mirror or lens. (p. 72)

inertia The tendency of an object to continue in motion at the same speed and in the same direction unless acted upon by a force. (p. 35)

inflation Short period of unchecked cosmic expansion early in the history of the universe. During inflation, the universe swelled in size by a factor of about 10^{50}. Inflation provides a plausible solution to the horizon and flatness problems in cosmology. (p. 447)

infrared radiation Region of the electromagnetic spectrum, lying between the radio and visible ranges, corresponding to light of wavelength longer than that of red light. (p. 45)

infrared telescope Telescope designed to detect infrared radiation. Infrared telescopes are designed to be lightweight so that they can be carried above (most of) Earth's atmosphere by balloons, airplanes, or satellites. (p. 88)

inner core The central part of Earth's core, believed to be solid, composed mainly of nickel and iron. (p. 137)

instability strip Region in the Hertzsprung–Russell diagram where stars become pulsationally unstable, leading to RR Lyrae and Cepheid variables. (p. 355)

intensity A basic property of electromagnetic radiation that specifies the amount or strength of the radiation. (p. 51)

intercrater plains Relatively smooth regions on the surface of Mercury that do not show extensive cratering. (p. 161)

interferometer Collection of two or more telescopes that work together to observe the same object at the same time and at the same wavelength. The effective diameter of an interferometer is equal to the distance between its outermost telescopes. (p. 86)

interferometry Technique in widespread use to dramatically improve the resolution of radio and other telescopes. Several telescopes observe an object simultaneously and a computer analyzes how the signals interfere with one another to reconstruct a detailed image of the field of view. (p. 86)

interstellar medium The matter between stars composed of two components, gas and dust, intermixed throughout all of space. (p. 278)

inverse-square law The law that a field follows if its strength decreases with the square of the distance. Fields that follow the inverse-square law rapidly decrease in strength as the distance increases, but never quite reach zero. (p. 37)

ion An atom that has gained or lost one or more electrons and so has a net electrical charge. (p. 59)

ionosphere A layer of Earth's atmosphere above about 100 km, where the air is significantly ionized and conducts electricity. (p. 132)

ion tail Thin stream of ionized gas that is pushed away from the head of a comet by the solar wind. It extends directly away from the Sun. Often referred to as a plasma tail. (p. 107)

irregular galaxy A galaxy that does not fit into any of the other major categories in the Hubble classification scheme. (p. 381)

isotropic The same in all directions. In an isotropic universe, we count the same numbers of galaxies in any direction on the sky, no matter which direction we choose. (p. 432)

J

jovian planet One of the four giant outer planets of the solar system which resemble Jupiter in overall physical and chemical properties. (p. 102)

K

Kepler's laws See laws of planetary motion. (p. 29)

Kirchhoff's laws Three rules governing the formation of different types of spectra. (p. 57)

Kuiper belt A region in the plane of the solar system beyond the orbit of Neptune where most short-period comets are thought to originate. (p. 109)

L

law of conservation of mass and energy A fundamental law of modern physics which states that the sum of mass and energy must always remain constant in any physical process. In fusion reactions, the lost mass is converted into energy, primarily in the form of electromagnetic radiation, according to Einstein's formula $E = mc^2$. (p. 245)

laws of planetary motion Three empirical laws derived by Johannes Kepler, based on precise observations of the motions of the planets by Tycho Brahe, summarizing the motions of the planets about the Sun. (p. 30)

light See electromagnetic radiation. (p. 52)

light curve A plot of the variation in brightness of a star with time. (p. 268)

light-year The distance that light, moving at a constant speed of 300,000 km/s, travels in one year. One light-year is about 10 trillion kilometers. (p. 2)

lighthouse model The leading explanation for pulsars. A small region of a neutron star, near one of the magnetic poles, emits a steady stream of radiation that sweeps past Earth as the star rotates. The pulse period is thus the neutron star's rotation period. (p. 329)

lobe radio galaxy An active galaxy whose radio emission comes mainly from two extended radio lobes lying well outside the visible part of the galaxy. (p. 410)

Local Group Small galaxy cluster that includes the Milky Way Galaxy. (p. 386)

look-back time The length of time since a distant object emitted the radiation we see today. (p. 418)

luminosity One of the basic properties used to characterize stars, luminosity is defined as the total energy radiated by a star each second, at all wavelengths, in all directions. (p. 233)

luminosity class A classification scheme that groups stars according to the widths of their spectral lines. For a group of stars having the same temperature, luminosity class differentiates between supergiants, giants, and main-sequence dwarfs. (p. 271)

lunar eclipse Celestial event during which all or part of the Moon passes through Earth's shadow, temporarily darkening it. (p. 11)

lunar phase The changing appearance of the Moon over the course of a month as it orbits Earth. (p. 11)

M

Magellanic Clouds Two small irregular galaxies that are gravitationally bound to the Milky Way Galaxy. (p. 382)

magnetic field Field that accompanies any changing electric field and governs the influence of magnetized objects on one another. (p. 47)

magnetosphere Zone of charged particles, trapped by a planet's magnetic field, lying above the planet's atmosphere. (p. 128)

magnitude scale A system that ranks stars by apparent brightness developed by the Greek astronomer Hipparchus. Originally, the brightest stars in the sky were categorized as being of first magnitude, while the faintest stars visible to the naked eye were classified as sixth magnitude. The scale has since been extended to cover stars and galaxies too faint to be seen by the unaided eye. Increasing magnitude means fainter stars and a difference of five magnitudes corresponds to a factor of 100 in apparent brightness. (p. 258)

main sequence A well-defined band in the Hertzsprung–Russell diagram, running from the top left of the diagram to the bottom right, on which most stars are found. (p. 265)

main-sequence turnoff The point on the Hertzsprung–Russell diagram for a star cluster where stars are just beginning to evolve away from the main sequence. If all stars in the cluster are plotted on the diagram, the lower-mass stars will trace out the main sequence up to the turnoff point. (p. 319)

mantle Layer of Earth just interior to the crust and surrounding the core. (p. 128)

mare Relatively dark-colored, smooth region on the surface of the Moon (plural: maria). (p. 142)

mass A measure of the total amount of matter contained within an object. (p. 35)

matter-dominated universe A universe in which the density of matter exceeds the equivalent density of radiation. The present-day universe is matter-dominated. (p. 443)

mesosphere Layer of Earth's atmosphere lying between the stratosphere and the ionosphere, 50–80 km above Earth's surface. (p. 132)

meteor Bright streak in the sky, often referred to as a "shooting star," resulting from a small piece of interplanetary debris entering Earth's atmosphere and heating air molecules which subsequently emit light as they return to their ground states. (p. 110)

meteor shower Event caused by Earth's yearly passage through the debris spread along the orbit of a comet during which many meteors can be seen each hour. (p. 110)

meteorite Any part of a meteoroid that survives passage through the atmosphere and lands on Earth's surface. (p. 110)

meteoroid Chunk of interplanetary debris before it encounters Earth's atmosphere. (p. 110)

meteoroid swarm Pebble-sized fragments, dislodged from the main body of a comet, moving in nearly the same orbit as the parent comet. (p. 110)

micrometeoroids Relatively small chunks of interplanetary debris ranging from dust-particle-sized to pebble-sized fragments. (p. 110)

Milky Way Galaxy The spiral galaxy in which the Sun resides. The disk of our Galaxy is visible in the night sky as the faint band of light known as the Milky Way. (p. 352)

millisecond pulsar A pulsar whose period indicates that the neutron star responsible for the pulses is rotating nearly 1000 times each second. The most likely explanation for these rapid rotators is that they have been spun up by drawing in matter from a companion star. (p. 334)

molecular cloud A cold, dense interstellar cloud containing a high fraction of molecules. It is widely believed that the relatively high density of dust particles in these clouds plays an important role in both the formation and the protection of the molecules. (p. 285)

molecular-cloud complex Collection of molecular clouds, spanning as much as 50 parsecs, possibly containing enough material to make millions of Sun-sized stars. (p. 286)

molecule A tightly bound collection of atoms held together by the electromagnetic fields of those atoms. Molecules, like atoms, emit and absorb photons at specific wavelengths, but molecular spectra generally bear little resemblance to the spectra of their component atoms. (p. 63)

N

nebula General term for any "fuzzy" patch on the sky, whether light or dark. (p. 116)

nebular theory One of the earliest models of solar system formation, dating back to Descartes, in which a large cloud of gas was envisaged to collapse under its own gravity to form the Sun and planets. (p. 116)

neutrino Virtually massless and chargeless particle that is one of the products of fusion reactions in the Sun. Neutrinos move at close to the speed of light and interact only very weakly with matter. (p. 246)

neutrino oscillations Possible solution to the solar neutrino problem, but only if the neutrino has a small, nonzero mass. In this view, some neutrinos can "oscillate," or become transformed into other particles en route to Earth from the solar core, and hence go undetected. (p. 248)

neutron An elementary particle having roughly the same mass as a proton, but which is electrically neutral. Along with protons, neutrons form the nuclei of atoms. (p. 62)

neutron star A dense ball of neutrons that may remain at the core of a star after a Type II supernova explosion has destroyed the rest of the star. Typical neutron stars are about 20 km across and contain more mass than the Sun. (p. 328)

new Moon Phase of the moon during which none of the lunar disk is visible. (p. 11)

Newtonian mechanics The basic laws of motion and gravity, postulated by Newton, which are sufficient to explain and quantify virtually all of the complex dynamical behavior found on Earth and elsewhere in the universe. They are superceded by Einstein's theories of relativity in circumstances where velocities approach the speed of light or gravitational fields become very strong (such as in the vicinity of a black hole). (p. 35)

Newtonian telescope A reflecting telescope in which incoming light is intercepted before it reaches the prime focus and is deflected into an eyepiece at the side of the instrument. (p. 77)

Newton's laws of motion Three laws that describe how objects move and interact with one another. (p. 35)

Newton's law of gravity Two massive objects exert a gravitational force on one another that is directly proportional to the product of their masses and inversely proportional to the square of the distance between them. (p. 36)

nonstellar radiation A radiation spectrum (for example, of an active galaxy) that cannot be explained simply as the combined spectra of many stars. (p. 406)

north celestial pole Point on the celestial sphere directly above Earth's north pole. (p. 6)

nova A star that suddenly increases in brightness, often by a factor of as much as 10,000, then slowly fades back to its original luminosity. A nova is the result of an explosion on the surface of a white dwarf star caused by hydrogen fusion in matter accreted onto its surface from a binary companion. (p. 310)

nuclear fusion Mechanism of energy generation in the core of the Sun in which light nuclei are combined (fused) into heavier ones, releasing energy in the process. (p. 245)

nucleotide base Organic molecules, the building blocks of genes, which pass hereditary characteristics from one generation of living creatures to the next. (p. 457)

nucleus Dense, central region of an atom containing both protons and neutrons, usually surrounded by one or more electrons. (p. 59)

nucleus The solid region of ice and dust that comprises the central part of the head of a comet. (p. 107)

O

Olbers's paradox A thought experiment suggesting that if the universe were homogeneous, infinite, and unchanging, then the entire night sky should be as bright as the surface of the Sun. (p. 433)

Oort Cloud Spherical halo of material surrounding the solar system out to a distance of about 50,000 A.U.; the region of the solar system where most comets originate. (p. 109)

opacity A measure of a material's ability to block electromagnetic radiation. Opacity is the opposite of transparency. (p. 51)

open cluster Loosely bound collection of hundreds to thousands of stars, a few parsecs across, generally found in the plane of the Milky Way. (p. 294)

open universe Geometry that the universe would have if the total density is less than the critical value. In the absence of a cosmological constant, an open universe does not contain enough matter to halt the present cosmic expansion. An open universe is infinite in extent. (p. 441)

outer core The outermost part of Earth's core believed to be liquid and composed mainly of nickel and iron. (p. 136)

outflow channel Surface features on Mars believed to be the relics of catastrophic flooding about three billion years ago providing evidence that liquid water once existed there in great quantities. Found only in the equatorial regions of the planet. (p. 170)

ozone layer Layer of Earth's atmosphere at an altitude of 20 to 50 km where incoming ultraviolet solar radiation is absorbed by oxygen, ozone, and nitrogen in the atmosphere. (p. 134)

P

parallax The apparent motion of a foreground object with respect to the background as the location of the observer changes. (p. 18)

parsec The distance at which a star must lie in order that its measured parallax be exactly one arc second; equal to 206,000 A.U. (p. 254)

partial eclipse Celestial event during which only a part of the occulted body is obscured from view. (p. 11)

penumbra Region of the shadow cast by an eclipsing object in which the eclipse is seen as partial. (p. 14)

penumbra Outer region of a sunspot, not as dark nor as cool as the central umbra. (p. 240)

perihelion A planet's point of closest approach to the Sun. (p. 32)

period Time needed for an orbiting body to complete one revolution. (p. 33)

period–luminosity relationship Relation between the pulsation period of a Cepheid variable and its absolute brightness (luminosity). Using this relation, measurement of the pulsation period allows the distance of the star to be determined. (p. 356)

permafrost Layer of permanently frozen water ice believed to lie just under the Martian surface. (p. 171)

phase See lunar phase. (p. 11)

photon Packet of electromagnetic energy comprising electromagnetic radiation. (p. 60)

photosphere The visible surface of the Sun lying just above the uppermost layer of the Sun's interior and just below the chromosphere. (p. 233)

planet A large body (not itself a star) orbiting a star. (p. 100)

planetary nebula The ejected envelope of a red giant star, typically spread over a volume roughly the size of our solar system. (p. 308)

planetesimal Object in the early solar system that had grown to the size of a small moon, at which point its gravitational field was strong enough to begin to influence its neighbors. (p. 118)

plate tectonics Motion of large regions of Earth's crust (plates), which drift with respect to one another as a result of convection in the upper mantle. Also known as continental drift. (p. 139)

polarity The direction of a magnetic field conventionally taken to run from S to N. (p. 241)

positron Subatomic particle with properties identical to those of a negatively charged electron except for its positive charge. The positron is the antiparticle of the electron. Positrons and electrons annihilate one another when they meet and produce pure energy in the form of gamma rays. (p. 247)

precession Slow change in the direction of the rotation axis of a spinning object caused by some external influence. (p. 10)

primary atmosphere The chemical components that would have surrounded Earth just after it formed. (p. 173)

prime focus Point on the axis of a reflecting telescope where the mirror focuses incoming light. (p. 72)

primordial nucleosynthesis Production of elements heavier than hydrogen by nuclear fusion in the high temperatures and densities found in the early universe. (p. 444)

prominence Loop or sheet of glowing gas ejected from an active region on the solar surface which then moves through the inner parts of the corona under the influence of the Sun's magnetic field. (p. 244)

proper motion Angular movement of a star across the sky, as seen from Earth, measured in seconds of arc per year. The result of the star's actual motion through space. (p. 255)

proton An elementary particle carrying a positive electric charge that is a component of all atomic nuclei. The number of protons in the nucleus of an atom determines what element it represents. (p. 47)

proton–proton chain The chain of fusion reactions leading from hydrogen to helium that powers most main-sequence stars. (p. 246)

protoplanet Clump of material, formed in the early stages of solar system formation, that was the building block of the planets we see today. (p. 116)

protostar Stage in star formation when the interior of a collapsing fragment of gas becomes opaque to its own radiation. The protostar is the hot, dense region at the center of the fragment. (p. 288)

protosun The central protostar during the early stages of solar system formation. (p. 118)

Ptolemaic model Solar system model developed by the second-century astronomer Claudius Ptolemy, perhaps the best geocentric model ever proposed. It predicted with great accuracy the positions of the then-known planets, using more than 80 circles (epicycles and deferents) to model them. (p. 25)

pulsar Object that emits electromagnetic radiation in the form of rapid pulses, with a characteristic pulse period and duration. Pulsars are believed to be rapidly rotating neutron stars where charged particles, accelerated by the star's magnetic field, flow along the field lines, producing radiation that beams outward as the star spins on its axis. See lighthouse model. (p. 329)

pulsating variable star A star whose luminosity varies in a predictable, periodic way. A pulsating variable is typically an otherwise unremarkable star which happens to be experiencing a temporary period of instability as a normal part of its evolution. (p. 355)

Q

quarter Moon Lunar phase in which the moon appears as a half disk. (p. 11)

quasar Starlike radio source with an observed redshift that indicates an extremely large distance from Earth. A member of the most energetic class of active galaxies. (p. 419)

quasi-stellar object (QSO) See quasar. (p. 419)

R

radar Acronym for *ra*dio *d*etection *a*nd *r*anging. Radio waves are bounced off an object and the time taken for the echo to return indicates its distance. (p. 34)

radiation-dominated universe Early epoch in the universe when the equivalent density of radiation in the cosmos exceeded the density of matter. (p. 444)

radiation zone Region of the Sun's interior where extremely high temperatures guarantee that the gas is completely ionized. Photons are only occasionally diverted by electrons and travel through this region with relative ease. (p. 233)

radioactivity The release of energy by rare, heavy elements when their nuclei decay into lighter nuclei. (p. 184)

radio galaxy Type of active galaxy that emits most of its energy in the form of long-wavelength radiation. (p. 408)

radio lobe Roundish region of radio-emitting gas lying well beyond the center of a radio galaxy. (p. 410)

radio telescope Large instrument designed to detect radiation from space at radio wavelengths. (p. 89)

radio waves Region of the electromagnetic spectrum, to the long-wavelength side of the infrared range, corresponding to radiation of the longest wavelengths. (p. 45)

radius–luminosity–temperature relation A mathematical proportionality arising from simple geometry and Stefan's law which allows astronomers to determine the radius of a star indirectly once its luminosity and temperature are known. (p. 263)

reddening Dimming of starlight by interstellar matter, which tends to scatter higher-frequency (blue) radiation more efficiently than the lower-frequency (red) components, leading to a distinct reddish appearance in the light we receive. (p. 279)

red dwarf Small, cool, faint star at the lower-right end of the main sequence on the Hertzsprung Russell diagram. Its color and size give it its name. (p. 266)

red giant A giant star whose surface temperature is relatively low, so it appears red. (p. 263)

red giant branch The portion of the evolutionary track of a star corresponding to hydrogen shell burning which drives the expansion and cooling of the star's outer envelope. (p. 304)

red giant region The upper-right-hand corner of the Hertzsprung–Russell diagram where red giant stars are found. (p. 267)

redshift Any Doppler shift of radiation toward longer wavelengths (lower frequencies); the result of relative motion of the observer away from the source. (p. 64)

red supergiant An extremely luminous red star. Often found on the asymptotic giant branch of the Hertzsprung–Russell diagram. (p. 306)

reflecting telescope A telescope that uses a mirror to gather and focus light from a distant object. (p. 72)

refracting telescope A telescope that uses a lens to gather and focus light from a distant object. (p. 72)

refraction The tendency of a wave to bend as it passes from one transparent medium to another. (p. 72)

retrograde motion A backward (westward) loop traced out by a planet with respect to the fixed stars as Earth overtakes it in its orbit. (p. 24)

revolution Orbital motion of one body about another, such as Earth about the Sun. (p. 8)

right ascension Celestial coordinate used to measure "longitude" on the celestial sphere. The zero point is the position of the Sun at the vernal equinox. (p. 6)

ringlet Narrow region in Saturn's ring system where the density of ring particles is particularly high. *Voyager 1* discovered that the rings visible from Earth are actually composed of tens of thousands of ringlets. (p. 219)

Roche limit Often called the tidal stability limit, the Roche limit gives the distance from a planet at which the planet's tidal force on a moon exceeds the gravitational forces holding the moon together. Objects within the Roche limit are unlikely to accumulate into larger objects. The rings of the jovian planets occupy the regions within the planets' Roche limits. (p. 219)

rotation Spinning motion of a body about an axis. (p. 5)

rotation curve Plot of the orbital speed of disk material in a galaxy against distance from the galactic center. Analysis of rotation curves of spiral galaxies indicates the existence of dark matter. (p. 366)

RR Lyrae variable Variable star whose luminosity changes in a characteristic way. All RR Lyrae stars have more or less the same luminosity. (p. 355)

runaway greenhouse effect Process in which the heating of a planet leads to an increase in its atmosphere's ability to retain heat, thus increasing the temperature still further. Ultimately, this causes extreme changes in the planet's surface temperature and atmospheric composition. (p. 177)

runoff channel River-like surface feature in the southern highlands of Mars providing evidence that liquid water once existed there in great quantities. Runoff channels are thought to have been formed by water that flowed nearly four billion years ago. (p. 170)

S

S0 galaxy Galaxy that shows evidence of a thin disk and a bulge, but which has no spiral arms and contains little or no gas. (p. 381)

SB0 galaxy S0-type galaxy whose disk shows evidence of a bar. (p. 381)

scarp Surface feature on Mercury, believed to be the result of cooling and shrinking of the crust, forming a wrinkle on the face of the planet. (p. 162)

Schwarzschild radius The distance from the center of an object at which, if all the mass were compressed within that region, the escape velocity would equal the speed of light. Once a stellar remnant collapses within this radius, light cannot escape and the object is no longer visible—it forms a black hole. (p. 337)

seasons Annual changes in average temperature and length of day that result from the tilt of Earth's (or any planet's) axis with respect to the plane of its orbit. (p. 9)

secondary atmosphere The chemicals that comprised Earth's atmosphere shortly after the planet's formation, after volcanic activity had outgassed chemicals from the interior. (p. 173)

seeing The ease with which good telescopic observations can be made from Earth's surface given the blurring effects of atmospheric turbulence. (p. 79)

seeing disk Roughly circular region on a detector over which a star's pointlike image is spread due to atmospheric turbulence. (p. 79)

seismic wave A wave that travels outward through Earth's interior from the site of an earthquake. (p. 135)

self-propogating star formation The theory that the formation and eventual explosion of massive stars can create shock waves that trigger the next round of star formation, possibly accounting for the spiral structure observed in many galaxies. (p. 364)

semi-major axis Half the length of an ellipse's major axis. The conventional measure of the "size" of an ellipse. (p. 32)

Seyfert galaxy Type of active galaxy whose emission comes from a very small region within the nucleus of an otherwise normal-looking spiral galaxy. (p. 407)

shepherd satellite Satellite whose gravitational effects are responsible for maintaining a ring. Examples are Prometheus and Pandora, the two satellites of Saturn whose orbits lie on either side of the F ring. (p. 220)

shield volcano Volcano produced by repeated nonexplosive eruptions of lava, creating a gradually sloping, shield-shaped low dome. There is often a caldera (crater) at its summit. (p. 165)

sidereal day Time required for Earth to rotate exactly once, relative to the stars. (p. 7)

sidereal month Time required for the Moon to complete one orbit around Earth, relative to the stars. (p. 11)

sidereal year The time required for the constellations to complete one cycle around the sky and return to their starting points, as seen from a given point on Earth; the time required for Earth to complete one orbit around the Sun, relative to the stars. (p. 10)

singularity A point in the universe where the density of matter and the gravitational field become infinite, such as at the center of a black hole. (p. 344)

solar constant The amount of solar energy reaching Earth (above the atmosphere) per unit area per unit time, approximately 1400 W/m^2. (p. 233)

solar cycle The 22-year period needed for both the average number of spots and the Sun's magnetic polarity to repeat themselves. The Sun's polarity reverses on each new 11-year sunspot cycle. (p. 242)

solar day Period of time between noon one day until noon the next day. (p. 7)

solar eclipse Celestial event during which the new Moon passes directly between Earth and the Sun, temporarily blocking the Sun's light. (p. 13)

solar nebula The swirling gas surrounding the early Sun during the epoch of solar system formation; also referred to as the primitive solar system. (p. 116)

solar neutrino problem The discrepancy between the theoretically predicted numbers of neutrinos streaming from the Sun as a result of nuclear fusion reactions in the core and the numbers actually observed. The observed number of neutrinos is about half the predicted number. (p. 248)

solar system The Sun and all the planets that orbit the Sun—Mercury, Venus, Earth, Mars, Jupiter, Saturn, Uranus, Neptune, and Pluto. (p. 100)

solar wind An outward stream of fast-moving charged particles emanating from the Sun. (p. 107)

south celestial pole Point on the celestial sphere directly above Earth's south pole. (p. 6)

special theory of relativity Einstein's reformulation of Newton's laws of motion to incorporate the fact that the speed of light is constant, regardless of the motion of the observer. (p. 340)

spectral class Classification scheme based on the strength of stellar spectral lines and an indication of the temperature of a star. (p. 262)

spectrometer Instrument used to produce detailed spectra of stars. Usually, a spectrograph records a spectrum on a photographic plate or CCD, or more recently, in electronic form on a computer. (p. 82)

spectroscope Instrument used to split a light source into its component colors. (p. 55)

spectroscopic binary A binary-star system which from Earth appears as a single star, but whose spectral lines show back-and-forth Doppler shifts as two stars orbit one another. (p. 267)

spectroscopic parallax Method of determining the distance to a star by measuring its temperature and then determining its absolute brightness by comparing with a standard Hertzsprung–Russell diagram. The absolute and apparent brightness of the star give its distance from Earth by using the inverse-square law. (p. 271)

spectroscopy The study of the way in which atoms absorb and emit electromagnetic radiation. Spectroscopy allows astronomers to determine the chemical composition of distant objects. (p. 57)

speed of light The fastest speed possible, according to the currently known laws of physics. Electromagnetic radiation exists in the form of waves or photons moving at the speed of light. (p. 48)

spiral arm Observed distribution of material in a disk galaxy in a pinwheel-shaped pattern apparently emanating from near the galactic center. (p. 361)

spiral density wave One proposed explanation for the existence of galactic spiral arms in which coiled waves of gas compression move through the galactic disk, triggering star formation as they pass. (p. 362)

spiral galaxy Galaxy containing a flattened, star-forming disk which may have prominent spiral arms and a large central galactic bulge. (pp. 353, 379)

standard candle Any object with an easily recognizable appearance and known luminosity that can be used in estimating distances. Type I supernovae, which all have roughly the same peak luminosity, are particularly useful standard candles used to determine distances to other galaxies and to probe the large-scale structure of the universe. (p. 384)

Standard Solar Model A self-consistent mathematical model of the Sun obtained by incorporating into a computer program all physical processes believed to be important in determining the Sun's internal structure and adjusting the calculation until its predictions (solar radius, luminosity, temperature, etc.) agree with observations. (p. 234)

star A glowing ball of gas held together by its own gravity and powered by nuclear fusion in its core. (p. 232)

star cluster A group of anywhere from a few dozen to a few million stars that formed at the same time from the same cloud of interstellar gas. Stars in clusters are useful in aiding our understanding of stellar evolution, because they all have roughly the same age and chemical composition and lie at roughly the same distance from Earth. (p. 294)

starburst galaxy Galaxy in which a violent event, such as near-collision, has caused a sudden, intense burst of star formation in the recent past. (p. 393)

Stefan's law Relation giving the total blackbody energy emitted per square meter per second by an object of a given temperature. The energy emitted increases proportional to the fourth power of the temperature. (p. 52)

stellar occultation The dimming of starlight produced when a solar system object, such as a planet, moon, or ring, passes directly in front of a distant star, as seen from Earth. (p. 222)

stratosphere Layer of Earth's atmosphere lying above the troposphere, extending up to an altitude of 40–50 km. (p. 132)

subgiant branch Portion of the evolutionary track of a star occurring just after core hydrogen burning ceases. Shell hydrogen burning heats the outer layers of the star, causing a general expansion of the stellar envelope. (p. 304)

summer solstice Point on the ecliptic where the Sun is at its northernmost point above the celestial equator, occurring on or near June 21. (p. 9)

sunspot An Earth-sized dark blemish on the surface of the Sun. The sunspot is darker than its surroundings because its temperature is lower than that of the rest of the solar photosphere. (p. 240)

sunspot cycle The variation with time of the numbers and distribution of sunspots. The average number of spots reaches a maximum roughly every 11 years, then falls off to almost zero. Solar activity is greatest at times of sunspot maximum. (p. 242)

supercluster Grouping of several clusters of galaxies into a larger, but not necessarily gravitationally bound, unit. (p. 386)

supergiant Star with a radius between 100 and 1000 times that of the Sun. (p. 263)

supergranulation Large-scale flow pattern on the surface of the Sun, consisting of cells measuring up to 30,000 km across, believed to be the imprint of large convective cells deep in the solar interior. (p. 236)

supernova Explosive death of a star that is one of the most energetic events of the universe. A supernova may temporarily outshine the rest of the galaxy in which it resides. (p. 314)

supernova remnant The scattered glowing remains from a supernova that occurred in the past. The Crab Nebula is one of the best-studied supernova remnants. (p. 315)

surface gravity The strength of the gravitational field at a body's surface. (p. 128)

synchronous orbit State of an object when its rotation period is exactly equal to its orbital period around some other body. The Moon is in a synchronous orbit around Earth, and so presents the same face to Earth at all times. (p. 132)

synchrotron radiation Type of nonthermal radiation produced by high-speed charged particles, such as electrons, as they accelerate in a strong magnetic field. (p. 417)

synodic month Time required for the Moon to complete a full cycle of phases, as seen from Earth. (p. 11)

T

T Tauri phase Protostar in the late stages of formation, often exhibiting violent surface activity and strong stellar winds. T Tauri stars have been observed to brighten noticeably in short periods of time. (p. 291)

tail Component of a comet consisting of material streaming away from the main body, sometimes spanning hundreds of millions of kilometers. May be composed of dust or ionized gases. (p. 107)

telescope Instrument used to capture as many photons as possible from a given region of the sky and concentrate them into a focused beam for analysis. (p. 72)

temperature A measure of the speed of the atoms and molecules comprising an object. The higher a body's temperature, the faster the random motion of its constituent particles. (p. 51)

terrestrial planet The four innermost planets of the solar system, resembling Earth in general physical and chemical properties. (p. 102)

tidal bulge Elongation of Earth caused by differences in the Moon's gravitational force between the side of Earth nearest the Moon and the side farthest from the Moon. The bulge is roughly aligned with the line joining Earth to the Moon. More generally, the deformation of any body produced by the tidal effect of a nearby gravitating object. (p. 129)

tidal force The variation in one body's gravitational force from place to place across another body—for example, the variation of the Moon's gravity from one part of Earth to another. (p. 130)

tides Rising and falling of bodies of water on Earth exhibiting daily, monthly, and yearly cycles. Ocean tides on Earth are caused by the competing tidal influences of the Moon and Sun on our planet. (p. 129)

time dilation A prediction of the general theory of relativity that is closely related to the gravitational redshift. To an outside observer, a clock lowered into a strong gravitational field will appear to run more slowly. (p. 342)

total eclipse Celestial event during which one body is completely blocked from view by another. (p. 11)

transition zone Region in the solar atmosphere where the temperature increases rapidly. The transition zone lies above the (4500 K) chromosphere and below the (1 million K) corona. (p. 233)

triangulation Method of determining distance based on simple geometry. A distant object is sighted from two well-separated locations. The distance between the two locations (the baseline) and the angles between the line joining them and the sight lines to the distant object are all that are necessary to infer the object's distance. (p. 17)

troposphere Lowest layer of Earth's atmosphere, extending from the surface to an altitude of about 15 km, where convection occurs. (p. 132)

Trojan asteroids One of two groups of asteroids orbiting at the same distance from the Sun as Jupiter, 60 degrees ahead of and behind the planet. (p. 103)

tropical year The time interval between one vernal equinox and the next. (p. 16)

Tully–Fisher relation An empirical connection between a spiral galaxy's total luminosity and its rotational velocity (usually determined spectroscopically, from measurements of line broadening). (p. 384)

21-centimeter line Radio energy emitted by a hydrogen atom when the spin of an electron in the ground state suddenly reverses direction, becoming parallel to the spin of the proton in the nucleus. (p. 285)

Type I supernova Explosive death of a star in which a white dwarf in a binary system accretes so much mass that it cannot support its own weight. The star collapses and temperatures become high enough for carbon fusion to occur. Fusion begins throughout the white dwarf almost simultaneously and an explosion results. (p. 314)

Type II supernova Explosive death of a high-mass star in which the star's highly evolved iron core rapidly implodes and then explodes, destroying the surrounding star in the process. (p. 314)

U

ultraviolet radiation Region of the electromagnetic spectrum, lying between the visible and X-ray ranges, corresponding to light of wavelength shorter than that of blue light. (p. 45)

ultraviolet telescope A telescope designed to collect radiation in the ultraviolet part of the spectrum. Earth's atmosphere is partially opaque to ultraviolet wavelengths, so ultraviolet telescopes are placed on rockets, balloons, or satellites to get high above most or all of the atmosphere. (p. 90)

umbra Central region of the shadow cast by an eclipsing body. (p. 14)

umbra Central region of a sunspot; the darkest and coolest part of the spot. (p. 240)

universe The totality of all space, time, matter, and energy. (p. 2)

V

Van Allen belts At least two doughnut-shaped regions of magnetically trapped charged particles high above Earth's atmosphere. (p. 146)

variable star A star whose luminosity changes with time. (p. 354)

vernal equinox Date on which the Sun crosses the celestial equator moving northward, occurring on or near March 21. (p. 9)

visible light The small range of the electromagnetic spectrum that human eyes can see. The visible spectrum ranges from about 400 to 700 nm, corresponding to blue through red light. (p. 45)

visual binary A binary-star system in which both components can be seen from Earth as separate stars. (p. 267)

void Large (tens of megaparsecs across), relatively empty region of the universe surrounded by galaxy clusters, superclusters, and filaments. (p. 399)

volcano Upwelling of hot lava from below a planet's crust to its surface. (p. 138)

W

water hole The radio interval between 18 cm and 21 cm, the wavelengths at which hydroxyl (OH) and hydrogen (H) radiate (respectively) which intelligent civilizations might conceivably use for communications. (p. 466)

wave A way in which energy and information may be transferred from place to place without physical movement of material from one location to another. The energy is carried by a disturbance of some sort that moves in a distinctive, repeating pattern. A wave is characterized by its velocity, frequency, and wavelength. (p. 45)

wave period The time required for a wave to repeat itself at a specific point in space. (p. 46)

wavelength The distance from one wave crest to the next at any instant in time. (p. 46)

white dwarf A faint star with a sufficiently high surface temperature that appears white or blue-white in color. Typical white dwarfs are similar in size to Earth, but have masses comparable to the mass of the Sun. White dwarfs are the carbon cores of low-mass stars that have shed the rest of their envelopes as planetary nebulae. (p. 263)

white dwarf region The bottom left-hand corner of the Hertzsprung–Russell diagram where white dwarf stars are found. (p. 267)

white oval Light-colored storm system, smaller than the Great Red Spot, in Jupiter's atmosphere. (p. 192)

Wien's law Relation between the wavelength at which a blackbody curve peaks and the temperature of the emitter. The peak temperature is inversely proportional to the peak wavelength; the hotter the object, the bluer its radiation. (p. 52)

winter solstice Point on the ecliptic where the Sun is at its southernmost point below the celestial equator, occurring on or near December 21. (p. 9)

X

X ray Region of the electromagnetic spectrum between the ultraviolet and gamma-ray ranges, corresponding to radiation of high frequency and short wavelengths. (p. 45)

X-ray burster X-ray source that radiates thousands of times more energy than the Sun in bursts that last only a few seconds. The source is most likely a neutron star in a binary-star system, which accretes matter until its surface temperature reaches the point at which hydrogen fusion begins. The result is a sudden period of rapid nuclear burning and the release of energy in the form of X-rays. (p. 332)

Z

zero-age main sequence The region on the Hertzsprung–Russell diagram where stars lie at the onset of nuclear burning in their cores. (p. 292)

zodiac The twelve constellations through which the Sun moves as it traverses the ecliptic. (p. 9)

zonal flow Alternating regions of westward and eastward flow, roughly symmetrical about the equator, associated with the belts and zones in the atmosphere of a jovian planet. (p. 191)

zone Bright, high-pressure region in the atmosphere of a jovian planet where gas flows upward (away from the planet's center). (p. 190)

ANSWERS TO CONCEPT CHECK QUESTIONS

Prologue

1. (p. 7) Because the celestial sphere provides a natural means of specifying the locations of stars on the sky. Celestial coordinates are directly related to Earth's orientation in space, but are independent of Earth's rotation. 2. (p. 11) Precession is a slow shift in Earth's rotation axis due to the gravitational pulls of the Moon and the Sun. Over the course of one 26,000-year cycle, the north and south celestial poles will describe circles on the celestial sphere, while the celestial equator "wobbles" relative to the ecliptic (but always maintaining a constant angle to it). Equivalently, the familiar constellations will drift through the four seasons on Earth, eventually returning to their starting points. 3. (p. 14) If the distance doubles, the angular size of the Sun will be halved, making it easier for the Moon to eclipse the Sun. If the distance is halved, the angular size of the Sun will double, and total solar eclipses will not be possible. 4. (p. 18) The Sun is very far away, is very bright, and is somewhat "fuzzy" in appearance.

Chapter 1

1. (p. 28) In the geocentric view, retrograde motion is a real backward motion of a planet as it moves on its epicycle. In the heliocentric view, the backward motion is only apparent, caused by Earth "overtaking" the planet in its orbit. 2. (p. 30) Galileo was an experimentalist; Kepler was a theorist. Galileo found observational evidence supporting the Copernican theory; Kepler found an empirical description of the motions of the planets within the Copernican picture that greatly simplified the theoretical view of the solar system. 3. (p. 33) Yes to both. Kepler's laws apply to any body whose orbit around another is governed by the force of gravity. 4. (p. 34) Because the overall geometry of the solar system was determined by triangulation using Earth's orbit as a baseline, so all distances are known only relative to the scale of Earth's orbit—the astronomical unit. 5. (p. 39) In the absence of any force, a planet would move in a straight line with constant velocity (Newton's First Law) and therefore tends to move along the tangent to its orbital path. The Sun's gravity causes the planet to accelerate toward the Sun (Newton's Second Law), bending its trajectory into the orbit we observe.

Chapter 2

1. (p. 48) Light is an electromagnetic wave produced by accelerating charged particles. All the waves on the list are characterized by their velocities, frequencies, and wavelengths, and all carry energy and information from one location to another. Unlike waves in water or air, however, light waves require no physical medium in which to propagate. 2. (p. 51) All are electromagnetic radiation and travel at the speed of light; they differ only in wavelength (frequency). 3. (p. 54) As the switch is turned and the temperature of the filament rises, the bulb's brightness increases rapidly, by Stefan's law, and its color shifts, by Wien's law, from invisible infrared to red to yellow to white. 4. (p. 58) They are characteristic wavelengths (frequencies) at which matter absorbs or emits photons of electromagnetic radiation. They are unique to each atom or molecule, and thus provide a means of identifying the gas producing them. 5. (p. 59) Electron orbits can occur only at certain specific energies, and there is a ground state which has the lowest possible energy. Planetary orbits can have any energy. Planets can stay in any orbit indefinitely; in an atom, the electron must eventually fall to the ground state, emitting electromagnetic radiation in the process. 6. (p. 62) Spectral lines correspond to transitions between specific orbitals within an atom. The structure of an atom determines the energies of these orbitals, hence the possible transitions, and hence the energies (colors) of the photons involved.

Chapter 3

1. (p. 76) Because reflecting telescopes are easier to design, build, and maintain than refracting instruments. 2. (p. 79) The need to gather as much light as possible, and the need to achieve the highest possible angular resolution. 3. (p. 84) To reduce or overcome the effects of atmospheric absorption, instruments are placed on high mountains or in space. To compensate for atmospheric turbulence, adaptive optics techniques probe the air above the observing site and adjust the mirror surface accordingly to recover the undistorted image. 4. (p. 88) The long wavelength of radio radiation. Very large radio telescopes and interferometry, which combines the signals from two or more separate telescopes to create the effect of a single instrument of much larger diameter. 5. (p. 92) Benefits: they are above the atmosphere, so they are unaffected by seeing or absorption; they can also make round-the-clock observations. Drawbacks: cost/smaller size, inaccessibility, vulnerability to damage by radiation and cosmic rays.

Chapter 4

1. (p. 103) Because the two classes of planet differ in almost every physical property—orbit, mass, radius, composition, existence of rings, and number of moons. 2. (p. 106) Similarities: orbit in the inner solar system, solid bodies of generally "terrestrial" composition. Differences: asteroids are much smaller than the terrestrial planets, and their orbits are much less regular. 3. (p. 110) Most comets never come close enough to the Sun that we can see them. 4. (p. 113) Because it is much less evolved than material in planets, and hence gives a better indication of conditions in the early solar system. 5. (p. 116) Because the solar system has many notable irregularities in its properties, and the theory must be able to account for these too. 6. (p. 122) Yes, if the condensation theory is the correct description of planet formation, then the leftover building blocks—asteroids and comets—should be found in any planetary system.

Chapter 5

1. (p. 132) A tidal force is the variation in one body's gravitational force from place to place across another. Tidal forces tend to deform a body, rather than causing an overall acceleration, and they decrease proportional to the inverse cube of the distance from the other body, rather than as the inverse square. 2. (p. 135) It raises Earth's average surface temperature above the freezing point of water, which was important for the development of life on our planet. Should the greenhouse effect continue to strengthen, however, it may conceivably cause catastrophic climate changes on Earth. 3. (p. 139) Without volcanoes or earthquakes, we would have far less detailed direct (from volcanoes) or indirect (from seismology) information on our planet's interior. 4. (p. 142) Convection currents in the upper mantle cause portions of Earth's crust to slide around on the surface. As the plates move and interact, they are responsible for volcanism, earthquakes, the formation of mountain ranges and ocean trenches, and the creation and destruction of oceans and continents. 5. (p. 146) The maria are younger and are much less heavily cratered than the highlands. Both observations are consistent with current theories of lunar and planetary formation. 6. (p. 148) It tells us that the planet has a conducting, liquid core in which the magnetic field is continuously generated. 7. (p. 150) The Moon's small size, which is the main reason that the Moon (1) has lost its initial heat, making it geologically dead today, and (2) has insufficient gravity to retain an atmosphere.

Chapter 6

1. (p. 159) It has caused the rotation period to become exactly 2/3 the orbital period, and the rotation axis to be perpendicular to the planet's orbit plane. 2. (p. 161) Venus's atmosphere is much more massive, hotter, and denser than that of Earth, and is composed almost entirely of carbon dioxide. 3. (p. 163) They are thought to have been formed when the planet's iron core cooled and contracted, causing the crust to crack. Faults on Earth are the result of tectonic activity. 4. (p. 166) No, they are mainly shield volcanoes formed by lava upwelling through "hot spots" in the crust, and are not associated with plate tectonics (there is none on Venus). 5. (p. 172) No, because it wasn't created by running water. It is a huge crack in the planet's crust thought to be associated with upwelling in the mantle that never developed into convection. 6. (p. 177) Both might have climates quite similar to that on Earth, as the greenhouse effect would probably not have run away on Venus, and the water in the Martian atmosphere might not have frozen and become permafrost.

Chapter 7

1. (p. 186) Uranus's orbit was observed to deviate from a perfect ellipse, leading astronomers to try to compute the mass and location of the body responsible for those discrepancies. That body was Neptune. 2. (p. 190) Because the magnetic field is generated by the motion of electrically conducting liquid in the planet's deep interior. 3. (p. 193) Like weather systems on Earth, they are regions of high and low pressure and are associated with convective motion. However, unlike storms on Earth, they wrap all the way around the planet, because of Jupiter's rapid rotation. In addition, the clouds are arranged in three distinct layers, and the bright colors are the result of cloud chemistry unlike anything operating in Earth's atmosphere. 4. (p. 197) Uranus's low temperature means that the planet's clouds lie deeper in the atmosphere, making them harder to distinguish from Earth. 5. (p. 201) The dense cores are the result of accretion, as described in Sec. 4.3. They are larger than the terrestrial planets mainly because there was more "raw material" available in the outer part of the solar nebula. Eventually, the cores' gravitational pulls became strong enough to capture gas from the solar nebula, forming the extended envelopes.

Chapter 8

1. (p. 213) The tidal effect of Jupiter's gravitational field. 2. (p. 214) The moon's atmosphere is opaque to visible light. 3. (p. 216) They are all very heavily cratered. 4. (p. 223) The distance from a planet within which the planet's tidal forces would destroy a moon. Planetary rings are all found inside the planet's Roche limit. 5. (p. 225) They have similar masses, radii, composition, and perhaps origins in the Kuiper belt.

Chapter 9
1. (p. 234) Because we simply multiply the solar constant by the total area, we are implicitly assuming that the same amount of energy reaches every square meter of the large sphere in Figure 9.3. 2. (p. 236) (1) Radiation, where energy travels in the form of light, and (2) convection, where energy is carried by physical motion of upwelling solar gas. 3. (p. 239) (1) Emission lines of (2) highly ionized elements, implying high temperature. 4. (p. 245) Strong field, with a well-defined east–west organization, just below the surface. The field direction in the Southern Hemisphere is opposite to that in the north.
5. (p. 248) Because the sun shines by nuclear fusion, which converts mass into energy as hydrogen is turned into helium.

Chapter 10
1. (p. 255) Because stars are so far away that their parallaxes relative to any baseline on Earth are too small to measure. 2. (p. 259) Nothing—we need to know their distances before the luminosities (or absolute magnitudes) can be determined. 3. (p. 262) Because temperature controls which excited states the star's atoms and ions are in, and hence which atomic transitions are possible. 4. (p. 263) Yes, using the radius–luminosity–temperature relationship, but only if we can find a method of determining the luminosity that doesn't depend on the inverse-square law (see Sec. 10.7). 5. (p. 267) Because giants are intrinsically very luminous and can be seen to much greater distances than the more common main-sequence or white dwarfs. 6. (p. 270) We don't—we assume that their masses are the same as similar stars found in binaries. 7. (p. 273) All stars would be further away, but their measured spectral types and apparent brightnesses would be unchanged, so their luminosities would be greater than previously thought. The main sequence would therefore move vertically upward in the H–R diagram. (We would therefore use larger luminosities in the method of spectroscopic parallax, so distances inferred by that method would also increase.)

Chapter 11
1. (p. 280) Because the scale of interstellar space is so large that even low densities amount to a large amount of obscuring matter. 2. (p. 286) Because the UV light is absorbed by hydrogen gas in the surrounding cloud, ionizing it to the emission nebulae. The red light is part of the visible hydrogen spectrum emitted as electrons and protons recombine to form hydrogen atoms. 3. (p. 292) (1) The existence of a photosphere, within which the protostar is opague to its own radiation. (2) Nuclear fusion in the core.
4. (p. 293) No—different parts of the main sequence correspond to stars of different masses. A typical star stays at roughly the same location on the main sequence of most of its lifetime. 5. (p. 295) Because the most massive stars form most rapidly, and reach the main sequence long before low-mass stars do.

Chapter 12
1. (p. 306) Because the nonburning inner core, unsupported by fusion, begins to shrink, releasing gravitational energy, heating the overlying layers, and causing them to burn more vigorously. 2. (p. 311) No, because it is not a member of a binary system.
3. (p. 313) Because iron does not fuse to produce energy. As a result, no further nuclear burning is possible, and the core's equilibrium cannot be restored.
4. (p. 318) Because they are responsible for creating and dispersing all the heavy elements out of which we are made. In addition, they (or their progenitor high-mass star) may also have played a role in triggering the collapse of an interstellar cloud to form our solar system. 5. (p. 321) Because a star cluster gives us a "snapshot" of stars of many different masses, but of the same age and initial composition allowing us to directly test the predictions of the theory.

Chapter 13
1. (p. 329) No—only Type-II supernova. According to theory, the rebounding central core of the original star becomes a neutron star. 2. (p. 330) Because (1) not all supernovae form neutron stars, (2) the pulses are beamed, so not all pulsing neutron stars are visible from Earth, and (3) pulsars spin down and become too faint to observe after a few tens of millions of years.
3. (p. 336) The X-ray sources are binaries containing accreting neutron stars, which may be in the process of being spun up to form millisecond pulsars. 4. (p. 338) Its gravity becomes so strong that even light cannot escape, and it becomes a black hole. 5. (p. 339) Newton's theory describes gravity as a force produced by a massive object that influences all other massive objects. Einstein's relativity describes gravity as a curvature of spacetime produced by a massive object; that curvature then determines the trajectories of all particles—matter or radiation—in the universe.
6. (p. 344) Because the object would appear to take infinitely long to reach the event horizon, and its light would be infinitely redshifted by the time it got there.
7. (p. 346) By observing their gravitational effects on other objects, and from the X-rays emitted when matter falls into them.

Chapter 14
1. (p. 353) Because the Milky Way is the thin plane of the Galactic disk, seen from within. When our line of sight lies in the plane of the Galaxy, we see many stars blurring into a continuous band. In other directions, we see darkness. 2. (p. 357) No, because even the brightest Cepheids are unobservable at distances of more than a kiloparsec or so through the obscuration of interstellar dust. 3. (p. 361) Because all the gas and dust in the halo fell to the Galactic disk billions of years ago, so star formation has long since ceased.
4. (p. 364) Because differential rotation would destroy the spiral structure within a few hundred million years.
5. (p. 367) Its emission has not been observed at any electromagnetic wavelength. Its presence is inferred from its gravitational effect on stars and gas orbiting the Galactic center. 6. (p. 371) Observations of rapidly swirling gas and the variability of the radiation emitted suggest the presence of a million-solar-mass black hole.

Chapter 15
1. (p. 384) Spirals are neither the only nor the most common type of galaxy. Dwarf elliptical and irregular galaxies are by far the most numerous. 2. (p. 388) Because calibration errors in the cosmic distance ladder compound as we move farther out, and because different techniques lead to different and inconsistent results. 3. (p. 390) Because the amount of luminous matter seen in galaxies and galaxy clusters, even including the hot intracluster gas, falls short, by a factor of at least 10, of the mass inferred from studies of galaxy rotation curves and galaxy orbits. 4. (p. 393) Galaxies are thought to form and grow by collisions and mergers of smaller objects, while stars via the collapse and fragmentation of a large interstellar cloud.
5. (p. 399) Hubble's law, which describes the expansion, provides a direct connection between redshift (recession velocity) and distance.

Chapter 16
1. (p. 407) They are more luminous, emit a large fraction of their energy at radio and infrared wavelengths, and, in the case of quasars, all lie at great distances from Earth. 2. (p. 408) Both show rapid internal motion and nonthermal energy emission, but the Seyfert emits thousands of times more energy than the Galactic Center. 3. (p. 410) Many radio galaxies have high-speed jets of material originating in the nucleus and aligned with the radio lobes. 4. (p. 418) The energy is generated in an accretion disk in the galactic nucleus and then transported by jets into the lobes, where it is eventually emitted by the synchrotron process in the form of radio waves. 5. (p. 424) Because quasars generally appear starlike, as seen from Earth, and their redshifts were unexpectedly large for objects thought to be members of our own Galaxy. 6. (p. 426) Perhaps. Our Galaxy has a massive black hole at its center, which might possibly have supported a very high luminosity in the past. However, our central black hole has a mass only a few million times that of the Sun, so our Galaxy probably wasn't extremely luminous in its youth. Many nearby galaxies may well have been very bright quasars in the distant past.

Chapter 17
1. (p. 433) On very large scales—more than 300 Mpc—the distribution of galaxies seems to be roughly the same everywhere and in all directions.
2. (p. 435) Because, tracing the motion backwards in time, it implies that all galaxies, and in fact the universe itself, were located at a single point at the same instant in the past. 3. (p. 439) There doesn't seem to be enough matter to halt the collapse and, in addition, the observed cosmic acceleration suggests the existence of a large-scale repulsive force in the cosmos that also opposes recollapse. 4. (p. 442) A low-density universe has negative curvature, a critical density universe is spatially flat (Euclidean), and a high-density universe has positive curvature (and is finite in extent). 5. (p. 444) At the time of the Big Bang. It is the electromagnetic remnant of the primeval fireball. 6. (p. 445) Because the amount of deuterium observed in the universe today implies that the present density of normal matter is at most a few percent of the critical value—much less than the density of dark matter inferred from dynamical studies. 7. (p. 448) Inflation implies that the universe is flat, and hence that the total cosmic density equals the critical value. However, the matter density is only about 30 percent of critical, and the density of electromagnetic radiation (the microwave background) is only a tiny fraction of the critical density. The remaining density may be in the form of the "dark energy" thought to be powering the accelerating cosmic expansion (Section 17.3).
8. (p. 450) Because the dark matter clumped to form the "skeleton" of the large-scale structure in the universe. Normal matter subsequently accumulated in the densest regions, forming galaxies there.

Chapter 18
1. (p. 459) The formation of complex molecules from simple ingredients by non-biological processes has been repeatedly demonstrated, but no living cell or self-replicating molecule has ever been created.
2. (p. 460) Mars remains the most likely site, although Europa and Titan also have properties that might have been conducive to the emergence of living organisms.
3. (p. 464) It breaks a complex problem up into similar "astronomical," "biochemical," "anthropological," and "cultural" pieces, which can be analyzed separately. It also identifies the types of stars where a search might be most fruitful. 4. (p. 466) It is in the radio part of the spectrum, where Galactic absorption is least, at a region where natural Galactic background "static" is minimized, and in a portion of the spectrum characterized by lines of hydrogen and hydroxyl, both of which would likely have significance to a technological civilization.

ANSWERS TO SELF-TEST QUESTIONS

Prologue
True or False? 1. T 2. F 3. T 4. F 5. F 6. F 7. T 8. F 9. F 10. F
Fill in the Blank 1. Galaxy 2. ecliptic 3. winter solstice; lowest 4. stars 5. celestial sphere 6. vernal equinox 7. celestial equator 8. lunar 9. parallax 10. angular diameter
Problems 1. (a) 10^3, 10^{-6}, 1.001×10^3, 10^{15}, 1.23×10^5, 4.56×10^{-4}, (b) 31,600,000, 299,800, 0.0000000000667, 2, (c) 2000.01, 3.33×10^5, 9.47×10^{12} 2. (c) 3. It would decrease by roughly 8 minutes 4. Leo 5. (a) 33 arc minutes, (b) 33 arc seconds, (c) 0.55 arc seconds; 56.4 minutes 6. 144.3 m 7. (a) 57,300 km, (b) 3.44×10^6 km, (c) 2.06×10^8 km 8. 12,000 km 9. 391 10. 1.08×10^5 km, 2.58×10^6 km

Chapter 1
True or False? 1. F 2. T 3. F 4. F 5. T 6. F 7. T 8. T 9. F 10. F 11. T 12. T
Fill in the Blank 1. retrograde 2. Earth 3. scientific method 4. moons, phases, sunspots 5. Brahe 6. ellipse, circle 7. focus 8. square, cube 9. radar ranging 10. 150,000,000 11. force 12. product, square
Problems 1. (a) 110 km, (b) 44,000 km, (c) 370,000 km 2. 27 arc minutes, retrograde 3. 700 s 4. 3 A.U., 0.333, 5.2 yr 5. 35 A.U. 6. Pluto perihelion = 29.65 A.U., Neptune perihelion = 29.80 A.U. 7. 9.42 (10^{-4} solar = 1.88×10^{27} kg 8. 9.50, 7.33, 1.49 m/s^2 9. For a 70 kg person, force (weight) = 684 N = 154 pounds 10. 1.7 km/s

Chapter 2
True or False? 1. T 2. F 3. F 4. T 5. F 6. F 7. F 8. T 9. F 10. T 11. T 12. F 13. T 14. F 15. F
Fill in the Blank 1. wavelength 2. frequency 3. electric, magnetic 4. 400, 700 5. radio, infrared, visible 6. temperature 7. the 1200 K object 8. continuous 9. the Sun 10. dense 11. cool 12. positive, negative 13. adsorbs 14. emits 15. difference
Problems 1. 1480 m/s 2. 3 m 3. 23 Hz; radio 4. 3.25; 112 5. 310, 9.4 microns, infrared 6. 6.4×10^7 W/m^2; 3.9 (10^{26} W 7. 10^{10} 8. 3, 6 9. 137 km/s, approaching 10. 1.94×10^{27} kg

Chapter 3
True or False? 1. F 2. T 3. F 4. F 5. F 6. T 7. T 8. F 9. T 10. T 11. F 12. F 13. T 14. F 15. F
Fill in the Blank 1. refracting 2. reflecting 3. reflecting 4. area 5. diameter, wavelength 6. turbulent atmosphere 7. 1 8. digital 9. angular resolution 10. reflecting 11. interferometer 12. X-ray 13. infrared 14. infrared 15. images
Problems 1. 0.6 arc seconds; 6.8 pixels 2. 4 percent, taking the cage to be 1 m across 3. 6.7 minutes; 1.7 minutes 4. (a) 0.25″, (b) 0.01″ 5. 4 minutes 6. roughly 50 times 7. 0.7 light years, 0.014 light years 8. 14.1 m, 16 m 9. (a) 0.003 arc seconds, (b) 0.005 arc seconds 10. The X-ray measurement, since the angular resolution of the X-ray image is around 1″ (the equation in the text gives a diffraction-limited resolution of 0.0005″, but the effective resolution is complicated by the X-ray mirror arrangement, and is actually determined by other design factors), much less than the 1° resolution of *GRO*

Chapter 4
True or False? 1. T 2. F 3. F 4. F 5. F 6. F 7. T 8. F 9. F 10. T 11. T 12. T 13. T 14. F 15. F
Fill in the Blank 1. Mercury, Pluto 2. rocky, icy 3. Mars, Jupiter 4. 100–1000, 100 5. Jupiter 6. 10, 1 A.U. 7. meteoroid swarm 8. meteor 9. 4.6 billion 10. gravity 11. gas 12. icy 13. comets 14. Jupiter 15. spin
Problems 1. Mercury: 0.307, 0.467 A.U.; Mars: 1.382, 1.666 A.U.; Pluto: 29.65, 49.31 A.U. 2. Total mass = 10^{21} kg = 0.00017 Earth masses; 40 km 3. Semi-major axis = 0.67 A.U., aphelion = 1.13 A.U. 4. 3 kg equivalent 5. 1.6×10^{21} kg, 1.6×10^{20} kg, 1.6×10^{19} kg, 1.6×10^{18} kg 6. (a) 11.1 million years, (b) 68.4 A.U. 7. (a) 1.94×10^{43}, 7.83×10^{42}, 2.66×10^{40} SI units, (b) 10^{31} SI units 8. Increases by a factor of 2 9. 477,000 10. 2×10^8 comets; 1 per 2.5 years

Chapter 5
True or False? 1. F 2. F 3. F 4. T 5. T 6. T 7. F 8. F 9. T 10. F 11. F 12. T 13. F 14. F 15. F
Fill in the Blank 1. 1/4 2. crust 3. liquid water 4. variation 5. maria 6. meteoritic impact 7. mantle 8. nitrogen, oxygen 9. convection 10. infrared 11. increase 12. aurora 13. solid, liquid 14. molten 15. plate tectonics
Problems 1. Surface gravity = 0.55 the current value (5.3 m/s^2); escape speed = 8.3 km/s 2. 30 kg equivalent 3. 21 m 4. 1.1×10^{-7} 5. Smaller by a factor of 6.0×10^{-6}. No! 6. 5.0×10^{18} kg = 8.4×10^{-7} times Earth's mass 7. 45 percent 8. 43 minutes 9. 1.2 trillion years 10. 93

Chapter 6
True or False? 1. F 2. T 3. T 4. F 5. F 6. F 7. T 8. T 9. F 10. T 11. F 12. F 13. F 14. T 15. F
Fill in the Blank 1. larger 2. radar 3. poles 4. atmosphere 5. slow and retrograde 6. carbon dioxide 7. closer 8. radar 9. volcanism 10. temperature, atmospheric pressure 11. cratered 12. bulge 13. gravity 14. water ice 15. water
Problems 1. 8.8 minutes 2. Roughly 20 percent greater than it is now 3. 96 minutes 4. 4.9×10^{20} kg, 98 more massive than Earth's atmosphere, or 10^{-4} the mass of Venus 5. 400 km/h, 250 mph 6. 6600 km 7. GM/R^2 = 3.7 m/s^2, or 40 percent the Earth value 8. 10 m/s 9. 2.9×10^{17} kg, or 5.9×10^{-4} the mass of Venus's atmosphere 10. 6.4×10^{23} kg

Chapter 7
True or False? 1. F 2. T 3. F 4. F 5. F 6. T 7. F 8. T 9. F 10. T 11. T 12. F 13. T 14. F 15. T
Fill in the Blank 1. average density 2. Uranus 3. Neptune 4. methane 5. hydrogen, helium 6. zones, belts 7. hurricanes 8. ammonia 9. gravity 10. Neptune 11. rotation 12. parallel 13. 2 14. hydrogen 15. helium precipitation
Problems 1. (a) 150 km, (b) 1100 km 2. 2.2×10^{17} newton; 1.4×10^{21} newton 3. 1.17 hours; 84,000 km 4. 1.2×10^{23} kg = 2.1 percent the mass of Earth 5. GM/R^2 for Jupiter is 24.8 m/s^2; for Saturn, it is 10.4 m/s^2 6. 1.5 percent the present mass 7. 10.5 days 8. 74 K 9. 1.7×10^{25} kg (2.8 Earth masses); 81 percent 10. 49 μm; infrared

Chapter 8
True or False? 1. T 2. T 3. T 4. T 5. F 6. F 7. T 8. F 9. T 10. T 11. T 12. T 13. T 14. F 15. F
Fill in the Blank 1. Ganymede 2. water ice 3. volcanoes 4. 3 5. A, B 6. Roche limit 7. shepherd 8. nitrogen 9. water 10. Miranda 11. retrograde 12. nitrogen 13. Triton 14. Triton 15. eclipses
Problems 1. 2.3 planetary radii 2. 3.4×10^{-3}; 3.8×10^{-3} 3. Io: 36′, Europa: 18′, Ganymede: 18′, Callisto: 9′, Sun: 6.1′; Yes 4. 38 km radius 5. GM/R^2 for Titan is 1.35 m/s^2 (Earth = 9.80); escape speed = 2.64 km/s 6. 20.3 km/s; Earth orbit: 7.6 km/s 7. 1.1×10^{18} 8. 6.8 times farther out 9. For a weight of 70 kg: 4.7 kg on Pluto, 2.1 kg on Charon 10. 11.1 hours; 20,000 km

Chapter 9
True or False? 1. T 2. F 3. T 4. F 5. F 6. F 7. F 8. T 9. F 10. T 11. T 12. F 13. F 14. F 15. T
Fill in the Blank 1. photosphere 2. chromosphere, transition zone, corona 3. convection, radiation, core 4. granulation 5. photosphere 6. hydrogen 7. helium 8. 99.9 9. solar wind 10. cooler 11. 11, twice 12. flare 13. core 14. 6, helium−4, neutrinos, gamma-ray radiation 15. few
Problems 1. 52 W/m^2 2. 3000 km; orbital period = 167 minutes ≈ 30 times wave period 3. 0.29 nm (gamma ray), 29 nm (UV), 290 nm (near-UV) 4. 1000 s = 17 minutes ≈ 2 × granule lifetime 5. Sunspot emits 36 percent, or 64 percent less 6. Mass loss by radiation = 4.3 million ton/s = 4.7 × solar wind 7. 151 days 8. 310,000 years 9. 74 billion years 10. 4.1×10^{28}

Chapter 10
True or False? 1. T 2. T 3. F 4. T 5. F 6. T 7. F 8. F 9. F 10. F 11. F 12. T 13. F 14. T 15. T
Fill in the Blank 1. Earth's orbit 2. distance, proper motion 3. temperature, luminosity 4. white dwarfs 5. smaller 6. temperature 7. ionized 8. in the ground state 9. G2 10. temperature (color), luminosity (magnitude) 11. Main Sequence 12. red giants 13. white dwarfs 14. binary 15. decrease
Problems 1. 77 pc, ϕ.39″ 2. 47 km/s 3. (a) 80 solar luminosities, (b) 2 solar radii 4. B is 3 times farther away 5. A is 10 times farther away 6. 3.3×10^{-10} W/m^2, 2.3×10^{-13} times the solar constant 7. factor of 4 million 8. 100 pc 9. 3.5, 2.3 solar masses 10. (a) 200 billion years, (b) 1 billion years, (c) 100 million years

Chapter 11
True or False? 1. F 2. T 3. T 4. F 5. F 6. F 7. F 8. T 9. F 10. F 11. T 12. T 13. F 14. F 15. F
Fill in the Blank 1. gas, dust 2. similar 3. spin, hydrogen 4. 20 5. radio 6. hydrogen 7. 1,000,000 8. 8,000 9. radio 10. evolutionary track 11. increase 12. infrared 13. (zero-age) main sequence 14. 50 15. rapidly
Problems 1. 1.1 grams 2. 1.43×10^9 Hz; frequency range (relative to unshifted frequency): -3.6×10^5 Hz to $+2.3 \times 10^5$ Hz; wavelength range: -35 μm to $+53$ μm 3. 3500 A.U. 4. factor of 9.5×10^3; magnitude increases by 9.9 5. escape speeds (km/s): 1.3, 0.81, 0.79, 0.57; average speeds (km/s): 13.6, 14.0, 14.6, 14.2; No 6. Yes—escape speed = 29 km/s, average molecular speed = 0.35 km/s 7. (a) 190, (b) 2.3 solar radii 8. 8 9. 10^{-6} 10. 17 pc

Chapter 12
True or False? 1. T 2. T 3. F 4. T 5. F 6. T 7. F 8. F 9. F 10. T 11. T 12. T 13. F 14. T 15. F
Fill in the Blank 1. pressure 2. 5 billion 3. hydrogen, helium 4. 100 million K 5. contract 6. star clusters 7. 10,000 km 8. high, low 9. decrease 10. accretion

11. hydrogen, surface 12. neutron, neutrino 13. Type-I/carbon-detonation 14. light curves 15. decreases
Problems 1. 3.5×10^7 kg/m^3 (25,000 times solar value of 1400 kg/m^3); 5.7×10^{-4} kg/m^3 (4×10^{-7} times solar) 2. 220 pc 3. 5 solar masses 4. 2.9 yr; 25,000 yr 5. 5400 km/s; 4.9×10^6 m/s^2 (Earth gravity = 9.8 m/s^2) 6. 3,2 million pc; 1 billion pc 7. 0.48 pc; no—there are no O or B stars (in fact, no stars at all) within that distance of the Sun 8. apparent magnitude = −11.5; Moon: −12.5, Venus −4.4, so slightly fainter than the full Moon, much brighter (by a factor of about 2600) than Venus 9. roughly 1000 km/s; not too bad an assumption—the Nebula is moving too fast to be affected much by gravity, although it is probably slowing down as it runs into the interstellar medium 10. 1.2×10^{44} joules; 1.4×10^{27} kg = 230 Earth masses

Chapter 13

True or False? 1. T 2. F 3. T 4. T 5. F 6. F 7. T 8. F 9. T 10. F 11. F 12. T 13. F 14. F 15. T
Fill in the Blank 1. Type I (carbon detonation) 2. 20 3. high, strong 4. radio 5. a few milliseconds, a few seconds 6. rotation period 7. binary 8. neutron 9. gamma rays 10. gravity 11. lose 12. singularity 13. X-rays 14. binary 15. variability
Problems 1. 1.1 revolutions per second 2. a mass of 80 kg becomes 2.4×10^{16} kg = 3.3×10^{-7} Moon masses, or 15,000 asteroid masses (assuming an asteroid density of 3000 kg/m^2) 3. surface gravity = 1.9×10^{12} m/s^2 = 1.9×10^{11} Earth gravities; escape speed = 193,000 km/s = 0.47 c; escape speed = 300,000 km/s = c 4. 1.6×10^{-5} solar, 1.6×10^3 solar, 1.6×10^{11} solar; they might be very bright immediately after formation, but become very faint as they cool; the coolest could be plotted at the bottom right of the diagram (very faint, very hot O-type "star") 5. 2.4×10^{44} J; 2.4×10^{34} J; 1.4×10^{25} J 6. 63,000 km/s = 0.21 c; 137,000 km/s 7. 3 million km = 4.3 solar radii; 3 billion km = 20 A.U. = half the size of Pluto's orbit 8. 9.9×10^9 m/s^2 = 10^9 g; 10^{-3} g; 10^{-9} g 9. 1400 km 10. 1.2×10^6 km

Chapter 14

True or False? 1. F 2. T 3. F 4. F 5. F 6. F 7. T 8. F 9. T 10. T 11. F 12. T 13. F 14. T 15. F
Fill in the Blank 1. Galactic disk 2. low 3. disk 4. halo 5. luminosities 6. 1 to 100 days 7. instability strip 8. higher 9. RR Lyrae stars 10. 8,000 11. 220 12. random 13. halo 14. larger; dark matter 15. high
Problems 1. 1″; 950 A.U. 2. 10 kpc 3. 10 times farther 4. 1.7 Mpc 5. 0.014″; yes—proper motions have been measured for many relatively nearby globular clusters 6. 2.7×10^{11} solar masses 7. 530 million years (estimating 240 km/s at 15 kpc); 490 million years (estimating 200 km/s at 5 kpc) 8. 18 (18.7 rounded down) 9. 1 kpc 10. 2.2 million solar masses

Chapter 15

True or False? 1. T 2. T 3. F 4. T 5. T 6. F 7. F 8. T 9. F 10. F 11. F 12. F 13. F 14. T 15. T
Fill in the Blank 1. Hubble 2. large 3. E (elliptical); Irr (irregular) 4. flattened/elongated 5. Local Group 6. width 7. superclusters 8. increases 9. gravity 10. X-rays 11. dark matter 12. mergers; fragmentation 13. distances 14. 50; 80 15. 100 Mpc
Problems 1. 360 Mpc 2. 17 Mpc (estimating the star's luminosity = 10^4 solar) 3. 3.4 billion years 4. 500 km/s 5. 2.6×10^{11} solar masses 6. 3.7×10^{13} solar masses 7. 130 Myr 8. 13000 km/s, 62 Mpc; 10000 km/s, 80 Mpc; 16000 km/s, 50 Mpc 9. 15 billion years 10. 18.5

Chapter 16

True or False? 1. T 2. F 3. F 4. F 5. F 6. T 7. T 8. F 9. F 10. T 11. F 12. T 13. F 14. F 15. T
Fill in the Blank 1. long/radio/infrared 2. nucleus 3. core 4. larger 5. center 6. nucleus 7. 1 parsec 8. 3 billion 9. 10 solar masses per year 10. starlike 11. redshifted 12. expansion 13. short 14. galaxy 15. fainter
Problems 1. 2.3×10^8 solar masses 2. 14° = 28 × Moon's diameter 3. 2.2 million years 4. 1.1×10^{35} W = 2.8×10^8 solar luminosities 5. 5000 km/s 6. −27; 5×10^{12} solar luminosities; roughly the same apparent brightness 7. 60 million years 8. 23 Mpc 9. 33″ 10. 3.4×10^{14} solar masses

Chapter 17

True or False? 1. F 2. T 3. F 4. T 5. T 6. T 7. F 8. T 9. T 10. T 11. F 12. F 13. F 14. F 15. T
Fill in the Blank 1. 200 2. everywhere 3. directions 4. cosmological principle 5. homogeneity 6. isotropy 7. 15 8. 3 9. closed 10. 3 11. matter 12. deuterium, helium 13. cool 14. equal to 15. temperature; COBE
Problems 1. 1000 Mpc 2. 400 million 3. 1100 km/s along diagonal 4. 19.5 billion years, 15.0 billion years, 12.2 billion years 5. (a) 270,000 tons, (b) 2.9 pc on a side 6. 3.1×10^{15} solar masses; 1200 km/s 7. (a) 8×10^{53} kg, (b) 4×10^{23} solar masses, (c) 4×10^{11} galaxies 8. 8×10^{-18} kg/m^3, 5×10^{-19} kg/m^3 9. 166 10. 1 Mpc

Chapter 18

True or False? 1. T 2. F 3. T 4. T 5. F 6. F 7. T 8. F 9. T 10. T 11. T 12. T 13. T 14. F 15. T
Fill in the Blank 1. proteins 2. methane, ammonia 3. radioactivity, lightning, volcanism, meteoritic bombardment, solar radiation 4. 3.5 billion 5. 1 billion 6. amino acids 7. technical civilizations 8. unstable 9. low 10. distances 11. light 12. dust 13. 70 14. water hole 15. 200
Problems 1. 6.3 seconds (for a 20 year old reader); 16 seconds (in 2001); 71 seconds; 161 seconds; 5.2 days 2. 0.27–0.64 A.U. 3. 5 billion 4. 4.6 A.U. 5. (a) 0.2, (b) 20, (c) 2000 6. 3100 years; 1 million years 7. 32,000 km/s 8. 5×10^8 W (500 times the emission of the Sun) 9. 1.43×10^9 to 1.67×10^9 Hz; 2.4×10^6 channels 10. 2.3 years; 55 years

INDEX

These star maps show the brighter stars and the prominent constellations as they appear on the dates and at the times indicated. To use these maps, face the south and hold the book overhead with top of the map toward the north and the right-hand edge toward the west. The brightest stars are indicated by the star symbol (☆) and the names are indicated. (Star maps courtesy of Robert Dixon, *Dynamic Astronomy*, 6th ed., Prentice Hall, 1992.)

WINTER

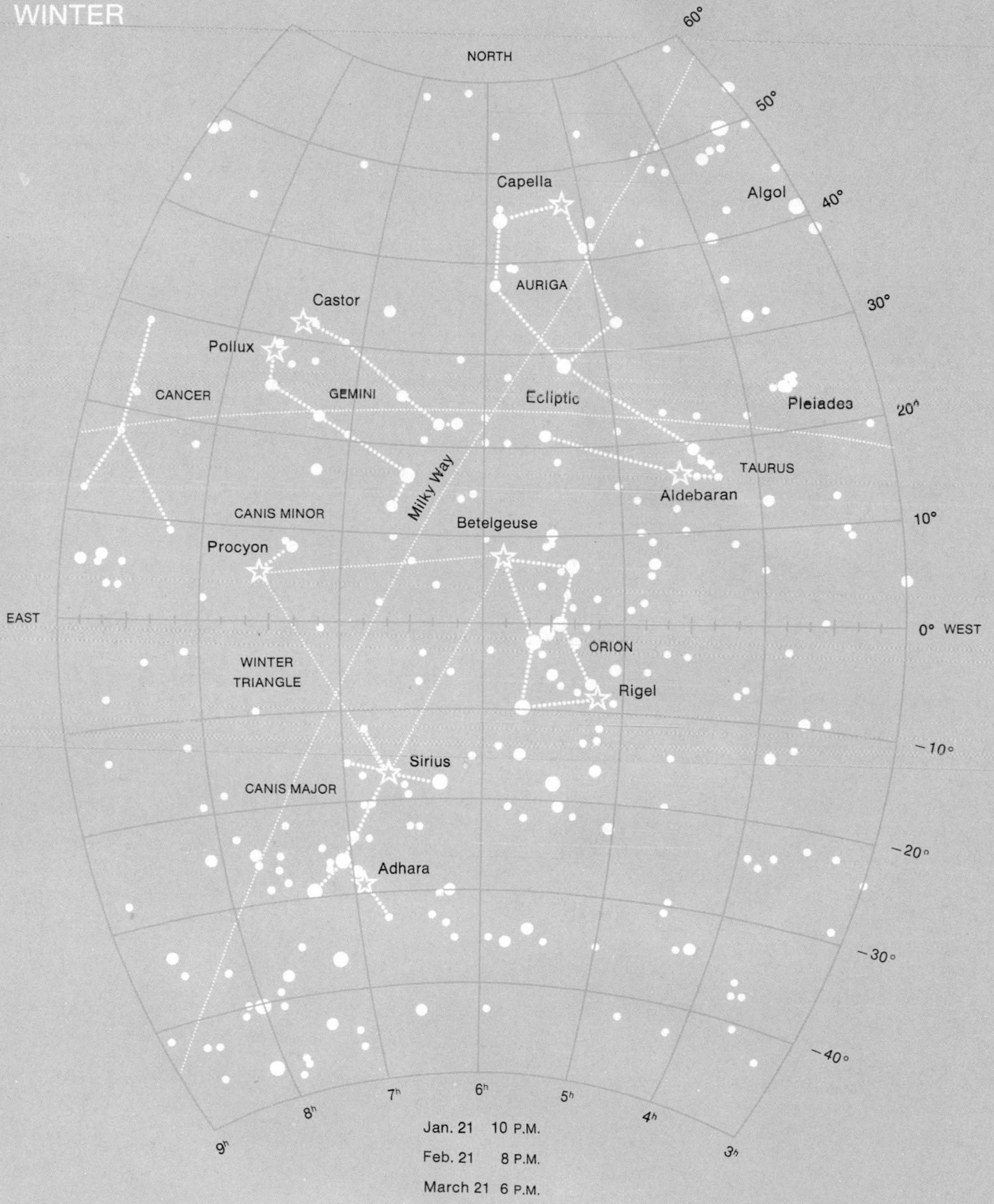

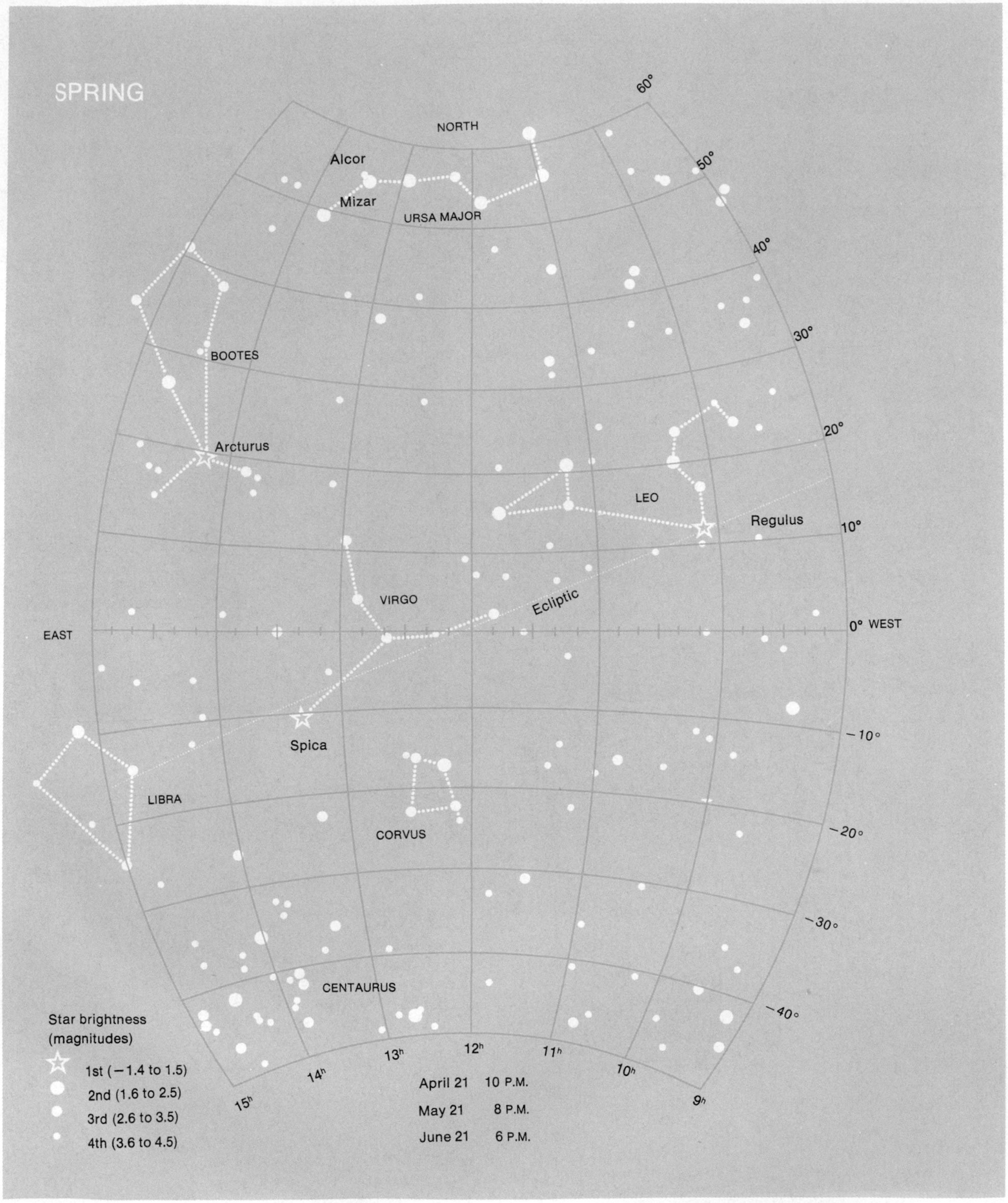
SPRING
NORTH
60°
50°
40°
30°
20°
10°
0° WEST
−10°
−20°
−30°
−40°
EAST
Alcor
Mizar
URSA MAJOR
BOOTES
Arcturus
LEO
Regulus
VIRGO
Ecliptic
Spica
LIBRA
CORVUS
CENTAURUS
15h
14h
13h
12h
11h
10h
9h
Star brightness
(magnitudes)
1st (−1.4 to 1.5)
2nd (1.6 to 2.5)
3rd (2.6 to 3.5)
4th (3.6 to 4.5)
April 21 10 P.M.
May 21 8 P.M.
June 21 6 P.M.

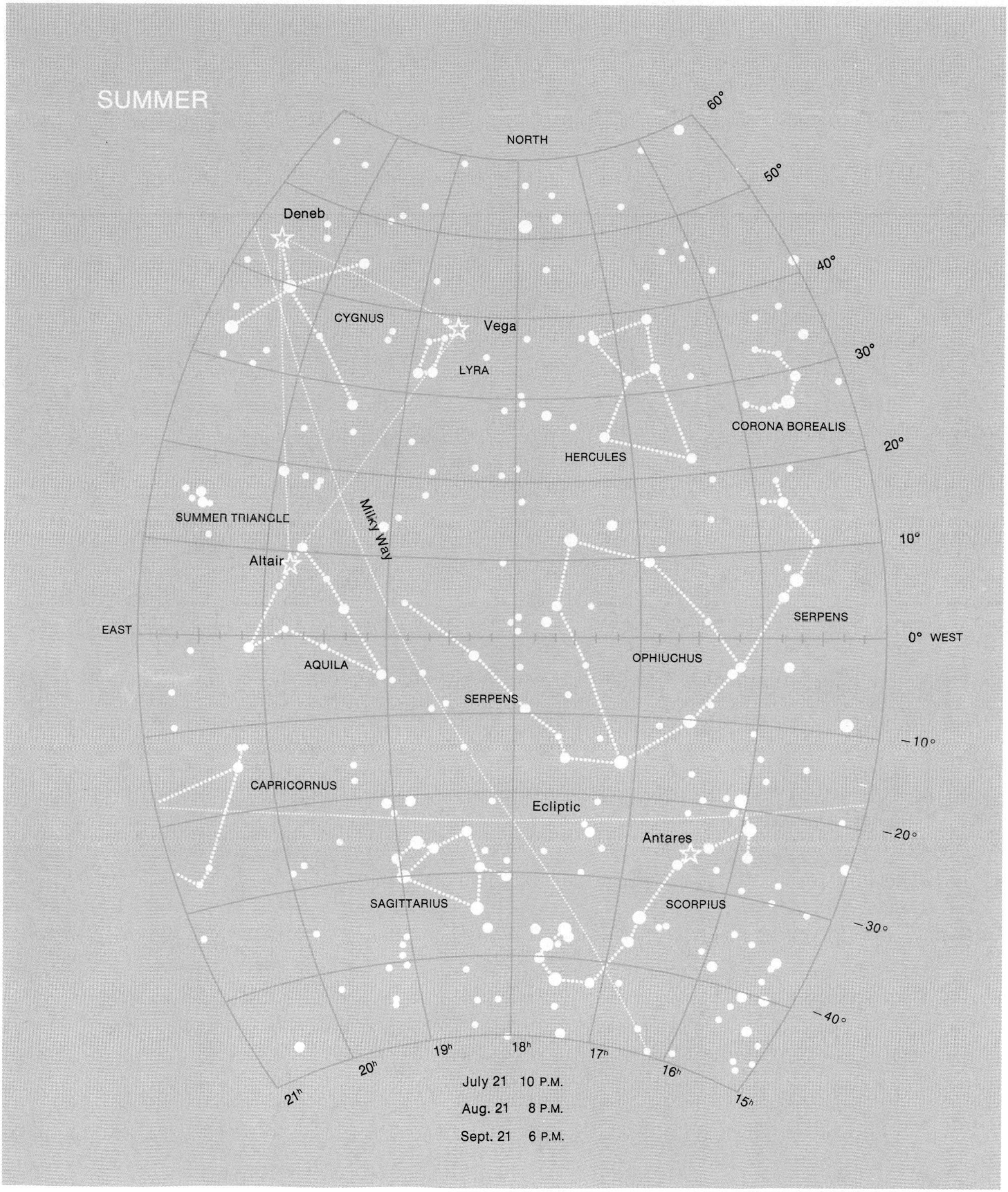
SUMMER
NORTH
60°
50°
40°
30°
20°
10°
0° WEST
-10°
-20°
-30°
-40°
EAST
Deneb
CYGNUS
Vega
LYRA
HERCULES
CORONA BOREALIS
SUMMER TRIANGLE
Milky Way
Altair
SERPENS
AQUILA
OPHIUCHUS
SERPENS
CAPRICORNUS
Ecliptic
Antares
SAGITTARIUS
SCORPIUS
21h
20h
19h
18h
17h
16h
15h
July 21 10 P.M.
Aug. 21 8 P.M.
Sept. 21 6 P.M.

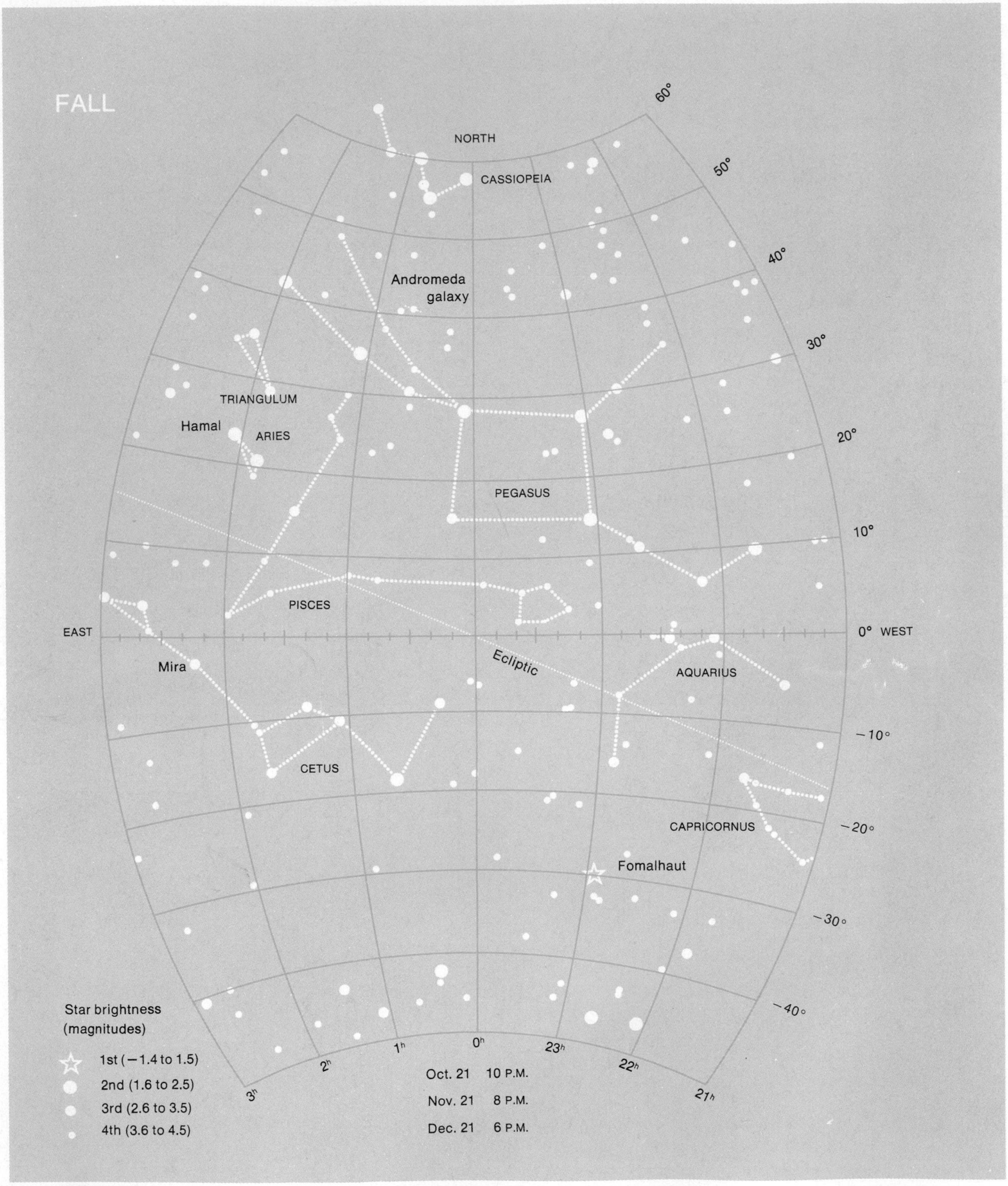
FALL
NORTH
CASSIOPEIA
Andromeda galaxy
TRIANGULUM
Hamal
ARIES
PEGASUS
PISCES
EAST
0° WEST
Mira
Ecliptic
AQUARIUS
CETUS
CAPRICORNUS
Fomalhaut
60°
50°
40°
30°
20°
10°
−10°
−20°
−30°
−40°
3h
2h
1h
0h
23h
22h
21h
Star brightness (magnitudes)
1st (−1.4 to 1.5)
2nd (1.6 to 2.5)
3rd (2.6 to 3.5)
4th (3.6 to 4.5)
Oct. 21 10 P.M.
Nov. 21 8 P.M.
Dec. 21 6 P.M.

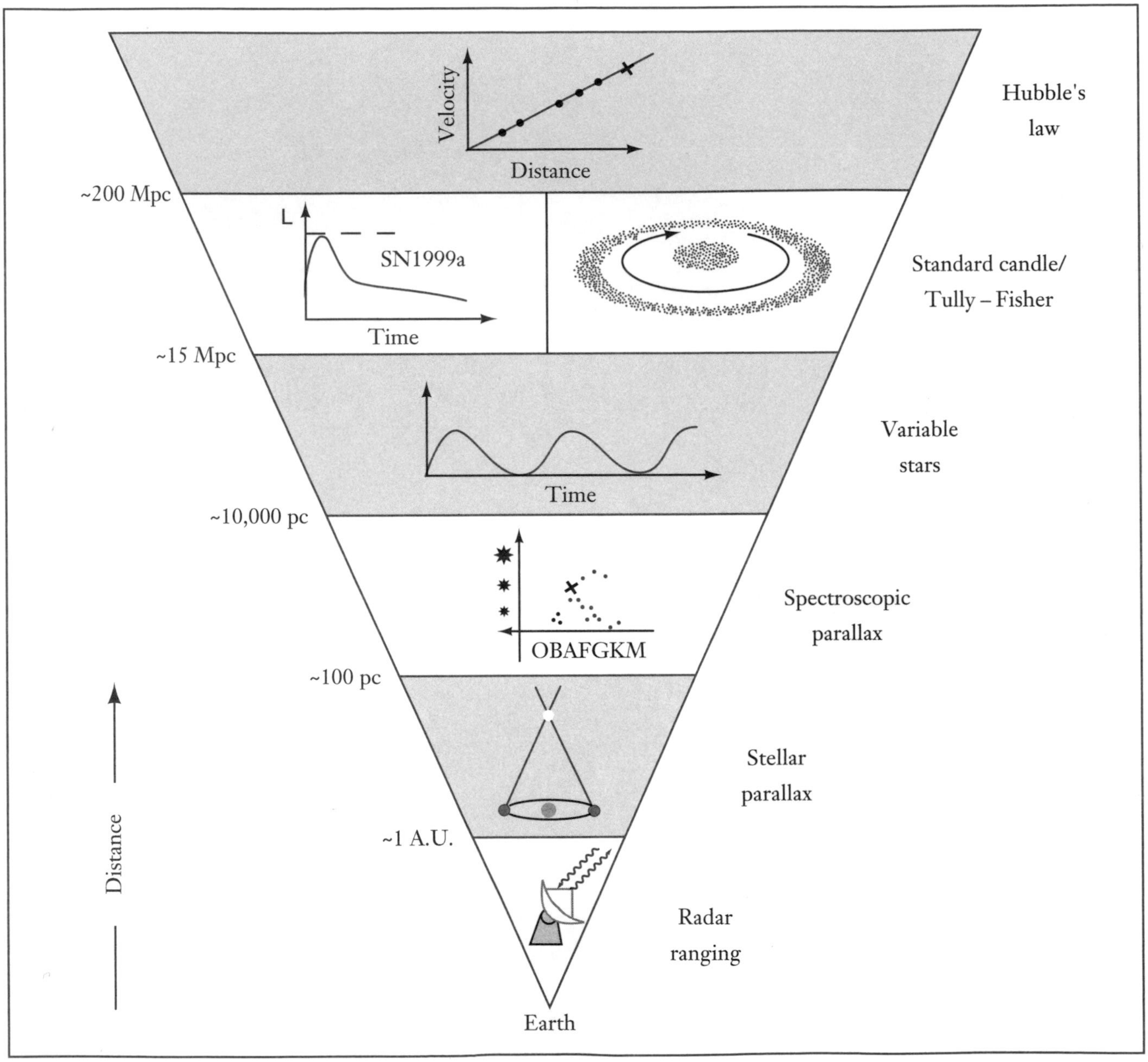
Velocity
Distance
Hubble's law
~200 Mpc
L
SN1999a
Time
Standard candle/ Tully – Fisher
~15 Mpc
Time
Variable stars
~10,000 pc
OBAFGKM
Spectroscopic parallax
~100 pc
Stellar parallax
~1 A.U.
Radar ranging
Earth
Distance

Main-Sequence Stellar Properties by Spectral Class

Spectral Class	Typical Surface Temperature (K)	Color	Mass* ($M_\odot$)	Luminosity* ($L_\odot$)	Lifetime* (10^6 yr)	Familiar Examples
O	>30,000	Electric Blue	>20	>100,000	<2	
B	20,000	Blue	8	3000	30	Spica (B1)
A	10,000	White	3	75	400	Vega (A0) Sirius (A1)
F	7,000	Yellow - white	1.5	4	4000	Procyon (F5)
G	6,000	Yellow	1.0	1.5	9000	Sun (G2) Alpha Centauri (G2)
K	4,000	Orange	0.5	0.1	60,000	Epsilon Eridani (K2)
M	3,000	Red	0.1	0.005	200,000	Proxima Centauri (M5) Barnard's Star (M5)

***Approximate values for stars of solar composition**

ASTRONOMY: A Beginner's Guide to the Universe, 3/E

(0-13-087307-1)
Prentice-Hall, Inc.

INSTALLATION INSTRUCTIONS
Please refer to the Read Me file on the CD-ROM for information regarding program launch instructions.

MINIMUM SYSTEM REQUIREMENTS:

WINDOWS/PC

Pentium processor
Windows 95/98/NT 4.0x
640x480 pixel screen resolution (1024x768 recommended)
16 MB RAM, 48 MB recommended
4X CD-ROM drive
Netscape Navigator 4.08 (included on this CD-ROM)
Apple QuickTime 4.0 (included on this CD-ROM)
Color monitor set to 'High Color' (16 bit)
Soundcard and Speakers

MACINTOSH

PowerPC
System 7.6.1 or above
640x480 pixel screen resolution (1024x768 recommended)
16 MB RAM, 48 MB recommended
4X CD-ROM drive
Netscape Navigator 4.08 (included on this CD-ROM)
Apple QuickTime 4.1.2 (included on this CD-ROM)
Color monitor set to 'Thousands of colors' (16 bit)